THE ELEMENTS OF PHYSICS

THE ELEMENTS OF
PHYSICS

I. S. GRANT
W. R. PHILLIPS

OXFORD
UNIVERSITY PRESS

OXFORD
UNIVERSITY PRESS

Great Clarendon Street, Oxford OX2 6DP

Oxford University Press is a department of the University of Oxford.
It furthers the University's objective of excellence in research, scholarship,
and education by publishing worldwide in

Oxford New York

Athens Auckland Bangkok Bogotá Buenos Aires Cape Town
Chennai Dar es Salaam Delhi Florence Hong Kong Istanbul Karachi
Kolkata Kuala Lumpur Madrid Melbourne Mexico City Mumbai Nairobi
Paris São Paulo Shanghai Singapore Taipei Tokyo Toronto Warsaw

with associated companies in Berlin Ibadan

Oxford is a registered trade mark of Oxford University Press
in the UK and in certain other countries

Published in the United States
by Oxford University Press Inc., New York

© Oxford University Press, 2001

British Library Cataloguing in Publication Data

Data available

Library of Congress Cataloging in Publication Data

Data available

ISBN 0-19-851878-1

Typeset by Newgen Imaging Systems (P) Ltd., Chennai, India
Printed in Italy
by Giunti Industrie Grafiche, Florence

Preface

This book explains the basic principles and concepts of physics to those taking their first university courses in the subject. The text is intended as a complete course for students taking physics as their chief subject in the first year of their studies. The subject is treated at a suitable level to enable students to progress to higher level courses. The book is written in independent parts covering distinct branches of the subject. Each part contains some more difficult sections, without encroaching much on material that we consider should be covered in more advanced courses. The book is thus kept to a manageable size.

The physics knowledge assumed is that covered in physics courses during the final year of studies in science at school. Students often have difficulty with the level of mathematics needed in first-year physics courses. In its final chapter the book summarizes the mathematics needed to follow the text. This includes some discussion of vectors, with elementary trigonometry, and with differential and integral calculus. Much of the mathematics will have been taught at school, and ancillary mathematics courses in the first year at university will cover any outstanding topics.

Mathematical skills are needed to practise physics and the mathematics in the summary is an aid to acquiring those skills. However, detailed mathematical rigour is not always sought or maintained in the text, since our chief concern is that the student can reason from basic physical principles, and use these principles to discuss their consequences and predictions in a numerate fashion. Mathematical explanations and proofs are sometimes inserted in the text. These inserts may be omitted without compromising the understanding of what follows. Comments on applications or advanced topics related to the subject under consideration, are also separated from the main text.

The book is written in six parts covering different branches of first year physics. These are Dynamics, Vibrations and waves, Quantum physics, Properties of matter, Electricity and magnetism, and Relativity. The parts are largely independent and may be used in any order as texts for appropriate courses. There are inevitably some places in each part that impinge on material covered more extensively in other parts. Cross-references are given where necessary. At the end of each part there are summaries of the material presented, highlighting the most important ideas and equations.

A proper understanding of the basic principles of physics requires that a student should be able to apply the laws of physics as well as to state them and define the variables that occur in the equations. For this reason the book contains several worked examples in each chapter and exercises to which answers are given without derivation.

At the end of each chapter there are about 30 problems set at three levels. The first level consists of exercises that are mostly simple applications of laws and relations explained in the text. The second level contains problems that are harder and may require connections to be made with related topics. At the third level, there are problems of a more challenging nature. The problems and exercises at the end of each chapter are an integral part of the design of the book and students should examine and attempt them. Worked solutions to about one-third of the problems are given at the end of each chapter, as are answers to about a further one-third.

When first met, the definitions of physical quantities are given in bold-faced type, and the page number indicated in the index italicized to distinguish that reference from others. Equations that are considered to be more important than most are made distinctive by being coloured blue. Important physical laws, when expressed in words, are given in bold-faced type and also printed in blue.

We would like to acknowledge the helpful suggestions and constructive criticism of many friends and colleagues; any errors that remain are our responsibility. The contents of the book have been strongly influenced by the first-year courses given in the Department of Physics of the University of Manchester over several years. Feedback from these courses, both from students and from staff, has been of great help in determining which topics need a more extensive treatment than others.

I. S. Grant
W. R. Phillips

Symbols and notation

Physics describes the behaviour of many observable quantities. When the relationships betwen physical quantities are expressed in mathematical formulae it is necessary to use symbols to represent the different quantities. There is a difficulty in finding sufficient simple symbols to cope with the demand. Sometimes different physical observables are represented by the same symbol. This does not occur often, but it would not be helpful to try to ensure that it never happened because there are several symbols universally used for more than one quantity. There should be no confusion over the meaning of a symbol in any particular section. Our usage of symbols is, almost without exception, that found in other modern texts, and we restrict symbols to letters of the Roman and Greek alphabets.

When quoting numbers for fundamental constants or answers to problems and exercises, only sufficient significant figures appropriate to the context are given. For example, if an exercise asks for an estimate, the answer is usually given to two significant figures; fundamental constants, when introduced, are given to four significant figures. The accuracies on experimental numbers are given when the quantities are first introduced and are the standard errors on the means.

We finally give the meanings of some symbols used in the text that are shorthand for words in mathematical expressions. The equals sign means 'identically equal to' subject to the approximations and limitations within which the discussion is taking place. The $\sim$ sign means 'roughly equal to', and the $\approx$ sign is used when two sides of an equation are nearly equal. The $\equiv$ sign means 'is equivalent to' and is used, for example, when we indicate how many basic units of a quantity like seconds there are in composite units. For example, one microsecond $\equiv 10^{-6}$ seconds. The $>$ and $<$ signs mean 'greater than' and 'less than', respectively. We use a bar above a symbol to denote the average of the quantity the symbol represents. The average of the square of a quantity denoted by x is written $\overline{x^2}$. If a quantity appears between vertical lines, it means 'take its absolute value', which is always positive.

Contents

Part 2 **Vibrations and waves**

Chapter 1

Introduction

This chapter introduces the subject of physics, its methodology, and applications. The units and dimensions of physical quantities are described, and the errors on measurements discussed.

Physics is the subject that underpins science. It lays the foundations of all scientific investigations. Physics is the vehicle for understanding natural phenomena in a framework of logic and order, and determining the rules that govern how the Universe behaves.

Physics is the study of causes and effects, from the extraordinarily small to the unimaginably large. It seeks to describe observed phenomena within a framework of scientific laws that involve abstract concepts developed from experiments and the interpretation of experimental results. Physics is the art of putting together the results and conclusions of experiments within the framework of such physical laws. Abstract concepts are given shape and form in mathematical language, and natural laws, once deduced, are expressed mathematically. When laws have been deduced from observations on measurable quantities, they are used to make predictions about situations to which they have not already been applied. If new observations fail to reproduce the predictions of a particular law, the law is false and has to be modified.

1.1 Physics and the scientific method

The scientific method is the way in which laws are deduced from observation. Laws tested and found true become part of the accepted structure of science. The scientific method relies on the belief that, if one observer obtains a set of experimental results under given experimental conditions, then a second will obtain the same results (subject to the uncertainties of measurement discussed below) if the same conditions apply and similar procedures are undertaken. Of course, erroneous results can be obtained in experiments, but as time goes by further experiments highlight the errors and allow an understanding of their source. When sufficient trust in the validity of experiments has been established, the relationships determined between defined quantities are then deduced from the accepted data and expressed in laws and mathematical relationships. A consensus develops among the scientific community that the laws and principles deduced from the experiments have validity. As new results are accepted, they are placed within the framework of existing theories. If this cannot be done, new theories and laws have to be formulated or the old ones modified.

Many of the phenomena allowed by the laws of physics are not obvious. A mathematical law can hide many physical consequences that remain undiscovered. New phenomena are continually being discovered, both through theoretical examination of the accepted theories and from experimental tests of their predictions under new conditions. When new observations do require explanation and cannot be described within the conventional wisdom, it often requires the radical thinking of a genius to establish a new system of order. This happened when Albert Einstein formulated the theory of special relativity to give a cohesive description of high-speed phenomena. Sometimes such radical thinking comes in advance of the experimental observations. For example, it was postulated that moving masses have wave-like properties, in addition to the familiar properties of classical mechanics, before it was experimentally verified.

1.2 Physics and technology

During the last 100 years or so, the way of life of people in developed countries has been transformed by the systematic application of scientific discoveries. Since the basic laws of physics underpin each different science, physics contributes at some level to all the benefits of the application of science. Here we highlight a few examples where the role of discoveries in physics has been particularly important.

One of the keys to economic advancement is *power*. The ability to do work at a much greater rate than can be achieved by men or animals frees people to do other things. Almost all electrical power is produced in generators based on the principle of *induction*, which is discussed in Part 5. Electrical power now provides almost all the lighting without which modern life is unimaginable. Electrical power is clean and easy to transport and to adapt for different purposes. For this reason more and more new applications use electricity as their source of power.

An understanding of wave motion—the subject of Part 2 of this book—underlies all kinds of observations and measurements and is the basis of almost all communications. Electromagnetic waves are transmitted directly or via satellites for radio, television, and telephones, or in the form of infrared light along optic fibres. Optic fibres are also used with visible light to observe places that are otherwise inaccessible, for example by providing close-up views for diagnosis and key-hole surgery in cardiac and orthopaedic medicine. High-freqency sound waves are also used in medicine to observe the behaviour of internal organs in real time. Medical physics has now become a recognized branch of physics, with particular application to imaging the interior structure of parts of the body with a variety of physical techniques.

The study of both the large-scale and the microscopic properties of materials is an important part of physics. Part 4 of this book is devoted to introducing some of the concepts needed to understand the behaviour of solids, liquids, and gases. The understanding of solids in particular is of enormous technological importance. Safe engineering depends on the ability to predict how structures will respond to the stresses to which they will be subjected in the course of their useful life. As a result of understanding their atomic structure, materials that are strong or can withstand high temperatures or particular types of corrosive environment may be developed.

Experimental physicists are continuously striving to increase the precision of measurements. The measuring techniques they have developed have been essential for the advancement of science and technology. Engineering tolerances are much smaller than they used to be, and as a result both the performance and the reliability of products has greatly improved. For centuries the difficulty of measuring time accurately enough to determine longitude gave great problems for navigators. Now a quartz watch with a timing mechanism that costs less than the watch strap is accurate to a few seconds a month, or about one part in a million, easily good enough for navigation.

1.3 Physics and its branches

The behaviour of interacting objects, from the largest to the smallest, reveals that there are four different types of forces in the physical world. Gravity is responsible for the interactions between massive objects like planets and the Sun. Electric and magnetic forces are responsible at the atomic level for the existence of the varied worlds of chemistry and biology. The remaining two types of force are known as the strong force and the weak force and are responsible for the behaviour of matter at the subatomic level.

In this book we are taking the first step in physics and introducing the basic principles and concepts that are used in the description of everyday phenomena or of observations that may be made in laboratory experiments. It is necessary to understand the elements of physics before going on to more advanced topics.

The teaching of first courses in the subject is usually done in separate sections covering subfields with headings that identify the major areas of attention. The names of the different branches have become standard and here we follow the traditional headings in the separate parts in which the book is written.

The first part deals with *dynamics*, which is the motion of objects when under the influence of forces. Classically, this topic is associated with Newton's laws, which are used to predict trajectories, and with conservation principles, which must be obeyed by an object isolated from outside effects.

The second part deals with *vibrations and waves*. Objects can vibrate either after being disturbed or when under the influence of a continuous oscillating force. If we have very many adjacent particles or objects constituting a continuous medium, the vibration of one object may be passed on the the next and a wave may travel through the medium.

Electromagnetic waves, in which there are related and oscillating electric and magnetic fields, may travel through free space as well as through continuous media.

The third part deals with *quantum physics*. It introduces the ideas required for the description of microscopic, atomic phenomena, where quantum theory replaces classical theories and the motions of microscopic particles are described by wave functions that have a probabilistic interpretation. The allowed energies of interacting particles do not always form a continuous set but may also have discrete values.

Certain *properties of matter* in bulk form the subject of the next part. Heat and temperature are introduced, and the dependence on temperature of the way the total energy of a composite system of particles is distributed between the constituents is deduced. Gases, liquids, and solids are discussed, and the way they respond to temperature changes, for example, is investigated.

The fifth part considers *electricity and magnetism*. Electric charges give rise to forces on other charges that can be described in terms of electric and magnetic fields. Stationary charges produce an electric field only; both electric and magnetic fields result when charges are in motion. Accelerated charges produce electromagnetic radiation, which consists of related, oscillating electric and magnetic fields.

The final part considers the reformulations of classical ideas about time and space needed to develop theories valid for objects moving relative to each other at speeds comparable with the speed of light. This is the subject of *special relativity*, which introduces formulae giving the equivalence of mass and energy and the variation with speed of the mass of an object.

1.4 Units and dimensions

The standard units

Numbers are essential in conveying any sort of quantitative information. Usually a number on its own is not sufficient. For example, to specify someone's height requires a system of *units* as well as a number. The answer to the question 'How tall is that man?' will depend on who gives the answer. In the USA and the UK the height is most likely to be given in feet and inches, and the answer might be 'Five foot ten inches tall'. In Europe the answer would be '1.78 metres tall'. The two answers are, of course, the same, but they refer to two different standards of length. At one time the standards of length were metal bars. The standard metre was a platinum bar kept near Paris. Nowadays the *standard metre* is defined differently. However, one metre corresponds as nearly as possible to the same length as the old platinum bar, but the new standard can be measured with much greater precision and reproduced in different laboratories around the world. Measuring instruments may be calibrated by comparison with the standard, and in turn used to calibrate others. Millions of measuring instruments, from those used in the most exact engineering to simple tapes and rules, are thus referred back to the standard with an appropriate precision.

The metre is one of the key standards in science, and it is given the symbol 'm'. Thus the height of a person is written, for example, as 1.78 m. Depending on the lengths involved, the metre may be inconveniently large or small. Multiples and submultiples of the metre are given their own special symbols, such as km for *kilometre* (1000 metres) and μm for *micrometre* (10^{-6} m). Feet and inches are called *imperial units* and are now subsidiary to the metre. One inch is defined to be *exactly* 0.0254 metres. Expressed in imperial units 1.78 m is $1.78/0.0254 = 70.08\ldots$ inches to as many decimal places as is required. For measuring someone's height, about a half an inch is usually a good enough precision, and we can round the height to 70 inches, which is five feet and ten inches.

Not all measurements need to be referred to their own standard, even when the measurements are made in terms of a special unit. For example, the volume of beer is measured in litres or in pints. But the volume of liquid does not depend on the shape of its container. If the liquid is placed in a rectangular container, then its volume is determined by the length and width of the container and the depth of the liquid. These are three lengths, each of which may be measured in metres, and the volume has then the unit (metre × metre × metre) or (metre cubed), represented by the symbol m^3. The litre and the pint are not independent units, but can be expressed in m^3. In fact, 1000 litres is exactly the same as 1 m^3. (The pint is not so conveniently related to the scientific standard, and is not even the same in the USA and the UK: one litre is roughly the same as 1.6 (UK) pints.)

Besides the standard of length, there are other standards that have world-wide recognition. Two particularly important ones are standards of mass and of time. The standard unit of mass is the *kilogram*, denoted by the symbol kg, and the standard unit of time is the *second*, denoted by the symbol s. Once the standard metre, kilogram, and second have been set up, many other measurements may be expressed in terms of these units.

The system of units used throughout this book is the Système Internationale, or SI for short. The SI units of length, mass, and time are the metre, the kilogram, and the second, and are the generally accepted units in scientific publications. The SI units for some quantities are not common in everyday usage: for example, pressure is measured in *pascals*, a unit that you are not likely to find in a list of tyre pressures. A great advantage of SI units is that they are consistent, and any calculation carried out completely in SI units gives the answer in SI units, without the need to multiply by a conversion factor. In this book the SI unit for each quantity is defined as it arises, and a summary of all the units is given in Appendix A.

Many quantities are *defined* in terms of the standards of length, mass, and time. Consider speed as an example. The speed of an object is defined to be the distance it travels in unit time. Writing this definition as an equation, if a distance d is travelled at a constant speed v in a time t, $v = d/t$. Putting $d = 1$ metre and $t = 1$ second, we find $v = 1$ metre per second, and the unit of speed in SI units is thus one metre per second. This procedure applies to any system of units, not just to SI. If we choose to measure distance in miles and time in hours, the speed is given in miles per hour. In general the unit of speed for any choice of standard units is (unit of distance) $\times$ (unit of time)$^{-1}$. Using symbols for SI units, a speed of three metres per second is written as $3 \, \mathrm{m \, s^{-1}}$. All the other quantities that occur in physics are, like speed, each defined by just one equation, and this equation determines the unit for that quantity.

Temperature and electric current have SI units—*kelvins* and *amperes*, respectively—that are not defined in terms of metres, kilograms, and seconds. Their definitions are given in Appendix A, and the units are discussed more fully in the chapters where they arise. A final basic unit is the *mole*, which defines an *amount* of material as distinct from its mass. One mole of any pure substance contains the same number of atoms or molecules.

Dimensional analysis

Most of the equations met in physics do not define new quantities, but are relations between quantities measured in units that have already been defined. Clearly, both sides of an equation must have the same units. As an example consider a satellite in a circular orbit around the Earth. How to calculate the time T taken to complete one orbit is explained in Chapter 5. The result is given by the equation

$$T = 2\pi \sqrt{\frac{r}{g}} \tag{1.1}$$

where r is the radius of the orbit and g is the acceleration of gravity. The left-hand side of this equation is a time, and it makes no sense unless the right-hand side is also a time. If the orbit is not very far from the Earth's surface the radius r is about 6000 km, or in standard SI units 6×10^6 m. We know that g, the acceleration of gravity, is the same for all objects, whatever their mass. Acceleration is the change in speed per unit time, so it is measured in metres per second per second, and its value is about

9.8 m s^{-2}. Substituting these values for r and g into the equation gives the numerical result that T is about 5000. This calculation has been carried out in SI units, and the answer must also be in SI units of time, which are seconds. The time for a satellite to complete one circular orbit not too far from the Earth's surface is thus about 5000 seconds, or about 83 minutes.

We found the units of speed to be m s^{-1} by the cumbersome procedure of writing out the units of the defining equation in full. A convenient method for working out the units of an expression is to determine its dimensions. A **dimension** is a basic independent physical quantity in terms of which other physical quantities may be defined. Length, mass, and time are dimensions, and length is represented by the symbol $[L]$, mass by $[M]$, and time by $[T]$. In this notation the dimensions of speed are $[L][T]^{-1}$. To illustrate the use of dimensions, we examine the right-hand side of eqn (1.1). The radius of the orbit r is a length, that is, its dimensions are represented by $[L]$. The acceleration of gravity is a length per unit time per unit time and its dimensions are $[L][T]^{-2}$. Hence the dimensions of r/g are $[L]/([L][T]^{-2}) = [T]^{2}$. The time taken for the orbit is proportional to the square root of r/g, which has dimensions $([T]^{2})^{1/2} = [T]$. The right-hand side of the equation is indeed a time. Notice that this analysis has made no reference to SI units; it is applicable to whatever system of units you may be using.

In our example the left-hand side of eqn (1.1) was a single symbol representing a time. However, in any equation, both sides must have the same dimensions, even if both sides are complicated algebraic expressions. Everyone makes mistakes, and it is a good practice after completing the derivation of an equation to make sure that the dimensions are correct by evaluating the dimensions of both sides. The problems in this book often require several mathematical steps, and it is easy to make a slip, particularly in adding up the powers of different quantities. Such a slip is usually uncovered if it is found that the dimensions are wrong at the end of the calculation. For this reason it is advisable to complete a calculation algebraically whenever possible, and only to insert numerical values at the end. Once numerical values are used the dimensions are no longer visible.

Sometimes the possible form of an equation is limited by the requirement that the dimensions are the same on both sides. This is best illustrated by a simple example, again involving gravity. Suppose we want to know how far an object will fall in a given time when it is dropped in a vacuum. This is a simple calculation, which is done in Chapter 2, but we can find the answer, apart from a numerical factor, simply by considering dimensions. The distance fallen must depend on the acceleration of gravity g as well as the time t. The dimensions occurring in the equation do not include mass, since we know that g has the same value whatever the mass of the object. In a vacuum there are no frictional forces, and there are no other factors that might affect the distance fallen.

Let us call the distance fallen z, and suppose that the equation relating it to g and t has the form

$$z = kg^{\alpha}t^{\beta} \tag{1.2}$$

where k, α, and β are numerical constants. The dimensions of g have already been given above as $[L][T]^{-2}$. The dimensions of z and t are $[L]$ and $[T]$, respectively. Now write down the dimensions of both sides of eqn (1.2),

$$[L] = [L]^{\alpha}[T]^{-2\alpha} \times [T]^{\beta}.$$

Comparing powers of $[L]$ on both sides leads immediately to $\alpha = 1$. Comparing powers of $[T]$, $0 = -2\alpha + \beta$ and $\beta = 2$. This analysis does not give any information about the constant k, but it has limited the form of the equation to

$$z = kgt^2.$$

The value of k is in fact $\frac{1}{2}$, and it is, of course, important to know its value in order to perform quantitative calculations. Nevertheless, it is a good start to know the functional form of the equation. Some less obvious examples, which include the use of $[M]$ as well as $[L]$ and $[T]$, are included in problems later in the book.

1.5 Measurement errors

Quantitative experiments require all measurements to be expressed numerically. Numerical measurements always have uncertainties. There is no such thing as an exact measurement and to pass on sensible and useful information to someone you must state the degree of confidence you have in the number you quote.

Detecting instruments always have errors; the responses given to identical signals usually differ. Even if the numerical accuracy of the instrument is very high, it will rarely give the same response. For example, if the instrument is a timer that measures intervals between the ticks of a hypothetical perfect clock, the measurements have errors because of imperfections in the recording system. Many measurements of the interval between ticks are distributed around a most probable value. The deviations of the measurements from the 'true' value occur randomly and are referred to as *random errors*. The occurrence of a given deviation is independent of the previous history of the deviations, and the deviations are equally as likely to be positive as to be negative.

If the timing sytem that measures the clock intervals is good, the distribution of the measured time intervals is narrow and centred on the most likely observed interval. If the system is poor, the distribution is wide. The width of the distribution is clearly related to the confidence we can have in giving the most probable value of a limited number of measurements of the interval. The answer quoted as the best result of an experiment is the *mean* of the measured values. The error quoted to give the information on the accuracy to which the mean has been measured is called the *standard error of the mean*. The way random errors are treated in the analysis of experimental data is discussed further in Appendix B.

The above remarks on errors considered random errors only. Experiments also have *systematic errors*. These are uncertainties that affect all the data points in the same way. For example, voltage and current measurements may be used to determine the resistance of a wire at different temperatures. The devices that give the voltages and currents have to be calibrated against standards, and there will be errors in the calibrations. These errors are common to all the resistance measurements and have to be taken into account when quoting values for the resistances. Systematic errors of this kind do not change the error on *relative* values of resistance at different temperatures.

The scientific theories that describe and crystallize the results of experiments may also be approximate. Theories, to be accepted, have to give predictions that are correct to within the accuracy of existing experiments. In many applications of physics, we often have no need to be concerned about the approximate nature of a theory or of a particular formula; the validity of a procedure is already established to be sufficiently accurate for the purposes at hand. When a formula is used to calculate the value of a physical quantity there is often no need for the answer to be known to better than a given accuracy. In such situations, there are often terms in the expressions used to give the answers that may be neglected without affecting the results to the required accuracy. Part of the art of physics is to make suitable approximations in determining quantities using the accepted formulae.

Part 1

Dynamics

Chapter 2

Newton's laws of motion

*The laws governing
motion are explained in
this chapter and applied
to some simple
examples.*

Dynamics is the science of motion: it is the study of how objects move under the action of forces. For ordinary everyday objects Newton's three laws of motion sum up the whole of dynamics. In this chapter we shall be discussing Newton's laws and explaining how to use them, answering such questions as 'How fast is a rocket moving after it has burnt all its fuel?'

Newton's laws do not apply to very small systems or to objects moving at speeds close to the speed of light. The motion of atomic constituents over distances comparable with the size of an atom must be described by *quantum mechanics*, and the *special theory of relativity* is needed to explain the phenomena observed when objects are moving very fast. Quantum mechanics and special relativity are the subjects of Parts 3 and 6, respectively. For now we shall only be concerned with *classical dynamics*, which is the motion of bodies of such a size and moving at such speeds that Newton's laws are to all intents and purposes exact.

2.1 Forces and Newton's first law

Everyone has a commonsense idea of what is meant by the word force. When you push an object with your arms, you exert a force on it. What happens as a result depends on the circumstances. Just by thinking carefully about some simple examples, it is possible to pick out several properties of forces that must be incorporated into a quantitative theory of dynamics.

Imagine two arm-wrestlers sitting opposite each other with their elbows on a table each trying to force the other's arm down. If one is stronger than the other, he will succeed in moving his opponent's arm. However, if they are a well-matched pair, neither gives way, and both arms remain upright: the forces exerted by the two wrestlers balance and no movement occurs. For this balance to be maintained the forces exerted by the wrestlers must be of equal strength and in opposite directions.

This example shows that, if a force is to be specified completely, not only must its magnitude be known, but also the direction in which it acts. In other words, a force is a **vector** quantity, and the net force obtained by adding the two equal and opposite forces is zero.

Equilibrium

If an object is initially at rest and the net force acting on it is zero, it will remain at rest. The object is said to be in **equilibrium**. In the case of the two wrestlers, they exert equal and opposite forces on one another, as shown in Fig 2.1. These forces cancel out, which is expressed in vector language by saying that their vector sum is zero. The wrestlers' arms are in equilibrium.

The equilibrium will not last very long, since eventually one wrestler tires and his arm will be pushed down against the table by his opponent. But looking around you will see that almost all stationary objects are in a position of **stable** equilibrium, i.e. they will remain stationary indefinitely unless disturbed by a substantial force. Everything experiences a gravitational force pulling downwards, so for a chair to remain standing on the floor it must be held up by an upwards force that exactly cancels its weight and ensures that equilibrium is maintained. Here again there are two forces—the weight of the chair and the reaction of the floor—which act in opposite directions and directly cancel out. The chair will only move if it is pushed by a big enough horizontal force to overcome friction or if it is lifted up. In contrast, a pencil standing on its end is in **unstable** equilibrium—a very small disturbance will cause it to fall over.

Often, more than two forces are acting on a stationary object, and then no two of them need be oppositely directed. For instance, Fig 2.2 shows a metal bar hanging from a fixed support. The total weight of the bar is balanced by an upward force exerted by the support, but the two strings are not pulling vertically upwards on the bar. The strings are taut and each one exerts a force on the bar: this force is called the **tension** in the string.

The bar is in stable equilibrium and it remains stationary. This means that the net force acting on it is zero, even though the three forces are all acting in different directions. To find the tensions in the strings in terms of the weight of the bar, we must combine the forces using the rules for vector addition, which are summarized in Chapter 20.

Vectors may be represented in a diagram as lines with lengths proportional to the magnitudes of the vectors, pointing in the same directions of the vectors, as indicated by arrows. The vector diagram of the forces on the bar is shown in Fig 2.3. Bold letters are used for vector quantities, and in the diagram the line labelled $\mathbf{F}_1$ represents the weight of the bar. The tensions in the strings are represented by $\mathbf{F}_2$ and $\mathbf{F}_3$. The combined effect of $\mathbf{F}_2$ and $\mathbf{F}_3$, written as $(\mathbf{F}_2 + \mathbf{F}_3)$, is represented by the diagonal of the parallelogram formed by the vectors $\mathbf{F}_2$ and $\mathbf{F}_3$. The combined force $(\mathbf{F}_2 + \mathbf{F}_3)$ is equal in magnitude and in the opposite direction to the weight of the bar $\mathbf{F}_1$. The three forces $\mathbf{F}_1$, $\mathbf{F}_2$, and $\mathbf{F}_3$ can make up a triangle with the arrows following on from one vertex to the next.

The net force on the bar is the vector sum of the three forces, and the vector equation

$$\mathbf{F}_1 + \mathbf{F}_2 + \mathbf{F}_3 = 0$$

● Stable and unstable equilibrium

Fig. 2.1 If neither arm is moving, each is exerting a force of the same magnitude, but in the opposite direction, as the other.

● The net force on a stationary object is zero

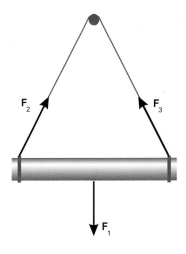

Fig. 2.2 Three different forces are acting on the bar, but the net force is zero.

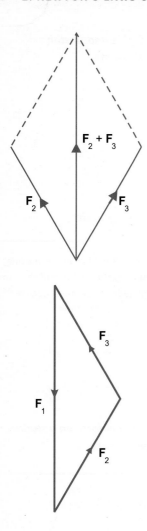

Fig. 2.3 The vector diagrams for three forces in equilibrium.

is the mathematical way of expressing the statement that the net force is zero.

Equation (20.8) expresses the sum of two vectors in terms of components along coordinate axes: the sum of the components of the separate vectors along any coordinate axes equals the component of the summed vector in the same direction. This statement then clearly extends to apply to the components of any number of vectors added together. In the example of the suspended bar, the net force on the bar is zero and both the vertical and the horizontal components of the separate forces on the bar must cancel out. The force due to the weight of the bar acts vertically downwards, and this equals the sum of the upward vertical components of the tensions in the strings. The strings pull inwards as well as upwards, and their horizontal components are equal and opposite. Notice that the horizontal components of the tension are trying to compress the bar. The metal is not very compressible, and it can react against the string with scarcely any change of length.

We can express the fact that the net force on a stationary object is zero in vector notation, without referring to particular coordinate axes. Suppose that a number of forces $\mathbf{F}_1$, $\mathbf{F}_2 \dots \mathbf{F}_i \dots$ keep the object in equilibrium. Then the vector sum of all these forces must be zero. This general statement is written as

$$\sum_i \mathbf{F}_i = 0. \tag{2.1}$$

The summation symbol $\sum_i$ indicates the sum over all values of i, that is, over all the forces acting on the object. If the vectors $\mathbf{F}_i$ are represented in a vector diagram, the net force is found by repeated application of the procedure for adding two vectors described above. When the net force is zero, the diagram is a closed polygon with the arrows all in the same sense around the polygon.

Statics and dynamics

The bar in Fig 2.2 is a very simple example of equilibrium. In large structures it is often of great importance to know the forces acting everywhere in the structure. For example, in a suspension bridge the forces must not be large enough for there to be a risk of damaging the materials of which the bridge is made. Calculation of these forces is an essential part of engineering design. Such calculations, which may be extremely lengthy, depend on a knowledge of the properties of the materials as well as the shape of the bridge. The study of forces in stationary structures is called *statics*, in contrast to dynamics, which deals with moving objects. Statics and dynamics together are referred to as *mechanics*. The subject of statics is not included in this book, but the properties of solid materials are considered in Chapter 14.

Worked Example 2.1 The strings holding the bar in Fig 2.2 are at an angle of 30° to the vertical. Compare the tension in each string with the weight of the bar.

Answer Write the magnitudes of the forces $\mathbf{F}_1$, $\mathbf{F}_2$, and $\mathbf{F}_3$ as F_1, F_2, and F_3. The strings are both at an angle 60° to the horizontal, and the magnitudes of the horizontal components of $\mathbf{F}_2$ and $\mathbf{F}_3$ are $F_2 \cos 60°$ and $F_3 \cos 60°$. Because the bar is at rest, there is no net horizontal force on it and, since the horizontal components of $\mathbf{F}_1$ and $\mathbf{F}_2$ are in opposite directions, $F_2 \cos 60° = F_3 \cos 60°$, i.e. $F_2 = F_3$. Similarly, the vertical component of the net force is zero, or $F_1 = (F_2 + F_3) \cos 30° = \frac{\sqrt{3}}{2}(F_2 + F_3) = \sqrt{3}F_2$. The tension in each string is thus equal to $1/\sqrt{3}$ times the weight of the bar. The compressional forces are $F_2 \cos 60° = \frac{1}{2}F_2$ and $F_3 \cos 60° = \frac{1}{2}F_3$, each $1/2\sqrt{3}$ times the weight of the bar.

Velocity

Our main concern in this chapter is with objects that are moving. To specify the motion of any object we need to state its direction as well as its speed, that is to say, the motion must be described by a vector. The vector in the direction of motion and with magnitude equal to the speed is called **velocity**.

Velocity is a relative quantity. Usually people talk about velocities relative to the Earth's surface. When you say 'the car was going at a hundred miles an hour' it is not necessary to add that you mean relative to the Earth—that is understood. But for a satellite docking with a space station what is more important is their velocity relative to each other. This can be worked out if the velocity of each relative to the Earth is known. Because velocity is a vector quantity, the velocities must be combined according to the rules of vector addition, the same rules that apply to the addition of forces.

The addition of velocities can be illustrated with the simple example of a man moving on board ship. Suppose that the ship is sailing in a calm sea at a speed V. Let us choose a coordinate system fixed relative to the water, with x- and y-axes along the direction of the ship's motion and perpendicular to it. If the man now walks at a speed v' (relative to the ship), in the y-direction, the velocity diagram is as shown in Fig 2.4. The man's velocity $\mathbf{v}$ relative to the water is the vector sum of the ship's velocity $\mathbf{V}$ relative to the water and his velocity $\mathbf{v}'$

● *Relative velocity*

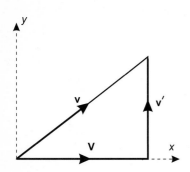

Fig. 2.4 In this velocity diagram, $\mathbf{v}'$ represents the velocity of the man who is walking across the ship, perpendicular to the ship's velocity $\mathbf{V}$ relative to the water. The vector sum $\mathbf{v} = \mathbf{v}' + \mathbf{V}$ is the velocity of the man relative to the water.

relative to the ship;

$$\mathbf{v} = \mathbf{v}' + \mathbf{V}.$$

The x- and y-components of $\mathbf{v}$ are $v_x = |\mathbf{V}|$ and $v_y = |\mathbf{v}'|$, using the notation $|\mathbf{a}|$ for the magnitude or **modulus** of a vector $\mathbf{a}$. For example, if V is eight kilometres per hour (8 kph) and v' is 6 kph, by applying Pythagoras' theorem to the triangle of velocities, we find that the man's speed relative to the water is $|\mathbf{v}| = 10$ kph.

Worked Example 2.2 A motorboat sails WNW at 10 kilometres per hour from a harbour on an east–west coastline. After what time is the boat 1 km from the shore?

Answer The angle between the boat's direction and the coastline is $22\frac{1}{2}^{\circ}$ (Fig 2.5). The component of velocity perpendicular to the coastline is thus $10\sin(22\frac{1}{2}^{\circ}) = 3.8$ kph. In about a quarter of an hour the boat is 1 km from the coast.

Notice that we have not given exact numerical answers to this example. Your pocket calculator will give you $\sin(22\frac{1}{2}^{\circ})$ with ten figure accuracy. However, it makes no sense to work out the mathematics with great precision when the physics is more approximate. No boat ever sails at an exactly constant speed, nor can it keep a precise bearing. Two figure precision is the most that can be justified in this problem. Many of the examples and exercises in this book deal with realistic problems, and the answers are given with appropriate precision.

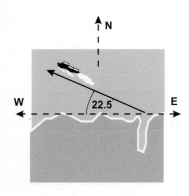

Fig. 2.5 The motorboat sails out of the harbour on a fixed bearing. If there are currents flowing the skipper must take account of them to work out his position.

Exercise 2.1 The boat in the worked example above meets a current moving northwards at 6 kph. If the skipper takes no action, in what direction is the boat now sailing relative to the land?

Answer N 43.2° W, i.e. close to NW.

Now that we have set out the notation for describing moving bodies, we can consider how motion is affected by forces. The earlier examples showed that, whenever a body remains at rest, there is no net force acting on it. What about objects that are already moving? How do they behave when there is no net force pushing them one way or the other? Of course, we must always balance out the gravitational force, but it is not difficult to arrange that any other forces are negligible, or at least very small. For example, if you throw a stone across a smooth frozen pond, the ice sustains the weight of the stone, but since it is slippery it cannot exert a large force horizontally. The stone slithers a long way before slowing down, and you will notice that it moves in a straight line. The smoother the stone and the more slippery the ice, the farther the stone will slide. This suggests that, if the ice were perfectly frictionless, the stone would go on sliding in the same direction indefinitely, without slowing down at all. This situation can be described more concisely by saying that the stone has a constant velocity, since the velocity is a vector specified by both the speed and the direction of motion.

A nearly frictionless system is shown in Fig 2.6—this system is often seen in science museums and in teaching laboratories. A heavy block is kept floating just above an inverted V-shaped beam by an array of air jets. There is almost no force acting on the block along the length of the beam. If it is given a gentle push, it will move at a constant speed to the other end of the beam.

The steady motion of the stone on the ice, and of the block on the air cushion are in accordance with **Newton's first law of motion**, which states:

Any body moves with constant velocity (or remains at rest) if the net force acting on it is zero.

This section started by recognizing that, for a body to remain stationary, the forces acting on it must cancel out. The same applies for an object moving in a straight line with a steady speed. This means that the converse of Newton's first law is also true, and we can deduce that there is no net force acting on any body that is stationary or is moving with constant velocity.

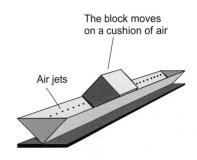

The block moves on a cushion of air

Air jets

Fig. 2.6 A stream of tiny air jets keeps the block floating just above the trough. It can move along the trough almost without friction.

2.2 Momentum and Newton's second law

When the net force on a body is not zero, there is a change in its velocity. For example, if a supermarket trolley is at rest, a push will set it moving in the direction in which it is pushed (if you are lucky enough to find one with smoothly moving wheels!). Both the force and the change in velocity that it causes are vectors, and they point in the same direction. The harder you push, the faster the trolley will move but, if you are looking for a relationship between force and velocity, it is pretty obvious that we must also consider how heavy the trolley is. To cause the same motion in a full trolley as in an empty one, a larger force needs to be applied to the full trolley.

The **mass** of the trolley is the quantity that measures how difficult it is to make the trolley move fast. The difference between mass and weight is explained later in this chapter: for the moment we simply note that mass is measured by comparison with standard masses on a balance and that the unit of mass used in science is the kilogram. The argument about mass and speed also applies to objects which are not initially at rest: the force required to stop a moving trolley depends both on how heavily it is loaded and on how fast it is moving.

A quantity that depends on both mass and velocity is **momentum**.

The momentum of a body of mass m moving with a velocity v is defined to be the product mv.

We shall denote momentum by the symbol **p**:

$$\mathbf{p} = m\mathbf{v}. \tag{2.2}$$

Momentum is a vector, pointing in the same direction as **v**. It is the momentum of a body (such as a moving trolley) that is changed by the force acting on it. Clearly, the change in momentum also depends on the length of time during which a force has been applied and, in fact, the momentum change is directly proportional to both the force and the time. This is the basis of Newton's second law, which we shall now express in a mathematical form.

● *Forces change momentum*

Suppose a steady force **F** is applied to a stationary body during a short time interval δt. The body acquires a momentum proportional to **F** and to δt, and, because the initial momentum was zero, this momentum represents the *change* of momentum occurring in δt. Since δt is small, the momentum change is also small and we shall call it δ**p**, writing

$$\delta\mathbf{p} \propto \mathbf{F}\delta t. \tag{2.3}$$

Generally speaking, when two quantities are proportional to one another a constant of proportionality has to be included in the equation relating them. The numerical value of the constant of proportionality depends on the units that are used to measure the quantities occurring in the equation. In this book we use the SI units, which are nowadays almost universally chosen by physicists. The SI system of units, which is explained in Chapter 1 and Appendix A, is internationally recognized, and, indeed, SI simply stands for the French words 'Système Internationale'.

● *SI units*

In the SI system distance is measured in metres, mass in kilograms, and time in seconds. These units are abbreviated as m, kg, and s, respectively. Derived units must also be based on the same units of length, mass, and time. The speed of an object moving at constant velocity is (*distance divided by time*) and the SI unit of velocity is thus metres per second (m s^{-1}). The SI unit of force is called the **newton**, abbreviated as N. The

Isaac Newton 1642–1727

Isaac Newton's father, like all the Newtons before him, was illiterate and could not write his own name. Isaac's father died before his only child was born. The baby was premature, 'so little they could put him in a quart pot and so weakly that he was forced to have a bolster all round his neck to keep it on his shoulders'. Fortunately he survived. When Isaac was three years old an elderly clergyman married 'the widow Newton', but disagreeably refused to take Isaac into his house. The boy was left with his mother's parents, who seemed not to care for him either—at least his grandfather excluded Isaac from his will.

Isaac's mother was widowed again when he was ten, and she returned to their original home, now with three more children from the second marriage. But two years later he was sent away to a grammar school. His mother's family were higher in the social order than the Newtons; one of her brothers was rector of a nearby church, and it may have been due to his influence that Isaac was given a formal education. At school he learned to be fluent in Latin. This was almost all he did learn, but it was later to be a useful accomplishment, as Latin was the language of scholarly communication in the seventeenth century. Young Isaac, who had been lonely at home and now lived in lodgings, did not make friends with his school contemporaries. It is perhaps not surprising that he remained reclusive and prickly all his life. While at school he spent his money on tools and his spare time making doll's furniture for his landlord's daughters or models such as windmills and sundials for his own amusement. Isaac left school when he was seventeen and went home, where his mother wanted him to run their farm. This was not a success, as he was always daydreaming or making his models. His uncle the rector and his old schoolmaster recognized that Isaac was an exceptional individual, and they eventually persuaded his mother to let him return to school to prepare for university.

In 1661 Newton went to Trinity College in Cambridge as a subsizar. A subsizar earned his keep by acting as a servant for other students, and paid to attend lectures. This lowly status again cut Newton off from his fellows, quite unnecessarily, since his mother was now wealthy from the estate of her second husband. However, Newton now had access to books and the opportunity for concentrated study. Although in the corrupt world of the seventeenth century advancement owed more to patronage than to talent, Newton obtained a scholarship in 1664 (worth his keep and a little more than one pound a year), was elected a fellow of the college in 1667 and became Lucasian Professor in 1669. The professorship, which was handsomely endowed, was the only one in the university with any connection with mathematics or natural philosophy: classical studies and theology still formed the main part of the curriculum.

Newton stayed in Cambridge until 1696, living a solitary existence, scarcely speaking to the other fellows. During that time he made an extraordinary contribution to science. He laid the foundations of modern mechanics, celestial dynamics, and optics. Newton's fame spread as some of his results became known through correspondence, but he was reluctant to publish his work, and was paranoid when he thought he had been plagiarized. A particularly bitter controversy arose when the German mathematician Leibniz published a paper on the calculus and Newton claimed that his own results had been used without acknowledgement. The truth seems to be that Newton had certainly devised his method first, but that Leibniz probably invented calculus independently. After cajoling on the part of those who wanted Newton's work to be widely available, and also because Newton wanted to justify his reputation, his work eventually appeared in two great books, the *Philosophiae naturalis principia mathematica* in 1687 and *Opticks* in 1704. By 1704 Newton's pre-eminence was acknowledged. He was Master of the Mint (a post that had usually been regarded as a sinecure, but that Newton undertook with great energy and efficiency) and

he was the President of the Royal Society. Newton's name deservedly lives on. His ideas revolutionized science and his papers in the *Philosophical Transactions of the Royal Society* stand out for the modernity of their outlook. Yet shortly before he died, this great man wrote to an acquaintance, 'I don't know what I may seem to the world, but, as to myself, I seem to have been only like a boy playing on the sea shore, and diverting myself in now and then finding a smoother pebble or a prettier shell than ordinary, whilst the great ocean of truth lay all undiscovered before me.'

● **The SI unit of force**

newton is defined so that the constant of proportionality in the relation (2.3) is one, i.e.

$$\delta \mathbf{p} = \mathbf{F} \delta t,$$

$$\text{or} \quad \mathbf{F} = \frac{\delta \mathbf{p}}{\delta t}. \tag{2.4}$$

It is important to remember that eqn (2.4) only applies if the force, the momentum, and the time are all in SI units. Momentum is (*mass × velocity*) and, if, for example, the velocity is given in kilometres per hour, it must be converted to metres per second before applying eqn (2.4).

Worked Example 2.3 Very heavy objects can be easily moved if they are standing on a platform supported a few mm above the ground on an air cushion maintained by air jets directed downwards from the underside of the platform. There is then almost no friction, and even a small force is sufficient to move the heavy object. A steel block weighing one tonne (one tonne $\equiv$ 1000 kg) is supported on such a platform. A horizontal force of 10 newtons is applied to the block, which is initially at rest. Neglecting the mass of the platform, estimate how fast the block is moving after 10 s.

Answer From eqn (2.4) a change of momentum is *force × time*. The force is 10 N and after 10 s the momentum Δp of the initially stationary block is $10 \times 10 = 100 \, \text{N} \, \text{s}$. There is no special SI unit for momentum and the result is expressed here in newton seconds (N s). Momentum can also be written as mv, and the momentum of the block can equivalently be expressed as $100 \, \text{kg} \, \text{m} \, \text{s}^{-1}$. Since $\Delta p = mv$ the speed is $\Delta p / m$. In this equation the mass of one tonne must be expressed in SI units as 1000 kg leading to $v = 10 \times 10 / 1000 = 0.1 \, \text{m} \, \text{s}^{-1}$. A force of 10 N is just sufficient to lift a mass of 1 kg, yet after 10 s the steel block is moving at $0.1\,\mathrm{m\,s^{-1}}$, or $10\,\mathrm{cm\,s^{-1}}$, quite fast for such a massive load!

According to eqn (2.3), since **F** is constant, $\delta\mathbf{p}$ increases at a constant rate as δt increases. The right-hand side of eqn (2.4) is just this rate, and we can express eqn (2.4) in words by saying that the force is equal to the rate of change of momentum.

If the time interval δt is made to tend to zero, the right-hand side now becomes $\mathrm{d}\mathbf{p}/\mathrm{d}t$, the derivative of **p** with respect to time. Don't worry if the derivative of a vector is unfamiliar. The derivative $\mathrm{d}\mathbf{p}/\mathrm{d}t$ is just a vector whose components along the coordinate axes x, y, and z are $\mathrm{d}p_x/\mathrm{d}t$, $\mathrm{d}p_y/\mathrm{d}t$, and $\mathrm{d}p_z/\mathrm{d}t$, the derivatives with respect to time of the corresponding components p_x, p_y, and p_z of the vector **p**; $\mathrm{d}\mathbf{p}/\mathrm{d}t$ is still described in words as the rate of change of momentum.

Although we introduced eqn (2.4) by discussing a body that was initially at rest, this restriction is not necessary, and the equation applies whatever the initial momentum may be. Nor when we go to the limit as δt tends to zero need **F** be a constant: **F** is now the instantaneous value of the force at the time when $\mathrm{d}\mathbf{p}/\mathrm{d}t$ is evaluated. This relation between force and momentum is **Newton's second law of motion**.

In words the second law states that

The net force acting on a body equals its rate of change of momentum.

The mathematical expression of the law is

$$\mathbf{F} = \frac{\mathrm{d}\mathbf{p}}{\mathrm{d}t}. \tag{2.5}$$

Once again we emphasize that the units of force are *defined* by this equation. In effect, the second law constitutes the definition of force: if the momentum of an object is changing, we can say that it is experiencing a force given by the second law, even if the force is not otherwise measurable.

● *Forces are defined by Newton's second law*

Gravitational and electrical forces

Now that we have defined forces through Newton's first and second laws, it is natural to ask what different forces we can observe in the world around us. Any object falls unless it is supported. This tells us that the force of gravity is universal, attracting everything towards the Earth. Gravitational forces and their role in astronomy are discussed in Chapter 5. Here we need to point out only that gravitational forces are always

● *Gravitational forces are weak*

attractive—there are no materials that are pushed away from the Earth by the gravitational force—and also that they are not very strong. This may seem a strange statement if you have just been trying to lift a large sack of flour, but the sack is only so heavy because it is being attracted by the whole Earth. Gravitational forces are also acting between other pairs of objects, such as sacks of flour or lumps of lead, but, as we shall find in Chapter 5, these forces are so small that they can normally be completely neglected.

It is easy to think of many situations where forces are causing a change of motion or resisting another force. In the example of a bar hanging on strings, the weight of the bar is balanced by the tensions in the strings. Why are the strings able to withstand the pull of the bar? What is the origin of the tension that they exert? The same sort of questions can be asked about all kinds of other forces. A spring pulls when it is stretched and pushes when it is compressed. Similarly, the air in your bicycle pump resists with a progressively stronger force as you compress it. There is a force of buoyancy keeping you afloat when you swim, and you have to push hard against the water to make progress. The forces in these examples may seem all to be very different, but remarkably they are all electrical in origin.

Electrical forces only act between objects that carry electric charge. The important differences between electrical and gravitational forces are that electrical forces may be either repulsive or attractive and that, by comparison with gravitation, electrical forces are immensely strong. Most large objects carry little or no electrical charge, and we do not notice any force between them. However, all matter is made of atoms built up out of minute charged particles carrying two different kinds of charge, called positive and negative. Unlike charges attract one another while like charges repel, according to the laws governing electrical forces, which are explained in Chapters 15 to 17. A spring, which is made up of equal numbers of positively and negatively charged particles, balances the attraction and repulsion between them if no external force is exerted. We can understand that, if this balance is disturbed, the spring may pull when it is stretched and push when it is compressed.

There are two other fundamental types of force, called the strong force and the weak force. These forces act within atomic nuclei, and they are briefly discussed in Chapter 11. The strong and weak forces do not contribute to the forces between pieces of ordinary matter. The only forces that are relevant when applying Newton's laws to the motion of large-scale objects are gravity and forces that have electrical origins.

To illustrate how Newton's second law is applied, we shall discuss a few simple examples that are one-dimensional, i.e. in which the motion is in a straight line. In these examples the forces only act in the

● *Electrical forces are very powerful, but appear to be weak because the effects of positive and negative charges tend to cancel out*

direction of this straight line, and there is no need to use the vector notation: we shall go on to deal with three-dimensional motion in Section 2.4.

Free fall

Firstly, let us consider the motion of a body falling freely under gravity. For example, think about what happens when an object is dropped on to the floor. Provided that it is compact, and dense enough for air resistance to be unimportant, the time it takes to reach the floor does not depend on its size and shape, nor on how heavy it is. As it falls it picks up speed, and careful measurements have shown that all objects falling in vacuum, and acted on by no force other than gravity, have a constant acceleration, that is to say, the rate of increase of speed is the same whether the object is moving fast or slowly.

The constant acceleration is called **acceleration of gravity**. Its magnitude, which is denoted by g, is about $9.8 \, \mathrm{m \, s^{-2}}$, though the exact value depends on where you are on the Earth's surface. Small local variations in g are determined by the geology of the neighbourhood. Gravity surveys (i.e. contour maps showing the small changes in g) give information about what lies beneath the Earth's surface, and are used along with other evidence in the search for oil fields.

● *The acceleration of gravity*

If an object is at rest before being dropped, it moves in a straight line—downwards. Only the distance z it has fallen is needed to specify where it has reached in its downward journey (Fig 2.7). For a falling body of mass m, let us write the magnitude of the (downward) gravitational force as F and the magnitude of its (downward) speed as v. The downward momentum is mv, so that Newton's second law becomes

$$F = \frac{\mathrm{d}(mv)}{\mathrm{d}t}$$

or, since m is constant,

$$F = m\frac{\mathrm{d}v}{\mathrm{d}t}. \tag{2.6}$$

Now $\mathrm{d}v/\mathrm{d}t$ is the rate of increase of speed, i.e. the acceleration, which we know is equal to g, whatever the value of m. To be consistent with Newton's second law the gravitational force acting on a body of mass m must be mg. This force is called the **weight** of the body.

Notice that, although the mass is always the same, the weight varies slightly from place to place according to the local value of g. If you weigh out a kilogram of sugar on an old-fashioned pair of scales with balance pans, you compare the weight of sugar with the weight of a standard one kilogram

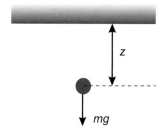

Fig. 2.7 Provided that we can neglect friction, the falling body experiences no force other than the gravitational force mg acting vertically downwards. In the diagram the distance z is the distance fallen since release from rest at $z = 0$.

● *The distinction between mass and weight*

Scales

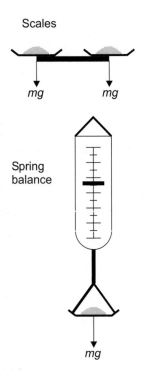

mg mg

Spring
balance

mg

Fig. 2.8 Scales measure *mass*, but a spring balance measures *weight*.

● *Solving a simple differential equation*

mass. This is illustrated in Fig 2.8. At balance the mass in each pan must be the same, whatever is the local value of g, and what you are measuring really is the mass of the sugar. A spring balance, on the other hand, measures the weight by comparing it with the force exerted by a compressed spring. The force exerted by the spring is independent of gravity, and slightly different masses are needed to compress it by the same amount when the balance is moved from one place to another—though for most purposes these changes are far too small to be of any significance.

A falling object is subjected only to the gravitational force mg, and it has a constant acceleration g. In each unit of time its speed increases by an equal amount, and at a time t after being dropped from rest it has reached a speed gt, i.e.

$$v = gt. \tag{2.7}$$

There is no need to do any mathematics to find that the speed is gt after a time t: this is simply a statement of what is meant by constant acceleration. However, to illustrate the method used to determine the motion in more complicated examples, we shall derive the same result starting from eqn (2.6). Equation (2.6) is a first-order differential equation, and its solution must be obtained by integration. To do this, multiply each side of the equation by the infinitesimal time dt, rewriting it as

$$F\,dt = mg\,dt = m\,dv,$$

or

$$g\,dt = dv.$$

Integrating both sides,

$$\int g\,dt = \int dv.$$

The integral on the right-hand side is v, and we may write

$$v = \int g\,dt.$$

Since g is constant, the result of the integration is

$$v = gt + C$$

where C is a constant of integration. Whatever the value of C, this equation is a solution of the differential equation (2.6), because the result of differentiating a constant is zero. For this reason a **constant of integration** always occurs in the solution of a differential equation.

To find the solution for our particular problem we must also specify the **initial conditions**. In this case if we choose to measure time from the moment when the object is dropped, the initial condition is that $v = 0$ at $t = 0$. Substituting in the above equation leads to $C = 0$ and to the correct solution

$$v = gt.$$

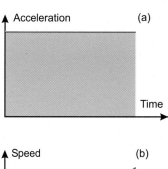

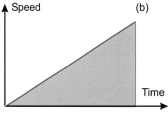

The speed v in eqn (2.7) can also be represented as an integral:

$$v = \int g \, dt. \tag{2.8}$$

This is explained in the reminder above for readers who are unsure about the mathematics. The result of the integration can be visualized by remembering that an integral represents the area under a curve. In this case g remains constant with time and is represented in Fig 2.9(a) by a horizontal straight line. Between the times 0 and t the area under this line is gt.

Now we know the speed of the object after it has been falling for a time t, but we still do not know its position. Let us call the distance it has fallen z. The z-axis points downwards and the downwards speed is $v = dz/dt$. Equation (2.7) becomes

$$\frac{dz}{dt} = gt.$$

Integration of this equation gives an expression for z:

$$z = \int dz = \int gt \, dt = \tfrac{1}{2}gt^2 + C \tag{2.9}$$

where C is again a constant of integration. However, $z = 0$ at $t = 0$ so that the constant is zero, and finally

$$z = \tfrac{1}{2}gt^2. \tag{2.10}$$

The steadily increasing speed is shown in Fig 2.9(b) as a straight line passing through the origin, and the integral leading to eqn (2.10) represents the area under this line between times 0 and t. The curve in Fig 2.9(c) shows the distance increasing as the square of the time: this curve is called a *parabola*.

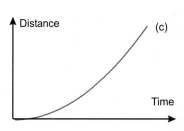

Fig. 2.9 Speed is the area under the graph of acceleration against time, and distance is the area under the speed graph.

Exercise 2.2 A splash is heard 1.5 s after a stone is dropped into a well. At what depth is the water?

Answer About 11 metres.

Up and down motion

If a ball is thrown vertically upwards it goes up and comes down again to land on the spot where it started. If the mass of the ball is m, the force on it has a magnitude mg, and Newton's law gives us exactly the same equation of motion as for free fall. The only difference is that the initial condition has changed—instead of starting from rest the ball begins with an upward speed v_0, say. How long does it take before the ball comes down again? How high up does it get?

Not much more work has to be done to answer these questions, but it is necessary to be careful about the sign of the speed of the ball. Although the motion is in a straight line, the velocity and the gravitational force are still vector quantities, so, of course, the speed is reduced when the force is opposite to the direction in which the ball is moving. Now it is convenient to choose the z-axis to point upwards so that the z-coordinate is the height above the point where the ball was released. The initial speed is then $+v_0$, since the ball is moving upwards, in the direction of increasing z.

The gravitational force mg is downwards, in the direction of decreasing z, and Newton's law becomes

$$mg = -m\,\frac{dv}{dt}. \tag{2.11}$$

The speed decreases with a deceleration g until the ball reaches its greatest height, when it is momentarily at rest. Since the rate of change of speed is always $-g$, the speed after a time t is

$$v = v_0 - gt. \tag{2.12}$$

The time taken for the upward speed to change from $+v_0$ to zero is thus v_0/g and from eqn (2.10) we find that the height the ball reaches is $\frac{1}{2}v_0^2/g$. It then retraces its path exactly, and the time taken to come down again is also v_0/g, reaching the starting point at a time $2v_0/g$. Putting this time into eqn (2.12) gives the *upwards* speed $v = -v_0$, which means that the ball is now moving downwards at the same speed as that at which it was originally thrown upwards.

The mathematical procedure for solving the problem of the ball thrown upwards is the same as that used for free fall. The only difference (apart from the sign change of the position coordinate) is in the initial conditions. Equation (2.11) leads to

$$\mathrm{d}v = -g\,\mathrm{d}t.$$

Integrating both sides,

$$\int \mathrm{d}v = -\int g\,\mathrm{d}t.$$

The solution, including a constant of integration C, is

$$v = -gt + C.$$

The initial condition in this problem is that, at $t = 0$, $v = +v_0$, requiring that $C = v_0$, i.e. $v = v_0 - gt$, the result given in eqn (2.12).

As for free fall, another integration is needed to find the position as a function of time. Rewrite eqn (2.12) as

$$v = \frac{\mathrm{d}z}{\mathrm{d}t} = v_0 - gt.$$

Integrating this equation, and introducing another constant of integration D,

$$\int \mathrm{d}z = \int v_0\,\mathrm{d}t - \int gt\,\mathrm{d}t, \text{ leading to } z = v_0 t - \tfrac{1}{2}gt^2 + D.$$

The constant of integration D is zero, since we have chosen the origin $z = 0$ to be at the point where the ball is thrown upwards. Finally, the solution is

$$z = v_0 t - \tfrac{1}{2}gt^2. \tag{2.13}$$

Equation (2.13) satisfies both the initial conditions that, at $t = 0$, the position is $z = 0$ and the speed $v = v_0$. Figure 2.10 shows the upward speed of the ball, and the upward distance it has moved, both as a function of time. The distance is again a parabola, but it is upside down when compared with the free fall diagram in Fig 2.9 because of the choice of an upward- instead of a downward-pointing z-axis. The upward speed changes from a positive to a negative at the time when the upward distance is maximum.

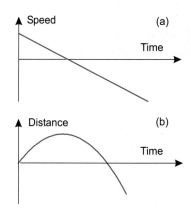

Fig. 2.10 The speed of object thrown upwards is zero when it reaches its highest point. The time taken to fall down again is the same as the time taken to reach the highest point.

Worked Example 2.4 When a hosepipe is pointed vertically upwards the water reaches a height of 20 m above the end of the hose. The hosepipe is then pointed vertically downwards from a point 20 m above the ground. How fast is the water moving when it reaches the ground?

Answer We must be careful with the signs in this example. The time taken for the water to reach its greatest height z_0 when the hose is pointing upwards is the same as the time taken for water to fall a distance z_0. From eqn (2.10) this time is $t = \sqrt{2z_0/g}$. The speed needed to reach z_0 is given by eqn (2.12) as $v_0 = gt = \sqrt{2z_0g}$. When the hosepipe is pointed downwards the initial speed, the distance, and the acceleration are all in the downward direction. A *positive* sign is needed in place of the negative sign in eqn (2.13). The initial speed is again $v_0 = \sqrt{2z_0g}$ and eqn (2.13) becomes $z_0 = \sqrt{2z_0g}\, t + \frac{1}{2}gt^2$ which may be rewritten as $\frac{1}{2}t^2 + \sqrt{2z_0/g}\, t - z_0/g = 0$.

The solution of the quadratic is $t = -\sqrt{2z_0/g} \pm \sqrt{(2z_0/g) + (2z_0/g)}$. The positive solution is the one representing the time for the water to reach the ground and this is $(2 - \sqrt{2})\sqrt{z_0/g}$. The water speed at this time is $v_0 + gt = \sqrt{2z_0g} + (2 - \sqrt{2})\sqrt{z_0g} = 2\sqrt{z_0g}$. Putting $g = 9.8 \text{ m s}^{-2}$ and $z_0 = 20$ m, the water speed is $2 \times \sqrt{196} = 28 \text{ m s}^{-1}$.

Notice that the final water speed is given by a simple algebraic expression. The intermediate quantities v_0 and t are not required, and it is better not to insert numerical values until the calculation is complete.

● *Most problems are best solved algebraically before putting numbers in*

An equation of the same form as eqn (2.13) gives the position of *any* object moving with a constant acceleration in a straight line, whether or not the acceleration is caused by gravity. The only change is that g is replaced by a constant a which is the magnitude of the acceleration. The equation does not refer to the mass of the object or to the magnitude of the force acting on it, and is therefore independent of Newton's second law. The calculation of distances in terms of known or measured speeds and accelerations, without reference to force, is called **kinematics**.

● *Kinematics*

Remember to be careful about the *sign* of the acceleration in calculations of straight line motion. Equation (2.13) applies when the acceleration is *opposite* to the z-axis, and g appears with a negative sign, representing deceleration. When the acceleration is in the *same* direction as the z-axis, a positive sign is required. A practical example of uniform deceleration that is not caused by gravity follows.

● *Be careful of the signs of speeds and accelerations*

Worked Example 2.5 The brakes on an aeroplane are controlled so that they exert very nearly the maximum braking force that can occur without skidding. A plane touches down on the runway at 180 kph, and the braking force remains constant at 10% of the weight of the plane. How long does it take for the plane to come to rest?

Answer If the mass of the plane is m, then the braking force is $0.1mg$ and the deceleration is $0.1g \approx 0.98\,\mathrm{m\,s^{-2}}$. The initial speed, in SI units, is $180 \times 1000/3600 = 50\,\mathrm{m\,s^{-1}}$. The plane comes to rest after $50/0.98 \approx 51\,\mathrm{s}$.

Exercise 2.3 What distance does the plane travel on the runway before stopping?

Answer 1.28 km

The effect of air resistance

The solution for free fall given by eqn (2.10) is a good approximation for objects falling from a small height, but very poor for a large drop. As a falling body speeds up, the air resistance becomes more and more important, and eventually the body reaches a constant **terminal velocity** when the weight and the air resistance are equal and opposite. The amount of air resistance depends on the shape of the falling body and its orientation to the vertical. Sky divers can to some extent control their speed by the attitude they adopt, as shown by some amazing mid-air rescues when sky divers have been able to overtake colleagues whose parachutes have failed to open.

The forces on irregularly shaped objects are rather complicated, but in some circumstances the air resistance is roughly proportional to speed. In order to get an understanding of the effect of air resistance, let us assume that it is exactly proportional to speed. The net force on the falling body includes air resistance as well as its weight mg. The air resistance is, of course, in the opposite direction to the weight—it is slowing the rate of fall—and we shall write it as $(-bv)$, where b is a constant. The equation of motion is the one-dimensional version of Newton's second law (eqn (2.6):

$$mg - bv = m\,\frac{\mathrm{d}v}{\mathrm{d}t}. \tag{2.14}$$

We will see that eqn (2.14) does give a reasonable description of the behaviour of the object falling from rest, namely that it accelerates to begin with and gradually builds up speed towards the terminal velocity. What is the initial acceleration? Just at the moment when the object is released, its speed v is zero, and then according to eqn (2.14) the acceleration $\mathrm{d}v/\mathrm{d}t$ is just the gravitational acceleration g. This makes

● *Initial acceleration and terminal speed*

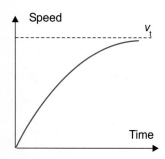

Speed

v_t

Time

Fig. 2.11 When there is air resistance, the speed of a falling object gradually approaches its terminal value v_t.

sense, because there is no air resistance to begin with, and the acceleration should be the same as it is in vacuum. After a long time, however, the weight is balanced by the air resistance and there is no longer any acceleration, i.e. $\mathrm{d}v/\mathrm{d}t$ is zero. The body has then reached its terminal speed, which from eqn (2.14) is $v_t = mg/b$.

The speed actually approaches its terminal value exponentially: starting the fall from rest at $t = 0$, v is given by

$$v = \frac{mg}{b}(1 - e^{-bt/m}) \tag{2.15}$$

where e is the base of natural logarithms. The result of differentiating $e^{-bt/m}$ with respect to time is $-(b/m)e^{-bt/m}$, and you will find by differentiating eqn (2.15) and substituting for $\mathrm{d}v/\mathrm{d}t$ in eqn (2.14) that it is indeed a solution. It also has the correct properties that $\mathrm{d}v/\mathrm{d}t$ is g at $t = 0$, and that v approaches the terminal velocity mg/b as t reaches large values, when the exponential term becomes very small. The variation of speed with time is sketched in Fig 2.11.

Exercise 2.4 A light ball reaches a terminal speed of $20\,\mathrm{m\,s^{-1}}$ when it is falling vertically in air. If it is dropped from rest, how long does it take for the ball to reach a speed of $10\,\mathrm{m\,s^{-1}}$?

Answer About $1.4\,\mathrm{s}$.

The solution of eqn (2.14) is obtained by separating the variables v and t, in just the same way as for solving an ordinary algebraic equation. Gathering the terms in v to one side of the equation and rearranging the constants we find

$$\int \frac{\mathrm{d}v}{(mg/b) - v} = \frac{b}{m}\int \mathrm{d}t.$$

The left-hand side can be integrated by putting it in the form of the standard integral

$$\int \frac{\mathrm{d}x}{x} = \ln x + \text{constant}$$

with the change of variable $x = (mg/b - v)$. The result is

$$-\ln\left(\frac{mg}{b} - v\right) = \frac{bt}{m} + C.$$

The speed v is zero at $t = 0$, requiring the constant of integration to be $C = -\ln(mg/b)$, and

$$-\frac{bt}{m} = \ln\left(\frac{mg}{b} - v\right) - \ln\left(\frac{mg}{b}\right) = \ln\left(1 - \frac{bv}{mg}\right).$$

Hence $1 - \dfrac{bv}{mg} = \mathrm{e}^{-bt/m}$

or

$$v = \frac{mg}{b}\left(1 - \mathrm{e}^{-bt/m}\right),$$

the result quoted in eqn (2.15).

2.3 Balancing forces: Newton's third law

At the beginning of this chapter we discussed static equilibrium, a phrase describing a body that is not moving and is in a stable position, so that it remains at rest. An example is a chair standing on the floor. If the mass of the chair is m, its weight is mg, i.e. the chair experiences a downward gravitational force mg. The chair does not move, which means that the net force acting on it is zero. The floor must react to the presence of the chair by exerting an upward force mg that balances the downward gravitational force.

This is rather obvious when applied to objects standing on the floor. What is not so obvious is that, when there are forces acting at a distance between objects that are not in contact, there is also a balance of forces. How can there be a balance when something is falling, for example, if you drop a chair? As we have seen, an object falling freely accelerates all the time, so there does not seem to be any balance here. What force is balancing the gravitational force on the falling chair? The answer is that the chair is exerting an upwards force on the whole of the Earth that is exactly equal and opposite to its own weight, as indicated in Fig 2.12.

Of course, you don't notice this, because the mass of the Earth is so enormous compared to that of the chair that it scarcely moves. Indeed, it would be silly to try to take into account the Earth's motion as a result of the reaction to the weight of the chair. Many other forces are being exerted on the Earth at the moment when you drop your chair but, for all practical purposes, the Earth remains at rest. There are always other forces that you also forget about because they are too small to have a noticeable

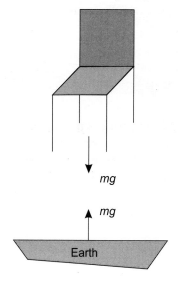

Fig. 2.12 The falling chair exerts the same force on the Earth as the Earth exerts on the chair.

effect—the gravitational pull of the Sun and the moon, for example. Although the Sun is much more massive than the Earth, it is so far away that it has practically no effect. When trying to drop something on to a particular point on the ground you don't have to allow for the direction of the Sun, which varies with the time of day.

Scientists are always making approximations of this kind. The world is too complicated for it to be possible to take everything into account. The judgement of what is important and what may be neglected is not always easy, and it depends on circumstances. For example, although the gravitational attraction of the moon can be safely neglected when considering a falling object on the Earth, it is the main force causing tides in the oceans. Similarly, the gravitational attraction of the Sun is responsible for maintaining the planets in their orbits. Just as there is a reaction on the Earth equal and opposite to the weight of a falling object, so the attraction of a planet to the Sun is balanced by an equal and opposite force on the Sun. If the Sun were viewed from a great distance, it would be seen to move as a result of the net effect of the forces exerted by its planets. It is not possible to get far away enough to make this observation, but in recent years tiny movements of some stars have been interpreted as being due to the existence of an orbiting planet—the Sun proves not to be the only star with a planetary system.

The properties of planetary orbits will be discussed in more detail in Chapter 5. For the moment we only need to state the rule that, whenever two bodies are attracted by gravity, the force on each body has the same strength and is directed towards the other. Making use of this rule, and knowing the strength of the gravitational force, it is possible to calculate the positions of the planets in our own solar system with great precision. Astronomical observations give a very strong proof that, for gravitational forces, action and reaction are always equal and opposite.

Like gravitational forces, electric and magnetic forces also act at a distance. If two balloons are charged up by rubbing them against wool, they will repel one another. Similarly, if current is passed through two coils of wire, they will attract or repel one another depending on whether the coils are wound in the same direction or in opposite directions. A pair of permanent magnets, too, attract or repel, and in this case when holding the magnets it is easy to feel that there are forces in the opposite direction on the two magnets (Fig 2.13).

The forces between atoms and molecules are electric and magnetic in origin and, together with gravitation, these are the only forces experienced by everyday objects. In our study of dynamics, we may be confident that in all circumstances the forces between two objects balance. This important fact, which is called **Newton's third law of motion**, is illustrated schematically for two objects, A and B, in Fig 2.14. Bearing in

● *Forces that have little effect on results should be neglected*

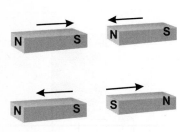

Fig. 2.13 Whether they are attracting or repelling one another, the forces between two magnets are equal and opposite.

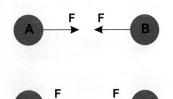

Fig. 2.14 The forces between two objects A and B are equal and opposite, whether they are attractive or repulsive.

Newton's third law can be deduced from very simple observations, without having to know the laws governing gravitational and magnetic forces. Think again about two magnets attracting one another (Fig. 2.13). If the magnets are in contact and at rest, the forces between them must cancel, otherwise the pair of magnets would spontaneously accelerate. This does not happen, so the third law must apply to the magnets. The argument is valid whatever may be the separation of the magnets, since they may be kept at rest by a block of nonmagnetic material. Any object is held together by internal forces between its constituent atoms. These internal forces must balance, and only net external forces can cause an object to move.

mind that forces are vectors, the third law may be stated in the form:

If an object A exerts a force F on an object B, then B exerts an equal and opposite force (−F) on A.

Newton's third law is often summarized as 'Action and reaction are equal and opposite'. In terms of the fuller statement above, if the force exerted by A on B is called 'action', then the 'reaction' is the force on B exerted by A. It is fine to refer to action and reaction provided it is remembered that they are forces on *different* objects. The forces on an individual object need not balance and, of course, acceleration only occurs when there is a nonzero net force. The application of the third law to an accelerating system is illustrated in the following worked example.

Worked Example 2.6 A railway engine pulls two wagons each of mass 1000 kg (Fig 2.15). The train has an acceleration of 1 m s^{-2}. What are the forces between the engine and the first wagon and between the two wagons?

Answer The forces are shown separately in the lower part of the figure. Writing the mass of each wagon as m, the equation of motion for the second wagon at an acceleration a is $F_2 = ma$. For the first wagon the net force is $F_1 - F_2 = ma$, and $F_1 = F_2 + ma = 2ma$. For $m = 1000$ kg and $a = 1$ m s^{-2}, $F_1 = 2000$ N and $F_2 = 1000$ N. The engine is also accelerating, and if its mass is M, the forward force on it is $Ma + F_1$. This force is exerted on the engine by the rail (the action), which by Newton's third law experiences an equal and opposite force from the engine (the reaction).

Interplanetary vehicles

There are not many circumstances in which objects experience no external force at all. It would be possible if an interplanetary vehicle

● *Action and reaction*

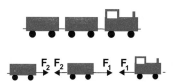

Fig. 2.15 Newton's third law is needed to work out the forces on the couplings on the accelerating train.

happened to be at a position where the gravitational attractions of the Sun and the planets cancelled out. Imagine that a vehicle is in such a position and that it is slightly off course. The mission controller decides to fire one of the small rocket motors on board the vehicle to get it back on to the correct course. We can use Newton's third law to find the change of velocity caused by firing the rocket. During the rocket burn, hot gases are expelled from the motor at a speed v_e, say, with respect to the vehicle. Suppose that the rocket is fired for a short time δt, and in that time the mass of expelled gas is δm. Then the change in momentum of the gas is

$$\delta p = \delta m \times v_e.$$

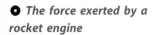
● *The force exerted by a rocket engine*

Now by Newton's second law the force required to cause this change in momentum is

$$F = \frac{\delta p}{\delta t} = v_e \frac{\delta m}{\delta t}. \tag{2.16}$$

This force is in the direction in which the hot gas is expelled. Since action and reaction are equal and opposite, there must be an equal and opposite force on the vehicle itself. This force also acts for a time δt, and causes a change of speed δV in the vehicle's speed in the direction opposite to the motion of the hot gas, as shown in Fig 2.16. If the mass of the vehicle M is very much bigger than δm, we can neglect the change in M, and the change in momentum is $M\delta V$, so that

$$F = v_e \frac{\delta m}{\delta t} = M \frac{\delta V}{\delta t}$$

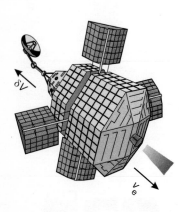

or $\quad \delta V = \dfrac{\delta m \times v_e}{M}. \tag{2.17}$

The mission controller knows the rate at which fuel is burnt and the speed v_e, and by using the above equation he is able to decide how long to fire the rocket to put the vehicle back on its proper course.

Fig. 2.16 A rocket motor has to exert a force on the exhaust gases to expel them from the engine nozzle, but the gases exert an equal and opposite force on the rocket.

The rocket equation

Usually rockets are not in positions where there is no gravitational force. The effect of gravity must be included in working out a rocket's motion. For simplicity let us assume that the rocket is being fired vertically upwards from the Earth's surface, as is the case for NASA's launchings of the Space Shuttle. The rocket initially travels in a straight line, and only its vertical height z above the ground changes.

Before it is launched, a large fraction of the weight of the rocket is fuel. While the rocket motors are being fired, the change in mass of the rocket cannot be ignored, and we must think about both the mass and the speed as functions of time. At a time t after launch, write the mass of the rocket and its remaining fuel as M and its upward speed relative to the Earth as V: the upward momentum is then MV. Suppose that a mass δm of fuel has just been burnt in the short time δt between $t - \delta t$ and t, and that the exhaust gases are expelled downwards at a speed v_e relative to the rocket, as shown in Fig 2.17. The speed of this mass of gas relative to the Earth changed from V to $(V - v_e)$, so that its momentum change is $-v_e \delta m$. Since the time taken for the change is δt, the force exerted by the rocket motor on the gas is $-v_e \delta m / \delta t$. The minus sign indicates that the force is downwards.

Newton's third law tells us that an equal and opposite force $v_e \delta m / \delta t$ acts on the rocket. During the time δt this force increased the speed of the rocket by δV, from $V - \delta V$ to V, while the mass of the rocket decreased from $M + \delta m$ to M. We have already accounted for the momentum change of the mass δm, and the rate of change of momentum for the remaining mass M is $M \delta V / \delta t$. This equals the net force on the rocket, which is the difference between the upward force generated by the motor and the weight of the rocket:

$$M \frac{\delta V}{\delta t} = v_e \frac{\delta m}{\delta t} - Mg. \qquad (2.18)$$

Notice that the upward force does not depend on V. The motor continues to accelerate the rocket even when it is travelling faster than the speed v_e of the exhaust gases. Consumption of a mass δm of fuel *reduces* the mass M of the rocket by δm, so $\delta M = -\delta m$, and in the limit as $\delta t \to 0$, eqn (2.19) becomes

$$M \frac{dV}{dt} = -v_e \frac{dM}{dt} - Mg. \qquad (2.19)$$

This is the **rocket equation** for this example of a rocket moving in a straight line against gravity. Remember that the first term on the right-hand side is the upward force from the rocket motor and is positive; dM/dt is negative because the rocket is getting lighter as it uses fuel. The rocket equation is only valid if this term is initially bigger than Mg—otherwise the rocket stays on the ground!

The rocket equation is a differential equation in which both the quantities M and V are functions of time. Generally speaking it is not possible to solve such a differential equation, but in this case it turns out that the variables M, V, and t can be separated and each term integrated separately. This is achieved by multiplying eqn (2.19) by dt and dividing

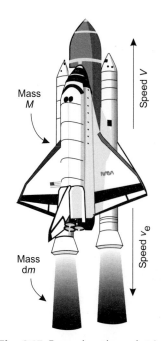

Fig. **2.17** Even when the rocket is moving upwards faster than the speed at which it is expelling the exhaust gas, there is an upward force on the rocket and it continues to accelerate.

⬤ *A rocket can travel faster than the speed of the exhaust gases that drive it* by M, leading to

$$-v_e \frac{dM}{M} - g\,dt = dV. \tag{2.20}$$

Choose $t = 0$ at the moment of lift-off, when $V = 0$. Suppose that the mass of the rocket is initially M_i, falling to M_f when all the fuel is burnt, at a time t_f. At this time the speed has reached V_f. Integrating eqn (2.20),

$$-v_e \int_{M_i}^{M_f} \frac{dM}{M} - g\int_0^{t_f} dt = \int_0^{V_f} dV. \tag{2.21}$$

Performing the integration we find

$$-v_e [\ln M]_{M_i}^{M_f} - g[t]_0^{t_f} = [V]_0^{V_f}$$

$$\text{or } V_f = -v_e \ln M_f + v_e \ln M_i - gt_f = v_e \ln \frac{M_i}{M_f} - gt_f. \tag{2.22}$$

The rocket was initially stationary, and V_f is the boost in speed it has received after burning the fuel. For this boost in speed to be large the speed v_e of the expelled gas must be high and the payload must be made as light as possible so that the ratio M_i/M_f is considerably larger than one. Also t_f must be small, which means that the fuel must be burnt as quickly as possible to give the largest possible vertical thrust. If gt_f is very small, the rocket's upward motion is almost the same as if it were launched from the ground at a speed $v_e \ln(M_i/M_f)$—not much energy is wasted in lifting fuel to a great height.

In this discussion we have assumed that the value of g is constant. This will not be true for a rocket being delivered to an orbit far above the Earth. We shall return to the rocket problem in Chapter 5 in connection with orbits, and find the conditions to launch into any orbit.

Worked Example 2.7 The speed of the exhaust gas from a modern rocket is about 2.5 km s^{-1}. The motor and the body of the rocket are necessarily heavy, and the ratio of the initial mass to the final mass, including the payload, is typically 5. Estimate the speed of such a rocket if all the fuel is burnt in 20 seconds.

Answer From eqn (2.22), the speed V_f of the rocket at the end of the fuel burn is nearly 4.0 km s^{-1}. This is certainly very fast, but not in fact fast enough to put the rocket in orbit. Because of the logarithmic dependence of the mass ratio, increasing the size of the rocket is not a good way to solve the problem—increasing the mass ratio to 20 increases V by less than 90%. Satellite launches are actually carried out by multistage rockets. At the end of the first burn the motor and support structure of the first stage are jettisoned. One or more remaining stages are each capable of providing a boost in speed similar to that of the first stage, allowing launch into orbit or interplanetary space.

The initial conditions before the rocket is fired have been treated by a different method from the one used in the earlier section on free fall (p. 00). In that section **indefinite integrals** were solved, which are integrated to give the most general solution of the differential equation, including constants of integration. The values of the constants of integration were then found by requiring that the solutions satisfy the initial conditions. In contrast, eqn (2.22) has no constant of integration. This is because **definite integrals** were used in eqn (2.21) to solve the differential eqn (2.20). The initial conditions—that $M = M_i$ and $V = 0$ at $t = 0$—are included as the lower limits of the integrals in eqn (2.21). The definite integral is just the difference between the values of the indefinite integral evaluated at the arguments given by the upper and lower limits of the integral. The constant of integration drops out in the subtraction, and the initial conditions are already automatically included.

● *Definite integrals*

2.4 Motion in three dimensions

For simplicity all the examples of the application of Newton's laws have so far been of motion in a straight line. These examples give an insight into the way forces cause changes in speed and position that are easy to visualize and that do not require complicated mathematics. However the real world has three dimensions, and it is important to know how to apply Newton's laws to motion that is not constrained to lie on a straight line.

Newton's laws as they have been stated in this chapter need no modification since they already refer to forces, velocities, and momenta that are vectors, and may point in any direction in three-dimensional space. The usual rules for vector addition apply, and can be used, for example, to work out the net force on an object subjected to forces acting

● *Vector differential equations*

in different directions, as in the example of a bar hanging on a hook in Section 2.1. Let us write Newton's second law (eqn (2.5)) again in its vector form:

$$\mathbf{F} = \frac{d\mathbf{p}}{dt}.$$

What may be new or unfamiliar about the law in this form is that it is a *differential equation* involving vector quantities. If we use Cartesian coordinates, and choose the x-, y-, and z-axes to be in some particular directions, then eqn (2.5) applies to each component separately. For the x-component $F_x = dp_x/dt$, and so on.

● *Vector equations do not depend on a particular set of axes*

It is an advantage of writing the law in terms of vectors that it makes no reference to any particular set of coordinate axes. We are free to choose the x-axis to be in any direction we please, and the y- and z-axes in any directions that ensure that all the axes are mutually perpendicular. The vector equation gives the correct relation between the components whatever the choice of axes. But when we wish to solve the vector differential equation in a particular case, then a definite choice of axes must be made. The components along the x-, y-, and z-axes each obey an equation of the same form as eqn (2.5), and in general there will be three differential equations to solve. Often the amount of mathematics can be reduced by a sensible choice for the directions of the axes. This is illustrated in the following examples of electrons in an oscilloscope and of non-vertical motion under gravity.

Deflection of an electron in an electric field

In an oscilloscope, electrons are fired from an electron gun and accelerated towards the screen by a high voltage V_a. Although, as was mentioned in the introduction to this chapter, quantum mechanics is needed to explain the behaviour of electrons over distances comparable to atomic dimensions, Newton's laws are adequate to describe their motion in the oscilloscope. After acceleration, the electrons are deflected by applying a voltage between a pair of plates as shown in Fig 2.18. Electrical forces are discussed in detail in Part 5. Here we shall simply quote the relevant formulae and use them in the equations of motion for the electrons.

For the moment let us accept that the electrons are accelerated towards the screen by the voltage V_a and that they are deflected when a voltage is applied to the plates. Choose the z-axis to be in the direction from the electron gun to the screen and the x-axis perpendicular to the plates, as in Fig 2.18. Before entering the region between the plates an electron moves along the z-axis at a speed v_z, determined by the accelerating voltage V_a.

The relation between v_z and V_a is:

$$\tfrac{1}{2}m_e v_z^2 = eV_a \tag{2.23}$$

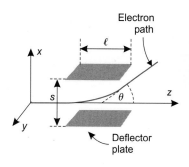

Fig. 2.18 The blue line shows the path of the electron beam as it passes through the deflector plates.

where m_e is the mass of the electron and e the magnitude of the negative charge it carries. (This equation equates the kinetic energy of the electron to the work done by the accelerating voltage: kinetic energy and work are discussed in the next chapter.)

As in Fig 2.18, the separation of the plates is s, their length is ℓ, and the voltage between them is V_p. The direction of deflection depends on which plate has a positive voltage with respect to the other: let us connect the plates so that if V_p is positive the electron is deflected upwards in the diagram, in the x-direction.

The deflecting force on the electron actually begins before the electron reaches the region between the plates, and then builds up to its maximum value inside this region. However, we can get a very good approximation to the electron's path by assuming that the force has a uniform value $F_x = eV_p/s$ between the plates and is zero outside. Now we need to look at the components of Newton's second law. There is no force in the z-direction, i.e. $F_z = 0$, and the z-component of the vector form of Newton's law is

$$0 = \frac{\mathrm{d}p_z}{\mathrm{d}t}$$

or $p_z = m_e v_z = $ constant.

The velocity component v_z in the z-direction thus remains constant as the electron travels through the plates. In the x-direction the force $F_x = eV_p/s$ causes an acceleration

● *The speeds in the x- and z-directions are independent of one another*

$$\frac{F_x}{m_e} = \frac{eV_p}{sm_e} \tag{2.24}$$

for the time ℓ/v_z during which the electron is between the plates. Before entering the plates the x-component of velocity is zero, and it is therefore accelerated to a final value

$$v_x = \frac{eV_p}{sm_e} \times \frac{\ell}{v_z}.$$

There is no motion in the y-direction, and after passing through the plates, the electron has components of velocity v_x and v_z in the x- and

z-directions, so that it is now moving at an angle θ given by

$$\tan\theta = \frac{v_x}{v_z} = \frac{eV_p\ell}{sm_e v_z^2}$$

or, using eqn (2.23),

$$\tan\theta = \frac{V_p\ell}{2V_a s}. \tag{2.25}$$

By varying the deflection voltage V_p the electron may be scanned back and forth on the screen.

A second pair of plates deflects the beam in the y-direction, and so the electron beam spot may be swept over the whole screen simply by varying the voltages across the two pairs of plates.

Exercise 2.5 The dimensions of the deflector plates in an oscilloscope are $\ell = 20\,\text{mm}$, $s = 5\,\text{mm}$, $V_p = 100$ volts, $V_a = 2000$ volts, and the distance from the end of the plates to the screen is 100 mm. An electron is moving along the axis when it enters the deflector plates. How far from the axis does the electron strike the screen?

Answer Assuming a uniform force within the plates and no force outside, the electron leaves the plates at a distance of 1 mm from the axis and moves a further 10 mm from the axis before reaching the screen.

Three-dimensional motion including gravity

If a stone is thrown horizontally off a cliff at the seaside, it can go a long way before it splashes into the sea. What is the trajectory of the stone, and how does the distance it goes depend on how hard it is thrown? There is again an obvious choice of axes. Put the origin at the point where the stone is released, and call the vertical axis z, with increasing z being downwards. The other two axes are horizontal, and we can choose the x-axis to be in the direction in which the stone is thrown. There is then no motion in the y-direction; $y = 0$ at all times. If we neglect air resistance, there is no force acting on the stone in the x-direction after it has been thrown, and its velocity component v_x remains constant at its initial value, v_0, say. In a time t it will travel a horizontal distance

$$x = v_0 t.$$

The gravitational force mg acts along the z-axis, and the equation of motion that applies to the z-components is

$$F_z = mg = m\frac{\mathrm{d}v_z}{\mathrm{d}t}.$$

This is the same as the equation for a body falling from rest, and the relation between time and the distance fallen has already been given in eqn (2.10):

$$z = \tfrac{1}{2}gt^2.$$

Now we have equations for the distance travelled in both the x- and z-directions as a function of time, and we can combine them to find the relation between x and z:

$$z = \tfrac{1}{2}gt^2 = \tfrac{1}{2}g\left(\frac{x}{v_0}\right)^2. \tag{2.26}$$

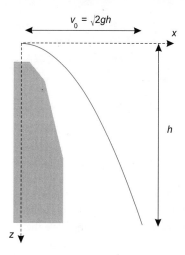

The trajectory of the stone is a parabola, as shown in Fig 2.19.

If the height of the cliff is h, the time taken for the stone to reach sea-level is $\sqrt{2h/g}$ from equation (2.10). This time does not depend on how fast the stone is thrown, and the horizontal distance it travels before falling into the sea is simply $v_0\sqrt{2h/g}$, proportional to the speed v_0 with which it is thrown.

When a ball is being thrown from one person to another on flat ground, a practised thrower knows instinctively the right angle at which to throw. He has found out by experience how to throw the longest distance, without having to solve any equations. Again, neglecting air resistance, we can make an estimate of how steeply he should in fact throw. We have already found that, if something is thrown vertically upwards with an initial speed v_0, it takes a time $2v_0/g$ to reach the ground again, since it has a constant downward acceleration g, and returns with the same speed v_0 downwards. If the ball is now thrown at an angle θ to the horizontal, it is in the air for a time $(2v_0 \sin\theta)/g$. Since its horizontal speed is constant in the absence of air resistance and has the value $v_0 \cos\theta$, in this time the ball travels a horizontal distance

$$2v_0^2 \cos\theta \sin\theta/g = v_0^2 \sin 2\theta/g.$$

The largest possible value of $\sin 2\theta$ is one, when $\theta = 45°$. If air resistance is negligible, $45°$ is therefore the best angle at which to throw.

Fig. 2.19 A stone thrown from a cliff top follows a parabolic path to the sea.

⬤ *How to throw a ball a long way*

Fig. 2.20 This picture, obtained with a scanning tunneling microscope, shows a surface of nickel on which there are some gold atoms. The gold atoms have clumped together to make little triangular islands, so that on this scale the surface is bumpy. The length of the picture is less than 5 nm, and individual atoms are clearly seen.

● *Static and dynamic friction*

Fig. 2.21 Another picture of a nickel surface. The cliff-like structures are ledges one atom high, where an extra layer of nickel has formed.

2.5 **Friction**

When an object is moving on a surface without leaving it, then the component of its velocity perpendicular to the surface is always zero. It follows that, if there are any forces, such as gravitational forces, that have components normal to the surface, these components must always be balanced by a reaction from the surface. The same applies to objects moving along a line: the component of a force in any direction perpendicular to the line must be balanced by a reaction from the line. However, the total reactive force need not necessarily be in the perpendicular direction. There may also be a component along the surface or line. Such non-perpendicular components are called **frictional forces**. Frictional forces always act so as to slow down or prevent motion. They depend sensitively on the materials of the surfaces in contact, and on their state—whether they are rough or smooth, for example. Surfaces are rarely perfectly flat on an atomic scale; the high magnification photographs in Figs 2.20 and 2.21 show how they are covered with bumps and ledges. The actual area of contact between surfaces is usually very small, as indicated in Fig 2.22. If the two surfaces in this figure are dragged sideways across one another, there will be a force opposing the motion as the bumps collide.

The size of the frictional force F_f is related to the size of the normal reaction N—it is much easier to push an empty box across the floor than the same box filled with something heavy like a pile of books. It is often a useful approximation to assume that F_f and N are proportional to one another. The constant of proportionality is called the **coefficient of friction**, usually denoted by μ. The value of μ is not well-defined, and it is worthwhile to distinguish between **static** friction and **dynamic** friction. Once an object has started to move it generally experiences a smaller frictional force than when it is at rest. The **static coefficient of friction** μ_s is therefore larger than the **dynamic coefficient of friction** μ_d.

Suppose a box with a mass $M = 50\,\text{kg}$ is standing on the floor, and the static coefficient of friction between box and floor is 0.2. The gravitational force on the box is Mg, and this is balanced by an equal and opposite vertical reaction N from the floor (Fig 2.23). The horizontal force needed to move the box is then $\mu_s N = \mu_s Mg$, in this case about 100 newtons. Once the box is sliding on the floor, a smaller force is sufficient to keep it moving.

Worked Example 2.8 A block is placed on a slope as shown in Fig 2.24. The static and dynamic coefficients of friction between the block and the plane are $\mu_s = 0.25$ and $\mu_d = 0.2$, respectively. What is the steepest angle θ for which the block will remain stationary? If θ is just greater than the critical value, what is the acceleration of the block down the slope once it has started sliding?

Answer The normal reaction N ensures that there is no motion of the block perpendicularly to the slope, i.e. $N = mg \cos \theta$. For the block to start moving, the force $mg \sin \theta$ down the slope must be bigger than the frictional force $\mu_s N = \mu_s mg \cos \theta$, that is to say, the critical angle for sliding to start is $\tan^{-1}(\mu_s) = 14°$. Once moving, there is a net force down the slope equal to $mg(\sin \theta - \mu_d \cos \theta) = mg \sin \theta (1 - \mu_d / \tan \theta) = mg \sin \theta (1 - \mu_d / \mu_s) \approx 0.05\, mg$. The block therefore slides down the slope with a constant acceleration $0.05g$.

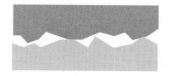

Fig. 2.22 Even surfaces that seem very smooth to the touch are covered with microscopic bumps and ledges, and the area of contact between two surfaces is always small. Neighbouring parts of the surface will snag if the surfaces are slid across one another.

It must be emphasized that the treatment of friction given above is approximate. The velocity dependence of friction is usually more complicated than the simple division between static and dynamic friction would suggest. It is, in fact, not possible to formulate exact laws to describe the effects of friction between solid surfaces. Since the frictional forces depend on the precise shape of surfaces on an atomic scale, they will vary from one sample to another even if all the materials are produced in the same way.

Friction also occurs when objects move through fluids as well as when solid surfaces are in contact. We have already met this in considering the air resistance to a falling body. There we assumed that the air resistance is proportional to speed, and were then able to solve the equation of motion. Again the solution is not exact—air resistance is only proportional to speed for low speeds. For fast-moving objects higher powers of speed are important and there are also large changes in air resistance when turbulence occurs.

It does not follow that, because the treatment of friction is approximate, it is of no use. As we pointed out when discussing action and reaction, exact answers are not available for almost all of practical science and engineering. Real systems are usually too complicated for us to know precisely all the forces which are acting. Friction is a good example where this is so. Nevertheless, it is valuable to solve equations of motion using approximate frictional forces with values based on theory, observation, and experiment. The solutions give insight into the behaviour of mechanical systems and provide a much better basis for design than purely empirical rules.

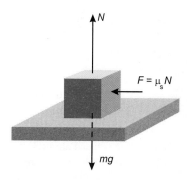

Fig. 2.23 The normal reaction of the floor on the box is N. The force F needed to move the box gets bigger as N gets bigger, and F is roughly proportional to N.

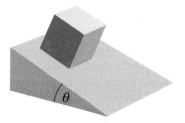

Fig. 2.24

Problems

Level 1

2.1 A sign of mass 4 kg is fixed to a wall by a bracket as shown in Fig 2.25. Calculate the tension or compression in the two arms of the bracket.

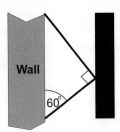

Fig. 2.25

2.2 The cable above and below a cable-car of mass M is at angles θ_1 and θ_2 to the horizontal. What are the tensions in the upper and lower parts of the cable?

2.3 A telephone call from London to New York is made via a satellite situated about 37 000 km above the Earth's surface. Estimate the minimum time the Londoner must wait for a reply after finishing speaking. (The speed of light is close to 3.0×10^8 m s^{-1}).

2.4 A boat sailing up and down a river takes m minutes to go one kilometre upstream and n minutes to go one kilometre downstream. Find the speed of the current and of the boat relative to it.

2.5 A cyclist pedals up a hill for 1 km and coasts down the other side at 40 kph for another 1 km. The journey up and down takes $6\frac{1}{2}$ minutes. What was the cyclist's average speed when going uphill?

2.6 A boat that moves at 12 km per hour in still water is sailing across an estuary. A current flows at 9 km per hour relative to land. How long does it take the boat to move 1 km relative to land in a direction perpendicular to the current?

2.7 Neglecting air resistance, calculate the time taken for an object to reach the ground from the top of the Eiffel tower, which is 300 m high.

2.8 A cylindrical rod, starting from rest with its axis horizontal, rolls down an inclined plank. It accelerates uniformly and after 2 s it has moved a distance of 1 m down the plank. How fast is it going at this point?

2.9 Electrons in a cathode ray tube, which are initially at rest, are accelerated uniformly through a voltage difference of 1000 V over a distance of 1 cm. The electrons then coast a further distance of 20 cm to a screen. What is the total time for an electron to travel through the tube? (The mass m_e of an electron is 9.1×10^{-31} kg and its charge e has a magnitude 1.6×10^{-19} coulombs—the coulomb is the SI unit of charge).

2.10 The mass of a Space Shuttle at lift-off is about 2000 tonnes (i.e. 2×10^6 kg). The maximum thrust of its two booster rockets and the three main engines exceeds 3×10^7 N. The booster rockets have consumed their propellant and are discarded about two minutes after lift-off. The actual launch profile of the Shuttle is complicated: to get some idea of the magnitudes involved, calculate the height and speed of the Shuttle after 2 minutes, assuming a constant thrust of 3×10^7 N and neglecting the mass of the propellant used up to this time.

2.11 Two trains set off 5 minutes apart. If each accelerates uniformly to its maximum speed of 150 km hr^{-1} in 5 minutes, find how far the first train has gone when the other starts, and how far apart they are when both are at full speed.

2.12 A woman who weighs 50 kg is standing in a lift. What force does the floor of the lift exert on her (a) if the lift is accelerating upwards at 1 m s^{-2}; (b) if the lift is accelerating downwards at 1 m s^{-2}?

2.13 Two equal masses are connected by a light string passing over a light pulley as shown in Fig 2.26. Calculate their acceleration, neglecting friction for the mass on the wedge.

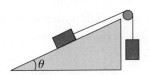

Fig. 2.26

2.14 A planetary probe is on its way from Earth to Venus. During the flight it needs a mid-course adjustment of its path. At the time of the adjustment the probe is travelling at a speed of 30 000 m s^{-1} relative to the Sun, and the direction it is moving (also relative to the Sun) needs to be changed by 0.1°. The probe has a mass of 2000 kg, and the exhaust velocity from its correction rocket is 50 000 m s^{-1}. What mass of fuel must be expended to put it on course?

2.15 A climber on an overhanging cliff falls 10 m and is held by a rope fixed near the point of fall. Estimate the force on the rope, assuming that the climber weighs 50 kg and that a constant force acts over a period of 0.5 s. (A better assumption, that the force exerted by the rope increases as it is stretched, is used in Problem 3.10.)

2.16 A man finds that the greatest distance he can throw a cricket ball is 100 m. Estimate the maximum height of the trajectory in his 100 m throw (again neglect air resistance).

Level 2

2.17 Two jet planes are flying at the same altitude. One is 10 km due south of the other and is flying in a northeasterly direction at 800 kph. The second plane is heading due east at 600 kph. In which direction is the relative motion of the planes? If they continue on their courses, what time elapses before they reach their distance of closest approach? Will the pilots have to record a 'near miss', i.e a closest approach of less than 1 km?

2.18 A hose pipe with a diameter of 13 mm is directing water horizontally at a vertical wall at a rate of 1 litre per second. Estimate the force exerted on the wall. (The density of water is 1000 kg m^{-3} ≡ 1 g cm^{-3}.)

2.19 The fastest sports car you can buy is capable of reaching a speed of 100 kph in 4.0 s, starting from rest. Estimate how far it has travelled in that time, assuming constant acceleration. Draw the graph of speed against time, and verify that the area under the graph is equal to the distance.

In practice the acceleration decreases as the car approaches 100 kph. Sketch a more realistic speed−time curve. Did your calculation overestimate or underestimate the distance travelled?

2.20 A truck moving at a steady speed of 10 m s^{-1} passes a stationary car. When the truck is 100 m ahead

the car starts moving with a uniform acceleration 1.5 m s^{-2} in the same direction as the truck. How far has the truck moved when it is overtaken by the car, and what is the speed of the car when it overtakes?

2.21 A water-skier is pulled over a ramp at an angle of 15° to the horizontal. The top of the ramp is 1 m above the water. The skier's rope is slack and he lands 20 m from the ramp. How fast was the skier moving when he reached the ramp?

2.22 A sphere is dropped down the centre of an evacuated tube. Three horizontal laser beams cross the centre of the tube and are placed so that each pair is separated by exactly 50 cm. The times at which the sphere interrupts the laser beams are recorded by photodetectors. In one measurement the time differences for the upper and lower pairs are 175.9 ms and 116.9 ms respectively. (1 ms ≡ 1 millisecond, i.e. 10^{-3} s.) Use these data to calculate the acceleration of gravity.

2.23 An empty rail truck of mass M starts from rest and is pulled by a constant force F. At the same time that the truck begins to move, coal from a hopper, which is stationary with respect to the rail track, starts to fall vertically into the truck. The mass of coal in the truck increases at a constant rate $b = \mathrm{d}m/\mathrm{d}t$. If the speed of the truck is v, show that

$$F = bv + (M + bt)\frac{\mathrm{d}v}{\mathrm{d}t},$$

and hence, by integrating this equation with respect to time, that

$$v(t) = \frac{Ft}{M + bt}.$$

2.24 A rocket in a firework display is fired vertically upwards. The rocket burns its fuel in 1 s and then rises a further 80 m. The fuel makes up three-quarters of the weight of the rocket. Use these data to estimate the speed of ejection of the exhaust gas from the rocket.

2.25 In the oscilloscope described in Exercise 2.5, not all the electrons arrive at the deflecting plates moving exactly along the z-axis. Assuming the same dimensions and voltages as in Exercise 2.5, calculate the trajectory of an electron arriving at the plates in the xz-plane, at $z = 0$ and at an angle of 1° to the z-axis. When this electron reaches the screen, how far away is it from an electron arriving at the plates on axis? (This problem illustrates the need for focusing even for narrow electron beams.)

2.26 A block slides down a slope as in Fig 2.27. The angle θ of the slope to the horizontal is 20°. Starting from rest, the block slides 2 m in 2 s. Estimate the dynamic coefficient of friction between the block and the slope.

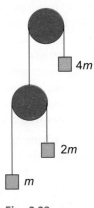

Fig. 2.28

Fig. 2.27

Level 3

2.27 A ferry sails between two jetties directly opposite one another on the banks of a river of width d. The water is stationary at the banks and the water speed increases linearly with distance from the nearest bank to a maximum value u in the middle of the river. The flow of water is everywhere perpendicular to the line joining the jetties. In still water the ferry sails at v m s^{-1}. Derive an expression for the minimum time needed for the ferry to cross from one jetty to the other. You will need the standard integral

$$\int \frac{dx}{\sqrt{a^2 - x^2}} = \sin^{-1} \frac{x}{a}.$$

2.28 Strings pass over two light pulleys as shown in Fig 2.28. The pulleys are frictionless so that the tensions are the same throughout the length of each string. Calculate the accelerations of the masses.

2.29 Due to air resistance a rifle bullet has a deceleration bv^3, where v is its speed and b is a constant. Show that v, the time t, and the distance z travelled by the bullet are related by the equations

$$t = \tfrac{1}{2}bz^2 + z/u \quad \text{and} \quad v = u/(1 + bzu)$$

where u is the initial speed of the bullet. You will need to use the standard integral

$$\int \frac{dx}{\sqrt{a + bx}} = \frac{2\sqrt{a + bx}}{b}.$$

For a rifle with muzzle velocity 1000 m s^{-1}, the speed of the bullet has reduced to 980 m s^{-1} after it has travelled 100 m. Calculate the time needed to traverse 1000 m.

2.30 A box standing on a smooth floor is pushed by a force just strong enough to start it moving. The force is maintained at a constant value for 2 seconds after movement begins, and is then removed. If the coefficients of static and dynamic friction are 0.2 and 0.15, respectively, calculate how far the box moves when it is accelerating and when it is slowing down.

2.31 A chain is placed on a table so that one-fifth of the chain is hanging vertically over the edge and the other four-fifths is lying horizontally in a straight line perpendicular to the table's edge. The chain is released and starts to slide off the table. Calculate the speed at which it is moving at the moment when all the chain has just left the table and it starts to fall freely. Assume that the dynamic coefficient of friction is 0.1.

2.32 The string holding the bar shown in Fig 2.2 slides over the hook so that the bar hangs at an angle. The bar will remain stationary at angles up to 30° to the horizontal, but slips if it is placed more steeply. Show that the string attached to the upper end of the bar is at 56.1° to the horizontal when the bar is at 30°. What are the tensions in the string attached to the upper and lower ends of the bar when it is on the point of slipping? Assuming that the static coefficient of friction μ_s between the string and the pulley is constant, what is the value of μ_s?

Some solutions and answers to Chapter 2 problems

2.1 The tension in the upper arm is 19.6 N and the compressing force on the lower arm is 34.0 N.

2.2 Call the tensions in the upper and lower parts of the cable T_1 and T_2 respectively. Provided that the cable-car is not accelerating, the net horizontal force on it is zero: $T_1 \cos\theta_1 = T_2 \cos\theta_2$. The vertical component of the tension in the upper part balances the vertical component of the lower part and sustains the weight of the car: $T_1 \sin\theta_1 = Mg + T_2 \sin\theta_2$. Hence

$$T_1\left(\sin\theta_1 - \frac{\cos\theta_1 \sin\theta_2}{\cos\theta_2}\right) = Mg, \text{ leading to}$$

$$T_1 = \frac{Mg \cos\theta_2}{\sin(\theta_1 - \theta_2)} \quad \text{and} \quad T_2 = \frac{Mg \cos\theta_1}{\sin(\theta_1 - \theta_2)}.$$

2.4 Call the speed of the current u kph, and the speed of the boat relative to the water v kph. The number of minutes taken to go 1 km upstream is then

$$m = \frac{60}{v - u} \text{ and the number to go 1 km downstream}$$
is $n = \dfrac{60}{v + u}.$

Solving for u and v, $u = 30\left(\dfrac{1}{n} - \dfrac{1}{m}\right);$

$$v = 30\left(\frac{1}{n} + \frac{1}{m}\right).$$

2.6 About $7\frac{1}{2}$ minutes.

2.7 7.8 s.

2.9 According to equation (2.24), the acceleration experienced by an electron caused by a voltage V applied across a distance s is $a = eV/sm_e$. For an electron starting from rest, the time taken to move a distance s with a uniform acceleration a is $\frac{1}{2}at^2$, leading to

$$t = \sqrt{\frac{2s}{a}} = \sqrt{\frac{2s^2 m_e}{eV}}.$$

The speed at t is $v = at = \sqrt{\dfrac{2eV}{m_e}}.$

In the problem $s = 10^{-2}$ m and $V = 1000$ volts. Hence

$$t = \sqrt{\frac{2 \times 10^{-4} \times 9.1 \times 10^{-31}}{1.6 \times 10^{-19} \times 1000}} = 1.07 \times 10^{-9} \text{ s.}$$

The time taken to coast a further distance ℓ is

$$\frac{\ell}{v} = 0.2\sqrt{\frac{9.1 \times 10^{-31}}{2 \times 1.6 \times 10^{-19} \times 1000}} = 1.08 \times 10^{-8} \text{ s,}$$

and the total time is 1.18×10^{-8} s.

2.10 36 km.

2.14 2.1 kg.

2.15 From eqn (2.13) the time taken for the climber to fall a distance z is $t = \sqrt{2z/g}$. The speed after this time is $v = gt = \sqrt{2gz}$. For a 10 m fall, v is $\sqrt{2 \times 9.8 \times 10} \sim 14 \text{ m s}^{-1}$. The momentum is mv, where m is the mass of the climber. If the climber is brought to rest by a steady force F acting for a time Δt, the force is the rate of change of momentum $= mv/\Delta t = 50 \times 14/0.5 = 1400$ N.

2.16 Suppose that the man can throw the ball at a speed v_0. If air resistance can be neglected, the longest throw is when the horizontal and vertical components of the initial speed are both $v_0/\sqrt{2}$, i.e. the angle is 45° to the horizontal. When the ball returns to the ground, the downward speed is the same as its initial upward speed. At a deceleration g, the time taken for the upward speed to change from $v_0/\sqrt{2}$ to zero is $t = v_0/\sqrt{2}g$, and the height it reaches is $\frac{1}{2}gt^2 = v_0^2/4g$. The ball has a constant horizontal speed $v_0/\sqrt{2}$ and in a time $2t$ it travels a horizontal distance $2v_0/\sqrt{2} \times v_0/\sqrt{2}g = v_0^2/g$, four times the maximum height reached. For the 100 m throw, the maximum height of the ball is therefore 25 m.

2.18 The area of the hosepipe is $\pi/4 \times 0.013^2 = 1.33 \times 10^{-4}$ m². For a flow of one litre ($\equiv 10^{-3}$ m³) per second through this area the speed of the water is $10^{-3}/(1.33 \times 10^{-4}) = 7.5 \text{ m s}^{-1}$. The mass of water leaving the hose is 1 kg s⁻¹, and the momentum delivered per second is 7.5 m kg s⁻². The force exerted when it is stopped by the wall equals the rate of change of momentum, and is 7.5 newtons.

2.20 The truck has moved another 200 m when it is overtaken by the car, which is then moving at 30 m s^{-1}.

2.21 If the skier's speed on take-off is v, the time taken to move 20 m horizontally is $t = 20/v \cos 15°$, neglecting air resistance. The upward speed at take-off is $v \sin 15°$, and if the time taken to land one metre below the ramp is t, then after a time t the vertical distance from the top of the ramp is (eqn (2.13)) $z = v \sin(15°) \times t - \frac{1}{2}gt^2 = -1$. This leads to $v = 15.3$ m s^{-1}, or 55 kph.

2.22 The sphere is moving at a speed u when it reaches the first laser beam at $t = 0$, say. It then falls a distance $s_1 = 0.5$ m to the second beam and arrives at $t_1 = 0.1759$ s. The distance from the first to the third beam is $s_2 = 1$ m and the time of arrival at the third beam is $t_2 = 0.1759 + 0.1169 = 0.2928$ s. The distances are given in terms of speeds and times by

$$s_1 = ut_1 + \tfrac{1}{2}gt^2 \quad \text{and} \quad s_2 = ut_2 + \tfrac{1}{2}gt_2^2.$$

Multiplying the first equation by t_2 and the second by t_1:

$$s_1 t_2 = ut_1 t_2 + \tfrac{1}{2}gt_1^2 t_2; \quad s_2 t_1 = ut_1 t_2 + \tfrac{1}{2}gt_2^2 t_1.$$

Subtracting and rearranging,

$$g = \left(\frac{2(s_2 t_1 - s_1 t_2)}{t_1 t_2 (t_2 - t_1)} \right)$$

$$= \frac{2(0.1759 - 0.5 \times 0.2928)}{0.1759 \times 0.2928 \times 0.1169} = 9.80 \text{ m s}^{-2}.$$

2.23 At the time t the mass of the truck is $(M + bt)$ and its momentum $(M + bt)v$. The force F is rate of change of momentum, i.e.

$$F = \frac{d}{dt}(M + bt)v = M\frac{dv}{dt} + bv + bt\frac{dv}{dt}$$

or $F \, dt = M \, dv + bv \, dt + bt \, dv = M \, dv + d(tv)$.

Integrating from $t = 0$ to t, $Ft = Mv + btv$,

and $v = \dfrac{Ft}{M + bt}$.

The result was obtained in this way to emphasize that when the mass is changing the force contains a term depending on the rate of change of mass—here b. The final result can be obtained more simply by noting that, for a constant force F acting for a time t, the change in momentum is Ft. The mass at t is $(M + bt)$ and so the momentum is $(M + bt)v$, leading directly to $Ft = (M + bt)v$, the same expression as before.

2.25 2.1 mm.

2.26 0.255.

2.27 At a distance x from the river bank the water speed is $2xu/d$. At this point, if the ferry is steered directly towards the opposite jetty, its speed towards the jetty is

$$\frac{dx}{dt} = \sqrt{v^2 - \frac{4x^2 u^2}{d^2}}.$$

If the time taken to cross the river is T, the time to reach the half-way point is

$$\int_0^{T/2} dt = \frac{T}{2} = \int_0^{d/2} \frac{dx}{\sqrt{v^2 - 4x^2 u^2/d^2}}$$

$$= \frac{d}{2u} \sin^{-1}\left(\frac{u}{v}\right)$$

from the standard integral. The minimum time for crossing is T: if $u \ll v$, $\sin^{-1}(u/v) \sim u/v$ and T reduces to d/v.

2.31 2.63 m s^{-1}.

2.32 The tension in the string attached to the upper end of the bar is 0.433 mg and the tension in the string attached to the lower end is 0.684 mg.

Chapter 3

Momentum and energy

In this chapter, the energy associated with motion is introduced, and the role of momentum and energy in describing the behaviour of moving systems is discussed.

When material objects are split up into parts, their total amount is unaltered. For example, if a bag of apples is shared among several people, the total number of apples is the same, however many are given to each individual. Some more abstract quantities, such as the momentum of a moving body, also have the property that they can be split into parts without affecting the total. Particularly when speaking of an abstract quantity, we then say that the quantity is *conserved* or that it obeys a **conservation law**.

This language, referring to laws, sounds formal and difficult, but the idea is really perfectly simple. Money is a good example of a conserved quantity, which can be used to illustrate conservation laws in general. If I give you some money, you have more money and I have less, but the total amount of money we hold between us as notes and coins is unchanged. You may choose to use a credit card to buy some goods. You take away the goods, but nothing else changes hands. Nevertheless, you have less money and the storekeeper has more, so that again the total amount of money is unchanged. The same applies no matter how many people are involved in buying and selling—making money transfers. The total amount of money in the whole group of people is constant. The group of people can carry on their monetary transactions among themselves indefinitely. They form a closed system in which the total amount of money is conserved. Money obeys a conservation law. This state of affairs persists until someone buys or sells outside the group. In this case the system is no longer closed, and the outsiders may change the amount of money within the system.

The same idea applies to momentum. In the next section we shall show that, when changes in momentum occur in parts of an isolated system, the momentum of the whole system is unaltered. This fact is expressed by saying that momentum is a conserved quantity. Energy is also a conserved quantity. We shall find that, by making use of the conservation laws for momentum and energy, many dynamical problems can be solved easily, without the need for setting up and solving differential equations.

3.1 Conservation of momentum

Correcting the orbit of an interplanetary vehicle was discussed in Chapter 2. This example is a good one to use in introducing momentum conservation. The correction is applied by firing a rocket motor for a short time δt. While the motor is on, it exerts a force of magnitude F on the expelled rocket gas, and the gas acquires a momentum δp. Newton's third law tells us that the force on the vehicle has the same magnitude as the force on the expelled gas, and we used eqn (2.16) to calculate the change in momentum of the vehicle. Equation (2.16) may be rewritten as

$$\delta p = F\delta t. \tag{3.1}$$

The quantity $F\delta t$ has a special name in mechanics. It is called **impulse**. Equation (3.1) can be expressed in words by saying that the change in momentum of the vehicle is equal to the impulse given to it by the motor.

The name impulse suggests something that has happened quickly. Here we are assuming that the motor is fired for a short enough time that we do not have to worry about how far the vehicle moves during the firing. Its path is just turned into a slightly different direction. In working out the impulse we assumed that the rocket motor exerts a constant force while it is firing. However, the impulsive force may not be constant during the short time over which it is applied—the force may build up from zero after the motor is turned on. The force is then a function of time and may be written as $\mathbf{F}(t)$. The impulse is now defined as $\int \mathbf{F}(t)\,dt$, where the integral is evaluated over the short time during which the force is applied. If no other forces are acting, the impulse equals the change in momentum.

⊙ *An impulse is the result of a force acting for a very short time*

Because the forces on the interplanetary vehicle and on the expelled gas are equal and opposite, the impulses acting on them, and hence the momentum changes they experience, are also equal and opposite. The total momentum change of the system as a whole (vehicle plus expelled gas) is zero, since it is the vector sum of the equal and opposite contributions of each part separately. In other words, the momentum of this isolated system is conserved.

It is easy to see that momentum is conserved in any isolated system, no matter how many parts it contains. If there is a force acting between any pair of parts, the argument given above proves that momentum is conserved for this pair. The same applies to all other pairs, so momentum must be conserved for the system as a whole, provided no net external force is acting on it.

A mathematical proof of the conservation of momentum, based on Newton's laws, goes as follows. Suppose a closed system is made up of a number of particles. Label the particles with a suffix i. The ith particle has a mass m_i and at a particular moment it is moving with a velocity $\mathbf{v}_i$. Here we must use a vector symbol for the velocity, since each particle may be moving in any direction. The momentum of the particle labelled i is $m_i\mathbf{v}_i$ and the momentum $\mathbf{p}$ of the whole system is

⊙ *The total momentum of a system of particles*

$$\mathbf{p} = \sum_i m_i\mathbf{v}_i. \tag{3.2}$$

Here the subscript i on its own, without specifying the largest and smallest value of i, indicates that the summation is over all the particles in the system.

There may be internal forces acting within the system. Call the net force on the ith particle $\mathbf{F}_i$. If there is no net external force acting on the system, the sum of the forces on all the particles must be zero:

$$\sum_i \mathbf{F}_i = 0.$$

Newton's second law tells us that the rate of change of momentum of the ith particle is

$$\frac{d}{dt}(m_i\mathbf{v}_i) = \mathbf{F}_i,$$

and the rate of change of momentum of the whole system is

$$\frac{d\mathbf{p}}{dt} = \frac{d}{dt}\left(\sum_i m_i\mathbf{v}_i\right) = \sum_i \frac{d}{dt}(m_i\mathbf{v}_i) = \sum_i \mathbf{F}_i = 0. \tag{3.3}$$

Expressing this equation in words, the **law of conservation of momentum** states that

The momentum of any system remains constant unless it is acted on by a net external force.

Worked Example 3.1 A shunting engine gives a push to an empty railway wagon, which rolls by itself into a siding and couples on to two similar stationary wagons, as shown in Fig 3.1. If the first wagon is going at 30 m s^{-1} when it strikes the others, how fast do the three wagons start moving after they are coupled together?

Answer It is not necessary to know anything about the forces on the buffers or couplings to work this out—conservation of momentum is all that is required.

If the mass of each wagon is m and the initial speed of the first one is v, then the momentum before impact is mv. After coupling, the mass of the wagons that are moving together is $3m$, and the momentum is still mv. Their speed must be $v/3$, i.e. 10 m s^{-1}.

Exercise 3.1 If the wagons do not couple together, but the single one and the pair move independently, more information is needed to work out the speeds after impact. For example, what is the speed of the two wagons if they remain coupled together and the single wagon is observed to be at rest immediately after the impact?

Answer $15\,\text{m s}^{-1}$.

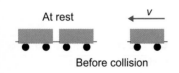

At rest $\xleftarrow{v}$

Before collision

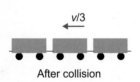

$\xleftarrow{v/3}$

After collision

Fig. 3.1 The moving wagon collides with two stationary wagons. After the collision the three wagons move together. Their speed is determined by conservation of momentum.

More general collisions, in which the colliding objects do not move on the same straight line, are discussed in Section 3.4, after we have dealt with conservation of energy.

3.2 Transformations between frames of reference

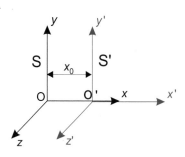

The position vector **r** and the velocity **v** of a moving particle are defined with respect to a coordinate system. This coordinate system is called the **frame of reference** for the moving particle. The actual values of the coordinates depend on the choice of frame of reference. For example, consider a frame which we shall call S′ with origin O′ lying on the x-axis of a frame called S with origin at O, as in Fig 3.2 The distance between the origins O′ and O is x_0. Clearly the difference between the coordinates $x′$ in S′ and x in S is x_0 for *any* point. The y- and z-axes are in the same direction as the y′- and z′-axes, so that the other coordinates are the same in S and S′, and we can write

$$x′ = x - x_0; \quad y′ = y; \quad z′ = z. \tag{3.4}$$

Fig. 3.2 The frames of reference S and S′ have different origins.

These equations relating the coordinates in the two frames are referred to as the **transformation** of coordinates between S and S′.

The origin O′ of a frame of reference S′ need not lie on an axis of S. A fixed origin O′ may be at any position with respect to the origin of S and the axes of S and S′ may point in different directions. Each different pair of frames S and S′ has its own transformation linking the coordinates $x′$, $y′$, and $z′$ to x, y, and z. The actual position described by a set of coordinates does not depend on the choice of the frame of reference. Different frames label positions equally well and the choice between them is just a matter of convenience: often the mathematics can be expressed more simply with a particular choice of coordinate system.

● *Transformation of coordinates*

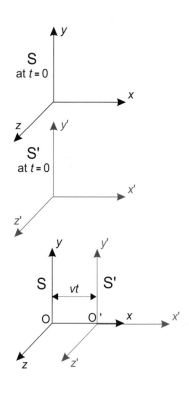

The Galilean transformation

So far we have considered frames of reference that are both fixed. Two different frames of reference may also be moving with respect to one another. The transformation between them then depends on time as well as on the position coordinates. In this section we are concerned with frames of reference that are moving at a uniform speed with respect to one another. Firstly, we shall obtain the transformation of coordinates that applies in these circumstances and then investigate the transformation of velocity and momentum.

To be definite we shall again place the origin of S′ on the x-axis of S. The origin O′ is no longer fixed in the frame S—let it move at a constant speed V along the x-axis. The position of O′ depends on the time t, and again for simplicity let S and S′ coincide at $t = 0$. Figure 3.3 shows a small ball, which is stationary in S′, placed at the origin in both S and S′ at

Fig. 3.3 The frame S′ is moving at a steady speed with respect to S. For clarity S and S′ have been separated in the diagram for $t = 0$, when they coincide with one another.

$t = 0$. At the time t, the ball is still at the origin of S′, but O′ has moved a distance Vt with respect to O. The two sets of coordinates at time t are shown together in the lower part of Fig 3.3. This is exactly the same as Fig 3.2 except that x_0 has been replaced by Vt. The transformation of coordinates between the frames S and S′ for any point is therefore given by putting Vt in place of x_0 in eqn (3.4):

● *Transformation of coordinates for a moving frame of reference*

$$x' = x - Vt; \quad y' = y; \quad z' = z. \tag{3.5}$$

This transformation may be written in vector notation as

$$\mathbf{r}' = \mathbf{r} - \mathbf{V}t \tag{3.6}$$

where $\mathbf{V}$ is a vector with magnitude V pointing along the x-axis. The x'-, y'-, and z'-axes are in the same directions as the x-, y-, and z-axes, and the components of eqn (3.6) along these axes are simply the three separate eqns (3.5).

Equation (3.6) is not restricted to relative motion along the x-axis. Consideration of its x-, y-, and z-components in turn show that it applies to a constant velocity $\mathbf{V}$ pointing in any direction, provided that the origins of the frames S and S′ coincide at $t = 0$. This transformation of coordinates between two frames of reference moving at a uniform speed with respect to one another is called the **Galilean transformation**.

Transformation of velocity and momentum

● *Relative velocities*

As a reminder of how velocities change when they are referred to different frames of reference, look again at the example, discussed in Section 2.1, of a man walking across a ship sailing in calm water at a steady velocity $\mathbf{V}$. The man moves with velocity $\mathbf{v}'$ with respect to the ship, perpendicular to the direction in which the ship is sailing. The man's velocity $\mathbf{v}$ with respect to the water is the vector sum of $\mathbf{v}'$ and $\mathbf{V}$ as shown in Fig 3.4:

$$\mathbf{v} = \mathbf{v}' + \mathbf{V}. \tag{3.7}$$

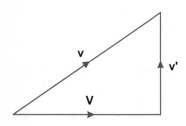

Fig. 3.4 A man on board a ship walks at a velocity $\mathbf{v}'$ across the ship. The ship is moving at a velocity $\mathbf{V}$ through the water. The man's velocity with respect to the water is given by the vector $\mathbf{v}$ in the triangle of velocities: $\mathbf{v}$ represents the vector sum of $\mathbf{v}'$ and $\mathbf{V}$.

Now let us say the same thing in a more abstract way by referring the velocities to the frames of reference S and S′ in Fig 3.5. The frame S is fixed with respect to the water; the x-axis is chosen to be in the direction in which the ship is moving. The other frame, S′, is fixed with respect to the ship—this is the frame of reference that applies to measurements on board ship.

Suppose that at the time $t = 0$ the axes of the two frames coincide with their origins at the position of the man. After a time t the man's position in the two frames is shown in the lower part of Fig 3.5. The man's z-coordinate is unchanged in both S and S′ and the z-axes are omitted from the diagrams. The velocities $\mathbf{v}$ and $\mathbf{v}'$ in S and S′ are simply the rates of change of the man's position in the two coordinate systems. The rates of change are written in vector notation as $\mathbf{v} = d\mathbf{r}/dt$ and $\mathbf{v}' = d\mathbf{r}'/dt$.

By differentiating eqn (3.6) with respect to time, it follows that, in going from the S coordinate system to the S′ system, the velocities are related by the equation

$$\mathbf{v}' = \mathbf{v} - \mathbf{V}. \tag{3.8}$$

This is, of course, the same as the result given in eqn (3.7), which has simply been rearranged.

The vector eqn (3.8) makes no mention of a particular set of axes in either S or S′, and does not require that the x-axis should be in the direction of motion of the ship, nor that the man should walk at right angles to this direction. Figure 3.6 is the velocity diagram for the general case in which $\mathbf{v}$ and $\mathbf{V}$ can be at any angle to one another. The relation between velocities in two frames of reference with a constant relative velocity $\mathbf{V}$ is always given by eqn (3.8).

What happens to momentum when it is referred to a new coordinate system? The momentum of a particle labelled i is $m_i\mathbf{v}_i$, where m_i is its mass and $\mathbf{v}_i$ its velocity. The velocity $\mathbf{v}_i$ of each particle must now be transformed from S to S′ according to eqn (3.8):

$$\mathbf{v}'_i = \mathbf{v}_i - \mathbf{V}.$$

The momentum of the whole system of particles, which eqn (3.2) gives as $\mathbf{p} = \sum_i m_i\mathbf{v}_i$ in coordinate system S, changes to

$$\mathbf{p}' = \sum_i m_i\mathbf{v}'_i = \sum_i m_i(\mathbf{v}_i - \mathbf{V})$$

or

$$\mathbf{p}' = \mathbf{p} - M\mathbf{V} \tag{3.9}$$

where $M = \sum_i m_i$ is the total mass of all the particles in the system.

The change in momentum in switching from coordinate system S to S′ is $M\mathbf{V}$, the same as the momentum of a single body of mass M moving at the relative velocity $\mathbf{V}$ of the two frames of reference. This result, like the Galilean transformation of coordinates in eqn (3.6), applies only to coordinate systems that are moving at a uniform speed with respect to one another.

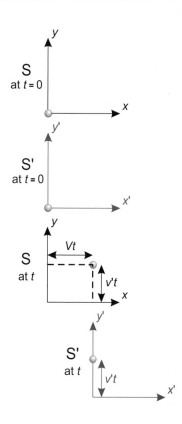

Fig. 3.5 The velocity v′ in Fig 3.4 is now referred to a coordinate system S′ moving along the x-axis at a speed V: S′ is at rest with respect to the ship. The vector equation relating the speeds v and v′ can be regarded as a transformation between the coordinate systems S (at rest with respect to the ship) and S′ (at rest with respect to the water). The man's position is represented in both frames by a blue ball.

⬤ *Transformation of momentum*

Exercise 3.2 A car of mass 1 tonne (1000 kg) is going at $10\,\mathrm{m\,s^{-1}}$ and is approaching a lorry of mass 5 tonnes moving at $10\,\mathrm{m\,s^{-1}}$ in the opposite direction. What is the momentum of the two vehicles (a) in the frame of reference fixed with respect to the road and (b) in the frame in which the car is stationary?

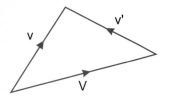

Fig. 3.6 The transformation of velocities between the S and S′ coordinate systems is given by eqn (3.5) even when the velocities **V** and **v**′ are not pointing along the axes.

Answer (a) 4×10^4 N s, (b) 10^5 N s. Unfortunately for the driver of the car, if the two vehicles collide the initial impact is the same as if he were stationary and were hit by a lorry travelling at $20\,\mathrm{m\,s^{-1}}$.

If **V** changes with time, so does **p**′, just as though an external force were acting. This fairly obvious statement leads on to the ideas of General Relativity, which is outside the scope of this book. The relation between different frames of reference that are moving with a constant relative velocity is the province of the Special Theory of Relativity, which is discussed in Chapter 19. Here we should point out that, according to this theory, the Galilean transformation, given by eqns (3.5), is incorrect. However, it is a very good approximation for speeds that are small compared with the speed of light. As we explained at the beginning of Chapter 2, this part of the book is concerned with classical mechanics, i.e. with large-scale objects that always move at speeds much less than the speed of light. Relativistic corrections are then extremely small.

The centre of mass frame

A frame of reference that is particularly useful in discussing the motion of a number of particles is the **centre of mass frame**. The origin of the centre of mass frame is defined so that the positions $\mathbf{r}'_i$ of the particles in the system satisfy the equation

$$\sum_i m_i \mathbf{r}'_i = 0. \tag{3.10}$$

The origin of this frame is itself called the **centre of mass** or the **centre of gravity** of the system of particles. For a symmetrical object made of a uniform material, the centre of mass is the same as the actual centre: each particle m_i with position vector $\mathbf{r}'_i$ is balanced by an opposite one at the same distance from the origin, that is, with position vector $-\mathbf{r}'_i$. Thus the centre of mass of a sphere is at the centre of the sphere, and the centre of mass of a bar with a uniform cross-section is at the centre of the bar.

The mass m_i of each particle is constant, having the same value whether the particle is moving or stationary. The result of differentiating eqn (3.10) is therefore

$$\frac{\mathrm{d}}{\mathrm{d}t} \sum_i m_i \mathbf{r}'_i = \sum_i m_i \frac{\mathrm{d}\mathbf{r}'_i}{\mathrm{d}t} = \sum_i m_i \mathbf{v}'_i = 0,$$

● *The momentum is zero in the centre of mass frame*

so that the momentum $\mathbf{p}' = \sum_i m_i \mathbf{v}_i'$ is automatically zero in the centre of mass frame. The momentum is also zero in any other coordinate system that is not moving with respect to the centre of mass frame, but normally the centre of mass frame itself is the most convenient one to use. Since $\mathbf{p}' = 0$ in the centre of mass frame, the velocity $\mathbf{V}_{cm}$ of the centre of mass frame with respect to the original coordinate system S is given by eqn (3.9) as

$$\mathbf{V}_{cm} = \frac{\mathbf{p}}{M}. \tag{3.11}$$

Since the momentum in the centre of mass frame is always zero, it follows from eqn (3.11) that the momentum $\mathbf{p}$ of a system in any frame is its total mass M times the velocity $\mathbf{V}_{cm}$ of its centre of mass in that frame. This is an important result that we often use without thinking about it. Even in a rigid body the atoms that make it up are moving about, and move faster if the body is heated up. But we can apply Newton's second law to the body as a whole, since the internal motion does not affect its momentum.

The result of any calculation in mechanics must be independent of the choice of coordinate system. Sometimes a calculation can be simplified by using centre of mass coordinates; such a case is discussed in Section 3.5. For the moment we shall illustrate how centre of mass coordinates are applied to the simple problem already solved in Worked example 3.1, in which a moving railway wagon collides with two stationary ones.

With respect to the Earth the moving wagon, of mass m, has momentum mv pointing in the direction in which it is moving; the other two wagons have zero momentum. The system of three wagons has mass $M = 3m$ and momentum $\mathbf{p} = m\mathbf{v}$. Now the velocity $\mathbf{V}_{cm}$ of the centre of mass can be found from eqn (3.11) to be

$$\mathbf{V}_{cm} = \frac{\mathbf{p}}{M} = \frac{\mathbf{v}}{3}.$$

After the wagons are coupled together they are all at rest in the centre of mass frame, i.e. they move at a speed $V = \frac{1}{3}v$ along the rails. Figure 3.7 shows the wagons before and after coupling together in both sets of coordinates. The diagrams make it clear that the speed of the single wagon relative to the other two is the same in both frames. This speed is v before the wagons are coupled together and is zero after coupling.

Worked Example 3.2 A man is standing on a platform supported on an air cushion so that there is negligible friction with the floor. The length of the platform is 4 m and its mass is 20 kg. The man, whose mass is 80 kg, walks from one end of the platform to the other. How far does the platform move?

● *Internal motion does not affect the momentum of the centre of mass*

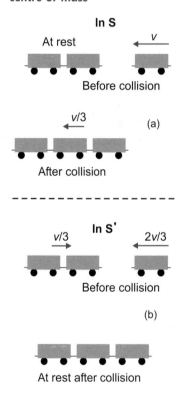

Fig. 3.7 The collision between railway wagons can be described with respect to the ground, as in diagram (a). Alternatively, as in diagram (b), the velocities may be referred to the centre of mass coordinate system, which moves at the speed $v/3$ with respect to the ground. In this system the net momentum is zero before and after the collision.

Fig. 3.8 The man starts walking on the frictionless platform. When he reaches the other end, the centre of mass of the platform plus man has not moved.

Answer Both the man and the platform are at rest before he moves, and their centre of mass is therefore also at rest. No external horizontal forces are acting, and the centre of mass stays in the same place as the man walks from one end to the other. If the platform moves a distance x metres with respect to the floor, as shown in Fig 3.8, the man moves $(4 - x)\,\mathrm{m}$ in the opposite direction. Since the centre of mass does not move, $80(4 - x) - 20x = 0$, i.e. $x = 3.2\,\mathrm{m}$: relative to the ground the man has only moved 80 cm.

3.3 Conservation of energy

Up to now we have been investigating how forces cause motion without worrying very much about how the forces are generated. But forces cannot make something move without expending *energy*. When an object starts to move it uses up energy in one form or another. Anything that falls acquires energy, at the expense of gravitational energy; gravitational energy is discussed later in Section 3.6. A variety of other primary sources of energy is available. There are fuels used for combustion—oil, gas, coal and also wood or waste materials. Nuclear, water, and wind power are used almost exclusively for electricity generation. Solar power may be harnessed nowadays, particularly for some small-scale specialist uses.

Occasionally, the primary source of energy is used directly, as in an old-fashioned waterwheel where the primary source—the moving water in a river—exerts a force on the moving wheel. More usually, the energy from the primary source is applied indirectly. For example, aviation fuel drives the jet engine that propels the hot gases to provide the force on an aeroplane in just the same way as in the rocket motors we have already discussed. In an electric locomotive the primary source of energy is far away in an electricity-generating station; electrical energy is transported on overhead lines and extracted by the locomotive. To a limited extent some of the energy used by the locomotive may be recovered. By using the engine as a brake some electrical energy may be returned to the overhead lines. But at the end of the journey the train is stationary, as are the passengers, and the train company has an electricity bill to pay.

At first sight it seems that after energy has been used to cause motion it just disappears. However, there is one very important form of energy that has not yet been mentioned, namely, heat. When you apply the brakes on a car, they get hot. The energy the car possessed by virtue of its motion has been transformed into heat. Heat as a form of energy is discussed in more detail in Chapter 12. Here all we need to know is that, when heat is taken into account, nothing is lost when different kinds of energy are transformed into one another.

Work

If we are to set up physical laws relating to energy, a *unit of energy* must be defined. The same unit must be able to be applied to all the various different forms energy may take. It was noted above that forces must be expending energy when they cause motion—the force is the agency for transforming another kind of energy into motion. This energy expended by a force is called **work**. No motion occurs in static equilibrium: in order to do work a force has to push an object through some distance in the direction it is acting. This leads to the definition of work in its scientific sense:

Work done = force × distance moved in the direction of the force.

(3.12)

We already have units for force and distance. Force is measured in newtons, and distance in metres. The unit of work is the **joule**, named after James Prescott Joule, the British physicist who demonstrated that, when work is converted into heat, equivalent amounts of work always give the same amount of heat. From eqn (3.12) we see that one joule of work is the amount done by a force of one newton moving through one metre. This is true whatever the source of the force. If a piston in an engine and the core of a solenoid exert the same force, as illustrated in Fig 3.9, each does the same amount of work in moving through the same distance. The source of energy is quite different in the two cases. The piston is driven by the chemical energy released in burning the fuel, whereas the solenoid is powered electrically.

● *When a force moves an object it does work*

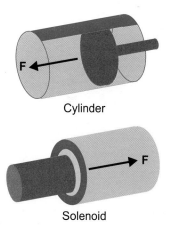

Fig. 3.9 Work is done whenever a force moves an object. The piston in a car engine compresses the gas in the cylinder, and the solenoid pulls an iron rod inwards. If both the piston and the core exert forces of the same magnitude, both do the same amount of work in moving through the same distance.

Worked Example 3.3 Estimate the amount of energy needed: (a) to open a sliding door; (b) to walk upstairs to the next floor.

Answer (a) If the mechanism of the sliding door is in good order, the force needed to move it should be around 10 N, similar to the force exerted in lifting a bag of sugar. Moving the door one metre thus uses about 10 joules, or in the abbreviated form, 10 J. This is a very small amount of energy: the running cost is not a matter of much concern for automatic doors.
(b) The upward force needed to walk upstairs must just exceed the gravitational force mg. For a person of mass 60 kg and a 3 m distance between floors, the work done is $60 \times 9.8 \times 3 \approx 1800$ J.

Work is energy expended, so the joule is the unit of energy, applied to all forms of energy, including heat. Now that we have a unit for

● *The joule is the unit for all forms of energy*

energy we can express the fact that energy is not 'lost' when it is changed from one form to another, i.e. that energy obeys a conservation law. Provided a system made up of any number of parts does not exchange energy with anything outside, the total amount of energy inside remains constant. The energy referred to includes heat: the system must be perfectly insulated so that no heat moves in or out even if there are temperature changes. Such a system is effectively isolated from the outside world. We can now state the **law of conservation of energy** as follows:

The total energy of an isolated system is constant.

The law of conservation of energy is a universal law of physics and it is not restricted to the domain where Newtonian mechanics is valid. It also applies to microscopic systems that can only be described by quantum mechanics and to systems including light and objects moving at speeds close to the speed of light. No single experiment proves the validity of the law, but all the calculations that assume conservation of energy always turn out to be correct. From now on we shall accept that the law of conservation of energy is correct: whenever energy is expended to do work, we know that the energy is still present in some other form.

3.4 **Kinetic energy**

We can now use the law of conservation of energy to find out how much energy is associated with motion, i.e. how much work must be done in order to make an object move at a particular speed. The energy associated with motion is called **kinetic energy**. To obtain an expression for kinetic energy we shall go back to the example of an object falling from rest that was considered in Section 2.2. Equations (2.7) and (2.10), which were derived in that section, give the speed v of the object and the distance z it has fallen as functions of time

$$v = gt$$

and

$$z = \tfrac{1}{2}gt^2.$$

This is not quite what is needed here; since work is (force × distance) the equations need to be rearranged to give us the speed as a function of distance instead of as a function of time. This is easily done by eliminating time from the above equations to give $t = v/g$ and

$$z = \frac{v^2}{2g}.$$

A constant gravitational force mg acts on the falling object, so the total amount of work done on it is

$$(\text{force} \times \text{distance}) = mgz = \tfrac{1}{2}mv^2.$$

Applying the law of conservation of energy we can now say that, because it is moving at a speed v, the falling object has acquired a kinetic energy $\tfrac{1}{2}mv^2$ equal to the work done on it.

Although we have only worked out the kinetic energy for the particular case of a falling object, the result is quite general. The whole idea of conservation of energy is that different forms of energy transform into one another. Irrespective of the agency doing work to make something move, the amount of energy gained by a stationary object of mass m accelerated to a speed v is *always* $\tfrac{1}{2}mv^2$. We shall use the notation E_K for kinetic energy, and we can write

$$E_K = \tfrac{1}{2}mv^2. \tag{3.13}$$

In words,

The kinetic energy of a mass m moving at a speed v is $\tfrac{1}{2}mv^2$.

Exercise 3.3 Calculate the kinetic energies of (a) a car of mass 1 tonne ($\equiv 1000\,\text{kg}$) moving at $90\,\text{km}\,\text{h}^{-1}$; (b) an aeroplane of mass 100 tonnes moving at $800\,\text{km}\,\text{h}^{-1}$; (c) an oil tanker of mass 200 000 tonnes moving at $20\,\text{km}\,\text{h}^{-1}$.

Answer (a) $3.1 \times 10^5\,\text{J}$; (b) $2.5 \times 10^9\,\text{J}$; (c) $3.1 \times 10^9\,\text{J}$. These answers have been given to two significant figures. This is sufficient precision for comparing the kinetic energies in an example like this, where the data are approximate. Here the masses are simply rough estimates for a vehicle of each kind.

● *Kinetic energy of a falling object*

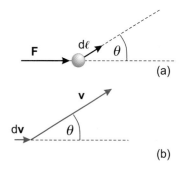

Fig. 3.10 An object is not always moving in the direction in which it is being pushed. Diagram (a) shows an object moving through a small distance $d\ell$ while a force **F** is acting. The velocity **v** is in the same direction as $d\ell$, but the *change* in velocity $d\mathbf{v}$ must be in the direction in which the force is acting, as shown in (b).

Here is a proof that the kinetic energy is always $\tfrac{1}{2}mv^2$, whatever the origin of the force causing motion, and whether or not the force is constant. Figure 3.10(a) shows an object of mass m that moves a distance $d\ell$ while it is acted on by a force F. The force is at an angle θ to the direction of motion. The work done by the force while the object moves through $d\ell$ is $F\,d\ell\cos\theta$, since $d\ell\cos\theta$ is the distance moved in the direction of the force.

In the vector notation, $F\,d\ell\cos\theta = \mathbf{F}\cdot d\boldsymbol{\ell}$, the scalar product of the vectors **F** and $d\boldsymbol{\ell}$ (see Section 20.2 for a definition of scalar product).

● *The vector expression for work*

Here $\mathrm{d}\boldsymbol{\ell}$ is an infinitesimal vector of length $\mathrm{d}\ell$ pointing in the direction of motion. The corresponding velocity diagram is shown in Fig 3.10(b). The velocity

$$\mathbf{v} = \frac{\mathrm{d}\boldsymbol{\ell}}{\mathrm{d}t}$$

is in the direction of motion, i.e. in the same direction as $\mathrm{d}\boldsymbol{\ell}$, while the *change* in $\mathbf{v}$, which is labelled $\mathrm{d}\mathbf{v}$, is in the same direction as the force. We may write

$$\mathbf{F} \cdot \mathrm{d}\boldsymbol{\ell} = \mathbf{F} \cdot \frac{\mathrm{d}\boldsymbol{\ell}}{\mathrm{d}t}\mathrm{d}t = \mathbf{F} \cdot \mathbf{v}\mathrm{d}t. \tag{3.14}$$

According to Newton's second law,

$$\mathbf{F} = \frac{\mathrm{d}\mathbf{p}}{\mathrm{d}t} = m\frac{\mathrm{d}\mathbf{v}}{\mathrm{d}t}$$

for an object of mass m, so eqn (3.14) becomes

$$\mathbf{F} \cdot \mathrm{d}\boldsymbol{\ell} = m\mathbf{v} \cdot \frac{\mathrm{d}\mathbf{v}}{\mathrm{d}t}\mathrm{d}t = m\mathbf{v} \cdot \mathrm{d}\mathbf{v}. \tag{3.15}$$

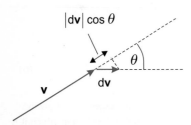

Fig. 3.11 The change in the magnitude of the velocity $\mathbf{v}$ is the component of the small change $\mathrm{d}\mathbf{v}$ in the direction of $\mathbf{v}$.

The right-hand side of eqn (3.15) can be expressed in terms of the speed v, without needing to know the direction of $\mathbf{v}$. Referring to Fig 3.11, we see that, if the change $\mathrm{d}\mathbf{v}$ is at an angle θ to $\mathbf{v}$, then the scalar product $\mathbf{v} \cdot \mathrm{d}\mathbf{v} = v|\mathrm{d}\mathbf{v}|\cos\theta$. The figure shows that the change in the magnitude of $\mathbf{v}$ is $\mathrm{d}v = |\mathrm{d}\mathbf{v}|\cos\theta$. The scalar product $\mathbf{v} \cdot \mathrm{d}\mathbf{v}$ is just $v\mathrm{d}v$. (But note that $\mathrm{d}v$ is *not* the magnitude of the vector $\mathrm{d}\mathbf{v}$). Equation (3.15) for the work done as the object moves through the distance $\mathrm{d}\ell$ becomes

$$\mathbf{F} \cdot \mathrm{d}\boldsymbol{\ell} = m v\,\mathrm{d}v. \tag{3.16}$$

The total work done when the object is accelerated by a force is the integral of this quantity. The particle does not have to be moving in a straight line but, in each little segment $\mathrm{d}\boldsymbol{\ell}$ of its path, work is done on it according to eqn (3.16).

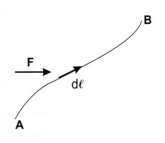

Fig. 3.12 A particle moves along the blue line from A to B and a force acts on it as it moves. The total work done by the force $\mathbf{F}$ between A and B is the line integral $\int_{A}^{B}\mathbf{F} \cdot \mathrm{d}\boldsymbol{\ell}$.

The total work done over a path such as the one shown in Fig 3.12 is called a **line integral** and is written as $\int \mathbf{F} \cdot \mathrm{d}\boldsymbol{\ell}$. If the particle is accelerated from rest to a speed v, then the total amount of work done, which is equal to the kinetic energy of the particle, is

$$\int \mathbf{F} \cdot \mathrm{d}\boldsymbol{\ell} = \int_{\Gamma}^{v} mv\,\mathrm{d}v = \tfrac{1}{2}mv^2.$$

● *The work done by a force is given by a line integral*

This equation holds even if the force $\mathbf{F}$ changes in magnitude and direction while the particle is accelerating. Kinetic energy is always $\tfrac{1}{2}mv^2$, no matter how the particle reaches the speed v.

Kinetic energy in the centre of mass frame of reference

Kinetic energy is a scalar quantity, which has a magnitude but no direction. The kinetic energy of the object of mass m is $\frac{1}{2}mv^2$ whatever may be its direction of motion. The actual value given to the kinetic energy does, however, depend on the frame of reference used to measure it. Going back to our man on a ship, if he is relaxing in a deck chair he considers himself to be at rest, since the convenient frame of reference for him moves along with the ship. For an observer ashore, on the other hand, the man in his chair is moving at the speed of the ship and he has kinetic energy. Note that the law of conservation of energy applies irrespective of the frame of reference. The man on the ship and the observer ashore may assign different values to the total energy of an isolated system, but they agree that the energy remains constant so long as no external force acts on the system.

How does kinetic energy change from one frame to another? Suppose in a particular frame of reference S that a number of particles with masses m_i have velocities $\mathbf{v}_i$. Then their total kinetic energy, which we shall call E_K, is $\sum_i \frac{1}{2}m_i v_i^2$. The squares of the speeds v_i can be written as $\mathbf{v}_i \cdot \mathbf{v}_i$, since the scalar product of any vector with itself is equal to the square of its magnitude. Now if we choose a new frame of reference S' moving with a velocity $\mathbf{V}$ with respect to the first, each velocity changes, according to eqn (3.8), to $\mathbf{v}'_i = \mathbf{v}_i - \mathbf{V}$. The total kinetic energy in S' is

● *Transformation of kinetic energy*

$$E'_K = \sum_i \tfrac{1}{2}m_i v_i'^2 = \sum_i \tfrac{1}{2}m_i \mathbf{v}'_i \cdot \mathbf{v}'_i = \sum_i \tfrac{1}{2}m_i(\mathbf{v}_i - \mathbf{V}) \cdot (\mathbf{v}_i - \mathbf{V})$$

or, writing the scalar products on the right-hand side in terms of the x-, y-, and z-components of the vectors $\mathbf{v}_i$ and $\mathbf{V}$,

$$E'_K = \sum_i \tfrac{1}{2}m_i(v_{xi} - V_x)^2 + \sum_i \tfrac{1}{2}m_i(v_{yi} - V_y)^2 + \sum_i \tfrac{1}{2}m_i(v_{zi} - V_z)^2.$$

$$(3.17)$$

This kinetic energy can clearly be made as big as you please by choosing any of the components V_x, V_y, or V_z large enough. But each term in the sum in eqn (3.17) is the square of a magnitude, so the whole expression is positive and there must be some minimum value for E'_K.

Which frame of reference S' will be the one with minimum kinetic energy? To answer this question we have to minimize the sums over the x-, y-, and z-components separately. First look at the x-components. At the minimum, the rate of change of E'_K with V_x must be zero:

$$\frac{dE'_K}{dV_x} = -\sum_i m_i(v_{xi} - V_x) = 0,$$

leading to

$$V_x = \frac{\sum_i m_i v_{xi}}{\sum_i m_i} = \frac{1}{M}\sum_i m_i v_{xi},$$

● *The kinetic energy is least in the centre of mass frame*

where $M = \sum_i m_i$ is the total mass of the system of particles. Similar expressions hold for the y- and z-components, leading to

$$\mathbf{V} = \frac{1}{M}\sum_i m_i \mathbf{v}_i = \frac{1}{M}\mathbf{p}. \tag{3.18}$$

In S′ the total momentum is

$$\mathbf{p}' = \sum_i m_i \mathbf{v}'_i = \sum_i m_i \mathbf{v}_i - \sum_i m_i \mathbf{V} = \mathbf{p} - M\mathbf{V} = 0$$

from eqn (3.18). In other words, the frame of reference in which the kinetic energy has a minimum value is the one in which the total momentum of the system is zero. This is the centre of mass frame, which was discussed in Section 3.2.

The centre of mass frame of reference turns out to be special for kinetic energy as well as for momentum. How is the kinetic energy in the original frame S related to its value in the centre of mass frame? We have to go back to frame S now and find the kinetic energy E_K in terms of E'_K. In the S frame $E_K = \sum_i \frac{1}{2}m_i v_i^2$. The velocities $\mathbf{v}_i$ can be expressed in terms of the velocities $\mathbf{v}'_i$ in a frame S′ moving at a velocity $\mathbf{V}$ with respect to S using eqn (3.7):

$$\mathbf{v}_i = \mathbf{v}'_i + \mathbf{V}.$$

Now, putting $\mathbf{V} = \mathbf{V}_{cm}$ for the special case when S′ is the centre of mass frame,

$$E_K = \sum_i \frac{1}{2}m_i v_i^2 = \sum_i \frac{1}{2}m_i \mathbf{v}_i \cdot \mathbf{v}_i = \sum_i \frac{1}{2}m_i (\mathbf{v}'_i + \mathbf{V}_{cm}) \cdot (\mathbf{v}'_i + \mathbf{V}_{cm})$$

$$= \sum_i \frac{1}{2}m_i v_i'^2 + \sum_i m_i \mathbf{v}'_i \cdot \mathbf{V}_{cm} + \sum_i \frac{1}{2}m_i V_{cm}^2.$$

● *Total kinetic energy is the sum of the kinetic energy of the centre of mass and internal energy*

The middle term is zero since the momentum $\sum_i m_i \mathbf{v}'_i$ in the centre of mass frame is zero, and all that is left is

$$E_K = \sum_i \frac{1}{2}m_i v_i'^2 + \frac{1}{2}MV_{cm}^2 = E'_K + \frac{1}{2}MV_{cm}^2. \tag{3.19}$$

Now $\frac{1}{2}MV_{cm}^2$ is just the kinetic energy of an object of mass M moving at a speed V. The rest of the total kinetic energy is the kinetic energy in the centre of mass frame. This is an **internal energy** for the system, which is unaffected by the motion of the system as a whole.

Sometimes internal energy may be easy to see, as for example when an aeroplane propeller rotates about its hub. As the plane gathers speed for take-off, the propeller rotates faster and it also moves forwards. According to eqn (3.19) the total kinetic energy is just the sum of the energies due to rotation and forward motion taken separately. The rotational energy is the same as the energy of the rotating propeller if the plane were to remain at rest: to this we must add the kinetic energy associated with the motion of the propeller's centre of mass, which is simply the energy the propeller would have if the plane were moving forwards without the propellers turning.

Very often the internal energy of an object remains constant. When considering its motion you then need only consider the term $\frac{1}{2}MV_{cm}^2$. Together with the earlier result that the momentum of the object is $M\mathbf{V}_{cm}$, this is an enormous simplification. When thinking about the motion of large bodies, there is no need to worry about what is going on inside them.

What to include and what to ignore depends on the scale of the problem. When calculating the Earth's orbit about the Sun, the motions of all the objects on the Earth are irrelevant. The momentum of the Earth and everything on it is zero in their centre of mass frame. They contribute a constant internal energy, which does not affect the additional kinetic energy in the coordinate system fixed on the Sun. Similarly, ordinary macroscopic objects are made up of atoms that are in constant motion. Common sense suggests that you can ignore the atomic motion when working out how a football or a car or a railway wagon will move. The commonsense view is justified whenever changes in internal energy are negligible.

When the internal energy of a system does change, some of this change may have to be accounted for as heat. Heat is energy that is distributed randomly among all the atoms and molecules that make up the system. Even a very small piece of matter like a grain of sand contains such an enormous number of atoms that it would be impossible to list the energy of each one at a particular moment. Impossible and also pointless, since the distribution of energy among the atoms is always changing on a short time-scale.

3.5 Collisions

Both the laws of conservation of momentum and conservation of energy must hold when objects collide. Conservation of momentum has already been invoked to find out what happens in a collision of railway wagons in Section 3.1. A single wagon of mass m moving at speed v collides with two stationary wagons each also of mass m, and the three wagons then move off coupled together, as in Fig 3.13. It immediately follows from conservation of momentum that the speed of the three wagons is $v/3$.

Before collision

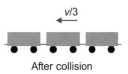

After collision

Fig. 3.13 Conservation of momentum requires that, after coupling together, the three wagons move at one-third the initial speed of the single wagon.

● *Kinetic energy may be lost in a collision*

Before the impact the kinetic energy of the single wagon is $\frac{1}{2}mv^2$. Afterwards the kinetic energy of all three is $\frac{1}{2} \times 3m \times \left(\frac{1}{3}v\right)^2 = \frac{1}{6}mv^2$, only a third of the initial kinetic energy. The remaining two-thirds is now internal energy of the system of three wagons, and it has been converted into heat. As a result of the collision some parts of the wagons are hotter than they were before.

The heat is energy that is distributed randomly among the atoms of the hot parts of the wagons, and it cannot easily be converted into other forms of energy. Although the amount of heat generated in the collision is just what is required to satisfy conservation of energy, the collision cannot be reversed—if the first wagon is uncoupled from the other two it does not recover its initial speed. (A fuller discussion of heat energy is given in Chapter 12 on thermal physics.)

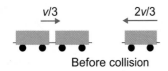

v/3   2v/3

Before collision

At rest

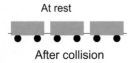

After collision

Fig. 3.14 The same collision as the one in Fig 3.13 with speeds referred to centre of mass coordinates. In this frame of reference the trucks are stationary after the collision, and all their initial kinetic energy has turned into heat.

Exercise 3.4 What is the speed relative to the ground of the centre of mass of the three wagons discussed in the paragraph above? Calculate their kinetic energy in the centre of mass frame of reference before the collision. This kinetic energy is the internal energy of the system of three wagons. Check that this internal energy is the same as the amount of heat generated in the collision.

Answer Figure 3.14 is a diagram of the collision in centre of mass coordinates. The speed of the centre of mass relative to the ground is $v/3$ and the internal kinetic energy is $\frac{1}{3}mv^2$.

Elastic and inelastic collisions

● *No kinetic energy is lost in elastic collisions*

If the kinetic energy is the same before and after a collision, the collision is called **elastic**. A collision like the one between the railway wagons in which some kinetic energy is converted into heat is an **inelastic** collision. Most collisions are inelastic, and often it is easy to see that this is so. For example, if a tennis ball is dropped on to the ground, it bounces up again but does not reach its original height. Successive bounces are lower and lower, showing that kinetic energy is being transferred into heat at each bounce. In contrast the 'bouncy balls' that children often play with bound back almost to their original height at each bounce off a hard surface. The kinetic energy the ball has just before reaching the ground is almost restored when it flies up again. No collisions between macroscopic objects are perfectly elastic, but very little kinetic energy is lost in the collision between a bouncy ball and the ground; it is quite a good approximation to assume that the collision is elastic.

If the bounce is perfectly elastic, no kinetic energy is lost, and the upwards speed of the ball after bouncing is the same as the downwards speed beforehand. The ratio of speeds before and after bouncing is a measure of how elastic is the collision that has occurred. For a perfectly elastic collision the ratio is one, and it is less than one for any other collision. The ratio is called the **coefficient of restitution**. Remembering that speeds are relative quantities, the coefficient of restitution is strictly defined as the ratio of the relative speeds of colliding objects before and after collision. For the bouncy ball the speed is relative to the Earth, which is not affected by the collision. For the colliding railway wagons the relative speed is v before collision and zero after, since all the wagons are then moving at the same speed. The coefficient of restitution is thus zero, and this collision is as inelastic as possible—the maximum possible amount of kinetic energy has been transferred into heat.

As these examples show, the value of the coefficient of restitution depends on the nature of the colliding objects, and it can vary from zero to very nearly one. For collisions of the same objects, the coefficient is often roughly the same at different relative speeds, though in practice it is never exactly constant. To illustrate how the result of a collision can be calculated if the coefficient of restitution is known, we shall consider the bouncing tennis ball in more detail in the following worked example.

Worked Example 3.4 A tennis ball is dropped from a height of 2 m and bounces back to 1.5 m above the ground. What height will it reach on the second bounce, assuming that the coefficient of restitution is the same for each bounce?

Answer Neglecting air resistance, eqn (2.10) gives the time taken for the ball to fall a distance z as $\sqrt{2z/g}$. At a uniform acceleration g, it reaches a speed $\sqrt{2gz}$ in this time. Calling the height from which it is dropped h_1 and the height of the first bounce h_2, as shown in Fig 3.15, the speeds just before the first and second bounces are $v_1 = \sqrt{2gh_1}$ and $v_2 = \sqrt{2gh_2}$. The upwards speed of the ball just after the first bounce is the same as its downwards speed just before the second bounce. Hence the coefficient of restitution is

$$\frac{v_2}{v_1} = \frac{\sqrt{2gh_2}}{\sqrt{2gh_1}} = \sqrt{\frac{h_2}{h_1}}.$$

The same coefficient applies for the second bounce as for the first, so if the ball reaches a height h_3 after the second bounce, then $\sqrt{h_3/h_2} = \sqrt{h_2/h_1}$, or $h_3 = h_2^2/h_1$. In this example $h_3 = (1.5)^2/2 = 1.125$ m. So long as the coefficient of restitution maintains the same value, each succeeding bounce will reduce the height by the same factor 0.75.

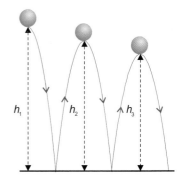

Fig. 3.15 A tennis ball bouncing on the ground reaches a smaller height at each successive bounce.

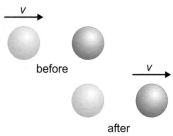

Fig. 3.16 The moving white ball collides head-on with the stationary blue ball. After the collision the white ball is stationary and all its energy has been transferred to the blue one.

⊙ *Conservation of both energy and momentum are invoked to find the result of a collision*

Collisions between pool or snooker balls are also nearly elastic. These are more interesting than balls bouncing on the ground because energy is now transferred from one ball to another, and because the balls may move off at different angles. First consider the simplest case of a head-on collision when the balls do travel in a straight line (Fig 3.16). The balls have the same mass m. Before the collision the cue ball—the white one—is moving at a speed v across the table and the blue ball is stationary. Afterwards, call the speeds relative to the table v_1 for the white ball and v_2 for the blue ball. Conservation of momentum requires

$$mv = mv_1 + mv_2. \tag{3.20}$$

Provided that the collision is elastic, conservation of energy requires

$$\tfrac{1}{2}mv^2 = \tfrac{1}{2}mv_1^2 + \tfrac{1}{2}mv_2^2. \tag{3.21}$$

Eliminating v_1 from eqns (3.20) and (3.21), $v_1 = v - v_2$ and $v^2 = (v - v_2)^2 + v_2^2$, which has the solution $v_2 = v$, and thus $v_1 = 0$. The cue ball stops and the other ball moves forward with the same kinetic energy as the cue ball had before the collision.

The calculation has ignored the fact that the balls are rolling as well as moving across the table. The balls rotate about their centres of mass, and the energy of rolling is internal motion. Rotations are the subject of the next chapter, and it is shown in Section 4.6 that, if the balls roll without slipping, the energy of rotation is unchanged by the collision. Under these conditions it is justified to use eqn (3.21), which includes only the straight line motion of the balls across the table.

Exercise 3.5 In the game of bowls the object is to roll a heavy ball, called a 'wood', so that it stops as close as possible to a small ball called a 'jack'. A player wishes to strike the jack hard to move it away from his opponents' balls. His wood collides head on with the jack at a speed of $1\,\mathrm{m\,s^{-1}}$. If his ball weighs five times as much as the jack, what speed does the jack have just after the collision, assuming the collision to be elastic?

Answer $\tfrac{5}{3}\,\mathrm{m\,s^{-1}}$.

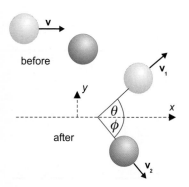

Fig. 3.17 If the collision is not head-on, the balls move off in different directions.

Two-dimensional collisions

Pool and snooker balls do not often line up with the pockets on the edge of the table and, in order to pocket the struck ball, the player must usually make it move off at an angle. What happens in the collision if the white cue ball is deflected through an angle θ, as shown in Fig 3.17?

As well as the speeds v_1 and v_2, there is now a third unknown, the recoil angle ϕ of the struck ball. Conservation of momentum is illustrated by the vector diagram in Fig 3.18. In the collision between the balls the velocities before and after collision all lie in the plane of the table, so by choosing x- and y-coordinate axes in this plane as shown in the diagram, two equations are sufficient to specify momentum conservation.

Conservation of momentum in the x-direction requires

$$mv = mv_1 \cos\theta + mv_2 \cos\phi; \tag{3.22}$$

conservation of momentum in the y-direction requires

$$mv_1 \sin\theta = mv_2 \sin\phi. \tag{3.23}$$

For elastic collisions, conservation of energy requires:

$$\tfrac{1}{2}mv^2 = \tfrac{1}{2}mv_1^2 + \tfrac{1}{2}mv_2^2. \tag{3.24}$$

This equation is simply Pythagoras's theorem applied to the vector triangle in Fig 3.18, and it follows that the angle between the vectors $\mathbf{v}_1$ and $\mathbf{v}_2$ is 90°. This means that the two balls move away after the collision at right angles to one another. The blue ball moves in the direction of the force that moves it, which is along the line through the point of contact and the centres of the two balls, as shown in Fig 3.19. The point of contact must also be on the face of the ball visible from the cue ball direction: the blue ball can be made to go nearly sideways by a glancing blow, but always acquires some forward momentum.

Describing how to pocket the ball like this makes it sound very easy, but a lot of practice is needed to be able to do it! The top players' skill goes far beyond the application of momentum and energy conservation: they greatly extend the possible outcomes of collisions by imparting spin to the cue ball and making use of the friction between the balls and the table.

Collisions in the centre of mass frame

We have looked at the collision of snooker balls in a coordinate system fixed with respect to the snooker table. Let us repeat the calculation using the centre of mass coordinate system. All of the kinetic energy will now be designated as internal energy. The centre of mass of the two identical balls is at the point half-way between them. It moves at a speed $v/2$ with respect to the table.

Assuming that the two balls do not lose energy through friction on the table during their collision, their internal energy stays the same, and so does the kinetic energy of the centre of mass. After the collision the centre of mass continues at the speed $v/2$ as though nothing had happened. Figure 3.20 represents the collision in the centre of mass frame, which is

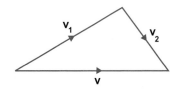

Fig. 3.18 The vector diagram for the velocities contains a right angle, since $\mathbf{v}$, $\mathbf{v}_1$, and $\mathbf{v}_2$ must satisfy eqn (3.24).

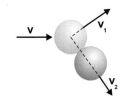

Fig. 3.19 The force between the two balls acts along the line joining their centres.

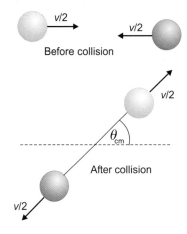

Fig. 3.20 In the centre of mass frame of reference both balls have the same speed before and after the collision.

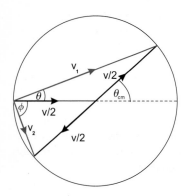

Fig. 3.21 This diagram shows the transformation from the centre of mass frame to the frame fixed with respect to the table. The horizontal line of length $v/2$ represents the velocity of the centre of mass with respect to the table.

the one in which the total momentum is zero. This means that the balls must be approaching one another at the same speed. Since their relative speed is v, each has a speed $v/2$. The kinetic energy in the centre of mass coordinates, which is the internal energy of the system of two balls, is $2 \times \frac{1}{2}m(v/2)^2 = \frac{1}{4}mv^2$.

The collision is elastic, so the coefficient of restitution is one, and the balls have the same relative speed after the collision, i.e. they move *away* from the centre of mass, each at a speed $v/2$. Energy is thus conserved in the centre of mass frame, in which the momentum remains zero.

Energy and momentum are conserved in the frame fixed with respect to the table as well as in the centre of mass frame. The velocity of the centre of mass is unaltered by the collision: it is still in the x-direction and has magnitude $v/2$. To find the speeds and directions on the table after the collision, we must add the velocity of the centre of mass to the velocities of the outgoing balls. The vector diagram in Fig 3.21 does this, and the blue arrows represent the velocities on the table after the collision.

The result of the calculation is, of course, the same whether it is done in laboratory or in centre of mass coordinates. In this simple example, where the laboratory angles θ and ϕ satisfy $\theta + \phi = \pi/2$, the speeds v_1 and v_2 and the angle ϕ are easily found from the geometry of Fig 3.21. Collision calculations are generally simpler in centre of mass coordinates.

Exercise 3.6 A cue ball collides with a stationary snooker ball and is deflected through 30°. What fraction of its energy is transferred to the other ball in the collision?

Answer One-quarter.

3.6 Potential energy

We have seen that, when kinetic energy is dissipated into heat, the kinetic energy cannot be immediately recovered by making the process go backwards. After railway wagons have collided and coupled together, their kinetic energy is reduced and it cannot be increased again without doing more work on the wagons.

Sometimes, however, the kinetic energy of an object increases and decreases without any change in its internal energy. A good example is a bouncy ball making elastic collisions with the ground. The ball is moving fast when it hits the ground, i.e. it has a large kinetic energy. It bounces up and at the top of the bounce it once more has no kinetic energy. How can this be squared with the conservation of energy? As the ball is falling, work is done on it by the gravitational force. For a ball of mass m falling through a height h, the work done is mgh, and this amount of work is equal to the kinetic energy of the ball as it hits the ground. When the ball was lifted up to a height h before release, exactly the same amount of work—mgh—was done against the gravitational force. Gravitational energy has been stored by lifting the ball. This energy is called **potential energy**, because the ball has the potential to convert the stored gravitational energy into kinetic energy by falling under the action of the gravitational force.

○ *Work done against gravity is stored as potential energy*

Kinetic energy and gravitational energy are exchanged as the ball bounces up and down. If the ball's collisions with the ground were elastic and there were no air resistance the ball would bounce forever. The total amount of energy, shared between kinetic energy and gravitational, would remain constant as required by the law of conservation of energy. In practice, of course, this energy is gradually dissipated as heat because the collisions with the ground are not perfectly elastic and because of friction in the air. As well as the kinetic and gravitational energy, heat must be included to maintain energy conservation.

Conservative and non-conservative forces

An important feature of the bouncing ball example is that the gravitational potential energy is recovered after the bounce. Exactly as much work is done *by* the gravitational force on a ball falling through a height h as is done *against* it in lifting a ball through h. Nor does it make any difference if the ball is lifted at an angle instead of vertically; horizontal movement does not require any work to be done (see Fig 3.22). The height alone determines the potential energy. It follows that, if an object moves along any path whatsoever, so long as it comes back to its original position there is no change in potential energy. The potential energy depends only on the present position and not on the history of how the object got into that position. A force that has this property, that the associated potential energy of an object has the same value whenever it returns to the same position, is called a **conservative** force. A force remains conservative even if there are different forces acting at the same time that are not conservative. The gravitational potential energy of the ball at a height h is the same when it is returned to this height even if it has got warm by bouncing in the meantime. In this case the collisions with the ground are inelastic and non-conservative, and work has to be done to recover the gravitational energy.

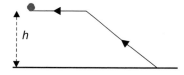

Fig. 3.22 The ball follows the path given by the arrows. No work is done against gravity when the ball is moved horizontally. The total work done against gravity is determined only by the height difference h.

○ *The gravitational force is conservative*

Frictional forces are always non-conservative. They act in a direction to oppose motion and the work they do on a moving object always reduces the kinetic energy. It is characteristic of frictional forces that their effects are not reversible. When a falling ball loses energy because of the frictional force of the air, this energy is not recovered when the ball bounces up again; on the contrary still more energy is lost. The energy has been dissipated as heat.

Worked Example 3.5 A sledge of mass m slides down a slope at an angle θ to the horizontal for a distance d and then runs along a flat area at the bottom of the hill for another distance d before coming to rest (Fig 3.23). Assuming that the coefficient of friction μ is constant throughout the sledge's run, obtain an expression for μ. What is the value of μ if $\theta = 15°$?

Answer The source of energy moving the sledge is the change in gravitational potential energy in dropping through a height $d \sin \theta$. This change is $mgd \sin \theta$, no matter how far the sledge has moved horizontally. Friction opposes the motion, and the frictional force is $mg\mu \cos \theta$ on the slope and $mg\mu$ on the flat. The change in potential energy equals the total amount of work done by friction, i.e. $mgd \sin \theta = mg\mu(d \cos \theta + d)$ or $\mu = \sin \theta (1 + \cos \theta)^{-1}$. Putting $\theta = 15°$ we find $\mu = 0.13$.

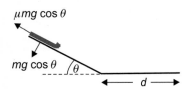

Fig. 3.23 A sledge runs down a hill and then continues for some distance along the flat before it stops. The work done by friction can be evaluated simply by using conservation of energy.

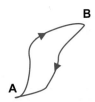

Fig. 3.24 The work done against a conservative force along a path from A to B is the same for any path. This work may be recovered by returning from B to A along any other path. No work is done by a conservative force in going around a closed loop.

The condition for a force to be conservative is expressed in a concise way in vector notation. The work done by a force $\mathbf{F}$ on an object moving through an infinitesimal distance $\mathrm{d}\boldsymbol{\ell}$ is $\mathbf{F} \cdot \mathrm{d}\boldsymbol{\ell}$, as in eqn (3.14). Work done *against* the force is thus $-\mathbf{F} \cdot \mathrm{d}\boldsymbol{\ell}$. This work is positive if $\mathbf{F}$ and $\mathrm{d}\boldsymbol{\ell}$ are in opposite directions—as they are when you lift a ball against the gravitational attraction of the Earth. The work done in pushing against the force $\mathbf{F}$ along the path in Fig 3.24 between points A and B is given by the line integral $-\int_A^B \mathbf{F} \cdot \mathrm{d}\boldsymbol{\ell}$ (line integrals are explained in Section 3.4).

For a conservative force this integral has the same value for any path joining the points A and B, and it represents the difference in potential energy between these points:

$$\text{Potential energy difference between A and B} = -\int_A^B \mathbf{F} \cdot \mathrm{d}\boldsymbol{\ell}. \qquad (3.25)$$

Reversing the path from B to A changes the sign since the sign of each infinitesimal section $\mathrm{d}\boldsymbol{\ell}$ of the path is changed. It follows that for any path from A to B and back again the integral is zero. An integral around such a closed path is written as $\oint$. Hence

$$\oint \mathbf{F} \cdot \mathrm{d}\boldsymbol{\ell} = 0$$

for a conservative force.

Potential energy in a spring

The coiled spring in Fig 3.25 is fixed at the top. If you attach a mass m to the spring it is stretched, and the spring exerts an upward *restoring force*. The more the spring is stretched, the stronger the restoring force. There is an equilibrium position where the restoring force F exactly balances the downward gravitational force mg on the mass. In Fig 3.25 equilibrium occurs when the spring is stretched a distance x beyond its length when no mass is attached.

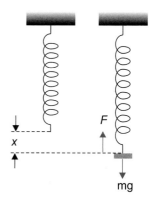

Fig. 3.25 When the spring is stretched, the restoring force F balances the weight mg.

Work has been done by the gravitational force in stretching the spring, and the energy expended in this work is now stored in the spring: the spring has acquired potential energy. That the spring has potential energy can easily be demonstrated by removing the mass—the spring flies back and the potential energy is converted into kinetic energy.

It is often a very good approximation to assume that the restoring force F exerted by the spring is proportional to the amount x by which it is extended.

This statement is written mathematically as

$$F = -kx. \tag{3.26}$$

The proportionality constant k is called the **spring constant**, and the minus sign indicates that the restoring force is in the opposite direction to the displacement x from the resting position of the spring. Equation (3.26) is known as **Hooke's law**.

Hooke's law is applicable to many other materials besides coiled springs. A thin elastic band or a rope will obey Hooke's law provided you do not stretch them too much. The spring constant k is small for the weak material—the elastic band—and large for a strong rope. Materials such as steel that do not seem to stretch or compress at all also obey Hooke's law for small distortions, and for them the spring constant is very large indeed.

We can easily find the potential energy of a stretched spring by equating it to the work done against the restoring force. The work done in extending the spring from x to $(x + dx)$ is $kx\,dx$, since work done is (force × distance). The total work done in extending the spring from its resting position at $x = 0$ to a displacement x is

● **Work done in extending a spring**

$$\int_0^x kx\,dx = \tfrac{1}{2}kx^2. \tag{3.27}$$

The potential energy $\tfrac{1}{2}kx^2$ does not depend on the sign of x. If Hooke's law is obeyed, the potential energy is the same in compression as in tension for the same displacement.

Like the gravitational force, the force exerted by a spring that obeys Hooke's law is conservative: the potential energy depends only on x. The potential energy is $\tfrac{1}{2}kx^2$ when the extension is x, whether the spring was at

● **Potential energy functions**

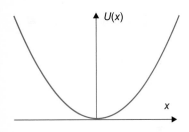

Fig. 3.26 The potential energy function for a spring is a parabola. The force for positive or negative x points down the slope towards the origin.

a larger or smaller value of x beforehand. The potential energy is a *function of position* which we shall denote by $U(x) = \frac{1}{2}kx^2$. The potential function $U(x)$ is a parabola with a minimum at $x = 0$ as sketched in Fig 3.26.

Since $U(x)$ was obtained by integrating the force exerted against the spring in eqn (3.27), the force may be found by differentiating the potential energy function. Bearing in mind that the force F generated *by* the spring has the opposite sign to the force needed to extend the spring, we find

$$F = -\frac{\mathrm{d}U}{\mathrm{d}x}. \tag{3.28}$$

Equation (3.28) applies to the potential energy function for any conservative force. Putting $U(x) = \frac{1}{2}kx^2$ for the spring leads to $F = -kx$, in agreement with Hooke's law.

Potential energy functions are very important in physics, and we shall meet them again many times in this book. In particular, Schrödinger's equation, which is the fundamental equation of quantum mechanics, is expressed in terms of potential energy rather than force. Schrödinger's equation is discussed in Chapters 10 and 11.

Worked Example 3.6 A spring is extended by 5 cm when a mass of 1 kg is hung on it. The mass is pulled down a further 2 cm and then released. How fast is the mass moving when it has risen 2 cm, back to its point of release?

Answer The restoring force equals the weight of 1 kg for an extension $x = 0.05\,\mathrm{m}$. From eqn (3.26) the spring constant is $k = mg/x \approx 9.8/0.05 \approx 200\,\mathrm{N\,m^{-1}}$. The potential energy in extending 2 cm is $\frac{1}{2}kx^2 = \frac{1}{2} \times 200 \times 0.02^2 = 0.04\,\mathrm{J}$. This equals the kinetic energy $\frac{1}{2}mv^2$ when the mass is back at its equilibrium position, leading to $v = 0.28\,\mathrm{m\,s^{-1}}$.

Exercise 3.7 Note that in the solution above the potential energy was calculated using the displacement from the equilibrium position with the mass in place, when the spring is already storing potential energy. By considering the gravitational energy of the mass and the total potential energy of the spring, check that this procedure is valid.

In Worked example 3.6, the potential energy of the spring has been converted into kinetic energy of the mass. If undisturbed, the mass will come to rest above the equilibrium position and then start down again.

It will continue to oscillate up and down until all the initial potential energy is dissipated as heat. Oscillations of this kind are very important in physics, and they are discussed in Chapter 6.

The advantages of using conservation laws

In this chapter energy and momentum conservation have been used to find the answers to a number of problems involving motion without having to work out any forces or apply Newton's laws. In coupling railway trucks together the conservation laws alone were enough to find the speed after collision and the amount of heat generated; in the elastic collision of snooker balls, the angles and energies of the cue ball and the struck ball were evaluated. We were able to calculate the coefficient of friction of the sledge in Worked example 3.5 without having to calculate how fast the sledge had been moving at any stage, nor the time-scale of its run. If you have a problem in dynamics to solve, first look at the conservation laws. Sometimes you find your solution without going any further, and save a lot of mathematics.

● *Conservation laws can lead to simple solutions of problems in mechanics*

The laws of conservation of momentum and energy are not limited to large objects, but also apply on the atomic scale. This is illustrated in Worked example 3.7 and some of the problems at the end of this chapter. The speeds of the particles in these examples are small compared to the speed of light. The expressions we have derived in this chapter for momentum and kinetic energy are not valid for objects moving at speeds near to the speed of light: energy and momentum at such speeds are discussed in Chapter 19. Although momentum and energy conservation apply to very small particles, not all the results of classical mechanics, derived from Newton's laws, are valid on the atomic scale. The circumstances in which classical ideas do not apply are described in Chapter 9.

● *Energy and momentum are conserved even when Newton's laws do not apply*

Worked Example 3.7 The atomic nuclei of helium, which are called α-particles, are emitted in one type of radioactive decay. In 1911 Rutherford, Geiger, and Marsden found that, when α-particles from a radioactive source impinged on a gold target, most passed through the target, but a few were scattered backwards. This led Rutherford to propose the nuclear picture of the atom, in which nearly all of the mass of the atom is concentrated in a very small nucleus at its centre. The speed of the α-particles in these pioneering experiments was about $1.5 \times 10^7 \, \mathrm{m \, s^{-1}}$, and the mass of the gold nucleus is about 50 times that of the α-particle. When an α-particle is scattered directly backwards, through $180°$, what is the speed of the gold nucleus after the collision?

Answer Write the mass of the α-particle as m, its speed before and after head-on collision with a gold nucleus as v and v_1, respectively, and the speed of the gold nucleus after collision as v_2. The momentum before the collision is mv and conservation of momentum leads to

$$mv = 50mv_2 - mv_1.$$

This is similar to eqn (3.20) for the head-on collision between two equal masses except for the occurrence of the minus sign. The difference arises because, even in a head-on collision, we must remember that momentum is a vector quantity: after the collision the α-particle and the gold nucleus are moving in opposite directions. Kinetic energy is a scalar, and there is no need to worry about signs in the equation expressing conservation of energy, which is

$$\tfrac{1}{2}mv^2 = \tfrac{1}{2}mv_1^2 + 25mv_2^2.$$

Eliminating v_1 we find $v_2 = v/25.5$ and, for $v = 1.5 \times 10^7\,\mathrm{m\,s^{-1}}$, $v_2 \approx 6 \times 10^5\,\mathrm{m\,s^{-1}}$.

The speed of the incident α-particles in this example is about 5% of the speed of light. Although the corrections required by relativity are not completely negligible, the classical result we have calculated differs from the exact result by only about 1 part in a thousand.

3.7 Power

Conservation of energy implies that you never get energy for nothing. If a certain amount of kinetic energy is supplied to an object, the source of energy necessarily loses at least an equivalent amount of energy. Usually the source loses more than the minimum, since energy is unavoidably dissipated as heat. The total amount of energy transferred is limited by the capacity of the source. A battery, for example, will keep a light burning for only a limited time before it is flat and requires to be recharged. The distance a car can travel before needing more petrol depends on the way it is driven—high speeds and frequent acceleration are uneconomical, but it is still true that the engine cannot deliver more energy than is stored in the petrol.

● *Power is the rate of using energy*

Energy sources are limited not only by the total amount of energy available, but also by the rate at which it can be delivered. The rate of work done by a source of energy is called its **power**. Since energy is measured in joules and time in seconds, the units of power are **joules per second** ($\mathrm{J\,s^{-1}}$). Power is such an important quantity that it has a special unit of its own, called the **watt**: one watt is the power corresponding to a rate of energy transfer of one joule per second.

The watt is named after James Watt, the engineer famous for supposedly watching the lid moving up and down as a kettle boiled. Watt was a successful businessman as well as an inventor, and his factory in Birmingham supplied many steam engines in the early days of the industrial revolution.

One watt is a small amount of power. Certainly, if it were converted into heat, one watt would not do much to keep you warm—an electric kettle usually consumes about 3000 watts (three kilowatts or 3 kW). The examples below indicate the power required in different kinds of transport.

Worked Example 3.8 (a) The most powerful sports car you can buy can reach 100 kph from a standing start in about 4 seconds. What is its average power output, assuming the mass of the car to be about one tonne. (b) In 4 seconds a 60 kg sprinter might reach $10 \, \text{m s}^{-1}$. What is his power output?

Answer The average power is the kinetic energy $\frac{1}{2}mv^2$ divided by the time taken. (a) For the car, the top speed of 100 kph $\approx 28 \, \text{m s}^{-1}$. The kinetic energy for a mass of 1000 kg is nearly 400 kJ (one kilojoule $\equiv 10^3 \, \text{J}$), and the average power is about 100 kW. (b) the power output of the runner is 750 W. This may seem to be a small power output, but humans can only keep up this work rate for a very short time.

Exercise 3.8 (a) For electric motors it is often a good approximation to assume a constant power output over a wide range of speeds. A 200 tonne train has a 1 MW electrical power unit (one megawatt $\equiv 10^6 \, \text{W}$). Estimate the time taken for the train to reach 180 kph, neglecting air resistance. (b) The aircraft in Worked example 2.6 weighs 100 tonnes, lands at 180 kph, and comes to rest in 50 s under the action of a constant braking force. What is the maximum power dissipated by the brakes?

Answer (a) 250 s, (b) 5 MW, which is twice the average power.

There are always losses in energy transfer. This is so even if the type of energy is not changed in the transfer. For instance, as we have seen in the collision of railway wagons that are being coupled together, the kinetic energy before and after collision is not the same. Kinetic energy has been dissipated as heat. Transfer of electrical energy can be done with little loss. This is why it is worthwhile to distribute electrical energy over large distances on high-voltage transmission lines. But even here, the cables are heated by the current they are carrying and the amount of energy received at the end of the transmission line is less than the amount delivered at the beginning.

The **efficiency** of any energy transfer system is the ratio of energy received to the energy delivered by the source. Efficiency is usually quoted in per cent, and for an electrical transmission line it is very high—over 99%. Electric motors also have very high efficiencies, typically over 90%. However, for systems using heat as part of the energy transformation process, efficiencies are much lower. For example, in an internal combustion engine less than half of the energy of combustion is converted into useful mechanical energy. This is not because of friction or bad design, but because of a fundamental property of heat engines that limits their efficiency to a maximum value depending on their operating temperatures. This topic is discussed in Chapter 12.

● *The efficiency of engines relying on heat is limited by temperature*

When discussing power, it is necessary to be careful to define whether you are talking about power delivered from a source, or power received. The power of a car engine, for example, is normally quoted in **brake horse power**. This is an unfortunate unit derived from the amount of continuous work a horse was supposed to be able to do! One horse power is equivalent to 746 watts. For the car engine it is the rate at which the engine can do work against a stationary force, i.e. a brake exerting a force on the wheels. This amount is not a constant but depends on the speed at which the engine is running, and you will not be surprised to learn that the value given by the car manufacturers is the maximum power that the engine can deliver.

● *Power delivered by gravity*

The variable output power of the internal combustion engine is one of the reasons for having gears in a car. The gear ratios are chosen so that the engine is running at a speed not too far from the speed at which it can deliver peak power, whatever may be the speed of the car. This not only gives the driver the opportunity of high acceleration, but is also economical, since the engine is then most efficient at converting the energy of the fuel into mechanical work.

Work is done on a falling object by the gravitational force, which therefore delivers power to the object. Let us call the power P. What is the value of P? Since work done is force × distance, if an object of mass m falls a distance δz in a time δt, the work done is $mg\delta z$ and the rate of work $mg\delta z/\delta t$. In the limit when δt tends to zero, $\delta z/\delta t$ is the speed $v = dz/dt$, so the rate of work done by the gravitational force, which is the power delivered, is $P = mgv$. In the absence of any air resistance, all the work is converted into kinetic energy. The kinetic energy is $\frac{1}{2}mv^2$, and its rate of increase is

$$\frac{d}{dt}\left(\tfrac{1}{2}mv^2\right) = mv\frac{dv}{dt} = v\frac{dp}{dt}, \tag{3.29}$$

since $p = mv$ is the momentum. Newton's second law tells us that dp/dt is the force mg, and eqn (3.29) may be rewritten as

$$P = mgv = \frac{d}{dt}\left(\tfrac{1}{2}mv^2\right). \tag{3.30}$$

This equation is simply a statement of the conservation of energy applied to the falling object.

Equation (3.29) can easily be generalized to apply to forces that are not necessarily in the same direction as the velocity. If the net force on an object is $\mathbf{F}$, the work done by $\mathbf{F}$ in moving through an infinitesimal distance $d\boldsymbol{\ell}$ is $\mathbf{F} \cdot d\boldsymbol{\ell}$. This work is related to the change in speed dv for an object of mass m by eqn (3.16):

$$\mathbf{F} \cdot d\boldsymbol{\ell} = mv\,dv.$$

The power is the work per unit time, i.e.

$$P = \mathbf{F} \cdot \frac{d\boldsymbol{\ell}}{dt} = mv\frac{dv}{dt} = \frac{d}{dt}\left(\tfrac{1}{2}mv^2\right),$$

the same result as eqn (3.30), which was derived for the special case of an object falling with no air resistance.

In practice, for an object falling towards the earth through the atmosphere, there is always some air resistance. The power generated by the gravitational force approaches a limiting value mgv_t, where v_t is the terminal speed of the object. Notice that the power received by the object at first increases and then decreases to *zero*, since at the terminal speed the *net* force acting on the object is zero, because the gravitational force is exactly balanced by the air resistance. Calculation of the total energy supplied by gravity in this case is left as an example for you to solve.

Enormous amounts of power can sometimes be delivered by gravitational forces. An avalanche in high mountains is a vivid and frightening example of this. On the cosmic scale huge explosions are powered by gravity. Figure 3.27 shows an image of a radio source outside our galaxy. The image was made by the Hubble telescope using visible light and the contours are from the radio emissions measured with an array of radio telescopes. Analysis of images made at different times show that after the explosion the matter ejected from the source was moving at 0.98 of the speed of light and reached a temperature of about 100 000 million degrees.

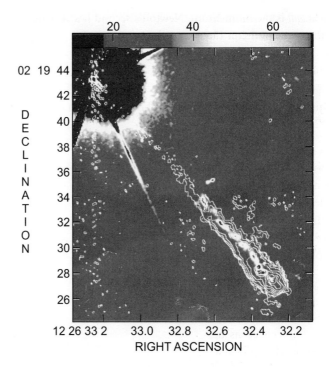

Fig. 3.27 Gravitational energy contributes to the enormous power observed in astronomical events. The jet of luminous matter in this image is emitting about 10^{40} watts.

Problems

Level 1

3.1 A ball of mass m_1 moving at a speed v hits a stationary ball of mass m_2. What is the maximum possible speed of the ball of mass m_2 after the impact?

3.2 At a fairground an apple is placed at a height of 1.5 m above the ground on a model head representing William Tell's son. The arrow is fired at 50 m s^{-1} and has a mass one-third of that of the apple. The arrow hits the apple and sticks to it. How far behind the head does the apple hit the ground?

3.3 A poacher shoots a duck with a rifle as it is approaching him. It is flying at 20 m s^{-1}, and the bullet hits it at an angle of 45° to its flight direction. The bird weighs 20 times as much as the bullet, which moves at 150 m s^{-1}. Through what angle is the bird deflected by the bullet?

3.4 A simple pendulum, consisting of a small mass suspended on a light rod 50 cm long, is released from rest at an angle of 30° to the vertical. Calculate the speed of the small mass when the rod is vertical, neglecting air resistance.

3.5 A rail wagon of mass 5000 kg is shunted in a

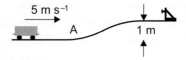

Fig. 3.28

goods yard. It rolls freely and reaches the point A at the bottom of the incline shown in Fig 3.28 at a speed of 5 m s^{-1}. How fast is the wagon moving at the top of the incline, and what is its energy when it hits the buffers? You may neglect the energy of rotation of the wheels.

3.6 A block of mass 1 kg, starting from rest, slides down a ramp at an angle of 30° to the horizontal with negligible friction. What is its kinetic energy when it has moved 2 m down the ramp? (You only need conservation of energy to solve this part of the question.) Compare the time to slide 2 m with the time taken for the block to fall freely through the same vertical distance.

3.7 In Worked example 3.5 the coefficient of friction was calculated for a sledge sliding down a hill and then coming to rest on a flat surface. Assuming the angle θ to be 15° and the coefficient of friction 0.13 as before, and taking the distance d to be 100 m, find out how fast the sledge is going at the bottom of the hill.

3.8 Five steel balls are strung from a bar as shown in Fig 3.29. A ball at one end is pulled aside, in the plane including the other balls, and then released. What happens if the collisions between the balls are elastic? How is the result modified if the coefficient of restitution between the balls is 0.95?

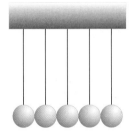

Fig. 3.29

3.9 A fire hose has a diameter of 2.5 cm and is required to direct a jet of water to a height of at least 40 m. What is the minimum power of the pump needed for this hose?

3.10 The climber in Problem 2.15 falls off the cliff again, once again falling 10 m vertically until held by a rope fixed near the point where she fell. The rope stretches 1 m in bringing the climber to rest. Calculate the maximum tension in the rope, assuming that it obeys Hooke's law.

3.11 The propeller of a wind turbine has a diameter of 20 m. Estimate the maximum power it can deliver when the wind speed is 50 kph.

Level 2

3.12 A missile travelling north at 500 m s^{-1} at a height of 500 m explodes into three pieces, which later hit the ground simultaneously. One piece, of mass 150 kg, lands directly below the explosion point, while the second, of mass 150 kg, lands 5 km north and 3 km west of the explosion point. The third piece has a mass of 200 kg. Where does it land? (You may neglect air resistance.)

3.13 Nuclear power is derived from the fission of uranium nuclei, which split into two unequal pieces with the release of an enormous amount of kinetic energy. The mass of the fissioning uranium nucleus is 235 times that of a hydrogen nucleus. In a particular fission event, the nucleus splits into parts with masses 138 and 97 times the mass of the hydrogen nucleus. What fraction of the kinetic energy is carried by the lighter fragment?

3.14 Four particles, each of mass m, are positioned at the corners of a square. All are moving in the plane of the square at the same speed v, in the directions shown in Fig 3.30. What is the momentum of the centre of mass of the particles? What is the internal energy of the system?

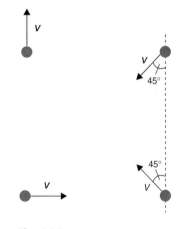

Fig. 3.30

3.15 A bullet is fired along a rifle barrel at a speed v relative to the barrel. The mass of the rifle is 50 times that of the bullet. What is the speed of the centre of mass (a) if the rifle is fixed and (b) if it is free to move. Draw diagrams showing the velocities of the bullet and of the rifle with respect to the centre of mass.

3.16 An object falling vertically at a speed v experiences a retarding force kv due to air resistance. By integrating over the power delivered to it as it falls, show that the total amount of work done on the object by gravity is $\frac{1}{2}mv_t^2$, where v_t is its terminal speed.

3.17 A wedge has an angle of 30° and a height of 20 cm. It is placed at rest on a horizontal air cushion so that it can move almost without friction. A puck of mass m is at the top of the wedge and is initially also at rest. The puck then slides down the wedge onto the air cushion. If the mass of the wedge is $4m$, at what speed does the puck move over the air cushion after leaving the wedge?

3.18 A ball is thrown downwards from a height of 2 m above the ground and bounces up to 4 m. At what speed was the ball thrown? How long after it is released does it reach the top of the first bounce? Assume that the bounce is elastic.

3.19 Two balls, one of mass m and the other of mass $3m$, are approaching one another moving at the same speed, and the angle between their tracks is 90°. They collide elastically, and the lighter ball is deflected through 90°. What is the angle of deflection of the other ball?

3.20 If the speed of the balls before collision in Problem 3.19 is v, what is the speed of the centre of mass? Calculate the energies of the two balls after collision in centre of mass coordinates, and the energy of the centre of mass in laboratory coordinates.

3.21 The force between two small bar magnets separated by a distance r varies as $1/r^4$. If the force is 1 N at 1 cm, what is the potential energy of the magnets at this distance? (Two magnets attracting one another are often used to keep refrigerator doors closed—you may have noticed that, although it can be quite hard to begin to open the door, the attractive force quickly gets smaller as the magnets get further apart.)

3.22 A large oil drum is 5 m high and has a diameter of 2 m. A pipe at the bottom of the drum has a diameter of 5 cm. The drum is full, but the bung comes out of the pipe and oil gushes out so that the level falls by 10 cm in

5 s. If the drum were able to recoil frictionlessly, how fast would it be going after this time?

3.23 The fission of a uranium nucleus is described in Problem 3.13. The mass of a uranium nucleus undergoing fission is about 4×10^{-25} kg and about 3×10^{-11} J of energy is released as kinetic energy. The kinetic energy is dissipated as heat. Estimate the mass of uranium consumed in one year by a nuclear reactor that has a thermal power output of 3000 MW.

Level 3

3.24 A wire makes up an equilateral triangle of side 50 cm as shown in Fig 3.31. Beads of masses m and $3m$ are placed at the corners so that the top two can slide downwards. The whole triangle is suspended so that it is free to move horizontally. Friction is negligible. How fast is the triangle moving horizontally at the moment when the top bead has reached the bottom of the triangle?

3.25 Two weights of masses m and $3m$ are connected by a light string passing over a pulley. The system is initially at rest. After falling 3 m, the heavier weight hits the ground and stays at rest. It is then pulled up again by the other mass. What is its initial upward speed when it leaves the ground? How far does it now move before coming to rest again?

3.26 In a nuclear reactor the fission of uranium is induced by neutrons, which are uncharged particles with a mass nearly the same as that of a hydrogen atom. In the splitting of each uranium nucleus, a number of new neutrons are released, which are able to maintain a chain reaction. In a certain type of nuclear reactor, fast neutrons are slowed down by collisions with the carbon atoms in graphite.

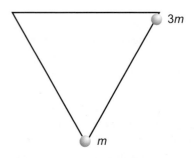

Fig. 3.31

These collisions are perfectly elastic. What is the maximum fraction of its initial energy that a neutron can lose in a single collision? You may assume that the mass of a carbon atom is exactly 12 times that of a neutron.

3.27 An accelerator generates a beam of nuclear particles moving at a speed v which strike a stationary target. The mass of the nuclei in the beam is twice that of the nuclei in the target. In an elastic collision with one of the particles in the beam, a target nucleus moves forward at an angle of 30° to the beam direction. Draw the velocities of the beam and target nuclei in the centre of mass coordinate system after the collision, and calculate the speed of the knocked-on target nucleus in the laboratory frame of reference.

3.28 A puck sliding along a horizontal surface strikes a stationary puck of the same mass. After the collision both pucks are moving at an angle of 26.6° to the direction of the incoming puck before the collision. Calculate the coefficient of restitution in the collision and the fraction of the incident kinetic energy that is lost.

3.29 A platform is supported by a spring that is free to expand and contract in a vertical direction, as shown in Fig 3.32. When two masses, one of 400 g and the other of 1600 g, are placed on the platform, it is depressed by 5 cm. The 1600 g mass is suddenly removed. What is the upward speed of the 400 g mass when it leaves the

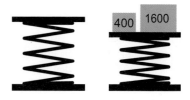

Fig. 3.32

platform? Assume that none of the potential energy of the spring is lost as heat, and that it obeys Hooke's law.

3.30 A rocket motor fires exhaust gases at a speed $3 \times 10^4 \, \mathrm{m\,s^{-1}}$. The fuel for the motor has a mass of 80 tonnes and it is all used in burning at a uniform rate for 15 seconds while the motor is held at rest in a test firing. What is the power delivered to the exhaust gases in this test firing? The motor is now used for the first stage in the launch of a rocket with mass 100 tonnes at lift-off. Calculate the kinetic energy delivered to the rocket's payload, and the efficiency of the rocket, i.e. the ratio of this kinetic energy to the energy you calculated for the test firing.

3.31 In Problem 2.29 a bullet is slowed down by air resistance. Calculate the energy dissipated by integrating the power expended against air resistance, and show that your result is consistent with conservation of energy.

Some solutions and answers to Chapter 3 problems

3.1 The maximum speed of the struck ball occurs in an elastic head-on collision. The momentum in the direction of the initial motion of the ball with mass m_1 is the same before and after the collision, and the momentum is zero perendicular to this direction. If the speeds of m_1 and m_2 after the collision are v_1 and v_2, respectively, then $m_1v = m_1v_1 + m_2v_2$. For an elastic collision, the relative velocity of the two balls is the same before and after the collision: $v_2 - v_1 = v$. Hence $m_1v = m_1(v_2 - v) + m_2v_2$, giving the result

$$v_2 = \frac{2m_1v}{m_1 + m_2}.$$

If $m_1 \gg m_2$, v_2 approaches $2v$.

3.3 The bird is deflected through an angle of 20°.

3.4 The speed of the bob is $1.15 \, \mathrm{m\,s^{-1}}$ when the rod is vertical.

3.7 The components of the gravitational force on the sledge are $mg \sin\theta$ down the slope and $mg \cos\theta$ perpendicular to the slope. If the coefficient of friction is μ_d, the net force down the slope is $(mg \sin\theta - \mu_d mg \cos\theta)$, and the acceleration is $a = g(\sin\theta - \mu_d \cos\theta)$. Putting $\theta = 15°$ and $\mu_d = 0.13$ gives $a = 1.306 \, \mathrm{m\,s^{-2}}$, and the speed of the sledge after moving 100 m down the slope is $\sqrt{2 \times 1.306 \times 100} = 16.2 \, \mathrm{m\,s^{-1}}$.

3.9 The speed of the water leaving the hose must be $\sqrt{2gh}$ if it is to reach a height h when directed vertically.

If the diameter of the hose is d, the volume of water ejected at this speed is $\frac{1}{4}\pi d^2 \times \sqrt{2gh}$ m³ s⁻¹, and for a water density ρ the mass ejected is $\frac{1}{4}\pi\rho d^2 \times \sqrt{2gh}$ kg s⁻¹. The kinetic energy of this water leaving the hose is $\frac{1}{2}mv^2 = \frac{1}{8}\pi d^2 \times (2gh)^{3/2}$. The density of water is 10^3 kg m⁻³, and the power is $\frac{1}{8} \times 10^3 \times \pi \times (5 \times 10^{-2})^2 \times (2 \times 9.8 \times 40)^{3/2} = 21\,500$ W, or 21.5 kW.

3.10 $10\,000$ N, about the same as the weight of 1000 kg or one tonne.

3.11 An optimistic upper limit to the power obtainable from the windmill is obtained by the assumption that all the kinetic energy of the air passing through the area swept out by the propeller is converted into useful energy. If the radius of the propeller is r and the wind speed is v, a column of air of volume $\pi r^2 v$ crosses this area each second. For an air density ρ, the kinetic energy of this column is $\frac{1}{2}\rho \times \pi r^2 v \times v^2$. In this example the wind speed is 50 km h⁻¹ = 13.9 m s⁻¹ and the power is therefore $\frac{1}{2} \times 1.2 \times 10^2 \times 13.9^3 \approx \frac{1}{2}$ MW. A single fossil-fuelled power station may generate as much as 2000 MW of electrical power. Only very large 'farms' containing many windmills can make a significant contribution to electricity production.

3.12 The third fragment of the missile lands 8.9 km north and 2.25 km west of the explosion point.

3.14 The momentum of the centre of mass of a set of particles with masses m_i and velocities $\mathbf{v}_i$ is $\sum_i m_i\mathbf{v}_i$. The upward components of the two masses on the right of the square in Fig 3.30 are equal and opposite. The net upward momentum is therefore mv. The net horizontal momentum is $(2mv\cos 45° - mv) = 0.414mv$ from right to left. The total momentum of the centre of mass thus has a magnitude $\sqrt{(1 + 0.414^2)} \times mv = 1.08mv$ at an angle $\tan^{-1}(0.414) = 22.5°$ to the upward direction.

3.17 Let the mass of the puck be m and its speed be v with respect to the wedge when it reaches the bottom of the wedge. The wedge recoils horizontally at a speed v_w, say. The horizontal component of the puck's momentum relative to the air cushion is then $(mv\cos\theta - mv_w)$ and this equals the opposite momentum $4mv_w$ of the recoiling wedge. Hence $v_w = \frac{1}{5}v\cos\theta$ and the horizontal component of the puck's velocity relative to the air cushion is $\frac{4}{5}v\cos\theta$. Since the vertical component of its velocity is $v\sin\theta$, the total kinetic energy of the puck plus wedge is $\frac{1}{2}m(v^2\sin^2\theta + \frac{16}{25}\cos^2\theta) + 2mv_w^2 = \frac{17}{40}mv^2$

for $\theta = 30°$. The total kinetic energy equals the initial potential energy mgh of the puck; hence $v = 2.15$ m s⁻¹ and the horizontal component of the puck's speed is 1.47 m s⁻¹. The air jets on the cushion stop the puck's vertical motion and after leaving the wedge it moves along the cushion at 1.57 m s⁻¹.

3.19 Suppose that the ball of mass $3m$ is moving in the x-direction, the ball of mass m in the y-direction, and that each ball has a speed v, as shown in the diagram. After the collision the balls move off at speeds v_3 and v_1, at angles θ and $90°$ to their initial directions. The momentum in the x-direction is the same before and after the collision: $3mv = mv_1 + 3mv_3\cos\theta$; in the y-direction the momentum is $mv = 3mv_3\sin\theta$.

Conservation of energy gives the equation $2mv^2 = \frac{1}{2}mv_1^2 + \frac{3}{2}mv_3^2$. Hence

$$v_3 = \frac{v}{3\sin\theta}; \quad v_1 = 3(v - v_3\cos\theta) = 3v(1 - \tfrac{1}{3}\cot\theta).$$

Substituting in the energy conservation equation and dividing by $\frac{1}{2}mv^2$

$$4 = 9(1 - \tfrac{1}{3}\cot\theta)^2 + \frac{1}{3\sin^2\theta}.$$

Now $\dfrac{1}{\sin^2\theta} = \dfrac{\cos^2\theta + \sin^2\theta}{\sin^2\theta} = 1 + \cot^2\theta,$

so that $4 = 9(1 - \tfrac{1}{3}\cot\theta)^2 + \tfrac{1}{3}(1 + \cot^2\theta).$

This equation simplifies to $2\cot^2\theta - 9\cot\theta + 8 = 0$, with the solution

$$\cot\theta = \frac{8 \pm \sqrt{9^2 - 8^2}}{4} = \frac{9 \pm 4.12}{4}.$$

The two solutions correspond to $\theta = 17°$ and $\theta = 39°$. Both solutions are possible, depending on the point of impact between the two balls.

3.20 The speed of the centre of mass is $v_{cm} = \sqrt{10/16}\,v = 0.791v$. The energies of the heavier and lighter balls in centre of mass coordinates are $\frac{9}{16}mv^2$ and $\frac{3}{16}mv^2$ respectively. The energy of the centre of mass in laboratory coordinates is $\frac{1}{2} \times 4m \times v_{cm}^2 = \frac{5}{4}mv^2$. The sum of these three energies is $2mv^2$.

3.22 After 5 s the drum is moving at 2.5 cm s⁻¹.

3.23 The mass of ²³⁵U used in a year is about 1260 kg $\approx 1\frac{1}{4}$ tonnes.

3.25 The initial upward speed of the weight is 2.7 m s⁻¹ and it rises 50 cm before coming to rest.

3.27 If the speed of the beam particles is v, the speed of the centre of mass is $2v/3$. The target nucleus moves towards the centre of mass at a speed $2v/3$, with a momentum equal and opposite to that of the beam particle. After the collision the target moves away from the centre of mass at a speed $2v/3$. To find its speed in laboratory coordinates its velocity is added vectorially to the centre of mass velocity. If the target nucleus moves away at 30° to the beam in laboratory coordinates, the diagram shows that its speed in the laboratory is $2 \times \frac{2}{3}v \times \cos 30° = 1154v$.

3.28 The coefficient of restitution is 0.5 and $\frac{3}{8}$ths of the initial kinetic energy is lost in the collision.

3.30 The total energy of the exhaust gas is $\frac{1}{2} \times 80 \times 1000 \times 9 \times 10^8 = 3.6 \times 10^{13}$ J. This energy is expended uniformly in 15 s, so the power is $3.6 \times 10^{13}/15 = 2.4 \times 10^{12}$ W.

The rocket equation (2.18) leads to the following result for the speed V_f of the rocket after the launch:

$$V_f = v_e \ln \frac{M_i}{M_f} - gt_f. \tag{2.22}$$

In this example $V_f = 3 \times 10^4 \ln 5 - 9.8 \times 15 \approx 48 \text{ km s}^{-1}$. The kinetic energy of the 20 tonne payload now moving at this speed is $\frac{1}{2} \times 20 \times 1000 \times (4.8 \times 10^4)^2 = 2.3 \times 10^{13}$ J. The 'efficiency' of the rocket is the ratio of this energy to the energy released in the static firing, i.e. 64%.

3.31 The net force acting on the object when it is moving at a speed v is $F = mg - kv = m\,dv/dt$. The time taken to reach the speed v is

$$\int_0^t dt = \int_0^v \frac{m\,dv}{mg - kv},$$

$$\text{giving} \quad t = -\frac{m}{k}\left[\ln\left(\frac{mg}{k} - v\right)\right]_0^v$$

and leading to $v = \frac{mg}{k}\left(1 - e^{-kt/m}\right)$.

The power delivered by the force F is $Fv = (mg - kv)v$ and the work done in the time t is

$$\int_0^t (mgv - kv^2)\,dt = \int_0^t \frac{m^2 g^2}{k}$$

$$\times \left\{\left(1 - e^{-kt/m}\right) - \left(1 - e^{-kt/m}\right)^2\right\}dt$$

$$= \frac{m^2 g^2}{k}\int_0^t \left\{e^{-kt/m} - e^{-2kt/m}\right\}dt$$

$$= \frac{m^2 g^2}{k}\left\{\frac{m}{k} - \frac{m}{2k}\right\} = \frac{1}{2}m\left(\frac{mg}{k}\right)^2 = \frac{1}{2}mv_t^2,$$

since the terminal speed v_t satisfies $mg - kv_t = 0$.

Chapter 4

Rotation

Rotation is a special kind of motion that requires some new concepts. These concepts are introduced in this chapter and their use in calculating the motion of rotating objects is explained.

Rotations, slow or fast, can be seen in all kinds of different contexts. The Earth goes round the Sun once a year and spins on its axis to complete one revolution each day. Nearly all land transport depends on rotation—bicycles, motor vehicles, and trains all run on wheels. Ships are driven by rotating propellers. Aeroplanes are powered by jet engines that have as their main components rotating compressors and rotating turbines.

Rotational motion is clearly of great practical importance, and the whole of this chapter is devoted to considering it. We shall find that no new laws of physics are needed to understand rotation—Newton's laws are sufficient. But in Section 4.3 we shall develop new equations, based on Newton's laws, that are particularly suited to describe rotational motion.

4.1 Rotational energy

Any object rotating about an axis is in motion and possesses kinetic energy. The total kinetic energy of the whole object is the sum of the kinetic energies of its different parts, which are not usually all moving at the same speed. The purpose of this section is to work out the total kinetic energy, but to begin with we must learn how to calculate the speeds of the different parts.

Angular speed

Fig. 4.1 The bicycle is standing upside down, and its front wheel can spin freely.

A wheel makes a good starting point for discussing rotational motion. Imagine that you have turned a bicycle upside down and it is now standing on the floor (Fig 4.1). Each of the wheels can now rotate about a fixed axis perpendicular to the plane of the wheel and passing through its hub. Spin the front wheel at such a speed that the valve for inflating the tyre is at the top of the wheel once every second. The valve traces out a circle every second, and so does every other part of the wheel. The rotational speed of the wheel is described by the number of revolutions it makes in unit time. Engineers often use revolutions per minute (rpm) as the unit of rotational speed. The wheel going round once a second therefore has a rotational speed of 60 rpm.

When setting up equations of rotational motion, instead of counting revolutions, it is more convenient to look at the angle through which the wheel turns and to talk about its **angular speed**. In one complete revolution the wheel turns through an angle of 2π radians or, equivalently, $360°$. We shall measure angles in radians, and thus the wheel with a rotational speed of 1 rpm has an angular speed equal to 2π radians per second (rad s^{-1}). At this angular speed the wheel will take $\frac{1}{4}$ s to move $\pi/2$ radians (i.e. $90°$) between the positions shown in Fig 4.2(a) and (b). The Greek symbol ω (omega) is used to represent angular speed. Since

$\omega = 2\pi$ rad s^{-1} at one revolution per second, the time taken to complete one revolution at an angular speed ω rad s^{-1} is $2\pi/\omega$ s and the time to rotate through an angle θ is θ/ω.

Look at what happens to one of the spokes as the wheel turns through a small angle $\delta\theta$ as in Fig 4.3. The time taken is $\delta t = \delta\theta/\omega$ and hence $\omega = \delta\theta/\delta t$. Taking the limit as $\delta\theta$ and δt tend to zero the angular speed becomes

$$\omega = \frac{\mathrm{d}\theta}{\mathrm{d}t}. \tag{4.1}$$

The outside parts of the wheel move faster than the parts near to the hub. Consider a point on the spoke at a distance r from the centre of rotation at the hub. In the time δt this point moves through a circular arc of length $r\delta\theta$, and its speed along the arc is $v = r\delta\theta/\delta t = r\omega$ and, in the limit as $\delta\theta$ and δt tend to zero,

$$v = r\frac{\mathrm{d}\theta}{\mathrm{d}t} = r\omega. \tag{4.2}$$

Any point on the wheel has a fixed value of r, and the speed $v = r\omega$ is constant for a wheel rotating with a constant angular speed ω. For varying angular speed, v represents the instantaneous speed of a point at a distance r from the hub at a moment when the angular speed is ω.

Worked Example 4.1 The diameter of a bicycle wheel is 24 inches. What is the angular speed of the wheel when the bicycle is being ridden at 20 miles per hour?

Answer There are 5280 feet in one mile and 12 inches in one foot, so the speed of the bike can be rewritten as

$$\frac{20 \times 5280 \times 12}{3600} = 352 \text{ inches per second.}$$

From eqn (4.2) the angular speed ω is $v/r = 352/12 = 29.3$ rad s^{-1}. Here the speed and the radius have been expressed in terms of the same Imperial unit—the inch (one inch is defined to be *exactly* 2.54 cm). This gives the angular speed in rad s^{-1}, the same as a calculation in SI units, because the conversion factor from inches to metres cancels out in the ratio v/r. It is necessary to be familiar with Imperial units, since they are still widely used in engineering in the USA and in the UK. This is a very simple example and it is clearly safe to stick with Imperial units here. But if you are faced with other units in more complicated problems, the best advice is: when in doubt, convert all your units to SI.

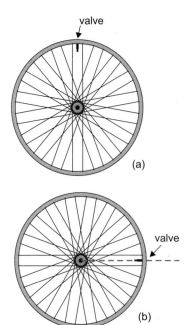

Fig. 4.2 The position of the wheel can be followed by looking at the valve. The wheel has moved through 90° between (a) and (b).

● *The relation between speed and angular speed*

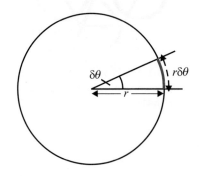

Fig. 4.3 As the wheel turns through a small angle $\delta\theta$, a point at a distance r from the centre advances a distance $r\delta\theta$.

Rotational energy and moment of inertia

The centre of mass of a balanced wheel is located at the hub. If the wheel as a whole is not moving, as in the upside-down bicycle, the centre of mass is stationary and the wheel's momentum is zero. The energy due to the rotation is part of what was called *internal energy* in Section 3.4.

If the wheel has a mass M and its centre of mass starts to move at a velocity **v**, then according to eqn (3.17) the additional energy acquired is $\frac{1}{2}Mv^2$, however fast or slowly it may be rotating. The internal energy of rotation has no effect on its centre of mass motion, which can be calculated using Newton's laws as they apply to any object of mass M. We shall call the motion of the centre of mass *translational motion* to distinguish it from the rotation about the axis.

How can we calculate the kinetic energy of a wheel rotating at an angular speed ω? Suppose for the moment that all the mass of the wheel is concentrated in its rim, at a radius R from the hub. Divide the rim into many small pieces and call the mass of the ith piece δm_i. The total mass M of the rim is then $\sum_i \delta m_i$. The ith piece contributes $\frac{1}{2}\delta m_i \times v^2$ to the kinetic energy of the wheel. We can use eqn (4.2) to work out the kinetic energy E_K of the whole wheel in terms of the angular speed ω:

$$E_K = \sum_i \tfrac{1}{2}\delta m_i \times v^2 = \sum_i \tfrac{1}{2}\delta m_i \times R^2\omega^2 = \tfrac{1}{2}R^2\omega^2 \sum_i \delta m_i$$
$$= \tfrac{1}{2}MR^2\omega^2. \tag{4.3}$$

The quantity $I_{\text{rim}} = MR^2$ is called the **moment of inertia** of the rim about the axis through the hub of the wheel. The moment of inertia has dimensions (mass $\times$ length^2) and in SI units is measured in $\text{kg}\,\text{m}^2$.

The rotational kinetic energy is proportional to the square of the angular speed, and in terms of the moment of inertia is

$$E_K = \tfrac{1}{2} I_{\text{rim}}\omega^2. \tag{4.4}$$

● *Rotational energy depends on the moment of inertia and on angular speed*

Worked Example 4.2 Estimate the energy of a bicycle wheel with radius 30 cm and mass about 0.5 kg, when it is rotating at one revolution per second.

Answer Most of the mass of the wheel is concentrated around the rim, so the moment of inertia is about $0.5 \times 0.3^2 \simeq 0.05\,\text{kg}\,\text{m}^2$. The angular speed is $\omega = 2\pi\,\text{rad}\,\text{s}^{-1}$, and the rotational energy $\frac{1}{2}I\omega^2$ is about 1 joule. This is roughly the same as the translational kinetic energy of the wheel after being dropped from a height of 20 cm.

The expression for the rotational kinetic energy given in the previous paragraph applies only to a wheel with all its mass concentrated on the rim. However, eqn (4.4) can be generalized to apply to more realistic rotating objects with mass distributed at different distances from the axis. As a simple example, consider a disc of radius R and uniform thickness h made of a material with constant density ρ. Like the bicycle wheel, the disc is rotating at an angular speed ω about an axis through its centre and perpendicular to its plane. An annulus of the disc centred on the axis with inner and outer radii r and $(r + dr)$, respectively, has an area $2\pi r \times dr$ and a mass $\rho h \times 2\pi r dr$. According to eqn (4.2) the speed of this part of the disc is $v = r\omega$ and its kinetic energy is $\frac{1}{2}\rho h \times 2\pi r dr \times r^2\omega^2$. The total rotational kinetic energy is found by integrating over all distances from the centre at $r = 0$ to the outer radius of the disc at R:

$$E_K = \tfrac{1}{2}\rho h \int_0^R 2\pi r \times r^2\omega^2 dr = \tfrac{1}{2}\rho h \omega^2 \int_0^R 2\pi r^3 dr = \tfrac{1}{4}\pi \rho h R^4 \omega^2.$$

The total mass of the disc is $M = \rho h \times \pi R^2$, and the rotational kinetic energy may be put in the same form as eqn (4.4) by writing

$$E_K = \tfrac{1}{4}MR^2\omega^2 = \tfrac{1}{2}I_{\text{disc}}\omega^2$$

where the moment of inertia of the disc is

$$I_{\text{disc}} = \tfrac{1}{2}MR^2. \tag{4.5}$$

Exercise 4.1 A flywheel consists of a solid steel cylinder of radius 20 cm, thickness 5 cm, and density 8 g cm^{-3}. What is its kinetic energy when it is rotating about its axis at 5000 rpm?

Answer 137 kJ. The moment of inertia of a disc varies very rapidly with its size. The mass of an object varies as the cube of its dimensions, and there is also a factor R^2 in eqn (4.5). This means that, if the radius and the width of the disc in this exercise are both reduced by a factor of ten, the moment of inertia falls by a factor 10^5. Only a little energy is stored in the rotating parts of small electric motors, and they can be brought up to full speed very quickly.

4.2 The moment of inertia of a rigid body

In eqn (4.5) the kinetic energy of a disc rotating about an axis through its hub and perpendicular to the plane of the disc is expressed in terms of the moment of inertia I_{disc}. At an angular speed ω the rotational kinetic

● *The moment of inertia of a disc*

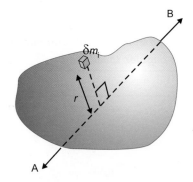

Fig. 4.4 The object in the diagram is divided into small elements like the one with mass δm_i shown in the diagram. Each element contributes to the moment of inertia about the axis of rotation AB.

energy is $\frac{1}{2}I_{\text{disc}}\omega^2$. An expression of the same form applies to any rigid body rotating about a fixed axis. The axis does not have to pass through the centre of mass—for example, a swinging gate rotates about its hinges, which are at one end of the gate, not in the middle. Whatever the shape of an object, and whatever the axis about which it is rotating, the rotational kinetic energy at an angular speed ω may always be written as $\frac{1}{2}I\omega^2$.

What is the moment of inertia of an object like the one in Fig 4.4 for rotation about the axis AB? The argument used to find the moment of inertia of a wheel still applies. Imagine that the object is divided up into many very small pieces, with the ith piece having a mass δm_i. If the perpendicular distance of this piece from the axis of rotation is r_i, then its speed as it rotates is $v_i = r_i\omega$ and its kinetic energy $\frac{1}{2}\delta m_i \times v_i^2 = \frac{1}{2}\delta m_i \times r_i^2\omega^2$. The rotational kinetic energy of the whole object is just the sum of the energies of all the pieces:

$$E_K = \sum_i \tfrac{1}{2}\delta m_i \times r_i^2\omega^2 = \tfrac{1}{2}I\omega^2,$$

with the moment of inertia

$$I = \sum_i \delta m_i \times r_i^2. \qquad (4.6)$$

Calculation of the moment of inertia

The rotational kinetic energy $\frac{1}{2}I\omega^2$ looks just like the kinetic energy $\frac{1}{2}mv^2$ for translational motion, with the speed v replaced by the angular speed ω, and the mass m replaced by the moment of inertia I. When discussing rotations we need to know the moment of inertia, which is a quantity more difficult to find out than the mass. To find the mass, you weigh the object, and that is the end of it—you don't need to know anything about its shape. In contrast, the moment of inertia depends on the position of the different parts of the object as well as its mass, and on the axis about which it rotates.

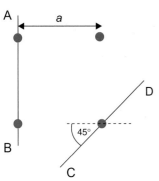

Fig. 4.5 The blue circles represent masses at the corners of a square of side a. Moments of inertia are required about the axes AB and CD.

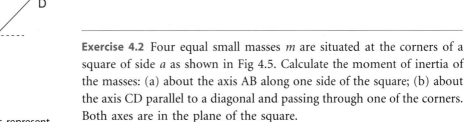

Exercise 4.2 Four equal small masses m are situated at the corners of a square of side a as shown in Fig 4.5. Calculate the moment of inertia of the masses: (a) about the axis AB along one side of the square; (b) about the axis CD parallel to a diagonal and passing through one of the corners. Both axes are in the plane of the square.

Answer (a) $2ma^2$; (b) $3ma^2$.

The expression for moment of inertia was given in eqn (4.6) as a sum over all the parts of a composite object. This is fine for an object made up of a number of small masses, as in Exercise 4.2. However, it is much more common to need the moment of inertia of an object made of a continuous material. The way to calculate moments of inertia for continuous bodies is to perform an integration. We shall go to the limit in which the masses δm_i in eqn (4.6) are infinitesimal and express the moment of inertia as an integral rather than as a sum. If the object is made of a uniform and continuous material—like a steel bar, for example—then each of the small pieces with mass δm_i will be squeezed up against its neighbours with no spaces in between. The small pieces fill the whole volume of the object. Call the volume of the ith piece δV_i. The material is uniform, and, if its density has the constant value ρ, then

$$\delta m_i = \rho \delta V_i,$$

and eqn (4.6) becomes

$$I = \rho \sum_i r_i^2 \delta V_i.$$

Now we can take the limit as all the volumes δV_i tend to zero and write

$$I = \rho \int_V r^2 \mathrm{d}V \qquad (4.7)$$

where r is the perpendicular distance of the volume element $\mathrm{d}V$ from the rotation axis.

The integral in eqn (4.7) is a **volume integral**, i.e. it is an integral in three dimensions. The suffix V on the integral sign indicates that the integral is over the whole of the volume occupied by the object.

Worked Example 4.3 Calculate the moment of inertia of a cube with side a about an axis passing through its centre and through the centre of two of its six faces, as shown in Fig 4.6. The cube is made of a uniform material of density ρ.

Answer To find the moment of inertia we must evaluate the volume integral in eqn (4.7). How can such a three-dimensional integral be worked out? To do this we have to choose a coordinate system and express the volumes of the infinitesimal boxes $\mathrm{d}V$ in terms of the coordinates. For the cube, Cartesian coordinates x, y, and z are clearly suitable. The origin O is placed at the centre of the cube, and the moment of inertia is calculated about the z-axis. The infinitesimal boxes $\mathrm{d}V$ become cubes with volume $\mathrm{d}x\,\mathrm{d}y\,\mathrm{d}z$. Integrations must be performed over x, y, and z separately, and each variable lies within the limits $\pm\frac{1}{2}a$. For the box inside the cube with the position coordinates x, y, and z, the square of the distance from the z-axis is $r^2 = x^2 + y^2$ (Fig 4.7). Writing out the integrations in eqn (4.7)

● *The moment of inertia for an object of any size and shape made of a uniform material*

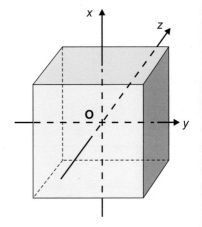

Fig. 4.6 The axes x, y, and z lie parallel to the sides of the cube. The moment of inertia of the cube can be evaluated by integrating over each of the variables x, y, and z in turn.

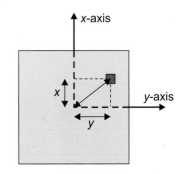

Fig. 4.7 A cross-section through the cube passing through the box of volume $\mathrm{d}V = \mathrm{d}x\,\mathrm{d}y\,\mathrm{d}z$. The axis of rotation coincides with the z-axis. The distance r of the box from the axis of rotation is $r = \sqrt{x^2 + y^2}$.

● *Multidimensional integrals*

separately for each of the three variables x, y, and z,

$$I = \int_{z=-a/2}^{+a/2} \int_{y=-a/2}^{+a/2} \int_{x=-a/2}^{+a/2} \rho(x^2 + y^2)\, dx\, dy\, dz. \tag{4.8}$$

The order in which the integrations are performed makes no difference to the final answer, but we shall choose to do the x-integration first, then the y- and z-integrations. The x-integral is $\int_{-a/2}^{+a/2} \rho(x^2 + y^2)dx$. This looks awkward because y occurs in the expression as well as x. However, in this example y is completely independent of x, since the limits of x are $\pm\frac{1}{2}a$ for any value of y that lies between $\pm\frac{1}{2}a$. While integrating over x, the variable y may be treated as a constant, and

$$\int_{-a/2}^{+a/2} \rho(x^2 + y^2)dx = \rho\left[\tfrac{1}{3}x^3 + y^2 x\right]_{-a/2}^{+a/2} = \tfrac{1}{12}\rho a^3 + \rho a y^2.$$

The integral over y becomes

$$\int_{-a/2}^{+a/2} (\tfrac{1}{12}\rho a^3 + \rho a y^2)dy = \rho\left[\tfrac{1}{12}a^3 y + \tfrac{1}{3}ay^3\right]_{-a/2}^{+a/2} = \tfrac{1}{6}\rho a^4$$

and, finally,

$$I = \int_{-a/2}^{+a/2} \tfrac{1}{6}\rho a^4 dz = \tfrac{1}{6}\rho a^5.$$

● *The moment of inertia of a cube*

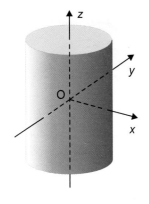

Fig. 4.8 To work out the moment of inertia of a solid cylinder about an axis along its centre, we have again chosen the z-axis to coincide with the axis of rotation.

The mass of the cube is $M = \rho a^3$ and we can write the moment of inertia as $I = \tfrac{1}{6}Ma^2$.

The moment of inertia about a particular axis is sometimes described in terms of the **radius of gyration** R_g, which is defined as the distance from the axis at which the whole of the mass would have to be placed in order to have the same moment of inertia. The moment of inertia may then be written as $I = MR_g^2$. In Worked example 4.3, the moment of inertia about the axis Oz is $I = \tfrac{1}{6}Ma^2 = MR_g^2$, and $R_g = \sqrt{\tfrac{1}{6}}a$.

Cylindrical polar coordinates and cylindrical symmetry

Rotating objects are often cylindrical in shape, rotating about the axis of the cylinder. For cylinders it is better to work out the volume integral in a different way. Figure 4.8 represents a uniform cylinder of radius R and height h. The origin has been chosen to be at the centre of the cylinder and the z-axis to coincide with its axis. The circle in Fig 4.9 represents a section through the cylinder lying in the xy-plane at a height z above the base. The shaded strips in the figure are each of width dy, and the x-integration is made along the length of each strip. The limits of the x-integration therefore depend on y, and the integration becomes a bit messy.

A better way of dividing up the cylindrical section into little boxes is to define the position of a point in the circle by its distance r from the centre and its angle θ from the x-axis, as shown in Fig 4.10. Any point in the cylinder can now be specified by the three coordinates r, θ, and z. These coordinates are called **cylindrical polar coordinates**. The coordinate r, which is the distance from the axis of the cylinder, is just what we want for calculating the moment of inertia.

In cylindrical polar coordinates, the volume of an infinitesimal box must be expressed in terms of r, θ, and z. The area of the box shaded dark in Fig 4.10 is $dr \times r d\theta$. If the thickness of the section is dz, the volume element dV is $r\, dr\, d\theta\, dz$. For a cylinder made of a uniform material with density ρ, the mass of the infinitesimal volume element is then $\rho dV = \rho r\, dr\, d\theta\, dz$. Substituting this value into eqn (4.7), the moment of inertia of a cylinder of height h lying between $z = -\frac{1}{2}h$ and $z = +\frac{1}{2}h$ now becomes

$$I = \int_{z=-h/2}^{h/2} \int_{\theta=0}^{2\pi} \int_{r=0}^{R} \rho r^3 dr\, d\theta\, dz. \tag{4.9}$$

The integrand depends only on r, so the θ and z integrations may be done at once;

$$\int_{\theta=0}^{2\pi} d\theta = 2\pi, \qquad \int_{z=-h/2}^{h/2} dz = h,$$

leading to

$$I = 2\pi\rho h \int_{r=0}^{R} r^3 dr = \tfrac{1}{2}\pi\rho h R^4$$

or, in terms of the mass $M = \pi\rho h R^2$ of the cylinder,

$$I = \tfrac{1}{2}MR^2,$$

the result already obtained for a uniform disc in Section 4.1.

In this integration nothing depends on the variable θ, which does not occur in the integrand or in any of the integration limits. The integration over θ covers the whole of its possible range from 0 to 2π. If the cylinder is rotated through some angle about the z-axis there is no way of telling what the angle is. The cylinder looks just the same after the rotation as

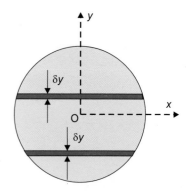

Fig. 4.9 Slices through the cylinder at different distances from the axis are all of different lengths. Cartesian coordinates are not the most convenient choice for calculating the moment of inertia.

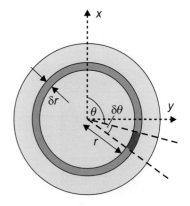

Fig. 4.10 For a uniform cylinder the mass contained in a height dz and within the dark shaded area does not depend on the angular variable θ. The integration over θ then becomes very easy.

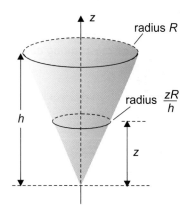

radius R

radius $\dfrac{zR}{h}$

h

z

z

Fig. 4.11 Cylindrical polar coordinates are the best choice for any object that possesses cylindrical symmetry like this cone. A section of the cone perpendicular to the z-axis is circular at any height z.

before. This state of affairs is described by saying that the cylinder possesses **cylindrical symmetry**. Rotating objects often have cylindrical symmetry about their axes of rotation. Whenever this occurs the integration over θ is simply 2π, and all that needs to be worked out to find the moment of inertia is a two-dimensional integral over r and z. This is why cylindrical polar coordinates are frequently the best choice for working out moments of inertia.

Worked Example 4.4 Calculate the moment of inertia of the cone shown in Fig 4.11, about the axis Oz.

Answer The algebra is made easier by choosing the origin O of coordinates to be at the apex of the cone rather than at its base. At a distance z from the apex of the cone its radius is zR/h and the limits on the integration over r therefore depend on z. It is easiest to do the integration over r first. Integrating over r and then over z, we find

$$I = 2\pi \int_{z=0}^{h} \int_{r=0}^{zR/h} \rho r^3 \, \mathrm{d}r \, \mathrm{d}z = 2\pi\rho \int_{z=0}^{h} \frac{1}{4}\left(\frac{zR}{h}\right)^4 \mathrm{d}z = \frac{\pi\rho R^4 h}{10}.$$

The mass of the cone is given by the same integral except that r^3 is replaced by r. This works out to be $M = \frac{1}{3}\pi\rho R^2 h$, so that $I = \frac{3}{10}MR^2$, i.e. the radius of gyration of a cone about its symmetry axis is $\sqrt{\frac{3}{10}}R$ whatever the height from the base to the apex.

Symmetry

Symmetry is a very important concept in physics. The cylindrical symmetry we have described represents the equivalence of all directions perpendicular to a *line*. There can also be symmetry with respect to *points* or *planes*. For example, the Sun is nearly spherical and, although sunspots come and go, on average the whole of its surface is rather uniform. It follows that in our solar system the intensity of radiation received from the Sun depends only on the distance from the centre of the Sun, and not on direction. This is an example of **spherical symmetry**. Similarly, **planar symmetry** occurs when all points on a plane surface are equivalent. Planar symmetry is strictly only valid for infinite planes, but in practice it is often a good approximation to assume that it applies to slabs with small thickness compared to their other dimensions. For example, in a window made of a uniform pane of glass, the temperature inside the glass depends only on the perpendicular distance from the surface of the glass. The temperatures of the inside and outside surfaces are the same, and the temperature must also be constant (except near the edges) on any parallel plane within the glass.

More subtle effects of symmetry are found in solid state physics and in quantum mechanics. These effects are not easy to explain in a few words and, indeed, some are not explicable at all in terms of classical physics. Their consequences are far-reaching, and the study of symmetries in Nature has led to a deeper understanding of the structure of the world around us.

So far we have been considering rotating objects made up of a uniform material with a constant density. The moment of inertia then depends only on the total mass and shape of the objects. These conditions do not always apply. A rotating star, for example, has a density that varies with the distance from the centre. Writing the density of the material in the volume element dV as ρ_V, the volume integral in eqn (4.7) becomes

$$I = \int_V \rho_V r^2 dV. \tag{4.10}$$

Although it is easy to write down this equation, performing the integration may be far from easy. If the integration cannot be done in terms of well-known functions, then it is necessary to return to eqn (4.6) and evaluate the moment of inertia numerically by adding the contributions from small pieces.

Numerical methods are also likely to be the most efficient way of calculating moments of inertia for objects that do not have regular shapes. Components in rotating machinery, for example, may consist of many parts made of different materials and having complicated shapes. The reader should also remember that the concept of moment of inertia is only useful for *rigid* bodies. A mass of fluid may rotate about a common axis, but with different parts of the fluid having different angular speeds.

The theorem of parallel axes

A rotating object does not necessarily spin about an axis that passes through its centre of mass.

For example, if a circular disc rotates about an axis perpendicular to the disc and passing through its circumference, the moment of inertia about that axis is needed to work out its rotational energy. Figure 4.12 is a plan view of such a disc, with an element of mass δm_i at a distance d_i from the centre at O and at a distance r_i from the axis of rotation passing through the point A at the edge of the disc.

The **theorem of parallel axes** relates the moment of inertia about the axis through A to the moment of inertia about an axis parallel to the first but passing through the centre of mass at O. The theorem is most easily derived by writing eqn (4.6) in vector form. The vector representing the displacement from A to the element δm_i is $\mathbf{r}_i = \mathbf{d}_i - \mathbf{R}$. Summing over all the elements of the disc, the moment of inertia about the axis through A is

$$I_A = \sum_i \delta m_i r_i^2 = \sum_i \delta m_i \mathbf{r}_i \cdot \mathbf{r}_i = \sum_i \delta m_i (\mathbf{d}_i - \mathbf{R}) \cdot (\mathbf{d}_i - \mathbf{R})$$

$$= \sum_i \delta m_i d_i^2 + \sum_i \delta m_i R^2 - 2 \sum_i \delta m_i \mathbf{d}_i \cdot \mathbf{R}. \tag{4.11}$$

● *A completely general expression for the moment of inertia of a body rotating about a fixed axis*

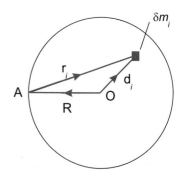

Fig. 4.12 The moment of inertia about an axis through A can be related in a simple way to the moment of inertia about a parallel axis passing through the centre of mass at O.

Now the term $\sum_i \delta m_i d_i^2$ is the moment of inertia I_C about the centre of mass of the disc, and $\sum_i \delta m_i$ is its total mass M. From eqn (3.10), since the vector $\mathbf{d}_i$ is measured from the centre of mass, $\sum_i \delta m_i \mathbf{d}_i = 0$, and the last term on the right of eqn (4.11) is zero. The equation simplifies to

$$I_A = I_C + MR^2. \tag{4.12}$$

This is the theorem of parallel axes. The argument that led to the theorem was general and did not depend on the fact that we were using the circular disc as an example. In words, the theorem is expressed as

The moment of inertia of a rigid object about any axis is equal to the moment of inertia about a parallel axis passing through the centre of mass, plus the total mass times R^2, where R is the perpendicular distance between the two axes.

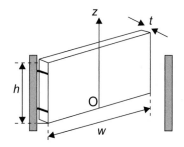

Fig. 4.13 The z-axis passes through the centre of the gate, which swings about hinges at one end.

Worked Example 4.5 The gate in Fig 4.13 is made with a rectangular piece of wood 2 m wide and hinged at one end. The mass of the gate is 60 kg. Estimate its moment of inertia about an axis through the hinges.

Answer Call the width of the gate w, its height h, thickness t, and density ρ. The axis Oz passes through the centre of the gate. The mass of a vertical section of width dx is $\rho h t dx$ and the mass of the whole gate is $M = \rho h t w$. If t is much less than w, it is a good approximation when estimating the moment of inertia to assume that all the mass of the gate lies in the xz-plane. The moment of inertia about the vertical axis Oz of the gate is then

$$I_C = \int_{-w/2}^{w/2} \rho h t x^2 \, dx = \left[\tfrac{1}{3} \rho h t x^3 \right]_{-w/2}^{w/2} = \tfrac{1}{12} \rho h t w^3 = \tfrac{1}{12} M w^2.$$

Now we can use the theorem of parallel axes to calculate the moment of inertia about the hinges at a distance $w/2$ from the centre of mass:

$$I = \tfrac{1}{12} M w^2 + M(w/2)^2 = \tfrac{1}{3} M w^2.$$

Since M is 60 kg and w is 2 m, the moment of inertia is 80 kg m^2.

In this example the moment of inertia about the hinges can be easily evaluated directly by choosing different limits of integration:

$$I = \int_0^w \rho h t x^2 \, dx = \tfrac{1}{3} \rho h t w^3 = \tfrac{1}{3} M w^2,$$

in agreement with the answer obtained using the theorem of parallel axes. Frequently, as in the following exercise, a direct calculation is not straightforward, and the problem is greatly simplified by use of the theorem.

Exercise 4.3 Evaluate I_A in eqn (4.12) for the circular disc shown in Fig 4.12, and hence find the radius of gyration of the disc about the axis through A and perpendicular to the plane of the disc.

Answer The radius of gyration is $R_g = \sqrt{\frac{3}{2}}R$.

It has already been shown in Section 3.4 that the centre of mass is a point of special importance when discussing kinetic energy. Kinetic energy can be divided into two parts: the internal energy of motion with respect to the centre of mass, and the remainder which is the same as for a point mass located at the centre of mass and moving with the same velocity as the centre of mass. We can see how this applies to rotation about a fixed axis by using eqn (4.12). For any object rotating at an angular speed ω about an axis that passes through a point A, the total kinetic energy is

$$\tfrac{1}{2}I_A\omega^2 = \tfrac{1}{2}I_C\omega^2 + \tfrac{1}{2}MR^2\omega^2.$$

Here I_A is the moment of inertia about the axis of rotation and I_C the moment of inertia about a parallel axis passing through the centre of mass. The first term on the right-hand side is the internal rotational energy. The perpendicular distance between the centre of mass and the axis through A is R, and the second term is the rotational energy associated with the rotation of the centre of mass.

The theorem of perpendicular axes

Another theorem that is occasionally useful applies to the moments of inertia of thin discs. The moment of inertia of the disc in Fig 4.14 about an axis perpendicular to the disc and passing through the origin at O is $\sum_i \delta m_i r_i^2$. The sum of the moments about the axes x and y lying in the plane of the disc is $\sum_i \delta m_i(y_i^2 + x_i^2) = \sum_i \delta m_i r_i^2$. This equation is correct for a disc of any shape, whatever the orientation of the axes x and y, provided that these axes are in the plane of the disc and are perpendicular to one another. This is the **theorem of perpendicular axes**, which is stated in words as:

The moment of inertia of a flat disc about an axis perpendicular to the disc equals the sum of the moments of inertia about two perpendicular axes lying in the plane of the disc and both meeting the first axis.

● *Kinetic energy for rotation about an axis that does not pass through the centre of mass*

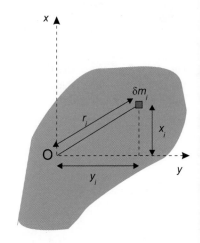

Fig. 4.14 The diagram represents a thin disc that is rotating about an axis perpendicular to the plane of the disc and passing through O. The moment of inertia about this axis is the sum of the moments of inertia about the axes x and y in the diagram.

Exercise 4.4 Calculate the moment of inertia of a thin circular disc of radius R and mass M about an axis in the plane of the disc and tangential to it.

Answer $\frac{5}{4}MR^2$.

4.3 The equation of motion for rotations

● *A torque changes the angular speed of a rotating object*

What is needed to start a wheel turning is to twist it about its hub. The strength of twist is measured by a quantity called a *torque*. A torque plays the same role in rotational motion as force does in translational motion. We shall set up an equation for rotational motion similar in form to Newton's second law, allowing us to calculate how changes in rotational speed are related to the applied torque. Just as the speed of an object can only be increased or decreased by the application of a force, so changes in angular speed only occur when a torque is applied. Before giving a full definition of torque and deriving the equation for rotational motion, we must explain the meaning of *angular velocity*, *angular acceleration*, and *angular momentum*. These quantities are analogous in rotational motion to the velocity, acceleration, and momentum that occur in translational motion.

Angular velocity

The speed alone is not enough to specify the translational motion of an object. A velocity vector, indicating direction as well as speed, is required. Similarly, there is a direction associated with rotation, and we need to define an *angular velocity vector* as well as angular speed.

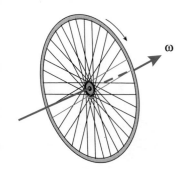

Fig. 4.15 The angular velocity of a rotating object is a vector with magnitude equal to the angular speed about the axis. When viewed in the direction of the angular velocity vector, the object is rotating in a clockwise direction.

Moving the handlebars of a bicycle alters the direction of the axle at the hub of the wheel. The direction associated with the rotation can be aligned along the axle, but this is not enough information to define the angular velocity vector. If you look at the axle from the side of the bicycle, is the vector pointing towards you or away from you?

The convention is that, if you look along the direction in which the angular velocity vector is pointing, the wheel is turning clockwise, as illustrated in Fig 4.15. Equivalently, you can think of the direction of the vector as the direction in which a right-hand threaded screw moves when it is rotating in the same sense as the wheel. With this convention it is now possible to define the **angular velocity vector ω**.

● *The right-hand rule for angular velocity*

The angular velocity is the vector with length equal to the angular speed ω measured in rad s^{-1} and pointing along the axis of rotation in the direction defined by this right-hand screw rule.

According to this definition, when you are pedalling forwards on your bike, the angular velocity of the wheels is directed horizontally to your left.

Angular acceleration and deceleration

Let us return to the example of the upside-down bicycle that we used at the start of this chapter. Suppose you spin the front wheel. If the bearings at the hub are of good quality, the wheel will keep turning for a long time. It is reasonable to expect that, if the bearings were frictionless and there were no air resistance, the wheel would not slow down at all. This is reminiscent of the stone sliding across ice that was considered in Section 2.1, leading to the idea that the momentum of the stone can only be changed by the application of an external force. Similarly, a torque must be applied to change the angular velocity of a wheel.

Imagine that the wheel in Fig 4.16, which has a radius r, is already rotating at an angular speed ω and that it is slowed down by applying a brake at the rim. The figure shows a braking force of magnitude F acting tangentially at the rim, and an opposite force F at the hub—for the hub to remain stationary there must be a reaction that ensures that the net force on the wheel is zero.

As the wheel slows, it is doing work against the brake. Work was defined in eqn (3.12) as 'force×distance moved'. Here the brake is stationary and you may worry that no work can be done. However, as each section of the wheel rim reaches the brake shoe, it pushes against the braking force. When the rim moves a distance $d\ell$, an amount of work $F\,d\ell$ is done by the section in contact with the brake shoe. The power, which is the rate at which work is done, as discussed in Section 3.7, is $F\,d\ell/dt = Fv$, where v is the speed of the rim. This power is dissipated in heating the brake shoe. Conservation of energy requires that the power consumed by the brake equals the rate of change of kinetic energy of the wheel. The kinetic energy E_K is given by eqn (4.3), so the rate at which it is changing is

$$\frac{dE_K}{dt} = \frac{d}{dt}\left(\tfrac{1}{2}I\omega^2\right).$$

The moment of inertia I is constant and, since

$$\frac{d}{dt}\omega^2 = \frac{d\omega^2}{d\omega}\frac{d\omega}{dt} = 2\omega\frac{d\omega}{dt},$$

we may write

$$\frac{dE_K}{dt} = I\omega\frac{d\omega}{dt}.$$

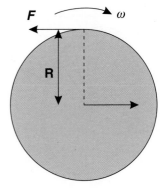

Fig. 4.16 For clarity, the wheel in this figure is represented simply by the blue disc. A braking force applied to the rim of the wheel slows its rotation, but the hub remains at rest. The reaction at the hub is equal and opposite to the braking force: since the wheel as a whole does not move, the net force acting on it must be zero.

⊙ *A wheel loses rotational kinetic energy when it does work on a brake*

Because the wheel is slowing down, $d\omega/dt$ is negative. The magnitude of $d\omega/dt$ is the **angular deceleration** of the wheel. The rate of work done by the wheel on the brake is Fv. This must equal the rate of change of kinetic energy; remembering the negative sign of $d\omega/dt$, we find that

A wheel slowing down

$$Fv = Fr\omega = -I\omega\frac{d\omega}{dt}$$

or

$$Fr = -I\frac{d\omega}{dt}. \tag{4.13}$$

An accelerating wheel

If the force on the wheel were in the opposite direction to the braking force, it would increase the speed of the wheel. The magnitude of $d\omega/dt$ is then positive and it represents an **angular acceleration**. The sign changes in eqn (4.13), which becomes

$$Fr = +I\frac{d\omega}{dt}. \tag{4.14}$$

The left-hand side of eqns (4.13) and (4.14) measures the strength of the torque that causes the change of angular speed. We can now define the magnitude of the torque as the product of the force exerted by the brake and the distance r between the brake and the axis.

To change angular speed quickly, a strong force is needed, but it is also helpful to increase the torque by applying the force at a large distance from the axis. The same applies for making the wheel turn faster as well as for braking. On a bicycle the low gears have a large cog on the back wheel, and it is the low gears that give fast acceleration from rest. Similarly, when turning a nut stuck tightly on its thread, a long-handled wrench is much easier to use than a short-handled one because the long handle helps you to exert a large torque.

The magnitude of a torque

The wheel is shown again in Fig 4.17 with a set of Cartesian coordinates having their origin at the centre of the wheel and the z-axis along the axis of rotation. If a point P on the wheel makes an angle θ to the x-axis, as shown in the figure, then the angular speed $\omega = d\theta/dt$. The force is in the plane of the wheel, in the direction required to cause an angular acceleration. Equation (4.14) applies, and it can be rewritten as

$$Fr = I\frac{d\omega}{dt} = I\frac{d^2\theta}{dt^2}.$$

Here $d^2\theta/dt^2$ is another way of writing the angular acceleration $d\omega/dt$.

This equation corresponds, for rotational motion about a fixed axis, to the familiar equation $F = ma$ which applies to motion in a straight line. The force F has been replaced by the torque Fr, the mass m by the moment of inertia I, and the translational acceleration $a = d^2x/dt^2$ by the angular acceleration $d^2\theta/dt^2$.

Worked Example 4.6 A solid pulley wheel, of thickness 1 cm, rotates about an axis through its centre and perpendicular to the plane of the pulley. A cord wound tightly around the pulley is pulled with a steady force of 1 N. Calculate the angular acceleration of the pulley (a) if its radius is 10 cm, (b) if its radius is 20 cm. (The density of steel is $8 \, g \, cm^{-3}$.)

Answer The moment of inertia of a disc of radius R, thickness h, and density ρ is $\frac{1}{2} MR^2 = \frac{1}{2} \pi \rho h R^4$ (see eqn (4.5)). The cord is *accelerating* the pulley and $d\omega/dt$ is therefore positive. Using eqn (4.14) we have

$$FR = \tfrac{1}{2} \pi \rho h R^4 \frac{d\omega}{dt}, \quad \text{and the angular acceleration} \quad \frac{d\omega}{dt} = \frac{2F}{\pi \rho h R^3}.$$

(a) Using SI units for all the quantities in this expression

$$\frac{d\omega}{dt} = \frac{2}{\pi \times 8000 \times 0.10 \times 10^{-3}} \approx 8 \ \text{rad} \, s^{-2}.$$

(b) The moment of inertia of the pulley varies as R^4. Although the torque exerted by the cord is twice as large for the 20 cm radius wheel as for the 10 cm wheel, the angular acceleration is reduced by the factor 2^3 to $1 \ \text{rad} \, s^{-2}$.

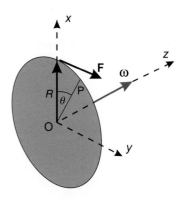

Fig. 4.17 A perspective view of the wheel as it is being braked shows the braking force **F**, the position vector **r** from the axis to the point where the brake is applied, and the angular velocity **ω**. These three vectors are perpendicular to one another.

Angular momentum

The way in which rotating wheels respond to torques acting on them was worked out above. But the discussion was only concerned with magnitudes and not directions. Equation (4.14) needs to be generalized by expressing it in terms of vectors. The angular velocity vector ω has already been defined, and the wheel in Fig 4.17 rotates clockwise when viewed in the direction of ω, or in other words the angular velocity vector is described by the right-hand rule. The z-axis in the figure is chosen to be in the same direction as ω, so that the z-component of ω is $\omega_z = \omega$, the total angular speed, and $\omega_x = \omega_y = 0$.

A torque is also associated with a direction. In this case the force **F** in Fig 4.17 is trying to turn the wheel clockwise about its axis. It is therefore natural to describe the torque as pointing into the page, in the same direction as the angular speed ω. The force **F** is applied at the point on the wheel with position vector **r** with respect to an origin at the centre of the wheel. The vectors **F** and **r** both lie in the plane of the wheel. The vector product **r** × **F** (see Section 20.2 in the mathematical review) is perpendicular to **r** and to **F** and also points into the page. Since **r** and **F** are also perpendicular to each other, the magnitude of the vector product **r** × **F** is Fr, the same as the magnitude of the torque acting as a brake on the wheel. We now define the **torque**, as a vector quantity that

● *The definition of the vector torque*

we shall call $\boldsymbol{\tau}$, to be equal to this vector product:

$$\boldsymbol{\tau} = \mathbf{r} \times \mathbf{F}. \tag{4.15}$$

The vector $\mathbf{r}$ is the position vector, with respect to an origin at O, of the point at which the force is acting. In our example of a brake speeding up a wheel, the origin at O is at the hub of the wheel. However, the same equation for a torque applies for an origin at *any* point O, whether or not it lies on a rotation axis. In general,

If a force F acts at a point with position vector r with respect to an origin at O, then the torque exerted by F about O is given by the vector product r × F.

Let us now apply the vector definition of torque to the problem of the accelerating wheel considered before. In Fig 4.17 the torque is pointing in the z-direction and, by using the right-hand rule, the z-component of $\boldsymbol{\tau}$ is found to be $\tau_z = Fr$. The angular velocity $\boldsymbol{\omega}$ points in the positive z-direction and the z-component of its rate of change is $\mathrm{d}\omega_z/\mathrm{d}t$. The x- and y-components of both $\boldsymbol{\tau}$ and $\boldsymbol{\omega}$ are zero, so their magnitudes are equal to their z-components. Equation (4.14) can be written as

$$\tau_z = I\frac{\mathrm{d}\omega_z}{\mathrm{d}t}. \tag{4.16}$$

Similarly, for a torque that is decelerating the wheel, the z-component of torque is $\tau_z = -Fr$, and eqn (4.16) correctly includes the signs occurring in eqns (4.13) and (4.14).

Equation (4.16) can be written directly in vector notation because the x- and y-components of the torque and of the angular momentum are zero:

$$\boldsymbol{\tau} = I\frac{\mathrm{d}\boldsymbol{\omega}}{\mathrm{d}t}. \tag{4.17}$$

We now define the **angular momentum** of the wheel to be the vector $I\boldsymbol{\omega}$, which is denoted by the symbol $\mathbf{L}$:

⬤ Angular momentum for rotation about a fixed axis

$$\mathbf{L} = I\boldsymbol{\omega}. \tag{4.18}$$

Differentiating eqn (4.18) with respect to time,

$$\frac{\mathrm{d}\mathbf{L}}{\mathrm{d}t} = \frac{\mathrm{d}(I\boldsymbol{\omega})}{\mathrm{d}t} = I\frac{\mathrm{d}\boldsymbol{\omega}}{\mathrm{d}t}$$

since for the rotating wheel the moment of inertia is constant. In terms of the angular momentum, eqn (4.17) is written as

$$\boldsymbol{\tau} = \frac{\mathrm{d}\mathbf{L}}{\mathrm{d}t}. \tag{4.19}$$

Expressing this result in words;

For a wheel rotating about its centre, the torque applied about the centre equals the rate of change of angular momentum.

Although we have only proved eqn (4.19) for the special case of a wheel rotating about its centre, it will be shown later that the equation applies to any object. This is the equation for rotational motion, which corresponds to Newton's second law for translational motion.

Newton's second law was given in eqn (2.5) as

$$\mathbf{F} = \frac{\mathrm{d}\mathbf{p}}{\mathrm{d}t}$$

where the translational momentum $\mathbf{p}$ is (mass × velocity): $\mathbf{p} = m\mathbf{v}$. The rotational equation of motion has a similar form, with torque taking the place of force and angular momentum replacing translational momentum.

Rotation of a rigid body about a fixed axis

Rotation is often constrained to occur only about a fixed axis—doors swing on their hinges, the axles of machines are held in fixed bearings, and so on. When a rigid body rotates about a fixed axis, every particle in the body follows a circular path and has the same angular velocity. The vector notation is not needed for the equation of motion and, if we choose the z-axis to be along the axis of rotation, the vector eqn (4.19) always reduces to eqn (4.16), which refers only to the z-component of an applied torque. There may be components of the torque in directions other than z but, although these may cause stresses in the body or forces on the bearings, they can have no effect on the rotation.

● *Torques in rigid body rotation*

The z-component of the torque due to a force applied at a point is calculated in the mathematical insert. The result can be stated in words as follows:

The z-component of the torque exerted by a force on any body rotating about the z-axis equals the component of the force in the direction of motion of the body at the point of application of the force times the perpendicular distance of that point from the axis.

The proof of the statement above is as follows. The torque exerted by a force is defined by eqn (4.15). The value of the torque depends on the choice of origin, since $\mathbf{r}$ is the position vector of the point of application of the force with respect to that origin. We are now interested in rotation about the z-axis, and the origin must lie on this axis. Figure 4.18 shows a force $\mathbf{F}$ acting on a rigid body at a point P with position vector $\mathbf{r}$. The part of the rigid body now at P is moving in a circle around the z-axis. The origin of coordinates O lies on the z-axis, but it is not in the plane of the circle.

Equation (20.12) defines the vector product of two vectors $\mathbf{a}$ and $\mathbf{b}$ to be

$$\mathbf{a} \times \mathbf{b} = (a_y b_z - a_z b_y)\mathbf{i} + (a_z b_x - a_x b_z)\mathbf{j} + (a_x b_y - a_y b_x)\mathbf{k},$$

where $\mathbf{i}$, $\mathbf{j}$, and $\mathbf{k}$ are vectors of unit length pointing along the positive x-, y-, and z-axes. The z-component of the vector $\mathbf{a} \times \mathbf{b}$ is the coefficient of the unit vector $\mathbf{k}$. In Fig 4.18 the y-axis is chosen to be perpendicular to the position vector $\mathbf{r}$. The y-component of $\mathbf{r}$ is therefore zero, and the z-component of the torque $\mathbf{r} \times \mathbf{F}$ is xF_y. The perpendicular distance from P to the axis is x and the part of the rigid body at P is moving in the y-direction.

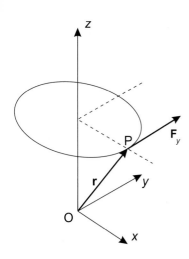

Fig. 4.18 The point P has a position vector $\mathbf{r}$ with respect to the origin at O. The point P is located on a rigid body that is rotating about the axis Oz. A part of the rigid body that passes through P moves in a circle in the xy-plane. A force acting at P has a y-component F_y which causes a change in the angular momentum about the axis.

Knowing the z-component of the torque allows us to calculate the motion of any rigid body about a fixed axis. The method of calculation is illustrated for rotation about an axis that does not pass through the centre of mass of the rigid body in the following worked example.

Worked Example 4.7 The moment of inertia of the hinged gate in Worked example 4.5 is 80 kg m². If a force of 10 N is applied to the end of the gate opposite to the hinges, and perpendicular to the surface of the gate, what is the gate's angular acceleration about the hinges?

Answer The force of 10 N is applied at a perpendicular distance of 2 m. The torque about the hinges is thus 20 N m, and this is also the rate of change of angular momentum about the hinges. Since the moment of inertia of the gate about the hinges has the constant value 80 kg m², the angular acceleration is $20/80 = 0.25$ rad s^{-2}.

Newton's second law requires conservation of translational momentum for systems for which the net external force is zero—this was proved in Section 3.1. In just the same way eqn (4.19) leads to conservation of the angular momentum of any rotating object unless it is acted on by a net external torque. We shall consider conservation of angular momentum more thoroughly in Section 4.5. Before doing this we need to define angular momentum in a more general way to include objects that are not moving in a circle.

Angular momentum in non-circular motion

When an athlete throws the hammer he whirls round in the turning circle. The hammer follows a roughly circular path, accelerating as the athlete pulls it round. After a few turns he releases the hammer, which flies off, eventually falling to the ground due to gravity. The gravitational force cannot exert a horizontal force, however. We are interested in the horizontal motion of the hammer: for the present discussion we will forget about gravity and neglect air resistance, so that we may assume that after release the hammer travels horizontally at a constant speed in a straight line. A plan view is shown in Fig 4.19.

Just before release, the hammer, which has a mass M, is moving horizontally in a circular path of radius R centred at the point O. Its motion at this time is the same as if it were attached to the edge of a horizontal wheel rotating about a vertical axis passing through O. Hammer throwers usually twist in the direction shown in Fig 4.19, anticlockwise when looking down on them as in the plan view: the angular momentum of the hammer is therefore pointing vertically upwards. Its moment of inertia about the vertical axis passing through O is $I = MR^2$. If its angular speed just before release is ω, the speed at that time is $v = R\omega$ and the magnitude of the angular momentum is

$$L = I\omega = MR^2\omega = MvR.$$

The coordinate system in Fig 4.19 has the origin at O and the z-axis vertically upwards, out of the page. In the notation used in eqn (20.7) of the mathematical review, $\mathbf{k}$ is the vector of unit length along the positive direction of the z-axis, which is vertically upwards in this case. The vector representing the angular momentum of the hammer at the point of release is therefore

$$\mathbf{L} = MvR\mathbf{k}. \tag{4.20}$$

After release there is no torque acting on the hammer and it follows from eqn (4.19) that $\mathbf{L}$ remains constant at this value.

After the hammer is released, no force acts on it and its translational momentum also remains constant. The magnitude of the momentum is Mv, and choosing the y-axis to be in the direction of flight as in the figure, the vector representing the momentum is

$$\mathbf{p} = Mv\mathbf{j}, \tag{4.21}$$

where $\mathbf{j}$ is the vector of unit length pointing along the positive y-axis. Figure 4.20 shows the position of the hammer when it has moved a distance y after being released. Its position vector is now

$$\mathbf{r} = R\mathbf{i} + y\mathbf{j}, \tag{4.22}$$

where $\mathbf{i}$ is the vector of unit length along the positive x-axis. The vectors $\mathbf{p}$ and $\mathbf{r}$ both lie in the horizontal xy-plane while $\mathbf{L}$ is directed vertically upwards. Comparing eqns (4.21) and (4.22) with the definition given in

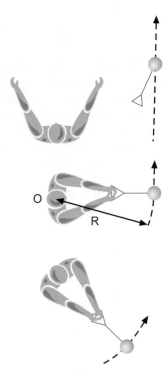

Fig. 4.19 Looking down on a hammer thrower as he turns on an axis and releases the hammer. After he has let go, no horizontal force is acting on the hammer, and the angular momentum about O must remain unchanged.

● *The definition of angular momentum about a point*

eqn (20.12) for a vector product, we find that $\mathbf{r} \times \mathbf{p} = MvR\mathbf{k}$ and hence from eqn (4.20)

$$L = \mathbf{r} \times \mathbf{p}. \qquad (4.23)$$

At all points on the straight line trajectory after release, the magnitude of the angular momentum of the hammer about O is the product of the magnitude of its momentum $p = Mv$ with the perpendicular distance R of the straight line from O.

Equation (4.23) is not limited to objects moving horizontally. It is the completely general expression for angular momentum that we were seeking.

The angular momentum of a moving object about *any* point O is $\mathbf{r} \times \mathbf{p}$ where $\mathbf{p}$ is the object's momentum and $\mathbf{r}$ is the position vector of the object with respect to an origin at O.

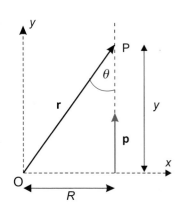

Fig. 4.20 The angular momentum of any moving object about a point O is the vector product of the position vector of the object and its translational momentum.

The total angular momentum of a system of particles

The angular momentum about a point O of a system of particles with momenta $\mathbf{p}_i$ and positions $\mathbf{r}_i$ with respect to an origin at O is simply the vector sum of all the individual angular momenta with respect to the same point O:

$$L = \sum_i \mathbf{r}_i \times \mathbf{p}_i. \qquad (4.24)$$

Using this expression it is now straightforward to derive a general equation for the rate of change of angular momentum. The same rule for differentiating products of scalar quantities also applies to vector products—the vector equations are just a shorthand way of writing separate equations for the three components. Differentiating eqn (4.24) with respect to time, we find

$$\frac{d\mathbf{L}}{dt} = \sum_i \left(\frac{d\mathbf{r}_i}{dt} \times \mathbf{p}_i + \mathbf{r}_i \times \frac{d\mathbf{p}_i}{dt} \right).$$

● *Derivation of the general relation between angular momentum and torque*

Now $d\mathbf{r}_i/dt$ is the velocity of particle i and is in the same direction as the momentum $\mathbf{p}_i$. The vector product of two vectors pointing in the same direction is zero, so the first term on the right-hand side vanishes. By Newton's second law, $d\mathbf{p}_i/dt$ is the net force $\mathbf{F}_i$ on the ith particle, and the equation reduces to

$$\frac{d\mathbf{L}}{dt} = \sum_i \mathbf{r}_i \times \mathbf{F}_i = \sum_i \tau_i.$$

The vector sum $\tau = \sum_i \tau_i$ is the net torque about the origin for the whole system of particles, and we can write

$$\tau = \frac{dL}{dt}.$$ (4.25)

Equation (4.25) is the general equation needed to describe all rotational motion: it is the same as eqn (4.19), which was derived for the special case of the rotation of a wheel.

The derivation of eqn (4.25) makes it clear that no new laws of physics are required to explain rotational motion. Newton's laws are all that is needed, but because rotations and angular momentum are so important it is worthwhile to write the equation of motion for rotation in the special form of eqn (4.25). The equation is expressed in words by stating that

For any system of objects or particles, the net torque acting about any point is equal to the rate of change of angular momentum about the same point.

An important consequence of this statement is that, if the net torque about a particular point is zero, the angular momentum about that point is constant. In other words, the angular momentum about that point is *conserved*. Conservation of angular momentum is considered in more detail in Section 4.5.

Equation (4.25) had already been derived for the special case of a wheel rotating about its hub—eqn (4.19) is exactly the same as eqn (4.25). Here we remind the reader how this equation of motion applies to rotations of a rigid body about a fixed axis.

When a rigid body rotates about a fixed axis the angular momentum is given by eqn (4.18) as $L = I\omega$, where ω is the angular velocity and I the moment of inertia about the axis of rotation. The moment of inertia about the fixed axis is a constant quantity, and eqn (4.25) then reduces to

$$\tau = \frac{dL}{dt} = I\frac{d\omega}{dt}$$

where τ is the torque about an origin placed anywhere on the axis of rotation.

As was explained in the earlier discussion of rotation of a rigid body about a fixed axis, since ω has a fixed direction, the net torque τ must be in the same direction. Any external torques applied in other directions are cancelled out by forces on the bearings or internal forces in the rigid body that ensure that the rotation remains about the fixed axis.

● *The definition of a couple*

Whenever the net force on an object is zero, the net torque is called a **couple**. A wheel rotating about a fixed hub, for example, has no net force acting on it, because the centre of mass is stationary, but its angular speed may be increased or decreased by a couple. The magnitude and direction of a couple do not depend on the point about which the net torque is acting. Suppose that forces $\mathbf{F}_i$ are applied at points on an object with position vectors $\mathbf{r}_i$ with respect to an origin at O, giving a net torque $\sum_i \mathbf{r}_i \times \mathbf{F}_i$ about O. Now choose a new origin O′ with position vector $\mathbf{d}$ with respect to O. The net torque about O′ is $\sum_i(\mathbf{r}_i - \mathbf{d}) \times \mathbf{F}_i = \sum_i \mathbf{r}_i \times \mathbf{F}_i - \mathbf{d} \times \sum_i \mathbf{F}_i = \sum_i \mathbf{r}_i \times \mathbf{F}_i$ since the net force $\sum_i \mathbf{F}_i$ is zero. This is the same as the net torque about O, and there is no need to specify the origin when discussing a couple.

4.4 Centrifugal and centripetal forces

If you stand on a rotating platform like the ones found in children's playgrounds you have to hang on tightly to avoid falling off the edge. You are exerting an *outwards* pull on the platform away from its axis. Similarly, when going round a corner at high speed in a car you push against the door on the outside of the bend. This force outwards from the centre of rotation is called a **centrifugal force**. Another way of describing the same phenomenon is to say that the platform, or the door of the car, is exerting an *inward* force on you to keep you moving on a circular path. A force towards the axis of rotation is called a **centripetal force**.

● *Force is needed to maintain circular motion*

The force occurs because an object moving in a circle is being continuously accelerated. Even if the *speed* in the circle is constant, the *velocity* is continuously changing direction and is therefore not constant. A changing velocity implies acceleration and hence the circular motion can only be maintained in the presence of a force.

Figure 4.21 shows an object of mass m moving in a circle of radius r in the xy-plane. It is attached to the axis of rotation by a rod of negligible mass that keeps the mass m at the same distance r from the axis at all times. The mass has a constant angular speed ω and in a time t it sweeps out an angle ωt. If we choose the zero of time to be at a moment when the mass is on the x-axis, then ωt represents the angle between the x-axis and the rod (multiples of 2π may be added to the angle without changing the position of the rod, so our diagram is really referring to the first revolution after $t = 0$).

In order to find the acceleration of the rotating mass, it is easiest to express its position as a vector. At time t the mass is at a point with position coordinates

$$x = r \cos \omega t; \quad y = r \sin \omega t; \quad z = 0.$$

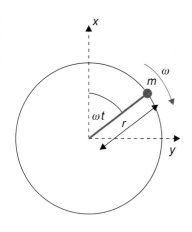

Fig. 4.21 The velocity of an object moving in a circle is continually changing. The change in velocity implies that a force is acting: this force acts along the line joining the object to the centre of the circle.

Using the conventional vector notation outlined in the mathematical review (eqn (20.7)), the position vector **r** is written as

$$\mathbf{r} = r\cos\omega t\mathbf{i} + r\sin\omega t\mathbf{j} \qquad (4.26)$$

where **i** and **j** are vectors of unit length pointing along the *x*- and *y*-axes. The velocity **v** of the mass is given by differentiation of eqn (4.26) with respect to time:

$$\mathbf{v} = \frac{d\mathbf{r}}{dt} = -r\omega\sin\omega t\mathbf{i} + r\omega\cos\omega t\mathbf{j}. \qquad (4.27)$$

The velocity **v** is a vector with magnitude $r\omega$, which is always at right angles to the position vector. You should convince yourself that this is so by showing that $\mathbf{v}\cdot\mathbf{r}=0$. To find the acceleration **a**, differentiate eqn (4.27) with respect to time:

● *Acceleration in circular motion*

$$\mathbf{a} = \frac{d\mathbf{v}}{dt} = \frac{d^2\mathbf{r}}{dt^2} = -r\omega^2\cos\omega t\,\mathbf{i} - r\omega^2\sin\omega t\,\mathbf{j} = -\omega^2\mathbf{r}$$

and the final result is

$$\mathbf{a} = -\omega^2\mathbf{r}. \qquad (4.28)$$

The acceleration of an object moving in a circle at an angular speed ω has a magnitude $r\omega^2$ and is in the same direction as the vector $-\mathbf{r}$, i.e. towards the centre of the circle. Since the speed round the circle is $v=r\omega$, the acceleration may also be written as v^2/r:

In circular motion at constant speed the acceleration is towards the centre of the circle with a magnitude $r\omega^2 = v^2/r$.

The *inwards* centripetal force that keeps the mass *m* on its circular path has a magnitude $ma = m\omega^2 r$. The rod is in tension and it is also exerting an *outwards* centrifugal force $m\omega^2 r$ on the bearing at the centre.

Worked Example 4.8 Suppose that, when an athlete throws a hammer, the hammer turns through a full circle with radius about one metre in about one second before it is released. What is the magnitude of the centripetal force?

Answer The angular speed of the hammer is $\omega = 2\pi\,\text{rad s}^{-1}$ and the acceleration $\omega^2 r = 4\pi^2 r \approx 40\,\text{m s}^{-2}$. For the hammer the acceleration *a* is four times bigger than $g(\approx 10\,\text{m s}^{-2})$ so that the athlete has to pull it inwards with a force about four times its weight.

Centripetal forces occur whenever objects are constrained to follow a curved path, whether or not the paths form complete circles. Over a short distance a curved path coincides with the arc of a circle with a definite centre and radius. This is illustrated in Fig 4.22. For the short section of

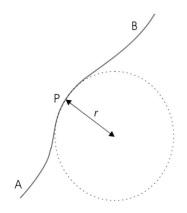

Fig. 4.22 An object moves along the continuous curved line from A to B. When it is at P, a circle of radius r centred on O has an arc that follows the path for a short distance through P. The radius r is called the radius of curvature of the path at P.

● *Centripetal force on a car*

the path lying on the circle in the diagram, the radius r is called its **radius of curvature**. The centripetal force acts along this radius, which is always perpendicular to the path of the moving object. Notice that the centripetal force, which is the force needed to keep the object moving at constant speed on its curved path, can therefore never do any work.

Centripetal forces can be very large. In highly manoeuverable military aircraft the minimum radius of curvature when coming out of a dive may be limited by the ability of the pilot to remain conscious. Accelerations several times larger than the gravitational acceleration g can occur, so that the forces are several times greater than the pilot's body weight. Quite large forces may occur even at more modest speeds. How large is the centripetal force for a car travelling at 150 km per hour around a bend with a 500 m radius of curvature? The speed of 150 kph $\approx 40 \, \mathrm{m \, s^{-1}}$, so the radial acceleration is $40^2/500 \approx 3 \, \mathrm{m \, s^{-2}}$. Since g is a little less than $10 \, \mathrm{m \, s^{-2}}$, this means that the tyres must sustain a sideways force nearly equal to a third of the weight of the car without skidding.

Exercise 4.5 The cylinder of a washing machine has a diameter of 50 cm and it spins at 1000 rpm. What is the maximum ratio of the centrifugal force to the weight for clothes in the machine?

Answer 280.

4.5 Conservation of angular momentum

From eqn (4.25) we see that

If the net torque acting on any system is zero, the angular momentum remains constant.

This is a statement of the **conservation of angular momentum**. When discussing conservation of momentum and energy, we showed that the answers to some problems can be found by using the conservation laws alone, without having to solve any differential equations. The same applies to conservation of angular momentum, as is illustrated in Worked example 4.9.

Worked Example 4.9 A girl jumps down through a height of 1 m on to the end of a see-saw. The see-saw consists of a uniform plank 3 m long pivoted at its centre, and it is horizontal and at rest before the girl jumps. The mass of the see-saw is twice the mass of the girl. What is the angular speed of the see-saw immediately after the girl has landed on it?

Answer After jumping down through a height h, the girl's speed is $\sqrt{2gh}$. Calling her mass m and the length of the plank ℓ, her angular momentum about the pivot when landing on the end of the see-saw is $\frac{1}{2}m\ell\sqrt{2gh}$. In Worked example 4.5 the moment of inertia of a plank about an axis through its centre was calculated: for a plank of mass $2m$ and length ℓ it is $\frac{1}{6}m\ell^2$. If the angular speed of the plank is ω, its angular momentum is $\frac{1}{6}m\ell^2\omega$. The girl is moving at a speed $\frac{1}{2}\ell\omega$ at a distance $\frac{1}{2}\ell$ from the pivot, and the combined angular momentum of the girl plus the see-saw is now $\frac{1}{6}m\ell^2\omega + \frac{1}{4}m\ell^2\omega = \frac{5}{12}m\ell^2\omega$. Since angular momentum is conserved, this equals the initial angular momentum $\frac{1}{2}m\ell\sqrt{2gh}$, i.e. $\omega = \frac{6}{5}\sqrt{2gh}/\ell$. For $\ell = 3$ m and $h = 1$ m, $\omega \approx \frac{2}{5} \times \sqrt{20} \approx 1.8$ rad s^{-1}.

⬤ *Varying rotational energy at constant angular momentum*

Conservation of angular momentum does not necessarily imply that the rotational *energy* stays constant. Rotational energy and other forms of internal energy may be interchanged even though the angular momentum must not change.

For example, imagine a spoked wheel with weights on each spoke that are free to slide without friction inwards or outwards along the spoke, as shown in Fig 4.23. To make the calculation simpler, suppose that the mass of the wheel is negligible, that each weight has the same mass m, and that each is at the same distance r_0 from the axis. The wheel is given an impulse that sets it rotating at an angular speed ω_0. If there are N weights, their combined moment of inertia is $I_0 = Nmr_0^2$, the angular momentum is $L = Nmr_0^2\omega_0$, and the rotational energy, which we shall call $\mathcal{R}_0$, is $\frac{1}{2}I_0\omega_0^2 = \frac{1}{2}Nmr_0^2\omega_0^2$.

There is no frictional force to oppose the centrifugal forces and the weights begin to slide outwards on their spokes. Since there is no external torque acting on the wheel we can invoke angular momentum conservation to work out its angular speed ω when the weights have moved out to a radius r. The moment of inertia has become $I = Nmr^2$ and, since the angular momentum L has not changed,

$$L = Nmr^2\omega = Nmr_0^2\omega_0,$$

leading to

$$\omega = \left(\frac{r_0^2}{r^2}\right)\omega_0.$$

The rotational energy is now

$$\mathcal{R} = \frac{1}{2}Nmr^2\omega^2 = \left(\frac{r_0^2}{r^2}\right)\mathcal{R}_0. \tag{4.29}$$

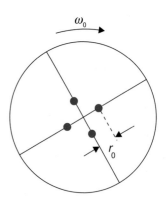

Fig. 4.23 The weights sliding along the spokes of the wheel are initially all at a distance r_0 from the centre.

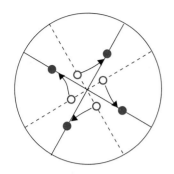

Fig. 4.24 As the wheel spins the weights move out from the centre. The angular speed of the wheel changes so that the total angular momentum remains constant.

Both ω and $\mathcal{R}$ are smaller than their initial values: the wheel slows down as the weights move outwards and its rotational energy decreases. The total kinetic energy has not changed, however. The weights are also moving radially outwards with the speed required to satisfy energy conservation. By the time the radius is twice its initial vaue r_0, the angular speed is only $\frac{1}{4}\omega_0$ and the rotational energy is $\frac{1}{4}\mathcal{R}_0$. The remaining three-quarters of the kinetic energy is in the radial motion.

The weights spiral outwards following the path shown in Fig 4.24. At all points on each path the frictionless reaction of the spoke on the weight is perpendicular to the path and no work is done on the weights. The weights have the same speed as when the rotation started, but the direction of their motion is continually bending outwards.

In this example some of the kinetic energy of rotation has been converted into kinetic energy of radial motion. But other forms of internal energy are also possible. When the weights reach the edge of the wheel they will crash into the rim. If they do not bounce, but stay on the rim, all their radial kinetic energy is absorbed as heat. The wheel then continues rotating at the speed required to conserve angular momentum.

Another possibility is that each weight is attached to the hub by a spring. In that case the outward radial motion is slowed down by the spring, which does do work on the weights, storing potential energy. The potential energy may later be released if the spring pulls the weight inwards, thus increasing the angular speed and the rotational kinetic energy. This is similar to what happens when a skater spinning on the ice with outstretched arms pulls her arms inwards. A dramatic end to a free-skating programme is often made in this way as the skater can make a big reduction to her moment of inertia and spin extremely fast. She has done work in pulling her arms inwards and this work has been converted into rotational energy.

4.6 Combined translational and rotational motion

This chapter started by looking at a turning wheel on an upside-down bicycle. Let's now set the bicycle the right way up and consider what happens when someone pedals it. The wheels go round and the bicycle moves forward—we have both rotational and translational motion. Watch a good cyclist at red traffic lights. He can balance the machine so that it remains stationary. When the lights turn green he pedals away. What are the forces and torques that move the bicycle and turn its wheels?

Forces and torques acting together

It is easiest to start by thinking about forces and torques acting on a one-wheeled cycle that has no chains or cogwheels. The performer on the one-wheeled cycle in the circus ring does the same thing as the cyclist at the traffic lights. When the cycle is stationary his weight Mg is equally divided between the front and back pedals as in Fig 4.25. The length of both crankshafts is ℓ so that equal and opposite torques are acting on the hub. If the pedals are horizontal and the cyclist transfers the whole of his weight to the front pedal, there is now a torque $Mg\ell$ about the hub of the wheel (Fig 4.26).

If there were no friction between the wheel and the ground, the wheel would slip and turn without moving the rest of the cycle. But, usually, there is enough friction to prevent slipping and the wheel rolls forward. The frictional force F opposes the rotation. Calling the radius of the wheel R and its moment of inertia I, the net torque about the centre of the wheel is initially $(Mg\ell - FR)$ and from eqn (4.17) its angular acceleration $d\omega/dt$ is given by

$$Mg\ell - FR = I\frac{d\omega}{dt}. \tag{4.30}$$

Since the wheel is rolling but not slipping, its speed v along the ground is the same as the speed of the rim with respect to the hub; $v = R\omega$. Its initial *translational* acceleration is $a = dv/dt = Rd\omega/dt$. The mass of the rider is M and, if m_c is the mass of the whole cycle including the wheel, the total mass that is moving forward is $(M + m_c)$. The only external force acting on the wheel is the frictional force F. Now we can apply Newton's second law to the forward motion to obtain

$$F = (M + m_c)R\frac{d\omega}{dt}. \tag{4.31}$$

Eliminating F from eqns (4.30) and (4.31), we find the initial forward acceleration to be

$$a = R\frac{d\omega}{dt} = \frac{Mg\ell R}{(M + m_c)R^2 + I}. \tag{4.32}$$

If the mass of the wheel is m_w and is mostly concentrated near the rim, $I \approx m_w R^2$, and

$$a \approx \frac{Mg\ell}{(M + m_c + m_w)R} \approx g\frac{\ell}{R} \tag{4.33}$$

for a cycle that is much lighter than its rider.

Almost the same equation as (4.32) applies to a bicycle with two wheels. There is just enough friction to turn the front wheel, but the main

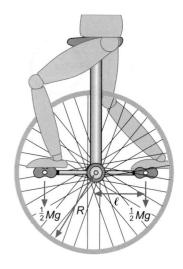

Fig. 4.25 The one-wheeled cycle can be balanced with the rider's weight divided equally between the two pedals.

Fig. 4.26 When the rider puts all his weight on the front pedal there is a torque about the hub of the wheel.

⬤ *The acceleration of a bicycle*

Fig. 4.27 When the bicycle is accelerating there is a couple on the wheels and also a net force on the whole bicycle from the frictional reaction of the ground. It is this frictional force that makes the bicycle move faster.

driving force is from the reaction F of the ground to the back wheel, shown in Fig 4.27. It is surprising but true that all wheel-driven vehicles are being pushed along by the frictional reaction of the ground!

For a bicycle the moments of inertia of both wheels occur in the denominator of eqn (4.32), and there is a small additional term to allow for the rotation of the chain and pedals. But, provided the bicycle is still light compared with the rider, eqn (4.33) will still be roughly correct. The length of the crankshaft on a full-size bicycle is about 18 cm and the radius of the wheels about 33 cm. When the rider puts his weight on the front pedal of a stationary bicycle, his initial acceleration is almost $g/2$.

This large initial acceleration suggests that the bicycle is a remarkably efficient machine. We can estimate its efficiency without having to solve the equations of motion, simply by considering conservation of energy. This method works when there is rotation, just as it did for translational motion alone. Since the purpose of the bicycle is to carry its rider forwards, we can define its efficiency to be the ratio of his forward kinetic energy to the total work he has done.

Initially, before there is any appreciable air resistance, all the work is converted into translational or rotational kinetic energy. The rotational energy of the two wheels, if each has a moment of inertia I, is $I\omega^2 = Iv^2/R^2$. Neglecting the small amount of rotational energy of the chain and the pedals, the total kinetic energy is then $\frac{1}{2}(M+m_c + 2I/R^2)v^2$ where, as before, M is the mass of the rider and m_c the mass of the whole bicycle. Finally,

$$\text{Efficiency} = \frac{\text{kinetic energy of rider}}{\text{total work done}} = \frac{M}{M + m_c + 2I/R^2}.$$

Of course, the efficiency of the bicycle defined in this way declines as air resistance increases, and indeed is zero when the cyclist is moving at a steady speed. However, on any measure it is the most efficient type of wheeled transport, since so little energy is used in moving anything other than the rider.

Exercise 4.6 A bar of length ℓ and mass m is lying on a smooth surface with negligible friction. A horizontal force F, in a direction perpendicular to the length of the bar, is applied to one end for a short time δt. Calculate the speed of the centre of mass and of the end of the bar to which the force was applied after the time δt.

Answer The speed of the centre of mass is $F\delta t/m$ and the speed of the end of the bar is $4F\delta t/m$.

Conservation of both translational and angular momentum

In the absence of external forces or torques, translational and angular momentum are conserved. The translational and angular momenta are both vector quantities, and conservation must apply to the x-, y-, and z-components separately. This means that, in any collision between two objects, the net value of each component of both the translational momentum and the angular momentum must be the same before and after the collision. The directions of both the translational and angular momenta of the colliding objects may change, and calculating the result of a collision may require rather complicated mathematics. To illustrate the principles of the method of calculation for collisions including rotation, we shall examine a collision between a moving pool or snooker ball and a stationary one.

The collision is illustrated in Fig 4.28. The balls each have a mass M, radius R, and a moment of inertia I about an axis through their centres. The collision is not head-on, and the moving ball is deflected through 45°. We shall assume that the collision is perfectly elastic—which means that no energy is lost as heat—and that there is friction large enough to ensure that the balls roll without slipping but small enough for us to neglect frictional energy losses just before and after the collision. Under these assumptions, we shall use conservation laws alone to calculate the direction of the struck ball and the speeds of both balls after the collision.

To calculate angular momenta before and after the collision we must again remember that the internal motion of the balls and the motion of their centres of mass are independent, as explained at the end of Section 3.4. For the rolling balls the internal motion is the rotation about their centres of mass. To consider the angular momentum of both balls, we need to know how to work out the angular momentum of a rolling ball about a point that is not on the rotation axis. This is done by using the following theorem, which is proved in the mathematical box.

The angular momentum of an object about a particular point in space is the sum of the angular momentum about its centre of mass and the angular momentum of the centre of mass about the same point.

Here is the proof that the total angular momentum of a group of particles about a point in space is the sum of the internal angular momentum about the centre of mass and the angular momentum of the centre of mass about the same point. In Fig 4.29 the centre of mass C has a position vector **R** with respect to an origin at O. A particle of mass m_i is situated at the point P with position vector $\mathbf{r}_i$ with respect to O and $\mathbf{r}_{iC}$ with respect

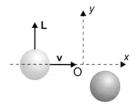

● *A collision between rolling balls*

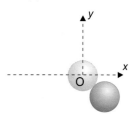

(a) Before collision

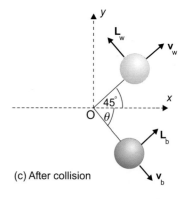

(b) At the moment of collision

(c) After collision

Fig. 4.28 In an elastic collision between rolling balls both energy and angular momentum must be conserved.

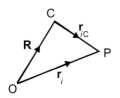

Fig. 4.29 The angular momentum of an object about any point O can be related to the angular momentum about its centre of mass at C.

to C:

$$\mathbf{r}_i = \mathbf{r}_{iC} + \mathbf{R}.$$

The angular momentum about O is

$$L = \sum_i \mathbf{r}_i \times \mathbf{p}_i = \sum_i m_i \mathbf{r}_i \times \frac{d\mathbf{r}_i}{dt} =$$

$$\sum_i m_i(\mathbf{r}_{iC} + \mathbf{R}) \times \left(\frac{d\mathbf{r}_{iC}}{dt} + \frac{d\mathbf{R}}{dt}\right) = \sum_i m_i \mathbf{r}_{iC} \times \frac{d\mathbf{r}_{iC}}{dt}$$

$$+ \sum_i m_i \mathbf{r}_{iC} \times \frac{d\mathbf{R}}{dt} + \sum_i m_i \mathbf{R} \times \frac{d\mathbf{r}_{iC}}{dt} + \sum_i m_i \mathbf{R} \times \frac{d\mathbf{R}}{dt}. \qquad (4.34)$$

The centre of mass is defined by the condition $\sum_i m_i \mathbf{r}_{iC} = 0$, and the second term in the last expression vanishes. The third term can be rearranged as $\mathbf{R} \times d/dt(\sum_i m_i \mathbf{r}_{iC})$, and it also vanishes. Now write $\mathbf{P} = M\, d\mathbf{R}/dt = \sum_i m_i\, d\mathbf{R}/dt$ for the momentum of a mass M moving at the velocity of the centre of mass, and $\mathbf{L}_{cm} = \sum_i m_i \mathbf{r}_{iC} \times d\mathbf{r}_{iC}/dt$ for the angular momentum with respect to C. Equation (4.34) becomes

$$\mathbf{L} = \mathbf{L}_{cm} + \mathbf{R} \times \mathbf{P}. \qquad (4.35)$$

This is the result stated in words earlier.

A special case of interest is when the centre of mass is stationary. The momentum $\mathbf{P}$ is then zero, and $\mathbf{L} = \mathbf{L}_{cm}$ for any choice of $\mathbf{R}$, i.e. the angular momentum about any point whatsoever for an object with a stationary centre of mass equals its internal angular momentum. This is important when dealing with a stationary object made up of a number of parts, such as an atom or a molecule. One can talk about 'the angular momentum' or 'the z-component of angular momentum', without having to specify a particular origin.

Now we choose to calculate the angular momenta with respect to the origin at O in Fig 4.28(b), which is the centre of the white ball just at the moment of collision. Since the white ball was moving towards O, the angular momentum of its centre of mass about O is zero. Similarly, after the collision both balls are moving away from O and there is again no contribution from the centre of mass motion of either ball to the angular momentum. We only need to consider the internal angular momentum of each ball due to its rolling.

Before the collision the moving ball—the white one—has a speed v in the x-direction. It has an angular speed $\omega = v/R$ and the angular momentum $L = Iv/R$ points in the y-direction), perpendicular to the momentum vector $m\mathbf{v}$. The rotational kinetic energy $\frac{1}{2}I\omega^2 = \frac{1}{2}Iv^2/R^2$.

After the collision the velocities and angular momenta are as shown in Fig 4.28(c). The conservation laws now lead to the following equations.

• Conservation of kinetic energy E_K

Since the collision is elastic, the kinetic energies are the same before and after the collision:

$$E_K = \tfrac{1}{2}Mv^2 + \tfrac{1}{2}I\frac{v^2}{R^2} = \tfrac{1}{2}M(v_w^2 + v_b^2) + \tfrac{1}{2}I\left(\frac{v_w^2}{R^2} + \frac{v_b^2}{R^2}\right) \qquad (4.36a)$$

where v_w and v_b are the speeds of the white and the blue balls after the collision.

• Conservation of angular momentum **L**

Similarly, the components of the angular momentum **L** are unchanged,

$$L_x = 0 = -\frac{Iv_w}{\sqrt{2}R} + \frac{Iv_b \sin\theta}{R}, \qquad (4.36b)$$

$$L_y = \frac{Iv}{R} = \frac{Iv_w}{\sqrt{2}R} + \frac{Iv_b \cos\theta}{R}. \qquad (4.36c)$$

Solving for $v_w v_b$, and θ leads to $v_w = v_b = v/\sqrt{2}$ and $\theta = 45°$.

The problem has been solved without having to invoke conservation of translational momentum. This is because, since the balls are rolling without slipping, once we know their velocities we know both their translational and angular momenta. If the balls may slip, translational and angular momenta are not linked in this way, and they become independent variables. Two more equations, representing conservation of translational momentum in the x- and y-directions, are then needed.

4.7 Tops and gyroscopes

Tops and gyroscopes provide very good illustrations of the way in which rotating systems respond to forces. One of the little tops that play a tune when you spin it with your fingers starts with its axis vertical and appears to be almost motionless—it is said to be 'sleeping'. It will remain sleeping for long enough to play 'Happy Birthday to You' two or three times while it is slowing down because of friction. But it does not just fall down and roll away. When it has slowed down sufficiently some roughness on the table will cause the axis to tilt. The top of the spindle then traces a circle about the vertical. The circle gradually gets bigger until the angle of the spindle is such that the body of the top touches the table, and only then does it fall down. Why does it behave in this way?

Toy gyroscopes do something very similar. If the gyroscope in Fig 4.30 is spinning horizontally and one end is placed on a bearing as in the

Fig. 4.30 The force on the spinning gyroscope is downwards, yet it moves horizontally, with the flywheel describing a circle around the bearing at the top of the tower.

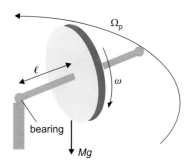

Fig. 4.31 The torque on the gyroscope is acting in the horizontal plane, and the rate of change of angular momentum is in the same direction.

● *Precessional frequency*

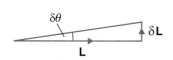

Fig. 4.32 Looking down on the gyroscope we can see that the small change in angular momentum $\delta \mathbf{L}$ is perpendicular to the angular momentum of the flywheel.

figure, the axis turns around the bearing while remaining horizontal (or wobbling up and down only a small amount).

The turning of the axis of a top or a gyroscope is called **precession**. This precession can be explained by eqn (4.25). The gyroscope in Fig 4.31 has a mass M and the distance from the centre of mass to the bearing is ℓ. It is spinning at an angular speed ω and the moment of inertia about its axis is I. This axis is not fixed, and we must consider the effect of torques in all directions, using the vector equation of motion (4.25) to describe the changing angular momentum.

When the axis is horizontal, the weight of the gyroscope exerts a torque $Mg\ell$ about the bearing. This torque is perpendicular to the axis and hence to the gyroscope's angular momentum $L = I\omega$. According to eqn (4.25), the rate of change of angular momentum equals the torque. In a short time δt the angular momentum changes by $\delta \mathbf{L} = \tau \delta t$, and hence $\delta \mathbf{L}$, like τ, is perpendicular to $\mathbf{L}$. This is shown in Fig 4.32, looking down on the gyroscope. The vector addition of $\mathbf{L}$ and $\delta \mathbf{L}$ represents a shift of the axis by an angle

$$\delta\theta = \frac{\delta L}{L} = \frac{\tau \delta t}{L}.$$

In the limit as $\delta t \to 0$, $\delta\theta/\delta t$ is the precessional angular frequency, which we shall call Ω_{p}:

$$\frac{\mathrm{d}\theta}{\mathrm{d}t} = \Omega_{\mathrm{p}} = \frac{\tau}{L} = \frac{\tau}{I\omega}$$

or

$$\tau = I\omega\Omega_{\mathrm{p}}. \tag{4.37}$$

Equation (4.37) is sometimes called the **gyroscope equation**. It tells us that, for a given torque, the faster the gyroscope is spinning on its axis (ω large), the slower it precesses (Ω_{p} small). This means that it is harder to change the direction of the axis when the angular momentum about that axis is high.

Worked Example 4.10 Let us use eqn (4.37) to estimate the precession frequency of the horizontal toy gyroscope in Fig 4.30, which is started by pulling a string wrapped around its spindle. For a typical toy gyroscope $M = 200$ g, $R = 2.5$ cm, and $\ell = 3$ cm. The string wrapped around the spindle is pulled with a steady force of 40 N over 50 cm. What is the period for precession, i.e. the time taken for the gyroscope to complete one horizontal revolution?

Answer Assume that nearly all the mass M of the gyroscope is in the flywheel, which is a uniform disc of radius R. Then the moment of inertia of the gyroscope is $\frac{1}{2}MR^2$ and the torque about the bearing is $Mg\ell$. The gyroscope equation becomes $Mg\ell = \frac{1}{2}MR^2\omega\Omega_p$, leading to

$$\Omega_p = 2g\ell/\omega R^2. \tag{4.38}$$

The angular speed ω is calculated by equating the rotational energy to the work done in pulling the string. If the gyroscope is started with a steady pull by a force F over a distance d, then the work done is Fd:

$$\frac{1}{2}I\omega^2 = \frac{1}{4}MR^2\omega^2 = Fd. \tag{4.39}$$

From eqn (4.39) $\omega = 800$ rad s^{-1} and from eqn (4.38) $\Omega_p = 1.18$ rad s^{-1}. The period for a complete revolution of the precessional motion is $2\pi/\Omega_p$—about 5 seconds.

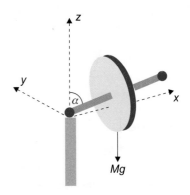

Fig. 4.33 A gyroscope can precess steadily with its axis at a constant angle α to the vertical.

Even if a gyroscope is not horizontal it may exhibit steady precession. Figure 4.33 shows a gyroscope precessing with its axis at an angle α to the vertical. It is spinning at an angular speed ω about its own axis and precessing at an angular speed Ω_p. The angular momentum about the gyroscope's axis and the precessional angular momentum now both have horizontal and vertical components. Choose the origin O at the bearing, and, at the moment when Fig 4.33 applies, let the gyroscope axis lie in the xz-plane.

The gyroscope has a mass M and the moment of inertia about its axis is I. The gyroscope is spinning about its axis at an angular speed ω and the centre of mass has a position vector ℓ. We shall use the usual trick and split the angular momentum of the gyroscope into two parts, namely, the angular momentum of the gyroscope about its centre of mass and the angular momentum of the centre of mass about O. The centre of mass at ℓ is moving perpendicularly to the xz-plane at a speed $\ell\Omega_p \sin \alpha$ and has a momentum **P** in this direction of magnitude $M\ell\Omega_p \sin \alpha$.

The angular momentum of the centre of mass about the origin is $\mathbf{L}_{cm} = \ell \times \mathbf{P}$. This vector product is perpendicular to ℓ and to **P**, i.e. $\mathbf{L}_{cm}$ lies in the xz-plane at right angles to the gyroscope axis and has a magnitude $M\ell^2\Omega_p \sin \alpha$. The internal motion contributes an angular momentum $I\omega$ directed along the axis, but the whole gyroscope is precessing, and there is also an internal angular momentum in the direction of $\mathbf{L}_{cm}$. The theorem of parallel axes tells us that the total angular momentum in this direction has a magnitude $(M\ell^2 + I_\perp)\Omega_p \sin \alpha$, where $I_\perp$ is the moment of inertia of the gyroscope about an axis through the centre of mass perpendicular to ω. The horizontal component of angular momentum, including the contributions from spinning about

● *Precession of a gyroscope when its axis is not horizontal*

the gyroscope axis and from precession, is

$$L_x = I\omega \sin\alpha - (M\ell^2 + I_\perp)\Omega_p \sin\alpha \cos\alpha.$$

The rate of change of angular momentum is found in the same way as for the horizontal gyroscope, and equals the torque $Mg\ell \sin\alpha$ exerted about O by the weight Mg. The equation of motion for the precession becomes

$$Mg\ell \sin\alpha = \Omega_p L_x = I\omega\Omega_p \sin\alpha - (M\ell^2 + I_\perp)\Omega_p^2 \sin\alpha \cos\alpha. \quad (4.40)$$

Since this equation is quadratic in Ω_p, there are two solutions, which are

$$\Omega_p = \frac{I\omega \pm \sqrt{I^2\omega^2 - 4(M\ell^2 + I_\perp)Mg\ell \cos\alpha}}{2(M\ell^2 + I_\perp)\cos\alpha}. \quad (4.41)$$

If the second term under the square root is small compared to the first, i.e. if the gyroscope is spinning very fast or is not far from horizontal, one solution is very close to the value $Mg\ell/I\omega$ that applies to horizontal precession. Under these conditions (spinning fast or nearly horizontal), the other solution gives a very large value of Ω_p, which is not likely to be realized in practice.

● *'Sleeping' tops*

The same equation for precession applies to a top, which has its base in contact with the surface on which it is spinning. The base is not perfectly sharp and, if the top is not vertical, the point of contact is moving on the surface. As well as precessing, the frictional reaction of the surface causes the top to move horizontally, and also exerts a small torque about its spinning axis. This torque twists the top towards an upright position. This explains the observation that a top precesses and moves horizontally until it is vertical, when it 'sleeps', remaining vertical and with its base stationary on the surface. When the top is displaced slightly from the vertical, eqn (4.41) applies. If $4I_\perp Mg\ell \cos\alpha$ becomes larger than $I\omega$, there is no real solution for Ω_p when $\cos\alpha$ is near one. This means that the vertical position is no longer stable; the top tips and precesses faster and faster as ω decreases until it finally falls over.

● *Gyroscopes in navigation*

Gyroscopes are not only toys—they also have practical applications. Gyroscopes are sometimes installed in ships to give increased stability against rolling. A ship rolls from side to side under the influence of torques pointing along the length of the ship, from bow to stern and vice versa. A vertical gyroscope senses the rolling motion and the resulting precession is used to drive fins at the sides of the ship to counteract the rolling.

Gyroscopes are also used for navigation in **inertial guidance systems**. In combination with accelerometers that measure components of acceleration in three directions, the responses of three gyroscopes can be used to measure the components of angular acceleration in three directions. These translational and angular accelerations can be converted into differences in position by integrating twice. It seems almost incredible that such measurements can be made accurately enough to be useful over any great distance. Yet inertial navigation is used by nuclear submarines, which may spend long periods under ice when they are unable to receive outside signals.

Aircraft also use inertial guidance systems, which have a precision of better than a mile after a crossing of the Atlantic. The measurements are independent of wind speed and, although not as accurate as global positioning using satellite signals, provide an on-board system that is available if there is a breakdown in the satellite navigation. Inertial navigation also has military use in the guidance of missiles. Although it has been developed to reach this remarkable precision, gyroscopic inertial navigation is rapidly becoming an obsolete technology. Angular movements can now be measured more accurately with laser 'gyroscopes' that use the interference between two counter-rotating laser beams.

Problems

Level 1

4.1 The diameter of a car wheel is 60 cm. What is its angular speed when the car is travelling at 90 km h^{-1}? If the car takes 2 seconds to accelerate uniformly from 30 to 40 km h^{-1}, what is the angular acceleration of the wheel during this acceleration?

4.2 An electric motor is controlled to deliver a constant power of 1 kW as it drives a flywheel up to a speed of 5000 rpm. The flywheel has a diameter of 20 cm, is 2.5 cm thick, and is made of steel of density 7.5 g cm^{-3}. Neglecting friction, calculate how long it takes for the flywheel, starting from rest, to reach 5000 rpm? What is the angular acceleration just before the flywheel reaches maximum speed?

4.3 A boy is sitting on one end of a see-saw that consists of a uniform plank pivoted at its centre at a height of 1 m above the ground. The plank and the boy have the same mass. Another boy, with a mass of twice that of the first boy, climbs on to the other end of the see-saw. What is his speed when his end of the plank hits the ground?

4.4 A Catherine wheel is a firework consisting of a tube rolled into a disc that ejects hot gas. The total mass of a Catherine wheel is 0.1 kg and it is an approximately uniform disc of radius 5 cm, emitting 0.01 kg of hot gas per second at a speed of 20 m s^{-1}. Neglect friction and estimate its angular speed 0.5 s after it starts to rotate.

4.5 A bicycle wheel is spinning freely at one revolution per second. A steady braking force of 0.025 N (roughly $\frac{1}{4}$ kg wt) is applied to the rim. The wheel has a radius 30 cm and a mass 0.5 kg: most of the mass is concentrated around the rim. Estimate the time needed to stop the wheel.

4.6 A thin-walled steel pipe of radius 0.1 m and mass 300 kg rolls without slipping down a slope inclined at 15° to the horizontal. What is the speed and the kinetic energy of the pipe after it has rolled 3 m along the length of the slope?

4.7 A boy weighing 27 kg is on the edge of a playground roundabout rotating once every 6 seconds. The roundabout platform, which has a radius of 1.5 m and weighs 70 kg, is of uniform thickness and is much heavier than the rails the boy holds on to. The boy pulls himself in to a distance of 0.75 m from the centre of the roundabout. How fast is it now turning?

4.8 The rectangular end of a gate-legged table is 0.5 m wide and is hinged along its one metre length. The gate leg is carelessly folded away, and the end of the table swings on its hinge. How fast is the edge of the table moving when it hits the table leg?

4.9 Two cylinders have the same radius and each has a mass of 10 kg. One is solid and the other is hollow. Both roll down a slope without slipping. Which one rolls faster? What is the acceleration of the solid cylinder down a slope at 20° to the horizontal?

4.10 A solid spherical ball is slipping without rolling on a smooth horizontal surface at a speed v. It reaches a rougher surface where it rotates without slipping. What is now the speed of the ball? (The moment of inertia of a sphere of mass M and radius R about an axis through its centre is $\frac{2}{5}MR^2$.)

Level 2

4.11 A weight is attached to a string wound round a pulley so that it does not slip. The pulley, which has a radius of 3 cm, may be taken to be a uniform disc of the same mass as the weight. The weight is initially at rest one metre above the floor. When it is released, how long does it take to reach the floor?

4.12 A constant radial force of 10 N is applied to the rim of a uniform flywheel of mass 10 kg and radius 50 cm. If the wheel comes to rest after 10 s from an initial angular speed of one revolution per second, what is the coefficient of friction between the wheel and the brake?

4.13 A cylindrical bar is rotating about an axis through its centre and perpendicular to its length at 10 000 rpm. The radius of the bar is very much less than its length, which is 40 cm, and it is made of steel of density 7.5 g cm^{-3}. The bar is in tension because of the centrifugal force. Calculate the stress (i.e. the tensional force per unit area) at the centre of the bar.

4.14 The gate in Worked examples 4.5 and 4.7 was made with a rectangular piece of wood 2 m long and with a mass of 60 kg. It is hinged at one end and struck at the other end with a force of 10 N. What is the force on the hinge? If the force is applied to some point along its length, the reaction on the hinge changes. Find the point where the force may be applied so that there is no reaction at the hinge.

This point is called the centre of percussion. For impulsive forces that may be very large for a short time, selection of the centre of percussion greatly reduces the stress at the axis of rotation. For example, tennis rackets are designed so that a player is swinging the racket about an axis close to his or her elbow; the force on the elbow joint is minimized.

4.15 A bar of mass m and length ℓ is lying on a smooth table. One end of the bar is struck by a puck of mass $\frac{1}{4}m$ moving at a velocity v in a direction at right angles to the length of the bar. Calculate the speed of the centre of

mass of the bar and its rate of rotation after the collision, assuming the collision to be elastic.

4.16 Calculate the translational and rotational speeds of the bar in Problem 4.15 if the puck sticks to the end of the bar after the collision. How much energy is dissipated in this collision?

4.17 A thin circular ring of mass M lies on a smooth and frictionless horizontal surface. It is free to pivot about a fixed point on its circumference. A bead of mass m slides smoothly round the ring. Initially the ring is at rest and the bead is at the pivot, moving with a speed v. Show that when the bead is diagonally opposite the pivot its speed is $v(1 + 2m/M)^{-1/2}$.

4.18 By carrying out the integral in eqn (4.7) prove that the moment of inertia of a uniform sphere of mass M and radius R about an axis through its centre is $\frac{2}{5}MR^2$.

4.19 The 'ballistic pendulum' shown in Fig 4.34 is used to measure the speed of a bullet. The pendulum consists of a target free to swing about a horizontal axis perpendicular to the path of the bullet and thick enough to stop it. The mass of the pendulum is M, its moment of inertia about the axis is I, and the distance of its centre of mass from the axis is h. A bullet of mass $m \ll M$ strikes the pendulum at a distance ℓ from the axis and causes it to swing through an angle α before momentarily coming to rest. Show that the speed v of the bullet is given by

$$v = \frac{\sqrt{2IMgh(1 - \cos\alpha)}}{m\ell}.$$

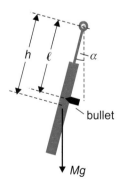

Fig. 4.34 The ballistic pendulum swings through an angle α after it has been hit by the bullet.

4.20 A small mass m is located at the rim of a uniform disc of mass $2m$, which is rotating without friction at an angular speed ω_0. The mass m is pulled slowly in to the axis by a light cord. Show that the angular speed is $2\omega_0$ when the small mass reaches the axis. Obtain an expression for the tension in the cord as a function of distance from the axis. Calculate the total work done by the cord in pulling the small mass from the rim to the centre of the disc, and demonstrate that this work equals the change in rotational kinetic energy of the system.

4.21 A bicycle has wheels of radius 0.3 m and is moving at 10 m s^{-1}. Each wheel has a moment of inertia about its axis of 0.1 kg m^2. The total mass of the rider and the bicycle is 70 kg and the centre of mass is one metre above the ground. By what angle must the line from the centre of mass to the tyres be tilted from the vertical to keep the bicycle on a turning circle of radius 5 m? (You may assume that this angle is small, so that the solution to eqn (4.38) is almost the same as if the angle were zero. Remember that the bicycle has two wheels!)

Level 3

4.22 A cross-section through a diameter of a pulley has the dimensions shown in Fig 4.35. The groove in which a cord runs has a semicircular shape. Calculate the moment of inertia of the pulley about the axis perpendicular to the plane of the pulley passing through its centre.

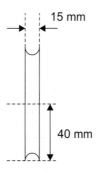

Fig. 4.35 The cross-section through the diameter of a pulley.

4.23 A wheel can rotate in a vertical plane about a horizontal axis through its centre O. The wheel, which is 40 cm in diameter, is unbalanced by a 0.1 kg weight at a point P on its rim, as shown in Fig 4.36, The wheel is held with OP inclined at 30° above the horizontal and then released. Owing to a frictional torque of constant magnitude τ at the bearing, the first swing carries OP to

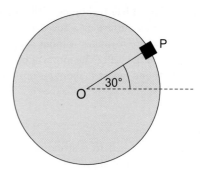

Fig. 4.36 The weight on the rim of the wheel causes it to swing.

a position only 45° beyond the vertical. Determine the value of τ and prove that in the next swing OP will come to rest before reaching the vertical.

4.24 A snooker ball moving at a speed v collides with a stationary ball and is deflected through 30°. Assuming that the collision is elastic and that the balls roll without slipping before and after the collision, calculate the speed and the direction of the struck ball. Demonstrate that angular momentum as well as translational momentum is conserved in such an elastic collision.

4.25 A solid uniform wheel of mass M and radius R is free to rotate without friction about a fixed axis through its centre. A particle of mass m falls through a height h and sticks to the rim of the wheel at a point making an angle θ to the vertical, as shown in Fig 4.37.

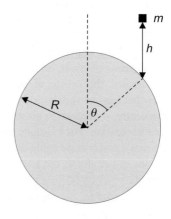

Fig. 4.37 The wheel is made to rotate by the impact of a weight falling on to its rim.

Show that the maximum subsequent angular speed of the wheel is

$$\omega = \frac{2}{R}\left\{\frac{mg}{M+2m}\left\{\frac{2mh\sin^2\theta}{M+2m} + R(1+\cos\theta)\right\}\right\}^{1/2}.$$

What is the minimum value of h such that the wheel will rotate continuously?

4.26 A door knocker consists of a metal ring of mass M and radius R pivoted at the top, as illustrated in Fig 4.38. When the ring is vertical it is in contact with a stud in the door. Estimate the moment of inertia of the ring about the pivot, assuming that the thickness of the ring is much less than R. The ring is raised and then released; on impact with the door it immediately comes to rest. Show that the impulse at the stud (A) is three times that at the pivot (B).

Fig. 4.38 When the door-knocker is released it raps the stud at A.

4.27 A small solid sphere of radius r rolls without slipping on the outside surface of a large fixed hemisphere of radius $R(R \gg r)$. It starts at the top and rolls down under gravity. Find the speed of the centre of mass of the sphere after it has rolled through the angle β shown in Fig 4.39. Show that the sphere loses contact with the hemisphere when $\cos\beta = 10/17$.

4.28 The gyroscope in Worked example 4.10 is precessing steadily with its spindle at an angle of 45° to the vertical. Calculate the precessional frequency when the angular frequency ω of the gyroscope about it own axis is $800\ \mathrm{rad\,s^{-1}}$. When calculating the moment of inertia about an axis in the plane of the gyroscope's disc, neglect the thickness of the disc.

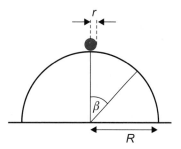

Fig. 4.39

4.29 A gyroscope will retain its orientation with respect to the stars when it is mounted in frictionless bearings turning about three perpendicular axes. If such a gyroscope is mounted horizontally on a north–south axis at the equator, it will remain stationary with respect to the Earth if its axis lies north–south, but will rotate in the vertical plane with a period of one day if its axis is east–west. Show that, if a weight is hung on the mounting as shown in Fig 4.40, there is a torque tending to cause the gyroscope to align along a north–south axis. (This arrangement is the basis of the gyrocompass, which was an important instrument before the advent of satellite and inertial navigation.)

4.30 A gyroscope is spinning about a horizontal axis and precessing in a horizontal plane. One end of the spindle is held in a fixed bearing, which allows the axis to point

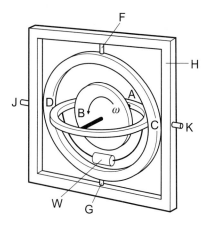

Fig. 4.40 The gyrocompass is free to rotate about the bearings CD, FG, and JK. The weight W exerts a torque to ensure that the compass always points in a north–south direction.

in any direction. A frictionless vertical surface is now placed against the other end of the spindle, preventing the precession. Describe the subsequent motion of the gyroscope immediately after the impact and calculate the force it exerts on the vertical surface. (Remember that, as soon as the precession stops, the torque causing the precession must also become zero.)

Some solutions and answers to Chapter 4 problems

4.2 The angular speed of the flywheel is 5000 rpm after 4 s, and its angular acceleration just before reaching maximum speed is $65 \, \text{rad s}^{-1}$.

4.3 Call the length of the plank 2ℓ, the height of the pivot above the ground h, and the masses of the boys M and $2M$. The moment of inertia of the plank plus the boys about the pivot is $I = \frac{1}{3}M\ell^2 + 3M\ell^2 = \frac{10}{3}M\ell^2$. The boy of mass $2M$ falls a distance $2h$ and the other boy is lifted up a distance $2h$, so the net loss of potential energy is $2Mgh$. This equals the rotational energy $\frac{1}{2}I\omega^2$ at the moment when the end of the plank hits the ground; hence $\omega = \sqrt{4Mgh/I}$. At this moment the speed of the end of the plank is $\ell\omega = \ell\sqrt{4Mgh/I} = \sqrt{\frac{12}{10}gh}$. For $h = 1$ m, this speed is $\sqrt{1.2 \times 9.81} = 3.4 \, \text{m s}^{-1}$.

4.4 The force F exerted by the gas equals its rate of change of momentum, in this case $0.01 \times 20 = 0.2$ N. If the radius of the Catherine wheel is R, the torque about its centre is $FR = I \mathrm{d}\omega/\mathrm{d}t$. Assuming that the moment of inertia I is constant in the first half second, during which the mass M of the wheel reduces by only 5%,

$$\frac{\mathrm{d}\omega}{\mathrm{d}t} = \frac{2FR}{MR^2} = \frac{0.4}{0.1 \times 0.05} = 80 \, \text{rad s}^{-1}.$$

4.7 The roundabout rotates once in about 4 s.

4.8 If the width of the end of the table is w, its moment of inertia about the hinge is

$$\int_0^w \rho x^2 \mathrm{d}x = \left[\frac{1}{3}\rho x^3\right]_0^w = \frac{1}{3}\rho w^2.$$

The speed of the edge of the table is $\ell\omega = \sqrt{3g\ell}$. This is $3.8 \, \text{m s}^{-1}$ for $\ell = 0.5$ m.

4.9 The acceleration of the solid cylinder is $2.23 \, \text{m s}^{-2}$.

4.10 The speed when the ball is rolling without slipping is $\sqrt{\frac{5}{7}}v$.

4.12 The coefficient of friction is about 0.15.

4.13 The bar is rotating about its centre at an angular speed ω and the acceleration is $\omega^2 x$ at a distance x from the axis. The mass of a length $\mathrm{d}x$ of the bar is $\pi r^2 \rho \mathrm{d}x$, where r is the radius and ρ the density. The centripetal force needed to keep the segment $\mathrm{d}x$ moving in a circular path is $\pi r^2 \rho \omega^2 x \mathrm{d}x$ and the total force at the centre of the bar is

$$\int_0^{\ell/2} \pi r^2 \rho \omega^2 x \mathrm{d}x = \frac{1}{8}\pi r^2 \rho \omega^2 \ell^2.$$

The stress, which is the force per unit area, is

$$\frac{1}{8}\rho\omega^2\ell^2 = \frac{7500 \times (2\pi \times 10000)^2 \times 0.4^2}{8 \times 60^2}$$

$$= 1.6 \times 10^8 \, \text{N m}^{-2}.$$

The force divided by the weight of the bar is $\omega^2\ell/8g \approx 5600$. The stress is easily sustained by a steel bar, and causes scarcely any extension of its length, even though the force is nearly 6000 times the weight of the bar.

4.14 The force on the hinge is 5 N in the same direction as the force on the end of the gate. The centre of percussion is two-thirds of the distance from the hinge to the end of the gate, namely, 1.33 m from the hinge.

4.15 After the collision the speed of the bar is v_b and the speed of the puck v_p. Conservation of translational momentum gives

$$mv = mv_b + mv_p. \tag{1}$$

The moment of inertia of the bar about an axis through its centre is $\frac{1}{12}m\ell^2$. Conservation of angular momentum about the centre gives

$$\frac{1}{2}mv\ell = \frac{1}{2}mv_p + \frac{1}{12}m\ell^2\omega. \tag{2}$$

Kinetic energy is conserved in the elastic collision:

$$\tfrac{1}{2}mv^2 = \tfrac{1}{2}mv_b^2 + \tfrac{1}{24}m\ell^2\omega^2. \tag{3}$$

From (1), $v_b = v - v_p$. Substituting for v_b in (2) gives $\ell\omega = 6v$. (3) now leads to $v_p = \tfrac{2}{5}v$ and $v_b = \tfrac{3}{5}v$.

4.16 The centre of mass speed is $\tfrac{1}{5}v$ and the bar rotates at an angular speed $\tfrac{6}{5}v/\ell$. 52% of the initial kinetic energy is retained as translational and rotational kinetic energy and the remaining 48% is dissipated in the collision.

4.17 Initially, the angular momentum about the pivot is zero. The moment of inertia of the ring about the pivot (using parallel axes) is $2MR^2$. When the bead is opposite the pivot let its speed be v': the ring is rotating at an angular speed ω, say, in the opposite direction. The angular momentum is now $2MR^2\omega - mv' \times 2R = 0$. Hence $\omega = mv'/MR$. The equation for conservation of energy is $\tfrac{1}{2}mv^2 = \tfrac{1}{2}mv'^2 + \tfrac{1}{2} \times 2MR^2\omega^2$. Substituting for ω gives the required expression for v'.

4.20 The moment of inertia of the disc about its centre is $\tfrac{1}{2} \times 2m \times R^2 = mR^2$. When the small mass is at a distance r from the centre, its moment of inertia is mr^2 and the disc rotates at an angular speed ω, say, and the angular momentum about the axis is $m(R^2 + r^2)\omega$. Initially m is at R, the angular speed is ω_0, and the angular momentum is $2mR^2\omega_0$. Since angular momentum is conserved, $m(R^2 + r^2)\omega = 2mR^2\omega_0$ and

$$\omega = \frac{2R^2\omega_0}{R^2 + r^2}.$$

Hence $\omega = 2\omega_0$ at $r = 0$.

The centripetal force on the small mass is $mr\omega^2$ and the work done to move it from $r = R$ to $r = 0$ is

$$\int_0^R mr\omega^2 dr = \int_0^R \frac{4mR^4\omega_0^2 r dr}{(R^2 + r^2)^2}.$$

Putting $y = R^2 + r^2$, $dy = 2rdr$ and the integral becomes

$$2mR^4\omega_0^2 \int_{R^2}^{2R^2} \frac{dy}{y^2} = 2mR^4\omega_0^2 \left[-\frac{1}{y} \right]_{R^2}^{2R^2} = mR^4\omega_0^2.$$

The initial kinetic energy of the disc plus the small mass is $\tfrac{1}{2}mR^2\omega_0^2 + \tfrac{1}{2}mR^2\omega_0^2 = mR^2\omega_0^2$, and the final kinetic energy when the small mass is at $r = 0$ is

$\tfrac{1}{2}mR^2 \times (2\omega_0)^2 = 2mR^2\omega_0^2$. The difference between initial and final kinetic energies is $mR^2\omega_0^2$, just the amount of work done to pull the mass into the centre.

4.21 The angle is about 1°.

4.22 To solve this problem you will have to look up and evaluate integrals that may be expressed in the form $\int x^n \sqrt{a^2 - x^2} dx$, for $n = 0$ to $n = 3$. The moment of a solid disc with the outside dimensions of the pulley is 1.03×10^{-3} kg m^2, and this is reduced by about 15% by the presence of the groove.

4.26 The moment of inertia of the ring about the pivot is $\tfrac{3}{2}MR^2$. Suppose that the angular speed about the pivot is ω at the moment just before the ring is stopped by impulse of magnitude α at the stud. The moment of this impulse about the pivot is equal to the change in angular momentum: $\alpha \times 2R = \tfrac{3}{2}MR^2\omega$, $\alpha = \tfrac{3}{4}MR\omega$. The speed of the centre of mass on impact is ωR. The net impulse required to stop the translational motion of the ring is $\alpha + \beta = M \times \omega R$, where β is the impulse at the pivot. Thus $\beta = \tfrac{1}{4}M\omega R = \tfrac{1}{3}\alpha$.

4.28 The moment of inertia of the gyroscope about its axis is $\tfrac{1}{2}MR^2$, where M is the mass of the disc and R its radius. If the disc is thin, the moment of inertia about an axis in the plane of the disc and through its centre is given by the theorem of perpendicular axes as $\tfrac{1}{4}MR^2$. Equation (4.40) becomes

$$\Omega_p = \frac{\tfrac{1}{2}MR^2\omega \pm \sqrt{\tfrac{1}{4}M^2R^2\omega^2 - M^2R^2 g\ell\cos\alpha}}{\tfrac{1}{2}MR^2\cos\alpha}.$$

Dividing by $\tfrac{1}{2}MR^2$,

$$\Omega_p = \frac{\omega \pm \sqrt{\omega^2 - 4g\ell\cos\alpha/R^2}}{\cos\alpha}.$$

For the gyroscope in the problem, rotating at 800 rad s^{-1},

$$\Omega_p = \frac{800 \pm \sqrt{800^2 - 4 \times 9.8 \times 0.03 \times \cos 45°/0.025^2}}{\cos 45°}$$

$$= 1.178,$$

scarcely different from the value 1.176 that applies to horizontal precession.

4.30 When the gyroscope strikes the surface the precession stops and the disc starts to fall with acceleration g. The precession is now downwards with angular frequency initially $\Omega_d = 0$, increasing at the rate $d\Omega_d/dt = g/\ell$. The rate of change of torque is

$$\frac{d\tau}{dt} = \frac{d}{dt}(I\omega\Omega_d) = I\omega\frac{g}{\ell}.$$

The torque is $\tau = F \times 2\ell$, where F is the force on the stopping surface. This force is initially increasing at a rate $dF/dt = \frac{1}{2}I\omega g/\ell$.

Chapter 5

Gravity and orbital motion

This chapter discusses the nature of gravity and describes how gravitational forces govern the motion of objects in the solar system.

The explanation of astronomical observations has been one of the great successes of classical physics. From the earliest times, men have been interested in looking at the heavens. Every day the Sun moves in a great arc across the sky, and the variation with the season of its rising and setting must have aroused the curiosity of primitive people. No instruments are needed to see the changing phase of the Moon and the movement of constellations of stars across the night sky. Artefacts such as Stonehenge show that these phenomena had been carefully observed thousands of years ago.

The motion of the planets is much more puzzling than that of the other heavenly bodies. The positions of all stars are fixed relative to one another, but the planets wander about among them—indeed the name planet is derived from the Greek word for 'wanderer'. Nor are the planets fixed relative to the Sun. Venus, for example, sometimes rises before dawn (when it is called the 'morning star'), and sometimes sets after the Sun (the 'evening star'). Ancient astronomers devised a complicated system of cyclical motions to describe the planets, which they pictured as moving on 'celestial spheres' centred on the Earth.

It was not until the sixteenth century that Copernicus advanced his heretical heliocentric theory, proposing that the planets are following orbits around the Sun. Demonstrating that this theory is correct by observing the positions of planets in the sky is no easy task. Only after many years of numerical work was Kepler able to deduce that the orbits of the planets around the Sun are *ellipses*. In an elliptical orbit the distance between a planet and the Sun is changing, unlike the constant distance in the simpler circular motion considered in Chapter 4. What force between a planet and the Sun can keep it on its elliptical path? Isaac Newton proved that an elliptical orbit is only possible with a force which varies as the inverse of the square of the distance between the planet and the Sun. This force is the universal force of gravitation, which acts between any pair of objects, the force that pulls objects vertically downwards on Earth. In this chapter we shall discuss Newton's law of gravity and use it to explain the orbits of the planets. Gravitational forces also act on satellites, comets, and space probes, and we shall outline the principles used to determine the paths of these objects.

5.1 Gravitational forces

Forces acting at a distance

There are many examples of motion caused by forces arising from objects in contact. You have to get hold of the supermarket trolley to push it, a bicycle does not move unless your feet are on the pedals, nothing happens in a collision between railway wagons until they are in contact, and so on. The force of gravity is not at all like this: it can act between objects that are not in contact, and always attracts them towards each other. Gravity is not affected by the presence of other forces: the force you experience from the Earth's gravity is the same whether you are supported by the ground or falling towards the water after diving from a springboard.

Wherever you are on the Earth's surface, gravity is a force that acts vertically, in a downwards direction. The usual way to find the vertical direction is with a plumb-line consisting of a small weight—the plumb-bob—hanging on a flexible cord: the vertical direction coincides with the cord. The plumb-line is a simple and practical piece of apparatus, but it hides the fact that the gravitational force at a particular point is actually built up in a complicated way. There is an attractive gravitational force—a vector with a direction and magnitude—between any two objects. The gravitational force acting on the plumb-bob is the net effect of the attraction between the bob and the whole of the rest of the world. Fortunately, the effect of the rest of the world on the plumb-bob can be approximated in a very simple way, as we shall explain by imagining a rather idealized Earth.

When the plumb-line is suspended over an expanse of still water, the water surface must be perpendicular to the cord (Fig 5.1). If the surface were not perpendicular to the cord there would be a component of the gravitational force along the surface, causing the water to run downhill. If the Earth were covered with water and had a perfectly spherical shape, then all vertical lines through its surface would pass through the centre of the sphere.

It turns out that the gravitational force outside a spherically symmetrical Earth would be exactly the same as if all the mass were concentrated at the centre. Of course, the real Earth is not at all smooth and there are small variations in gravity because of the presence of mountains and buildings and because of changes from place to place in the nature of geological formations below ground. But the scale of these variations is small compared to the size of the Earth, which has a radius of about 6400 kilometres. The gravitational force on the plumb-bob is almost the same wherever it is placed on the Earth's surface, and vertical lines do in fact all pass close to the centre of the Earth.

Moving out into space, the roughness of the Earth's surface becomes progressively less important. For astronomical calculations, it is a very good approximation to represent the gravitational effect of the Earth by a point object with the same mass. The other planets, their moons, and the Sun can similarly all be regarded as massive point objects. The gravitational attraction between any pair of these objects acts across vast distances of empty space and is unaffected by the presence of any other objects in between.

Fig. 5.1 The plumb-line is hanging vertically above the water: the water surface is exactly perpendicular to the plumb-line.

● *The Earth's gravity behaves as if all the Earth's mass were at its centre*

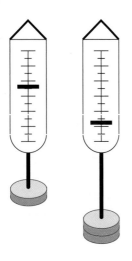

Fig. 5.2 Two equal masses move the pointer on the spring balance twice as far as one pointer on its own

but on the Moon the pointer does not move nearly so far.

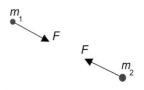

Fig. 5.3 The gravitational forces attracting the two masses are of equal strength and act in opposite directions. The net force on the pair of masses is zero.

Newton's law of gravity

The weight of any object is the force acting on it because of the Earth's gravity. In Section 2.2 it was explained that, because the acceleration of a freely falling body is independent of its size and shape, the weight of the body must be proportional to its mass. Weights can be measured with a spring balance, and indeed two equal masses extend a spring balance twice as much as one on its own (Fig 5.2). The weight also depends on the mass of the Earth, the other partner in the gravitational attraction. On the Moon, which has a mass much less than that of the Earth, weights are less than they are on Earth. The same mass will deflect a spring balance by about seven times as much on Earth as it does on the Moon. When astronauts landed on the Moon, they had great fun jumping about even though they were wearing cumbersome space suits—their suits did not weigh much on the Moon.

Gravitational forces fall off rapidly with distance. The distance to the Moon is about 60 times the Earth's radius, and when you are on the Moon the attraction of the Earth has become very small compared with the attraction of the Moon—even though the Moon's mass is much less than the Earth's, a plumb-bob on the Moon points nearly towards the centre of the Moon.

The gravitational force between two objects is thus dependent on the mass of each and on the distance between them. This statement is given precise form by **Newton's law of gravity**. More than 300 years ago Newton proved that the elliptical orbits of the planets could only be explained if the force of gravity decreases as the inverse square of the distance between a planet and the Sun. Apart from some tiny discrepancies that can be explained by General Relativity theory, Newton's law has proved to be consistent with all astronomical observations and all laboratory experiments. The law of gravity states that:

The attractive gravitational force between two small bodies acts along the line joining them: its magnitude is proportional to the product of their masses and inversely proportional to the square of the distance between them.

The law applies in the limit that the dimensions of the bodies are vanishingly small compared to their distance apart. The equal and opposite gravitational forces between two bodies are illustrated in Fig 5.3. In symbols, if the masses are m_1 and m_2 and their separation is r, the magnitude of the gravitational force F attracting them is

$$F \propto \frac{m_1 m_2}{r^2}.$$

The units of force, mass, and distance are all defined already, so a constant of proportionality is required to give the actual magnitude of the force. The constant is called the **gravitational constant** and is denoted by G:

$$F = G \frac{m_1 m_2}{r^2}. \tag{5.1}$$

Force is a vector quantity and it must be remembered that the force with magnitude given by eqn (5.1) acts along the line joining the masses m_1 and m_2. When two or more masses are attracting another one, the forces must be added according to the rules of vector addition. Equation (5.1) can be written in vector form but, since we shall be concentrating in this chapter on examples in which the gravitational attraction of a single mass is predominant, the expression for the magnitude of the force is sufficient.

Although gravitational forces are so important in everyday life and in astronomy, the constant G is still only known with rather poor accuracy. This is because gravitational forces are extremely weak in the sense that the attraction between objects on a human scale is scarcely discernible. Whereas most fundamental constants are known to better than one part per million, G has an uncertainty of more than one part in ten thousand. The currently accepted best value is

$$G = 6.6726 \pm 0.0009 \times 10^{-11} \, \text{N} \, \text{m}^2 \, \text{kg}^{-2}. \tag{5.2}$$

Writing the units of G in this way makes it clear that the right-hand side of eqn (5.1) is a force, measured in newtons. Since the dimensions of force are (mass × acceleration), 1 newton is $1 \, \text{kg} \, \text{m} \, \text{s}^{-2}$, and the units of G can equivalently be expressed as $\text{m}^3 \, \text{kg}^{-1} \, \text{s}^{-2}$. The uncertainty $\pm 0.0009 \times 10^{-11} \, \text{N} \, \text{m}^2 \, \text{kg}^{-2}$ in the value for G is its *standard deviation,* the statistical measure of error which is defined in Appendix B.

The fact that G is a small number when expressed in the standard SI units reflects the weakness of the gravitational attraction between objects that people or machines can move about. For example, a sphere of lead weighing 1000 kg has a radius of about 30 cm. (A mass of 1000 kg is often called one tonne or one metric ton. One tonne is about $1\frac{1}{2}\%$ less than one Imperial ton; confusingly both units have the same pronunciation.) Two such spheres with their centres one metre apart experience a gravitational attraction of about $6.7 \times 10^{-5} \, \text{N}$. This is less than the weight of a mass of one milligram, which is only one billionth of the mass of each sphere. The vertical gravitational force on the spheres is so large simply because the mass of the Earth is enormous.

Now that we know G we can actually work out the Earth's mass M_e in terms of its radius R_e and the acceleration of gravity g. The weight of an object of mass m on the surface of the Earth is

$$mg = G \frac{m M_e}{R_e^2},$$

● *Gravitational forces are weak*

The variation of the gravitational force with distance appears to follow the inverse square law *exactly*. There has been speculation that there might be an additional component in the attraction between masses that falls off more rapidly with distance. However, attempts to discover this component—called the *fifth force*—have not been successful and it is now generally accepted that it does not exist. An inverse square law of force has some very special properties. One property, which we have already used and which applies only to inverse square law forces, is that the gravitational effect outside a sphere is the same as for a point mass at the centre. Electrical forces also obey an inverse square law, and the corresponding result for a spherical distribution of electrical charge is proved in Chapter 15.

---

making the assumption that the gravitational force is the same as for two point masses separated by a distance R_e. This leads to

$$M_e = \frac{gR_e^2}{G}.$$ (5.3)

Exercise 5.1 The radius of the Earth is about 6400 km, and g is about 9.8 m s^{-1}. Estimate the mass of the Earth.

Answer $M_e \approx 6.0 \times 10^{24}$ kg.

Worked Example 5.1 What is the average density of the Earth?

Answer The volume of the Earth is $\frac{4}{3}\pi R_e^3$. Denoting its average density by ρ_e, we have $M_e = \frac{4}{3}\pi R_e^3 \rho_e$, and

$$\rho_e = \frac{6.0 \times 10^{24}}{1.333\pi \times (6.4 \times 10^6)^3} \approx 5500 \text{ kg m}^{-3}$$

or equivalently, 5.5 g cm^{-3}, i.e. 5.5 times the density of water.

(a)

(b)

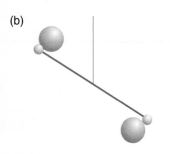

Fig. 5.4 The beam carrying the small spheres is deflected when the large spheres are moved from their position in (a) to their position in (b).

Measurement of G

Accurate values of G are obtained by measuring the small force between two spheres whose mass and position are known precisely. This method dates from 1798, when Cavendish achieved a result within 1% of the best value. Two spheres at the ends of a 6 foot long beam were suspended from a torsion wire as shown in Fig 5.4. Two huge lead spheres, each 1 foot in diameter, were placed close to the suspended spheres, first on one side and then on the other. The deflection of the beam is determined by the gravitational attraction of the spheres and the torsional strength of the suspension. The period of horizontal oscillations of the beam also depends on this torsional strength, which can be eliminated to give G in terms of this period, the masses of the spheres, and the dimensions of the apparatus.

Oscillations are discussed in Chapter 6, and the calculation of G is given as an example at the end of that chapter. A modification of the method devised by Cavendish is still used today, and the factor of about 100 in accuracy arises simply from the greater sensitivity of modern measuring techniques.

Gravitational potential energy

Gravitational potential energy was discussed in Section 3.6 for changes in the height of objects near the surface of the Earth. It was shown that the work done in lifting an object is stored as potential energy, which can be recovered by lowering it again. The gravitational force was described as *conservative*, which means that no work is done by gravity when the object is moved away and then returned to its starting place. Differences in gravitational potential energy between two points are determined only by the change in height, independently of the path between the points or of the action of other, non-gravitational forces. Now we need to extend the argument used in Section 3.6 to account for the variation of the gravitational force with distance.

First, consider a stationary object of mass m at a distance r_1 from the centre of the Earth, as shown in Fig 5.5. How much work is needed to raise it vertically to a distance r_2? From eqn (5.1) the gravitational force $F(r)$ on the object when it is at a distance r is

$$F(r) = \frac{GmM_e}{r^2}.$$

The work done in increasing the distance by a small amount dr is $F(r)dr$, and the total work done from r_1 to r_2 is

$$W = \int_{r_1}^{r_2} F(r)\,dr = \int_{r_1}^{r_2} \frac{GmM_e}{r^2}\,dr = GmM_e\left[\frac{1}{r}\right]_{r_1}^{r_2} = GmM_e\left(\frac{1}{r_1} - \frac{1}{r_2}\right).$$

$$(5.4)$$

The work done is equal to the difference in potential energy. The potential energy is largest when the mass is a long way from the Earth, since energy can be recovered by allowing it fall back towards the Earth. But only differences in potential energy can be calculated and, if an arbitrary constant is added to the potential energy at both distances r_1 and r_2, there is no change in W. It is obviously reasonable in this example to choose the potential energy to be zero when the mass is at a great distance from the Earth where the gravitational force is negligible. As the mass approaches the Earth, its potential energy then becomes *negative*. The work needed to move the mass from a distance r_1 to an infinite

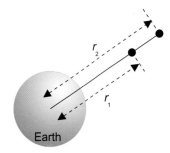

Fig. 5.5 Work must be done against the gravitational force to move a mass further away from the Earth.

● *The potential energy is inversely proportional to the distance from the centre of the Earth*

distance ($r_2 = \infty$) is

$$W_\infty(r_1) = \frac{GmM_e}{r_1}.$$

The potential energy at a distance r_1 from the centre of the Earth is $U = -W_\infty(r_1)$, and thus

$$U = -\frac{GmM_e}{r_1}. \tag{5.5}$$

The negative potential energy indicates that an amount of energy $W_\infty(r_1)$ is required to remove the mass from the influence of the Earth's gravity: $W_\infty(r_1)$ is called the **binding energy** of the stationary mass—it is as though the mass were tied to the Earth and must struggle to escape. Keeping up the metaphor, the *escape velocity* is defined to be the minimum velocity needed for the object to leave the influence of the Earth's gravity.

Worked Example 5.2 What is the escape velocity of an object on the Earth's surface?

Answer The minimum velocity v_{esc} occurs when the object moves vertically upwards with kinetic energy equal to $W_\infty(R_e)$—the kinetic energy then tends to zero far from Earth. Using eqns (5.5) and (5.3), the kinetic energy is

$$\tfrac{1}{2}mv_{esc}^2 = W_\infty(R_e) = \frac{GmM_e}{R_e} = mgR_e,$$

or $v_{esc} = \sqrt{2gR_e}$. Now $g = 9.81\,\mathrm{m\,s^{-2}}$ and $R_e = 6.4 \times 10^6\,\mathrm{m}$, leading to $v_{esc} = 11.2\,\mathrm{km\,s^{-1}}$. In practice this is an underestimate, since no allowance has been made for frictional energy losses as the object passes through the atmosphere. The high value of the escape velocity explains why such huge rockets are needed to launch space probes.

5.2 Circular orbits

At all times of year the rising or setting Sun appears to be roughly the same size. This observation means that the distance from the Earth to the Sun is nearly constant. In other words the Earth's orbit around the Sun must be roughly circular. In fact the orbit is not quite a perfect circle, and the Earth–Sun distance does vary by a few per cent during each year. But you can learn a lot about gravitational orbits by starting with circles. The effects of varying distance between the attracting objects will be left for Section 5.3.

The box on p. 140 has a schematic picture of the solar system. The system is held together by gravity: the Sun and all the planets have a gravitational attraction for one another. All these different attractions between moving objects make it seem that it will be very difficult to work out the Earth's orbit. Fortunately, the Sun is so big that the effect of the other planets can be neglected, just as we neglect nearby objects when considering gravity at the surface of the Earth. The mass of the Sun is 300 000 times as much as that of the Earth, and even a thousand times the mass of Jupiter, which is the largest planet. For most purposes we can think of the motion of the Earth around the Sun as if nothing else in the solar system existed.

⬤ *The Earth's attraction towards the Sun far outweighs the effects of the other planets*

Radius, period, and energy in circular orbits

Suppose for the moment that the Earth's orbit is a perfect circle. In Section 4.4 it was shown that a centripetal force towards the centre of the circle is required to maintain the circular motion. The centripetal force on the Earth comes from the gravitational attraction of the Sun, pulling like an invisible string. According to eqn (4.28), the acceleration of the Earth towards the Sun is $a = \omega^2 R_{\text{orbit}}$, where ω is the angular speed of the Earth around the Sun and R_{orbit} the radius of its orbit (Fig 5.6). The angular speed is related to the period T of the rotation by $\omega = 2\pi/T$, since there are 2π radians in one revolution—in this case T is, of course, 1 year. Calling the masses of the Earth and the Sun M_e and M_s, the gravitational attraction between them is given by eqn (5.1):

⬤ *The relation between period and radius for circular orbits*

$$G\frac{M_e M_s}{R_{\text{orbit}}^2} = M_e a = M_e \omega^2 R_{\text{orbit}} = \frac{4\pi^2 M_e R_{\text{orbit}}}{T^2}.$$

The mass of the Earth cancels out in this equation, which simplifies to

$$T^2 = \frac{4\pi^2 R_{\text{orbit}}^3}{GM_s}. \tag{5.6}$$

For any planet moving in a circular orbit the period is determined only by the distance between the planet and the Sun, independently of its mass. Planets close to the Sun have shorter periods than those of more distant ones. Venus, which has a nearly circular orbit smaller than that of the Earth, completes one revolution in about 225 of our days.

We shall find in the next section that eqn (5.6) requires only a small modification to be applicable to planetary orbits that are non-circular: it is still almost exactly correct provided R_{orbit} is taken to be the mean radius. The mean radius of the Earth's orbit around the Sun is called the **astronomical unit**, denoted by AU. It is a very important quantity in astronomy, since many other measurements depend on it. The

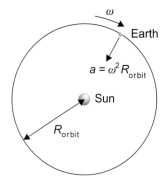

Fig. 5.6 The Earth rotates around the Sun at an angular speed ω. To stay in its circular orbit, the Earth must experience acceleration towards the centre of the circle.

The solar system

The solar system of planets orbiting the Sun is shown schematically in the diagram. All the planets except Pluto, the outermost, have their orbits lying nearly in the same plane. The angle of Pluto's plane is inclined at about 15° to the others. All the orbits are nearly circular, though the Sun is not always at the centre of the circle: here we imagine we are looking at the solar system from outside at an angle, so that the circular orbits appear to be oval. The radii of the orbits vary over a huge range, and are drawn on a logarithmic scale. The shaded orbit in the gap between Mars and Jupiter represent the *asteroids*, thousands of small bodies independently orbiting the Sun. If we count the asteroids as though they were a single planet, the planetary orbits are rather evenly spaced on the logarithmic scale, suggesting that a simple numerical expression might reproduce the radii. This was noticed in the eighteenth century, and *Bode's law* gives the planetary radii in units of the radius of the Earth's orbit. In these units, the radius of Mercury's orbit is 0.4, and according to Bode's law the radius of the nth planet (including the asteroids as $n = 3$) is $R = 0.4 + 0.3 \times 2^n$, with $n = 0$ for Venus, 1 for the Earth, and so on. Bode's law is accurate to within a few per cent except for Pluto, which has an orbit nearly twice the size given by the law. But Pluto is a bit odd anyway, since its orbit is not in the same plane as the others.

On the diagram the large planets are bigger than the small ones, but again are not to scale. The planets' orbital radii, and their own radii and masses are listed in the table. The values are relative to the Earth, for which the orbital radius is 1.5×10^{11} m, the radius 6.4×10^6 m, and the mass 6.0×10^{24} kg.

Planet	Orbital radius	Planet's radius	Planetary mass
Mercury	0.39	0.38	0.05
Venus	0.72	0.96	0.82
Earth	1.00	1.00	1.00
Mars	1.52	0.53	0.11
Jupiter	5.20	11.19	317.8
Saturn	9.54	9.47	95.2
Uranus	19.2	3.73	14.5
Neptune	30.1	3.49	17.2
Pluto	39.4	0.47	0.8

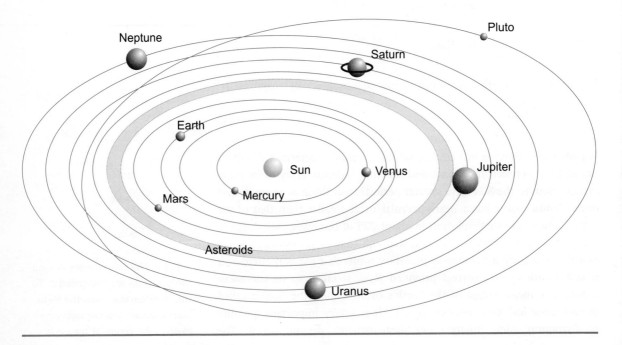

astronomical unit is known with great precision, and to four significant figures its value is

1 astronomical unit (AU) $= 1.496 \times 10^8$ km,

very close to 150 million kilometres or about 93 million miles.

Exercise 5.2 Use eqn (5.6) to calculate the mass of the Sun.

Answer 1.99×10^{30} kg.

A good example of a circular orbit is provided by a geostationary satellite. Geostationary satellites are in orbits such that they remain vertically above the same point on the Earth's equator the whole time. This is very convenient for communications and television broadcasting, since transmitting and receiving equipment can be fixed, always pointing in the same direction. The radius R_g of the geostationary orbit satisfies an equation of the same form as eqn (5.6), with the mass of the Sun replaced by the mass of the Earth, and the period T_{sat} equal to one day instead of one year:

● *Geostationary satellites*

$$T_{sat}^2 = \frac{4\pi^2 R_g^3}{GM_e}. \tag{5.7}$$

The radius R_g is about 42 000 km, seven times the radius of the Earth (Fig 5.7).

The distance from Earth to satellite and back again is so great that signals travelling at the speed of light are delayed by nearly a quarter of a second. Fortunately, this is a short enough time that people whose calls are routed through geostationary satellites are scarcely aware of the delay. Communicating in this way is very inefficient in the sense that only a small fraction of the energy beamed up to the satellite is intercepted, and a large fraction of the Earth's surface is covered by the return signal. Fortunately, what matters is not the absolute magnitude of the signal, but how much bigger it is than background noise. At the very high frequencies used for satellite communication, background noise is small from directions near to the vertical, and the satellites only use a few kilowatts in retransmitting the signals.

If a geostationary satellite were not quite at the correct radius, it would have a period slightly different from one day, and its position in the sky would drift. Suppose the satellite were a little too near the Earth. Then its period would be too short—it would be going too fast. If the position is corrected by moving it further from the Earth, work must be done against gravity and its potential energy is increased. The speed in the corrected orbit is less, so that the kinetic energy has been reduced.

Satellite

Earth

Fig. 5.7 A geostationary satellite beams signals back to Earth, covering nearly half the Earth's surface.

● *Kinetic and potential
energies in circular orbits*

Let us compare the difference in kinetic and potential energies when the satellite changes orbit. From eqn (5.7) the speed v_g of the satellite at the radius R_g is

$$v_g = \omega_g R_g = \frac{2\pi R_g}{T_{sat}} = \sqrt{\frac{GM_e}{R_g}}.$$

If the mass of the satellite is m_{sat} the kinetic energy is

$$E_K(R_g) = \tfrac{1}{2}m_{sat}v_g^2 = \frac{Gm_{sat}M_e}{2R_g}.$$

The potential energy of the satellite at the geostationary radius R_g is given by eqn (5.5) as

$$E_P(R_g) = -\frac{Gm_{sat}M_e}{R_g}$$

and the total energy

$$E(R_g) = E_K(R_g) + E_P(R_g) = -\frac{Gm_{sat}M_e}{2R_g}.$$

If the satellite were initially in orbit at a smaller radius R_1, the amount of work needed to move it out to the geostationary orbit at R_g is

$$\Delta W = E(R_g) - E(R_1) = \tfrac{1}{2}Gm_{sat}M_e\left(\frac{1}{R_1} - \frac{1}{R_g}\right).$$

As the satellite moves out from R_1 to R_g, the *increase* in its potential energy is twice the *decrease* in its kinetic energy. The binding energy $W_\infty(R_1)$ of the satellite at a radius R_1 (which is the work needed to move it infinitely far from the Earth) is thus

$$W_\infty(R_1) = \frac{Gm_{sat}M_e}{2R_1}, \tag{5.8}$$

equal to the orbital kinetic energy at this radius.

Worked Example 5.3 Compare the energy needed to launch a geostationary satellite with its escape energy.

Answer While it is stationary on the ground, the binding energy of the satellite is $Gm_{sat}M_e/R_e$ (eqn (5.5)). The binding energy in geostationary orbit is $\tfrac{1}{2}Gm_{sat}M_e/R_g$. The ratio of the two energies is $\tfrac{1}{2}R_e/R_g$. Putting $R_e = 6400\,\text{km}$ and $R_g = 42\,000\,\text{km}$, this ratio is 0.076. The energy needed to launch the geostationary satellite is thus over 90% of the energy needed for it to escape from the Earth altogether.

Reduced mass

So far we have been thinking about orbits for which one of the attracting objects has a much larger mass than the other. For example, the Sun is so much heavier than the Earth that the centre of mass of the Earth and the Sun is somewhere deep inside the Sun. We talk about 'the Earth rotating about the Sun', but strictly speaking the rotation is an internal motion of the Earth and Sun together. The Sun and the Earth are both rotating about their common centre of mass. From a great distance this would not be very obvious since the centre of mass is inside the Sun. However, there are many examples of 'double stars' consisting of two stars with similar masses rotating about their common centre of mass.

In a double star we cannot forget about the motion of one component and concentrate on the other—both are moving at similar speeds. There are two stars, and their motion can be specified by describing the velocities and positions of each one separately. But it is usually more convenient to separate the problem in a different way. Everything about the motion of the double star is given by the values of the velocity and the position of each star. However, the motion of the double star as a whole is described by giving the velocity and position of its centre of mass. Only one velocity and one position are enough to describe the orbital motion of the two components with respect to the centre of mass.

In Fig 5.8 the two stars are shown rotating about the origin O in the centre of mass coordinate system. Since the origin is at the centre of mass, the two position vectors $\mathbf{r}_1$ and $\mathbf{r}_2$ are not independent, but must satisfy the equation

$$m_1\mathbf{r}_1 + m_2\mathbf{r}_2 = 0.$$

(The centre of mass of a system of massive particles was defined in Section 3.2, and the above equation is simply eqn (3.10) applied to two

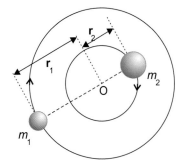

Fig. 5.8 The two components of a double star rotate about their centre of mass at O.

Detecting double stars

It is not usually possible to see the two components of a double star separately, since stars are so far away that most of them appear like point objects even when viewed by the best telescopes. The tell-tale signature of a double star is seen in the way the light emitted varies with wavelength. Single stars emit some light at sharply defined wavelengths determined by the properties of particular atoms in their surfaces. In double stars the wavelengths are shifted by the Doppler effect as the stars whirl round in their orbit: the light becomes bluer for a star coming towards you and redder for the companion going away. The Doppler effect is described in more detail in Section 8.5, and here we shall just take it for granted that by measuring wavelengths we can find the relative speeds of the two companions of a double star. If we are observing from a direction at a small angle to the plane of the orbit, the wavelength of the light from each star will oscillate with the period of the orbit. From the period and the size of the Doppler shifts the relative masses of the two stars can be calculated.

masses.) Since the position vectors point in exactly opposite directions, writing this equation in terms of magnitudes leads to $m_1 r_1 = m_2 r_2$. Now that we know the relation between r_1 and r_2, we can express the distances of each star from the origin in terms of the single distance $r = r_1 + r_2$:

$$r = r_1 + \frac{m_1 r_1}{m_2} \text{ or } r_1 = \frac{m_2 r}{m_1 + m_2}; \text{ similarly } r_2 = \frac{m_1 r}{m_1 + m_2}.$$

If the angular frequency about the centre of mass is ω, then the speeds of the stars in their orbits are $r_1\omega$ and $r_2\omega$, and the orbital kinetic energy of the two stars is

$$E_K = \tfrac{1}{2} m_1 (r_1 \omega)^2 + \tfrac{1}{2} m_2 (r_2 \omega)^2 = \tfrac{1}{2}\left[\frac{m_1 m_2}{m_1 + m_2}\right] r^2 \omega^2 = \tfrac{1}{2} m r^2 \omega^2 \tag{5.9}$$

where

$$m = \frac{m_1 m_2}{m_1 + m_2} \tag{5.10}$$

is called the **reduced mass** of the double star.

For the rest of this chapter, whenever the symbol m appears without a suffix, it stands for the reduced mass of two objects. If one mass, m_2 say, is very much greater than the other, $m_1 + m_2 \simeq m_2$, and m is almost the same as m_1—we are back to a light object orbiting around a heavy one. In the Earth–Sun system the reduced mass differs from the mass of the Earth by only about one part in 300 000. This is a negligible correction at the precision we have been working to, and it was justifiable to use the mass of the Earth instead of the reduced mass in our earlier calculations.

Equation (5.9) does not apply only to stars, but represents the rotational kinetic energy of *any* pair of masses rotating at an angular speed ω. Rotational kinetic energy was expressed in terms of the moment of inertia in eqn (4.4) as

⬤ *The kinetic energy and moment of inertia of two objects rotating about their centre of mass are determined by their reduced mass*

$$E_K = \tfrac{1}{2} I \omega^2.$$

Comparing this equation with eqn (5.9) we see that, for rotation about the centre of mass the moment of inertia of two masses m_1 and m_2 separated by a distance r depends only on the reduced mass, and is $I = m r^2$.

Exercise 5.3 The mass of the planet Jupiter is 1050 times smaller than that of the Sun, and the distance from Jupiter to the Sun is 778 million kilometres. Calculate the reduced mass of Jupiter and the Sun, and the distance of their centre of mass from the centre of the Sun. (Take the mass of the Sun as 1.99×10^{30} kg: this result is from Exercise 5.2.)

Answer The reduced mass is 1.89×10^{27} kg, to this precision not different from the actual mass. The centre of mass is at a distance 740 000 km from the centre of the Sun. This is only just outside the Sun.

We have found the expression for the reduced mass by considering circular orbits, but eqn (5.10) applies to any other example of the relative motion in a two-body system. The same reduced mass must be used for the momentum appropriate for relative motion, which is given by the expression $p = m\mathrm{d}r/\mathrm{d}t$. The gravitational force depends on the two masses separately, and Newton's second law (eqn (2.5)) for the relative motion becomes

● *The reduced mass must be used when Newton's second law is applied to relative motion*

$$F = \frac{Gm_1 m_2}{r^2} = -\frac{\mathrm{d}p}{\mathrm{d}t} = -m\frac{\mathrm{d}^2 r}{\mathrm{d}t^2}. \tag{5.11}$$

The force F acts along the direction between the two masses, and the minus sign indicates that the gravitational force is attractive, i.e. it is trying to make r smaller.

5.3 Kepler's laws and planetary orbits

It is not at all easy to deduce the orbits of the planets from astronomical observations. The platform from which we observe the planets, the Earth, is itself rotating about its axis and moving in its orbit around the Sun. Yet the planetary orbit puzzle was unravelled nearly 400 years ago. Near the end of the seventeenth century Tycho Brahe, an eccentric Danish nobleman, was given permission by the Danish King to build an observatory on an island in the strait between Denmark and Sweden.

Tycho Brahe made measurements of planetary directions with an accuracy of about one minute of arc (one minute is one-sixtieth of a degree); no distances were known. Tycho accumulated an enormous mass of data. Tycho himself did not believe in the heliocentric theory that the planets are orbiting the Sun, and it was left to Johannes Kepler to interpret the data. A little of the history of these two great astronomers is summarized in the box on p. 146. Both men must have been obsessive. Imagine recording so much data on primitive instruments, and performing endless calculations in spherical trigonometry without calculators or computers! Nevertheless, the task was completed, and Kepler enunciated three laws of planetary motion:

1. Each planet moves in an elliptical orbit with the Sun at a focal point.
2. The position vector of a planet relative to the Sun sweeps out a constant area per unit time within the ellipse.
3. The squares of the periods of the planets are proportional to the cubes of their mean distances from the Sun.

We shall show that Kepler's laws are consistent with Newton's law of gravity. First, we need to make a diversion to explain how to describe an ellipse in mathematical terms.

Tycho Brahe and Johannes Kepler

Tycho Brahe

Johannes Kepler

Tycho Brahe and Johannes Kepler between them laid the foundations of our understanding of the solar system. Tycho Brahe was born in 1546, 25 years before Kepler. Tycho's interest in astronomy was aroused by seeing a solar eclipse when he was a young man, and he was impressed by the fact that the eclipse occurred on the day when it was predicted. Later, in 1572, there appeared a new star (what we still call a *nova*) that was so bright that for some time it was visible in daylight. Tycho made careful observations of this nova until it died away, but he kept up his observing and found that the 'planetary tables' predicting the positions of the planets were not accurate. He resolved to improve the accuracy of astronomical measurements. Tycho gained the support of the King of Denmark, and built an observatory—the ruins of which are still preserved—at an isolated spot on an island in the strait between Denmark and Sweden. The measurements were made using long sighting arms: this was 30 years before Galileo built his astronomical telescope. Over a period of many years Tycho accumulated a huge mass of data, until he fell out with the King

of Denmark and moved to Prague. Here he was appointed Imperial Mathematician by the Austrian Emperor Rudolph, and with a team of assistants set to work to interpret planetary and lunar measurements.

As a small boy, Kepler was also attracted to astronomy by a spectacular event—a brilliant comet that appeared in 1577. Tycho Brahe had observed that this comet did not exhibit a daily parallax with the stars, and thus proved that it moved in celestial regions far beyond the moon. Kepler joined Tycho as one of his assistants in 1601, but in a few months Tycho died. Kepler was appointed his successor as Imperial Mathematician, charged with the task of completing Tycho's work on planetary tables. Tycho did not believe that the Earth revolved around the sun, but Kepler was forced by his analysis to support the Copernican theory. He published his first two laws of planetary motion by 1604; the third was not to follow until 1618. In the meantime, Kepler had seen a telescope and realized how the action of lenses is explained by the refraction of light. In 1611 he published *Dioptrics*, the first book on geometrical optics.

Properties of the ellipse

If you loop a piece of string around a peg you can draw a circle with a pencil passed through the loop. If the loop is then required to go round two pegs some distance apart, as in Fig 5.9, you draw an ellipse provided that the string is kept taut. The positions of the two pegs are called the **focal points** of the ellipse. Since the two pegs are a fixed distance apart and the string has a constant length, the sum of the distances between any point on the ellipse and the focal points is also constant—that is one way we can define an ellipse.

   The ellipse is drawn again in Fig 5.10. The line passing through the focal points cuts the ellipse where it is widest: this line is called the **major axis** of the ellipse. The distance labelled a, half the width of the ellipse, is the **semi-major axis**. The line through the centre of the ellipse and perpendicular to the major axis passes through the highest part of the ellipse: the height b is the **semi-minor axis**. The amount by which the ellipse departs from a circular shape is determined by the separation of the focal points. The distance from the centre to one of the focal points is denoted ae. The fraction e is called the **eccentricity** of the ellipse. For a circle, e is zero and, as e increases from 0 to 1, the ellipse becomes thinner and thinner.

   In Fig 5.11 an ellipse is shown as a mathematical diagram. With reference to the origin O at a focal point, the point P on the ellipse is labelled by the coordinate r (which is the length OP), and by θ (the angle between OP and the major axis). The coordinates (r, θ) are *polar coordinates* in two dimensions (cylindrical polar coordinates in three dimensions were described in Section 4.2 and used for the discussion of moments of inertia). Polar coordinates are a good choice for describing planetary motion, since the Sun is situated at a focus of the elliptical orbits. The equation giving r in terms of θ is

$$r = \frac{a(1 - e^2)}{1 - e \cos \theta} \tag{5.12}$$

where as before a is the semi-major axis and e the eccentricity.

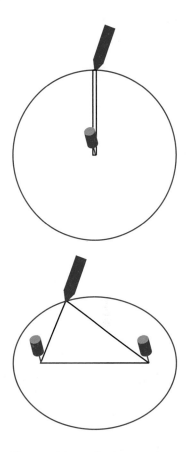

Fig. 5.9 A loop of string round a pencil and one peg allows you to draw a circle. With two pegs you draw an ellipse.

Let us check that eqn (5.12) does describe the shape that was drawn with the loop of string and two pegs. To do this we must show that $r + r'$ is constant, where r' is the distance from P to the second focus of the ellipse at O'. The ellipse is symmetrical with respect to O and O', and eqn (5.12) also applies to the coordinates r' and θ' referred to an origin at O'. The perpendicular from P meets the major axis at F; denote the distance FC from F to the centre of the ellipse at C by af. Now OC = O'C = ae, so that

$$\cos \theta = \frac{OF}{OP} = \frac{a(e - f)}{r} \quad \text{and} \quad \cos \theta' = \frac{O'F}{O'P} = \frac{a(e + f)}{r'}.$$

⬤ *The mathematical description of an ellipse*

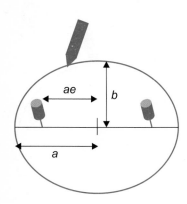

Fig. 5.10 The distances a and b are the semi-major and semi-minor axes of the ellipse.

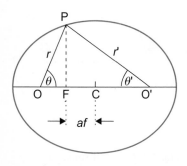

Fig. 5.11 The sum $r + r'$ of the distances from a point P on the ellipse to the two foci is $2a$, the length of the major axis.

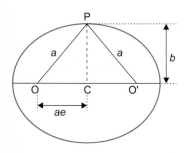

Fig. 5.12 The eccentricity of an ellipse determines the ratio of the minor and major axes.

Putting these values of $\cos\theta$ and $\cos\theta'$ into eqn (5.12),

$$r(1 - e\cos\theta) = r - ae(e - f) = a(1 - e^2),$$

giving

$$r = a - aef$$

for the coordinates r and θ with origin at O and

$$r'(1 - e\cos\theta') = r' - ae(e + f) = a(1 - e^2),$$

giving

$$r = a + aef$$

for the coordinates r' and θ' with origin at O'. Hence $r + r' = 2a$, a constant. This is the result we set out to prove—the locus of points satisfying eqn (5.12) and the construction with the loop of string are describing figures of the same shape.

In Fig 5.12, P lies on the minor axis of the ellipse and $OC = ae$. In this case $r = r' = a$ and the semi-minor axis $b = PC$ is given by applying Pythagoras's theorem to triangle OPC:

$$b = \sqrt{a^2 - (ae)^2} = a\sqrt{1 - e^2}. \tag{5.13}$$

There is not much difference between the major and minor axes unless the eccentricity e is near to one. For example, if the focus is half-way from the centre along the semi-major axis, i.e. $e = 0.5$, then the difference between the semi-minor and semi-major axes is only 13%. The eccentricity of the Earth's orbit is 0.017, and the closest and furthest distances to the Sun differ by about $3\frac{1}{2}$%. But the ratio of the minor and major axes is 0.99986, which means that the orbit is an almost perfect circle, though the focus is not at its centre.

Apart from the distant Pluto, the planet with the most eccentric orbit is Mercury, for which $e = 0.206$. Mercury's orbit around the Sun is shown in Fig 5.13. The orbit is an ellipse with the Sun at the focus but, although the Sun is obviously not at the centre of the orbit, the 2% difference between the major and minor axes is scarcely discernible to the eye.

Kepler's first law states that the planets move in elliptical orbits with the Sun at the focus. Calculation of a planet's orbit from measurements of its direction with respect to the Earth is not easy. Since the eccentricities of the orbits are rather small, it is remarkable that Kepler was able to deduce his first law from observations made before the invention of the telescope and with no mechanical aids to speed up his calculations.

Angular momentum and acceleration in elliptical orbits

The gravitational force between a planet and the Sun always lies on the line joining the two, and passes through their centre of mass. Such forces between two objects, pointing towards their centre of mass, are called **central** forces. Since the gravitational force is central, the component of the force acting perpendicularly to the position vector of the planet is zero: there is no torque acting about the centre of mass. According to eqn (4.19) the rate of change of angular momentum equals the net torque acting and the angular momentum about the centre of mass is constant.

Figure 5.14 shows the planet in an elliptical orbit—the eccentricity is greatly exaggerated in this diagram. In a time δt the angular coordinate changes by $\delta\theta$ with respect to the Sun at S as the planet travels from P to Q, moving a distance $r\delta\theta$ perpendicularly to the position vector $\mathbf{r}$. In the limit as δt tends to zero, $r\delta\theta/\delta t$ is the component of the planet's velocity $v_\perp$ perpendicular to $\mathbf{r}$:

$$v_\perp = r\frac{\mathrm{d}\theta}{\mathrm{d}t}.$$

The component of momentum perpendicular to $\mathbf{r}$ is $mv_\perp$, where $m = M_\mathrm{p}M_\mathrm{s}/(M_\mathrm{p} + M_\mathrm{s})$ is the reduced mass of the planet and the Sun. Since the mass of the planet M_p is much less than that of the Sun, the reduced mass is almost the same as M_p. Nevertheless, in order to be precise, we shall keep the distinction between m and M_p.

The angular momentum vector is given by eqn (4.23) as $\mathbf{r} \times \mathbf{p}$, perpendicular to the plane of the orbit, and of magnitude equal to the length of the position vector $\mathbf{r}$ multiplied by the component of momentum perpendicular to $\mathbf{r}$. The magnitude of the angular momentum about the centre of mass of the planet and the Sun is therefore

$$L_\mathrm{p} = mv_\perp r = mr^2\frac{\mathrm{d}\theta}{\mathrm{d}t}. \tag{5.14}$$

In the time δt an area δA, equal to the shaded area of the triangle SPQ in Fig 5.14 is swept out in the orbit. The area of the triangle is $\delta A = \frac{1}{2}r^2\delta\theta$. In the limit as $\delta t \to 0$, the rate at which the orbital area is swept is thus

$$\frac{\mathrm{d}A}{\mathrm{d}t} = \frac{1}{2}r^2\frac{\mathrm{d}\theta}{\mathrm{d}t}. \tag{5.15}$$

Since L_p and m are both constant it follows from eqns (5.14) and (5.15) that the area of the orbit swept out by the planet in unit time is constant. This is **Kepler's second law**, which we see is an immediate consequence of the conservation of angular momentum. Figure 5.14 makes it clear that

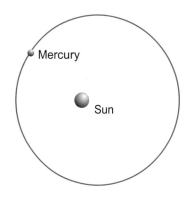

Fig. 5.13 The orbit of Mercury is very nearly circular, but the Sun is offset from the centre.

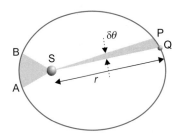

Fig. 5.14 A planet moves from A to B in the same time that it takes to move from P to Q. It must therefore move faster when it is nearer to the Sun.

● *Planets move fast when they are near to the Sun*

this law requires the planet to move faster when it is near the Sun than when it is far away.

Exercise 5.4 The largest and smallest distances from the Sun occur when a planet is at either end of the major axis of its orbit. In both positions it is moving perpendicularly to the major axis. The ratio of the fastest to the slowest speeds of Mars in its orbit around the Sun is 1.205. Calculate the eccentricity of Mars' orbit.

Answer 0.093.

● *The acceleration of a planet towards the Sun*

For an object moving in a circular orbit at angular speed ω, the centripetal acceleration towards the centre of the circle is $r\omega^2$ (eqn (4.28)). In the elliptical orbit of a planet the velocity is not everywhere perpendicular to the radius vector. There is a changing component of velocity along the radius vector which represents an acceleration in addition to the acceleration $r\omega^2$. It is proved in the box below that the total inward acceleration is

$$\alpha = \frac{L_{\rm P}^2}{m^2 a(1 - e^2)} \frac{1}{r^2} \tag{5.16}$$

at a moment when the planet is at a distance r from the Sun. Everything on the right-hand side of eqn (5.16) is constant except $1/r^2$.

The inward force needed to keep the planet in its elliptical orbit is $m\alpha$, and this force obeys an inverse square law. Conversely, the orbits of the planets orbiting around the Sun under the influence of the inverse square gravitational force must move in elliptical orbits, with the Sun at a focus. This is **Kepler's first law**.

The above argument may seem to be a backwards way of proving Kepler's first law, but it actually follows history. Kepler found by observation that the planetary orbits are elliptical, and only later did Newton show that an inverse square law of gravity is required to explain the observations.

The magnitude of the centripetal acceleration required to keep a rotating object at a constant distance from its axis is $r\omega^2$ (see eqn (4.28)). There is an additional term if the length r of the radius vector is changing, which can be found by writing the radius vector as $\mathbf{r} = r\cos\theta\,\mathbf{i} + r\sin\theta\,\mathbf{j}$ and differentiating twice. The result is that the total inward acceleration becomes

$$\alpha = r\omega^2 - \frac{\mathrm{d}^2 r}{\mathrm{d}t^2} = r\left(\frac{\mathrm{d}\theta}{\mathrm{d}t}\right)^2 - \frac{\mathrm{d}^2 r}{\mathrm{d}t^2}. \tag{5.17}$$

Write the equation of the ellipse, eqn (5.12), as

$$r(1 - e \cos \theta) = a(1 - e^2).$$

Differentiating this equation with respect to t,

$$\frac{dr}{dt}(1 - e \cos \theta) + er \sin \theta \frac{d\theta}{dt} = 0.$$

Using eqns (5.12) and (5.14) this may be written as

$$\frac{a(1 - e^2)}{r} \frac{dr}{dt} + \frac{eL_p}{mr} \sin \theta = 0$$

or

$$\frac{dr}{dt} = -\frac{eL_p \sin \theta}{ma(1 - e^2)}.$$

Differentiating with respect to t,

$$\frac{d^2 r}{dt^2} = -\frac{eL_p \cos \theta}{ma(1 - e^2)} \frac{d\theta}{dt}.$$

From eqn (5.14) $d\theta/dt = L_p/mr^2$ and hence

$$\frac{d^2 r}{dt^2} = -\frac{eL_p^2 \cos \theta}{m^2 r^2 a(1 - e^2)}.$$

The total inward acceleration, given by eqn (5.17), becomes

$$\alpha = r\left(\frac{d\theta}{dt}\right)^2 - \frac{d^2 r}{dt^2} = \frac{L_p^2}{m^2 r^3} + \frac{eL_p^2 \cos \theta}{m^2 r^2 a(1 - e^2)}.$$

Substituting for $e \cos \theta$ from eqn (5.12), the above equation reduces to

$$\alpha = \frac{L_p^2}{m^2 r^3} + \frac{L_p^2}{m^2 r^2 a(1 - e^2)} - \frac{L_p^2 a(1 - e^2)}{m^2 r^3 a(1 - e^2)} = \frac{L_p^2}{m^2 a(1 - e^2)} \frac{1}{r^2},$$

which is the result quoted in eqn (5.16)

To check **Kepler's third law** we need to work out the period. This follows from the rate at which the area of the ellipse is swept out, which is given by eqns (5.14) and (5.15):

$$\frac{dA}{dt} = \tfrac{1}{2}r^2\frac{d\theta}{dt} = \frac{L_p}{2m}.$$

We shall not formally calculate the area of an ellipse: it is actually $\pi \times (\text{major axis}) \times (\text{minor axis}) = \pi a^2 \sqrt{1-e^2}$. Bearing in mind that an ellipse looks like a squashed circle this result is not surprising—the ellipse is just a circle with the scale along the minor axis reduced by the ratio of minor to major axes. The period of the orbit is

$$T = (\text{Area of ellipse}) \div \left(\frac{dA}{dt}\right) = \frac{2\pi m a^2 \sqrt{1-e^2}}{L_p}. \tag{5.18}$$

Using eqn (5.16) we can relate the inward acceleration to the gravitational force. Newton's second law $F = ma$ for the internal motion of the Sun and the planet becomes

$$\frac{GM_s M_p}{r^2} = \frac{L_p^2}{ma(1-e^2)}\frac{1}{r^2},$$

and hence

$$L_p^2 = GM_s M_p ma(1-e^2). \tag{5.19}$$

Eliminating L_p from eqns (5.18) and (5.19),

$$T^2 = \frac{4\pi^2 m a^3}{GM_s M_p} = \frac{4\pi^2 a^3}{G(M_s + M_p)}, \tag{5.20}$$

since the reduced mass $m = M_s M_p/(M_s + M_p)$.

○ *The mean distance between a planet and the Sun is nearly the same as the semi-major axis of the orbit*

Equation (5.20) is nearly but not quite the same as Kepler's third law in the form stated at the beginning of this section. The third law stated that the square of the period is proportional to the cube of the mean distance from the Sun. Instead of the cube of the mean distance from the Sun, we have found a^3. But as we have seen, the difference between the major and minor axes is very small for the planets in the solar system. The points where the planet is closest to the Sun (called the **perihelion**) and most distant from it (the **aphelion**) lie on the major axis. The perihelion and aphelion distances are, respectively, $a(1-e)$ and $a(1+e)$. Their mean is indeed a, and the exact mean distance over the whole orbit differs from a only by an amount proportional to e^2.

The other differences from Kepler's third law in eqn (5.20) are that a is the distance to the centre of mass, not to the Sun, and the mass on the right-hand side is the sum of the masses of the Sun and the planet. The quantity multiplying a^3 on the right-hand side of eqn (5.20) is not constant, but includes variations depending on the ratio of the mass of the planet to the mass of the Sun. However, even for Jupiter, by far the most massive planet, the correction is only one part in a thousand. Kepler did a good job! His first two laws are exact within the framework of classical mechanics and, when applied to the solar system, only tiny corrections are needed to the third law.

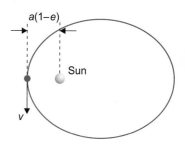

Worked Example 5.4 Obtain an expression for the binding energy of a planet in an elliptical orbit around the Sun.

Answer At the perihelion the planet is at a distance $a(1 - e)$ from the Sun, and it is moving perpendicularly to the semi-major axis, as shown in Fig 5.15. If the speed of the planet at this point is v, the angular momentum in the orbit is $L_p = mva(1 - e)$, where m is the reduced mass. The kinetic energy

Fig. 5.15 The angular momentum of the planet can be expressed in terms of the velocity at the perihelion.

$$E_K = \tfrac{1}{2}mv^2 = \frac{L_p^2}{2ma^2(1 - e)^2}.$$

The binding energy E_B of the moving planet is the binding energy if it were stationary at the perihelion *minus* E_K. From eqn (5.5),

$$E_B = \frac{GM_sM_p}{a(1 - e)} - \frac{L_p^2}{2ma^2(1 - e)^2}.$$

Substituting L_p^2 from eqn (5.18),

$$E_B = \frac{GM_sM_p}{a(1 - e)} - \frac{GM_sM_p(1 - e^2)}{2a(1 - e)^2}$$

$$= \frac{GM_sM_p}{a}\left(\frac{1}{1 - e} - \frac{1 + e}{2(1 - e)}\right) = \left(\frac{GM_sM_p}{2a}\right).$$

Here M_p is the actual mass of the planet, not its reduced mass. This expression for the binding energy has the same form as eqn (5.8) for a circular orbit, with the orbital radius replaced by the semi-major axis of the ellipse. For the planets in the solar system, there is very little difference between the mean radius and the semi-major axis: the energy as well as the shape of the orbit is hardly affected by the fact that the Sun is not at the centre of the orbit.

Launching satellites

So far we have discussed stable orbits for planets and satellites. Once in such an orbit, the planet or satellite will stay there, following the same path over and over again. How can we put a satellite into the particular stable orbit that we want it to occupy?

The most efficient way of achieving high speed for the payload in a rocket launch is to burn the fuel as fast as possible, to avoid wasting

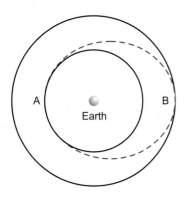

Fig. 5.16 Rockets fired at A and B transfer a satellite from the inner to the outer circular orbit.

● *Changing satellite orbits*

energy in lifting the fuel against gravity. (The speed achieved by the launch is given in eqn (2.22) for a vertical launch.) Unless further action is taken the payload will not reach a stable orbit. A satellite launched vertically will simply come straight down again, and one at an angle will follow an elliptical path that will also eventually strike the Earth's surface. At least one further velocity adjustment must be made to steer the satellite into a stable orbit.

Similar considerations apply to the transfer of a satellite from one orbit to another. If the two orbits do not touch, at least two velocity boosts are required. For example, Fig 5.16 shows how a satellite may be moved between circular orbits of different radii. At the point A on the inner orbit, the satellite fires a rocket directly backwards to increase speed without changing direction. For a velocity change of given magnitude, this leads to the largest increase in kinetic energy. Equation (2.17) shows that the change in velocity of the satellite is proportional to the amount of fuel burned, and increasing the speed in the direction of motion thus achieves the required energy increase with the use of the minimum amount of fuel.

After the rocket is fired, the satellite is still moving perpendicularly to the radius vector to the centre of the Earth, but is now going too fast for the circular orbit. It is at the **perigee** of a new elliptical orbit (i.e. the point nearest to the Earth, as the perihelion is the point nearest to the Sun for a solar orbit). The satellite follows this orbit, and at B, its **apogee** (when it is furthest from Earth), it can be given another velocity boost to put it into the outer circular orbit. This manoeuvre, using an elliptical orbit that is tangential to both of the circular orbits, burns the least possible amount of precious fuel in transferring the satellite between the two orbits.

This analysis suggests that the best strategy for launching a satellite is to fire it horizontally from the Earth's surface. In practice, of course, rockets are fired vertically to carry the payload as quickly as possible above the atmosphere where there is little air resistance. The launching trajectories are indeed made to be nearly horizontal as soon as the air resistance is small enough.

5.4 **Comets and space probes**

From time to time a comet is seen in the night sky. Comets are quite small objects, usually with a solid core only a few kilometres in diameter, carrying along with them a dust cloud of ice or solid carbon dioxide. It is the tiny dust particles in this cloud that make the comet visible. When a comet is at a distance from the Sun comparable with the size of the planetary orbits, some of the dust particles are blown away from the Sun to form the comet's tail. The wind that blows them is the 'solar wind' consisting of rapidly moving and electrically charged particles leaving the Sun. The pressure exerted by these particles, and by the light from the Sun, is large enough to overcome the Sun's gravitational attraction and waft the dust into the comet's tail. The collisions between the dust and the solar wind are violent enough to excite the dust particles so that they emit visible light.

Some comets are regular visitors to the solar system. The most famous comet is Halley's comet, which last passed by in 1986. Halley's comet has been known for hundreds of years, and records of comet sightings that probably refer to Halley's comet stretch back over 2000 years. Comets are always named after their discoverers: Halley's comet returns every 76 years, and from records of its three previous appearances, Halley correctly predicted that it would be seen in 1758.

● *Halley's comet*

The comet Hale–Bopp, photographed with its spectacular tail in Fig 5.17, was the brightest to be seen in recent years. For some time it was almost as bright as Sirius, which is the brightest of all the stars. Hale–Bopp has a period much longer than that of Halley's comet. It is predicted that Hale–Bopp will next return in about 2600 years.

Why do comets have such long orbiting periods? Recalling Kepler's third law, that the square of the period is proportional to the cube of the mean distance from the Sun, we see that long-period comets must have long thin elliptical orbits. Hale–Bopp will travel to a distance of about 370 astronomical units from the Sun at the far end of its elliptical orbit. Most comets have orbits with major axes that are hundreds or even thousands of astronomical units. Astronomers infer that there is a myriad of solid objects, called the Oort cloud, far outside the planets. Occasionally, an object in the Oort cloud may be deflected into an orbit passing close to the Sun, when for a brief period it is visible as a comet.

● *Long-period comets*

Worked Example 5.5 Halley's comet has a period of 76 years and its closest distance to the Sun is 0.587 AU. What is the eccentricity of its orbit and its greatest distance from the Sun?

Answer The mass of the comet is negligible compared to the Sun's mass, and from eqn (5.20), the semi-major axis of the orbit is given by

$$a^3 = \frac{T^2 \times GM_s}{4\pi^2}$$

$$= \frac{(76 \times 365 \times 24 \times 3600)^2 \times 6.67 \times 10^{-11} \times 2 \times 10^{30}}{4\pi^2}$$

$$\approx 1.94 \times 10^{37}\,\mathrm{m^3}, \text{ and } a \approx 2.69 \times 10^{12}\,\mathrm{m}, \text{ or } a \approx 17.9\,\mathrm{AU}.$$

Fig. 5.17 The comet Hale–Bopp with its bright tail was easily visible to the naked eye for more than a month.

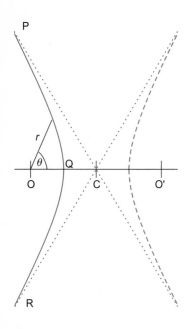

Fig. 5.18 The curve PQR is a hyperbola with focal points at O and O'.

● *The equation for a hyperbola*

The comet's closest approach to the Sun is $a(1 - e) = 0.587$ AU. Hence $e \approx 0.967$ and the greatest distance from the Sun is $a(1 + e) \approx 18$ AU.

Open orbits

So far we have been discussing planets and comets following closed orbits around the Sun, and satellites in orbits around the Earth. Space exploration is not limited to the Earth and the Moon, however, and space probes must be given enough energy to escape from the gravitational attraction of the Earth altogether. What trajectories does such a space probe follow as it leaves the Earth?

Since the same gravitational force is affecting a satellite whether it is in an elliptical orbit or escaping from Earth, it is not surprising that the mathematical description of the trajectory is similar in the two cases.

We have been representing an ellipse in polar coordinates, with origin at a focal point, by eqn (5.12):

$$r = \frac{a(1 - e^2)}{1 - e\cos\theta}.$$

The eccentricity e must lie between 0 and 1 to ensure that r is finite at all angles θ in the closed orbit. For comets with long thin orbits the value of e is very close to one.

What happens to an orbit if e is greater than one? When $e\cos\theta = 1$ the denominator of eqn (5.12) is zero, and r becomes infinite. The orbit is no longer closed. It is an **open orbit** that starts and finishes infinitely far away from the focal point. Such an orbit is called a **hyperbola**.

Equation (5.12) does represent a hyperbola for $e > 1$, but for our purpose it is more convenient to relabel the coordinates and to describe the hyperbola by the equation

$$r = \frac{a(e^2 - 1)}{1 + e\cos\theta}, \quad \text{for } e > 1 \text{ and } \cos\theta > -\left(\frac{1}{e}\right). \tag{5.21}$$

A hyperbola with eccentricity $e = 2$ is shown in Fig 5.18. The full line joining P, Q, and R is part of the hyperbola. At $e\cos\theta = -1$, which occurs at $\theta = \pm 120°$ for the hyperbola in the diagram, r is infinite. As the distance r from the origin at O becomes bigger and bigger, the angle θ approaches closer and closer to $\pm 120°$ without actually getting there. This is described in mathematical jargon by saying that θ asymptotically approaches $\pm 120°$ as r tends to infinity. At the same time the hyperbola asymptotically approaches the dotted lines crossing at C. The distance a occurring in eqn (5.21) is equal to QC. The shortest distance from hyperbola to the origin is OQ, which has the value $a(e - 1)$.

For angles $\theta > 120°$ the coordinate r in eqn (5.21) becomes negative. This may not seem to make much sense, but it actually represents another hyperbolic orbit, which is shown as a dashed line in Fig 5.18; this is a mirror image of the line PQR passing close to a second focal point O′, which is also at a distance a from C. We do not need this orbit, and will restrict discussion to the angular range specified in eqn (5.21).

Deflection in open orbits

Figure 5.19 shows a space probe following the orbit PQR around a planet, which is situated at the focal point O. If the probe were not deflected by the gravitational force, it would travel on a straight line through C, and its closest approach to the planet would be the perpendicular distance h. The distance h is called the **impact parameter** of the orbit. As a result of the attraction of the planet, the probe follows the hyperbola, and is deflected through the angle θ_d. It is left as an example (Problem 5.29) to prove that for a hyperbolic orbit the impact parameter h and the deflection angle θ_d are related to the mathematical parameters of the hyperbola by the equations

$$h^2 = a^2(e^2 - 1), \tag{5.22}$$

$$\cot(\theta_d/2) = h/a. \tag{5.23}$$

We can use these equations to find the angle through which a probe will be deflected if it is approaching the planet from a great distance at a speed v_s and with an impact parameter h. The mass of the space probe m_s is so much smaller than that of the planet that there is no need to make the distinction between the probe's mass and its reduced mass with respect to the planet. The total energy is $\frac{1}{2}m_s v_s^2$ and the angular momentum is $L_s = m_s v_s h$. The angular momentum is conserved in the orbit and, if the probe's speed is v_Q at Q, where, since $\cos\theta = 1$, it is at a distance $a(e-1)$ from the planet, then $L_s = m_s v_Q a(e-1)$, and the kinetic energy

$$E_K = \tfrac{1}{2}m_s v_Q^2 = \frac{L_s^2}{2m_s a^2(e-1)^2} = \frac{m_s v_s^2 h^2}{2a^2(e-1)^2}.$$

The total energy E of the probe is constant and, when the probe is at a great distance from the planet, $E = \frac{1}{2}m_s v_s^2$. At Q the total energy is the sum of the kinetic and (negative) gravitational energies, i.e.

$$E = \tfrac{1}{2}m_s v_s^2 = \frac{m_s v_s^2 h^2}{2a^2(e-1)^2} - \frac{GM_p m_s}{a(e-1)}.$$

Using eqn (5.22) and rearranging, this equation becomes

$$\frac{GM_p m_s}{a} = \tfrac{1}{2}m_s v_s^2 \left(\frac{e^2-1}{e-1} - (e-1) \right) = m_s v_s^2 \tag{5.24}$$

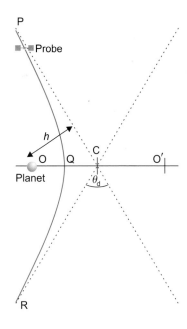

Fig. 5.19 The impact parameter h is the perpendicular distance from the asymptotic path to the focus at O.

and, substituting for a from eqn (5.24), eqn (5.23) becomes

$$\tan(\theta_d/2) = \frac{GM_p}{hv_s^2}. \tag{5.25}$$

The deflection angle θ_d of a space probe or of a comet is determined by the speed v_s and the impact parameter h, but does not depend at all on the mass m_s. For this reason the masses of comets are not well known: by observing their trajectories nothing can be learned about the masses.

Exercise 5.5 A space probe is approaching Jupiter on a path with an impact parameter of 2.4×10^8 m at a speed of $23\,\mathrm{km\,s^{-1}}$ with respect to the planet. What is the angle through which it is deflected, and what is its distance of closest approach to the planet? Take the mass of Jupiter to be 1.9×10^{27} kg. (*Hint.* Use eqn (5.24) to find a for the hyperbolic orbit.)

Answer The total deflection angle is about $90°$. The distance of closest approach to the centre of Jupiter is about 1.0×10^8 m. The radius of Jupiter is about 7.3×10^7 m, so the probe passes within about 3×10^7 km of the surface of the planet. Not much of the planet is seen at close range in such a fly-past, and rockets must be fired to place the probe in a closed orbit to view the whole planet.

● *The deflection of comets by planets*

The deflection of comets by planets is often significant. The planets themselves all move in roughly circular orbits, which lie in planes that are only at small angles to one another, but their orbits are at widely different radii. The planets are always a long distance apart from one another, and their orbits are scarcely disturbed by their mutual gravitational attraction. Comets, on the other hand, come whizzing in from any direction and at any distance from the Sun. It is quite possible for a comet to pass close enough to a planet for a small change to be made to its smooth orbit around the Sun. Comets are like grapeshot flying around in the sky in random directions and following uncertain paths. Fortunately, there are not very many comets and there is an awful lot of sky. However, in 1995, the comet Shoemaker–Levy crashed into Jupiter. Before the collision the comet broke up into a number of fragments, all of which struck the planet on the far side, which was not visible from Earth. But the effect of the crash landings was visible in the Jovian atmosphere when the planet had turned sufficiently far on its own axis. Enormous amounts of energy were dissipated in these collisions, and if a similar comet were to crash on Earth it would cause great devastation. There is now a world-wide programme of search for comets, and a not altogether fanciful discussion of what might be done to deflect one on a collision course with Earth.

The trajectories of charged particles

Everything we have proved about orbits governed by gravitational forces also applies to other systems with an inverse square law of force. The force between electric charges obeys an inverse square law, and the orbit PQR in Fig. 5.19 would be followed, on an appropriately smaller scale, by two charges of opposite sign. Masses always attract each other, whereas an electric charge is repelled by another charge of the same sign. Like charges can never move in a closed orbit because their potential energy gets bigger when they approach close together. The orbit is always hyperbolic but, of course, charges of different sign do not follow the same path. Looking again at Fig. 5.19, you see that the incoming object at P has the same impact parameter h with both the foci at O and O'. By comparison with the undeflected path, the object is being pushed away from the focal point O'. This is indeed the trajectory of a charge moving past a fixed charge of the same sign located at O'. The angle of deflection is the same as for unlike charges with the same impact parameter h, and an equation like eqn (5.24) still applies, with GM_p replaced by a factor representing the strength of the interaction between the charges.

Planetary visits

A space probe visiting another planet must be able to escape from the gravitational force of the Earth. The Earth's 'sphere of influence' extends to the distance where the Earth's gravitational force greatly exceeds that of the Sun. When it has left the sphere of influence, the probe must still be moving away from Earth towards its destination. The probe is in an open orbit, following a hyperbolic path while it is near to Earth and eventually entering an orbit around the Sun. The direction and timing of the launch can be chosen so that the solar orbit brings the probe within the range of the gravitational force of another planet. It then approaches this planet, again on an open orbit, which may swing past in a close encounter, or even lead to collision.

The solar orbit is a transfer orbit between the two open planetary orbits and, just as for the transfer between Earth orbits described in Section 5.3, the most efficient transfer orbit is one that is tangential to both the hyperbolic planetary orbits. Most of the time spent in an interplanetary mission is spent in the solar transfer orbit, and it is not difficult to estimate the time taken to travel from Earth to any of the other planets on the minimum energy path. For Mars the time is about $8\frac{1}{2}$ months, but it would take more than 30 years to reach Neptune with this strategy. In practice it would not be possible to accomplish an interplanetary mission with a single velocity boost: the required precision of speed and direction are unattainable. Midflight corrections are needed, and these can also be used to reduce the mission time. It is also possible to play clever tricks by using encounters with other planets to speed the journey to the destination—but these are specialized topics beyond the scope of this book.

Problems

Where necessary, make use of the following data in these problems.

Gravitational constant $G = 6.67 \times 10^{-11}\,\mathrm{N\,m^2 kg^{-2}}$.
Acceleration of gravity at the surface of the Earth $=$ $9.8\,m\,s^{-2}$.
Radius of the Earth $= 6400\,\mathrm{km}$. Mass of the Earth $=$ $6.0 \times 10^{24}\,\mathrm{kg}$.
Earth−Sun distance $= 1.5 \times 10^8\,\mathrm{km}$.
Earth−Moon distance $= 3.8 \times 10^5\,\mathrm{km}$.
Mass of the Sun $= 2.0 \times 10^{30}\,\mathrm{kg}$.

Level 1

5.1 Estimate the distance from the Earth of the point where the gravitational attractions of the Earth and the Sun are equal and opposite.

5.2 A plumb-bob is hanging vertically. A 40 tonne truck pulls up 10 metres away. What is the resulting deflection of the plumb-bob?

5.3 A pinhole is made in one end of a shoebox that is 60 cm long. An image of the Sun, with a diameter of 2 mm, is observed on the end of the box opposite to the pinhole. Estimate the acceleration due to gravity on the surface of the Sun.

5.4 Estimate the percentage difference in the acceleration of gravity g at sea level and at the height of Mount Everest (8800 m).

5.5 A manned satellite is in a circular orbit at an altitude above the Earth that is much less than the radius of the Earth. How long does it take to complete one revolution?

5.6 The Moon makes a complete orbit around the Earth in 27.3 days. What is its speed with respect to the Earth?

5.7 Use eqn (2.22) to calculate the minimum mass of fuel needed to lift a 1000 kg load from the surface of Mars in a one-stage rocket. The speed of the ejected fuel is 2.5 km s^{-1}, radius of Mars $= 3380$ km, mass of Mars $=$ 6.6×10^{23} kg.

5.8 The period of Venus's orbit around the Sun is 225 days, and the orbit is very nearly circular. Use Kepler's third law to find the radius of the orbit of Venus.

5.9 The angle subtended by the Moon's diameter at the Earth is close to $\frac{1}{2}^{\circ}$. A spacecraft orbiting the Moon at a low altitude completes one revolution every 110 minutes. Estimate: (a) the radius of the Moon; (b) the acceleration of gravity on the Moon's surface; (c) the mass and mean density of the Moon.

5.10 The eccentricity of the Earth's orbit around the Sun is 0.017. Use Kepler's second law to calculate the maximum and minimum speeds of the Earth's orbital motion.

5.11 A comet has been deflected into a hyperbolic orbit around the Sun. It is observed to be moving at a speed of 25 km s^{-1} with respect to the Sun when it is at its perihelion at 10^6 km from the Sun. What is its speed when it is far away from the Sun?

Level 2

5.12 Two lead spheres, each of diameter 10 cm and density 11.3 g cm^{-3}, are placed on a balance as shown in Fig. 5.20. A third similar sphere is placed directly underneath one of the others. Estimate the extra mass that needs to be added to the other sphere to restore balance, considering only the gravitational attraction between the two close masses. (Poynting used this balance technique to measure G with a precision better than 1% in the nineteenth century.)

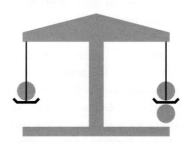

Fig. 5.20 A small mass must be added to the left-hand pan to balance the gravitational force due to the heavy sphere under the right-hand pan.

5.13 One of the techniques used in searching for oilfields is to survey the variation in the acceleration of gravity in the neighbourhood of the oil. To get some idea of the precision needed for such a survey, assume that a sphere of oil with a diameter of 1 km has its centre at a depth of 1 km. Taking the density of oil to be $1 \, \mathrm{g \, cm^{-3}}$ and the mean density of the Earth to be $5 \, \mathrm{g \, cm^{-3}}$, estimate the change in g just above the oilfield.

5.14 Compare the gravitational forces on an object at the surface of the Earth due to the Earth, the Moon, and the Sun. Use the mass of the Moon calculated in Problem 5.9.

5.15 (For this problem you again need the mass of the Moon as calculated in Problem 5.9.) Estimate the difference in the gravitational force per unit mass exerted by the Moon on the parts of the Earth that are closest to and most distant from the Moon. Express the difference as a fraction of g. (This differential gravitational pull of the Moon is responsible for the ocean tides.)

5.16 Calculate the rotational kinetic energy of the Earth and its translational kinetic energy in a frame of reference in which the Sun is at rest.

5.17 The planet Jupiter is the largest in the solar system, with a mass of 1.9×10^{27} kg. Its orbit has a radius of 780 million kilometres, and it completes one revolution around the Sun in 4330 days. Estimate the speed of rotation of the Sun about the centre of mass of the solar system, assuming that this rotation is entirely due to the presence of Jupiter.

5.18 A total eclipse of the Sun occurs over part of the Earth when the Sun's disc is completely obscured by the Moon. Estimate the largest distance between the Earth and the Moon for which a total eclipse is possible.

5.19 An object arrives from outer space and is deflected through $150°$ by the Sun. Its impact parameter with the Sun on arrival is 1.5×10^6 km. On its outward path it passes within 1.5×10^6 km of the Earth. Estimate the additional deflection due to the encounter with the Earth.

5.20 The comet Swift–Tuttle was visible in 1998. At its perihelion (closest approach), it was 0.963 AU from the Sun and moving at $4.3 \times 10^4 \, \mathrm{m \, s^{-1}}$ with respect to the Sun. When will this comet next be seen?

5.21 The comet Swift–Tuttle passes rather close to the Earth. Make a rough estimate of the energy released and the change in speed of the Earth if the comet were to make a head-on collision with the Earth. Assume that the mass of the comet is 10^{16} kg and that it is moving at $40 \, \mathrm{km \, s^{-1}}$ with respect to the Earth at the time of the collision. (The estimate is necessarily rough since the masses of comets are not known with any precision.)

Level 3

5.22 A lead hemisphere, of density $11.3 \, \mathrm{g \, cm^{-3}}$, has a diameter of 10 cm. Calculate the gravitational force between the hemisphere and a 1 g mass placed at the centre of the flat face of the hemisphere.

5.23 Observations on a star show that it is a binary system. The period of rotation is 100 days, and the maximum speeds of the two components are $13 \, \mathrm{km \, s^{-1}}$ and $26 \, \mathrm{km \, s^{-1}}$. Calculate the masses of the two components.

5.24 Starting from eqn (4.26) for the Cartesian co-ordinates of a position vector $\mathbf{r}$, demonstrate that the acceleration along $\mathbf{r}$ is given by eqn (5.17) and that the acceleration perpendicular to the direction of $\mathbf{r}$ is

$$\frac{1}{2r} \frac{\mathrm{d}}{\mathrm{d}t} \left\{ r^2 \left(\frac{\mathrm{d}\theta}{\mathrm{d}t} \right)^2 \right\}.$$

Equation (5.15) shows that this expression is zero, since the angular momentum L_p is constant for central forces.

5.25 An astronaut in a circular Earth orbit experiences no net force towards the Earth and floats inside his capsule. His orbit is then adjusted to be an ellipse with eccentricity 0.1 and perigee (the point nearest to the Earth) at a distance of 100 miles from the Earth's surface. Calculate the forces per unit mass he now experiences at the perigee and apogee of his orbit.

5.26 A rocket is fired from the surface of the Earth in such a way that it emerges from the atmosphere at an angle of $45°$ to the vertical and moving at half the escape speed. Calculate the greatest distance from the Earth reached by this rocket in its elliptical trajectory.

5.27 A satellite is in a circular orbit 3600 km above the Earth's surface. It is transferred to a circular orbit with a radius 40 000 km via an elliptical transfer orbit that is tangential to both the circular orbits. Calculate the velocity boosts needed in the transfer. Check that the additional kinetic energy given to the satellite during the two boosts is the correct amount needed to account for its change in binding energy.

5.28 Because of the Earth's rotation about its axis, the net force on an object on the Earth's surface is less than the gravitational force and is not quite in the same direction as the gravitational force. Assuming the

Earth to be spherical, calculate the latitude at which the deviation in direction caused by the rotation has its maximum value. Estimate the magnitude of the deviation.

5.29 Figure 5.21 represents a hyperbolic orbit described by the equation

$$r = \frac{a(e^2 - 1)}{1 + e \cos \theta}, \quad \text{for } e > 1. \tag{5.18}$$

The deflection of an object following the orbit is θ_d and the angle $\psi = (\pi/2 + \theta_d/2)$ is the asymptotic value of θ as r tends to infinity. The asymptotic value of p is the impact parameter h. By considering the properties of the triangle OPS in the limit as r becomes very large, show that $h = a \cot(\theta_d/2)$, and that $h^2 = a^2(e^2 - 1)$.

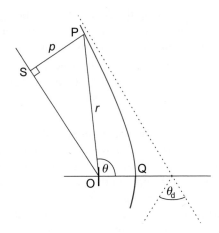

Fig. 5.21 PQ is a hyperbola with focal point at O. OS is parallel to the asymptotic trajectory (dashed line) and θ_d is the angle of deflection of the comet.

Some solutions and answers to Chapter 5 problems

5.1 If the distance from Earth is d, then the gravitational attractions of the Earth and the Sun are equal when

$$\frac{M_s}{(A - d)^2} = \frac{M_e}{d^2},$$

where M_e is the mass of the Earth and A is the distance from the Earth to the Sun (one astronomical unit). This equation is a quadratic in d: however, because M_s is so much larger than M_e, d is very small compared with A. To a good approximation

$$d = A\sqrt{\frac{M_e}{M_s}} = 1.5 \times 10^8 \sqrt{\frac{6 \times 10^{24}}{2 \times 10^{30}}}$$

$$= 2.6 \times 10^5 \text{ km from Earth.}$$

The error in this estimate is less than two parts in a thousand.

5.3 534 m s^{-2}.

5.4 0.275%.

5.5 Rearranging eqn (5.6),

$$T = 2\pi\sqrt{\frac{R^3}{GM_e}} = 5085 \text{ s}$$

$$\approx 85 \text{ minutes.}$$

5.7 From eqn (2.22) the speed of the spacecraft after firing the rocket is $v_f = v_e \ln(M_i/M_f) - g t_f$, where v_e is the speed of the gas ejected from the rocket, M_i and M_f are the masses of the rocket before and after firing, g the acceleration of gravity on Mars, and t_f the time taken to use all the rocket fuel. The minimum mass of fuel is used if t_f is small so that the second term may be neglected and if v_f is just greater than the escape velocity v_{esc} from Mars. Calling the mass of Mars M_M and its radius R_M, the escape velocity is given by equating the kinetic energy after firing the rocket with the potential energy at the surface of Mars:

$$\frac{1}{2} M_f v_{esc}^2 = \frac{GM_f M_M}{R_M}, \text{ leading to } v_{esc} = \sqrt{\frac{2GM_M}{R_M}}$$

$$= 5103 \text{ m s}^{-1}.$$

For $v_e = 2500\,\mathrm{m\,s^{-1}}$, $\ln(M_i/M_f) \geq (5103/2500)$ and $(M_i/M_f) \geq 7.7$. The minimum mass of fuel is therefore 6700 kg, or 6.7 tonnes.

5.9 (a) 1.7×10^3 km; (b) $1.5\,\mathrm{m\,s^{-2}}$; (c) 6.4×10^{22} kg and $3.25\,\mathrm{g\,cm^{-2}}$.

5.10 $15\,200\,\mathrm{m\,s^{-1}}$ and $14\,700\,\mathrm{m\,s^{-1}}$.

5.13 The change in g due to the presence of the oilfield is equivalent to the effect of a sphere of density $4\,\mathrm{g\,cm^{-3}}$. For a sphere with radius r and density ρ, the gravitational acceleration at a distance d is $\Delta g = \frac{4}{3}\pi\rho r^3 G/d^2 = 1.40 \times 10^{-4}\,\mathrm{m\,s^{-2}}$, and the fractional change in g is about 1.4×10^{-5}.

5.15 The change $\Delta g/g$ due to the Moon is about 2.0×10^{-7}.

5.16 The moment of inertia I of a sphere of mass M and radius R is $\frac{2}{5}MR^2$. The rotational energy of the Earth is therefore

$$\frac{1}{5}M_e R_e^2 \omega^2 = \frac{1}{5} \times 6.0 \times 10^{24} \times (6.4 \times 10^6)^2$$
$$\times\, 4\pi^2/(24 \times 3600)^2$$
$$= 2.6 \times 10^{29}\ \mathrm{J}.$$

The speed v_e of the Earth in its orbit is 2π AU per year, which converts to $30\,\mathrm{km\,s^{-1}}$. The translational kinetic energy $\frac{1}{2}M_e v_e^2$ is 2.7×10^{33} J, 10 000 times greater than the rotational energy.

5.17 The speed of Jupiter in its orbit is

$$\frac{2\pi \times 7.8 \times 10^{11}}{4330 \times 24 \times 3600} = 13\,100\,\mathrm{m\,s^{-1}}.$$

Both Jupiter and the Sun are rotating about their common centre of mass. The ratio of their masses is the same as the ratio of their distances to the centre of mass and hence of their speeds. The speed of the Sun is thus

$$\frac{13\,100 \times 1.90 \times 10^{27}}{2.0 \times 10^{30}} = 12.4\,\mathrm{m\,s^{-1}}.$$

5.19 1.3×10^{-3} degrees.

5.20 The mass M_c of the comet is very much less than the mass M_s of the Sun: the reduced mass is therefore almost exactly M_c. The perihelion distance is $a(1 - e)$, so the angular momentum of the comet about the Sun is $L_c = M_c v_c a(1 - e)$, where v_c is the speed of the comet at the perihelion. From eqn (5.19) $L_c^2 = GM_s\,M_c^2 a(1 - e^2) = M_c^2 v_c^2 a^2 (1 - e)^2$, which simplifies to $v_c^2 a(1 - e) = $

$GM_s(1 + e)$. Hence

$$1 + e = \frac{v_c^2 a(1 - e)}{GM_s}$$
$$= \frac{(4.26 \times 10^4)^2 \times 0.963 \times 1.496 \times 10^{11}}{6.67 \times 10^{-11} \times 2.0 \times 10^{30}} = 1.959$$

and $1 + e = 0.959$. Now $a(1 - e) = 0.963$ AU, leading to $a = 0.963 \times 1.496 \times 10^{11}/0.041 = 3.5 \times 10^{12}$. Using this value in eqn (5.20), the period of the comet is

$$T = 2\pi\sqrt{\frac{a^3}{GM_s}} = 2\pi\left\{\sqrt{\frac{(3.5 \times 10^{12})^3}{6.67 \times 10^{-11} \times 2.0 \times 10^{30}}}\right\}$$

$$= 3.56 \times 10^9\,\mathrm{s}\ \text{or}\ 113\ \text{years}.$$

Measurements must be very precise to determine the parameters of orbits with eccentricity close to one. In this example the data provided do not lead to an accurate value of the quantity $(1 - e)$ and hence of a. The actual period of Swift–Tuttle is 120 years and it is expected to return in the year 2118.

5.22 By symmetry the force is perpendicular to the flat face of the hemisphere. The mass of the annular element of thickness dr lying between θ and $\theta + d\theta$ shown in the figure is $2\pi r^2 \rho \sin\theta\, d\theta\, dr$, where ρ is the density of the lead. Each element of the annulus exerts a force at an angle θ to the net force, and the force on a mass m at the origin is $\cos\theta \times 2\pi r^2 \rho \sin\theta d\theta dr \times Gm/r^2 = 2\pi Gm\rho \sin\theta\, d(\sin\theta)\, dr$.

The integral over the angle is from $\sin\theta = 0$ to $\sin\theta = 1$ and the integral over r is from 0 to R. The total force is

$$\int_0^1 \int_0^R 2\pi Gm\rho \sin\theta\, d(\sin\theta)\, dr$$
$$= 2\pi Gm\rho R \left[\tfrac{1}{2}\sin^2\theta\right]_0^1 = \pi Gm\rho R.$$

Using the values given for m, ρ, and R, the force is 1.2×10^{-10} N.

5.23 8×10^{29} kg and 1.6×10^{31} kg.

5.26 The potential energy of a rocket of mass m on the surface of the Earth is GmM_e/R_e, and the escape velocity v_e is given by $\frac{1}{2}mv_e^2 = GmM_e/R_e$. The speed of the rocket after firing is $v = \frac{1}{2}v_e$ and the total (kinetic + potential) energy after launch is $\frac{1}{2}v^2 - GmM_e/R_e = -\frac{3}{4}GmM_e/R_e$. From Worked example 5.4, the total energy in the elliptical orbit is $-GmM_e/2a$ and thus $a = \frac{2}{3}R$.

The perpendicular distance from the rocket's launch trajectory to the centre of the Earth is $R_e/\cos\theta = R_e/\sqrt{2}$. The angular momentum L about the centre of the Earth is given by $L^2 = m^2 \times \frac{1}{4}v_e^2 \times \frac{1}{2}R_e^2 = \frac{1}{4}Gm^2 M_e R_e = Gm^2 M_e a(1 - e^2)$ from eqn (5.18). Since $a = \frac{2}{3}R$, this leads to $(1 - e^2) = \frac{3}{8}$, $e = 0.79$ and the greatest distance from the Earth $a(1 + e) = 1.19R_e$, or 7640 km. This is a height of 1240 km above the Earth's surface.

5.27 The first velocity boost is $1.68\,\text{km s}^{-1}$ and the second is $1.16\,\text{km s}^{-1}$.

5.29 The equation for the hyperbola is $r(1 + e\cos\theta) = a(e^2 - 1)$. The angle θ approaches its asymptotic value ψ as $r \to \infty$, so that $\cos\psi = 1/e$ or $e = \sec\psi$. Substituting for e in the equation for the hyperbola, and using $\tan^2\psi = \sec^2\psi - 1$, $r(1 - \sec\psi\cos\theta) = a\tan^2\psi$, which can be rearranged to give

$$\cos\theta = \cos\psi\left(1 - \frac{a}{r}\tan^2\psi\right)$$

Hence $\sin\theta = \sqrt{1 - \cos^2\theta}$

$$= \left\{1 - \cos^2\psi(1 - \frac{a}{r}\tan^2\psi)^2\right\}^{\frac{1}{2}}$$

$$= \sin\psi\left\{1 + \frac{a}{r} + \cdots\right\},$$

keeping only the term in $1/r$ in the expansion of $\sin\theta$. In the triangle OPS,

$$p = r\sin(\psi - \theta) = r\sin\psi\cos\theta - r\cos\psi\sin\theta$$
$$= r\sin\psi\cos\psi(1 - a/r\tan^2\psi)$$
$$\quad - r\cos\psi\sin\psi(1 + a/r)$$
$$= -a\sin\psi\cos\psi(1 + \tan^2\psi + \cdots)$$
$$= -a\sin\psi\cos\psi\sec^2\psi + \cdots$$
$$= -a\tan\psi + \text{terms in powers of } 1/r.$$

As $r \to \infty$, $p \to h$, the impact parameter, and $h = -a\tan\psi$. From above, $\tan^2\psi = \sec^2\psi - 1 = e^2 - 1$, giving $h^2 = a^2(e^2 - 1)$. The deflection angle $\theta_d = 2\psi - \pi$ leading to $\tan\psi = -\cot\theta_d/2$ and $h = a\cot(\theta_d/2)$.

Summary of Part 1
Dynamics

Important concepts and key equations from Part 1 are briefly explained in this summary

Chapter 2 **Newton's laws of motion**

Forces are vectors and the net force on an object is the vector sum of all the individual forces acting on it. An object that remains at rest under the action of a number of forces $\mathbf{F}_i$ is in **equilibrium** and the net force on it is zero:

$$\sum_i \mathbf{F}_i = 0. \tag{2.1}$$

The state of motion of a moving object is specified by its **velocity**. Velocity is also a vector, pointing in the direction of motion, and with a magnitude equal to the speed of the object. Velocity is a relative quantity; an object is always moving with respect to something else. For example, a person walking will usually be described as having a particular velocity $\mathbf{v}$, say, with respect to the Earth. However, if the person is walking on the deck of a ship, the walking speed and direction $\mathbf{v}'$ is given with respect to the deck. Velocities, like forces, obey the vector rule of addition and, if the velocity of the ship with respect to the water is $\mathbf{V}$, then

$$\mathbf{v} = \mathbf{v}' + \mathbf{V}.$$

The velocity of an object can only be changed by a net force acting on it. This fact is **Newton's first law of motion**, normally expressed in the form:

Any body moves with constant velocity (or remains at rest) if the net force acting on it is zero.

The momentum $\mathbf{p}$ of an object of mass m moving with a velocity $\mathbf{v}$ is defined to be the product $m\mathbf{v}$:

$$\mathbf{p} = m\mathbf{v}. \tag{2.2}$$

Newton's second law of motion states that

The net force acting on a body equals its rate of change of momentum,

which is expressed mathematically by the equation

$$\mathbf{F} = \frac{d\mathbf{p}}{dt}. \tag{2.5}$$

There is no constant of proportionality in this equation when SI units are used, because the unit of force (the *newton*) is defined so that eqn (2.5) is correct if mass is measured in kilograms, velocity in metres per second, and time in seconds.

The gravitational force on an object is constant and gravitational acceleration is downwards, with a constant value g. Integration of Newton's second law leads to the result that, for an object with initial upward speed v_0, the upwards distance moved under gravity in a time t is

$$z = v_0 t - \tfrac{1}{2} g t^2. \tag{2.13}$$

Newton's second law is a vector differential equation, applying to motion in three dimensions. This means that the law is equivalent to three different equations referring to three different coordinate axes. When forces and velocities are not in the same direction, the equations referring to each of these axes must be considered in turn.

The final law of classical mechanics is Newton's third law, which states that

If an object A exerts a force F on an object B, then B exerts an equal and opposite force $(-F)$ on A.

The third law is summarized as 'Action and reaction are equal and opposite': it must be remembered that the action and reaction referred to are forces on *different* objects.

Frictional forces act to slow down motion. Sometimes objects are constrained as to the directions in which they are able to move, as, for example, a box standing on a flat floor that can only move horizontally. The box is held in place by an upward force from the floor, counteracting the downward gravitational force. When an external horizontal force is applied, the external force is opposed by an opposite horizontal frictional force.

For a stationary object the frictional force has a maximum value and no motion occurs until the external force exceeds the maximum frictional force. The maximum frictional force is often roughly proportional to the weight of the object. The constant of proportionality μ_s is called the *static coefficient of friction*. Once the object is moving, the frictional force reduces to a value μ_d times the weight, where the coefficient μ_d is the dynamic coefficient of friction.

Frictional forces also oppose motion for objects moving through gases or liquids. At low speeds the frictional forces are then often approximately proportional to the speed through the fluid.

Chapter 3 **Momentum and energy**

For a system containing any number of different particles with masses m_i and velocities $\mathbf{v}_i$, the momentum of the whole system is

$$\mathbf{p} = \sum_i m_i \mathbf{v}_i. \tag{3.2}$$

Conservation of momentum follows from applying Newton's second law to the system.

The momentum of any system remains constant unless it is acted on by a net external force.

The mathematical statement of conservation of momentum is

$$\frac{d\mathbf{p}}{dt} = \sum \mathbf{F}_i = 0. \tag{3.3}$$

Positions and velocities depend on the coordinate system (the *frame of reference*) used to describe them. Consider two frames of reference S and S′, such that the origin of S′ is moving along the x-axis of S at a speed V. Suppose that the frames S and S′ coincide at the time $t = 0$. Then the coordinates in S and S′ are related by the *Galilean transformation*

$$x' = x - Vt;\ y' = y;\ z' = z. \tag{3.5}$$

Using the Galilean transformation we can find the relation between velocities and momenta in S and S′. If S′ moves at a constant velocity $\mathbf{V}$ with respect to S, these relations are

$$\mathbf{v}' = \mathbf{v} - \mathbf{V}. \tag{3.8}$$

and

$$\mathbf{p}' = \mathbf{p} - M\mathbf{V}, \tag{3.9}$$

whether or not $\mathbf{V}$ is in the direction of the x-axis.

The centre of mass frame is the special one for which $\mathbf{p}' = 0$. From eqn (3.9) it follows that the velocity $\mathbf{V}_{cm}$ of the centre of mass in the frame S is

$$\mathbf{V}_{cm} = \frac{\mathbf{p}}{M}. \tag{3.11}$$

Energy is expended by a force when it causes an object to move. The energy derived from mechanical forces is called *work*, and the work done by a force, measured in joules, is

Work done = force × distance moved in the

direction of the force. (3.12)

A conservation law applies to energy as well as to momentum. Conservation of energy may be stated as follows:

The total energy of any isolated system is constant.

The kinetic energy of a system is energy that it possesses by virtue of its motion. For a single particle of mass m moving at a speed v, the kinetic energy is

$$E_K = \tfrac{1}{2} m v^2. \tag{3.13}$$

In the centre of mass frame of reference the net momentum is zero. However, the total kinetic energy in this frame may be nonzero. If S′ is the centre of mass frame, the kinetic energy in another frame S can be written as

$$E_K = \sum_i \tfrac{1}{2} m_i v_i'^2 + \tfrac{1}{2} M V_{cm}^2 \tag{3.19}$$

where V_{cm} is the speed of the centre of mass in the frame S. The first term is an internal energy of the system, which has the same value whatever may be the velocity of its centre of mass.

Elastic collisions are those in which the total kinetic energy is the same before and after collision. Some kinetic energy is converted into internal energy in *inelastic* collisions. When a collision between two objects is inelastic, their relative speeds are smaller after the collision than before. The ratio of speeds, which can vary between zero and one, is called the *coefficient of restitution*.

The performance of mechanical work does not necessarily lead to an increase in kinetic energy equal to the work done. Work must be done against the gravitational force to raise an object, which may then remain at rest. The work done is stored as gravitational *potential energy*, which may be released and converted to kinetic energy by allowing the object to fall: work has been done *on* the gravitational force when the object is raised, and an equal amount is done *by* the gravitational force when it falls.

The work done by gravity in raising and lowering any object when it returns to its original starting point is zero, whatever path is followed by the object. A force that, like the gravitational force, has the property that no work is done by the force when an object moves around a closed path is called a *conservative force*. For a conservative force, the potential energy is always the same for an object at a particular point, whatever path it might have followed to reach that point.

The zero of potential energy may be chosen to be at any point but, once that is done, the potential energy is a definite *function of position*. For example, if the energy depends on a single coordinate x, as in the case of the gravitational force where height is all that counts, the potential energy can be written as a function of x only, $U(x)$, say. The force is then related to the potential energy $U(x)$ by the equation

$$F = -\frac{dU}{dx}.$$ (3.28)

Power is the rate of doing work, and is measured in watts. One watt is equivalent to one joule per second.

Chapter 4 Rotation

All parts of a rotating wheel have the same *angular speed*, measured in radians per second, about the axis of rotation. If the angular speed is $\omega = d\theta/dt$, a part of the wheel at a distance r from the axis is moving around the arc of a circle at a speed

$$v = r\frac{d\theta}{dt} = r\omega.$$ (4.2)

Rotating objects have kinetic energy as a result of their motion. The kinetic energy E_K depends on the angular speed ω and on the *moment of inertia* of the rotating object about the axis. For a uniform disc rotating about an axis through its centre and perpendicular to the plane of the disc, the moment of inertia is

$$I_{disc} = \tfrac{1}{2}MR^2.$$ (4.5)

The moment of inertia of a rigid body depends on its shape and the position and direction of the axis of rotation. To calculate the moment of inertia, an integral must be performed over the volume of the body. If the perpendicular distance of a volume element dV from the axis is r, and the rigid body has a constant density ρ, the moment of inertia is

$$I = \rho \int_V r^2 \, dV$$ (4.7)

where the suffix V on the integral sign indicates that the integral is over the whole of the volume occupied by the object.

Equation (4.7) is easily generalized to apply to bodies that do not have a uniform density. Writing the density within the volume dV as ρ_V, the moment of inertia may be written as

$$I = \int_V \rho_V r^2 \, dV$$ (4.10)

where as before r is the perpendicular distance from the axis of the volume element dV at the point with position vector $\mathbf{r}$.

It is often most convenient to calculate the moment of inertia about an axis passing through the centre of mass. The moment of inertia about any other axis parallel to this one can then be calculated by using the theorem of parallel axes. The theorem states that

The moment of inertia of a rigid object about any axis is equal to the moment of inertia about a parallel axis passing through the centre of mass, plus the total mass times R^2, where R is the perpendicular distance between the two axes.

The theorem of perpendicular axes applies only to thin discs, but it too can sometimes be used to simplify calculations. The theorem states that

The moment of inertia of a flat disc about an axis perpendicular to the disc equals the sum of the moments of inertia about two perpendicular axes lying in the plane of the disc and both meeting the first axis.

A rotation axis has a definite direction in space. To specify completely the rotational motion of a rigid body, the direction of the axis must be given as well as the angular speed. The *angular velocity* is a vector with magnitude equal to the angular speed and pointing in the same direction as the axis. A choice must be made between the two opposite directions along the axis. The convention is that the angular velocity vector is given by a right-hand rule. This means that, viewed in the direction of the angular velocity vector, the rotation is clockwise, the sense in which a right-handed screw turns.

Angular acceleration (or deceleration) is the rate of change of angular velocity. Angular acceleration or deceleration is effected by exerting a *torque*. The general definition of a torque is given by the statement that

If a force F acts at a point with postion vector r with respect to an origin at O, then the torque exerted by F about O is given by the vector product r × F.

The torque is thus a vector given by the equation

$$\tau = \mathbf{r} \times \mathbf{F}. \tag{4.15}$$

Note that a torque acts about a *point*, whereas in the discussion of rotation of rigid bodies we have been considering rotation about a fixed *axis*. When the axis is fixed, only the component of the torque in the direction of the axis can cause angular acceleration or deceleration. Other components may give rise to forces on the bearings or stresses within the rigid body, but they do not affect the rotational motion. The calculation of the component of torque along an axis, depends on the *perpendicular distance from the axis*. The result is that

The z-component of the torque exerted by a force on a body rotating about the z-axis equals the component of the force in the direction of motion of the body at the point of application of the force times the perpendicular distance of that point from the axis.

When a torque τ is applied in the direction of the axis of the angular velocity ω of a rigid body, the equation of motion for rotation is

$$\tau = I \frac{d\omega}{dt} \tag{4.17}$$

where I is the moment of inertia about the axis.

The *angular momentum* **L** is defined to be

$$\mathbf{L} = I\omega \tag{4.18}$$

and hence

$$\tau = \frac{d\mathbf{L}}{dt}. \tag{4.19}$$

Equation (4.19) has the same form as Newton's second law $\mathbf{F} = d\mathbf{p}/dt$, with the force **F** replaced by the torque τ and the linear momentum **p** replaced by the angular momentum **L**.

When there is no fixed axis of rotation, the angular momentum **L** is referred to a point rather than to an axis, with the following definition

The angular momentum of a moving particle about *any* point O is r×p where p is the object's momentum and r is the position vector of the particle with respect to an origin at O:

$$\mathbf{L} = \mathbf{r} \times \mathbf{p}. \tag{4.23}$$

The angular momentum of a system of particles with momenta $\mathbf{p}_i$ and positions $\mathbf{r}_i$ is simply the vector sum of all the individual angular momenta with respect to the same origin:

$$\mathbf{L} = \sum_i \mathbf{r}_i \times \mathbf{p}_i. \tag{4.24}$$

The rate of change of **L** is given by the same eqn (4.19) as was derived for the special case of a rigid body rotating about a fixed axis:

$$\tau = \frac{d\mathbf{L}}{dt}. \tag{4.25}$$

This is the general equation needed to describe all rotational motion. Expressing this equation in words:

For any system of objects or particles, the net torque acting about any point is equal to the rate of change of angular momentum about the same point.

In particular, if the net torque about a particular point is zero, the angular momentum about that point is constant. In other words, the angular momentum about that point is *conserved*.

The direction of motion is constantly changing for an object moving in a circle. A *centripetal* force on the object must act towards the centre of the circle to keep it on the circular path: the object exerts an equal and opposite *centrifugal* force on the body exerting the centripetal force. The acceleration towards the centre for an object moving in a circle of radius r at an angular speed ω is

$$\mathbf{a} = -\omega^2 \mathbf{r} \tag{4.28}$$

where the origin of the position vector $\mathbf{r}$ is at the centre of the circle. For objects moving in curved paths that are not circular, over a very short distance the path coincides with the arc of a circle that has a radius r equal to the *radius of curvature* of the path over the short distance. Over this section of the path the acceleration is given by eqn (4.28).

Consideration of the conservation of angular momentum often helps in the solution of problems involving rotation. Calculation of the angular momentum about a particular point may be simplified by making use of the following theorem.

The angular momentum of an object about a particular point in space is the sum of the angular momentum about its centre of mass and the angular momentum of the centre of mass about the same point.

In an obvious notation, this is expressed mathematically by the equation

$$\mathbf{L} = \mathbf{L}_{\text{cm}} + \mathbf{R} \times \mathbf{P}. \tag{4.35}$$

Tops and gyroscopes are rigid bodies that rotate about an axis that is not usually fixed. Their motion is described by the *gyroscope* equation

$$\tau = I\omega\Omega_{\text{p}} \tag{4.37}$$

where τ is a torque applied perpendicularly to the instantaneous axis of rotation, I and ω are, respectively, the moment of inertia and the angular speed about this axis, and Ω_{p} is the *precessional* angular speed about an axis perpendicular to both the torque and the spinning axis of the gyroscope.

Chapter 5 **Gravity and orbital motion**

Gravity is a universal force that attracts any two masses towards each other. The force obeys *Newton's law of gravity*, which states that:

The attractive gravitational force between two small bodies acts along the line joining them: its magnitude is proportional to the product of their masses and inversely proportional to the square of the distance between them.

The magnitude of the force between two particles of masses m_1 and m_2 separated by a distance r is given by the equation

$$F = G\frac{m_1 m_2}{r^2} \tag{5.1}$$

where G is the gravitational constant, which has the value

$$G = 6.6726 \pm 0.0009 \times 10^{-11} \, \text{N} \, \text{m}^2 \, \text{kg}^{-2}. \tag{5.2}$$

The gravitational attraction of a large spherically symmetrical object for masses outside it is the same as if all the mass were concentrated at its centre. The Earth, for example, is almost spherically symmetric. An object on the surface of the Earth experiences a gravitational force of almost the same magnitude at any position on the surface. Since the force is proportional to the mass of the object, its downwards acceleration g towards the centre of the Earth is the same whatever the magnitude of the mass, and is related to the mass M_{e} and the radius R_{e} of the Earth by the equation

$$M_{\text{e}} = \frac{gR_{\text{e}}^2}{G}. \tag{5.3}$$

Work must be done against the gravitational force to move an object further away from the Earth. The work done is stored as potential energy. Choosing the zero of potential energy to be at a great distance from the Earth, the potential energy of an object of mass m at a distance r_1 from the Earth's centre is negative, and has the value

$$U = -\frac{GmM_{\text{e}}}{r_1}. \tag{5.5}$$

The positive amount of work $W_\infty(r_1) = -U$ required to move the mass from r_1 to infinity is called its *binding energy*.

The period T of an object moving in a circular orbit around the Sun may be related to the orbital radius R_{orbit}

by considering the force required to maintain circular motion. The result is

$$T^2 = \frac{4\pi^2 R_{\text{orbit}}^3}{GM_s}. \tag{5.6}$$

The relative motion of two objects of masses m_1 and m_2 is independent of the motion of their centre of mass. The centre of mass motion is the same as for a single particle with the position coordinates of the centre of mass and mass equal to the total mass $M = m_1 + m_2$. For relative motion the position coordinates are given by a vector in the direction of the line separating the two objects and with magnitude equal to their separation, and the mass used in the equations of motion is the **reduced mass**

$$m = \frac{m_1 m_2}{m_1 + m_2}. \tag{5.10}$$

Planets move in *elliptical* rather than circular orbits. The equation that gives the locus of an *ellipse* in polar coordinates is

$$r = \frac{a(1 - e^2)}{1 - e\cos\theta}. \tag{5.12}$$

Here a is the *semi-major axis* of the ellipse, and e is a number less than one called its *eccentricity*.

The motion of planets is described by Kepler's laws, which are:

1. **Each planet moves in an elliptical orbit with the Sun at a focal point.**

2. **The position vector of a planet relative to the Sun sweeps out a constant area per unit time within the ellipse.**

3. **The squares of the periods of the planets are proportional to the cubes of their mean distances from the Sun.**

The first two laws are exact, and the second law is equivalent to the statement that the angular momentum of a planet in its orbit is constant. The third law is not quite exact. The precise equation that corresponds to the third law is

$$T^2 = \frac{4\pi^2 m a^3}{GM_s M_p} = \frac{4\pi^2 a^3}{G(M_s + M_p)}. \tag{5.20}$$

Here M_s is the mass of the Sun, M_p is the mass of the planet, and a is the mean distance of the planet from the centre of mass of the planet and the Sun. Since M_s is thousands of times greater than the mass of even the largest planet, Jupiter, the usual statement of the law given above is a very good approximation.

The motion of objects moving in space is affected by the gravitational attraction of stars and planets even if they are not trapped in closed orbits around a particular star or planet. When passing a centre of gravitational force, the path of an object is deflected. The locus of the path is a *hyperbola*, described by an equation, similar to that of an ellipse, with an eccentricity e greater than one:

$$r = \frac{a(e^2 - 1)}{1 + e\cos\theta}, \quad \text{for } e > 1 \text{ and } \cos\theta > -\left(\frac{1}{e}\right). \tag{5.21}$$

Part 2

Vibrations and waves

Chapter 6

Oscillations

This chapter discusses the oscillatory motion of bodies disturbed from equilibrium positions by momentarily applied or continuous forces.

In this chapter we discuss how an oscillatory motion is described as a function of time. Our considerations can readily be adapted to the discussion of the periodic variations with time of physical quantities other than position. A periodic motion is one that regularly repeats itself with each repetitive cycle taking a time T called the **period**. The **frequency** ν of the periodic motion is the number of cycles performed in one second.

$$\nu = \frac{1}{T}. \tag{6.1}$$

The units of ν are thus s^{-1}. However, a special name is given to the unit of frequency. It is called the **hertz** (Hz) after the discoverer of radio waves.

1 Hz ≡ 1 cycle per second.

A simple periodic motion is executed by a particle moving in a circle of radius a with constant speed v, as discussed in Chapter 4. The period is simply the time taken to travel around the circumference once and return to the same point: $T = 2\pi a/v$.

Figure 6.1 shows a particle moving clockwise around a circle. It passes the point A at time $t = 0$. After a period has elapsed, at time $t = T$, it has returned to the point A. At an intermediate time t the particle is at the point P and may be located on the circumference by drawing a radius from the centre of the circle at the point O to the point P. The line OP makes an angle ωt with the line OA, where ω is the angular speed of the particle's motion, introduced in Chapter 4, and is equal to the number of radians swept out by the line OP per second. The angular speed is measured in radians per second (rad s^{-1}). Hence

$$\omega = \frac{2\pi}{T} = 2\pi\nu \tag{6.2}$$

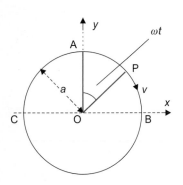

Fig. 6.1 A particle moving clockwise in a circle of radius a passes the point A at time $t = 0$. At time t it has reached the point P, with the angle AOP equal to ωt.

and in this example $\omega = v/a$.

If we make the point O on Fig 6.1 the origin of a Cartesian coordinate system, and let the line OA lie along the positive y-axis and OB lie along the positive x-axis, the function $x(t)$, which gives the dependence on time of the x-coordinate of the particle, takes the simple form

$$x(t) = a\sin(\omega t). \tag{6.3}$$

The x-coordinate varies sinusoidally with time and has a maximum value a when the particle on Fig 6.1 is at the point B and a minimum value $-a$ when the particle is at the point C. This is shown on Fig 6.2 where we have taken ω equal to 3 rad s^{-1}, when the period of the oscillation is $\frac{2}{3}\pi$ seconds.

The distance, or displacement, a is called the **amplitude** of the oscillations described by eqn (6.3). When discussing an oscillation, ω is referrred to as the **angular frequency** of the oscillation.

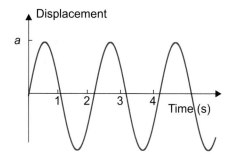

Fig. 6.2 The x-coordinate of a particle moving with constant speed in a circle varies sinusoidally with time. The figure is for a particle whose x-coordinate passes through the origin at $t = 0$ and whose angular speed in the circle is $\omega = 3 \text{ rad s}^{-1}$.

Normally, a periodic motion would have a more complicated time dependence than that given by eqn (6.3). However, as we shall see in Section 6.6, any periodic motion can be written as the sum of terms similar to the right-hand side of eqn (6.3) but with different amplitudes and angular frequencies, all of the latter being integral multiples of a fundamental angular frequency associated with the motion.

6.1 Simple harmonic motion

In this section we discuss the simplest type of periodic motion called **simple harmonic motion**, abbreviated SHM. Simple harmonic motion is important in practice because it is a good approximation to the free oscillations in many physical situations. Automobiles are usually suspended on springs, and an automobile would oscillate up and down in a way approximating SHM after the wheels hit a bump, unless steps were taken to stop it. The discussion of SHM is also important because it serves to introduce many features characteristic of more complicated oscillations.

Simple harmonic motion occurs when forces or torques that return an object to its equilibrium position are proportional to displacements from those positions. Many of the oscillatory motions in nature are SHM to a good approximation because, for a sufficiently small displacement of any body from equilibrium, the *restoring force* or *restoring torque* is proportional to the displacement.

Oscillating objects have energy and if energy is removed from the oscillations they die away. Pure SHM continues indefinitely when no energy is removed. Oscillations of real systems are invariably damped to some extent and thus motions are never exactly SHM. Damping of oscillations is considered in Section 6.2.

Examples of objects that oscillate with SHM to a very good approximation are provided by a mass on the end of a spring after the spring has been extended, a pendulum after it has been pulled sideways from its vertical equilibrium position, a ball making small excursions back and forth inside a smooth upturned bowl, or a fixed mass of liquid oscillating up and down in a U-shaped tube. The variation with time of many physical observables other than spatial or angular displacements is also well described within the same formalism as SHM. An example is that of charge oscillating back and forth between the plates of a capacitor in a circuit with an inductor of small resistance. This example is discussed in Chapter 18. Other examples occur in the exercises and problems of this chapter.

A good example of SHM in the microscopic world is provided within crystals such as quartz or diamond, and we will use this example to develop the discussion of simple harmonic motion. Crystals are made up of atoms making small oscillations about equilibrium positions that are points on a regular lattice structure. When an atom is slightly displaced, nearby atoms in the regular structure produce a net force that tends to restore the displacement to its equilibrium value. Although the possible motions are complicated and involve other atoms in the crystal, let us make the simple assumption that, when an atom is displaced by a distance x from its equilibrium position, the net restoring force is proportional to x and in the direction opposite to the displacement. In this simple picture the problem is a one-dimensional one, and Newton's second law relates the force on an atom to its acceleration. This law, eqn (2.6), can be written as

$$F(x) = -kx = m\frac{d^2x}{dt^2} \tag{6.4}$$

where k is the constant that gives the strength of the restoring force and m is the mass of the oscillating atom. The minus sign arises because with k positive the force acts to oppose x increasing.

Equation (6.4) may be rewritten in the form

$$\frac{d^2x}{dt^2} + \omega_0^2 x = 0 \tag{6.5}$$

with

$$\omega_0^2 = k/m. \tag{6.6}$$

It can readily be verified by substitution that the solution to eqn (6.5) is

$$x(t) = x_0\cos(\omega_0 t + \phi) \tag{6.7}$$

where the constant x_0 is the amplitude of the oscillations. The argument of the cosine term, $\omega_0 t + \phi$, is called the **phase** of the oscillation and the constant ϕ is a phase angle. The parameter ω_0 is called the **natural angular frequency** of the oscillations and eqn (6.4) was rewritten in the form of eqn (6.5) in order to exhibit clearly the relation between angular frequency, mass, and stiffness parameter k.

The solution of the second-order differential eqn (6.5) contains two constants as do the solutions to all such equations. These constants can be determined if the displacement and speed are both known at a given time. For example, if the mass is displaced by a certain amount x_0 and released from rest at time zero, the initial conditions that determine the two constants are that at $t = 0$ the speed dx/dt is zero and the displacement x is x_0. The speed is

$$\frac{dx}{dt} = -\omega_0 x_0 \sin(\omega_0 t + \phi). \tag{6.8}$$

Putting this equal to zero at time zero gives the angle ϕ equal to zero. The amplitude x_0 is then identified as the initial displacement. When the displacement is a maximum, at values of $\omega_0 t$ equal to $n\pi$ with n any integer, the speed is zero. When the speed is a maximum, at values of $\omega_0 t$ equal to $(n + \frac{1}{2})\pi$ with n any integer, the displacement is zero.

Energy in simple harmonic motion

The amplitude x_0 and the parameters k and m determine the energy of the oscillations. The force F is a conservative force (see Section 3.6) and so is expressed as the rate of change with x of a potential energy function $U(x)$ as in eqn (3.28).

$$F(x) = -\frac{dU(x)}{dx}$$

with

$$U(x) = -\int_0^x F(x')dx' = \frac{1}{2}kx^2$$

in our example. The total energy E of the oscillating atom at time t is the sum of potential and kinetic energies. Hence

$$E = U(x) + \frac{1}{2}m\left(\frac{dx}{dt}\right)^2$$

$$= \frac{1}{2}kx_0^2 \cos^2(\omega_0 t + \phi) + \frac{1}{2}m\omega_0^2 x_0^2 \sin^2(\omega_0 t + \phi),$$

or, using eqn (6.6),

$$E = \frac{1}{2}kx_0^2(\cos^2(\omega_0 t + \phi) + \sin^2(\omega_0 t + \phi))$$

$$= \frac{1}{2}m\omega_0^2 x_0^2(\cos^2(\omega_0 t + \phi) + \sin^2(\omega_0 t + \phi)).$$

Since the sum of the squares of the sines and cosines of any angle equals unity, as given in eqn (20.37),

$$E = \tfrac{1}{2}kx_0^2 = \tfrac{1}{2}m\omega_0^2 x_0^2. \tag{6.9}$$

We see that, as expected, E is constant for fixed m, k, and x_0, and during the motion potential and kinetic energy are interchanged continuously, the former being zero when the speed is at its maximum value, and the latter being zero when the speed is zero at the turning points of the oscillation.

Exercise 6.1 A mass of 25 g oscillates with simple harmonic motion on the end of a spring. The amplitude of the oscillations is 2 cm and the period is 0.3 s. What is the maximum kinetic energy of the mass? What is the spring constant?

Answer 2.19×10^{-3} J; 11.0 N m^{-1}.

Worked Example 6.1 A simple pendulum consists of a mass m on the end of a weightless string or rod of length ℓ. Show that the period of oscillations of small amplitude is $T = 2\pi\sqrt{\ell/g}$, where g is the acceleration due to gravity.

Answer Figure 6.3 shows the pendulum displaced from the vertical by the small angle θ. The mass oscillates back and forth in circular motion in a vertical plane along an arc of radius ℓ. The torque causing rotation about the point of suspension is the tangential component $mg\sin\theta$ of the gravitational force times the distance ℓ. The torque acts to oppose an increase in the angle θ. For small amplitude oscillations $\sin\theta \sim \theta$ and the equation of rotational motion is

$$I\frac{\mathrm{d}^2\theta}{\mathrm{d}t^2} = -mg\ell\theta$$

where I is the moment of inertia of the pendulum for rotation about the point of suspension. (Circular motion, torques, and moments of inertia are discussed in Chapter 4.)

The moment of inertia is $m\ell^2$, and the above equation becomes

$$\frac{\mathrm{d}^2\theta}{\mathrm{d}t^2} + \frac{g}{\ell}\theta = 0.$$

This represents SHM with angular frequency given by $\omega_0^2 = g/\ell$, and hence period $T = 2\pi/\omega_0 = 2\pi\sqrt{\ell/g}$.

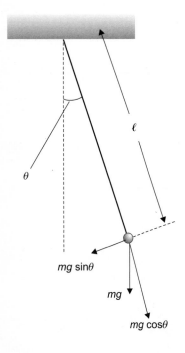

Fig. 6.3 A simple pendulum of length ℓ.

Exercise 6.2 A thin rod of mass M and length L is suspended horizontally by a wire attached to its midpoint. It makes small oscillations in the horizontal plane. For a small angular displacement θ from its equilibrium position the twisting of the wire results in a restoring torque of magnitude $C\theta$ where C is a constant. What is the period T of small oscillations?

Answer The period equals $2\pi\sqrt{ML^2/12C}$. You need to work out the appropriate moment of inertia of the rod to be $ML^2/12$.

6.2 Damped simple harmonic motion

We now proceed to a more realistic situation than that described by eqn (6.5) and include in the equation of motion forces that are always present to some extent in real physical situations. All oscillations are subject to non-conservative, **frictional** or **dissipative**, forces that remove energy from the oscillating system and transfer the removed energy to the surroundings. Examples of such dissipative processes are provided by a mass on the end of a spring in a viscous medium such as an oil bath, or a child on a swing subject to a retarding torque from the viscous drag of the air.

Consider a mass m on the end of a spring that has a constant k characterizing the size of the restoring force kx on the mass when it is displaced by a distance x from its equilibrium position. In many practical situations the drag force is proportional to the speed dx/dt. We limit our discussion to such situations, and include a dissipative force of this type in the equation of motion.

The equation of motion is now

$$m\frac{d^2x}{dt^2} = -kx - b\frac{dx}{dt} \tag{6.10}$$

where b is the constant of proportionality that gives the strength of the dissipative force. Equation (6.10) may be written in the form

$$\frac{d^2x}{dt^2} + \gamma\frac{dx}{dt} + \omega_0^2 x = 0 \tag{6.11}$$

with

$$\omega_0^2 = k/m$$

as before, and

$$\gamma = b/m.$$

The solution to eqn (6.11) gives the mathematical description of the motion for given values of the parameters γ and ω_0. We will approach the solution to the differential eqn (6.11) from a physics point of view, rather than pursue mathematical techniques for solving this type of equation.

Weak damping

The dissipative force removes energy from the motion and so, for small dissipation, we expect oscillations with amplitude decreasing with time. The oscillations are said to be **damped**. If the drag force is large enough, the mass may not oscillate at all if displaced, but simply fall back slowly to its equilibrium position. For a small dissipative force, however, we expect damped oscillations, and we will first assume that the solution corresponds to a function of time to be determined, multiplied by an exponential factor that ensures that at long times the motion has died away to nothing.

We thus try as a solution

$$x(t) = \exp(-\beta t)f(t) \tag{6.12}$$

with β a positive constant and $f(t)$ a function we have to determine. Substituting $x(t)$ and its derivatives into eqn (6.11) gives

$$\frac{d^2f}{dt^2} + (\gamma - 2\beta)\frac{df}{dt} + (\beta^2 + \omega_0^2 - \beta\gamma)f = 0. \tag{6.13}$$

This is still a complicated equation for the function f but we note that if we put γ equal to 2β it is an equation similar to eqn (6.5):

$$\frac{d^2f}{dt^2} + (\omega_0^2 - \gamma^2/4)f = 0. \tag{6.14}$$

● *An oscillating solution*

For positive values of $(\omega_0^2 - \gamma^2/4)$, the coefficient of f, eqn (6.14) describes SHM for f with angular frequency

$$\omega = (\omega_0^2 - \gamma^2/4)^{1/2}. \tag{6.15}$$

If the dissipative force is small, $\gamma^2/4$ is much less than ω_0^2 and ω is almost the same as the natural angular frequency ω_0.

Choosing the function f to have its maximum value x_0 at $t = 0$, we may write

$$f(t) = x_0 \cos \omega t. \tag{6.16}$$

Substitution of $f(t)$ from eqn (6.16) into eqn (6.12), with $\beta = \gamma/2$, gives the displacement

$$x(t) = x_0 \exp(-\gamma t/2) \cos \omega t. \tag{6.17}$$

The motion described by eqn (6.17) has a period $T = 2\pi/\omega$, and successive zeros of the displacement are separated by a time interval $\frac{1}{2}T$. However, this motion is not strictly periodic in the sense defined at the beginning of the chapter, since the amplitude of the oscillations decreases with time and $x(t)$ does not repeat the same values over and over again.

Note that for the motion described by eqn (6.17) the speed dx/dt is not zero at $t = 0$. The solution for a mass released from rest at a displacement x_0 has a different amplitude for $f(t)$ and a nonzero phase. The solution to this problem is cumbersome, and must be found by substituting $f(t) = A\cos(\omega t + \phi)$ into eqn (6.12) and deducing the values of A and ϕ from the conditions that $x(t) = x_0$ and $dx/dt = 0$ at $t = 0$. For weak damping the resulting solution differs very little from eqn (6.17).

Figure 6.4 shows the variation with time of the displacement of the mass for $x_0 = 0.08$ m, $\omega_0 = 10$ rad s^{-1}, and $\gamma = 0.5$ s^{-1}, i.e. $\omega_0/20$ s^{-1}. For these parameter values the angular frequency ω of the oscillations, given by eqn (6.15), differs from the natural frequency ω_0 by less than 0.04%.

The rate of fall-off of amplitude depends on γ and ω (see Problem 6.13); the ratio of successive swings on the same side is equal to $\exp(\pi\gamma/\omega)$. In the example shown in Fig 6.4 the amplitude falls to about one-half after about four swings, corresponding to a rather heavily damped situation. Many common phenomena exhibit much less damping. A struck tuning fork takes thousands of oscillations for its amplitude to fall away perceptibly, and its frequency is essentially that of the natural frequency ω_0.

Strong damping

As the damping force $b(dx/dt)$ increases in strength, i.e. γ increases above $\omega_0/20$, the oscillations die away more quickly. For $\gamma^2 > 4\omega_0^2$ there are no more oscillations. Equation (6.14) becomes

$$\frac{d^2f}{dt^2} - \alpha^2 f = 0, \tag{6.18}$$

with $\alpha^2 = (\gamma^2/4 - \omega_0^2)$. The functions $\exp(-\alpha t)$ and $\exp(+\alpha t)$ both satisfy eqn (6.18), and the general solution with two constants determined by the initial conditions is

$$f = A\exp(-\alpha t) + B\exp(+\alpha t), \tag{6.19}$$

giving the displacement

$$x = \exp(-\gamma t/2)(A\exp(-\alpha t) + B\exp(+\alpha t)). \tag{6.20}$$

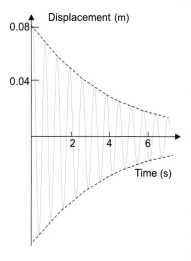

Fig. 6.4 The full line shows damped simple harmonic motion as given by eqn (6.17) with $x_0 = 0.08$ m, $\omega_0 = 10$ rad s^{-1}, and $\gamma = 0.5$ s^{-1}. The dotted lines give the loci of the peaks on either side of the equilibrium position.

● *The solution for strong damping*

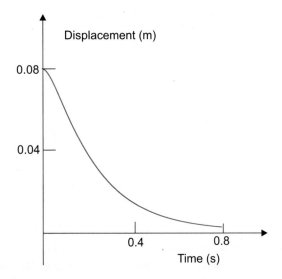

Fig. 6.5 The displacement as a function of time for a strongly damped oscillator with $\omega_0 = 10$ rad s^{-1} and $\gamma = 25$ s^{-1} after being released from rest at $x_0 = 0.08$ m.

The mass, once displaced to one side, will slowly move back towards its equilibrium position without oscillation. This situation, when the dissipative force with strength given by the parameter γ is large enough to prevent oscillatory behaviour, is called **strong damping**.

If the damping parameter γ for the oscillator in the above example with $\omega_0 = 10$ rad s^{-1} is increased to 25 rad s^{-1}, and the mass is again released from rest at an initial displacement of 0.08 m, the constants A and B are -0.0267 m and 0.1067 m, respectively. The curve showing how the displacement returns to the equilibrium position after release is shown in Fig 6.5.

Worked Example 6.2 If the damping in the example used with Fig 6.4 is increased so that γ becomes $\sqrt{800} = 28.3$ s^{-1}, the mass after being pulled from its equilibrium position will move back there without oscillating. If it is displaced by 8 cm and then let go at time $t = 0$, determine the expression giving its displacement as a function of time. How far back towards its equilibrium position has it gone after a time of 0.4 s?

Answer Using $\alpha^2 = (\gamma^2/4 - \omega_0^2)$, α is determined to be 10 s^{-1}. The values of A and B in eqn (6.20) are then given by setting x equal to 8 cm and the speed dx/dt equal to zero at $t = 0$. The results are $A = -1.66$ cm and $B = 9.66$ cm. The expression for x in cm is thus

$$x = e^{-14.14t}(9.66e^{10t} - 1.66e^{-10t}).$$

Evaluating x at $t = 0.4$ s, the last term can be ignored since it is very small and to a good approximation $x = 1.85$ cm. The mass has gone about 80% of the way back.

Critical damping

The special case when $\gamma = 2\omega_0$ corresponds to the mass returning towards its undisplaced position most quickly. For this case the function f is given by the equation $\mathrm{d}^2 f/\mathrm{d}t^2 = 0$, i.e. $f = (A + Bt)$. Hence

$$x = \exp(-\gamma t/2)(A + Bt). \tag{6.21}$$

This is called **critical damping**, and it is important to critically damp structures that are required to be steady but that may vibrate due to outside disturbances. The supporting springs of an automobile are typically damped so that the vehicle only just oscillates after hitting a bump. The floating platform on which a compact disc sits must be damped near to criticality so that the equilibrium position is very rapidly restored after a small displacement from a knock.

A value of the damping parameter γ of $20\,\mathrm{s}^{-1}$ critically damps an oscillator that has $\omega_0 = 10\,\mathrm{rad\,s}^{-1}$. Such an oscillator released from rest after an initial displacement of 0.08 m returns to its equilibrium position as shown in Fig 6.6. The return to a position very close to equilibrium is faster than when γ is $25\,\mathrm{s}^{-1}$.

Exercise 6.3 A mass on a spring is critically damped. The natural angular frequency is $10\,\mathrm{rad\,s}^{-1}$. The mass is displaced 8 cm from its equilibrium position and let go at $t = 0$. Determine the expression giving its displacement as a function of time. How far back to its equilibrium position has it gone after 0.4 s?

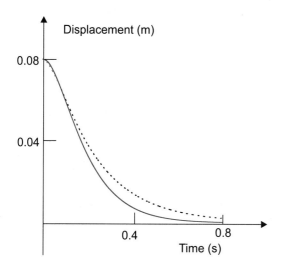

Fig. 6.6 The displacement as a function of time for a critically damped oscillator with $\omega_0 = 10\,\mathrm{rad\,s}^{-1}$ and $\gamma = 20\,\mathrm{s}^{-1}$ after being released from rest at $x_0 = 0.08$ m. The dotted curve gives the result for $\gamma = 25\,\mathrm{s}^{-1}$ shown on Fig 6.5.

Answer After calculating the constants, eqn (6.21) for the extension x in cm becomes $x = \mathrm{e}^{-10t}(8 + 80t)$. At $t = 0.4$ s, x is equal to 0.73 cm and the mass has gone about 91% of the way back.

6.3 Forced oscillations

In this section we consider a damped simple harmonic oscillator acted on by a harmonic force that varies sinusoidally with time. This is the simplest case mathematically and one used as the starting point for treatments of more complicated forces. We take the force to have constant amplitude F_0 and to be given by the expression $F_0 \cos \omega t$. It acts in the x-direction on the damped simple harmonic oscillator described in the last section.

As usual in the development of physics we are taking an idealized model. However, as well as acting as the starting point for more realistic problems it closely approximates several real examples and gives insight into the behaviour of all systems forced into oscillation by a periodic force.

The equation of motion describing the time variation of the forced oscillator is, using the notation of Section 6.2,

$$\frac{\mathrm{d}^2 x}{\mathrm{d}t^2} + \gamma \frac{\mathrm{d}x}{\mathrm{d}t} + \omega_0^2 x = \frac{F_0}{m} \cos \omega t. \tag{6.22}$$

This equation can be solved to give what is called the **particular integral**. However, we may add to the particular integral any solution of eqn (6.11). The resulting function still satisfies eqn (6.22). The solution to eqn (6.11) with two constants determined by the initial conditions of the problem is called the **complementary function**. It is always present immediately after the imposition of a force on a system although it dies away to insignificant levels after sufficient time has elapsed.

We will consider only the steady-state behaviour of the oscillator, which is given after a sufficiently long time by the form of the particular integral. Let us continue to use the model of a mass on the end of a spring. In the steady state the mass must oscillate at the same frequency as that of the impressed force and oscillate with a constant amplitude to which we will give the symbol x_0.

The oscillations, however, may differ from those of the applied force in that the peaks and troughs may occur at different times. In that case the oscillations and the driving force are said to be *out of phase*. Figure 6.7 shows how the applied force and the out-of-phase motion vary in an imaginary case. The peaks in the displacement come at a later time in the cycle than the peaks in the force, i.e. the displacement lags behind the force and the phase of the force leads the phase of the displacement.

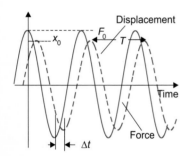

Fig. 6.7 The full line shows a force $F_0 \cos \omega t$. The dotted line shows a sinusoidal displacement $x_0 \cos(\omega t - \phi)$ lagging in phase by an angle $\phi \sim 60$ degrees with respect to the force. Its peaks occur at a time $\Delta t = \phi/\omega = T \times (\phi/2\pi) \sim T/6$ s later than those of the force.

● *Phase leads and lags*

The force $F_0 \cos \omega t$ has positive peaks at times t equal to 0, $2\pi/\omega$, $4\pi/\omega, \ldots$, etc. The positive peaks of the displacement, of size x_0, come at later times $\Delta t, (2\pi/\omega) + \Delta t, (4\pi/\omega) + \Delta t, \ldots$, etc., and hence the mathematical form of the displacement is

$$x = x_0 \cos(\omega t - \phi) \tag{6.23}$$

with $\phi = \omega \Delta t = 2\pi \Delta t / T$.

The phase of the force is ωt, the argument of the cosine term in the expression for the force; the phase of the displacement is $\omega t - \phi$, the argument of the cosine term in the expression for the displacement. Absolute phase has little meaning, since it depends on the choice of time zero; phase difference is a physically important quantity, representing the degree of synchronization of two sinusoidal signals.

● *Phase difference*

Equation (6.23) describes a displacement that has the same frequency as the driving force, has constant amplitude, and has a phase lag ϕ with respect to that of the force. It represents the steady-state solution we are seeking to eqn (6.22) and to determine the amplitude x_0 and phase difference ϕ we simply substitute (6.23) into (6.22). This results in the equation

$$(\omega_0^2 - \omega^2)x_0 \cos(\omega t - \phi) - \omega \gamma x_0 \sin(\omega t - \phi) = \frac{F_0}{m} \cos \omega t. \tag{6.24}$$

This equation must be true at all times t, and choosing times such that $(\omega t - \phi)$ is equal to zero and then equal to $\pi/2$ gives two simultaneous equations that can be solved for the unknowns x_0 and ϕ. These equations are

● *The amplitude and the phase lag of the displacement*

$$(\omega_0^2 - \omega^2)x_0 = \frac{F_0}{m} \cos \phi,$$

$$-\omega \gamma x_0 = \frac{F_0}{m} \cos\left(\frac{\pi}{2} + \phi\right).$$

When solving you will need to remember that $\cos(\pi/2 + \phi)$ is equal to $-\sin \phi$ and that $\cos^2 \phi + \sin^2 \phi = 1$, as given by the trigonometric relations in Chapter 20. The solutions are

$$x_0 = \frac{F_0/m}{\left\{(\omega_0^2 - \omega^2)^2 + \omega^2 \gamma^2\right\}^{1/2}} \tag{6.25}$$

and

$$\tan \phi = \frac{\omega \gamma}{(\omega_0^2 - \omega^2)}. \tag{6.26}$$

The speed and acceleration of the mass are obtained from the first and second differentials with respect to time of eqn (6.23). The speed differs in phase by $\pi/2$ from the displacement, and the acceleration differs in phase by π from the displacement.

Exercise 6.4 A mass of 25 g on the end of a spring with force constant $11 \ \mathrm{N\,m^{-1}}$ is subject to a harmonic force of amplitude 2 N. The viscous retarding force is given by $b\,dx/dt$ with $b = 0.3 \ \mathrm{kg\,s^{-1}}$. Determine the amplitude of the displacement and the phase difference between the force and the displacement in the steady-state motion when the angular frequency is (a) $5 \ \mathrm{rad\,s^{-1}}$, (b) $20 \ \mathrm{rad\,s^{-1}}$, and (c) $500 \ \mathrm{rad\,s^{-1}}$.

Answer (a) Amplitude 0.19 m, phase of displacement lags that of force by 8.2°; (b) 0.33 m, phase lags by 80.5°; (c) 3.2×10^{-4} m, phase lags by 178.6°.

The easiest way to solve eqn (6.22) mathematically, rather than finding the solution by arguments based on the physics of the situation, is to use a technique involving complex numbers, which are discussed in Section 20.6. We first note that the quantity $(F_0/m)\cos\omega t$ is the real part of $(F_0/m)\exp(\mathrm{j}\omega t)$. We thus allow the displacement x to be a complex number X and solve eqn (6.22) after substituting $(F_0/m)\cos\omega t$ by the complex expression $(F_0/m)\exp(\mathrm{j}\omega t)$. When the complex solution has been found, its real part gives the real displacement we seek. The complex equation requiring solution is

$$\frac{d^2 X}{dt^2} + \gamma \frac{dX}{dt} + \omega_0^2 X = \frac{F_0}{m} e^{\mathrm{j}\omega t}.$$

Let its solution be

$$X = X_0 e^{\mathrm{j}\omega t},$$

with X_0 complex. Substitution of this expression and its derivatives into the differential equation above gives

$$-\omega^2 X_0 + \mathrm{j}\omega\gamma X_0 + \omega_0^2 X_0 = 0.$$

Writing X_0 in its exponential form as

$$X_0 = r e^{-\mathrm{j}\phi}$$

we obtain an equation involving the quantities r and ϕ. This equation must be satisfied by both the real and imaginary components, and separating them gives two equations

$$-\omega^2 r \cos\phi + \omega\gamma r \sin\phi + \omega_0^2 r \cos\phi = \frac{F_0}{m}$$

and

$$+\mathrm{j}\omega^2 r \sin\phi + \mathrm{j}\omega\gamma r \cos\phi - \mathrm{j}\omega_0^2 r \sin\phi = 0.$$

After some algebra, r and ϕ are determined from these two equations to be

$$\tan \phi = \frac{\omega \gamma}{(\omega_0^2 - \omega^2)}$$

and

$$r = \frac{F_0/m}{\{(\omega_0^2 - \omega^2) + \omega^2 \gamma^2\}^{1/2}}.$$

The real part of

$$X = X_0 e^{j\omega t} = r e^{j(\omega t - \phi)}$$

is the displacement x, and is the same as that given by eqns (6.23), (6.25), and (6.26).

Resonance

We can now discuss how the motion of the mass on the spring varies as the amplitude F_0 and the angular frequency ω of the driving force are changed.

For our idealized system, eqn (6.25) shows that the amplitude of the displacement x_0 varies linearly with F_0. This is not precisely so for any real system, and as the amplitude of the driving force becomes large the equation of motion (6.22) becomes invalid. However, the assumptions made in setting up eqn (6.22) are good approximations for small amplitude motion.

The variations of x_0 and ϕ with the angular frequency ω of the impressed force demonstrate a very common phenomenon. The time variations of x_0 and ϕ depend on the degree of damping of the system, i.e. on the relative values of γ and ω_0. For a system that oscillates freely in its unforced state, the amplitude x_0, for fixed γ, ω_0, and F_0, goes through a maximum at a frequency near ω_0. This is shown in Fig 6.8, which is calculated for $F_0 = 0.5$ N, $m = 1$ kg, $\omega_0 = 10$ rad s^{-1}, and $\gamma = 0.5$ s^{-1}.

The amplitude is said to **resonate** at the frequency ω_m just below ω_0 and the curve of Fig 6.8 is called a **resonance curve**. Resonance behaviour is widespread, and ranges across all energy scales of natural phenomena, from the atomic to the macroscopic. It is observed at very high frequencies in the motions of electrons bound in atoms and subject to alternating electric fields of varying frequency, and it is felt when the front end of an automobile vibrates due to the motion of an unbalanced wheel at a certain speed of rotation.

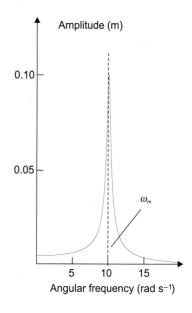

Fig. 6.8 The amplitude x_0 of the displacement of a forced oscillator as a function of the angular frequency ω of the applied force. The amplitude is given by eqn (6.25) with $F_0 = 0.5$ N, $m = 1$ kg, $\omega_0 = 10$ rad s^{-1}, and $\gamma = 0.5$ s^{-1}. The displacement reaches a maximum at an angular frequency ω_m just below ω_0.

Worked Example 6.3 If ω_{m} is the angular frequency at which the amplitude of a forced oscillator is a maximum, calculate the fractional difference $(\omega_0 - \omega_{\mathrm{m}})/\omega_0$ for a damped oscillator with natural resonant angular frequency $\omega_0 = 10 \text{ rad s}^{-1}$ as the damping factor γ increases from $\omega_0/10$ to ω_0. What happens when $\gamma = \omega_0 \times \sqrt{2}$?

Answer Differentiation of eqn (6.25) with respect to ω and putting the result equal to zero give maxima and minima of the amplitude. Zero value of the derivative requires

$$4(\omega_0^2 - \omega^2)\omega = 2\gamma^2\omega$$

and, by taking the second derivative of eqn (6.25), the above equation can be shown to give a maximum when

$$\omega_{\mathrm{m}} = \sqrt{(\omega_0^2 - \gamma^2/2)}.$$

Evaluating for $\omega_0 = 10 \text{ rad s}^{-1}$ gives $(\omega_0 - \omega_{\mathrm{m}})/\omega_0$ equal to 0.0025 when $\gamma = \omega_0/10$ and equal to 0.29 when $\gamma = \omega_0$.

When γ is equal to or greater than $\omega_0 \times \sqrt{2}$, x_0 no longer has a maximum and the amplitude as a function of frequency reduces smoothly from its value for a steady force F.

6.4 Power absorption

It is often as important to know the behaviour of the power absorbed by the system across a resonance, or how the phase of the oscillations changes with respect to the phase of the force, as it is to know the way in which the displacement responds to frequency changes.

Let us work out the average power absorbed by the oscillator as a function of frequency. Between times t and $t + \mathrm{d}t$ the mass moves $\mathrm{d}x$ and the work done on the mass by the applied force is $F\mathrm{d}x$. The instantaneous power absorbed is

$$P = F\frac{\mathrm{d}x}{\mathrm{d}t} = F_0 \cos \omega t \times \{-\omega x_0 \sin(\omega t - \phi)\}$$

$$= -\omega x_0 F_0 \cos \omega t \sin \omega t \cos \phi + \omega x_0 F_0 \cos^2 \omega t \sin \phi. \tag{6.27}$$

To obtain the time average power we must average the above over a period of the oscillations. The time average of the first term involving the product of a sine and a cosine is zero; the second term involves the square of a cosine and averages to one-half.

The average of $\sin \omega t \cos \omega t$ over a time interval of one period is

$$\frac{1}{T} \int_0^T \sin \omega t \cos \omega t \, dt = \frac{1}{\omega T} \int_0^T \sin \omega t \, d(\sin \omega t).$$

Integration gives

$$\frac{1}{2\omega T} [\sin^2 \omega t]_0^T = 0,$$

since ωt has the value 2π at the upper limit and zero at the lower.

The average of $\cos^2 \omega t$ over a time interval of one period is

$$\frac{1}{T} \int_0^T \cos^2 \omega t \, dt = \frac{1}{2T} \int_0^T (1 + \cos 2\omega t) \, dt.$$

The integral of $\cos 2\omega t$ can easily be shown to be zero, leaving one-half as the value of the right-hand side.

The average power $\bar{P}$ is given by

$$\bar{P} = \tfrac{1}{2} \omega x_0 F_0 \sin \phi$$

or

$$\bar{P} = \frac{1}{2m} \frac{\omega^2 F_0^2 \gamma}{\{(\omega_0{}^2 - \omega^2)^2 + \omega^2 \gamma^2\}} \tag{6.28}$$

after substituting for $\sin \phi$ from eqn (6.26) using the trigonometric identities of Chapter 20. The **frequency response curve** for $\bar{P}$ is shown on Fig 6.9 which is calculated for the parameters used for Fig 6.8. $\bar{P}$ is a maximum when $\sin \phi$ is a maximum, i.e. when $\sin \phi = 1$ and $\omega = \omega_0$.

At maximum power absorption, the phase of the force leads that of the displacement by $\pi/2$; when the force is a maximum the mass is at rest and about to move in the same direction as the force. The speed of the mass leads the displacement by $\pi/2$, and the speed synchronizes with the force. The way in which the phase difference between force and displacement varies with angular frequency can be determined from eqn (6.26). At low frequencies the force and displacement are in phase; at power resonance the force leads by $\pi/2$, and at very high frequencies the force and displacement are completely out of phase. When one is a maximum the other is a minimum and very little power is absorbed.

The Q factor

The sharpness of the power resonance may be defined quantitatively in terms of a parameter called the **Q factor**, or simply 'Q'. As the Q value increases the resonance becomes sharper.

● The frequency response

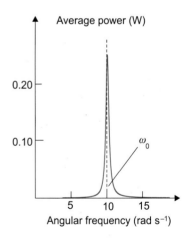

Fig. 6.9 The average power absorbed by a forced oscillator as a function of the angular frequency of the applied force. The curve is drawn for the same parameters as used for Fig 6.8. The power reaches a maximum at an angular frequency equal to ω_0.

For a given value of the natural frequency ω_0, Q depends on the parameter γ that determines the dissipative force and the rate of absorption of energy. As γ increases, the power needed to keep the oscillations going increases. The result is that Q decreases and the resonance becomes broader. For sufficiently large γ the resonance disappears. For very small γ the resonance becomes very sharp, as in a radio receiver tuned to receive a signal at a particular frequency only. (See Section 18.2.)

The connection between Q and the absorption of energy is made by defining Q in terms of the mean energy stored in the oscillating system and the energy dissipated in each cycle at resonance.

● *The definition of Q*

Q is equal to 2π times the energy stored at resonance divided by the energy dissipated per cycle at resonance.

The stored energy is the sum of kinetic and potential energy terms and, as in the simple harmonic oscillator discussed in Section 6.1, the stored energy is constant and equal to the maximum potential energy (and also equal to the maximum kinetic energy). The maximum potential energy is $kx_0^2/2$, and from eqn (6.25) and the relation $\omega_0^2 = k/m$, the energy stored at resonance is $F_0^2/2m\gamma^2$. This energy is maintained constant by the work done by the driving force. The energy absorbed per cycle at resonance is the average power performed by the driving force at resonance multiplied by the period $2\pi/\omega_0$, and equals $\pi F_0^2/m\omega_0\gamma$. Hence

$$Q = \frac{\omega_0}{\gamma}. \tag{6.29}$$

The sharpness of a curve that rises to a maximum and thereafter declines is often expressed in terms of the full-width of the peak, given by the points at which the curve has a height equal to one-half its maximum height, divided by the value at which the peak has its maximum. Using eqn (6.28) it may be shown, after some algebra, that for γ small compared with ω, the Q of the resonance curve showing the variation of power needed to drive a damped oscillator is approximately given by

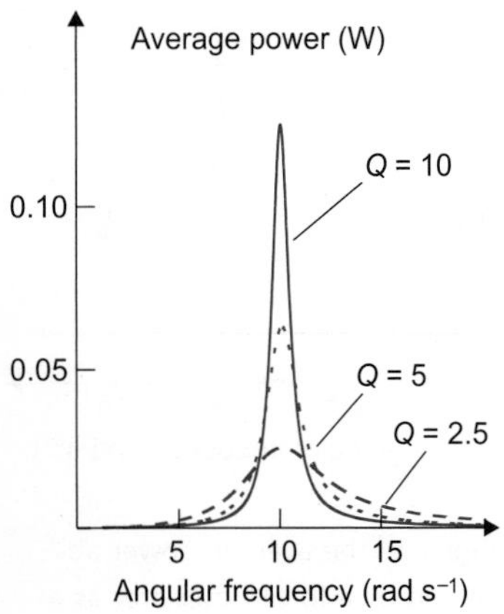

$$Q = \frac{\omega_0}{(\omega_2 - \omega_1)} \tag{6.30}$$

where ω_1 and ω_2 are the angular frequencies below and above ω_0 at which the power absorption falls to one-half of its maximum value, i.e. to $F_0^2/4m\gamma$.

Figure 6.10 shows the shapes of power resonance curves for three values of the parameter Q lower than the value 20 appropriate to Fig 6.9. This was effected by increasing the size of the damping parameter γ whilst maintaining other parameters at the values used in Figs 6.8 and 6.9.

Fig. 6.10 Power absorption curves for different Q-values. The parameter values used are the same as used for Figs 6.8 and 6.9 except for the increased sizes of γ.

Exercise 6.5 The power absorption of an oscillating mass of 0.5 kg subject to a sinusoidal force of amplitude 1 N goes through a maximum at an angular frequency of 10^3 rad s^{-1}. The Q factor is 100. What is the average power absorbed at the resonant frequency?

Answer 1/10 watt.

Gravitational waves

Theories of gravity describe the force between two masses in terms of gravitational waves, which travel at the speed of light in free space. These waves are exceedingly difficult to detect and so far have not been observed. It is predicted that waves that produce effects large enough to be detected by the most advanced techniques arise from rare cataclysmic events in our galaxy or in nearby galaxies. For example, the coalescence of binary stars may give a wave to which a suitable detector anywhere on Earth may respond.

One of the ways presently being used to search for gravity waves is to detect the motion of large metal cylinders that have very large Q factors. The cylindrical bars are supported by vibration-free suspensions and vibrate in response to a gravity wave. The effect of a gravitational wave passing over a bar is for the impulse given to the bar by the wave to set it in oscillation at a frequency given by eqn (6.15). Since Q is very large, this frequency is very close to the natural angular frequency ω_0 of the bar. The motion of the bar after the pasage of the wave is given by eqn (6.17), which can be used to show that the oscillations die away after a time of about $2/\gamma = 2Q/\omega_0$. The bar acts like a gong when hit by a gravity wave.

A gravity-wave detector at the University of Western Australia used a resonant bar isolated from local disturbances and kept at a very low temperature to minimize thermal noise effects. The bar has a natural frequency of 700 kHz and a Q factor of 2×10^8. The bar thus responds to a passing wave for about 10 minutes, the time for the induced oscillations to die away. A signal from the bar in Western Australia may be correlated with coincident signals from other detectors throughout the world to give information on the direction from whence the wave came.

The most advanced systems for the detection of gravity waves came into operation at the beginning of the millenium. These systems detect the change in travel time of light along two tubes, each a few kilometres long, arranged perpendicular to each other. Masses suspended at the ends of the tubes reflect laser light incident upon them. During the passage of a gravity wave, the masses change their dimensions by different amounts, and the times of travel of the light along the two tubes correspondingly change by different amounts. The minute difference is measured by very sensitive optical interference techniques. (Interference is discussed in Section 8.7.) The tubes have to be highly evacuated and the masses and reflectors have to be isolated from all sources of terrestrial disturbances.

6.5 Superposition

In this section we consider what happens when a damped harmonic oscillator is acted upon by a force that has two components oscillating at different frequencies rather than a force of a single frequency. We shall see that the resulting motion is the superposition of two motions that would arise if the two forces were acting separately.

Linear differential equations play a very important role in the description of many physical phenomena. For a differential equation determining the behaviour of a function y on a variable x to be linear, the function y or its derivatives must appear only in first-order in each term.

● *The definition of a linear equation*

The equations

$$a\frac{d^2y}{dx^2} + b\frac{d(y^2)}{dx} + cy = 0,$$

$$a\frac{d^2y}{dx^2} + b\frac{dy}{dx} + cy^{1/2} = 0,$$

and

$$a\frac{d^2y}{dx^2} + b\frac{dy}{dx} + cy^{1/2} = f(x),$$

with a, b, and c constants, are not linear.

The equation

$$a\frac{d^2y}{dx^2} + b\frac{dy}{dx} + cy = 0,$$

with a, b, and c constants, is linear. Hence, eqn (6.11) for the damped oscillator is linear.

Adding the driving force $F_0 \cos \omega t / m$ to eqn (6.11) gives eqn (6.22). It may easily be verified that, if $x_1(t)$ is the solution for a force $F_1(t)$ on the right-hand side and $x_2(t)$ is the solution for a force $F_2(t)$, then the solution for a combined force $F_1 + F_2$ is $x_1 + x_2$. Since

$$\frac{d^2}{dt^2}(x_1 + x_2) = \frac{d^2 x_1}{dt^2} + \frac{d^2 x_2}{dt^2}$$

and

$$\gamma \frac{d}{dt}(x_1 + x_2) = \gamma \frac{dx_1}{dt} + \gamma \frac{dx_2}{dt},$$

then the left-hand side of the equation

$$\frac{d^2 x}{dt^2} + \gamma \frac{dx}{dt} + \omega_0^2 x = F_1(t) + F_2(t) \tag{6.31}$$

separates into two parts, one of which equals $F_1(t)$ and the other $F_2(t)$.

The solution to eqn (6.22) when the force acting has two frequency components is thus given by

$$x = x_{10} \cos(\omega_1 t + \phi_1) + x_{20} \cos(\omega_2 t + \phi_2), \tag{6.32}$$

with x_{10} and x_{20} given by equations similar to eqn (6.25), and ϕ_1 and ϕ_2 given by equations similar to eqn (6.26). The resultant motion is the **superposition** of the two separate motions. The superposition is not harmonic and may not be periodic.

Worked Example 6.4 What conditions must be satisfied if the superposition of two harmonic signals is to be periodic?

Answer If the addition of the two signals is periodic with period τ, the displacements and speeds of the two components must be the same at times t and $t + \tau$. This requires that the components should execute an integral number of their individual cycles in the time interval τ. With reference to eqn (6.32), if the component with angular frequency ω_1 undergoes n_1 cycles and the other component n_2 cycles,

$$\omega_1 = n_1 \left(\frac{2\pi}{\tau} \right)$$

and

$$\omega_2 = n_2 \left(\frac{2\pi}{\tau} \right).$$

The resultant superposed motion will thus be periodic if

$$\frac{\omega_1}{\omega_2} = \frac{n_1}{n_2}$$

or

$$n_1 T_1 = n_2 T_2,$$

where T_1 and T_2 are the respective periods.

Beats

If the angular frequencies ω_1 and ω_2 are close in value we have the phenomenon of **beats**, commonly observed when two sound waves of close frequency are superimposed. The sensation of sound comes from vibrations of air hitting the ear drum, and the beat is heard as a regular variation in the intensity of the sound at a frequency equal to the difference in frequency of the two sound sources.

Equation (6.32) can be written in a way that makes it easier to visualize the motion resulting from two sinusoidal driving forces. The trigonometric relation (20.39) of Chapter 20 may be written

$$\cos(A + B) + \cos(A - B) = 2 \cos A \cos B.$$

● *Addition of two harmonic oscillations*

Using this identity, and putting x_{10} and x_{20} both equal to the value x_0 for simplicity, we find

$$x = 2x_0 \cos(\omega_0 t + \phi_0) \cos\left(\frac{\Delta\omega}{2} t + \frac{\Delta\phi}{2}\right) \tag{6.33}$$

where ω_0 and ϕ_0 are the averages of the two angular frequencies and phase angles, and $\Delta\omega$ and $\Delta\phi$ are their differences.

This mathematical expression represents an oscillation with an angular frequency ω_0. The oscillation does not have a constant amplitude but one that also varies with time. The amplitude of the oscillations at angular frequency ω_0 is **modulated** by the term $\cos(\Delta\omega t/2 + \Delta\phi/2)$, which results in the disturbance, averaged over a period $2\pi/\omega_0$, going through zeros at intervals of half the period of the modulating term, i.e. at intervals of $2\pi/\Delta\omega$.

The modulation is illustrated in Fig 6.11 which is calculated for $\omega_1 = 10$ rad s^{-1}, $\omega_2 = 10.5$ rad s^{-1} and where for simplicity we have put the phase angles ϕ equal and chosen the amplitudes x_{10} and x_{20} both equal to the same value x_0. The wave resulting from these simple choices of parameters still reveals many of the important features that arise from superposition. The modulation of the amplitude of the rapidly oscillating term gives rise to beats in whatever disturbance eqn (6.32) is describing. The beats correspond to the continued rise and fall of the amplitude of the underlying rapid oscillations, zeros occurring every time the modulating term $\cos(\Delta\omega t/2 + \Delta\phi/2)$ goes through zero, corresponding to a beat frequency of $\Delta\omega/2\pi$.

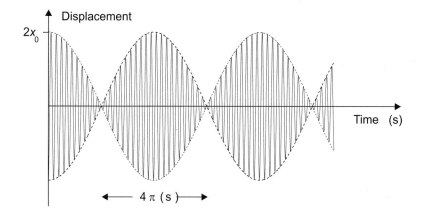

Fig. 6.11 The full line shows the resultant disturbance arising from the superposition of two sinusoidal displacements of equal amplitude x_0 and angular frequencies 10 rad s^{-1} and 10.5 rad s^{-1}. The envelope enclosing the signal, shown dotted, has zeros that occur at intervals of $2\pi/\Delta\omega \sim 12.6$ s.

For example, if two tuning forks are struck and held side by side, one hears a note at the average frequency of the two forks rising and falling in intensity at the beat frequency $\Delta\omega/2\pi$. If the two forks are tuned to 255 and 257 Hz and struck together, one hears a note at 256 Hz beating at a frequency of 2 Hz; the sound goes through minima twice every second.

6.6 **Fourier series**

In the last section we saw that, when two harmonic forces act at the same time on a system for which the free oscillations obey a linear differential equation, the resulting motion is the superposition of two motions corresponding to each force acting separately. This superposition remains valid for any number of simultaneously impressed harmonic forces; the resultant motion is the sum of motions each due to the separate harmonic contributions to the total force.

This suggests that we can determine the resultant motion for an arbitrary force if we can decompose the force into separate components of different angular frequencies. Such a decomposition can be made in the form of an infinite series if the force is periodic. This series representation is called a **Fourier series** after Jean Baptiste Fourier, a French mathematician who pioneered this approach in the eighteenth century. In this section we will develop the ideas of Fourier series and give an example of their application.

Suppose a periodic force $F(t)$, such as that shown on Fig 6.12, acts on a damped simple harmonic oscillator. The force is periodic with period T corresponding to an angular frequency $\omega = 2\pi/T$. It is possible to express the force as a sum of harmonic components with angular frequencies $n\omega$, with n an integer or zero. The different components have amplitudes A_n and phases $(n\omega t + \phi_n)$.

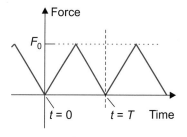

Fig. 6.12 A sawtooth periodic function with period T.

● *Series expansion of the force*

The mathematical formula for the Fourier series expansion of the force is

$$F(t) = \sum_{n=0}^{n=\infty} A_n \cos(n\omega t + \phi_n). \qquad (6.34)$$

● *An alternative expansion of the force*

Expanding the cosine term in the above equation gives the equivalent expression

$$F(t) = \sum_{n=0}^{n=\infty} (a_n \cos(n\omega t) + b_n \sin(n\omega t)), \qquad (6.35)$$

with $a_n = A_n \cos \phi_n$ and $b_n = A_n \sin \phi_n$. The coefficients a_n and b_n now determine the amplitudes and phases of the different frequency components, which are called **harmonics**. This name distinguishes them from the **fundamental** component, which oscillates at the basic angular frequency ω. Usually the amplitudes of the coefficients decrease as the integer index n increases.

Fourier series representations of periodic functions are useful in many situations. For example, a radio receiver may have a different response to different frequencies. In that case its output would not follow faithfully the form of an input signal, since the different harmonics of the input would be affected differently by the receiver; both the amplitudes of the various frequency components of the input, and their phases, could be changed resulting in distortion of the signal. For a given response function of the receiver, the output could be determined if the amplitudes and phases of the input were calculated by Fourier analysis.

If we are given the form of the periodic force, as in Fig 6.12 for example, how do we calculate the coefficients a_n and b_n? Let us concentrate on a specific coefficient a_p where p is the index of the required coefficient. Multiplying both sides of eqn (6.35) by $\cos(p\omega t)$ and integrating over the period T of the basic cycle gives the equation

$$\int_0^T F(t) \cos(p\omega t) \mathrm{d}t = \int_0^T \sum_{n=0}^{n=\infty} a_n \cos(p\omega t) \cos(n\omega t) \mathrm{d}t$$

$$+ \int_0^T \sum_{n=0}^{n=\infty} b_n \cos(p\omega t) \sin(n\omega t) \mathrm{d}t. \qquad (6.36)$$

Each of the integrals making up the second set on the right-hand side of the above equation is zero. This can be verified by rewriting the products of sine and cosine terms as sums of two sine terms, according to formulae given in Chapter 20, after which integration shows that each separate definite integral vanishes.

Here are some definite integrals that are very useful when making Fourier series expansions of a periodic function with period $T = 2\pi/\omega$. In the expressions below, p and n are any integers.

$$\int_0^T \cos(p\omega t)\cos(n\omega t)dt = 0, \quad \text{if } p \neq n,$$

$$\int_0^T \sin(p\omega t)\sin(n\omega t)dt = 0, \quad \text{if } p \neq n,$$

$$\int_0^T \cos(p\omega t)\cos(n\omega t)dt = T/2, \quad \text{if } p = n,$$

$$\int_0^T \sin(p\omega t)\sin(n\omega t)dt = T/2, \quad \text{if } p = n,$$

$$\int_0^T \cos(p\omega t)\sin(n\omega t)dt = 0, \quad \text{for all } p \text{ and } n.$$

The following indefinite integrals are also often useful.

$$\int t\cos t dt = \cos t + t\sin t,$$

$$\int t\sin t dt = \sin t - t\cos t.$$

If p and n are positive integers and p does not equal n,

$$\int \cos(p\omega t)\cos(n\omega t)dt = \frac{1}{2\omega(p+n)}\sin[(p+n)\omega t]$$
$$+ \frac{1}{2\omega(p-n)}\sin[(p-n)\omega t],$$

$$\int \sin(p\omega t)\cos(n\omega t)dt = \frac{1}{2\omega(n-p)}\cos[(n-p)\omega t]$$
$$- \frac{1}{2\omega(n+p)}\cos[(n+p)\omega t],$$

$$\int \sin(p\omega t)\sin(n\omega t)dt = \frac{1}{2\omega(p-n)}\sin[(p-n)\omega t]$$
$$- \frac{1}{2\omega(p+n)}\sin[(p+n)\omega t],$$

$$\int \cos^2(n\omega t)dt = \frac{1}{2}t + \frac{\sin(2n\omega t)}{4n\omega},$$

$$\int \cos(n\omega t)\sin(n\omega t)dt = \frac{1}{2}\frac{\sin^2(n\omega t)}{n\omega},$$

$$\int \sin^2(n\omega t)dt = \frac{1}{2}t - \frac{\sin(2n\omega t)}{4n\omega}.$$

Each of the integrals of the first set on the right-hand side of eqn (6.36) vanishes except that for which $n = p$. We thus find that, because of the properties of sines and cosines, the complicated eqn (6.36) reduces to the relatively simple expression

$$\int_0^T F(t)\cos(p\omega t)\mathrm{d}t = \int_0^T a_p\cos^2(p\omega t)\mathrm{d}t.$$

Evaluation of the integral involving $\cos^2(p\omega t)$ results in the value $Ta_p/2$, and so, finally, the coefficient a_p is given by the formula

$$a_p = \frac{2}{T}\int_0^T F(t)\cos(p\omega t)\mathrm{d}t. \tag{6.37}$$

The formula for the coefficient a_0, which gives the average value of $F(t)$, is

$$a_0 = \frac{1}{T}\int_0^T F(t)\mathrm{d}t. \tag{6.38}$$

● *Expressions for amplitudes of components*

By following a similar procedure it can easily be shown that the coefficients b_p are given by the expression

$$b_p = \frac{2}{T}\int_0^T F(t)\sin(p\omega t)\mathrm{d}t, \tag{6.39}$$

with $b_0 = 0$. We have now completed the Fourier series representation of the periodic function $F(t)$ by writing it in the form of eqn (6.35) with known coefficients a_n and b_n.

Even and odd functions

If a function $f(t)$ satisfies certain requirements, the Fourier series expansion takes on a simpler form than the general expression (6.35).

If the value of $f(t)$ for any time t is the same as its value for time $-t$,

$$f(t) = f(-t),$$

the function $f(t)$ is called an **even** function. For example, $\cos\omega t$ is an even function.

If the value of $f(t)$ for any time t is the negative of its value for time $-t$,

$$f(t) = -f(-t),$$

the function $f(t)$ is called an **odd** function. For example, $\sin \omega t$ is an odd function.

In the Fourier series expansion of even functions there are therefore only cosine terms (and also a possible average term a_0), and all coefficients b_n are zero. In the series for odd functions there are only sine terms and all the coefficients a_n are zero, including a_0.

The periodic function shown in Fig 6.12 is an even function of time and so its expansion has only cosine terms.

Worked Example 6.5 The negative half-cycles of a sinusoidal waveform $V = V_0 \sin \omega t$ provided by the mains electricity supply are removed by an electrical device called a rectifier in order to produce a nonzero average voltage, as shown in Fig 6.13. Show that the resulting wave is represented by the Fourier series

$$V = \frac{V_0}{\pi}\left(1 + \frac{\pi}{2}\sin \omega t - \frac{2}{3}\cos 2\omega t - \frac{2}{15}\cos 4\omega t - \cdots\right),$$

with successive coefficients of even harmonics of angular frequency $n\omega$ given by $-2V_0/\pi(n^2 - 1)$.

Answer In this answer we will use results for integrals given in the mathematical insert on p. 197. The function V is given by

$$V = V_0 \sin(\omega t) \quad \text{for } 0 < t < T/2$$

and

$$V = 0 \quad \text{for } T/2 < t < T.$$

This function is neither even nor odd and thus has both sine and cosine terms in its Fourier series expansion, which is

$$V = \sum_{n=0}^{n=\infty} a_n \cos(n\omega t) + \sum_{n=0}^{n=\infty} b_n \sin(n\omega t),$$

with the coefficients a_n for n not equal to zero, and b_n, given by eqns (6.37) and (6.39), respectively. The coefficient a_0 is given by eqn (6.38):

$$a_0 = \frac{V_0}{T}\int_0^{T/2} \sin(\omega t)\mathrm{d}t = \frac{V_0}{\pi}.$$

$$a_n = \frac{2V_0}{T}\int_0^{T/2} \sin(\omega t)\cos(n\omega t)\mathrm{d}t$$

$$= \frac{2V_0}{T}\left(\frac{\cos[(n-1)\omega t]}{2(n-1)\omega} - \frac{\cos[(n+1)\omega t]}{2\omega(n+1)}\right)_0^{T/2}.$$

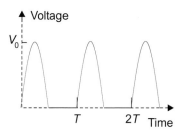

Voltage

V_0

T $2T$ Time

Fig. 6.13 A harmonic voltage signal of period T and amplitude V_0 with the negative-going parts removed.

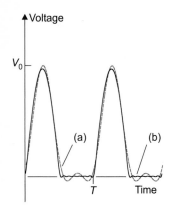

Fig. 6.14 Curves (a) and (b) show how the function shown in Fig 6.13 is reproduced more and more faithfully as successive harmonics are added to its Fourier series representation.

Evaluating this expression for different values of n gives $a_n = 0$ for all odd n and, $a_2 = -2V_0/3\pi$, $a_4 = -2V_0/15\pi$, $a_n = -2V_0/(n^2 - 1)\pi$.

$$b_n = \frac{2V_0}{T} \int_0^{T/2} \sin(\omega t) \sin(n\omega t) dt$$

and this integral is zero for all values of n not equal to one.

$$b_1 = \frac{2V_0}{T} \int_0^{T/2} \sin(\omega t) \sin(\omega t) dt$$

$$= \frac{2V_0}{T} \frac{1}{2} \int_0^{T/2} (1 - \cos(2\omega t)) dt = V_0/2.$$

The manner in which the original function $V(t)$ is reproduced more and more faithfully as successive harmonics are introduced into the series expansion is shown in Fig 6.14. The curve labelled (a) includes only the first three terms in the bracket on the right-hand side of the above equation for V; curve (b) shows the function obtained by using the first eight terms.

We can now answer the question posed at the beginning of this section about solving for the displacement of a damped oscillator under the influence of a periodic but non-harmonic force. If the force is expressed as a Fourier series, each component produces a displacement that can be obtained using eqns (6.25) and (6.26). The total displacement, using the principle of superposition, is the sum of the separate expressions, in the same way as eqn (6.32) describes the motion resulting from an impressed force consisting of a sum of only two harmonic components.

The series expression for a periodic function of time is made in terms of its different frequency components. However, the usefulness of Fourier series expansions is not confined to functions of time. A periodic function of a position coordinate, say x, can clearly be written as a Fourier series. For example, consider the function

● *Fourier series for function of position*

$$f(x) = \sum_{n=0}^{n=\infty} b_n \sin(2\pi nx/L). \tag{6.41}$$

The arguments of the sine functions change by integral multiples of 2π when x changes by L and $f(x) = -f(-x)$, so $f(x)$ is an odd function that is periodic in x over a distance L. The coefficients b_n are related to $f(x)$ by an equation similar to eqn (6.39):

$$b_n = \frac{2}{L} \int_0^L f(x) \sin(2\pi nx/L) dx. \tag{6.42}$$

The distance L is called the *wavelength* of the function $f(x)$.

Fourier expansions of a function of position x are useful in many situations. For example, they are used in analysing the motions of a string fixed at both ends, such as a guitar string. This is discussed further in Section 7.7.

6.7 Normal modes

So far in this chapter we have considered physical systems that can be described by the time dependence of one variable only, such as the current in an electrical circuit or the displacement of a mass on the end of a spring. However, in commonly occurring mechanical systems several oscillators interact with each other.

After the example of simple harmonic motion of one variable, the next simplest set of problems concerns systems that need two variables for their description, each of which may vary with time when the system oscillates freely or is forced to oscillate by an impressed force. An example of a system with two variables is provided by two simple pendulums connected by a spring and oscillating in a fixed vertical plane. The spring couples the two oscillations so that the oscillations of one affect the other. The two variables needed to describe the motions of the two pendulums are the angles they make with the vertical. A second example is that of two masses fixed to separate points by springs and oscillating in a straight line coupled by a third spring. Here we will use an example from atomic physics to illustrate the procedures adopted to determine the free oscillations of such **coupled oscillators**.

The carbon dioxide molecule consists of a central carbon atom of mass M in between two oxygen atoms each of mass m, the three forming a straight line. Electrons roam over the length of the chain and bind the oxygen atoms to the carbon atom. The bonding forces so produced can be likened to the effects of springs joining the outer atoms to the central one.

If the molecule is described in a coordinate system fixed in the laboratory, three variables x_1, X, and x_2 are needed to specify the displacements of the three atoms from their equilibrium positions in which the springs are unstretched. These are shown in Fig 6.15. However, we are interested in the motions within the molecule and uninterested in what is happening to the molecule's centre of mass. The latter is at rest or moving with uniform velocity, according to Newton's laws discussed in Chapter 2. We may impose the condition that it is at rest without losing any information about the internal motions. With the displacements shown on Fig 6.15, the displacement of the centre of mass is

$$\frac{mx_1}{M+2m} + \frac{MX}{M+2m} + \frac{mx_2}{M+2m}$$

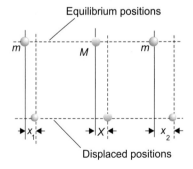

Fig. 6.15 The coordinates used to describe the displacements of oxygen and carbon atoms from their undisturbed positions in the carbon dioxide molecule.

and, with the condition that the centre of mass is at rest,

$$mx_1 + MX + mx_2 = 0. \tag{6.43}$$

The result is that only two of the displacements are independent variables and the discussion about internal motions within the molecule can be made using two variables only.

If we now write down the equations of motion of two of the masses, these expressions involve the three coordinates x_1, X, and x_2 but eqn (6.43) can be used to eliminate one. Consider the two oxygen atoms; their equations of motion are

$$m\frac{d^2x_1}{dt^2} = k(X - x_1)$$

and

$$m\frac{d^2x_2}{dt^2} = -k(x_2 - X).$$

The constant k in these equations gives the restoring forces provided by the springs. Substituting into these equations the value for X given by eqn (6.43) and rearranging gives

$$\frac{d^2x_1}{dt^2} = -kx_1\left(\frac{m + M}{mM}\right) - kx_2\frac{1}{M}$$

and

$$\frac{d^2x_2}{dt^2} = -kx_2\left(\frac{m + M}{mM}\right) - kx_1\frac{1}{M}.$$

● *Coupled equations for CO_2 molecule*

In a more concise notation we may write

$$\frac{d^2x_1}{dt^2} = -\alpha x_1 - \beta x_2 \tag{6.44}$$

and

$$\frac{d^2x_2}{dt^2} = -\alpha x_2 - \beta x_1 \tag{6.45}$$

where

$$\alpha = \frac{k}{m}\left(\frac{m}{M} + 1\right)$$

and $\beta = k/M$.

Equations (6.44) and (6.45) are called **coupled** differential equations because each involves both variables x_1 and x_2 and neither can be solved as they stand; the solution for x_1 involves knowledge of x_2 and vice versa. However, the equations can be **decoupled** by changing the variables in a way that produces two new equations each involving time and one of the new coordinates only. These equations can then be solved to determine how the new coordinates vary with time, after which transformation back to the original coordinates gives the possible vibrational patterns.

Mode frequencies

The decoupling procedure and the determination of possible vibrational motions in examples of coupled oscillations are best explained by doing them for the CO_2 molecule. Let us define new **mode coordinates** q_1 and q_2 by

$$q_1 = x_1 + x_2 \tag{6.46}$$

and

$$q_2 = x_1 - x_2. \tag{6.47}$$

In terms of these new coordinates, eqns (6.44) and (6.45) can be written

● *Decoupled equations in terms of mode coordinates*

$$\frac{d^2 q_1}{dt^2} + \frac{d^2 q_2}{dt^2} = -\alpha(q_1 + q_2) - \beta(q_1 - q_2)$$

and

$$\frac{d^2 q_1}{dt^2} - \frac{d^2 q_2}{dt^2} = -\alpha(q_1 - q_2) - \beta(q_1 + q_2).$$

Addition and subtraction of these equations leads respectively to

$$\frac{d^2 q_1}{dt^2} = -(\alpha + \beta)q_1 \tag{6.48}$$

and

$$\frac{d^2 q_2}{dt^2} = -(\alpha - \beta)q_2. \tag{6.49}$$

These equations are now decoupled, and describe simple harmonic motion for the variables q_1 and q_2. Their solutions are given by expressions similar to eqn (6.7):

$$q_1 = q_{10} \cos(\omega_1 t + \phi_1), \tag{6.50}$$

and

$$q_2 = q_{20} \cos(\omega_2 t + \phi_2) \tag{6.51}$$

where q_{10}, q_{20}, ϕ_1, and ϕ_2 are constants and the **mode angular frequencies**, ω_1 and ω_2, are given by

$$\omega_1^2 = \alpha + \beta = \frac{k}{m}\left(2\frac{m}{M} + 1\right) \tag{6.52}$$

and

$$\omega_2^2 = \alpha - \beta = \frac{k}{m}. \tag{6.53}$$

Substituting for q_1 and q_2 into eqns (6.46) and (6.47) we obtain the general formulae sought in the beginning for the displacements x_1 and x_2. They are

$$x_1 = x_{10} \cos(\omega_1 t + \phi_1) + x_{20} \cos(\omega_2 t + \phi_2) \tag{6.54}$$

and

$$x_2 = x_{10} \cos(\omega_1 t + \phi_1) - x_{20} \cos(\omega_2 t + \phi_2). \tag{6.55}$$

The constants x_{10}, x_{20}, ϕ_1, and ϕ_2 are determined by the initial conditions appropriate to the particular oscillation being considered.

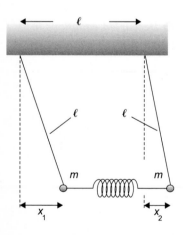

Fig. 6.16 Two identical pendulums coupled by a spring.

● *Coupled equations for pendulums*

Worked Example 6.6 Two simple pendulums, each consisting of a light string of length ℓ attached to a small bob of mass m, are suspended from rigid supports at the same height, as shown in Fig 6.16. The bobs are connected by an elastic spring of force constant k and natural length equal to the separation of the pendulum supports. Small oscillations of the system are set up by giving the left-hand mass an initial speed v to the left. Derive expressions for the displacements of the two masses at a later time t. At what times will the displacements of the two masses have equal magnitude and sign?

Answer Let us choose the x-axis along the line joining the masses, and positive displacements of the masses from their rest positions to be to the right. If the masses at time t have displacements x_1 and x_2, as shown in the figure, their equations of motion are

$$m\frac{d^2 x_1}{dt^2} = -\frac{mgx_1}{\ell} + k(x_2 - x_1)$$

and

$$m\frac{d^2 x_2}{dt^2} = -\frac{mgx_2}{\ell} - k(x_2 - x_1).$$

In the above we have assumed that the displacements are small so that the masses move along the horizontal and that the horizontal components of the gravitational forces on the masses, given by $mg \sin \theta_1$ and $mg \sin \theta_2$, with g the acceleration due to gravity, are approximately equal to mgx_1/ℓ and mgx_2/ℓ.

We now change the variables x_1 and x_2 to mode coordinates q_1 and q_2 with

$$q_1 = x_1 + x_2$$

and

$$q_2 = x_1 - x_2.$$

In terms of these coordinates, the equations of motion become

$$m \frac{d^2 q_1}{dt^2} = -\frac{mg}{\ell} q_1$$

and

$$m \frac{d^2 q_2}{dt^2} = -2kq_2 - \frac{mg}{\ell} q_2 = -(2k + mg/\ell)q_2.$$

These equations are solved to give simple harmonic motion for the variables q_1 and q_2, as in eqns (6.50) and (6.51), in terms of mode frequencies $\omega_1^2 = g/\ell$ and $\omega_2^2 = (2k/m + g/\ell)$. Reverting to real displacements gives equations similar to (6.54) and (6.55) where the four constants are determined by the initial conditions.

So far we have simply repeated material done in the text, but using a new example. Now we apply the given initial conditions to determine the actual motion of the pendulums. At $t = 0$, $x_1 = x_2 = 0$. Hence

$$0 = x_{10} \cos \phi_1 + x_{20} \cos \phi_2$$

and

$$0 = x_{10} \cos \phi_1 - x_{20} \cos \phi_2,$$

giving $\phi_1 = \phi_2 = \pi/2$, and

$$x_1 = -x_{10} \sin \omega_1 t - x_{20} \sin \omega_2 t,$$

$$x_2 = -x_{10} \sin \omega_1 t + x_{20} \sin \omega_2 t.$$

From these equations we derive expressions for the speeds of the masses and substituting in these expressions the conditions that at $t = 0$, $dx_1/dt = -v$ and $dx_2/dt = 0$ we obtain the results

$$x_1 = -(v/2\omega_1) \sin \omega_1 t - (v/2\omega_2) \sin \omega_2 t,$$

$$x_2 = -(v/2\omega_1) \sin \omega_1 t + (v/2\omega_2) \sin \omega_2 t.$$

The two displacements are equal when $x_1 = x_2$. This occurs when $\sin \omega_2 t = 0$ at $t = n\pi/\sqrt{2k/m + g/\ell}$, where n is an integer.

Exercise 6.6 The force constant k for the 'springs' in a CO_2 molecule may be taken to be $10^3 \, \mathrm{N \, m^{-1}}$. Calculate the normal mode frequencies. At a certain time the mode coordinates have the values $q_1 = 2 \times 10^{-10}$ cm and $q_2 = 1.4 \times 10^{-10}$ cm. What are the displacements x_1 and x_2 at that instant? The masses of the carbon and oxygen atoms may be taken to be 12 and 16 atomic mass units, respectively, with one atomic mass unit $= 1.66 \times 10^{-27}$ kg.

Answer The frequencies are about 3.1×10^{13} and 5.9×10^{13} Hz. The displacement $x_1 = 1.7 \times 10^{-10}$ cm, and $x_2 = 3.0 \times 10^{-11}$ cm.

In the above treatment of the CO_2 molecule we have shown that the displacements x_1 and x_2 of the oxygen atoms are given quite generally in terms of the sum of two harmonic oscillations with different frequencies and phases.

Two particularly simple motions result when the initial conditions require either x_{10} to be zero with $x_{20} = x_0$, or x_{20} zero with x_{10} equal to x_0. In the former case

$$x_1 = x_0 \cos(\omega_2 t + \phi_2), \tag{6.56}$$

$$x_2 = -x_0 \cos(\omega_2 t + \phi_2). \tag{6.57}$$

● *Normal modes of the CO_2 molecule*

The two oxygen atoms oscillate with the same amplitude x_0 and the same angular frequency ω_2, but are out of phase by π; when one is moving to the right the other is moving to the left. In the latter case

$$x_1 = x_0 \cos(\omega_1 t + \phi_1), \tag{6.58}$$

$$x_2 = x_0 \cos(\omega_1 t + \phi_1). \tag{6.59}$$

The two oxygen atoms are moving with the same amplitude x_0 and with the higher angular frequency ω_1, and are now moving in phase; when one is moving to the right the other is also moving to the right.

The two simple motions, each involving one frequency only, are called the **normal modes** of the coupled-oscillator system. The first can be excited by arranging (let us suppose we can do this with a carbon dioxide molecule!) that $x_1 = x_0 = -x_2$ at $t = 0$, and the second by making $x_1 = x_0 = x_2$ at $t = 0$. The first one corresponds to the molecule symmetrically vibrating, and successively stretching and compressing itself; the second to a mode in which the carbon atom vibrates between two oxygen atoms separated by a fixed distance.

The ratio of the frequency of the second mode to that of the first is given by eqns (6.52) and (6.53) as $(1 + 2m/M)^{1/2}$, which is about 1.91. A proper quantum mechanical treatment of the vibrations of the carbon dioxide molecule leads to the conclusion that allowed vibrations correspond to the classical normal modes discussed here, and two basic vibrations observed do indeed have a ratio of frequencies equal to about 1.91.

The fact that quantum mechanics allows only the normal modes to be observed and not mixtures can be made plausible by examination of the nature of the general solutions, eqns (6.54) and (6.55). Each of these has a structure similar to that of eqn (6.32) used in the discussion of beats, and, following Section 6.5, as time evolves they describe beats in the amplitudes of the displacements x_1 and x_2.

It can be shown that, when the envelope of the displacement x_1 is zero, the envelope of x_2 is a maximum, and vice versa. Hence the total energy is continually being exchanged between the two oscillating objects. Quantum theory allows measurements of the energy of a molecular vibrational state to give answers that correspond only to constant energy in a unique vibration, i.e. only when all the energy is concentrated in one mode.

Problems

Level 1

6.1 A particle vibrates with simple harmonic motion at a frequency of 50 Hz and with an amplitude of 2 mm. What is its maximum speed?

6.2 What is the steady speed of a particle executing circular motion with radius 20 cm when the projection of the particle's speed on an axis describes simple harmonic motion with a period of 4 s?

6.3 Assuming the average density of the Moon is half that of the Earth, and that its radius is one-quarter that of the Earth, calculate the length ℓ_1 of a simple pendulum that, on the Moon, has the same period as a simple pendulum of length ℓ_2 on the Earth.

6.4 If a tunnel could be drilled from one side of the Earth to the other passing through the centre, show that a mass freely falling between the two points on the surface executes simple harmonic motion. How long would it take for the mass to travel between the two points, given that the mean density of the Earth is 5500 kg m^{-3}? The value of the gravitational constant G is 6.67×10^{-11} m^3 kg^{-1} s^{-2}.

6.5 An object is suspended from a torsion wire that gives a restoring couple proportional to angular twist. The period of small oscillations is found to be 235 ms. A mass of 2.5 g, very small compared with the mass of the object, is attached to the body at a distance of 1 cm from the axis of revolution. This causes the period to increase by 0.1 ms. What is the moment of inertia of the object?

6.6 Assume that an automobile has all of its mass of 800 kg supported on springs whose total stiffness is 100 N m^{-1}. What is the natural frequency of vertical oscillations? Now, in order to assess how the vertical oscillations may be damped, assume that all the wheels are simultaneously raised by 1 cm. What will be the motion of the automobile if no dampers are fitted? Suggest a suitable strength of the dampers for a more comfortable ride.

6.7 The displacement x of a lightly damped simple harmonic oscillator at time t is given by

$$x = ae^{-0.01t} \sin 4t$$

where a is a constant. Determine the Q of the system.

6.8 Show that the power input required to maintain forced vibrations of amplitude x_0 and angular frequency ω is equal to $b\omega^2 x_0^2/2$, where b is the viscous retarding force for unit speed.

6.9 A thin, uniform rod of mass 1.4 kg and length 1.4 m is suspended vertically from a frictionless pivot 60 cm from one end. What is the period of small oscillations of the rod about the point of suspension?

6.10 Show that the period of oscillations of the rod in Problem 6.9 is unchanged if the pivot about which the rod swings gives a frictional torque of constant magnitude in a direction opposing the motion.

Level 2

6.11 A grandfather clock is adjusted so that the period of its simple pendulum is exactly 1 s at sea level. It is raised to a height of 1000 m above sea level and operates at the same temperature so that the length of the pendulum is unchanged. How many seconds does the clock gain or lose in a year? (Take the radius of the Earth to be 6.4×10^6 m.)

6.12 The potential energy for the force between two nuclei in a diatomic molecule has the approximate form

$$U(x) = -\frac{a}{x^6} + \frac{b}{x^{12}}$$

where x is the separation of the nuclei and a and b are constants. Find the force, the equilibrium separation, and the period for small oscillations about the equilibrium position. Take the mass associated with the oscillations to be the reduced mass m.

6.13 A mass m oscillating on the end of a spring of force constant k experiences a damping force bdx/dt, where b is a constant. The amplitude of the oscillations thus decreases with time. Determine the ratio of successive swings on the same side. (The natural logarithm of this ratio is sometimes called the **logarithmic decrement** of the damped oscillations.)

6.14 A platform executes simple harmonic motion in the vertical direction. Its amplitude is 5 cm. What is the maximum frequency with which it can oscillate if a mass sitting on the platform always maintains contact with the platform?

6.15 A mass m hangs vertically on the end of a light spring that has spring constant k and is attached to a fixed point. Vertical oscillations of the mass are damped by a force proportional to the speed of the mass. At time zero the mass is acted on by a vertical driving force F. If $m = 0.02$ kg, $k = 20$ N m^{-1}, the Q-factor is 80, and a driving force $F = 0.02 \cos(55t) \times \cos(5t)$ N is applied,

find the displacement of the mass as a function of time in the steady state.

6.16 A simple pendulum whose period in a vacuum is 2 seconds is placed in a resistive medium. Its amplitude on each swing is observed to be 2/3 that of the previous swing. What is its new period?

6.17 A mass of 10 g is suspended from a spring of force constant 1 N m^{-1}. The mass is set in motion and subject to a viscous damping force given by the parameter $b = 100$ g s^{-1}. What is the approximate percentage error made in calculating the frequency of the motion if the damping is neglected?

6.18 In Section 2.5, friction between dry bodies was represented by a force of size μN, where μ was the coefficient of friction and N the normal force. Friction on a body moving in a fluid needs a more complicated treatment; the frictional force may be assumed given by a power series involving powers of the speed of the body. If it is assumed that the retarding force on a mass m oscillating on the end of a spring of force constant k is proportional to the square of the speed, the equation of motion (6.11) requires modification, and the resulting motion differs from eqn (6.17). What qualitative changes would there be?

6.19 A mass of 0.01 kg hangs vertically on the end of a light elastic spring of force constant 20 N m^{-1}. The oscillations in the vertical direction are damped by a force proportional to the speed of the mass. The system has a Q-value of 50. A driving force $F = 0.02 \cos(60t) \cos(6t)$ N is applied. Find how the displacement of the mass varies with time after a steady state has been reached.

Level 3

6.20 A mass m hangs vertically on a spring attached to a support that moves when it experiences a force. The spring constant is k and the support moves downwards by an amount βF when it experiences a downward force F. Show that the angular frequency of vertical oscillations is

$$\omega = \sqrt{\frac{k}{m(1 + \beta k)}}.$$

6.21 Figure 6.17 is a diagram of an apparatus for measuring the gravitational constant G. A light beam of

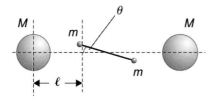

Fig. 6.17 Apparatus for measuring the gravitational constant.

length d carries two small masses m and is suspended by a thin torsion fibre. When the fibre is twisted through an angle θ, it exerts a couple $C\theta$ on the beam. Two large masses M are placed so that when the beam is at rest and the fibre untwisted the line joining the centres lies along the beam. Assuming that the small masses are point masses, show that the beam experiences a gravitational couple

$$C\theta = \tfrac{1}{2} GMmd^2\theta\{1/\ell^3 + 1/(d+\ell)^3\}$$

if the fibre is twisted through a small angle θ.

The torsion balance is constructed with $M = 10.5\,\text{kg}$, $\ell = 5.60\,\text{cm}$, and $d = 2.75\,\text{cm}$. The period of small amplitude torsional oscillations is found to be $316.10\,\text{s}$ when the large masses are removed, and $314.05\,\text{s}$ when they are present. Show that the difference between the squares of the angular frequencies of the torsional oscillations does not depend on C, and calculate the value of G.

The numbers used in this example are close to those of the apparatus used by the standards laboratory in the USA for the most accurate measurement of G to date. The 10.5 kg masses, the beam, and the small masses were all made of tungsten. In the real experiment the dimensions of the beam and the small masses are very important, and lengthy calculations are needed to calculate the gravitational couple and the moment of inertia of the oscillating beam. However, the result remains independent of the restoring couple of the fibre.

6.22 A uniform horizontal disc of mass m and radius a is suspended at its centre from a vertical torsion wire and performs simple harmonic oscillations. The restoring couple of the wire per unit twist is given by the constant C. A wire ring of mass m and radius a is dropped on to the disc and immediately sticks to it. Determine the new period and amplitude of the oscillations if the ring is dropped at the instant at which the disc is at the end of its swing and at rest. What are the new values if the ring is dropped when the disc is moving at its greatest angular speed?

6.23 Show that for light damping (values of the damping parameter very small so that γ^2 is much less than $4\omega_0^2$) the acceleration of a forced damped oscillator is a maximum when

$$\omega^2 = \frac{\omega_0^2}{(1 - \gamma^2/2\omega_0^2)}$$

and that the value at the maximum is equal to

$$\frac{QF_0}{m}\left(1 + \frac{1}{8Q^2}\right),$$

to first-order in $1/Q$.

6.24 Use the definition of Q as 2π times the energy stored at resonance divided by the energy dissipated per cycle, and eqn (6.9) for the total energy when the amplitude is x_0, to show that, for a lightly damped $(\gamma^2 \ll 4\omega_0^2)$ harmonically driven oscillator,

$$Q = \frac{1}{2}\left(1 + \left(\frac{\omega_0}{\omega}\right)^2\right)^{1/2}\left(\frac{\omega}{\gamma}\right)$$

where ω is the angular frequency at which the acceleration is a maximum.

6.25 Determine the Fourier series representation of the periodic force shown in Fig 6.12.

6.26 Show that a full-wave rectifier, which inverts the negative half-cycles of the sinusoidal voltage output from the electricity supply and adds them in real time to the unchanged positive half-cycles, as in Fig 6.18, has an output

$$V = \frac{2V_0}{\pi}\left(1 - \frac{2}{3}\cos 2\omega t - \frac{2}{15}\cos 4\omega t - \cdots\right).$$

6.27 Show that the Fourier series representation of the square-wave voltage signal of period 1 second and peak-to-peak voltage difference A shown in Fig 6.19 may be written in the form $\sum_{n=1}^{\infty} a_n \cos(2\pi nt)$, where $a_n = (-1)^{(n-1)/2}2A/\pi n$ if n is odd and $a_n = 0$ if n is even.

The signal is passed through a filter that transmits unchanged all frequencies between 2 and 6 Hz and rejects completely all other frequencies. Determine the shape of the filtered voltage.

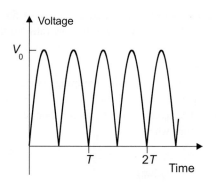

Fig. 6.18 A harmonic voltage signal of period T and amplitude V_0 with the negative-going parts inverted.

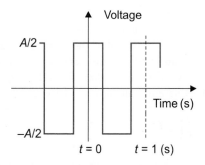

Fig. 6.19 A square-wave voltage signal. The peak-to-peak voltage difference is A and the period is one second.

A resonant system resonates at 5 Hz and has a Q-factor of 50. How does it respond to both the original voltage and the filtered voltage?

6.28 An underdamped harmonic oscillator is subjected to the periodic 'sawtooth' force shown in Fig 6.20. If the natural frequency of the oscillator is $\omega_0 = 4\pi/T$, calculate the amplitude of that part of the steady-state motion that is in resonance with ω_0.

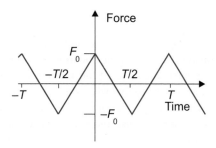

Fig. 6.20 The sawtooth force for Problem 6.28.

6.29 A coupled system consists of a ball of mass m suspended by a light spring of force constant k_1 supporting a mass m suspended by a light spring of force constant k_2. The system oscillates in a vertical line. Determine the ratio of the frequencies of the normal modes by using the fact that both masses can oscillate at the same time with the frequency of a normal mode.

Some solutions and answers to Chapter 6 problems

6.1 0.63 m s^{-1}.

6.4 2.53 × 10^3 s.

6.7 200.

6.9 The moment of inertia, I, for oscillations of the rod about the pivot is

$$I = \int_{-\ell_2}^{\ell_1} \rho x^2 \mathrm{d}x,$$

where ℓ_2 and ℓ_1 are the lengths of the rod below and above the pivot respectively and ρ is the mass per unit

length of the rod.

$$I = \tfrac{1}{3}\rho(\ell_2^3 + \ell_1^3) = \tfrac{1}{3}m(\ell_2^2 - \ell_2\ell_1 + \ell_1^2).$$

When the rod is displaced by a small angle θ from the vertical, the couple on the rod tends to restore the rod to the vertical and has magnitude

$$\rho g \ell_2 \left(\frac{\ell_2}{2}\right)\theta - \rho g \ell_1 \left(\frac{\ell_1}{2}\right)\theta,$$

where g is the acceleration due to gravity.

The equation governing small angular oscillations of the rod is thus

$$\frac{1}{3}m(\ell_2^2 - \ell_2\ell_1 + \ell_1^2)\left(\frac{d^2\theta}{dt^2}\right) = -\frac{1}{2}mg(\ell_2 - \ell_1)\theta.$$

This corresponds to SHM of angular frequency

$$\omega_0 = \sqrt{\frac{3}{2}\frac{(\ell_2 - \ell_1)g}{(\ell_2^2 - \ell_2\ell_1 + \ell_1^2)}}.$$

Inserting $\ell_2 = 0.8$ m, $\ell_1 = 0.6$ m, and $g = 9.8$ m s^{-2} into this equation gives $\omega_0 = 2.38$ rad s^{-1}.

6.11 The acceleration due to gravity at the surface of the Earth is given by

$$g_E = \frac{GM_E}{a_E^2}$$

where M_E is the mass of the Earth and a_E its radius. The period of a simple pendulum of length ℓ at the surface of the Earth is

$$T_E = 2\pi\sqrt{\frac{\ell}{g_E}} = 2\pi\sqrt{\frac{a_E^2\ell}{GM_E}}.$$

At a height a above the Earth's surface, similar reasoning gives

$$T = 2\pi\sqrt{\frac{\ell}{g}} = 2\pi\sqrt{\frac{(a_E + a)^2}{GM_E}}.$$

Hence,

$$\frac{T}{T_E} = \frac{(a_E + a)}{a_E} = 1 + 1.563 \times 10^{-4}.$$

A 'second' on the clock at 1000 m altitude is longer than a second on the clock at the surface, and the high clock loses 4928 seconds in one year.

6.12 The force is the negative of the differential of the potential with respect to x.

$$F = -\frac{dV}{dx} = -\frac{6a}{x^7} + \frac{12b}{x^{13}}.$$

The equilibrium separation x_0 is when the force is zero.

$$\frac{6a}{x_0^7} = \frac{12b}{x_0^{13}},$$

and

$$x_0 = \left(\frac{2b}{a}\right)^{1/6}.$$

For small displacements Δx from x_0, Taylor's theorem, eqn (20.60), gives

$$F(x + \Delta x) = F(x_0) + \left(\frac{dF}{dx}\right)_{x_0}\Delta x + \cdots.$$

The first term on the right-hand side is zero, and terms higher than the second can be neglected. Hence

$$F(x + \Delta x) = -36a\left(\frac{a}{2b}\right)^{4/3}\Delta x.$$

This force gives SHM of period

$$T = 2\pi\sqrt{\frac{m}{k}},$$

with

$$k = 36a\left(\frac{a}{2b}\right)^{4/3}.$$

6.13 Equation (6.17) applied at $t = 0$ and $t = 2\pi/\omega$, with ω given by eqn (6.15), can be used to determine the ratio of successive swings. The logarithmic decrement is $2\pi b/(4mk - b^2)^{1/2}$.

6.15 This is done by realizing that

$$0.02\cos(55t)\cos(5t) = 0.01\cos(60t) + 0.01\cos(50t).$$

6.17 15.5%. ($\omega_0/\omega = 1.155$.)

6.20 If the mass is displaced downwards by an amount y from its equilibrium position, the spring is extended by an amount $(y - x)$ where x is the extra downward displacement of the support from its equilibrium position. The restoring force on the mass due to the spring is $k(y - x)$ upwards and the total force on the mass is $mg - k(y - x)$. The equation of motion of the mass is thus

$$m\frac{d^2y}{dt^2} = mg - k(y - x).$$

The extra downward force on the support is due to the increased tension in the spring, and $x = \beta k(y - x)$. Solving this equation for x gives

$$x = \frac{\beta k y}{1 + \beta k},$$

and the equation of motion becomes

$$m \frac{d^2 y}{dt^2} = mg - y\left(k - \frac{k^2 \beta}{1 + \beta k}\right).$$

This describes simple harmonic motion with angular frequency

$$\omega = \sqrt{\frac{k}{m} \frac{1}{(1 + \beta k)}}.$$

6.21 For small angular displacements θ, we may approximate the separation of a small mass m from the nearest and farthest large masses by ℓ and $\ell + d$, respectively. The component in the direction perpendicular to the beam of the force on a small mass from the two large masses is thus

$$\frac{GMm}{\ell^2} \sin \phi + \frac{GMm}{(d + \ell)^2} \sin \psi,$$

with ϕ and ψ appropriate angles.

To the same approximation, the sine rule relating sides and opposite angles of a triangle, Problem 20.7, gives

$$\frac{d/2}{\sin \phi} = \frac{\ell}{\sin \theta},$$

and thus

$$\sin \phi = \theta d / 2\ell.$$

Similarly

$$\sin \psi = \frac{d}{2(d + \ell)} \theta,$$

and the component of force perpendicular to the beam is

$$\frac{GMm}{\ell^2} \frac{\theta d}{2\ell} + \frac{GMm}{(d + \ell)^2} \frac{d}{2(d + \ell)} \theta.$$

The couple on the beam from the perpendicular component on each small mass is

$$d\left(\frac{GMm\,\theta d}{\ell^2}\frac{1}{2\ell} + \frac{GMm}{(d + \ell)^2}\frac{d}{2(d + \ell)}\theta\right)$$

$$= \frac{1}{2} GMmd^2\theta\left(\frac{1}{\ell^3} + \frac{1}{(d + \ell)^3}\right).$$

6.23 The amplitude of the acceleration is $\omega^2 x_0$ with x_0 given by eqn (6.25). The maximum acceleration occurs at an angular frequency ω such that the derivative with respect to ω is zero. This gives the maximum acceleration at the value of ω given.

Substitution of this value into $\omega^2 x_0$ and using the binomial theorem to expand $(1 - \gamma^2/4\omega_0^2)^{-1/2}$, gives the maximum acceleration equal to the expression given.

6.25 The mathematical expression $F(t)$ that describes the given periodic force rises linearly to its maximum value F_0 during time $t = 0$ to $T = T/2$. It then falls linearly to zero over the next half-period. Hence

$$F(t) = \frac{2f_0}{T} t, \quad 0 < t < T/2,$$

and

$$F(t) = \frac{2f_0}{T}(T - t), \quad T/2 < t < T.$$

The function is even; hence only cosine terms appear in the expansion (6.35) and the coefficients a_n, other than a_0, are given by

$$a_n = \frac{4F_0}{T^2} \int_0^{T/2} t \cos(n\omega t)dt$$

$$+ \frac{4F_0}{T^2} \int_{T/2}^{T} (T - t) \cos(n\omega t)dt.$$

The integral involving $t \cos(n\omega t)$ can be evaluated by manipulation of the formula given in the mathematical insert on p. 000 for the integral of $t \cos t$, and the result is

$$\int t \cos(n\omega t)dt = \frac{t}{(n\omega)} \sin(n\omega t) + \frac{1}{(n\omega)^2} \cos(n\omega t)dt.$$

This allows the definite integrals to be evaluated, giving

$$a_n = -\frac{4F_0}{(n\pi)^2}$$

for n odd, and $a_n = 0$ for n even. We leave the algebra for this result to the reader. The term a_0 is nonzero and is the average value of the function over a period. The term a_0 equals $f_0/2$. The Fourier series for $F(t)$ is thus

$$F(t) = \frac{F_0}{2} - \frac{4F_0}{\pi^2}\cos(\omega t) - \frac{4F_0}{9\pi^2}\cos(3\omega t)$$
$$- \frac{4F_0}{25\pi^2}\cos(5\omega t) - \cdots . \tag{6.40}$$

6.29 The angular frequencies of the normal modes are given by

$$\omega^2 = \frac{k_1 + 2k_2}{2m} \pm \frac{1}{2}\sqrt{\frac{k_1^2 + 4k_2^2}{m^2}} .$$

Chapter 7

Waves

This chapter deals with the physics of wave motion, discussing the speeds of the waves and different possible wave patterns.

This chapter deals with the physics of wave motion. We are familiar with what is meant by a wave. It is a pattern of behaviour that a medium such as the ocean or a long rope under tension exhibits when it is more or less regularly disturbed. Winds are the usual cause of ocean waves, and up and down movement of one end causes waves on a rope. A common feature of the two disturbances is that energy can be transported through the medium from one part of it and deposited at another: ocean waves break on the shore and move the sand; a wave on a rope moves an object attached to the far end of the rope. Energy is transported in both situations but there is no bulk motion of the medium through which the energy passes. The energy transport in both situations is described mathematically in terms of waves.

If there is a periodic disturbance in the surface of a stretch of water, such as a stick moving up and down regularly in the middle of a pond, or periodic waggling of the end of a string under tension, the water surface or the string undergoes periodic displacements. If the oscillatory disturbance transmitting energy to the water or string is purely sinusoidal, it is a periodic disturbance of a single frequency and is called a **harmonic** disturbance.

Looking at *a fixed point in time*, when a sinusoidal disturbance has waggled the end of a string to the point *a* on Fig 7.1, a snapshot of the shape of the string looks like the sine wave shown. A short time later, when the disturbance has moved the end of the string to point *b*, the peak positions have moved forward. The shape of the string is still sinusoidal but the peak that was at point A is now at point B. As time goes on, the peaks appear to move forward horizontally while the rope as a whole stays with its end fixed to the vertical line in which the disturbance operates. The length interval after which the shape change repeats itself is called the **wavelength** and given the symbol λ. A wavelength has the dimensions of length and units of metres in the SI system.

Looking at *a fixed position along the string*, the vertical displacement changes with time and also exhibits a sinusoidal variation. This is shown in Fig 7.2. The time interval over which the pattern repeats itself is the period $T = 2\pi/\omega$ discussed at the beginning of Chapter 6, with ω the angular frequency of the wave which, in turn, is simply related to the frequency ν by eqn 6.2, $\omega = 2\pi\nu$.

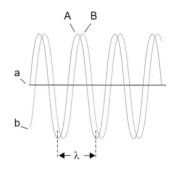

Fig. 7.1 A wave on a long string produced by regular sinusoidal up and down motion of one end. At any instant of time, the rope looks like a sine curve.

● *Wavelength*

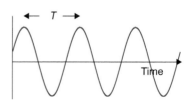

Fig. 7.2 At a fixed position along a string on which a harmonic wave of a single frequency is propagating horizontally, the displacement of the string in the vertical direction varies sinusoidally with time.

Exercise 7.1 A radio station transmits radio waves of frequency 102 MHz. Calculate the period and angular frequency of the waves.

Answer The period $T = 9.8 \times 10^{-9}$ s. The angular frequency $\omega = 6.409 \times 10^{8}$ s^{-1}.

One way to visualize what is happening as a wave progresses through a medium is to regard the medium as a large set of coupled oscillators similar to those discussed in Section 6.7. There we saw that, as time proceeded, the two masses of a coupled oscillating system could maintain the total energy constant with varying fractions taken up by each mass. At certain times all the energy could be in the motion of only one mass. Visualizing the constituents of the continuous medium as closely spaced coupled oscillators consisting of masses on springs, it can be imagined that as the first mass is made to vibrate at one side of a body it passes its energy on to the next mass, and so on until that energy reaches the far side. Continuous forced sinusoidal oscillation of the first mass will cause continuous sinusoidal oscillations throughout the medium and transfer of energy from one side to the other without overall motion of the medium itself.

It is interesting to note that for a long time the way in which transfer of energy in electromagnetic waves (to be introduced in Chapter 8) was visualized was through oscillations in an all-pervading continuous medium called the aether. However, no experiment was able to detect the existence of such a medium. Not until the late nineteenth century was a satisfactory description provided for transmission of electromagnetic waves through empty space.

In this chapter we discuss some basic properties of waves using the simple example of waves on stretched strings. Wave propagation is usually a three-dimensional problem, but we restrict attention to propagation in one dimension along a string to keep the mathematics simple. Most of the essential physics of waves can still be understood from the simplified treatment given here.

7.1 Waves on taut strings

Here we consider perhaps the simplest wave to visualize, that on a string or wire that is held under tension. The string is very long and lies stretched straight along the z-axis in the horizontal plane when at rest. One end is moved up and down sinusoidally in the vertical direction, which we take to be along the y-axis. We choose the origin to be at the point where this end crosses the z-axis.

To describe the shape the string adopts at any time as a wave passes down it we need to determine the vertical displacement y from the equilibrium rest position as a function of distance z along the string. We assume that the tension F in the string is constant throughout. We also assume that the forces on the string due to the tension are much larger than gravitational forces and neglect gravity. First, we shall set up an equation that applies to all possible waves on the string. This is a differential equation requiring constants of integration that are determined by the way the wave is maintained in a particular example.

The wave equation

Figure 7.3 shows a part of the string that includes a very small section of length δz with its centre at position z. The string is continuous, as is its slope to the horizontal, but the slopes at each end of the small section are different because of the changing shape of the string along its length. Figure 7.3 is a snapshot of the section of the string at a particular time t. The shape of the string also changes with time, and the vertical displacement is written as $y(z,t)$ to show that it is a function of both z and t.

For constant tension F, the horizontal force δF_z on the small section at position z is given by the difference between the horizontal forces on the right-hand end and the left-hand end. The horizontal force is

$$\delta F_z = F \cos \theta_2 - F \cos \theta_1 \tag{7.1}$$

where $\theta_1 = \theta(z - \frac{1}{2}\delta z, t)$ and $\theta_2 = \theta(z + \frac{1}{2}\delta z, t)$ are angles the string makes with the horizontal at time t at the points $(z - \frac{1}{2}\delta z)$ and $(z + \frac{1}{2}\delta z)$ respectively.

We now consider only small amplitude waves for which the maximum displacement in the y-direction is small compared with the wavelength. The angle θ the string makes with the horizontal is thus very small. The cosines of the two angles are both very nearly unity and the net horizontal force is negligible.

The force δF_y on the section in the vertical (y) direction is given by the tension multiplied by the difference between the sines of the two angles at

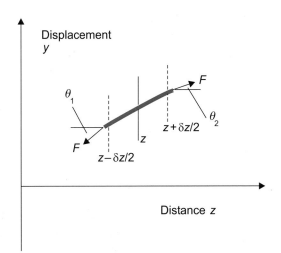

Displacement
y

Distance z

Fig. 7.3 A short length δz of string symmetrically placed around a point distant z from one end.

the ends of the small section

$$\delta F_y = F \sin[\theta(z + \tfrac{1}{2}\delta z, t)] - F \sin[\theta(z - \tfrac{1}{2}\delta z, t)]. \tag{7.2}$$

The sines of small angles are very nearly equal to the angles expressed in radians, and hence

$$\delta F_y = F[\theta(z + \tfrac{1}{2}\delta z, t)] - F[\theta(z - \tfrac{1}{2}\delta z, t)]. \tag{7.3}$$

Taylor's expansion, eqn (20.62), may be used to evaluate $\theta(z \pm \tfrac{1}{2}\delta z)$. The factor $(x - a)$ in eqn (20.62) becomes $\pm \tfrac{1}{2}\delta z$, and, since we will be interested in the limit as δz becomes vanishingly small, when setting up the differential equation there is no need to keep terms beyond the first order in δz. The Taylor expansion becomes

$$\theta\left(z \pm \frac{1}{2}\delta z, t\right) = \theta(z, t) \pm \tfrac{1}{2}\delta z \frac{\partial \theta}{\partial z}. \tag{7.4}$$

This is the first time we have used the notation $\partial y/\partial z$, which is called the **partial differential** with respect to z of the function y. The partial differential defines the rate of change of y with z alone, while the second variable t on which y depends is held constant. $\partial y/\partial z$ itself remains a function of the variables z and t and, for any reasonably complicated dependence of y on z and t, $\partial y/\partial z$ will also vary with both of these variables. The function $\partial y/\partial z$ gives the slope of the curve y at time t and position z. For a simple function $y = f(z, t)$, the partial differential of y with respect to z, $\partial y/\partial z$, although remaining a function of the two variables, may be independent of one or both of them. In a like manner, $\partial y/\partial t$ is the rate of change of y with t as z is held constant, and is simply the speed of the string in the y direction at a fixed point z. It is also, of course, a function of the two variables z and t. Similarly, $\partial^2 y/\partial t^2$, which equals

$$\frac{\partial}{\partial t}\left(\frac{\partial y}{\partial t}\right),$$

is the second (or double) partial differential of y with respect to t, namely the rate of change of the speed of the string in the y-direction, the acceleration of the string in that direction.

If the mass of the string per unit length is ρ, Newton's second law tells us that the mass of the section $\rho\delta z$ multiplied by its acceleration in the y-direction equals the force δF_y on the section in that direction. Hence,

$$\delta F_y = \rho \delta z \frac{\partial^2 y}{\partial t^2}. \tag{7.5}$$

Substituting eqns (7.4) and (7.5) into eqn (7.3), we find

$$\rho \frac{\partial^2 y}{\partial t^2} = F\left(\frac{\partial \theta}{\partial z}\right). \tag{7.6}$$

The angle θ in eqn (7.3) is an approximation for $\sin \theta$ and, since the sines and tangents of small angles are very nearly equal, we may replace θ by $\tan \theta$, the slope $\partial y/\partial z$ of the string. Equation (7.6) then becomes

$$\rho \frac{\partial^2 y}{\partial t^2} = F\left(\frac{\partial}{\partial z}\right)\left(\frac{\partial y}{\partial z}\right) = F\frac{\partial^2 y}{\partial z^2}$$

or

$$\frac{\partial^2 y}{\partial t^2} = \frac{F}{\rho}\frac{\partial^2 y}{\partial z^2}. \tag{7.7}$$

The above equation is the **wave equation** we have been seeking, which describes any wave on the string. It is a second-order **partial differential equation** and its solutions describe **transverse waves**, in which the oscillations are perpendicular to the direction of propagation of the wave.

This is the first time we have met a differential equation involving partial differentials. Previously, the differential equations encountered have related differentials of functions that depend on a single variable. When, as here, a function depends on more than one variable, differential equations may relate different partial differentials. A first-order partial differential equation involves only first partial differentials. Equation (7.7) gives the relationship between the double partial differential of y with respect to t and the double partial differential of y with respect to z. $\partial^2 y/\partial t^2$ is the acceleration of the string at time t and position z; $\partial^2 y/\partial z^2$ is the rate of change with position of the slope $\partial y/\partial z$ of the string at time t and position z. We are not interested here in techniques for solving partial differential equations. We simply wish the reader to appreciate that the relation (7.7) can be solved under certain conditions to give the displacement y of the string at time t and position z.

Simple solution

The solution to eqn (7.7) that fits the conditions externally imposed on the string, called the *boundary conditions*, gives the way in which the vertical displacement y of the string varies with time t and position z. There are standard mathematical techniques for solving such differential equations when given the boundary conditions and there are very many different waves that can exist. We are not interested at the moment in these techniques and merely want to extract from eqn (7.7) the physical properties of simple wave motions on the string.

Let us try as a solution the simple function

$$y(z,t) = y_0 \cos(\omega t - kz). \tag{7.8}$$

From now on we will omit the arguments (z,t) of the disturbance y and leave them to be understood. You can easily check that eqn (7.8) is indeed a solution to the wave equation as long as

$$\frac{\omega^2}{k^2} = \frac{F}{\rho}. \tag{7.9}$$

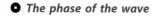
● *The phase of the wave*

The argument $(\omega t - kz)$ of the cosine function in eqn (7.8) is called the **phase of the wave** and we note that, if any constant is added to the argument, the new function remains a satisfactory solution to the wave equation. Addition of such a constant, call it ϕ, merely corresponds to a change in the origin of t or z. For example, the phase angle ϕ can be removed from the expression for the wave by moving the origin to the right of the previous origin by a distance Δz such that $k\Delta z = \phi$.

The relationship between the parameter ω and the period T of the wave can be determined by examining how eqn (7.8) varies with time at a fixed point. As in eqn 6.2 for simple harmonic motion,

$$T = \frac{2\pi}{\omega}, \tag{7.10}$$

and ω is thus the angular frequency. Similarly, the relation between the parameter k and the wavelength λ is determined by examining the variation with position at a fixed time, giving

$$\lambda = \frac{2\pi}{k}. \tag{7.11}$$

k is called the **wavenumber** of the wave with angular frequency ω, and has units m^{-1}.

The wave described by eqn (7.8) is called a **progressive** or **travelling wave** and it has a single frequency and wavelength.

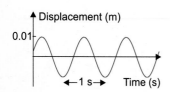

Fig. 7.4 The variation with time of the displacement y of a string from its undisturbed position at a fixed position z along the string, when a wave of fixed period and wavelength is propagating.

Figure 7.4 shows displacement against time for a wave with period T equal to 1 s, corresponding to an angular frequency of 2π rad s^{-1}. The time interval between successive crests is the period of the wave. The displacement is shown at a fixed point along the rope for a wave with amplitude 10^{-2} m. At a different position the shape of the curve remains the same but the pattern is shifted by an amount that depends on where the new point is with respect to the first.

Figure 7.5 shows displacement against distance for a wave with wavelength λ equal to 1.0 m, corresponding to wavenumber $k = 2\pi\,\mathrm{m}^{-1}$. The displacement is shown at a fixed time. The distance between successive crests is the wavelength of the wave, and the wave drawn has amplitude 10^{-2} m.

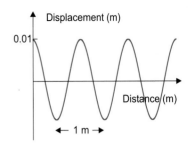

Fig. 7.5 The variation of the displacement y of a string from its undisturbed position against distance z along the string, at a fixed time t, when a wave of fixed period and wavelength is propagating.

Worked Example 7.2 A wave of frequency 5 Hz travels on a nylon string subject to a tension of 100 newtons and with a mass per unit length of 100 grams per metre. What is its wavelength? Calculate the wavelength of a wave of frequency 262 Hz (middle C) on a piano wire that has a mass per unit length of 40 grams per metre and a tension of 200 N.

Answer The wave of frequency 5 Hz has angular frequency ω equal to $10\pi\,\mathrm{rad\,s}^{-1}$. The wavenumber of the wave on the nylon string is given by eqn (7.9) as

$$k = \omega\sqrt{\frac{\rho}{F}},$$

and hence the wavelength is given by

$$\lambda = \frac{2\pi}{k} = \frac{2\pi}{\omega}\sqrt{\frac{F}{\rho}}.$$

Evaluating this expression gives $\lambda \sim 6.3$ m. Doing the same calculation for the piano wire gives $\lambda \sim 27$ cm.

7.2 Wave velocity

An important characteristic of a wave is the speed at which the crest of the wave moves. Figure 7.6 shows the waveform on the string discussed in the last section at time t. At time $t + \delta t$ the waveform has changed to have the shape of the curve shown dotted on Fig 7.6. The crest of the wave has moved forward by an amount $\delta t \times (\omega/k)$. This tells us that the wave described by eqn (7.8) is moving forward, in the positive z-direction, with speed v_p given by

$$v_p = \frac{\omega}{k}. \tag{7.12}$$

The speed v_p is called the **wave velocity** or **phase velocity** of the wave. A progressive wave moving backward with the same frequency, wavelength, and speed has the form

$$y = y_0 \cos(\omega t + kz). \tag{7.13}$$

As time progresses the maxima of this disturbance move in the negative z-direction.

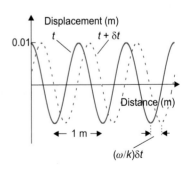

Fig. 7.6 This repeats Fig 7.5 but also shows the waveform at a later time $(t + \delta t)$ when the crest of the wave has moved forward by an amount $\delta t \times (\omega/k)$.

Wave velocity, frequency, and wavelength

Rewriting eqn (7.12) in terms of the wavelength λ and the frequency ν gives a fundamental relationship between these quantities and the wave velocity. This relation is applicable to all forms of wave motion.

$$\lambda\nu = v_p. \tag{7.14}$$

Exercise 7.2 A human ear can typically detect frequencies between 20 Hz and 15 000 Hz. Calculate the corresponding range of wavelengths. The speed of sound in air at normal temperature and pressure is $340\,\mathrm{m\,s^{-1}}$, which is about 750 miles per hour.

Answer The wavelength range is between $17.0\,\mathrm{m}$ and $2.27 \times 10^{-2}\,\mathrm{m}$.

The wave velocity on a taut string for small amplitude waves is given from eqns (7.9) and (7.12) as

● *The wave velocity on a string*

$$v_p = \sqrt{\frac{F}{\rho}}. \tag{7.15}$$

This accords with common experience: the tighter we make the string the faster the wave moves. For the nylon string used in Worked example 7.2 with a tension of 100 newtons and a mass per unit length of 100 grams per metre, the wave velocity is $31.6\,\mathrm{m\,s^{-1}}$. On the piano wire of the same example the velocity is $70.7\,\mathrm{m\,s^{-1}}$.

Equation (7.15) shows that the speed is independent of the frequency of the wave and waves of any wavelength or frequency travel at the same speed. The angular frequency ω is thus linearly related to the wavenumber k through eqn (7.12). When the speed is independent of the frequency there is said to be no *dispersion* of the waves. When the wave speed depends upon frequency the waves do suffer dispersion. Dispersion is treated in Section 7.5.

The wave given by eqn (7.8) represents an idealized situation. It is a wave of unique frequency. Naturally occurring waves never have unique frequencies. It is not possible for the source of the wave, the agency moving one end, to move the end perfectly sinusoidally. The movement is always a mixture of frequencies and the waveform propagating along the string is a group of waves rather than a single wave. With care the **wave group** may consist of a narrow band of frequencies centred around a mean value. A wave group retains its shape when there is no dispersion since all the components of different frequencies travel at the same speed.

7.3 **Energy in the wave**

The energy in a wave on a string depends on the amplitude, since if we increase the amplitude of vibration of the end of the string we expend more work per unit time. There is no dissipation of energy in the wave and we therefore expect that the amplitude remains constant as the wave progresses. In this section we determine an expression for the mean energy per unit length in a progressive wave in terms of the amplitude and frequency. The string, as before, has uniform tension F and mass per unit length ρ.

A very small section of length δz at position z and time t has both kinetic energy due to its speed in the y-direction and potential energy due to the work done by the force δF_y in displacing it a distance y from its equilibrium position. The kinetic energy is

$$\tfrac{1}{2}\rho\delta z\left(\frac{\partial y}{\partial t}\right)^2 = \tfrac{1}{2}\rho\delta z\omega^2 y_0^2 \sin^2(\omega t - kz), \tag{7.16}$$

and the potential energy is given by eqn (3.25) as $-\int_0^y \delta F_y\,\mathrm{d}y$. From eqns (7.3) and (7.4), $\delta F_y = F\delta z(\partial\theta/\partial z)$ and, since θ is approximately equal to $\partial y/\partial z$,

$$-\int_0^y \delta F_y\,\mathrm{d}y = -\int_0^y F\delta z\left(\frac{\partial^2 y}{\partial z^2}\right)\mathrm{d}y = \tfrac{1}{2}Fk^2\delta z y_0^2 \cos^2(\omega t - kz). \tag{7.17}$$

Because of the importance of the result, we show that the time average of $\sin^2(\omega t - kz)$ or $\cos^2(\omega t - kz)$ is equal to $\tfrac{1}{2}$.

The time average of a periodic function is equal to the integral of the function over the time T divided by T, and

$$\frac{1}{T}\int_0^T \sin^2(\omega t - kz)\,\mathrm{d}t = \frac{1}{2T}\int_0^T (1 - \cos 2(\omega t - kz))\,\mathrm{d}t.$$

The first term on the right-hand side gives the factor $\tfrac{1}{2}$. The second term is zero, since

$$\int_0^T \cos 2(\omega t - kz)\,\mathrm{d}t = \frac{1}{2\omega}\int_0^T \cos(2\omega t - 2kz)\,\mathrm{d}(2\omega t - 2kz)$$
$$= \frac{1}{2\omega}[\sin(2\omega t - 2kz)]_0^T = 0.$$

This result is valid for the time average of the square of any function that varies harmonically with time and does not change as the dependence of the function on position variables changes. As examples, the time average of the square of $\sin\omega t$ is one-half and the time average of the square of $\sin(\omega t - kz)$ is also one-half.

Using the above result to evaluate the time averages of the kinetic and potential energies, the time average of the kinetic energy is $\frac{1}{4}\rho\omega^2\delta z y_0^2$ and the time average of the potential energy is $\frac{1}{4}Fk^2\delta z y_0^2$. Since

$$\omega^2\rho = Fk^2, \tag{7.18}$$

from eqn (7.9), the two averages are equal, and the time average of the total energy in the wave per unit length of the string is independent of position and

● *Energy per unit length*   Average energy per unit length $= \frac{1}{2}\rho\omega^2 y_0^2 = \frac{1}{2}Fk^2 y_0^2$. (7.19)

The energy per unit length varies as the square of the amplitude of the wave and the square of the wave frequency. Equally, since the frequency is inversely proportional to the wavelength, the energy varies as the square of the amplitude and the square of the wavenumber.

Not only is the time-average energy per unit length constant for a wave with given amplitude and frequency but also, *at any time t*, the sum of kinetic and potential energies in a unit length is equal to the mean energy $\frac{1}{2}\rho\omega^2 y_0^2 = \frac{1}{2}Fk^2 y_0^2$, as can readily be seen from eqns (7.16) and (7.17). The average total energy of a small section of the string is constant but is continually exchanged between kinetic and potential energies, just like the energy of the simple harmonic oscillator discussed in Section 6.1.

Speed of energy flow

We now examine the rate at which energy in the wave flows down the string. The force moving the string to set up the wave is doing work against the tension in the string and the energy expended by the force is transmitted in the z-direction at a certain speed. If we call this speed v_1, the average amount of energy contained in a length v_1 of string is passed on each second and equals

$$\frac{1}{2}v_1 Fk^2 y_0^2, \tag{7.20}$$

from eqn (7.19).

The energy moving down the string per second is equal to the average power expended by the force which moves the string up and down in the vertical direction. If the location of this force is at the point with position coordinate z, the external force at time t must equal the vertical component of the tension in the string, $F\sin\theta$. As before, $\sin\theta$ is approximately equal to $\tan\theta$ for small angles θ and the external force in the z-direction is given by

$$-F\tan\theta = -F\left(\frac{\partial y}{\partial z}\right) = -Fky_0\sin(\omega t - kz). \tag{7.21}$$

The work done by this force in moving the string a small vertical distance δy equals the force times the distance, and the instantaneous power P expended by the force equals the force times the distance δy divided by the time δt to move that distance. As δy and δt become infinitesimally small, $\delta y/\delta t$ becomes the partial differential of y with respect to time and

$$P = -Fky_0 \sin(\omega t - kz)\left(\frac{\partial y}{\partial t}\right) = Fk\omega y_0^2 \sin^2(\omega t - kz). \tag{7.22}$$

Since the time average of $\sin^2(\omega t - kz)$ is $\frac{1}{2}$, the average power is

$$\bar{P} = \tfrac{1}{2} Fk\omega y_0^2. \tag{7.23}$$

Exercise 7.3 What power is needed to send waves of amplitude 1 cm and frequency 10 Hz down a string that has mass per unit length $\rho = 0.1 \, \text{kg m}^{-1}$ and tension F equal to 100 newtons?

Answer The power needed is about 0.6 watts.

Equating the average power given by eqn (7.23) to the energy flowing down the string each second, given by eqn (7.20), we find that the energy is moving at a speed

$$\frac{\omega}{k} = v_{\text{p}}. \tag{7.24}$$

As we might have expected, the speed with which energy moves in the progressive wave equals the phase velocity v_{p}. This result, as we will see in Section 7.5, is only true if the phase velocity v_{p} is independent of frequency, i.e. there is no dispersion. There is no dispersion for waves on strings that satisfy the assumptions made in deriving the wave eqn (7.7).

7.4 Superposition

Figure 7.7 shows two wave groups that constitute similarly shaped pulses going in opposite directions. First, the pulses are widely separated. As time increases they merge to form a complicated pattern. They then separate and recede from each other. On close inspection the complicated pattern seen during the overlap of the two waveforms decomposes into the sum of the two individual disturbances. *The resultant motion on a string due to two waves is the sum of the separate motions due to each.*

Similarly, if two stones are dropped at different points in a pond, waves spread out from each point, overlap, and move through each other. Figure 7.8 shows a photograph of overlapping ripples produced by raindrops falling on a pond. Ripples pass through each other and spread out over the two-dimensional surface of the water. The waves **superpose** where they overlap as discussed in Section 6.5; waves that obey a certain class of differential equations add when they come together. In this

● *Speed of energy flow equals phase velocity when there is no dispersion*

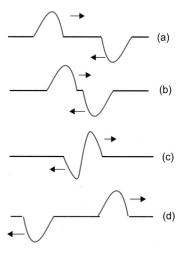

Fig. 7.7 Two disturbances travelling in opposite directions on a string meet, then pass through each other and move on independently, as shown by the sequence of time snapshots (a) through (d). The total disturbance is the sum of the two separate displacements superimposed on each other. When the two disturbances overlap they give a complicated pattern as in the snapshot (c). In region (d) they have passed through each other and proceed unchanged.

Fig. 7.8 Ripples caused by raindrops on a pond pass through each other and spread out over the two-dimensional surface of the water.

● *Linear differential equations*

section we discuss what type of equation waves have to obey in order to satisfy the principle and some mathematical consequences.

As stated in Section 6.5, a differential equation is linear if the function y, whose behaviour the equation governs, and the derivatives of y appear in first-order only.

The more general definition of a linear differential equation is that it must satisfy the condition that if y_1 and y_2 are each solutions, involving different constants of integration, the sum $y_1 + y_2$ is also a solution. An additional condition is that a constant multiplying any solution must also be a solution. Hence the function $Ay_1 + By_2$, with A and B constants, which is obtained by superposing two separate functions that satisfy a linear differential equation, is also a solution of the equation.

The equation

$$ax^2 \frac{d^2y}{dx^2} + bx \frac{dy}{dx} + cy = f_1(x),$$

with a, b, and c constants, is a linear equation. Its solution consists of a particular integral plus a complementary function containing two constants that depend on given conditions. Let its solution be y_1. If a different function of x, say f_2, were placed on the right-hand side instead of f_1, a new solution y_2 results. It can readily be verified that the solution to the equation

$$ax^2 \frac{d^2y}{dx^2} + bx \frac{dy}{dx} + cy = f_1(x) + f_2(x)$$

is the sum of the solutions to the separate equations involving f_1 and f_2 alone. This is an extension of the idea of superposition, and we have already used it in Section 6.5 in the discussion of the response of a damped oscillator to two impressed forces of different frequency.

The wave eqn (7.7), which we used to describe waves on strings, is a linear differential equation. Thus different solutions superpose. The two wave groups in Fig 7.7 separately satisfy the wave equation and so their sum, which is the resultant wave, also satisfies it.

● **The wave equation is linear**

The idea that the sum of any two solutions of a linear differential equation is also a solution can be extended to any number of waveforms that separately satisfy a wave equation. The function

$$y = \sum_{n=1}^{N} a_n y_n,$$

where a_n are constants, is an acceptable solution to a linear wave equation if all the separate functions y_n from $n = 1$ to $n = N$ are individually acceptable solutions.

We have already used this result at the end of Section 7.2 when introducing wave groups made up of many waves of the form of eqn (7.8) and reused it above when stating that the wave groups of Fig 7.7 separately satisfy the wave equation. The pulses going in opposite directions are made up of very many individual harmonic or purely sinusoidal waves with different amplitudes and phases, and the resultant disturbance when two such combinations occur together on the same string is simply the sum of all the individual components.

Superposition is widely used in the discussion of very many different types of waves met in physics. Many phenomena are described in terms of waves that obey linear equations such as (7.7). In the remainder of this chapter superposition will be regularly used to determine the mathematical form of various superposed waves so that the physical properties of the resultant waveforms can be more easily described.

7.5 Dispersion

Usually, real waves are dispersive. Wave groups and dispersion were introduced in Section 7.2. If the phase velocity depends upon frequency, different frequency components of a wave group travel at different speeds and the shape of the group changes.

Imagine a large pond on which ripples are formed by a non-harmonic but periodic disturbance in the middle. The disturbance produces waves with a spread of frequencies and we see ripples moving outwards from the centre with a speed roughly equal to the phase velocity at the mean frequency. However, the amplitudes of the ripples are contained within an

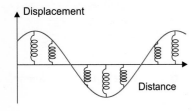

Displacement

Distance

Fig. 7.9 The stiffness of a string can be modelled by imagining the string connected to its undisturbed position by springs.

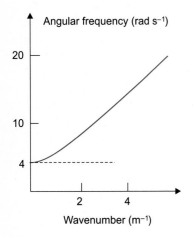

Angular frequency (rad s⁻¹)

Wavenumber (m⁻¹)

Fig. 7.10 The dispersion relation for a stiff string with $F/\rho = 16 \, m^2 \, s^{-2}$ and $\beta = 16 \, s^{-2}$. There is a low-frequency cut-off at $\omega_c = \sqrt{\beta} = 4 \, s^{-1}$, as given by eqn (7.26).

● *Cut-off frequency*

envelope that gradually loses its shape and moves at a different speed to that of the underlying oscillations.

Waves on real strings suffer dispersion. There are forces on sections of a real string or wire in addition to the resultant effect of the tension in the string calculated in Section 7.1. As an improved approximation to the equation governing the motion of an infinitesimally small section of a string we may add a term that resists the vertical displacement.

The string resists moving vertically from its equilibrium position. It is as if there were springs connecting it to its undisturbed shape and these springs pull the string back when it is displaced, as shown in Fig 7.9. This model gives an extra force proportional to displacement y and to the length of the section considered. The extra force is in the direction of y decreasing and, with its inclusion, eqn (7.7) becomes

$$\frac{\partial^2 y}{\partial t^2} = \frac{F}{\rho}\left(\frac{\partial^2 y}{\partial z^2}\right) - \beta y \tag{7.25}$$

where β is a constant depending on the strength of the resisting force and the mass per unit length of the string. This second-order partial differential equation still has a solution of the form of eqn (7.8), as can readily be checked, but subject to the condition that

$$\omega^2 = \frac{F}{\rho}k^2 + \beta. \tag{7.26}$$

The phase velocity ω/k now depends on k and the waves suffer dispersion.

Dispersion relations

A relation between the parameters ω and k for a particular wave motion is called a **dispersion relation**. The dispersion relation for a stiff string is given by eqn (7.26) and is shown in Fig 7.10. This relation has a low-frequency cut-off: no frequencies can propagate below the value $\sqrt{\beta}/2\pi$. If one tries to make a wave with lower frequency, the string as a whole simply moves up and down with the moving force with a displacement that dies away as distance from the end being moved increases. Sometimes this type of motion is called an **evanescent** wave.

The phase velocity of waves of angular frequency ω and wavenumber k on a stiff string is

$$v_p = \frac{\omega}{k} = \frac{\sqrt{Fk^2/\rho + \beta}}{k}. \tag{7.27}$$

The phase velocity is plotted against wavenumber k in Fig 7.11.

Figure 7.11 shows that the phase velocity of a wave increases rapidly as the wavenumber decreases. In theory it can exceed the speed of light in vacuum, which the theory of special relativity considered in Chapter 19 tells us is the maximum speed at which energy can travel. However, in nature the closest one can get to a unique frequency wave is a group of waves within a narrow frequency band. The speed at which the energy in such a group travels is always less than the speed of light.

7.6 Group velocity

Natural waveforms often consist of many waves with a small spread of frequencies, the whole making a wave group. The frequencies may be continuously distributed over a narrow band or the wave group may consist of discrete frequencies. For dispersive waves, the speeds at which the different frequencies in the wave group travel are different. The wave group changes its shape and the energy in the group travels at a speed that depends on the way in which the phase velocity varies with frequency. The speed at which the energy travels is called the **group velocity** and is considered in this section.

A group of waves susceptible to simple mathematical treatment is the one shown in Fig 7.12 consisting of two waves moving in the same direction with slightly different angular frequencies, $\omega - \delta\omega$ and $\omega + \delta\omega$, and thus slightly different wavenumbers $k - \delta k$ and $k + \delta k$. The resultant wave is the sum of the two.

For simplicity, let us assume that the two waves have equal amplitude y_0. The sum becomes

$$y = y_0 \cos\{(\omega - \delta\omega)t - (k - \delta k)z\} + y_0 \cos\{(\omega + \delta\omega)t - (k + \delta k)z\}.$$
$$(7.28)$$

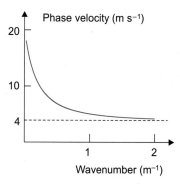

Phase velocity (m s^{-1})

Fig. 7.11 The variation of the phase velocity of waves on a stiff string with wavenumber k. The string has $F/\rho = 16\,\mathrm{m^2\,s^{-2}}$ and $\beta = 16\,\mathrm{s^{-2}}$. At low frequencies the phase velocity becomes very large. As the frequency increases, the phase velocity tends towards the value $(\omega/k) = \sqrt{F/\rho} = 4\,\mathrm{ms^{-1}}$ appropriate for a non-stiff string.

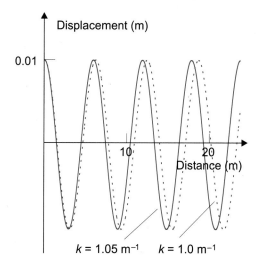

Fig. 7.12 The displacements in two forward-going waves with amplitudes both equal to $10^{-2}\,\mathrm{m}$ but slightly different frequency and wavelength are shown as a function of distance at the fixed time t equal to zero. One wave shown has wavenumber $1.00\,\mathrm{m^{-1}}$ and the other $1.05\,\mathrm{m^{-1}}$.

This equation can be rearranged with the aid of eqn (20.39), which may be written as

$$\cos(A + B) + \cos(A - B) = 2\cos A \cos B.$$

Using this identity, eqn (7.28) reduces to

$$y = 2y_0 \cos(\omega t - kz) \cos(\delta\omega t - \delta kz). \tag{7.29}$$

This represents a wave with frequency ω and wavenumber k equal to the respective means of the two waves making up the group. However, these waves are contained within an envelope given by the more slowly varying function

$$2y_0 \cos(\delta\omega t - \delta kz). \tag{7.30}$$

This function itself oscillates; at a fixed point it oscillates in time with the angular frequency $\delta\omega$ much lower than ω, and at a fixed time its variation with distance is given by the wavenumber δk much lower than k.

Figure 7.13 illustrates how the rapid oscillations are contained within the envelope given by eqn (7.30). From the form of eqn (7.30) it can be seen that the maxima of this envelope move forward with velocity $\delta\omega/\delta k$. This speed is the group velocity of the particular group of waves described by eqn (7.28). Equation (7.30) also shows that the maxima of the envelope at any time are separated by a distance $2\pi/\delta k$. If the wave group consists of a two very close frequencies, $\delta\omega/\delta k$ becomes $d\omega/dk$ and the

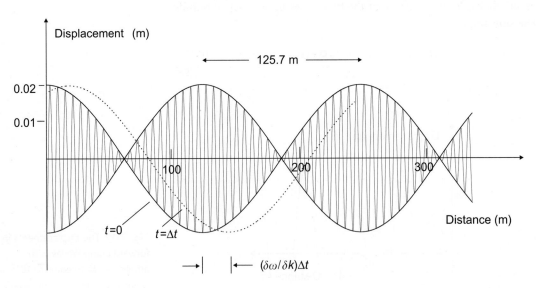

Fig. 7.13 The wave resulting from the addition of the two waves shown in Fig 7.12 is drawn as a function of distance at $t = 0$. At a short time Δt after $t = 0$ the maximum of the envelope within which the oscillations occur, shown dotted, has moved forward by an amount $(\delta\omega/\delta k)\Delta t$.

The distance between successive maxima of the envelope is 2π divided by the difference in wavenumbers of the two waves. In the example used here the distance beween successive maxima equals $(2\pi/0.05)$ m, about 125 m.

group velocity is

● *The expression for group velocity*

$$v_g = \frac{d\omega}{dk}. \tag{7.31}$$

It can be shown that, for the wave group discussed above consisting of two waves with closely spaced frequencies on a string, the energy in the group moves forward with the speed of the maxima of the envelope, namely, at the group velocity. The proof follows the same steps used in Section 7.3 to derive the speed of energy flow in a harmonic wave.

● *The energy in a group travels at the group velocity*

The result (7.31) for the group velocity, and the fact that the energy in a narrow band of waves travels with the group velocity, remain true for more complicated wave groups on a string made up of many waves within a narrow band of frequency. It is also true for other types of waves that occur in nature, and

The group velocity gives the speed of propagation of the energy in any narrow group of waves.

Calculation of group velocity

Returning to the stiff string, and using the dispersion relation (7.26), the group velocity can be calculated to be

$$v_g = \frac{Fk}{\rho} \frac{1}{\sqrt{\dfrac{Fk^2}{\rho} + \beta}}. \tag{7.32}$$

The group velocity tends to zero as k tends to zero, and the product of phase and group velocities is equal to F/ρ, the square of the constant phase velocity that waves would have on strings with no stiffness.

Worked Example 7.3 The dispersion relation for a string with no stiffness that carries beads of mass m, each separated by a distance a and with uniform tension F, neglecting the mass per unit length of the string, is

$$\omega = 2\left(\frac{F}{ma}\right)^{1/2} \sin(ka/2).$$

Deduce expressions for the phase and group velocities for long wavelengths. You will need to use the fact that $\sin x/x$ tends to unity as x tends to zero.

Answer The phase velocity is

$$v_p = \frac{\omega}{k} = 2\left(\frac{F}{ma}\right)^{1/2} \frac{\sin(ka/2)}{(ka/2)}\left(\frac{a}{2}\right).$$

As the wavelength becomes very large the wavenumber k tends to zero and $\sin(ka/2)/(ka/2)$ tends to unity, giving the result that the phase velocity tends to $(Fa/m)^{1/2}$.

The group velocity is

$$v_{\mathrm{g}} = \frac{\mathrm{d}\omega}{\mathrm{d}k} = 2\left(\frac{F}{ma}\right)^{1/2} \cos(ka/2)\left(\frac{a}{2}\right).$$

As k tends to zero the cosine term tends to unity and again v_{g} tends to the value $(Fa/m)^{1/2}$.

When a wave group has a broad spread of frequencies, or when the frequency varies very rapidly with wavenumber so that the approximation $\delta\omega/\delta k = \mathrm{d}\omega/\mathrm{d}k$ made in deriving eqn (7.31) is no longer valid, the concept of group velocity is no longer relevant. The behaviour of the resultant disturbance is complicated, as is the energy flow, and the speed at which energy propagates is no longer given by eqn (7.31).

7.7 Standing waves

In practice waves propagate on strings with finite length. If a continuous wave is begun at a certain time at one end, after a sufficient delay depending on the speed of the wave, there will be a wave reflected off the far end, unless something special is done to absorb all the energy. The resultant wave will be a combination of one going forward and one coming back. The two waves in opposite directions superimpose and combine to form what is called a **standing wave pattern** on the string.

Usually part of the forward-going energy is absorbed and the rest reflected, but in special circumstances all the energy is reflected and the backward-going wave has the same amplitude as the forward-going wave. For simplicity, let us consider this case, for which the resultant wave amplitude is

$$y = y_0 \cos(\omega t - kz) + y_0 \cos(\omega t + kz + \phi) \tag{7.33}$$

where the phase angle ϕ has been added to the reflected wave to account for a possible change of phase when the wave is reflected.

● *The sum of incident and reflected waves*

In the same way as with eqn (7.28), this expression for the resultant wave can be rearranged with the aid of eqn (20.39) to give

$$y = 2y_0 \cos(\omega t + \phi/2) \cos(kz + \phi/2). \tag{7.34}$$

This is not a progressive wave. There is no wave crest moving forward and carrying energy with it. The variables z and t are separated, and at any point z the string oscillates up and down with period $2\pi/\omega$ and with amplitude that depends on z and varies between zero and $2y_0$. This is shown in Fig 7.14.

Nodes and antinodes

At certain points such as the one labelled a on the string shown in Fig 7.14 the amplitude is always zero. These points are called **nodes** and occur when $\cos(kz + \phi/2)$ is zero, i.e. at points z given by

$$z = \frac{1}{k}\left(\left(n - \frac{1}{2}\right)\pi - \frac{\phi}{2}\right) \tag{7.35}$$

where n is a nonzero integer: $n = 1, 2, 3, \ldots$. The nodes are separated by a distance equal to π/k which is equal to half a wavelength. Points where the amplitude is a maximum are called **antinodes**.

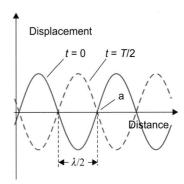

Fig. 7.14 Part of a standing wave pattern produced on a long string by the superposition of a forward-going wave and a reflected wave of equal amplitude and frequency. The full line gives the shape of the string at $t = 0$; the dotted line the shape at $t = T/2$, half a period later. At time $t = T/4$ the string has its undisturbed shape.

Exercise 7.4 The speed of waves on a stretched string is $2.0\,\mathrm{m\,s^{-1}}$. Standing waves are set up on the string with nodes 4.0 cm apart. What is the frequency of the waves? How many times per second does the string have its undisturbed shape and assume a straight line form? Contrast this with a progressive wave being transmitted on the string, when the whole string never has its undisturbed shape.

Answer The frequency is equal to 25 Hz; the string is straight twice every period of oscillation, i.e. 50 times per second.

The value of the phase change ϕ on reflection depends upon the motion of the string at the end of the string where the wave is reflected. If the end is fixed so that it can not move, there must be a node there. Other nodes occur at $\lambda/2$, λ, $3\lambda/2$, $\ldots$ from the end.

● *A string fixed at both ends*

If both ends of a string of length ℓ are fixed, at $z = 0$ and $z = \ell$, any wave on the string must have nodes at both ends. For this to be possible the length ℓ must be an integral number of half-wavelengths, since each node is separated by half a wavelength. Thus, only certain wavelengths, λ_n, given by

● *Allowed wavelengths on a fixed string*

$$\ell = \tfrac{1}{2}n\lambda_n, \tag{7.36}$$

where n is an integer, can exist on the length ℓ. The spatial part of expression (7.34) for a standing wave on a string fixed at both ends is thus $\sin(n\pi z/\ell)$, or equivalently $\sin(2\pi z/\lambda_n)$. The displacements of a string for the standing waves with n equal to 1, 2, and 3 are shown in Fig 7.15.

The fact that only certain wavelengths determined by the string length are allowed implies that the only standing waves that can exist on the length ℓ have frequencies ν_n given by

● *Allowed frequencies on a fixed string*

$$\nu_n = n\frac{\nu_\mathrm{p}}{2\ell} \tag{7.37}$$

where ν_p is the phase velocity of travelling waves on the string, assumed for simplicity to be constant. The time part of expression (7.34) for a standing wave on a string fixed at both ends is thus $\cos(2\pi\nu_n t)$ or equivalently $\cos(\omega_n t)$, where $\omega_n = 2\pi\nu_n$ and we have put the phase angle ϕ equal to zero.

If a string of length ℓ with its ends fixed at $z = 0$ and $z = \ell$ supports a standing wave with wavelength λ_n, eqn (7.34) for the wave reduces to

$$y = 2y_0 \cos(\omega_n t) \sin(2\pi z/\lambda_n). \tag{7.38}$$

If the fixed string can support all possible standing waves, the expression for the disturbance of the string becomes

$$y = \sum_{n=1}^{n=\infty} a_n(\cos(\omega_n t) \sin(2\pi z/\lambda_n)) \tag{7.39}$$

where the coefficients a_n are the amplitudes of the standing wave characterized by the integer n.

String instruments

Standing wave patterns on strings, and other standing wave patterns, are often met in real life. For example, musical notes are made with guitars and violins or other string instruments by setting up standing waves on metal strings fixed at both ends.

A guitar, when set in motion by plucking a string of fixed length, as in Fig 7.16, will support a complicated set of standing waves involving many different frequencies allowed by eqn (7.37). The lowest frequency, with $n = 1$, is called the *fundamental*. Standing waves with increasing values of

● *A general expression for a standing wave on a fixed string*

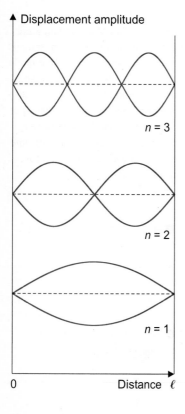

Displacement amplitude

$n = 3$

$n = 2$

$n = 1$

0   Distance ℓ

Fig. 7.15 The lowest three standing wave patterns on a string of length ℓ fixed at both ends.

Fig. 7.16 A guitar is made of six strings of different mass per unit length. The tensions in the strings and their masses per unit length are chosen to cover the required frequency range.

the integer n give *harmonics* at frequencies that are integral multiples of the frequency of the fundamental. However, the amplitude of the fundamental wave usually dominates. The waves on the string set the guitar body in motion, the air around the instrument is disturbed, and waves corresponding to compressions and rarefactions of the air travel outward to a listener.

If the only wave is that with the fundamental frequency, the listener hears a pure tone of the single frequency of the fundamental, as long as the frequency lies within the range 20 to 1.5×10^4 Hz, to which the human ear is normally sensitive. Harmonics are usually present, however, and the musical quality of the sound heard depends on the mix of harmonics and their relative amplitudes. The exact mix of frequencies is determined by how the string is set in motion by the guitar player and the construction of the guitar. Sound waves are discussed again in the next chapter.

If the initial shape of an isolated string is defined, the techniques of Fourier analysis discussed in Section 6.6 can be used to determine the relative amplitudes of the harmonics (see Problem 7.23).

Worked Example 7.4 A guitar string has a length of 60 cm and mass per unit length 5×10^{-4} kg m^{-1}. What should be the tension in the string if, when plucked, it is to produce a middle C fundamental note of frequency 262 Hz?

Answer Equation (7.37) with $n = 1$, $\nu = 262$ Hz, and $\ell = 60$ cm requires the speed of the waves to be 314.4 m s^{-1}. With a mass per unit length $\rho = 5 \times 10^{-4}$ kg m^{-1}, eqn (7.15) then requires the tension F to be about 49.4 N.

The standing waves discussed above have amplitudes that reduce to zero at nodes as a result of all the wave energy being reflected at either end. Often only part of the energy of a travelling wave is reflected at the termination of a string, or the wave is only partly reflected by other reflecting mechanisms, such as a join where there is a change in the mass per unit length of the string. The wave on the first string is now the sum of two waves, one being the original travelling wave and the other a reflected wave that has smaller amplitude and usually suffers a phase change ϕ on reflection, as written in eqn (7.33).

The resultant wave in these situations consists of a pattern similar to that shown in Fig 7.17. The amplitude of the disturbance on the string is not constant as in a progressive wave, nor does it ever reduce to zero as in the standing waves discussed above, but varies with position z on the string with maxima (and minima) separated by $\lambda/2$ as before. The

● *Partial reflection*

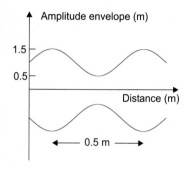

Fig. 7.17 The envelope defining the amplitude of oscillation of a string along its length when an incident wave of amplitude 0.1 m and wavenumber 1 m^{-1} is reflected with no phase change. The reflected wave has amplitude 0.05 m corresponding to a standing wave ratio of 3.

standing wave ratio of such a wave is the ratio of the displacement of the string at the maxima to the displacement at the minima.

Worked Example 7.5 A single-frequency wave on a string is a combination of one travelling to the right with amplitude y_0 and one to the left with amplitude αy_0. The backward-going wave also has an additional phase of ϕ with respect to the forward-going wave, and the wave has the form of eqn (7.33) appropriately modified by the amplitude of the second term becoming αy_0. Show that the resulting wave is given by

$$y = 2\alpha y_0 \left(\cos(\omega t + \phi/2) \cos(kz + \phi/2) \right) + (1 - \alpha) y_0 \cos(\omega t - kz),$$

and that the standing wave ratio is equal to

$$\frac{1 + \alpha}{1 - \alpha}.$$

Answer The resultant wave may be rewritten using the identity for the sum of two cosine terms. Equation (20.39) may be used to show that

$$\cos(A + B) + \cos(A - B) = 2 \cos A \cos B.$$

Hence

$$
\begin{aligned}
y &= y_0 \cos(\omega t - kz) + \alpha y_0 \cos(\omega t + kz + \phi) \\
&= (1 - \alpha) y_0 \cos(\omega t - kz) + \alpha y_0 \cos(\omega t + \phi/2 - kz - \phi/2) \\
&\quad + \alpha y_0 \cos(\omega t + \phi/2 + kz + \phi/2), \\
&= (1 - \alpha) y_0 \cos(\omega t - kz) + 2\alpha y_0 \cos(\omega t + \phi/2) \cos(kz + \phi/2).
\end{aligned}
$$

This represents a progressive wave of amplitude $(1 - \alpha) y_0$ and a standing wave that oscillates in time with angular frequency ω and with an amplitude that varies with position z as $2\alpha y_0 \cos(kz + \phi/2)$. The minimum amplitude $y_{\min}$ of the resultant is at a node of the standing wave and is $(1 - \alpha) y_0$. The maximum amplitude of the resultant is at an antinode of the standing wave and is $y_{\max} = (1 - \alpha) y_0 + 2\alpha y_0$. The standing wave ratio is thus

$$\frac{y_{\max}}{y_{\min}} = \frac{1 + \alpha}{1 - \alpha}.$$

7.8 Attenuation of waves

The vibrations of the air produced by the vibrating guitar string discussed in Section 7.7 remove small amounts of energy from the standing waves on the string. Some part of this energy reaches the listener's ear and causes the eardrum to vibrate. The small loss of energy to air vibrations and losses from other mechanisms reduce the amplitude of oscillation of the standing wave on the string as time goes on. The expression (7.34) for the standing waves takes no account of this effect. Similarly, there are energy losses when progressive waves travel on strings and the amplitudes of these waves decrease with distance from the source.

Energy loss has to be put into the wave equation for a realistic wave. The average rate of energy loss per unit length in a small section of string is often approximately proportional to the average energy per unit length at the position of the section. Let us use a model of a realistic string in which the string has no stiffness but in which there is a dissipative force. Let this force on an infinitesimal length dz be proportional to dz and to the vertical speed of the string. Equation (7.6) becomes

$$\rho \, dz \frac{\partial^2 y}{\partial t^2} = F \, dz \left(\frac{\partial^2 y}{\partial z^2} \right) - b \, dz \frac{\partial y}{\partial t} \tag{7.40}$$

where b is a constant depending on the strength of the dissipative force.

This approach to attenuation is analogous to the treatment of the damped simple, harmonic oscillator in Section 6.2, where the parameter b plays a similar role to that in eqn (7.40). With the addition of the dissipative force the wave eqn (7.7) becomes

$$\frac{\partial^2 y}{\partial t^2} = \frac{F}{\rho} \left(\frac{\partial^2 y}{\partial z^2} \right) - \gamma \frac{\partial y}{\partial t} \tag{7.41}$$

with $\gamma = b/\rho$.

In looking for a solution to this equation we consider a quantity that decays with distance such that its rate of decrease with distance at any point is proportional to its value at that point. As an example we determine how energy E decays with distance z when the energy loss $-dE$ over an infinitesimal element of length dz at position z is proportional to $E \, dz$. In that situation

● *Exponential decay in general*

$$-\frac{dE}{dz} = CE \tag{7.42}$$

where C is a constant. The solution to this equation is

$$E = E_0 \exp(-Cz) \tag{7.43}$$

where E_0 is the value of E at $z = 0$, and the energy decays exponentially with distance.

The energy in a wave is proportional to the square of the wave amplitude, as shown in eqn (7.19), and in view of the result (7.43) we expect the solution to eqn (7.41) to be a wave of the form of eqn (7.8) modified by a factor that makes the amplitude fall off exponentially with distance. With this in mind we try

$$y = y_0 \exp(-\alpha z) \cos(\omega t - kz) \tag{7.44}$$

as a solution, with α a constant. After working out the appropriate partial derivatives and substituting into eqn (7.41) we find that expression (7.44) does indeed satisfy the equation as long as

$$\omega^2 = (k^2 - \alpha^2)\frac{F}{\rho} \tag{7.45}$$

and

$$2\alpha k\left(\frac{F}{\rho}\right) = \omega\gamma. \tag{7.46}$$

If the parameter α is very much less than k, the above two equations reduce to

$$\frac{\omega}{k} = \left(\frac{F}{\rho}\right)^{1/2},$$

the same as without the damping force, and

$$\alpha = \frac{\gamma}{2}\left(\frac{F}{\rho}\right)^{-1/2}. \tag{7.47}$$

Exercise 7.5 A taut string such as might be used in a guitar has a tension $F = 50$ newtons and mass per unit length $\rho = 5 \times 10^{-4}\,\mathrm{kg\,m^{-1}}$. Determine the distance over which the amplitude of a progressive wave of frequency 262 Hz decreases by a factor of two if $\alpha/k = 10^{-2}$.

Answer The amplitude decreases by a factor of two over a distance of 13.6 m.

The above discussion of wave attenuation follows almost exactly the ideas introduced in Section 6.2 in the discussion of damped simple harmonic motion. Successive sections of the string on which the damped wave propagates oscillate with smaller and smaller amplitude as the wave energy is dissipated by the frictional force.

Problems

Level 1

7.1 The expression $y = y_0 \cos(\omega t - kz)$ represents a wave travelling in the positive x-direction. At position z equal to $5\lambda/8$, with λ the wavelength, is the disturbance y increasing or decreasing with time at time t equal to one-half of a period?

7.2 The equation of a transverse wave travelling on a string is

$$y = -0.2 \sin(\pi(0.5x - 200t)).$$

Find the amplitude, wavelength, frequency, period, and velocity of propagation. x and y are measured in cm, t in seconds.

7.3 If the mass per unit length of the string of Problem 7.2 is 5 g per cm find the tension in the string.

7.4 For the wave given in Problem 7.2, what are the distances between points on the string whose displacements have a phase difference of 45°?

7.5 For the wave given in Problem 7.2, what are the maximum speed and acceleration of points on the string?

7.6 A travelling wave of amplitude 1 mm is excited on a long string of mass per unit length $10^{-3}\,\mathrm{kg\,m^{-1}}$ maintained under a tension of 1 newton. Calculate the energy flowing per second if the frequency of vibration is 10 Hz.

7.7 The expression

$$y = A\cos(\omega t - kz) + B\sin(\omega t - kz)$$

represents a harmonic travelling wave of period $2\pi/\omega$. Calculate its amplitude.

7.8 A travelling wave of amplitude 1.5 mm propagates on a long string of mass $10^{-3}\,\mathrm{kg\,m^{-1}}$ maintained under a tension of 1 N. Calculate the energy flowing per second if the frequency of vibration is 12 Hz.

7.9 Equation (7.44) describes an attenuated wave that has displacement y_0 at position $z = 0$ and time $t = 0$. Show that the speed of the transverse motion on the string is zero when $z = t = 0$. Show that, if a phase angle

ϕ is added to the argument of the cosine in expression (7.44) for the displacement, the resulting expression is still a solution to eqn (7.41). What are the displacements and speed appropriate to the new expression at z and t equal to zero?

7.10 Show that eqn (7.41) is a linear differential equation, and thus that if two waves that each satisfy the equation exist at the same time on the string they superpose to give a resultant that is the sum of the two.

7.11 A wave on a string consists of two waves of different amplitude travelling in opposite directions,

$$y = A\cos(\omega t - kz) + B\cos(\omega t + kz).$$

Show that this may be expressed as

$$y = (A + B)\cos\omega t \cos kz + (A - B)\sin\omega t \sin kz.$$

7.12 A disturbance on the ocean bed creates a wave group on the surface with mean period 100 s. The wave properties are determined by gravity and the wave velocity is given by

$$v_{\mathrm{p}} = \sqrt{\frac{g\lambda}{2\pi}},$$

where g is the acceleration due to gravity at the Earth's surface. Calculate the group velocity of the waves. Take the acceleration due to gravity to be $9.8\,\mathrm{m\,s^{-2}}$.

7.13 A stretched string vibrates with a frequency of 20 Hz in the fundamental mode when tied to supports 40 cm apart. The string has a mass of 0.02 kg. If transverse waves are propagated on the string, what is their speed? What is the tension in the string?

Level 2

7.14 Equation (7.8) is a solution to the wave eqn (7.7) describing a travelling wave of a single frequency and wavelength related to ω and k, respectively. By substitution it can readily be shown that any function of the variable $(\omega t - kz)$, with ω and k any constants, satisfies the wave equation.

A pulse travelling along a wire under tension is given by the equation

$$y(z, t) = y_0 e^{-(az-bt)^2/2\sigma^2}.$$

Sketch the displacement y at time zero and at time t. What is the speed of the pulse?

Calculate how the transverse speed of the string varies over the length of the pulse at any fixed time.

7.15 A plane wave of wavelength λ propagates in a dispersive medium with phase velocity given by

$$v_p = a + b\lambda$$

where a and b are constants. What is the group velocity? By considering a pair of component waves separated in wavelength by a small amount $\delta\lambda$ (as in Section 7.6) show that any impulsive waveform will reproduce its shape at times separated by intervals of $1/b$, and at distance intervals of a/b.

7.16 Determine the relationship between the group velocity and the phase velocity v_p for the following types of waves, given the variation of v_p with wavelength λ:

(a) small-wavelength waves on the surface of water controlled by surface tension, for which $v_p = a\lambda^{-1/2}$;

(b) transverse waves on a rod for, which $v_p = a\lambda^{-1}$;

(c) radio waves in an ionized gas, for which $v_p = \sqrt{(c^2 + b^2\lambda^2)}$;

(d) attenuated waves on a string, with a dispersion relation given by eqn (7.45).

In these formulae, a, b, and c are constants.

7.17 The dispersion relation for waves on real wires may be expanded in the form

$$\omega^2 = a_0^2 k^2 (1 + a_2 k^2 + a_4 k^4 + \cdots)$$

where a_n are coefficients that decrease in size as n increases, and a_0 is given by eqn (7.15). If $1 \gg a_2 k^2 \gg a_4 k^4$, find approximate expressions involving terms in k no higher than k^6 for the phase and group velocities as a function of k.

7.18 Calculate the group velocity of light of wavelength 589 nm in glass for which the wave velocity at wavelength λ (in metres) is given by

$$\frac{c}{v_p} = 1.4 + \frac{3.5 \times 10^{-14}}{\lambda^2}$$

where c is the speed of light in free space.

7.19 A person who can hear sounds of frequency up to 12 kHz listens to the vibrations produced by plucking a stretched wire of length 0.8 m. If the speed of travelling waves on a similar wire is 4 km s^{-1}, how many different frequencies will the listener hear?

7.20 Show that, if the damping is very small, the amplitude of a travelling wave of frequency ν falls by a factor of 2 over a distance equal to

$$\frac{2 \ln 2}{\gamma} \left(\frac{F}{\rho}\right)^{1/2}$$

where the symbols have the meanings given to them in Section 7.8.

7.21 A standing wave of angular frequency ω exists on a stretched string of mass ρ per unit length and tension F. The amplitude of the standing wave is A. Show that the maximum kinetic energy per unit length at an antinode equals $\rho\omega^2 A^2/2$, and the maximum potential energy per unit length at a node due to the string stretching is $Fk^2 A^2/2$.

Level 3

7.22 The dispersion relation for waves in water of depth x depends on the acceleration due to gravity g, the surface tension Γ, and the water density ρ. The dispersion relation is

$$\omega^2 = \left(\frac{g + \Gamma k^2}{\rho}\right) k \left(\frac{e^{kx} - e^{-kx}}{e^{kx} + e^{-kx}}\right).$$

Show that the effect of surface tension is important only for wavelengths of around 20 mm or less.

Considering only the limits of (a) very deep and (b) shallow water, in each case calculate the phase velocity.

Calculate the minimum phase velocity in deep water. Take $g = 9.8 \text{ m s}^{-2}$, $\rho = 10^3 \text{ kg m}^{-3}$, and $\Gamma = 0.074 \text{ N m}^{-1}$.

7.23 As discussed in Section 7.7, Fourier analysis can be used to determine the mix of allowed frequencies on a string fixed at both ends and plucked in a defined way. Suppose that at time t equal to zero the shape of the string between $z = 0$ and $z = \ell$ is given by the function $y = f(z)$. The function $f(z)$ can be written in terms of its Fourier components using eqn (6.35) as

$$f(z) = \sum_{n=0}^{n=\infty} a_n \cos(k_n z + \phi_n),$$

where the wavenumber k_n of the nth harmonic is equal to $n\pi/\ell$. The harmonic with wavenumber k_n has frequency $vk_n/2\pi$, where v is the speed of waves on the string, and the vibrating string emits tones at frequencies given by the different values of the integer n. The energies or intensities in the notes with different frequencies produced by vibrating strings in practical string instruments are roughly proportional to the squares of the coefficients a_n. The determination of the coefficients thus gives an indication of the musical quality of the sound emitted.

If a string is plucked so that initially the midpoint has deflection y_0 with the two halves descending linearly to the fixed end points, find the harmonic mix of notes on the string.

You will need to use the result

$$\int_0^b x\sin(ax)\,\mathrm{d}x = \frac{1}{a^2}\left[\sin(ax) - ax\cos(ax)\right]_0^b$$

7.24 A taut wire of length 2ℓ is plucked by drawing it sideways a small distance at a point distant z from one end and releasing it. Discuss how the amplitudes of the first few harmonics depend on z.
For what values of z is the seventh harmonic absent?

7.25 A wire of mass ρ per unit length and length ℓ is fixed at the points $x = 0$ and $x = \ell$ and has a small mass M tied to its midpoint. Show that the frequencies ν_n of the normal modes are given by solutions to the equation

$$\frac{\pi\ell\nu_n}{v}\tan\left(\frac{\pi\ell\nu_n}{v}\right) = \frac{\rho\ell}{M}$$

where v_p is the speed of transverse waves on the wire.

7.26 An amplitude-modulated function of time has the form

$$A(t)\cos(\omega t).$$

An example is an amplitude-modulated radio wave with ω the angular frequency of the carrier wave and $A(t)$ an audio signal to be converted in a radio receiver. The frequency corresponding to ω could be $\sim 1\,\mathrm{MHz}$ and the

frequencies in the signal $A(t)$ would be in the range from about $20\,\mathrm{Hz}$ to $10\,\mathrm{kHz}$.

Suppose $A(t)$ is an even periodic function with a period $2\pi/\omega_0$ which is much greater than $2\pi/\omega$. Assume $\omega = N\omega_0$, with N a large integer. If the coefficients in the Fourier series expansion of $A(t)$ are a_n,

$$A(t) = \sum_{n=1}^{n=\infty} a_n\cos(n\omega_0 t).$$

Determine the coefficients a_m in the Fourier expansion of the modulated wave. All but the first few coefficients a_n in the expansion of $A(t)$ can be neglected.

7.27 An elastic string is stretched between two points separated by a distance ℓ. The tension in the string is F. Show that, for small transverse displacements of the string from equilibrium, the potential energy of a small element of the string of length δz situated a distance z from one end of the string is given by

$$\frac{1}{2}F\delta z\left(\frac{\partial\psi}{\partial z}\right)^2$$

where ψ is the transverse displacement at the point z.

Hence show that, if the string vibrates in its nth normal mode with amplitude a_n, the total energy of the string is

$$\frac{1}{4}n^2\pi^2\frac{F}{\ell}a_n^2.$$

Suppose the string vibrates so that

$$\psi(z, t) = a_1\sin\frac{\pi z}{\ell}\cos\frac{\pi t}{T} + a_2\sin\frac{2\pi z}{\ell}\cos\frac{2\pi t}{T}$$

where T is the period corresponding to a wavelength 2ℓ. Show that the total energy is now

$$\frac{1}{4}\pi^2\frac{T}{\ell}(a_1^2 + 4a_2^2),$$

i.e. the total energy is equal to the sum of the energies of each mode considered independently.

Some solutions and answers to Chapter 7 problems

7.2 Amplitude $= 0.002$ m; wavelength $= 0.04$ m; frequency $= 100$ Hz; period $= 0.01$ s; speed $= 4$ m s^{-1}.

7.3 Speed $v_p = \sqrt{F/\rho}$ and the tension $F = \rho v^2 = 8$ N.

7.5 Maximum speed $= 1.26$ m s^{-1}; maximum acceleration $= 790$ m s^{-2}.

7.6 Equation (7.23) gives the average energy flowing per second as

$$\bar{P} = \tfrac{1}{2} Fk\omega y_0^2.$$

Evaluation of this gives $\bar{P} = 6.24 \times 10^{-2}$ W.

7.9 The displacement is $y = y_0 \cos\phi$ and the speed is $\mathrm{d}y/\mathrm{d}t = -\omega y_0 \sin\phi$.

7.12

$$v_p = \frac{\omega}{k} = \sqrt{\frac{g\lambda}{2\pi}}.$$

The wavenumber k is $2\pi/\lambda$; hence

$$\omega = \sqrt{gk}$$

and

$$v_g = \frac{\mathrm{d}\omega}{\mathrm{d}k} = \frac{1}{2}\sqrt{\frac{g}{k}} = \frac{1}{2}\sqrt{\frac{g\lambda}{2\pi}}.$$

Since $\omega = \sqrt{gk}$, $\sqrt{k} = \omega/\sqrt{g}$ and

$$v_g = \frac{g}{2\omega} = \frac{gT}{4\pi},$$

giving $v_g = 78.0$ m s^{-1}.

7.14 As time progresses the pulse traves in the positive z-direction unchanged in form with speed b/a.

The transverse speed of the string, $\mathrm{d}y/\mathrm{d}t$, is

$$\frac{\mathrm{d}y}{\mathrm{d}t} = \frac{2b(az - bt)}{2\sigma^2} y.$$

At fixed time, say $t = 0$,

$$\frac{\mathrm{d}y}{\mathrm{d}t} = \left(\frac{baz}{\sigma^2}\right) \exp\left\{-\frac{a^2 z^2}{2\sigma^2}\right\}.$$

This is a skewed distribution, peaked slightly above the centre of the pulse and on average higher on the forward side than on the trailing side.

7.16 (a) $v_g = 3v_p/2$; (b) $v_g = 2v_p$; (c) $v_g = v_p(1 - (b_v^2\lambda_v^2/v_p^2))$; (d) $v_p v_g = F/\rho$.

7.22 In deep water $e^{-kx} \to 0$, and

$$\omega^2 = k\left(g + \frac{\Gamma k^2}{\rho}\right).$$

In very shallow water $kx \ll 1$, and

$$\left(\frac{e^{kx} - e^{-kx}}{e^{kx} + e^{-kx}}\right) \to kx$$

giving

$$\omega^2 = xk^2\left(g + \frac{\Gamma k^2}{\rho}\right).$$

The minimum phase velocity in deep water is 0.24 m s^{-1}.

7.23 This is the same problem as that done in Worked example 6.4. Position z and wavenumber k replace time and angular frequency, respectively.

7.25 The acceleration of the mass must always equal the product of M and the force on M provided by the tension in the string. The allowed standing waves on the string are given by

$$y = y_0 \sin\left(\frac{n\pi z}{\ell}\right) \cos(\omega_n t)$$

where n is an integer equal to the number of half-wavelengths in the standing wave and ω_n is the corresponding angular frequency. This may be rewritten in terms of the wavelength λ_n of the wave on the string characterized by the integer n. $\lambda_n = 2\ell/n$ and

$$y = y_0 \sin\left(\frac{2n\pi z}{\lambda_n}\right) \cos(\omega t).$$

From this we deduce the speed and acceleration of the string at its midpoint. The acceleration at $z = \ell/2$ is

$$\frac{\mathrm{d}^2 y}{\mathrm{d}t^2} = \omega_n^2 \sin\left(\frac{\pi\ell}{\lambda_n}\right) \cos(\omega_n t).$$

The standing waves are symmetrical about the midpoint of the string and the force on the mass M is $2F\mathrm{d}y/\mathrm{d}z$.

This equals the mass M times the acceleration, and

$$2Fy_0 \frac{2\pi}{\lambda_n} \cos\left(\frac{\pi\ell}{\lambda_n}\right) \cos(\omega_n t) = M \frac{d^2 y}{dt^2}.$$

Substitution of the acceleration gives the equation

$$\frac{\pi\ell}{\lambda_n} \tan\left(\frac{\pi\ell}{\lambda_n}\right) = \frac{\pi\ell}{\lambda_n} 2F\left(\frac{2\pi}{\lambda_n}\right) \frac{1}{M\omega_n^2}.$$

The speed v_p of travelling waves on the string is $v_p = \sqrt{F/\rho} = \lambda_n \omega_n / 2\pi$, and the right-hand side of the

above equation reduces to $\rho\ell/M$. Hence

$$\frac{\pi\ell}{\lambda_n} \tan\left(\frac{\pi\ell}{\lambda_n}\right) = \frac{\rho\ell}{M}$$

and

$$\frac{\pi\ell\nu_n}{v_p} \tan\left(\frac{\pi\ell\nu_n}{v_p}\right) = \frac{\rho\ell}{M}$$

where ν_n is the frequency of the mode characterized by n.

7.26 The coefficients a_m are closely grouped around a_N.

Chapter 8

Wave phenomena

This chapter discusses some common phenomena arising from different wave motions.

In the last chapter we outlined the mathematical formulae used to describe waves. Now that the formalism is in place, a good answer to the question 'What is a wave?' would be that it is a physical phenomenon that can be described to a good approximation by the mathematical treatment given in Chapter 7. There, the example of waves on strings was used to introduce the formalism and to determine mathematical expressions for waves on stretched strings or wires. In this chapter we discuss some aspects of other waves commonly met in the physical world.

The most obvious waves to which the senses of human beings respond are sound waves and light waves. Unlike waves on strings, these waves travel in three dimensions, and hence their mathematical description is more complicated than for waves on strings. We will not pursue the derivation and solution of wave equations that involve three spatial coordinates as well as time. Many of the important features of sound and light waves can be discussed and understood without recourse to a full mathematical treatment, and that is the approach adopted here.

Some waves require the presence of matter for their propagation. For these, the vibration of the microscopic constituents of the medium is the mechanism for the transfer of energy from one point to another. Transverse waves on strings, considered in the last chapter, are an obvious example. Sound waves, which correspond to longitudinal compressions and rarefactions of small sections of air or other media in which the sound is travelling, are another example. Electromagnetic waves, such as radio waves or light, and gravitational waves do not require the presence of a medium for their propagation, although they can transfer energy through media. Until about a century ago it was difficult for scientists to think in terms of waves and energy progressing in empty space. This reluctance was overcome when experiments were unable to detect the aether, the all-pervading medium through which it was thought light travelled. Nowadays, it is easily accepted that electromagnetic waves can be transmitted through a void.

● *Some waves need a medium for their propagation*

We are familiar with many of the wave phenomena treated in this chapter. We commonly experience the energy in reflected waves. Sound waves, light waves, and all waves suffer reflections when they meet a discontinuity in their path: sound waves reflected from the opposite side of a valley give echoes, and glare is produced when sunlight is reflected off windows and roads. Another familiar observation is that the whistle of a train has a higher pitch (frequency) when the train is approaching than when the train is receding. This is an example of the Doppler effect: the observed frequency depends on the relative motion of source and observer.

Waves superpose, and we have already discussed superposition of standing waves on strings in Section 7.7. Standing waves on strings are specific examples of the general phenomenon called interference, which happens whenever waves of the same frequency but with phases that have a fixed difference are added together. Interference is responsible for the colours seen when looking at an oil slick on a puddle, and for the beats heard when two tuning forks emitting slightly different frequencies are struck at the same time.

Another pattern of wave behaviour whose effects are familiar to us is that of diffraction. Sound waves travel round corners, as do radio waves, otherwise we would not be able to receive messages on mobile phones. Light waves are also diffracted and go round the edge of an obstacle. However, the extent to which light can turn corners is very small and for many purposes light can be considered to travel in straight lines.

8.1 Sound waves

The standing waves on a guitar string set up oscillations of the guitar, sending disturbances through the surrounding air. These propagate outwards and reach a listener's ear. There, a membrane vibrates in unison with the oscillations of the incoming air and gives the sensation of sound to the brain. In this section we discuss sound waves.

Sound waves require the presence of a medium for their propagation. If there were nothing between the guitar and you, you would hear nothing. Sound waves are **longitudinal waves**; they correspond to very small sections of the medium oscillating between compressions and rarefaction in the direction of travel of the waves. This is in contrast with the transverse waves on strings discussed in Chapter 7. In a gas, the progress of a sound wave may be pictured by imagining very small sections containing fixed numbers of molecules. These sections suffer compressions to slightly smaller volumes than occupied in the undisturbed state, followed by expansions to volumes slightly greater. A sound wave travelling down a long metal rod may be pictured in the same way. The two ends of a very small section cut perpendicular to the axis of the rod oscillate in the axial direction and give rise to compressions and expansions travelling down the rod. This is the model used below in the derivation of the wave equation.

● Sound waves are longitudinal waves

Sound waves in a metal rod

The problem of sound travelling down a long rod whose axis we take to be the z-axis can be treated to a good approximation in one dimension. Figure 8.1 shows part of such a rod and a section containing a small number of metal atoms which has length δz when the rod is undisturbed with no wave down it. With the rod undisturbed the section is situated between points distant z and $z + \delta z$ from the end of the rod, taken to be at $z = 0$.

When a sound wave moves down the rod, each end of the section is displaced in the z-direction under the influence of forces that are varying with both distance and time. We introduce a coordinate η that measures the displacement of the section from its undisturbed position. At a time t the left-hand end is at position $z + \eta$ and the right-hand end is at position $z + \delta z + \eta + \delta \eta$. The bottom part of Fig 8.1 shows the ends of the section when it is displaced from its undisturbed position. The volume or section containing a fixed number of atoms has expanded while a nearby volume containing a similar number of atoms has compressed. The expansion arises because the force on the left-hand edge is different from the force on the right-hand edge, giving a resultant force on the section.

The movement η and extension $\delta \eta$ are, in fact, very small compared with the width δz of the section; Fig 8.1 is drawn out of proportion for clarity. The force on the left-hand edge of the small volume being considered is the force $F(z)$ at distance z along the rod; the force on the right-hand edge is the force $F(z + \delta z)$ at distance $(z + \delta z)$ along the rod. The forces arise because a length δz has suffered an extension $\delta \eta$ under the influence of the tension in the rod. We can derive the net force on the section by considering the elastic properties of metal wires or rods.

An unstretched length ℓ of a metal wire or rod, when experiencing a force F stretches by an amount $\Delta \ell$ as shown in Fig 8.2. For small extensions the cross-sectional area S of the rod may be considered to remain constant and the fractional extension $\Delta \ell / \ell$ is proportional to the stress in the rod, given by the stretching force per unit area, F/S.

$$\frac{\Delta \ell}{\ell} \propto \frac{F}{S}.$$

The constant of proportionality is a property of the metal that depends upon the atomic constitution of the metal. We may write

$$\frac{\Delta \ell}{\ell} = \frac{F}{SY} \tag{8.1}$$

where Y is called **Young's modulus** for the metal.

A typical value of Young's modulus for a metal is about $10^{11}\,\mathrm{N\,m^{-2}}$. For steel, $Y = 2 \times 10^{11}\,\mathrm{N\,m^{-2}}$, so that a force of $10^4\,\mathrm{N}$, close to the weight of a mass of 1 kg, will stretch a steel wire of cross-sectional area $2\,\mathrm{mm^2}$ by about 2.5%.

We now assume that eqn (8.1), which applies to the static situation of a stretched wire, also applies to the rod down which a sound wave travels. If S is the cross-sectional area of the rod, eqn (8.1) gives

$$F(z) = SY \frac{\delta \eta}{\delta z}. \tag{8.2}$$

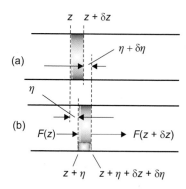

Fig. 8.1 (a) shows a small section of width δz at position z along an undisturbed solid rod or wire. (b) shows the section at time t, displaced to $z + \eta$ and with width changed to $\delta z + \delta \eta$ as longitudinal pressure variations are being propagated. The section changes its width and moves its position as the pressure oscillates with time.

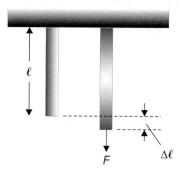

Fig. 8.2 The length ℓ of an unstretched rod or wire of cross-sectional area S increases by an amount $\Delta \ell = \ell F / YS$ when stretched by a force F. The parameter Y is a property of the material of the wire called Young's modulus.

As the section δz is made smaller and smaller, $\delta\eta/\delta z$ becomes the partial differential $\partial\eta/\partial z$ and, at the distance $(z + \delta z)$,

$$F(z + \delta z) = F(z) + SY\frac{\partial^2\eta}{\partial z^2}\delta z \tag{8.3}$$

where the second partial differential $\partial^2\eta/\partial z^2$ is evaluated at distance z. There is thus a net force on the small section equal to the difference of the forces given by eqns (8.3) and (8.2) and this force gives the section an acceleration $\partial^2\eta/\partial t^2$. If the density of the metal is ρ, the mass of the section is $\rho S\delta z$ and Newton's law tells us that

$$\rho S\delta z\frac{\partial^2\eta}{\partial t^2} = SY\frac{\partial^2\eta}{\partial z^2}\delta z. \tag{8.4}$$

● *The wave equation for sound in a rod*

This reduces to the equation

$$\frac{\partial^2\eta}{\partial t^2} = \frac{Y}{\rho}\frac{\partial^2\eta}{\partial z^2}, \tag{8.5}$$

which is the wave equation indicating how the displacement η varies with z and t.

Equation (8.5) is a similar equation to eqn (7.7) for waves on strings, and so has solutions of the form of eqn (7.8) for waves of a single frequency moving down the rod. The wave velocity of sound waves that obey eqn (8.5) is independent of frequency and is given by

● *The speed of sound in a metal rod*

$$v_p = \frac{\omega}{k} = \sqrt{\frac{Y}{\rho}}. \tag{8.6}$$

This equation is similar in form to eqn (7.15). The tension in the string is replaced by Young's modulus for the metal rod, and the mass per unit lengh of the string is replaced by the density of the metal.

Equation (8.6) is a good approximation for thin rods. Since it takes very large pressures to compress a metal, the cross-sectional area of a small section changes as the width changes. The degree to which a stretched wire or rod contracts in area as it is stretched is another property characteristic of a metal and is discussed in Section 14.4. It must be taken into account when sound waves are propagated in bulk media, when it typically changes the speed of the waves by a few per cent compared with the value predicted by eqn (8.6).

Exercise 8.1 Young's modulus for steel is $2 \times 10^{11}\,\mathrm{N\,m^{-2}}$ and the density of steel is $8 \times 10^3\,\mathrm{kg\,m^{-3}}$. Estimate the speed of sound in a steel rod.

Answer The speed is predicted from eqn (8.6) to be $5\,\mathrm{km\,s^{-1}}$.

Sound waves in air

A similar analysis to that given above for one-dimensional longitudinal waves in a metal rod can be made for the vibrations of air in pipes, provided the diameter of the pipe is small compared with its length. The sound wave now consists of density variations of small masses of gas as they are successively compressed and expanded. We will use the variable η to describe the departure of the gas density at position z and time t from its value when no disturbance is passing through the gas and derive a wave equation describing how η varies with time and position.

When deriving the wave equation for η, the resistance the gas presents to changes in its density takes on a similar role to that played by the resistance of the rod to stretching, and the density ρ of the gas at the mean pressure takes on the role of the metal density. The gas resistance is measured by the compressibility χ, which is the fractional decrease in volume V of the gas per unit pressure increase:

$$\chi = -\frac{1}{V}\left(\frac{\partial V}{\partial p}\right) \tag{8.7}$$

where p is the gas pressure. The value of the compressibility depends upon several factors, including the temperature and pressure of the gas. For now we take χ to be a compressibility whose value is given by observations on sound waves in air. With the introduction of χ, dimensional arguments tell us that the wave equation is

● *The wave equation for sound in air*

$$\frac{\partial^2 \eta}{\partial t^2} = \frac{1}{\rho\chi}\frac{\partial^2 \eta}{\partial z^2}, \tag{8.8}$$

since the constant on the right-hand side of the wave equation that multiplies the term $\partial^2\eta/\partial z^2$ must have the dimensions of the square of a velocity.

The wave velocity of sound is thus given by

● *The speed of sound in air*

$$v_p = \sqrt{\frac{1}{\rho\chi}}. \tag{8.9}$$

Sound waves suffer very little dispersion in air, and the speed of sound in air at sea level and the average temperate-zone temperature of 20 degrees centigrade is about $340\,\mathrm{m\,s^{-1}}$. Using a value of the density ρ equal to $1.29\,\mathrm{kg\,m^{-3}}$, we obtain a compressibility $\chi = 6.7 \times 10^{-6}\,\mathrm{m^2\,N^{-1}}$.

8.2 Electromagnetic waves

Electric and magnetic fields are discussed in Part 5 of this book. Here we indicate briefly how they arise in order to outline some properties of electromagnetic waves, which together with sound waves are the waves commonly experienced in normal life. An **electric field** is produced by electric charges: if a charge is present there is an associated electric field that produces forces on other charges. A **magnetic field** is produced by moving charges. Charges in uniform motion constitute a steady electric current and this current gives rise to a steady magnetic field. A magnetic field produces forces on other moving charges. Both electric and magnetic fields are vectors. The forces they impose upon a charge are determined by the magnitude of the charge and its velocity, and these forces are discussed in some detail in Part 5.

There may also be an electric field produced in the presence of changing magnetic fields, and a magnetic field produced in the presence of changing electric fields. This allows rapidly varying, related electric and magnetic fields, which have energy, to propagate outwards from their sources at a high speed characteristic of the medium in which the **electromagnetic waves** are being propagated.

Radio waves and the waves that bring television to our receivers are electromagnetic waves. The radar waves that detect planes and ships are electromagnetic waves. A hot object emits infrared waves, which produce the sensation of heat. Visible and ultraviolet light are electromagnetic waves, as are the X-rays that are used to image our bones in medical diagnosis.

● *The electromagnetic spectrum*

Electromagnetic waves are ubiquitous. Light, radar, and the other examples quoted all correspond to different intervals of frequency of the waves. The waves met in various situations are given different names according to the numerical value of the frequency interval they occupy: radio waves have frequencies around $10^4 - 10^8$ Hz; radar waves around 10^{10} Hz; light waves around 10^{14} Hz, and X-rays frequencies around $10^{17} - 10^{19}$ Hz. In a vacuum or in space, where there are very few atoms or molecules to influence the wave propagation, all the above waves travel at the same speed, c, the speed of light *in vacuo*, which is equal to 2.998×10^8 m s^{-1}, to four significant figures. They thus have different wavelengths, ranging from about 10^4 m for radio waves down to 10^{-8} m for X-rays. Figure 8.3 shows a schematic of the **electromagnetic spectrum**, covering the radio waves used at low frequency to gamma rays at the highest frequencies.

Electromagnetic waves are produced whenever charges are being accelerated and are thus in non-uniform motion. For example, when there is a high-frequency oscillating current in a wire, the charges constituting the current have an acceleration and radio waves are radiated. When electrons in atoms jump from one quantum level to another, the electrons are accelerated on very short time-scales and light waves or X-rays may be produced. The electromagnetic waves in both cases are made up of related electric and magnetic fields that

Wavelength (m)

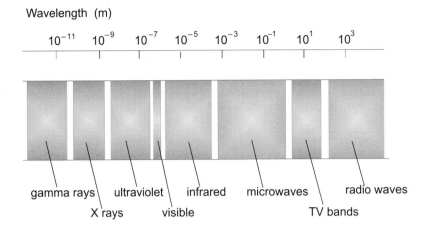

Fig. 8.3 The electromagnetic spectrum. The spectrum is divided roughly into wavelength intervals that correspond to different named parts. Gamma rays and X-rays are at the low-wavelength end and radio waves at the long-wavelength end.

Wavelength (m)

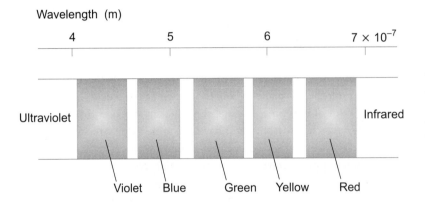

Fig. 8.4 The spectrum of visible light. The spectrum, which covers the wavelength range from about 400 nm to 700 nm, is divided roughly into wavelength intervals that correspond to different colours. The spectrum merges into the infrared at long wavelengths and into the ultraviolet at short wavelengths.

propagate outwards from the set of accelerated charges that generate them.

One of the most important regions of the electromagnetic spectrum for humans is the visible portion, since it is that region to which the human eye is sensitive. The visible region covers a range of wavelengths from about 4×10^{-7} m to about 7×10^{-7} m, as shown in Fig 8.4. The human eye interprets this range of wavelengths as colours, with violet at the short-wavelength end and red at the long-wavelength end. White light consists of a uniform mix of wavelengths distributed over the visible part of the spectrum. Removal of a narrow band of wavelengths from white light results in different sensations of colour depending on the waveband removed. The colours of objects thus depend on which wavelengths are absorbed and not reflected from incident white sunlight.

● *The visible part of the spectrum*

Part of the eye acts as a lens to focus light on to a mosaic of receptors at the rear. These receptors convert the light energy received, which is proportional to the square of the modulus of the electric field as for other types of waves discussed in Chapter 7, into electrical signals that are interpreted by the brain. The eye and the signal processing system constitute a brilliantly designed piece of scientific equipment with an amazing capability.

Plane waves

All electromagnetic waves in free space obey the same wave equation. The wave is a three-dimensional wave: the electric field **E** and the magnetic field **B** depend upon three position coordinates as well as upon time. We will not derive or discuss the equation obeyed by the waves in any detail. We can define what is meant by a plane wave without recourse to the wave equation.

Imagine a source of waves at the origin and look at waves coming towards you in the positive z-direction. If you are at very large distance z from the source there is no variation of the fields in the x- or y-directions. At large distances the electric field at any fixed time is the same over a large part of the xy-plane. Similarly, the magnetic field has its own constant value at that time. If a snapshot is taken at a later time, although the fields have changed, they still have the same value over much of the xy-plane. The wave observed is called a **plane wave**.

The simplest plane wave has a single angular frequency ω and its phase varies regularly with distance and time and is given by $(\omega t - kz)$. A wave of a single frequency is unattainable in practice, as indicated in Section 7.2. However, we will discuss monochromatic plane waves because waves whose properties closely approximate those of monochromatic plane waves do appear in nature and the mathematical description of a monochromatic plane wave is simple. It allows many features of electromagnetic waves to be explored in a simple way. The expression for a plane monochromatic wave is similar in form to that of eqn (7.8), and the electric field in a plane electromagnetic wave is

● *The mathematical form of a plane electromagnetic wave*

$$\mathbf{E} = \mathbf{E_0} \cos(\omega t - kz). \tag{8.10}$$

A **wavefront** of a wave is a surface over which the field **E** has constant amplitude and phase. The wavefronts of the plane wave (8.10) are planes perpendicular to the z-direction. Of course, the description (8.10) applies over a limited region of space only: for a sufficiently large excursion in the x- or y-direction, the wavefronts are no longer planes. Nevertheless, the idea of a plane wave is very useful and easy to think about.

The theory of electromagnetism shows that the vector **E** in a plane electromagnetic wave must be perpendicular to the direction of propagation, and a plane electromagnetic wave is thus a transverse wave. For the wave given by eqn (8.10) the vector amplitude $\mathbf{E_0}$ lies in the xy-plane. The plane wave (8.10) satisfies a wave equation similar to eqn (7.7) and, since we know that the speed of any electromagnetic wave in free space is c, the wave equation is

● *Electromagnetic waves are transverse waves*

$$\frac{\partial^2 \mathbf{E}}{\partial t^2} = c^2 \left(\frac{\partial^2 \mathbf{E}}{\partial z^2} \right). \tag{8.11}$$

● *The wave equation for electromagnetic waves in space*

If a wave travelling in free space enters a homogeneous medium, its speed is different from the speed c. The microscopic constituents of the medium oscillate at the wave frequency and make their own contributions to the fields and the result is a total field that travels with a phase velocity v_p different to c. The wave in the medium thus has a different wavelength to that of the wave in free space, and the wavenumber is changed from k *in vacuo* to k' in the medium. The mathematical form of a plane wave in a homogeneous medium is given by the expression

● *Electromagnetic waves in a medium*

$$\mathbf{E} = \mathbf{E_0} \cos(\omega t - k' z). \tag{8.12}$$

This function obeys the wave equation

$$\frac{\partial^2 \mathbf{E}}{\partial t^2} = v_p^2 \left(\frac{\partial^2 \mathbf{E}}{\partial z^2} \right) \tag{8.13}$$

with

$$v_p = \frac{\omega}{k'}. \tag{8.14}$$

When visible light travels in glass or most other homogeneous media, the phase velocity is lower than in free space. Moreover, the factor by which the phase velocity is reduced depends on frequency. The light waves suffer dispersion. Red light has a longer wavelength than blue light and travels in glass or water at a faster speed than the latter. This phenomenon is responsible for the colours of the rainbow and the colours seen when a beam of white light consisting of many wavelengths, and hence colours, is passed through a prism.

8.3 Refraction

When a wave crosses the boundary between two different media its phase velocity changes. For a wavefront incident at an angle to the boundary, which we will take to be a flat surface, this change of speed modifies the direction of propagation of the wave. The process is called **refraction**. Refraction occurs with waves of all types. However, for the rest of this section we restrict the discussion to light waves, those electromagnetic waves with frequencies in the visible part of the spectrum.

Plane waves at boundaries

We can calculate the change in direction of a plane wave with a simple construction that uses the wavefronts in the two media. In schematic drawings illustrating wave phenomena, **rays** are often used to picture the wave. The flow of energy in a wave at a point in space is in the direction of the ray at that point. At any point along a ray, the perpendicular to the wavefront at that point is in the direction of the ray. For plane waves, the rays are parallel straight lines perpendicular to the wavefront.

● *Representing waves by rays*

For a more complicated wave, wavefronts are continuous planes over which the waves have the same amplitude and phase, and at any two close points in space the ray may be constructed from the wavefronts by joining the corresponding perpendiculars. The wavefronts from a point source of light are spheres centred on the source. Rays used to define wavefronts and energy flows diverge from the source like spokes from the centre of a wheel.

Figure 8.5 shows a monochromatic plane wave incident at an angle to the surface between two media, medium 1 and medium 2, in which the phase velocities are v_1 and v_2. The wavefronts are planes perpendicular to the direction of travel of the wave and representative rays are shown by the lines AB and CD. The incident rays make an angle θ to the perpendicular to the surface, called the **angle of incidence**. The wavefront passing through the point B is plane and contains the line BD. The wavefront inside medium 2 passing through the point F contains the line FE. In the time interval $\Delta t = \mathrm{DE}/v_1$ the ray along the line CD has reached the boundary at E. In this time interval the ray along the line AB has progressed inside medium 2 to the point F, with $\Delta t = \mathrm{BF}/v_2$. The phases of the two different rays must be equal at F and E, and this requires that

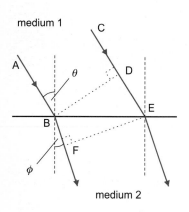

$$\mathrm{DE}/v_1 = \mathrm{BF}/v_2.$$

The direction of propagation in medium 2 is perpendicular to the line EF and makes an angle ϕ to the perpendicular to the boundary. This angle is called the **angle of refraction**. The above relation between the speeds in the two media and the lengths DE and BF may be used with the geometric properties of triangles BDE and BEF to show that

Fig. 8.5 A plane wave incident upon the boundary between two different media. The angle of incidence is θ and the angle of refraction is ϕ.

$$\frac{\sin \theta}{\sin \phi} = \frac{v_1}{v_2}. \tag{8.15}$$

If the speed v_2 is less than v_1, as for a light wave entering a block of glass from air, the angle ϕ is smaller than θ and rays are refracted towards the normal to the boundary. The ratio of speeds v_1/v_2 depends on the media and on the angular frequency of the wave. This ratio may be rewritten as

● Snell's law

$$\frac{v_1}{v_2} = \frac{v_1}{c}\frac{c}{v_2} = \frac{n_2}{n_1} \tag{8.16}$$

where $n_1 = c/v_1$ and $n_2 = c/v_2$. The factors n, which are the ratios of the speed of light *in vacuo* to the speed in the media, are called the **refractive indices** of the media, and eqn (8.15) becomes

$$\frac{\sin\theta}{\sin\phi} = \frac{n_2}{n_1}. \tag{8.17}$$

This equation is called **Snell's law of refraction**.

For a wave incident in air, $v \sim c$ and the refractive index of air at normal temperatures and pressure is unity to within a few parts in ten thousand. For that case

$$\frac{\sin\theta}{\sin\phi} = n_2. \tag{8.18}$$

Exercise 8.2 A light ray strikes a glass plate of thickness d equal to 1.5 cm at an angle of incidence θ equal to 50°. It emerges on the other side as shown in Fig 8.6. Show that the emerging ray is parallel to the incident ray and calculate the displacement D of the ray (measured along the side of the plate) if the refractive index of the glass for the light equals 1.5.

Answer 0.90 cm.

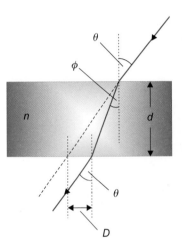

Fig. 8.6 A light ray passing through a slab of glass.

If a wave is entering a medium 2 in which the wave velocity is higher than in the first medium 1, for example, a wave going from glass to air, the wave is refracted away from the normal to the boundary. As the angle of incidence θ increases, the angle of refraction approaches 90° and, when $\sin\theta$ is equal to $n_2/n_1 \sim 1/n_1$, the wave in medium 2 is refracted along the boundary. For angles of incidence greater than the **critical angle** θ_c given by

$$\sin\theta_c = \frac{n_2}{n_1}, \tag{8.19}$$

the wave cannot emerge at all and is **totally internally reflected**. This topic is treated in more detail in the next section.

● Total internal reflection

Worked Example 8.1 Water waves in shallow water of depth less than the wavelength are non-dispersive, and the phase velocity is given approximately by $v_p = (gh)^{1/2}$, where g is the acceleration due to gravity and h is the depth. Water waves such as these are travelling in water of depth 1.2 m and meet a step where the depth changes to 2.1 m. If the angle of incidence of the wave on the step is 30°, what is the angle of refraction? What is the critical angle of incidence for which there is no wave crossing the step?

Answer If v_1 is the speed of the waves in water of depth 1.2 m and v_2 the speed at a depth of 2.1 m,

$$\frac{v_1}{v_2} = \sqrt{\frac{1.2}{2.1}}.$$

Equation (8.15) then gives the angle of refraction ϕ as

$$\sin\phi = \frac{v_2}{v_1}\sin\theta$$

with θ the angle of incidence. This gives $\sin\phi = 0.661$ and $\phi = 41.4°$.
  The critical angle is given by eqn (8.19) with

$$\frac{n_2}{n_1} = \frac{v_1}{v_2} = 0.756.$$

Thus the critical angle is 49.1°.

The refractive index of a material contains information about the material's composition, and instruments that measure the refractive index are called refractometers. These often work by determining the critical angle θ_c. Refractometers are part of the payload of the Cassini space mission to study Saturn and its rings, which left Earth in 1997. The Huygens probe on Cassini is scheduled to land on Titan, Saturn's largest moon, in the year 2004. Titan is thought to have a liquid surface and the refractometers will measure the liquid's refractive index at the point of landing. If the surface at landing is liquid, refractive index measurements will give valuable clues to its chemical and physical structure.

8.4 Reflection

When waves meet changes in the media through which they are propagating, for example, when a light wave hits a boundary between air and glass, some of the energy in the incident wave is reflected and some transmitted. The amplitude of the wave in the glass is less than that of the incident wave. There may also be phase differences between the reflected wave, the incident wave, and the transmitted wave. Both the amplitude changes and the phase changes depend upon the nature of the wave motion and the nature of the discontinuity the wave meets.

The amplitude and phase changes are determined by imposing appropriate conditions, called boundary conditions, on the wave at the boundary. These conditions ensure that the properties of the wave on the incident side match those on the exit side. The matching is performed by writing down an expression for the total wave, incident plus reflected, on the incident side of the boundary and an expression for the wave on the exit side. It is then required that the two expressions satisfy the required boundary conditions. This procedure determines the parameters describing the total waves on both sides of the boundary.

● *Boundary conditions*

The boundary conditions depend upon the physics of the problem. For example, when electromagnetic waves meet a boundary between two different media there are conditions that the fields **E** and **B** must satisfy; when sound waves meet the end of a rod there are conditions that the displacement and speed of the end of the rod must obey.

Waves on strings

The general ideas outlined above for matching waves on either side of a discontinuity are illustrated here with the simple one-dimensional example of a wave on two joined strings on which the wave has different speeds of propagation.

Let the undisplaced strings lie along the z-direction and the point where the strings are joined be at position z equal to zero. The string on the left-hand side propagates waves of angular frequency ω at speed v_1 corresponding to wavenumber k_1, and the string on the right-hand side of the join propagates waves at a speed v_2 corresponding to wavenumber k_2. We note the following.

1. The frequency of the waves on the two separate parts of the string must be the same; otherwise the join could not stay intact. At $z = 0$ the strings must go up and down together.

2. The tension in the strings must also be the same everywhere; otherwise the join would move in the z-direction.

3. The total disturbance on the left-hand string is the sum of the incident wave $y_0^i \cos(\omega t - k_1 z)$ and the reflected wave $y_0^r \cos(\omega t + k_1 z)$, where y_0^i and y_0^r are the respective amplitudes. The total wave on the left-hand string is thus given by

$$y_L = y_0^i \cos(\omega t - k_1 z) + y_0^r \cos(\omega t + k_1 z). \tag{8.20}$$

This is a progressive wave to the right plus a standing wave, as in Worked example 7.5.

4. The disturbance on the right-hand string is a progressive wave given by

$$y_R = y_0^t \cos(\omega t - k_2 z) \tag{8.21}$$

where y_0^t is the amplitude of the transmitted wave. We assume for the moment that there are no phase differences between the incident, reflected, and transmitted waves at $z = 0$.

Where the strings join at $z = 0$, the string is continuous; otherwise there is no join. Hence at all times the displacement on the left at $z = 0$ must equal the displacement on the right at $z = 0$. Also, in order to make the forces on a very small section at the join equal and opposite, the slopes of the string either side of the join must be the same: $(\partial y / \partial z)_L = (\partial y / \partial z)_R$ at $z = 0$. The conditions that the displacement and its derivative be continuous at the join are the boundary conditions that must be satisfied when the propagating wave meets the discontinuity.

● *Boundary conditions for waves on joined strings*

The boundary condition on string continuity applied at time $t = 0$ gives

$$y_0^i + y_0^r = y_0^t, \tag{8.22}$$

and the equal-slope condition applied at time $t = \pi / 2\omega$ gives

$$k_1 y_0^i - k_1 y_0^r = k_2 y_0^t. \tag{8.23}$$

● *Transmitted and reflected waves*

Manipulation of these two equations gives

$$\frac{y_0^r}{y_0^i} = \frac{k_1 - k_2}{k_1 + k_2} \tag{8.24}$$

and

$$\frac{y_0^t}{y_0^i} = \frac{2k_1}{k_1 + k_2}. \tag{8.25}$$

We have now determined the amplitudes of the transmitted and reflected waves in terms of the amplitude of the incident wave. This was done by assuming that the phases of all three waves were equal at $z = 0$. Equation (8.25) shows that the phase of the transmitted wave is equal to that of the incident wave; y_0^t has the same sign as y_0^i since their ratio is always positive. However, y_0^r may have a different sign to y_0^i depending on the values of k_1 and k_2. If $k_2 > k_1$, there is a phase difference of π between

● *The phase differences between incident and reflected waves*

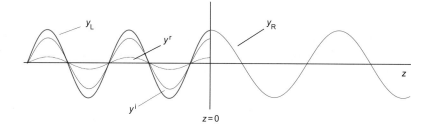

Fig. 8.7 A wave incident from the left joins, at $z = 0$, a string of smaller mass per unit length. Part of the incident wave is reflected with no phase change, giving the resultant wave y_L on the left; part of the incident wave is transmitted, giving the wave y_R on the right.

incident and reflected waves; adding π to the phase of the reflected wave in eqn (8.20) is equivalent to changing the sign of y_0^r.

Whether or not there is a phase change of π on reflection is determined by the sign of $(k_1 - k_2)$. The wavenumber on the right is greater than on the left if the speed of the waves on the right is less than on the left, from eqn (7.11). Thus, from eqn (7.11), the phase change is zero if the string on the right is lighter than on the left, or π if it is heavier.

Figure 8.7 shows at time $t = 0$ a wave of amplitude y_0^i incident on the join at $z = 0$ between two strings. The string on the right has a smaller value of mass per unit length ρ than the string on the left. The transmitted wave y_R has longer wavelength than the incident wave and greater amplitude y_0^t. It matches the total wave y_L on the left (composed of the incident plus the reflected) at $z = 0$.

If the string on the right is very heavy, the join does not move and k_2 tends to infinity. For this situation y_0^t is zero, the phase change is π and we have standing waves set up with infinite standing wave ratio as discussed in Section (7.7).

Energy flows

For two waves going in opposite directions, such as those represented by expression (8.20), the time average of the energy flow to the right is given by the average power expended by the force F. Following the procedure of Section 7.3, the average energy flow to the right may be shown to be equal to

$$\tfrac{1}{2} F \omega k_1 [(y_0^i)^2 - (y_0^r)^2]. \tag{8.26}$$

The power transmitted at the discontinuity and propagating to the right on the second string is given by eqn (7.20) to be

$$\tfrac{1}{2} v_2 F k_2^2 (y_0^t)^2.$$

It is straightforward to substitute for y_0^r and y_0^t from eqns (8.24) and (8.25), respectively, and to use the fact that the speed $v_2 = \omega/k_2$ to show that the above expression reduces to eqn (8.26). Thus the energy flow to the right on the left-hand string equals the energy flow to the right on the right-hand string. Energy is conserved, as it must be since we have introduced no energy dissipation into our discussion. The equality of the two expressions for the energy flow may be written in the form

● *Conservation of energy*

$$1 - \left(\frac{y_0^r}{y_0^i}\right)^2 = \frac{k_2}{k_1}\left(\frac{y_0^t}{y_0^i}\right)^2. \tag{8.27}$$

The above equation may be written in terms of energy reflection and transmission coefficients. If R is the fraction of energy reflected and T is the fraction transmitted, $R = (y_0^r/y_0^i)^2$ and $T = (y_0^t/y_0^i)^2 \times (k_2/k_1)$ and eqn (8.27) becomes

$$R + T = 1. \tag{8.28}$$

This is a more concise way of expressing the conservation of energy. Equation (8.28) is always obeyed when a wave is reflected at a boundary without energy loss.

Reflection of light

We have illustrated how the amplitude and phases of waves reflected and transmitted at a discontinuity can be determined by applying boundary conditions to the waves at the boundary. A similar technique is applicable to the reflection of other types of waves at different discontinuities. The resulting coefficients depend on whether the waves are sound waves, electromagnetic waves, or otherwise; on their frequency; and also on the properties of the media either side of the boundary.

If light waves are incident from air on to the plane surface of a nonconducting medium like glass, some of the energy is transmitted and some reflected. The ratio of the amplitude of the reflected wave to that of a wave incident in a direction perpendicular to the plane boundary may be anticipated from the above results for reflection at the junction of strings. For waves in air reaching a glass surface there is a phase change of π between the incident and reflected waves.

● *Waves reflected from glass have a phase change of π*

The reflection coefficient R is the ratio of the squares of the amplitudes of the incident and reflected waves. Using expression (8.24) for the ratio of the amplitudes and substituting for the wavenumbers of the light in media 1 and 2 we obtain

$$R = \left(\frac{n-1}{n+1}\right)^2 \qquad (8.29)$$

where n is the refractive index of the glass. For glass with a refractive index of about 1.5, the loss of intensity due to reflection at a surface is about 4%. For an optical system with several surfaces reflection could be responsible for a significant light loss unless remedial steps are taken.

If light waves are incident non-normally on a flat surface, the **law of reflection** tells us the direction of the reflected rays. It states that the angles of incidence and reflection are equal. The direction of the transmitted rays is given by the law of refraction, eqn (8.17). As discussed in the last section, total internal reflection can occur if the refractive index of the second medium is less than that of the first.

Worked Example 8.2 A glass slab of thickness $d = 5$ cm has a small object just underneath its bottom surface. The rays from the object leaving the top of the slab are contained within a circle of radius $r_0 = 4$ cm. What is the refractive index of the glass?

Answer Figure 8.8 shows how light from the object is totally internally reflected for angles of incidence θ greater than θ_c given by $\tan \theta_c = r_0/d$, with $r_0 = 4$ cm and $d = 5$ cm. Hence θ_c is equal to 38.7° and from eqn (8.19) the refractive index equals 1.60.

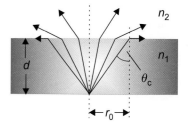

Fig. 8.8 Light rays from an object on the bottom surface of a block of glass of thickness d are totally internally reflected for angles of incidence greater than the critical angle θ_c.

Total internal reflection is responsible for the long distances over which infrared or visible waves are transmitted along optical fibres. An optical fibre is a thin thread of transparent material, with the diameter usually between a few and a few hundred microns, surrounded by a second material that has a refractive index smaller than that of the inner thread. Light travelling down the fibre at a small angle to the fibre axis is totally internally reflected each time it strikes the inner surface. See Problem 8.22.

Sound waves also provide an interesting example of total internal reflection. Unlike electromagnetic waves for which the speed in water is less than in air, the speed of sound in water is greater than in air. The speed in sea water at 25° centigrade can be estimated from eqn (8.9) using a value for the compressibility χ of 2.4×10^9 N m^{-2} and for the density ρ of 1.03×10^3 kg m^{-3}. The speed is ~ 1500 m s^{-1}, compared with ~ 340 m s^{-1} for sound in air. This difference is responsible for underwater swimmers being able to hear only sounds from a small region of the air above them. Sound from sources outside a cone that has its apex at the swimmer's position and a small half-angle are totally reflected at the water's surface.

Exercise 8.3 Calculate the half-angle of the cone from within which sounds can be heard by an underwater swimmer.

Answer   Using eqns (8.16) and (8.19), θ_c is determined to be about 13° for waves going into sea water from air.

The reflection coefficient of electromagnetic waves reflected off the surface of a conducting medium differs somewhat from that given by eqn (8.29). The electric field in the wave induces currents and these dissipate energy inside the conductor, reducing the amplitude of the transmitted wave to zero. Similarly, in the example of the joined strings shown in Fig 8.7, if the wave on the right-hand side of the join suffers attenuation because of energy loss processes, there would also be a different reflection coefficient to that given by eqn (8.24). Equation (7.42) would have to be used to describe the wave on the right, and the mathematics of matching the waves at the join would be different, resulting in different answers for R and T.

8.5 Doppler effects

Frequency and wavelength shifts occur when observer and source of waves are in relative motion. This is a commonly observed effect with sound waves. Aeroplanes, cars, and any moving sources of sound waves sound different when approaching to how they sound when receding. The notes are of higher frequency when the source is approaching than when the source is going away from the listener. These frequency changes are called **Doppler effects**.

We shall discuss Doppler effects in terms of how the frequency of a monochromatic wave is measured in two situations. The first is where the observer and the medium in which the wave is travelling are at rest while the source moves; the second where the observer is moving with respect to stationary source and medium. Most waves require a medium for their propagation; however, electromagnetic waves and gravitational waves do not need the presence of a medium and can travel in vacuum. If there is no medium present, the Doppler shifts are the same irrespective of whether the source moves or the observer moves, since there is no way of distinguishing the two situations.

Electromagnetic waves in free space travel with the speed of light in free space, c, and for these waves the arguments and results presented below no longer apply. The discussion of Doppler effects for electromagnetic waves is based on the principles of special relativity, and we leave electromagnetic waves until relativity is discussed in Chapter 19.

Moving source

Consider a source moving with speed V towards an observer who is stationary in a frame in which the medium is also at rest. Let the speed of a wave of angular frequency ω and period T_0, both values as measured by an observer at rest with respect to the source, be v. Let us imagine that the observer has a detector that counts every time a wave crest reaches it. At time $t = 0$ the source is at the point P_1, distance ℓ, as shown in Fig 8.9, and emits a wave crest. This reaches the observer at time $t_1 = \ell/v$. At time $t = T_0$ the source has moved a distance VT_0 towards the observer and is at point P_2 distance $(\ell - VT_0)$. At P_2, the source emits a second crest. This reaches the observer at time

$$t_2 = T_0 + \frac{\ell - VT_0}{v}.$$

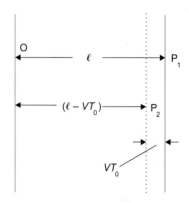

At time nT_0, the source emits its nth crest and this reaches the observer at time

$$t_n = nT_0 + \frac{\ell - nVT_0}{v}.$$

Hence in a time interval

$$\left(nT_0 + \frac{(\ell - nVT_0)}{v} - \frac{\ell}{v} \right)$$

the observer's detector counts n crests and the observer records the period of the wave as equal to T given by

$$T = T_0 - \frac{VT_0}{v} = T_0 \left(1 - \frac{V}{v} \right). \tag{8.30}$$

Fig. 8.9 A source moving with speed V towards a stationary observer at O emits a wave crest at the point P_1. It emits the next wave crest at P_2 after moving a distance VT_0 towards O.

This equation may be rewritten in terms of the frequency ν_0 that would be measured if source and observer were stationary, and the frequency ν observed when the source is moving, as

$$\nu = \nu_0 \left(1 - \frac{V}{v} \right)^{-1}. \tag{8.31}$$

If V is small compared with the wave speed v, this equation may be approximated by taking the binomial expansion to terms no higher than (V/v), giving

$$\nu \simeq \nu_0 \left(1 + \frac{V}{v} \right). \tag{8.32}$$

The observer thus measures a higher frequency when the source comes towards him than he does when it is at rest. For the source receding, the same procedure may be followed to show that the observed frequency is given by a formula similar to the above but with the sign of the speed V changed from plus to minus. If the source is moving with velocity $\mathbf{V}$ and the unit vector in the direction of the line joining the source to observer is denoted by $\mathbf{k_d}$, the observed frequency is given by

$$\nu = \nu_0\left(1 - \frac{\mathbf{V}\cdot\mathbf{k_d}}{\nu}\right)^{-1} \tag{8.33}$$

or approximately by

$$\nu \simeq \nu_0\left(1 + \frac{\mathbf{V}\cdot\mathbf{k_d}}{\nu}\right). \tag{8.34}$$

Exercise 8.4 The whistle on a French TGV train emits sound at a frequency of 900 Hz. What frequency will an observer down the track hear as the train approaches at a speed of 200 km per hour? What difference is made by using the approximate expression (8.34) instead of the more exact (8.33)? Take the speed of sound to be $340\,\mathrm{m\,s^{-1}}$.

Answer 1076 Hz using eqn (8.33) and 1047 Hz using eqn (8.34).

The above derivation of eqn (8.30), the new period when the source is moving towards the observer, depends upon integral numbers of crests reaching the observer before the source arrives. Clearly, there is a problem with the derivation if the time interval between crests approaches the time for the source to travel the distance ℓ. The wave crests become compressed closer and closer together as the speed V approaches the wave speed ν. Our derivation, and eqn (8.33), break down altogether when the speed of the source exceeds the speed of the wave in the medium. When this happens, all the energy in the compressed sound waves reaches us at the same time in the form of a **shock wave**. For example, when an aircraft approaches at a speed greater than the speed of sound in air, shock waves are produced that reach an observer after the plane has passed and manifest themselves as loud sonic bangs.

● *Shock waves*

Another example of a shock wave is provided by the bow wave a boat makes when it travels through the water at a speed greater than the speed of travel of the water waves, as illustrated in Fig 8.10. At a certain time $t = T$ the boat is at the position C. At time $t = 0$ the boat was at position A and at time $T/2$ it was at position B. The wave generated at time zero spreads out and at time T its wavefront is shown as the circle of radius $\nu_\mathrm{p}T$ centred on A, where ν_p is the phase velocity of the water waves. The wavefront generated at time $T/2$ is shown at time T as the circle of radius

$v_pT/2$ centred on B. The front edge of the bow wave, or shock wave, is the line drawn from C tangent to the two circles at D and E. This line makes an angle θ to the line ABC, the direction of travel of the boat. The angle θ can be obtained from the triangle AEC and is given by $\sin\theta = v_p/V$ where V is the speed of the boat.

Moving observer

The formula giving the Doppler change in frequency when the source is stationary in the propagating medium but the observer is moving is different to eqn (8.33), although for most practical purposes a single expression can be used to cover both cases.

An illuminating way of deriving the shift is to think in the reference frame in which the observer is at rest. In this frame the source and medium are approaching at speed V and the speed with which the wave approaches is $V + v$. By then following a similar procedure to that adopted before we find that the time interval between the arrival at the observer of the first and nth crests is

$$t_n - t_1 = nT_0 - \frac{nVT_0}{V+v}.$$

The observer thus measures the period of the wave to be

$$T = T_0 - \frac{VT_0}{V+v} = T_0\left(1 + \frac{V}{v}\right)^{-1},$$

giving

$$\nu = \nu_0\left(1 + \frac{V}{v}\right). \tag{8.35}$$

If V/v is small, the Doppler shift is almost the same whether it is the observer or the source moving, since the approximate formula (8.32) is the same as eqn (8.35).

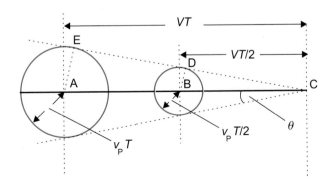

Fig. 8.10 A boat travelling at a speed V faster than the phase velocity v_p of the water waves generates a bow wave along the line CDE.

Exercise 8.5 If an observer is moving at 200 km per hour towards a stationary train emitting a sound at a frequency of 900 Hz, what frequency does he hear? Compare this with the answer to Exercise 8.4. Take the speed of sound to be 340 m s^{-1}.

Answer 1047 Hz.

Worked Example 8.3 A loudspeaker is generating sound at a frequency of 50 Hz when the amplitude of its motion is 1 mm. A listener sits directly in front of the speaker when it reflects a sound wave of frequency 1 kHz. What is the maximum Doppler shift in the reflected frequency the listener hears? Take the speed of sound to be 340 m s^{-1}.

Answer The diaphragm of the loudspeaker is executing simple harmonic motion of frequency 50 Hz when generating sound of that frequency. If the listener and loudspeaker lie along the x-axis, the displacement of the diaphragm from its normal position is

$$x_1 = x_{10} \cos \omega t,$$

where the amplitude, x_{10}, is 10^{-3} m and the angular frequency ω is 100π radians s^{-1}. The maximum speed of the diaphragm is thus $\omega x_0 = 0.1\pi$ m s^{-1}.

The maximum frequency observed at the diaphragm occurs when it is moving towards the source with its maximum speed. It is given by eqn (8.35) with $V = 0.1\pi$ m s^{-1} and $v = 340$ m s^{-1} and is equal to 1000.92 Hz. The maximum frequency heard by the listener is when the 1000.92 Hz wave is reflected from the diaphragm when it is moving at its maximum speed towards the listener. It is given by eqn (8.31) and is 1001.84 Hz. The maximum frequency shift is thus 1.84 Hz.

Doppler effects also occur when light sources and detection systems are in relative motion. As indicated earlier we need relativity for their proper description and this will be considered later. Here we simply state that if the relative velocity of source and observer, V, is small compared with the speed of the light waves v in the medium, eqn (8.34) can be used with very good accuracy to calculate the Doppler shift of light from a moving source. The fractional error involved in using the non-relativistic treatment is close to V^2/v^2. For a source in free space v is equal to c, the speed of light *in vacuo*.

8.6 Diffraction

A wave spreads out from its source, becoming a plane wave at large distances. Obstacles in the path of the wave affect the way it spreads out.

The simple ray-tracing procedure introduced in Section 8.3 for tracking the path of a wave suggests that light, for example, travels in straight lines. Often, this is a sufficiently good approximation. However, if the sizes of apertures or obstacles are comparable with the wavelength, all waves depart from straight-line propagation to some extent. If you are behind a wall that absorbs all sound waves incident on it without vibrating itself in response, you still hear noises from the other side. Similarly, if an obstacle prevents there being a direct line beween a light source and a detector, a sufficiently sensitive detector still records some light; if we look very carefully, shadows are not sharp.

The bending of waves round obstacles or the spreading out after passing through holes is called **diffraction**. In this section we discuss the diffraction of monochromatic plane waves. Most diffraction phenomena involve waves that are mathematically more complicated than plane waves. However, the physical ideas discussed here to work out diffractive effects with plane waves are applicable in principle to more complicated waves.

The spreading out of a wave after passing through an aperture depends on the ratio of the wavelength λ of the wave to the size of the aperture; the bigger the ratio the more the wave spreads out. Roughly speaking, if a plane wave passes through a circular aperture, the intensity on a screen placed downstream of the opening is smeared-out over an angular range $\sim \lambda/a$ where a is the radius of the aperture.

Plane wave at a slit

Consider the situation in which a plane wavefront moving in the z-direction meets a very long slit of width d cut in an otherwise opaque plane surface lying in the xy-plane. The plane wave can represent a disturbance of any kind; for example, it may be an electromagnetic wave or a sound wave. Choose axes of coordinates such that the long length of the slit is in the y-direction which is into the page on Fig 8.11. We may then ignore the variation of the wave with the coordinate y and treat the problem as if it were the two-dimensional situation shown in Fig 8.11. The narrow dimension of the slit is along the x-axis and at all points over the width d the incident waves have the same amplitude and phase. If the wave propagates by virtue of oscillations of the constituents of a medium, the resultant wave on the exit side of the slit is produced by wavelets that start out in phase over the whole width of the slit, since the incident wave at all points over the slit has the same phase. The principle of super-position can then be used to determine the resultant disturbance at any point on the exit side. If the wave is an electromagnetic wave *in vacuo*, it remains true that we can determine the radiation on the exit side of the slit in the same way. Using this procedure is known as following **Huygens' principle**.

● *Huygens' principle*

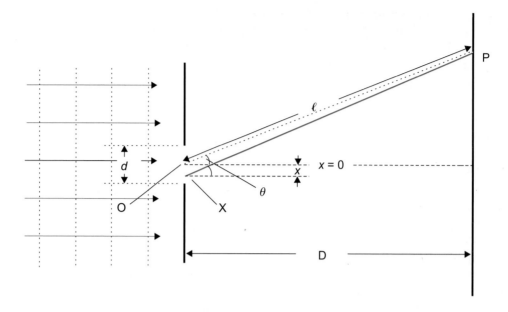

Fig. 8.11 A view looking down on a long, narrow slit of width d cut in an otherwise opaque plane surface. The slit is illuminated by a plane wave of monochromatic light incident perpendicularly to the surface. The wavefronts, parallel to the surface, are shown dotted.

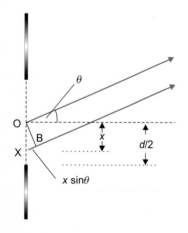

Fig. 8.12 The path difference of wavelets generated at strips of the slit with very small width dx at positions $x = 0$ and x. The resultant disturbance at the point P on the screen distant D shown in Fig 8.11 is the superposition of the amplitudes of all the wavelets originating over the slit.

We divide the slit up into infinitesimally small sections of width dx, from which the secondary wavelets originate. We take the origin of x at the midpoint O of the slit. Consider the wavelet from the small section dx at the point X shown on Fig 8.11 at position x along the slit. If the point P in the figure is far enough away from the slit, the wavelet originating at dx is approximately a plane wave when it reaches P, as are the wavelets from all over the slit.

Wavelets decay in amplitude as they spread out from their sources, since their initial energy is distributed over more space the farther they travel. However, since P is a distant point, we may make the approximation that all wavelets at P have travelled the same distance and so have the same amplitude dA. The amplitude dA is proportional to the elementary width dx over which it is generated, and $dA = \beta dx$, with β the constant of proportionality. Again, since P is distant, we make the approximation that the lines joining P to different parts of the slit are at the same angle θ to the x-axis.

Making the above approximations, the path difference to the point P between wavelets starting from strips of width dx at positions $x = 0$ and x is equal to $x \sin \theta$, where θ is the angle the mean ray from the slit to P makes with the normal to the slit. This is shown in Fig 8.12. The wave at P from the small element of the slit at x is thus given by

$$dA = \beta dx \cos(\omega t - k\ell - kx \sin \theta) \tag{8.37}$$

where ω and k are the angular frequency and wavenumber of the wave, and ℓ is the distance of P from the midpoint of the slit. Adding up the contributions from all the wavelets gives the resultant wave amplitude at P as

$$A = \int_{-d/2}^{d/2} \beta dx \cos(\omega t - k\ell - kx \sin \theta). \tag{8.38}$$

This integral can be evaluated to give

$$A = \frac{2\beta}{k \sin \theta} \sin(\tfrac{1}{2} kd \sin \theta) \cos(\omega t - k\ell). \tag{8.39}$$

● *The resultant amplitude at an angle θ*

The instantaneous intensity in the wave, I, is the energy crossing unit area per second and is proportional to the square of the amplitude, and thus

$$I(\theta) \propto \beta^2 d^2 \cos^2(\omega t - k\ell) \frac{\sin^2(\tfrac{1}{2} kd \sin \theta)}{(\tfrac{1}{2} kd \sin \theta)^2}. \tag{8.40}$$

Averaged over time, the intensity is

$$\bar{I} = I_0 \frac{\sin^2(\tfrac{1}{2} kd \sin \theta)}{(\tfrac{1}{2} kd \sin \theta)^2} \tag{8.41}$$

● *Average intensity*

where I_0, proportional to $\beta^2 d^2$, is the maximum of $\bar{I}$. This equation is the relation we have been seeking to describe how a diffracted plane wave of wavelength λ spreads out from an aperture that is a slit of width d.

There is a mathematically more elegant way of working out the resultant of all the wavelets. This is to represent a wavelet at P by the real part of the complex expression

$$\beta dx e^{-j(\omega t - k\ell - kx \sin \theta)}.$$

The real part is the same as eqn (8.37). Section 20.6 on complex numbers should be reviewed if there is any difficulty with this. The resultant wave at P is then the real part of the integral

$$\int_{-d/2}^{d/2} \beta e^{-j(\omega t - k\ell - kx \sin \theta)} dx = \beta e^{-j(\omega t - k\ell)} \int_{-d/2}^{d/2} e^{jkx \sin \theta} dx.$$

Evaluating the integral,

$$\int_{d/2}^{d/2} e^{jkx \sin \theta} dx = \frac{1}{jk \sin \theta} [e^{jkx \sin \theta}]_{-d/2}^{d/2}$$

$$= \frac{1}{jk \sin \theta} \left(e^{j(kd \sin \theta)/2} - e^{-j(kd \sin \theta)/2} \right)$$

$$= d \frac{\sin(\tfrac{1}{2} kd \sin \theta)}{\tfrac{1}{2} kd \sin \theta}.$$

The resultant amplitude A is thus the real part of

$$\beta d e^{-j(\omega t - k\ell)} \frac{\sin(\frac{1}{2} kd \sin \theta)}{\frac{1}{2} kd \sin \theta},$$

giving the same result as eqn (8.39).

The function

$$\frac{\sin^2(\frac{1}{2} kd \sin \theta)}{(\frac{1}{2} kd \sin \theta)^2} = \frac{\sin^2 \alpha}{\alpha^2} \tag{8.42}$$

● *Maxima and minima of the intensity distribution*

with $\alpha = (kd \sin \theta)/2$, has its maximum value unity when $\alpha = 0$. The maximum intensity I_0 thus occurs at θ equal to zero, and is the result of the in-phase addition of all the wavelets generated across the slit. The function (8.42) has zeros when $\sin \alpha = 0$ but $\alpha \neq 0$. The first zero thus occurs at

$$kd \sin \theta = 2\pi.$$

For light, λ is much less than d. Thus θ is very small and $\sin \theta$ is essentially equal to θ, giving the first zero at

$$\theta = \frac{\lambda}{d}. \tag{8.43}$$

The position of the first minimum is proportional to λ, and long wavelengths bend round corners more than short ones. The second zero occurs at $\theta = 2\lambda/d$ and the time average of the intensity in the diffracted wave, given by eqn (8.41), is drawn in Fig 8.13.

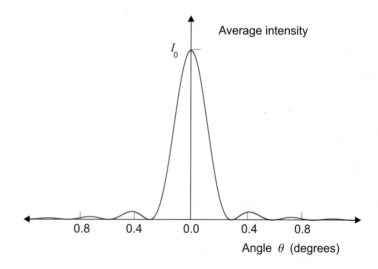

Fig. 8.13 The average intensity observed at angles θ shown on Figs 8.11 and 8.12 when a narrow slit of width 10^{-4} m is illuminated with a plane wave of monochromatic light of wavelength 500 nm.

Exercise 8.6 A plane wave of light of wavelength 690 nm is incident on a vertical slit of width 10^{-4} m. Sketch the intensity distribution on a screen 3 m from the slit placed parallel to the slit aperture. At what distances from the central maximum do the first two zeros occur?

A plane wave of sound of wavelength 10 cm is incident on an open door of width 1 m. Where are the first and second diffraction minima on a wall parallel to the door aperture and 10 m from it?

Answer The first minima occur on the screen at a distance of 2.1 cm either side of the straight-through maximum; the second occur at very nearly twice this distance.

The first minima on the wall occur at 1.00 m from the central maximum; the second at a distance of 2.04 m.

Circular apertures have a first zero at an angle given by an equation similar to (8.43) but with a multiplying factor close to unity. The first minimum when a plane wave falls normally on a very small circular aperture of diameter d is at

● *Circular apertures*

$$\theta = 1.22\frac{\lambda}{d}. \tag{8.44}$$

The pattern is a system of rings with circular symmetry. Plane waves fall on circular apertures in many practical situations and eqn (8.44) is an important result. For example, a circular telescope tube of diameter d defining the perimeter of a lens gives a ring pattern of a distant star in which the central bright disc has a total angular spread of $2.44\lambda/d$.

Limit to angular resolution

Diffraction influences many physical phenomena and one important consequence occurs with instruments that transmit or receive electromagnetic waves. If you are looking at two stars very close together the question arises as to what is the smallest angular separation of the stars for which your eye can distinguish them as separate. The light of wavelength λ from each star arrives at the eye as a plane wave. The pupil of the eye, of diameter d, diffracts the plane wave into a cone of half angle $\theta \sim \lambda/d$ which the eye detects as an image with this angular spread. The plane wave from the other star is similarly spread out and unless the angular separation of the stars is greater than $\sim \theta$, so that the maximum of the diffraction pattern from one star falls near the first minimum of the pattern due to the other, the two stars are indistinguishable and are registered as one object. This is referred to as the **Rayleigh criterion**. In the same way, the minimum angle of divergence of a radio beam of wavelength λ emitted from a radio transmitter of diameter d is about λ/d.

Exercise 8.7 The diameter of the dish in the Lovell radio telescope at Jodrell Bank, England, is 76 m. It is looking at electromagnetic radiation of frequency 1 GHz emitted from two stars that are very close together. Estimate the smallest angular separation of the pair that the telescope can resolve.

Answer $0.23°$.

8.7 Interference

When a wavefront is split up and the separate parts follow different paths before recombining, the principle of superposition tells us that the resultant disturbance is the sum of the separate disturbances. When there is a phase difference between two waves that combine, the magnitude of the resultant is less than the sum of their magnitudes. This phenomenon is called **interference**.

Consider the addition of two monochromatic waves with equal amplitude and with a fixed phase difference over the time during which an observation is made. If the phase difference is π, corresponding to a path difference of $N\lambda/2$ with N an odd integer, the resultant is zero and we have what is called **destructive interference**. If the phase difference over the observation time is zero, corresponding to a path difference of an integral number of whole wavelengths, the two disturbances add to give a maximum resultant amplitude and we have **constructive interference**.

Interference is illustrated in Fig 8.14, which shows waves resulting from two sources consisting of synchronized vibrators in a water surface. At certain points the two waves are always out of phase and there is zero disturbance; at others the amplitude of the resultant always has the maximum value.

The two-slit pattern

Let us consider the interference of light waves in order to be specific, and analyse the pattern produced when a plane wave of monochromatic light passes through two narrow slits in a screen. The analysis will be similar to that used in the discussion of diffraction.

Consider a wave propagating in the z-direction and passing through two narrow slits separated by a distance a in the x-direction as shown in

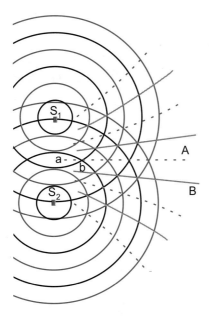

Fig. 8.14 A snapshot of water waves generated by synchronized vibrators in a water surface. Crests are shown as blue lines, troughs as black lines. In certain directions, shown on the diagram as black lines radiating outwards from the sources like the spokes on a wheel, crests overlap with troughs, the waves interfere destructively and there is no disturbance.

Fig 8.15. The Huygens' wavelets from each slit start out in phase. The resultant amplitude at the point P is the sum of the amplitudes of the two wavelets, which have traversed different paths. The wavelet from slit S_1 has gone a distance $\ell_1 = S_1 P$ and that from S_2 a distance $\ell_2 = S_2 P$. The wavelet from S_2 which at P combines with that from S_1 left S_2 at a previous time ℓ_2/c and that from S_1 at a previous time ℓ_1/c.

For interference to take place, the fixed phase relationship between the wavelets must be retained over a time interval of at least $(\ell_2 - \ell_1)/c$, the difference in the travel times to P of the wavelets from S_1 and S_2. If there are random changes of phase at time intervals smaller than this, the average wave energy seen at P is simply the sum of the separate averages, but if the phase relation is maintained the amplitudes add and the average energy depends on the average of the square of the resultant amplitude rather than the sum of the individual squares. The time interval over which the wavelets maintain a fixed phase relation is called the **coherence time** of the waves. If the coherence time is long compared with the time difference between the paths of the added wavelets the waves are said to be **coherent**. Coherence gives interference, when we see maxima and minima in the resultant light intensity; if the light is incoherent, we add separate intensities and obtain a smooth pattern of illumination. If slits S_1 and S_2 were illuminated by separate light sources of the same wavelength the wavelets from the two slits would usually be incoherent.

● *Coherence*

Fig. 8.15 A view looking down on two narrow slits S_1 and S_2 separated by a distance a and illuminated by a plane wave of monochromatic light. The plane wave is incident perpendicularly with wavefronts parallel to the plane containing the slits. The resultant disturbance at the point P on the screen at distance D is the superposition of the amplitudes of the wavelets originating from the two slits.

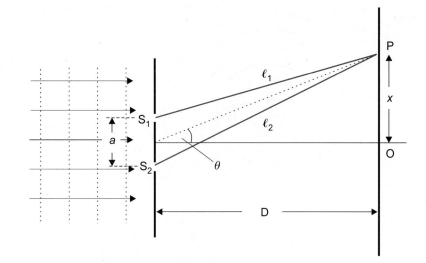

We will determine the light intensity on a screen distant D from the slits shown on Fig 8.15, where D is very large compared with a wavelength. If the separation a of the slits is small compared with D (the figure is drawn with greatly distorted scales for simplicity), the distance travelled by the separate waves from the two slits is very similar. Although the amplitudes of the waves from the slits fall off with distance travelled, if $D \gg a$ we may neglect any small difference in amplitudes on arrival at the screen and take the wavelets on arrival at P as having equal amplitudes A_1. The resultant amplitude is

$$A = A_1 \cos(\omega t - k\ell_1) + A_1 \cos(\omega t - k\ell_2) \tag{8.45}$$

$$= 2A_1 \cos(\omega t - k[\ell_2 + \ell_1]/2) \cos(k[\ell_2 - \ell_1]/2) \tag{8.46}$$

where ω and k are the angular frequency and wavenumber of the wave. Since the separation a is very small compared with the distance D, $(\ell_2 + \ell_1) \approx 2D/\cos\theta$. Also, since the angle θ is small, $\cos\theta \approx 1$, so that

● *The resultant amplitude from two slits*

$$A = 2A_1 \cos(\omega t - kD) \cos(k\Delta\ell/2) \tag{8.47}$$

where $\Delta\ell = (\ell_2 - \ell_1)$ is the path difference of the two wavelets and θ the angle shown.

The intensity I in the resultant wave at P is proportional to the square of this amplitude, and

$$I \propto A_1^2 \cos^2(\omega t - kD) \cos^2(k\Delta\ell/2).$$

● *The average intensity*

The time average of the intensity $\bar{I}$ is

$$\bar{I} = I_0 \cos^2(k\Delta\ell/2) \tag{8.48}$$

where I_0 is the average intensity observed at the maxima.

The intensity is zero whenever $\Delta\ell$ is an odd number of half-wavelengths and, when $\Delta\ell$ is an even number of half-wavelengths, has its maximum value of four times the intensity that would result from one slit alone. A pattern of bright and dark **interference fringes** is observed on the screen, the first bright fringe corresponding to straight-through light and $\Delta\ell \approx$ zero; the next bright fringe either side of the first corresponding to $k\Delta\ell/2 = \pi$, i.e. $\Delta\ell = \lambda$, and so on.

Since D is much greater than a, $\Delta\ell \approx a\sin\theta$ giving

$$\bar{I} = I_0 \cos^2(\tfrac{1}{2}ka\sin\theta)$$

or

$$\bar{I} = I_0 \cos^2\left(\frac{\pi a}{\lambda}\sin\theta\right). \tag{8.49}$$

Intensity maxima occur when $\pi a\sin\theta/\lambda = n\pi$, with n an integer or zero, and hence at angles θ given by

● *Maxima and minima of the intensity distribution*

$$\sin\theta = \frac{n\lambda}{a}. \tag{8.50}$$

For small values of θ, $\sin\theta \approx \theta \approx x/D$, and intensity maxima appear on the screen at distances x from the point O in Fig 8.15 given by

$$x = n\frac{\lambda D}{a}. \tag{8.51}$$

Minima, with zero intensity, occur at $\sin\theta = (n+\tfrac{1}{2})(\lambda/a)$, with n an integer or zero. This happens when

$$x = (n+\tfrac{1}{2})\frac{\lambda D}{a}.$$

The fringe pattern in a particular example is shown in Fig 8.16. Similar patterns result when two waves of any equal frequency from separate sources combine as long as they have fixed phase differences at the points of recombination. Exercise 8.8 is an example of electromagnetic waves from two radio antennas.

At small values of the angle θ shown in Fig 8.15, $\sin\theta \sim \theta$ and, if the slits are separated by 1 mm and the wavelength of the light is 500 nm, the maxima and minima are separated in angle by $\lambda/a = 5\times 10^{-4}$ radians $= 0.029$ degrees. If a screen is distant 1.5 m from the slits, the separation of the maxima and minima on the screen $\sim \lambda D/a = 0.75$ mm.

Exercise 8.8 Two antennas 200 m apart are fed by in-phase alternating currents and send out waves of wavelength 3 m distributed isotropically. Receivers are placed on a line parallel to the line joining the antennas. The

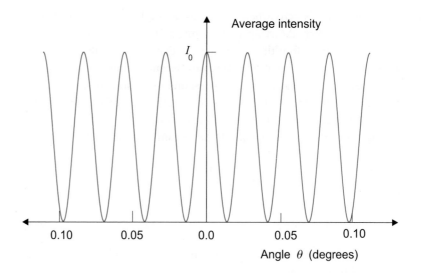

Fig. 8.16 The pattern of fringes resulting from the addition of wavelets from the two-slit system shown in Fig 8.15. The slits are separated by 10^{-3} m and are illuminated with light of wavelength 500 nm. The effects due to the nonzero widths of the slits are ignored and the curve is that predicted by the simplified treatment that results in eqn (8.49).

perpendicular distance between the two lines is 20 km. What is the separation of points that receive no signal? If the currents are out of phase by 30°, by how much will the position of a minimum change?

Answer The separation of minima is 300 m. When the currents are out of phase by 30° the waves from the two antennas start out 30° out of phase and the position of each minimum moves by 25 m.

Nonzero slit width

In the above derivation of the intensity pattern from two slits we ignored the widths of the slits and assumed that each gave out an in-phase disturbance of amplitude A_1. To be more exact we must regard each very small element across the widths of the slits as producing in-phase wavelets as we did in treating diffraction. Each of these wavelets now reaches the point P slightly out of phase. If the slits have a width d, the more exact treatment of the two-slit problem gives the result that the intensity given by eqn (8.49) is modulated by the intensity pattern due to diffraction of the incident plane wave at each slit. The more exact average intensity is given by

$$\bar{I} = I_0 \cos^2\left(\frac{\pi a}{\lambda}\sin\theta\right)\frac{\sin^2\left(\frac{1}{2}kd\sin\theta\right)}{\left(\frac{1}{2}kd\sin\theta\right)^2}. \qquad (8.52)$$

This intensity distribution is shown as a function of angle θ in Fig 8.17.

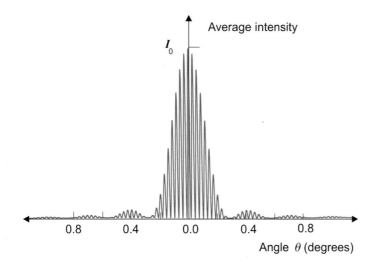

Fig. 8.17 The intensity distribution observed when two narrow slits of width 10^{-4} m separated by a distance of 10^{-3} m are illuminated by a plane wave of monochromatic light of wavelength 500 nm. The angle θ gives the departure from the straight-through beam measured in a direction perpendicular to the long sides of the slits, as shown on Fig 8.15.

Worked Example 8.4 A plane wave of monochromatic coherent light passes through two slits each of width 10^{-2} mm separated by a distance a. If the hundredth bright fringe away from the straight-through maximum of the two-slit interference pattern falls at the first minimum of the diffraction pattern due to each slit, what is a?

Answer The first minimum of the diffraction pattern from slits of width d occurs at an angle θ given by

$$\sin \theta = \frac{\lambda}{d}.$$

The hundredth maximum of the two-slit pattern, after that which occurs at zero degrees, occurs at the hundredth angle that makes eqn (8.49) a maximum after that at zero. This angle is given by

$$\sin \theta = \frac{100\lambda}{a}.$$

Hence the solution is obtained from

$$\frac{100\lambda}{a} = \frac{\lambda}{d}.$$

The distance a is thus 1 mm.

Diffraction gratings

A **diffraction grating** consists of a large number of parallel slits of width d, very small compared with the slits' length, ruled on a substrate. The slits allow the passage of light, while the space between the slits is opaque. Let there be a total of N slits with centres separated by a distance a. As in the previous discussions of diffraction and interference, if the long side of a slit is parallel to the y-axis and the perpendicular to the plane of the slits is in the z-direction the problem reduces to the two-dimensional one shown in Fig 8.18.

If a plane wave of monochromatic light falls normally on the grating, the transmitted intensity observed when rays emerging at the same angle θ to the normal are combined exhibits pronounced maxima as a function of θ. The separation and sharpness of the maxima depend upon the number of slits N and their separation.

Assuming that the width d of the slits is very small, we may ignore the effect of the nonzero slit width d that was taken into account when deriving eqn (8.52). Each of the N slits is the source of a wavelet of amplitude A_1 and the phase of each wavelet on leaving a slit is the same. The wavelets spread out, and rays emerging at the same angle θ superpose on a distant screen, or are focused by a lens, and combine to give a resultant amplitude A.

Combination at large distance corresponds to adding wavelets with the phases they have over the line CD shown. Over this line, wavelets from successive slits are out of phase by the amount equivalent to a path difference of $a \sin \theta$, as in the case of two slits. These path differences correspond to phase differences of $2\pi a \sin \theta / \lambda$, where λ is the wavelength. Hence the resultant amplitude is

● *The resultant amplitude from a diffraction grating*

$$A = A_1 \sum_{n=0}^{n=N-1} \cos(\omega t - k\ell - \phi_n)$$

where $\phi_n = 2\pi n a \sin \theta / \lambda$. This can most easily be evaluated if the waves are represented as the real parts of a complex expression, when the sum becomes a geometrical progression. The result is

$$A = A_1 \cos(\omega t - k\ell) \frac{\sin(N\pi a \sin \theta / \lambda)}{\sin(\pi a \sin \theta / \lambda)}.$$

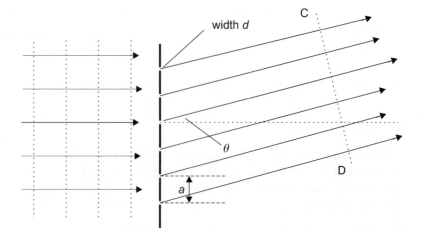

Fig. 8.18 A plane wave incident normally on a diffraction grating. The view looks down on the long length of the slits and shows, for clarity, a grating with only six slits.

The average intensity is

$$\bar{I} = I_0 \frac{\sin^2(N\pi a \sin\theta/\lambda)}{\sin^2(\pi a \sin\theta/\lambda)} \qquad (8.53)$$

● *The average intensity*

with I_0 the average intensity that would be seen if there were only one slit.

Figure 8.19 shows the intensity pattern observed when the number of slits N is ten and illustrates the properties of the function on the right-hand side of the above expression for the average transmitted intensity. Large intensity maxima, often called principal maxima, occur when

$$a \sin\theta = n\lambda \qquad (8.54)$$

● *Maxima and minima of the intensity distribution*

with n an integer or zero, as for two slits. The integer n defines the **order** of the principal maximum; $n = 1$ corresponds to the first order, $n = 2$ to the second, and so on. Smaller maxima occur when

$$a \sin\theta = \frac{(n' + \frac{1}{2})}{N}\lambda$$

where n' is an integer. There are $(N - 2)$ small maxima between each principal maximum and the next, and $(N - 1)$ minima.

The intensity at a principal maximum is $N^2 I_0$, the result of the in-phase addition of the amplitudes from each of the N slits.

For large N, the principal maxima are very sharp, since a small phase difference between wavelets from successive slits results in a small resultant amplitude even when the angle θ departs by only small amounts from those values which satisfy eqn (8.54). The intensities at the peaks of the principal maxima are proportional to N^2, while the intensities of the subsidiary maxima in between the principal maxima decrease as N increases.

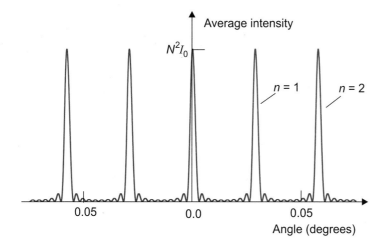

Fig. 8.19 The average intensity observed as a function of the angle θ when a monochromatic plane wave of wavelength 500 nm, incident normally, is transmitted through a diffraction grating that has 10 slits with centres separated by 10^{-3} m. The effect of the widths of the individual slits has been neglected. Compare this diagram with Fig 8.16. For practical gratings with a thousand slits or more with much smaller separation, the intensity pattern shows maxima with exceedingly narrow linewidths.

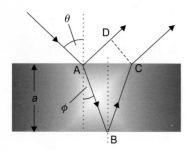

Fig. 8.20 A ray of light is partially reflected at the point A from the top surface of a thin film of oil of thickness a lying on water. A parallel ray emerges from the point C, after being reflected off the bottom surface of the oil, and interferes with the first ray.

● *Optical path difference*

Interference with thin films

A common example of interference observed with visible light is provided by the colours seen when a very thin layer of oil spreads over the top of a puddle or other smooth surface. White light of all colours irradiates the layer but some wavelengths experience destructive interference and others constructive interference due to path length differences between light reflected from the top and bottom of the oil film.

The mechanism can be illustrated with a simplified description of the process. Figure 8.20 shows a point source emitting light that strikes a thin film of oil on water. The eye looks at a small piece of the film and collects light reflected at a fixed angle θ. A ray with angle of incidence θ is partly reflected at the front face of the film at the point A and interferes with one transmitted into the film and reflected off the back face at the point B. This second ray emerges at the point C in a direction parallel to the first reflected ray, and both are focused by the eye and combined on the retina.

The geometry of the figure can be used to show that the difference in distances traversed by the two rays, the distance $AB + BC - AD$, is given by

$$\Delta \ell = \frac{2a}{\cos \phi} - 2a \tan \phi \sin \theta$$

where the first term is a distance travelled in oil and the second a distance in air.

If the oil has a refractive index n, a distance ℓ travelled in oil changes the phase of a wave by an amount $2\pi n\ell/\lambda$, while a similar distance in air changes the phase by an amount $2\pi \ell/\lambda$, taking the refractive index of air to be unity. The **optical path length** of a ray is defined as

$$\Delta \ell_o = \sum_i n_i \ell_i, \tag{8.55}$$

where ℓ_i is the distance travelled by the ray in medium i, which has refractive index n_i.

Using the law of refraction, eqn (8.17), the optical path difference between rays reflected off the front and rear surfaces of the oil film and recombined in the eye can be shown to be

$$\Delta \ell_o = 2na \cos \phi,$$

corresponding to a phase difference $(2\pi/\lambda) \times 2na \cos \phi$. This phase difference may result in destructive interference, in which case the corresponding wavelength is missing from the light observed and we see a colour. Defining the optical path difference to be the refractive index times the actual path difference gives the same phase difference for the same optical path difference in any medium.

In addition to the phase changes introduced by path differences, a phase change of π, equivalent to an optical path difference of $\lambda/2$ with λ the free-space wavelength, occurs when the light is reflected at the top surface of the oil at A, as given by eqn (8.29). The second wave reflected at B is incident in oil and is reflected at the boundary between oil and water. The refractive index of water for light waves is less than that of oil and so the second ray suffers no phase change. This phase difference must be considered when estimating which wavelengths are missing in the light reflected off films.

The same physical principles that give rise to interference are responsible for the colours of thin soap films, and are utilized in thin-film optical coatings, which are used to reduce reflections on camera lenses or spectacles.

● *Phase changes in reflection of light by thin films*

8.8 Wave focusing and ray optics

We have discussed how a wave spreads out from a source and how it is diffracted when passing through or round obstacles. Another property of certain waves is that their wavefronts can be altered in shape, and hence rays bent, by suitable choices of refracting media with appropriate shapes. A diverging wavefront emitted from a small source object can be altered so that it converges towards a small image spot. If a small source of radio waves with a wavelength of a few centimetres is placed a few metres on the left-hand side of a piece of paraffin wax shaped like a disc bulging at the centre, the source object can be focused to a small image spot on the right-hand side. Likewise, light from an object can be focused to an image by a lens of the right shape made of glass or other transparent material.

The focusing action of curved surfaces can be satisfactorily described by regarding light as waves. It can be shown that waves emitted from an object point converge to an image point such that the time taken for all rays to travel from object to image is the same. This is called **Fermat's principle** and can be proved using electromagnetic theory. The time taken is also the minimum time for any path between object and image.

In addition to focusing due to refraction of waves at boundaries between air and media, focusing can be effected by reflection of waves at surfaces. Light from an object can be brought to a focus by reflection off the curved surface of a mirror. Using Fermat's principle it is possible to determine the effect on wavefronts of arbitrarily shaped surfaces separating media with different refractive indices. However, in this chapter we adopt a simpler approach and treat the problem in terms of rays, which were introduced in Section 8.3. We discuss first-order optics and consider only rays within a narrow cone making a small angle with the line from object to image.

Single surface

Figure 8.21 shows part of a spherical boundary between two media of refractive indices n_1 and n_2. The centre of curvature of the surface is at C and the radius of curvature is R. A ray from an object at the point O strikes the surface at A at an angle of incidence θ. It is refracted at an angle of refraction ϕ and hits the axis containing the points O and C at the point I. This axis is called the **optical axis**. The surface cuts the optical axis at the point P.

To a first approximation, all rays that leave the object and are close to the axis converge on the point I which is thus an image of the object at O. The distance OP is the object distance and given the symbol u; the distance PI is the image distance and given the symbol v. Simple geometry and the law of refraction together give a relation between the image and object distances, the refractive indices, and the radius of curvature R.

The law of refraction gives

$$n_1 \sin \theta = n_2 \sin \phi$$

and, since we are considering only rays that are close to the axis, the angles θ and ϕ are small and $n_1\theta = n_2\phi$. Also, looking at the triangles in Fig 8.21 (which is distorted for clarity so that the ray OAI appears not to satisfy the 'near-axis' approximation), $\theta = \alpha + \beta$ and $\beta = \phi + \gamma$. Combining these relations gives

$$n_1(\alpha + \beta) = n_2(\beta - \gamma). \tag{8.56}$$

Since the angles α, β, and γ are small their tangents are approximately equal to the angles themselves and $\alpha = h/u$, $\beta = h/R$, and $\gamma = h/v$ where h is, to a good approximation, the vertical distance from A to the optical axis. Substitution of these data into eqn (8.56) and rearranging gives

$$\frac{n_1}{u} + \frac{n_2}{v} = (n_2 - n_1)\frac{1}{R}. \tag{8.57}$$

Fig. 8.21 A spherical surface with centre of curvature at the point C separates two media with refractive indices n_1 and n_2. To a good approximation, all rays from the point O close to the axis are deflected at the surface and focused on the axis at the image point I.

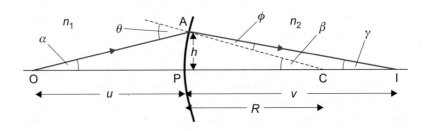

Fermat's principle gives a physical explanation of the image formation. The refractive index on the right is greater than that on the left and the wave travels more slowly in the medium on the right than it does on the left. A ray on the optic axis meets the medium with the higher refractive index at P at an earlier time than a ray such as OAI, which corresponds to a region of the wavefront off axis, meets the medium at A. The ray OAI is still travelling in the medium with the lower refractive index while the on-axis ray is travelling in the medium with the higher refractive index. The wavefront bends and adjusts itself so that the time taken for ray OAI to reach I is the same as the time taken for the ray on the axis to reach that point.

At this point a sign convention has to be adopted in order to obtain the correct results for surfaces that have different curvatures with respect to the object point. With an object O on the left, draw a line connecting O to the centre of curvature C of the spherical surface. This line is the optical axis and intersects the surface at the point P. The surface is *convex* to the point O if the centre of curvature is to the right of P, and *concave* if the centre of curvature is to the left. The sign of the radius R is positive if C is to the right of P and negative if it is to the left. The object distance u is positive if O is to the left of P.

● *The sign convention*

With this convention, eqn (8.57) gives the image distance v, and the sign of v may be positive or negative depending on the values of u and R. If v is positive the image point I lies to the right of P and the image is *real*, the rays converging and passing through the point I. A real image can be seen on a screen placed at P. If v is negative the image point I lies to the left of P and the image is *virtual*, the rays diverging at the surface and appearing to come from the point I. A virtual image cannot be seen on a screen placed at the image point and can only be seen looking through the refracting surface.

Thin lens

We now derive the formula relating object and image distances for a thin lens such as one used in spectacles to remedy eye defects. A lens for visible light is made of transparent material like glass and has two curved surfaces. Depending on the shape of the surfaces and the position of the object, the lens may either converge the rays from the object to produce a real image or may diverge them to give a virtual image.

Figure 8.22 shows an object in air a distance $u = OP$ from the point P where the axis intersects the lens, assuming the lens to be thin so that its width can be neglected. The first surface S_1 bends a ray OA towards the axis and would form an image at the point I_1 if the glass filled all the space to the right. In applying eqn (8.57) to this surface we thus have that $u = OP$, $v_1 = I_1P$, and $R_1 = PC_1$ with C_1 the centre of curvature of S_1.

Fig. 8.22 A thin lens with refractive index n has surfaces S_1 and S_2 with centres of curvature at C_1 and C_2, respectively. The optical axis of the lens is the line joining the two centres. The lens is assumed to be thin enough for distances to be measured from the point P on the axis.

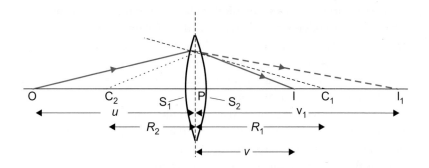

Equation (8.57), with n_1 put equal to unity for air and $n_2 = n_g$, becomes

$$\frac{1}{u} + \frac{n_g}{v_1} = (n_g - 1)\frac{1}{R_1}. \tag{8.58}$$

For the second lens surface S_2 the rays are entering in glass and appear to come from an object at I_1. This object is to the right of the point where the lens cuts the optic axis and so the object distance to be inserted into eqn (8.57) is $u = -v_1$. The rays exiting the second lens surface enter air and form a focus at the point I. Application of eqn (8.57) thus requires $n_1 = n_g$ and $n_2 = 1$, and the equation giving the final image distance v becomes

$$-\frac{n_g}{v_1} + \frac{1}{v} = (1 - n_g)\frac{1}{R_2}. \tag{8.59}$$

Adding eqns (8.58) and (8.59) results in the **lens maker's formula**

$$\frac{1}{u} + \frac{1}{v} = (n_g - 1)\left(\frac{1}{R_1} - \frac{1}{R_2}\right) \tag{8.60}$$

where u and v are the distances from the lens of the object and image distances, respectively, and R_1 and R_2 are the radii of curvature. The signs of all of these distances must be allocated using the sign convention adopted above. The factor $(n_g - 1)\left(\frac{1}{R_1} - \frac{1}{R_2}\right)$ in m^{-1} is called the **power** of the lens, and its reciprocal is called the **focal length** of the lens and given the symbol f. In terms of the focal length the lens maker's formula becomes

$$\frac{1}{u} + \frac{1}{v} = \frac{1}{f}. \tag{8.61}$$

If an object is very far away from a lens such that $1/u$ to a good approximation becomes zero, $v = f$, and the image is at the **focal point**, distance f from the lens.

Worked Example 8.5 A lens is made of material of refractive index 1.48. It is shaped as shown in Fig 8.23 and has surfaces with radii of curvature

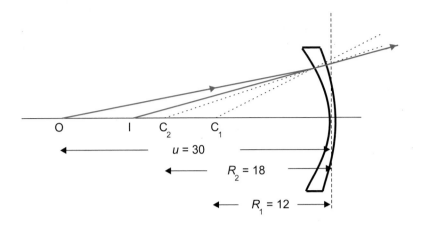

Fig. 8.23 A double-concave lens
gives a virtual image at the point I of
an object at the point O.

12 cm and 18 cm. The more curved surface of the lens, with radius 12 cm, faces an object on axis 30 cm away. Where is the image? Where is the image if the object is 3 m from the lens? Where are the images if the lens is turned round?

Answer In the first arrangement of the lens the surface on the left has radius of curvature -12 cm and the second surface has a radius of curvature -18 cm. The lens power is thus $0.48 \times (-\frac{1}{0.12} + \frac{1}{0.18})$, giving a focal length of -75 cm. For $u = 30$ cm the image is given by eqn (8.61) to be 21.4 cm to the left of the lens and is a virtual image. For $u = 3$ m the image remains virtual and is situated 60 cm to the left of the lens.

If the lens is turned round the power is now $0.48 \times (+\frac{1}{0.18} - \frac{1}{0.12})$. The power and the focal length are the same as before and the image position remains the same for a fixed object position.

A lens focuses rays incident parallel to the optical axis at the focal point. A ray through the centre of the lens is undeviated, because at the centre the lens is simply a thin parallel-sided slab. Using these properties of particular rays, it is easy to show using Fig 8.24 that, if we have an extended object, say a vertical post of height h, the transverse magnification equals the image distance over the object distance. With our sign convention, there is also a multiplicative factor of -1, since a real image is inverted and a virtual image is upright.

If two thin lenses are in close contact so that object and image distances may with sufficient accuracy be taken from the point where the lenses touch, it can be shown that the focal length of the combination is given by

$$\frac{1}{F} = \frac{1}{f_1} + \frac{1}{f_2}. \tag{8.62}$$

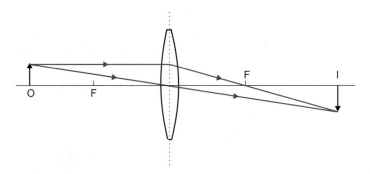

Fig. 8.24 A real image of an upright object is formed by a converging lens. The image is upside down and the magnification is the ratio of the image distance to the object distance.

The image position for two separated thin lens can be worked out using the image produced by the first lens as the object for the second together with the sign convention we have adopted.

Mirrors

Curved mirrors give focused images of objects by reflection, while the focusing action of lenses is due to refractive effects. The focal point of a mirror is the point where parallel rays from a distant object are brought to a focus, and this point may be determined by ray tracing, making the angles of incidence and reflection at the mirror equal. It is easily seen that the focal length for concave (centre of curvature to the left of the point where the mirror cuts the optical axis when the object is on the left as before) and convex (centre to the right) mirrors equals one-half the radius of curvature. If we adopt the same sign convention for radii of curvature and image and object distances, we can derive a formula relating these quantities in much the same way as the lens formula was derived.

Consider Fig 8.25, which shows a concave mirror cutting the optical axis at the point P and with its centre of curvature at C. An object at O, distant $u = OP$ from P, is imaged at I, distant $v = PI$ from P. Again we consider only rays close to the axis (the figure is out of proportion for clarity) so that all angles are small and the approximations may be used that their sines and tangents equal the angles themselves. In Fig 8.25, we see that $\beta = \alpha + \theta$ and $\gamma = \beta + \theta$, giving

$$\alpha + \gamma = 2\beta.$$

Since the angles are small, $\alpha = h/u$, $\gamma = h/v$, and $\beta = h/R$, where h is the vertical height of the point A above the axis and R is the radius of curvature of the mirror. Hence

$$\frac{1}{u} + \frac{1}{v} = \frac{2}{R}. \tag{8.63}$$

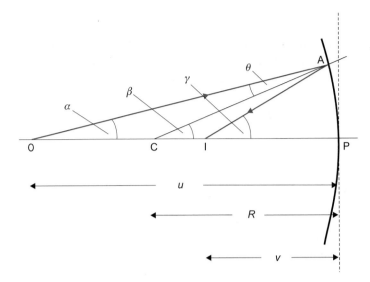

Fig. 8.25 A concave mirror with centre of curvature at the point C forms an image of a small object at the point O. The ray OAI, with angle of incidence θ, strikes the mirror at the point A and returns to the axis at the image point I.

In applying the above equation, the same sign convention as that adopted for lenses must be used. For example, in Fig 8.25 the centre of curvature of the mirror is to the left of the point P and the sign of R is negative. The image of a distant object is formed to the left of the mirror and v is negative as predicted by the negative sign of R. This image is real; the rays converge to the image, which can be seen on a small screen placed there. For mirrors, the sign convention we have adopted gives the correct positions of images but now a negative value for image distance v corresponds to a real image rather than a virtual one and a positive value for v corresponds to a virtual image. The situation with respect to real or virtual images is the opposite to that indicated with lenses.

Problems

Level 1

8.1 SONAR, which is an acronym for Sound Navigation and Ranging, is used at sea for determining the depth of water underneath a ship's keel. It works by measuring the time taken for a sound signal emitted from the bottom of the ship to be reflected off the sea bed and return back. If this time interval equals 0.1 s what is the water depth? The speed of sound in sea water is $\sim 1500\,\mathrm{m\,s^{-1}}$.

8.2 An underwater diver sees a distant object in the air above at 25° to the vertical. What is the real angle the object makes with the vertical? Take the refractive index of sea water for visible light to be 1.34.

8.3 A jar of depth d is half filled with a transparent medium that has refractive index n_1 and the top half is filled with a transparent medium that has refractive index n_2 with $n_2 < n_1$. Prove that the apparent depth of the jar, seen from above, is $(d/2)(1/n_1 + 1/n_2)$.

8.4 A train is travelling at 130 km per hour and blows a whistle of frequency 1500 Hz. What is the frequency heard by an observer by the side of the track as the train approaches? What is it when the train is past? The speed of sound is 340 m s^{-1}.

8.5 The phase velocity of water waves with wavelengths λ greater than a few cm in deep water is given by the relation

$$v_p = (g\lambda/2\pi)^{1/2}$$

where g is the acceleration due to gravity with value about 9.8 m s^{-2}. For a ship travelling at 15 nautical knots (1 knot is about 1.15 miles per hour, equivalent to about 0.51 m s^{-1}) calculate the angle the bow wave makes with the direction of travel of the ship on the assumption that the ship generates waves of wavelength 5 m.

8.6 What is the smallest angular resolution that an astronomical telescope of radius 8 m, working in the visible part of the spectrum at a wavelength of 0.55 μm, can, in principle, resolve?

In practice, fluctuations and turbulence in the atmosphere make the realizable resolution worse than the theoretical value.

8.7 Estimate the angular diameter of a celestial radio source that can be distinguished as other than a point when observed with a radio telescope 60 m in diameter used at a radiofrequency of 5×10^8 Hz.

8.8 Estimate the smallest possible angular divergence of a laser that has an aperture 1 cm wide when operating at a wavelength of 600 nm.

8.9 A plane acoustic wave meets a pair of parallel slits separated by 12 cm. The zeros in the interference pattern are spaced at angular intervals of approximately 10°. Estimate the wavelength of the sound waves.

8.10 If a diffraction grating illuminated by a plane wave incident normally gives the first maximum (away from zero degrees) of the transmitted light at an angle of 18° to the normal, what is the spacing of its slits? The incident light has a wavelength of 6×10^{-7} m.

8.11 A plane wave of light of wavelength 6×10^{-7} m falls normally on a diffraction grating that has narrow slits with centres separated by 10^{-5} m. At what angles to the normal are the first and second maxima (away from zero degrees) of the transmitted light intensity?

8.12 An object 1 cm high is 10 cm away from a convex lens on its principal axis. An image is formed 15 cm from the lens. What is the size of the image and the focal length of the lens?

8.13 Newton's form of the lens equation is $xy = f^2$, where x and y are the object and image distances measured from the lens foci. Show that Newton's formula is equivalent to eqn (8.61).

8.14 A lens with refractive index 1.48 is submerged in a liquid that has refractive index 1.61. How is its focal length changed?

Level 2

8.15 A prism is a wedge of transparent material that has a constant cross-sectional shape in the form of a triangle. A plane wave of nearly monochromatic light strikes one face of the equilateral glass prism shown in Fig 8.26 at an angle of incidence of 45° and emerges from the opposite face. What is the angular deviation of he plane wave if the refractive index of the glass is 1.55?

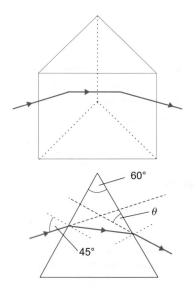

Fig. 8.26 A plane wave incident on one face of an

8.16 An aircraft flies at a constant height of 1 km at a speed of 0.7 km s^{-1}. A person on the Earth's surface hears a sonic bang. What is the distance between the aircraft and the individual when the sonic bang is heard? Take the speed of sound to be $340\,\mathrm{m\,s^{-1}}$.

8.17 Ocean waves of wavelength 10 m are diffracted by a gap of width 15 m in a harbour wall. Estimate the angular range within the harbour over which the amplitude of the waves is more than half the amplitude outside.

8.18 A radio telescope receiving a wavelength of 3 m is mounted on a cliff overlooking the sea 150 m below. If a source of radio waves rises over the horizon, at what angle of elevation will the received intensity first reach a maximum? Don't forget the phase change suffered by the ray reflected off the water. See the remarks at the end of Section 8.7.

8.19 A proposed design to ensure that aircraft coming in to land at an airport maintain the correct glide path uses a system of interference fringes produced by a radio transmitter working at a wavelength of 85 cm. The transmitter is at a height h above the ground, which acts as a reflecting plane for radiation that strikes it. Find the height h for a maximum signal to be received along a path at $3°$ elevation.

8.20 In the two-slit system of Fig. 8.15, where the slit separation is a, the wavelets from the two separate slits will start out of phase if the plane wave is incident on the slits at an angle ϕ. Show that the smallest value of ϕ that will give zero intensity in the direction perpendicular to the line joining the slits is approximately $\lambda/2a$.

8.21 Parallel light from a distant object passes from left to right through a converging lens of focal length 4 cm. This lens is 5 cm away from a second converging lens of focal length 4.5 cm. Where is the image of the object?

8.22 A ray of light enters a cylindrical transparent rod in a plane that contains the axis of the rod. If D is the diameter, L the length, and n the refractive index of the rod (which is in air), calculate the number of internal reflections made before the ray leaves the rod if it enters at an angle $\theta = 2°$ to the rod axis. Calculate this number for a ten-metre length of perspex ($n = 1.40$, $D = 10\,\mathrm{mm}$) and for a ten-metre length of glass fibre ($n = 1.55$, $D = 20\,\mathrm{\mu m}$). Estimate the transmitted intensity in both cases if the surfaces of the rods are such that only 99% of the intensity is internally reflected each time the ray strikes the surface of the rod.

Level 3

8.23 The lowest detectable power in a sound wave is about $10^{-12}\,\mathrm{W\,m^{-2}}$. Calculate the pressure change, particle velocity, and amplitude of a sound wave with this power in air for which the density is $1.2\,\mathrm{kg\,m^{-3}}$ and in which the speed of sound is $340\,\mathrm{m\,s^{-1}}$.

8.24 Light of frequency ν is scattered off a heavy particle moving in air with velocity **v**. The (dimensionless) unit vector in the direction of the incident wave is $\mathbf{k_i}$ and that in the direction of the scattered wave is $\mathbf{k_s}$. Show that the frequency of the scattered wave is given by

$$\nu' = \nu\left(1 + \frac{\mathbf{v}\cdot(\mathbf{k_i} - \mathbf{k_s})}{c}\right).$$

8.25 A radio transmitter emitting at 2 MHz free falls perpendicularly on to the surface of a distant planet from an infinitely large distance. The planet may be assumed to be spherical, homogeneous, and to have no atmosphere. Just before impact the signal received from the transmitter by a stationary detector situated on the line of fall is 200 Hz less than 2 MHz. Also, just before impact the rate of reduction in frequency of the received signal is 0.02 Hz per second. What are the mass and radius of the planet? The gravitational constant is $6.67 \times 10^{-11}\,\mathrm{m^3\,kg^{-1}\,s^{-2}}$.

8.26 A mirror is moving at speed v small compared with the speed of light. A ray of light falls on the miror surface at an angle of incidence θ as observed in the laboratory frame in which the mirror is moving. Calculate the angle of reflection in the laboratory frame. Remember that the law of reflection holds in the frame in which the mirror is at rest, and the angle of incidence θ' in the rest frame of the mirror equals the angle of reflection in that frame. The angle of incidence θ in the laboratory frame is different to θ' and the angle of reflection in the laboratory frame is different to both θ' and θ.

8.27 It is difficult to show diffraction or interference effects with X-rays with wavelengths shorter than about 5×10^{-9} m using ruled gratings or slits. However, interference can be produced by total internal reflection of X-rays at a mirror. At points on a detector distant from the source but on the same side of the mirror as the source, rays reflected at glancing angles off the mirror combine with direct rays from the source to give fringes. This wavefront-splitting interferometer is known as Lloyd's mirror.

A source of X-rays of wavelength 10^{-9} m is distant 1 m from a photographic plate and fringes are produced using reflection off a mirror whose surface is perpendicular to the plate. The mirror surface has a refractive index n of 0.9998 for the X-rays and this limits the angular range over which reflection occurs. What is the range of fringe spacing that might be observed?

8.28 A plane wave of monochromatic light is incident at an angle θ on a flat, thin film with parallel sides. The thickness of the film is d and the reflection coefficient for light incident from air is R. Derive expressions for the fractions of the incident intensity reflected and transmitted for small values of R.

8.29 A concave mirror of radius of curvature 2.5 m is positioned with its axis vertical and filled with a liquid of refractive index 1.3. The diameter of the mirror is small compared with its radius of curvature. Calculate the focal length of the system.

8.30 A converging and a diverging lens in contact produce an image of a distant object on a screen 25 cm away. If the diverging lens is moved 4 cm towards the screen, the screen has to be moved 6 cm to keep the image sharp. Which way must the screen be moved and what are the focal lengths of the lenses?

Some solutions and answers to Chapter 8 problems

8.1 75 m.

8.3 Figure 8.27 shows rays emerging from a point A on the bottom of the jar. These rays are collected by the eye of a viewer above A and appear to come from point C. (For clarity, the angles the rays make with the vertical are drawn much larger than they actually are.) From within medium 2 the rays appear to come from point B a vertical distance x_2 from the boundary between medium 2 and medium 1. The angles of incidence and refraction

at that boundary are related to n_1 and n_2. All angles of incidence and refraction are small and their tangents are closely equal to their sines and their cosines closely equal to unity. From the sines of the various angles in the figure it is straightforward to show that $x_2 = (d/2)(n_1/n_2)$ and then that $x_1 = DB/n_1 = (d/2) \times (1/n_1 + 1/n_2)$.

8.5 About 21°.

8.6 If the telescope looks at a small astronomical object, the maximum in intensity at zero degrees has zeros either side separated in angle by $\sim \lambda/d$, where d is the diameter of the telescope and λ the wavelength. A second object separated in angle from the first by less than λ/d will not be resolved. $\lambda/d = 20 \times 10^{-6}$ degrees.

8.8 3.4×10^{-3} degrees.

8.10 The first maximum away from zero when the grating is illuminated with a plane wave incident normally is given by eqn (8.54) with $n = 1$. This gives $a = 1.94 \times 10^{-6}$ m.

8.12 The inverted image is 1.5 cm high. The focal length of the lens is 6 cm.

8.14 The focal length of the lens is defined using eqn (8.60).

8.16 The aircraft is at point B on Fig 8.28 when the listener at C hears the sonic boom from the shock wave starting when the aircraft was at A. The angle α is given by $\sin \alpha = v_P/V$. But also $\sin \alpha = d/x$. Hence with $d = 1$ km, the distance $x = V/v_P = 2.06$ km.

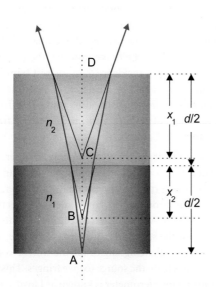

Fig. 8.27 To an observer above the jar, rays from a point A on the bottom appear to come from the point C.

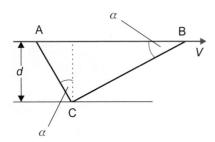

Fig. 8.28 A listener on Earth at the point C hears the sonic bang emitted by the aircraft when the aircraft is at the point B. The shock wave was actually emitted by the aircraft when it was at the point A.

8.18 Figure 8.29 shows a plane wave incident from over the horizon at an angle of elevation θ. The points B and C are in the wavefront and in phase. After reflection off the water the wave at C suffers a phase change of π equivalent to a path difference of $\lambda/2$. The rays passing through B and C, on recombining at the telescope at A, have a path difference $\Delta = \lambda/2 + \text{AC} - \text{AB}$. Hence

$$\Delta = \frac{\lambda}{2} + \frac{150}{\sin\theta} - \frac{150\cos 2\theta}{\sin\theta}$$

and, for the first maximum in the signal, $\Delta = \lambda$. This condition gives $\sin\theta = 1/200$ and $\theta = 0.29°$.

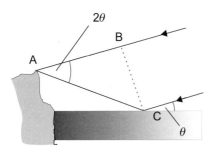

Fig. 8.29 The line BC is a wavefront of the plane wave at an angle of elevation θ. The transmitter at the point A receives the direct ray BA and the ray CA reflected off the water.

8.21 1.29 cm to the left of lens 2.

8.25 The speed v of the transmitter just before hitting is given by $\Delta\nu = \nu_0 v/c$ where $\Delta\nu$ is the frequency shift and $\nu_0 = 2 \times 10^6$ Hz. Hence $v = 3 \times 10^4$ m s^{-1}. Also

$$\frac{\mathrm{d}(\Delta\nu)}{\mathrm{d}t} = \frac{\nu_0}{c}\frac{\mathrm{d}v}{\mathrm{d}t},$$

and the acceleration just before hitting is 3.0 m s^{-2}. This is the acceleration g_s at the surface of the star, and thus

$$g_s = \frac{GM}{a^2} = 3.0$$

where G is the gravitational constant, a the star's radius, and M the mass of the star.

The kinetic energy just before hitting equals the loss of potential energy of the transmitter in the star's gravitational field and thus

$$\frac{1}{2}mv^2 = \frac{mGM}{a}$$

and

$$\frac{GM}{a} = \frac{1}{2}v^2 = 4.5 \times 10^8.$$

Combination of this with the expression above for GM/a^2 gives $a = 1.5 \times 10^8$ m, and then $M = 10^{27}$ kg.

8.27 Figure 8.30 shows the arrangement of the source S and the plate at a distance l. Since ℓ is large compared with the perpendicular distance $d/2$ of the source from the mirror, the rays that combine at P are rays that to a good approximation come from points S and S′ equidistant from the mirror and separated by a distance d. The fringe spacing at the plate is thus $\ell\lambda/d$. The smallest value of $d/2$ needed to accommodate an X-ray source we take to be about 10^{-5} m, and this value of $d/2$ gives the maximum fringe spacing to be about 10^{-4} m. The critical angle of reflection is given by $\sin\theta_c = n$, giving $(\pi/2 - \theta_c) = 0.02$ radians. Hence, the maximum value of d is about 0.02ℓ and the minimum fringe spacing is about 4×10^{-8} m.

8.29 0.96 m.

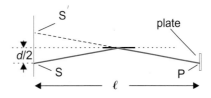

Fig. 8.30 The arrangement of source, mirror, and photographic plate.

Chapter 6 Oscillations

An object executes simple harmonic motion about a point when the force F on it is proportional to its displacement from the point and in a direction back towards the point. If x is the displacement,

$$F = -kx,$$

and the equation of motion of the object of mass m is

$$F(x) = -kx = m\frac{\mathrm{d}^2 x}{\mathrm{d}t^2}. \qquad (6.4)$$

The motion is periodic with angular frequency ω_0 satisfying

$$\omega_0^2 = k/m. \qquad (6.6)$$

The energy in the oscillations is constant and continuously exchanges between kinetic and potential, each with maximum value $m\omega_0^2 x_0^2/2 = kx_0^2/2$, where x_0 is the amplitude of the oscillations.

If there is a dissipative force involved in the motion, the energy in the oscillations is continually decreasing and the amplitude decreases with time. The equation of motion of a damped simple harmonic oscillator (when there is a dissipative force proportional to speed) becomes

$$m\frac{\mathrm{d}^2 x}{\mathrm{d}t^2} = -kx - b\frac{\mathrm{d}x}{\mathrm{d}t} \qquad (6.10)$$

where b gives the strength of the dissipative force. The solutions to this equation depend upon the relative magnitudes of the factors $\gamma^2 = (b/m)^2$ and $\omega_0^2 = k/m$. For strong damping, when $\gamma^2 > 4\omega_0^2$, the object does not oscillate after being displaced. If $\gamma^2 < 4\omega_0^2$, the object oscillates at angular frequency

$$\omega = (\omega_0^2 - \gamma^2/4)^{1/2}. \qquad (6.15)$$

The amplitude of the motion decays exponentially, with the displacement given by

$$x(t) = x_0 \exp(-\gamma t/2) \cos \omega t. \qquad (6.17)$$

If a damped simple harmonic oscillator is acted upon by a periodic force $F_0 \cos \omega t$, in the steady state, which is reached sooner or later depending upon frequency and damping, the object oscillates at angular frequency ω and with an amplitude

$$x_0 = \frac{F_0/m}{\{(\omega_0^2 - \omega^2)^2 + \omega^2\gamma^2\}^{1/2}}. \qquad (6.25)$$

The oscillations are out of phase with the applied force with phase lagging that of the force by an angle ϕ given by

$$\tan \phi = \frac{\omega\gamma}{(\omega_0^2 - \omega^2)}. \qquad (6.26)$$

The amplitude goes through a resonance when $\omega = \omega_0$, at which point the phase difference is 90° and the power absorbed by the oscillating object is a maximum. (See Fig 6.9.) For light damping, the resonance curve showing how the system behaves as a function of frequency is sharp. The sharpness of the resonance is given by the Q factor, which is approximately equal to ω_0/γ.

If a damped simple harmonic oscillator is acted upon by two periodic forces with different angular frequencies, the resultant motion is the sum of the two motions that would occur if the forces acted separately. The motions superpose, and the resultant is a displacement that oscillates at the average of the applied angular frequencies but with an amplitude that also oscillates in time with period $T = 4\pi/\Delta\omega$, where $\Delta\omega$ is the difference in the applied angular frequencies. This gives rise to beats since the amplitude goes through zeros at intervals of $2\pi/\Delta\omega$.

If a damped simple harmonic oscillator is acted upon by a force of any time dependence, as long as the force repeats itself regularly over time intervals T, it can be resolved by Fourier analysis into a sum of periodic forces of the type $F_n \cos(n\omega t)$ and $F_n \sin(n\omega t)$, with the angular frequency ω equal to $2\pi/T$. The resultant motion is then obtained by superposing the motions due to the separate resolved forces.

The least complicated system, other than a single oscillating object, is one consisting of two objects whose motions are coupled in some way. For linear restoring forces, as in SHM, the resultant motion of each object when either is displaced can be written in terms of two angular frequencies, called the mode frequencies. These are analogous to the angular frequency ω_0 for SHM of a single object. For certain initial conditions of the oscillating coupled system, the motions of the objects can be expressed in terms of one or the other of the mode frequencies only. These two states are called the normal modes of the system.

Chapter 7 **Waves**

A wave transmits energy between points without bodily displacement of any media between the points. For example, a wave on a string travelling in one dimension along the z-direction can be represented by solutions of a wave equation of the form

$$\frac{\partial^2 y}{\partial t^2} = \frac{F}{\rho}\left(\frac{\partial^2 y}{\partial z^2}\right) \tag{7.7}$$

where F is the tension in the string and ρ its mass per unit length. This equation applies if no energy is lost as the wave moves along the string. The simplest solution to wave eqn (7.7) is a plane wave of a single angular frequency ω moving in the positive z-direction. The mathematical representation of such a wave is

$$y(z, t) = y_0 \cos(\omega t - kz) \tag{7.8}$$

where k is the wavenumber; $k = 2\pi/\lambda$, with λ the wavelength. The wavelength and angular frequency are related to the wave velocity v_p of the wave by

$$v_p = \frac{\omega}{k}, \tag{7.12}$$

and the wavelength and frequency by

$$\lambda\nu = v_p. \tag{7.14}$$

Waves that obey the wave equation satisfy the principle of superposition: the addition of separate solutions of the wave equation is also an acceptable wave.

If the wave velocity v depends on frequency, the different components of a group of waves of different frequencies travel at different speeds and the shape of the group changes with distance. The wave is said to undergo dispersion, and the nonlinear relation between ω and k is called the dispersion relation. Most real waves suffer dispersion, and in that case the energy of a group of waves consisting of a narrow band of frequencies clustered around a mid-frequency travels with the group velocity v_g. The group velocity is given by

$$v_g = \frac{d\omega}{dk}. \tag{7.31}$$

If a wave is constrained between fixed points, at each of which it is reflected without loss of amplitude but perhaps with a phase change, standing waves are generated. Standing waves do not progress, and the function of z and t that describes them is of the form

$$y = 2y_0 \cos(\omega t + \phi/2)\cos(kz + \phi/2) \tag{7.34}$$

where y_0 is the amplitude of the forward- and backward-going waves, and ϕ is the phase angle between the forward- and backward-going waves at $z = 0$. Nodes where the amplitude is zero are separated by a distance of half a wavelength. Antinodes where the amplitude is greatest occur at regular spacings between the nodes. For a length ℓ between two fixed points at each of which the disturbance must be zero, as on a string fastened at $z = 0$ and $z = \ell$,

$$\ell = \tfrac{1}{2}n\lambda_n \tag{7.36}$$

where n is an integer. (See Fig. 7.15). Equation (7.36) requires that standing waves between the points must have frequencies

$$\nu_n = n\frac{v_p}{2\ell}. \tag{7.37}$$

If energy is lost from a wave as it travels or oscillates in a standing wave pattern, the amplitude of the wave decreases with time. The approximate expression for a wave in which the attenuation of the amplitude is small is

$$y = y_0 \exp(-\alpha z) \cos(\omega t - kz) \qquad (7.44)$$

where α gives the strength of the energy loss process.

Chapter 8 Wave phenomena

Sound waves and electromagnetic waves are those most commonly met. Sound waves consist of longitudinal oscillations of very small but macroscopic volumes of material under the influence of the pressure from preceding volumes. The speed of sound down a long thin rod of metal is given by

$$v_p = \sqrt{\frac{Y}{\rho}} \qquad (8.6)$$

where Y is Young's modulus and ρ is the density. The speed of sound in air is

$$v_p = \sqrt{\frac{1}{\rho \chi}} \qquad (8.9)$$

where χ is the compressibility of air and ρ its density. Electromagnetic waves consist of related electric and magnetic fields oscillating transversely to the direction of propagation of the wave. Electromagnetic waves need no medium for their propagation, and in free space travel at a fixed speed c. When travelling in a medium, electromagnetic waves travel at speed c divided by the refractive index n of the medium.

When a wave meets a discontinuity in the environment in which it is moving, the wave may be reflected and refracted. The law of refraction for a plane electromagnetic wave incident at an angle of incidence θ on a plane boundary between two media of refractive indices n_1 and n_2 is

$$\frac{\sin \theta}{\sin \phi} = \frac{n_2}{n_1} \qquad (8.17)$$

where ϕ is the angle of refraction. Some of a wave's energy is usually reflected at a boundary or discontinuity. Under certain conditions a wave may be totally internally reflected. For a wave on a string meeting a join past which the wavenumber is k_2, the ratio of the reflected amplitude to the incident amplitude is

$$\frac{y_0^r}{y_0^i} = \frac{k_1 - k_2}{k_1 + k_2} \qquad (8.24)$$

where k_1 is the wavenumber before the join. This expression can be applied to the reflection of an electromagnetic wave incident perpendicularly on a plane boundary separating a medium in which the wavelength is λ_1 from one in which the wavelength is λ_2. The ratio of reflected amplitude to incident amplitude is $(\lambda_2 - \lambda_1)/(\lambda_2 + \lambda_1) = (n_2 - n_1)/(n_2 + n_1)$, where n_1 and n_2 are the respective refractive indices.

If a source emitting a sound wave is moving with respect to an observer, there are Doppler shifts in the observed frequencies, which depend on the relative speed and on whether the source or observer is moving with respect to the air in which the wave propagates. For a source moving with speed V towards a stationary observer, the observed frequency is

$$\nu = \nu_0 \left(1 - \frac{V}{v}\right)^{-1} \qquad (8.31)$$

where v is the wave velocity of the wave. For an observer moving towards a stationary source of sound the observed frequency is

$$\nu = \nu_0 \left(1 + \frac{V}{v}\right). \qquad (8.35)$$

When a wave meets an obstacle whose dimensions are comparable with the wavelength, the wave suffers diffraction. It departs from straight-line propagation and spreads out with the time average of its intensity following a pattern characteristic of the size of the aperture or obstacle. If a plane, monochromatic, electromagnetic wave is incident upon a thin slit of width d, the intensity, averaged over time, at a distance from the slit very large compared with the wavelength is

$$\bar{I} = I_0 \frac{\sin^2(\frac{1}{2} kd \sin \theta)}{(\frac{1}{2} kd \sin \theta)^2} \qquad (8.41)$$

where θ is the angle in the plane perpendicular to the slit that the line from the point to the slit makes with the direction of the incident plane wave and k is the wavenumber of the wave. Diffraction limits the ability of instruments to resolve points or objects that are close together.

When two coherent waves are superimposed there are interference effects. Waves add coherently if they have the same frequency and there is a constant phase difference between them. Visible light usually consists of a multitude of wavelets of finite length with random phases. However, two coherent visible light waves can be produced if a wavefront is split, for example, by two narrow slits. Light waves from the two slits take different paths to points where they superpose and, if the individual wavelets are still overlapping in spite of the different paths taken, the waves add coherently. If a plane wave is incident perpendicularly to two slits each of width d separated by a distance a, and perpendicularly to the line joining the slits themselves, the average intensity observed beyond the slits is

$$\bar{I} = I_0 \cos^2\left(\frac{\pi a}{\lambda}\sin\theta\right)\frac{\sin^2\left(\frac{1}{2}kd\sin\theta\right)}{\left(\frac{1}{2}kd\sin\theta\right)^2}. \tag{8.52}$$

In this formula θ is the angle in the plane perpendicular to the slits between the direction of the incident wave and the direction along which the intensity is observed, and I_0 is the maximum intensity that would be observed from one slit alone. (See Fig 8.17.)

The focusing properties of lenses and mirrors may be comprehensively treated as problems in wave optics. However, they may also be treated at a certain level by consideration of light rays and their rectilinear propagation. The distance v from a thin lens of an image of an object at a distance u from a thin lens of focal length f is given by

$$\frac{1}{u} + \frac{1}{v} = \frac{1}{f}. \tag{8.61}$$

The corresponding distances from a mirror of radius of curvature R are given by

$$\frac{1}{u} + \frac{1}{v} = \frac{2}{R}. \tag{8.63}$$

These formulae must be used with the accompanying convention for the signs of the lengths they contain.

Part 3

Quantum physics

Chapter 9

The foundations of quantum physics

The observations that form the basis of quantum physics are described in this chapter.

In classical times, the Greeks considered what would happen if a lump of matter were successively divided into smaller and smaller pieces. Would the process ever come to an end, or would there be a smallest fragment that is indivisible? On philosophical grounds, they decided that there would be a limit to the division, and they called the smallest fragment an **atom**. We have kept the name atom for the smallest building blocks of ordinary materials, but we now know that atoms are themselves made up of even smaller constituents. In this chapter it is explained what atoms are and how they are made up from their constituent parts.

Many observations on atoms and other very small particles show that they behave in a quite different way from large-scale objects, and that in many ways they are similar to light. Very small particles do not follow the rules of classical mechanics, which are the subject of Part 1 of this book. Although Newton's laws are valid for everyday large-scale objects, a new theory is required to describe the behaviour of very small particles. The new theory is called **quantum theory** or **quantum mechanics**.

In this chapter some of the observations that cannot be explained by classical physics are discussed, and are used to illustrate the ideas of quantum mechanics. The equations used in quantum mechanics are set up and applied to atoms, molecules, and atomic nuclei in Chapters 10 and 11. Simple examples are chosen in which definite results can be obtained without using complicated mathematics. This should not be taken to imply that quantum mechanics is only relevant when discussing the behaviour of one atom or a few atoms, or that it is a specialized subject that few scientists need to know about. On the contrary, since matter is made up of atoms, the quantum mechanical properties of the individual atoms are usually important in determining the behaviour of matter on a large scale. The whole of solid state physics is based on quantum mechanics, and without it we should, for example, not be able to devise the electronic circuits that are so important in modern life. Chemistry too depends on quantum theory; the way different elements can combine to form chemical compounds is briefly explained in Chapter 11. It is just as important for all physicists to understand quantum mechanics as it is to know Newton's laws and the laws of electricity and magnetism.

9.1 The constituents of matter

At very low temperatures all materials are either solid or liquid. Solids and liquids are made up of atoms that are closely packed together and are described as **condensed matter** to distinguish them from matter in which the atoms are far apart. Nowadays we can look directly at the arrangement of atoms in solids. Figure 9.1 shows a picture of the atoms in a piece of aluminium, obtained with a scanning tunnel microscope. Each bright

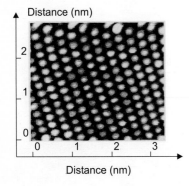

Fig. 9.1 A picture of the surface of an aluminium crystal. This picture was taken with a scanning tunnel microscope that is capable of making the individual aluminium atoms visible.

spot on the picture represents a single atom. The scale is labelled in **nanometres** (1 nanometre ≡ 1 nm ≡ 10^{-9} m). The distance between the centres of two neighbouring atoms is a few tenths of a nanometre. The atoms are stacked in rows like a box neatly packed with golf balls. The atoms have a well-defined size, and it would not be possible to pack them much more closely without squashing individual atoms. To this extent the analogy with golf balls is a good one, in that atoms are 'hard' in the sense that there is a strong force preventing them from being squashed. The difference between atoms and golf balls is that atoms attract one another when they are further apart—that is why solids and liquids form in the first place. The forces between atoms and how these forces affect the properties of matter are discussed in Part 4.

The structure of atoms

What happens to condensed matter when it is heated up? As the material gets hotter, the atoms move faster until eventually some of the atoms on the surface escape from the attraction of their neighbours. For a liquid this is called **evaporation**, for a solid it is **sublimation**. Often, a liquid contains atoms of different kinds, as for example water, which is made up of hydrogen and oxygen atoms bound together as water molecules. Water evaporates in molecular form—there is enough energy for the molecules to escape, but the affinity between the hydrogen and oxygen atoms is so strong that they remain bound together in each molecule.

The water molecules can themselves be broken up by giving them still more energy. This may be done by putting the molecules in a very high temperature environment so that they make energetic collisions. The molecules may also be dissociated into their constituent atoms by illuminating them with light of the right wavelength. We usually think of light as continuous, uniformly illuminating everything in its path. But, like matter, light is not indefinitely divisible, but comes along in packets called **photons**. When light is absorbed, each of the photons of which it is composed interacts with a single atom or molecule. If a photon is energetic enough it may, for example, dissociate a water molecule.

● *Molecules can be broken up by light*

Photons can also break up atoms—atoms are not indivisible as the Greeks imagined! The result of such a breakup is that the atom splits into two parts, one of which is much heavier than the other. The light particle, which is called an **electron**, carries a *negative electrical charge.* Electrical charges and the forces between charges are described in detail in Chapter 15. Here all we need to know is that there are two kinds of charge, called *positive* and *negative*, and that charges of the same sign repel one another while charges of opposite sign are attracted. Because electrons all carry a negative charge, they repel one another and they tend to move towards the positive terminal of a battery.

● *Electrical charges and forces*

Electrons are the same whichever atoms they have come from: all electrons have the same mass m_e and the same electrical charge $(-e)$. The SI unit of charge is the *coulomb*, which is denoted by the symbol C. The SI units are defined in terms of laboratory measurements on large-scale objects, and the magnitudes of the electron's mass and charge are extremely small in these units. To four significant figures, their values are

$$m_e = 9.109 \times 10^{-31} \text{ kilograms}$$

and

$$e = 1.602 \times 10^{-19} \text{ coulombs.}$$

Atoms are electrically *neutral*, which means that their net electrical charge is zero. When an electron carrying a charge $-e$ is removed from a neutral atom, the heavy particle that is left must have a charge $+e$. The heavy positively charged particle is called an **ion**. The positively charged ion and the negatively charged electron exert an attractive force on one another. Indeed, it is the electrical attraction that keeps electrons bound in a neutral atom. The unit of energy on the atomic scale is the **electron volt** (eV), which is the work needed to move an electron from one terminal to the other of a one volt battery. Electrical units are defined in such a way that one joule of work is needed to move one coulomb through one volt, and hence

One electron volt is equivalent to 1.602×10^{-19} joules.

The process of removing an electron from an atom is called **ionization**. Five to ten eV is sufficient energy to ionize most neutral atoms, leaving a **singly charged** ion with a net charge of $+e$. A second electron may be removed from a singly charged ion, leaving a doubly charged ion with charge $+2e$. Removal of each successive electron requires more and more energy, but this process can continue until all the electrons have been removed from the atom. The total number of electrons in a neutral atom is called its **atomic number**, usually designated by Z. The chemical properties of an atom are determined by its atomic number. Each chemical element has a different atomic number Z, and all atoms of a particular chemical element have the same atomic number.

● *Chemical elements are distinguished by the number of electrons in each atom*

When all the electrons have been removed from an atom, what is left behind is the **nucleus**, which includes nearly all the mass of the original neutral atom. The mass of the electron is almost two thousand times smaller than the mass of the lightest atom, hydrogen. For all other elements the mass of the Z electrons is an even smaller fraction of the atomic mass. Although it is heavy, the nucleus is extremely small, with a diameter only about one ten-thousandth of the distance between atoms in a solid.

The electrons in an atom move around the nucleus rather like planets in a solar system. This picture of an atom (see Fig 9.2) is quite good for reminding us that nuclei are small compared with atoms, but in other respects it is misleading. The electrons are not steadily moving in well-defined orbits like the planets. It is explained in Chapter 10 that it is not possible to predict the position of any one electron at a particular time—the electrons are more like a cloud of charge all round the nucleus. Nor can we add another electron with a kinetic energy of our own choosing as we do when launching a satellite to orbit the earth. The energy of electrons in an atom is restricted to a set of definite values, and only certain very particular states of motion are allowed.

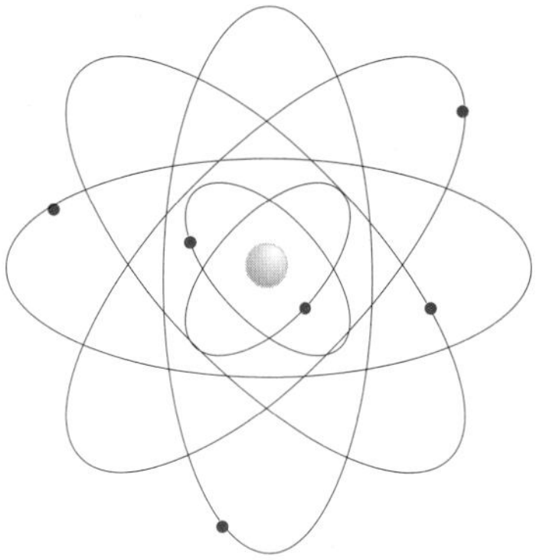

Fig. 9.2 A representation of the carbon atom, showing six electrons orbiting around the atomic nucleus. The picture of electrons in definite orbits was how atoms were imagined in the early days of quantum theory, but now we think of the electrons as though each one is smeared out into a cloud.

Worked Example 9.1 The ratio of the mass of the hydrogen atom to the mass of the electron is 1837, to the nearest integer. How many atoms are there in one microgram of hydrogen? (one microgram $\equiv 1\,\mu g \equiv 10^{-6}\,g \equiv 10^{-9}\,kg$.)

Answer The mass of the electron is given above as $m_e = 9.109 \times 10^{-31}\,kg$. The mass of the hydrogen atom is thus $1837 \times 9.109 \times 10^{-31} = 1.672 \times 10^{-27}\,kg$, and the number of atoms in one μg is $10^{-9}/1.673 \times 10^{-27} \approx 6 \times 10^{17}$. This result illustrates just how small atoms are. At ordinary pressures, one microgram of hydrogen gas occupies only a few cubic millimetres, yet it still contains an unimaginably large number of atoms.

The structure of nuclei

Electrons are regarded as one of the fundamental constituents of matter. They are all identical, and there is no evidence that they can be split up into smaller parts. Nuclei, however, have different charges for every element, and for each element there are many **isotopes**, which have the same atomic number (and hence the same chemical behaviour) but different masses. This diversity arises from the clustering together within the nucleus of two types of particle, called **protons** and **neutrons**. Protons carry a charge $+e$ and neutrons are uncharged, but otherwise protons and neutrons are rather similar. Since only the protons are charged, the number of protons is equal to the number of electrons in a neutral atom, i.e. to the atomic number Z. Just as electrons can be removed from atoms, so protons and neutrons can be removed from nuclei in energetic collisions. Nuclear forces are however much stronger than atomic forces: an energy of about seven MeV (millions of eV; $1\,\text{MeV} \equiv 10^6\,\text{eV}$) is required to remove a neutron or proton from the nucleus of a typical atom.

The **mass number** A of a nucleus is the sum of the number of protons and neutrons it contains. The masses of protons and neutrons are nearly the same, and hence the mass of a nucleus with mass number A is roughly A times the mass of a single proton. As an example to explain the

● *Nuclei are made up of protons and neutrons*

● *Masses on an atomic scale*

Energy scale

atoms	< 5 eV
electrons and ions	5 eV to 1 MeV
neutrons and protons	1 MeV to 1 GeV
quarks	> 1 GeV

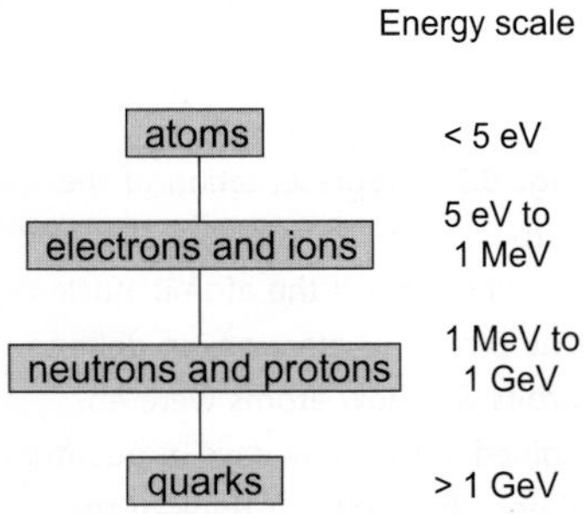

Fig. 9.3 The hierarchy of subatomic particles. Higher and higher energies are needed to investigate each step down in the hierarchy.

notation for describing isotopes, consider the element carbon, which has the chemical symbol C. About 99% of carbon is the isotope with mass number 12, which contains 6 protons and 6 neutrons. This isotope is referred to as carbon-12, and is written as ^{12}C. Sometimes Z is indicated by a subscript as in $^{12}_{6}$C, since it is not easy to remember the atomic numbers of all the elements. The mass of ^{12}C is roughly twelve times the mass of the proton. In fact the mass of ^{12}C is used as the standard of mass on the atomic scale. One **atomic mass unit** or **amu** is defined to be one-twelfth of the mass of a neutral carbon atom. In terms of kilograms

$$1 \text{ amu} = (1.661 \times 10^{-27}) \text{ kg},$$

to four significant figures.

Neutrons and protons are not themselves among the ultimate constituents of matter. When neutrons and protons are subjected to collisions with an energy of many GeV ($1 \text{ GeV} \equiv 10^{9} \text{ eV}$), their behaviour is explained by describing them as made up of **quarks**. Quarks, like electrons, are thought to be indivisible and to be among the fundamental building blocks of matter. There are six different kinds of quark, but they cannot be released in the same way as electrons from atoms or neutrons and protons from nuclei. No one has ever seen a free quark—they always go around in twos or threes. Figure 9.3 summarizes the hierarchy of particles that are constituents of matter and the energy appropriate to each level of the hierarchy.

9.2 Radioactivity

The simplest atom is hydrogen, which has the chemical symbol H. A neutral atom of hydrogen has one electron, so that its atomic number is one. Three different isotopes of hydrogen are known. Their nuclei contain zero, one, or two neutrons and each contains one proton (since $Z = 1$); their mass numbers are one, two, or three. The atom with a nucleus with mass number 2, containing one proton and one neutron, is called deuterium (chemical symbol D) or 'heavy hydrogen'. Although deuterium has the symbol D, its chemical behaviour is the same as that of ordinary hydrogen. Thus, just as the water molecule H_2O is formed by combining two atoms of H with one of oxygen (O), so deuterium can form the compounds HDO and D_2O ('heavy water'). The *physical* properties of D_2O are different from those of H_2O: it is about 10% denser than ordinary water and 'heavy ice' melts at 4°C, not at 0°C.

Both the proton and the deuterium nucleus are *stable*. This means that, so far as is known, a proton or a deuterium nucleus would remain unchanged forever if it were undisturbed. The third isotope of hydrogen, ^{3}H, is called tritium (chemical symbol T). Note that hydrogen is the only element that has special symbols for different isotopes. All the rest are

● *The isotopes of hydrogen*

labelled with the chemical symbol for the element together with the mass number, as described earlier for the isotope ^{12}C. Tritium is *unstable*. If tritium is kept for about 12 years, half of it is converted into ^{3}He, an isotope of the inert gas helium, which also has mass number three. The element helium has atomic number two, and its nucleus therefore contains two protons. In the transformation one of the neutrons in ^{3}H changes into a proton, as indicated schematically in Fig 9.4.

The phenomenon of the spontaneous transmutation of one nucleus into another, without the stimulus of any outside agency, is known as **radioactivity**. Radioactivity was discovered in 1896 by Antoine Becquerel, who found that photographic plates were darkened in the neighbourhood of salts of uranium, even though the plates were enclosed in opaque material. This unexpected observation that radiation of some kind is spontaneously emitted by the uranium salts caused great interest, and many other examples of radioactivity were soon discovered. The thickness of material that the radiation could penetrate was different from one source of radioactivity to another, and the radiation was divided into three categories; α-, β-, and γ-radiation. All three types of radiation cause ionization of the material through which they pass, and radioactivity is readily detected by observing this ionization. The names α-, β-, and γ-radiation are still used, although we now know the nature of each type of radiation. The properties of α-, β-, and γ-rays are summarized below.

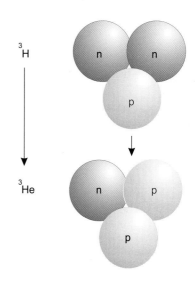

Fig. 9.4 Changing one neutron into a proton converts a ^{3}H nucleus into a ^{3}He nucleus.

● *α-, β-, and γ-radiation*

- α-rays are $^4_2$He nuclei, and α-rays from naturally occurring radioactivity are completely stopped by a sheet of paper or by a few cm of air. Emission of an $^4_2$He nucleus reduces the mass number of the radioactive nucleus by 4 and its atomic number by 2. For example, the polonium nucleus $^{208}_{84}$Po emits an α-particle, and is transmuted into the stable residual lead nucleus $^{204}_{82}$Pb.

- β-rays are electrons or particles called **positrons** that carry a charge $+e$ but are otherwise identical to electrons. The radioactivity is called β^- or β^+ for electrons or positrons, respectively; $\beta^\pm$-rays from natural radioactivity can penetrate several mm of plastic materials. After emission of a β-particle the mass number of the radioactive nucleus is unchanged but its atomic number is increased or decreased by one. For example, the long-lived naturally occurring isotope of potassium, $^{40}_{19}$K, emits an electron and is transmuted into the stable calcium isotope $^{40}_{20}$Ca, as is $^{40}_{21}$Sc (scandium) when it emits a positron.

- γ-rays are energetic photons, which may be able to pass through considerable thicknesses of lead. Since photons carry no charge or mass, emission of a γ-ray does not change the isotope. Frequently, γ-rays are emitted immediately after α- or β-radiation, but there are also long-lived radioactive nuclei that emit only γ-rays.

A large amount of energy is released in the radioactive transmutation of one nucleus into another. Nearly all of the released energy is carried away by the radiation, and only a small fraction is retained by the heavy residual nucleus. This is illustrated for α-decay in Exercise 9.1.

Exercise 9.1 When a nucleus of the isotope ^{238}U emits an α-particle the total energy released is 4.269 MeV. Conservation of momentum requires that the α-particle and the residual (^{234}Th) nucleus move in opposite directions to one another. Estimate the kinetic energy of the α-particle, assuming the masses of the residual nucleus and the α-particle to be 234 amu and 4 amu, respectively.

Answer 4.197 MeV.

Isotopes are known for nuclei with all atomic numbers between 1 and 112. Some of the 112 elements have no stable isotope. Stable isotopes occur for all atomic numbers between $Z = 1$ (hydrogen) and $Z = 83$ (bismuth: chemical symbol Bi) with the exception of $Z = 43$ (technetium: Tc) and $Z = 61$ (promethium: Pm). Isotopes of Tc and Pm can be made in the laboratory but they are radioactive. All the isotopes with atomic numbers from 84 to 112 are radioactive. The total number of known isotopes is over 2500, and the great majority are radioactive—there are only 266 stable isotopes.

Radioactive decay

For a sample of tritium gas in a closed vessel the transmutation into ^{3}He occurs smoothly, and the concentration of ^{3}He gradually builds up as the tritium disappears. After about 12 years, half of a sample of tritium is converted into ^{3}He. We can predict accurately the amount of tritium remaining at any time, since it follows an exponential law, which we shall derive below. But what happens if we have just one atom of tritium? When is this atom going to turn into a ^{3}He atom? No one knows: after some time has elapsed, maybe it will still be a tritium atom, or maybe it will have changed into a ^{3}He atom. All that can be said is that, after 12 years, half the tritium atoms in a large sample will have become ^{3}He atoms. For an individual atom the chances are the same that it will remain as tritium after 12 years or that it will have transmuted into ^{3}He.

This situation is very different from what happens in the classical physics discussed in Parts 1 and 2 of this book. In classical mechanics, if we know the state of a system now, we expect to be able to calculate its state at any time in the future. For example, when a stretched spring is released, it undergoes damped simple harmonic motion, and eqn (6.17) gives a definite value for its displacement at all subsequent times.

In radioactive decay we cannot work out the times when a particular atom will decay. We must use the language of probability. After a certain time has elapsed, we can calculate the probability that a tritium atom has transmuted into ^{3}He. It is a characteristic of quantum mechanics that predictions are only made about probabilities. Uncertainties are always present, but these uncertainties become negligible when we are dealing with large amounts of material, and classical physics then becomes applicable. Radioactivity is a very good example of this. Over a time comparable with the average time for a nucleus to transmute, an estimate of the time at which the change will occur for an individual atom is extremely uncertain. On the other hand, when dealing with, let us say, 10^{10} atoms— still less than a millionth of a millionth of a gram—the relative amounts of tritium and ^{3}He are known at all times with high precision.

● *Quantum mechanics can only predict probabilities*

Let us now generalize from the discussion of tritium and consider the radioactivity of any unstable isotope. The initial isotope is called the **parent** and, when it transmutes into another, it is said to **decay** into a **daughter** isotope. The probability per unit time that a parent atom will decay into a daughter atom is called the **decay constant**, usually denoted[1] by λ. The decay constant must be a constant for each radioactive isotope, independent of time, because there is no way of choosing a special origin of time. Provided that an atom has not already decayed, the probability that it will decay in the next short interval δt is $\lambda \delta t$, however long the atom may have previously existed.

Suppose that at a time t, there is a very large number N of parent atoms. Then the average number that will decay between t and $t + \delta t$ is $N\lambda\delta t$. The number of parent atoms has been reduced by this amount, and the change δN in N is

● *The equation for the radioactive decay of a large sample*

$$\delta N = -N\lambda\delta t.$$

Dividing by N and taking the limit as δt and δN become infinitesimal,

$$\frac{dN}{N} = -\lambda dt. \tag{9.1}$$

Integrating both sides of this equation gives

$$\int \frac{dN}{N} = -\lambda \int dt$$

and hence

$$\ln N = -\lambda t + C$$

---

[1] The symbol λ is also used in this chapter and elsewhere for wavelength. These notations are standard, and it is always clear from the context which quantity λ represents.

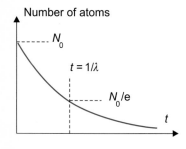

Fig. 9.5 The number of radioactive atoms in a sample decays exponentially from its initial value N_0.

● *The uncertainty in the number of atoms decaying*

where C is a constant of integration. If the number of atoms at $t = 0$ is N_0, then the value of C is $\ln N_0$, leading to

$$\ln N = -\lambda t + \ln N_0$$

or

$$\ln(N/N_0) = -\lambda t.$$

Finally, taking exponentials of both sides, the average number of parent atoms remaining at a time t is

$$N = N_0 e^{-\lambda t}. \tag{9.2}$$

The average number of parent atoms decays exponentially, as sketched in Fig 9.5. After a time $1/\lambda$, called the **decay time**, the number has reduced by a factor $e \approx 2.718$.

Equation (9.2) applies to the *average* number of parent nuclei remaining after a time t, but does not tell us anything about the uncertainty in N. What spread of numbers of remaining nuclei may we expect? This is actually a tricky problem in probability theory, because there is no uncertainty at $t = 0$ when we know that there are N_0 parent nuclei. Similarly, after very many decay times, it is almost certain that every parent nucleus will have decayed. However, at intermediate times there is an uncertainty and probability theory tells us that the greatest uncertainty occurs when half the atoms have decayed. At this time the *standard deviation* (a measure of uncertainty discussed in Appendix B) of the number of atoms is $\frac{1}{2}\sqrt{N_0}$.

For the example used above of 10^{10} tritium atoms, on average half of the sample is still tritium after 12 years, and the uncertainty in the number of tritium atoms is $\frac{1}{2}\sqrt{N_0} = 5 \times 10^4$. Now 5×10^4 is a large number, but nevertheless the *percentage* accuracy given by eqn (9.2) is very good—different measurements will vary from the average value by less than 10^5 parts in 10^{10}, or less than 0.001%. In this section we shall from now on refer to the 'number of atoms' in a radioactive decay when strictly the formulae apply to the average number. For practical purposes the spread about the average is negligible unless we are dealing with extremely small samples.

Worked Example 9.2 Use eqn (9.2) to show that the mean lifetime of a radioactive atom with decay constant λ is $1/\lambda$.

Answer From eqn (9.2), the probability that a single atom survives for a time t is $N/N_0 = \exp(-\lambda t)$. Since the decay probability per unit time is λ, the probability that it will decay between t and $t + \delta t$ is therefore $P(t)\delta t = \lambda \exp(-\lambda t)\delta t$: this expression satisfies the condition that the total probability of decaying at any time after $t = 0$ is one, since

$\int_0^\infty \lambda \exp(-\lambda t) \mathrm{d}t = 1$. The mean lifetime is thus $\int_0^\infty \lambda \exp(-\lambda t) t \mathrm{d}t$. Evaluation of the integral leads to the result $1/\lambda$.

The **half-life** $\tau_{1/2}$ of a radioactive isotope is the time taken for one-half of a sample of the isotope to decay. The half-life is usually the quantity listed in tables, and it is necessary to know the relation between half-life and decay constant. Starting with N_0 atoms, after a time $\tau_{1/2}$, $\frac{1}{2}N_0$ will remain. Hence $\frac{1}{2} = \exp(-\lambda \tau_{1/2})$, leading to

$$\lambda = \frac{\ln 2}{\tau_{1/2}} \approx \frac{0.693}{\tau_{1/2}}. \tag{9.3}$$

The half-lives of radioactive isotopes cover an enormous range. Many have half-lives much less than a second, and others have half-lives of many million years. Long-lived isotopes are found in nature: for example, thorium, the element with $Z = 90$, has an isotope, ^{232}Th, that is naturally occurring. Its half-life is 1.4×10^{10} years, roughly the same as the age of the universe.

The longest known radioactive half-life occurs in the α-decay of the isotope ^{144}Nd (neodymium), and is 2.1×10^{15} years. This is something like a million times longer than the age of the universe, so that only a very small fraction of the ^{144}Nd isotopes that have ever existed have decayed up to now. This raises the question of whether an isotope is radioactive or stable. Strictly speaking, an isotope is only stable if there is no way of breaking it into parts in such a way as to release energy. On this criterion, all nuclei with mass number greater than about 120 are unstable, since energy would be released if they split into two parts. They are *metastable*, in the same way that a stone in a hollow on top of a mountain is metastable. Energy would be released if the stone were to move to the bottom of the mountain. However, this will not occur since the stone cannot escape from the hollow. The same applies to the heavy 'stable' nuclei. They are held together by strong nuclear forces, which must be overcome if they are to reach the lower energy state when they split in two. In quantum mechanics there is an extremely small probability that the split will occur, but for practical purposes isotopes with half-lives much longer than the age of the universe may be regarded as stable. An indication of how small decay probabilities may be calculated for α-radioactivity is given in Section 10.5.

● *Isotopes with very long half-lives are effectively stable*

Large amounts of energy are carried by the ionizing radiation: for example, the energy of the α-particle released in the decay of ^{238}U was calculated in Exercise 9.1 to be 4.197 MeV. This is about a million times greater than the energy of a few eV needed to break up an atom or molecule to form an ion. When the α-particle slows down as it passes through matter, many ions are formed. For this reason the radiation emitted in radioactivity is referred to as **ionizing radiation**. The ionizing radiation damages material by breaking up atoms and molecules, and in particular it is harmful to living cells. Radioactive materials are very hazardous, and they must be used with great care.

The rate at which atoms decay in a sample of a radioactive isotope depends on the number N of atoms that are present and also on the half-life λ. Rearranging eqn (9.1) gives the rate of decay as

● **The definition of source strength**

$$\frac{dN}{dt} = -\lambda N. \tag{9.4}$$

The rate of decay is called the **source strength** of the isotope and it is measured in **becquerels** (symbol Bq), after the discoverer of radioactivity. One becquerel is one decay per second of the isotope in question. If N atoms are present and the decay constant is λ, then from eqn (9.4) the source strength is λN becquerels.

Although radioactive isotopes are hazardous, they may also be useful. Because the ionizing radiation from a single decay may be observed, the detection of radioactivity may be achieved with extremely high sensitivity. Radioactive isotopes have the same chemical behaviour as stable atoms of the same element and this sensitivity may be used to 'trace' the course of chemical and biological pathways. Tracer techniques are particularly important in medicine, where they may be used to learn about the reactions of biological molecules, which occur in minute quantities.

● **Radioactive clocks**

The half-life of a radioactive isotope is constant, and the change of its rate of decay according to eqn (9.2) can therefore be used to measure time intervals. For short times there is not much point in doing this—other methods of measuring time are generally more convenient. However, for very long periods, if there is some way of inferring the amount of a radioactive isotope that was present at an earlier time, the amount present now is a measure of the time elapsed since an event that occurred long ago. A number of different radioactive 'clocks' are based on this principle.

● **Carbon dating**

Carbon dating depends on the decay of the isotope ^{14}C. The element carbon has two stable isotopes, ^{12}C and ^{13}C. The unstable isotope ^{14}C is also found in nature, although it is very short-lived on a geological time-scale. It has a half-life of 5730 years, which means that in a sample of ^{14}C atoms, half will have decayed after a time of 5730 years. The reason that any ^{14}C is found in nature is that it is being continually produced by the bombardment of atmospheric nitrogen by cosmic rays. Carbon is an essential element in living things, and in both animals and plants carbon compounds pass to and from the atmosphere. As a result of this continual exchange of carbon, the small fraction of the isotope ^{14}C in carbon derived from living material is almost exactly the same in any sample from any part of the world. But, when the living object dies, there is no longer any exchange of carbon with the atmosphere. The ^{14}C decays and the proportion of radioactive ^{14}C to stable carbon gradually diminishes.

Over a range from a few hundred to more than 25 000 years, the age of bones, wooden objects, or fragments of woven materials can be estimated by measuring the remaining fraction of ^{14}C.

Worked Example 9.3 The Turin shroud carries the imprint of a man who, according to legend, was Jesus Christ. Three laboratories each made several measurements of the ^{14}C content of fragments of cloth from the shroud. The average value of the $^{14}C/^{12}C$ ratio was 92% of the value for modern material. Estimate the age of the shroud.

Answer In y years the ^{14}C content of a sample of dead material changes by a factor $x = 2^{-y/5730}$. Hence $y = -5730 \log x / \log 2$. Putting $x = 0.92$ gives the age y of the shroud to be about 690 years.

Exercise 9.2 The element uranium has two naturally occurring isotopes ^{235}U and ^{238}U. Their half-lives are 7.04×10^8 and 4.47×10^9 years, respectively. It is generally thought that the abundances of the two isotopes were the same when they were formed in a star. In samples of uranium found on Earth, 0.7% is ^{235}U and 99.3% is ^{238}U. Estimate how long ago terrestrial uranium was formed.

Answer 6×10^9 years.

Decay chains

As a radioactive parent isotope decays, so the daughter builds up. Generally, the parent decay is unique, although sometimes there may be alternative decay paths leading to different daughter isotopes. Consider first a parent isotope decaying with decay constant λ_1 to a single daughter isotope, which is stable. Starting with a sample of N_0 parent atoms, after a time t the number N_1 of parent atoms is given by eqn (9.2) as

$$N_1 = N_0 e^{-\lambda_1 t}.$$

Each parent atom is replaced by a daughter atom, so the number N_2 of daughter atoms is

$$N_2 = N_0 - N_1 = N_0\{1 - e^{-\lambda_1 t}\}. \tag{9.5}$$

The number of daughter atoms builds up from zero as shown in Fig 9.6.

The daughter may itself be radioactive, with a different decay constant λ_2 from its parent. The change δN_2 in the number N_2 occurring in a time δt is then the difference between the number formed from decay of the parent and the number decaying to the 'granddaughter':

$$\delta N_2 = \lambda_1 N_1 \delta t - \lambda_2 N_2 \delta t.$$

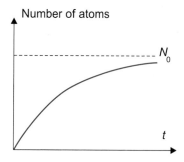

Fig. 9.6 When a radioactive source decays to a stable daughter, the number of daughter atoms builds up, eventually reaching the initial number of parent atoms N_0 after all the parent atoms have decayed.

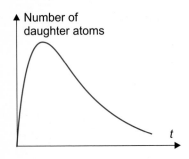

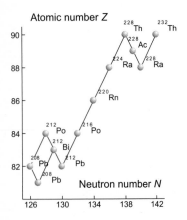

Fig. 9.7 If the daughter atoms of a radioactive source are themselves radioactive, the number of daughter atoms builds up and then decays.

● *Radioactive equilibrium*

Fig. 9.8 The decay chain of ^{232}Th.

In the limit as $\delta t \to 0$, this equation leads to the differential equation

$$\frac{dN_2}{dt} = \lambda_1 N_1 - \lambda_2 N_2 = \lambda_1 N_0 e^{-\lambda_1 t} - \lambda_2 N_2.$$

The solution to this equation, as may be verified by substitution, is

$$N_2 = \frac{\lambda_1 N_0}{\lambda_2 - \lambda_1} \left\{ e^{-\lambda_1 t} - e^{-\lambda_2 t} \right\}. \tag{9.6}$$

The growth and decay of the number N_2 is shown in Fig 9.7. At $t = 0$ there are no daughter atoms and the number N_2 begins to grow just as it did for the stable daughter. However, the daughter atoms themselves decay and, after a time such that both $\lambda_1 t$ and $\lambda_2 t$ are large, N_2 has fallen away to a small value.

If the half-life of the daughter is much shorter than that of the parent, i.e. $\lambda_1 \ll \lambda_2$, then the second term in eqn (9.6) rapidly becomes very small compared with the first. Neglecting λ_1 compared with λ_2 in the denominator, after a short time eqn (9.6) simplifies to

$$N_2(\lambda_1 \ll \lambda_2) = \frac{\lambda_1 N_0}{\lambda_2} e^{-\lambda_1 t}. \tag{9.7}$$

This limit is called radioactive equilibrium. A long-lived isotope may have a whole chain of daughters all decaying with the same long half-life, and each having a number of atoms inversely proportional to its own decay constant. This occurs naturally in the decay of the long-lived isotopes of uranium and of thorium (Th). The decay chains of ^{235}U, ^{238}U, and ^{232}Th all end at stable isotopes of lead (Pb). The decay chain of ^{232}Th is shown in Fig 9.8. All the radioactive isotopes in the chain have decay constants much larger than that of ^{232}Th, and all are decaying with the half-life of ^{232}Th, which is 1.4×10^{10} years.

9.3 Photons

As already mentioned in the introduction, light is not indefinitely divisible, but comes in a stream of particles called **photons**. Each photon carries a definite amount of energy. Such an indivisible amount of energy is called a **quantum** of energy and the photon itself is often described as a quantum of light. Quantum mechanics is the theory that allows for the discrete nature of quanta, and it gives very different results for the behaviour of very small objects from the classical theory, which deals only with continuous variables. Although light is made up of a stream of discrete photons, light nevertheless exhibits wave-like properties. Light provides an example of **wave–particle duality**, the name used to describe the puzzling fact that neither a wave nor a particle theory of light is adequate on its own.

Before discussing light in terms of photons, we must remind the reader of the evidence that light is propagated as waves. The wave nature of light is shown by the phenomena of diffraction and interference, which are discussed in Chapter 8. Diffraction and interference may easily be seen using a monochromatic plane wave of visible light, such as is provided by a small laser. If the light is passed through two narrow slits placed close together, the light reaches a screen behind the slits as light and dark bands spread out over a distance much wider than the separation of the slits.

A photograph of the pattern obtained in this way is shown in Fig 9.9. The light and dark bands show where constructive and destructive interference occur. The wavelength of the light can be calculated from the distance between the dark bands, as explained in Section 8.7. The wavelength of visible light is very short and is related to its colour: for yellow light the wavelength λ is about 500 nm. Equation (7.14) gives the relation between the wavelength λ and the frequency ν of a wave as $\nu = v/\lambda$, where v is the speed of the wave. Light of all wavelengths λ travel in a vacuum at the same speed $c \approx 3 \times 10^8 \, \mathrm{m\,s^{-1}}$, so the frequency for wavelength λ is $\nu = c/\lambda$. For $\lambda = 500$ nm this leads to $\nu = 6 \times 10^{14}$ Hz (the *hertz*, abbreviated as Hz, is the unit of frequency, which was defined at the beginning of Chapter 6; one Hz $\equiv$ 1 cycle per second.) A frequency of 6×10^{14} Hz is too high to measure directly but, since the wavelength of light can be measured accurately, the frequency is known equally accurately.

The frequency of light, as measured in interference experiments, determines the energy and momentum of the photons present in the light. The photon energy and momentum are established by observations on the photoelectric effect and on Compton scattering, both phenomena demonstrating the particulate nature of light. They are discussed in the remainder of this section.

The photoelectric effect

As mentioned in Section 9.1, neutral atoms may be ionized by being exposed to light. If the photon is absorbed by the atom, the process is called the **photoelectric effect**, and the electron released from the atom, leaving a charged ion behind, is a **photoelectron**. The photoelectric effect also occurs in solid materials, and may lead to the escape of photoelectrons from the surface.

The photoelectric effect at a solid surface is the basis for a very sensitive instrument for detecting light called a *photomultiplier*. A thin layer of the material that emits the photoelectron is placed on the inside of an evacuated glass container. A photon reaching this layer has a high probability of emitting a photoelectron into the vacuum. A strong electric field accelerates the photoelectron towards a metal surface with the property that an incoming electron can cause several electrons to leave the surface.

● *Wave properties of light*

Fig. 9.9 Interference fringes obtained with monochromatic light passing through two narrow slits.

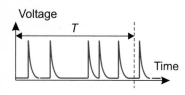

Fig. 9.10 The electrical pulses from a photomultiplier arrive at random times.

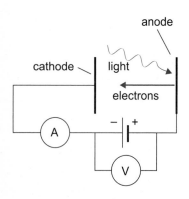

Fig. 9.11 A schematic diagram of the apparatus used to study the photoelectric effect. The space between the cathode and anode is a vacuum.

⬤ *No photoelectrons are ejected by light below a cut-off frequency*

These electrons are accelerated on to a further surface, and each generates several more electrons. This process is repeated many times, and the final result is that a single photoelectron leads to a huge number of electrons at the output of the photomultiplier.

The gain of a photomultiplier can be so large that a single photo-electron gives an output pulse that is easily observed on an oscilloscope. Figure 9.10 shows the response of a photomultiplier to a very weak light beam. The electrical pulses, and therefore the photoelectrons, arrive randomly, not at regular intervals. The *average* number of pulses in a fixed time interval, such as T in the figure, is proportional to the intensity of the light. However, after the arrival of one pulse, it is not possible to predict when the next one will arrive. Only the probability that a pulse occurs in any short interval δt can be known.

The explanation of this observation is straightforward when we regard the light as made up of photons. The light beam is not illuminating the photomultiplier uniformly as it would if it were a wave, but is arriving in discrete units—the photons—some of which have enough energy to liberate a photoelectron. The photons are arriving randomly because the emission of photons from the light source itself is random, just like the random emission of α-, β-, or γ-rays in radioactive decay.

How much energy is carried by a single photon? When photoelectrons are emitted from a metal surface, they have a range of different kinetic energies. However, for light of a particular frequency—or if you prefer, of a particular wavelength or colour—there is a well-defined maximum kinetic energy. This can be shown by measuring the current with the type of apparatus shown schematically in Fig 9.11. A voltage V is applied between two surfaces of the same metal, called the *cathode* and the *anode*, which are connected to the positive and negative terminals of a battery, respectively. There is an electric field between the anode and the cathode, which slows down photoelectrons emitted from the anode. Only those electrons with sufficient kinetic energy to overcome the decelerating voltage will reach the cathode and register a current in the ammeter A. No current at all is observed once the retarding voltage stops those electrons that have the maximum kinetic energy.

For light of any given frequency, the current measured at a fixed retarding voltage is proportional to the light intensity. But the voltage needed to stop the current is independent of intensity, and is determined only by the frequency of the light. Since one eV is the energy needed to move an electron through one volt, the voltage that just stops the current is the same as the maximum photoelectron energy in eV.

The variation of maximum kinetic energy with frequency is shown in Fig 9.12. The data points lie on a straight line which does not pass through the origin. For light below a certain minimum frequency, no photoelectrons are emitted. Extrapolating back to zero frequency, the straight line through the data intercepts the energy axis at a negative value of a few eV.

In the graph in Fig 9.12 the intercept (labelled ϕ in the diagram), is 2.2 eV and the slope $h = 4.1 \times 10^{-15}$ eV Hz$^{-1} \equiv 6.6 \times 10^{-34}$ J s. The intercept ϕ is different for different metals, but the slope always has the same value.

It helps in understanding the photoelectric effect to compare it with a simple gravitational analogy. Imagine a box containing a large number of identical balls, each with mass m, as in Fig 9.13. The box is not full, and the balls settle down to make a nearly flat surface at a distance d below the top of the box. The potential energy needed to raise a ball from the topmost layer to the top of the box is mgd. If a single ball is given a kinetic energy mgd and moves vertically upwards it just reaches the top of the box; mgd is the minimum amount of energy needed for a ball to appear at the top.

What happens to balls with kinetic energies greater than mgd? As an example let us assume that a number of balls is each given the same kinetic energy $3mgd$. Some of these balls will not appear above the top of the box either. They may be travelling at too shallow an angle, or may not start from the topmost layer, and lose energy in forcing a path through the higher layers. The maximum height above the box is for balls moving vertically upwards from the topmost layer. These balls rise a distance $2mgd$ above the topmost layer, reaching a height $2mgd$ above the box.

The maximum height above the box corresponds to the maximum electron energy in the experiment on photoelectrons, and the potential energy mgd corresponds to ϕ. The energy ϕ is thus the minimum energy needed to release a photoelectron from the surface of the metal: it is called the **work function** of the metal.

The energy of a photon

The observations on photoelectrons can be explained by making the following assumptions about light.

1. The energy of the light is carried in discrete packets called photons.

2. For a given frequency of the light, all the photons have the same energy.

3. The energy of a photon of frequency ν is proportional to ν and has the value

$$E = h\nu. \tag{9.8}$$

The universal constant h, which is called **Planck's constant**, occurs in many different contexts in quantum mechanics. To four significant figures, its value is $h = 6.626 \times 10^{-34}$ joule seconds. Equation (9.8) is an important relation between the particle and wave properties of light.

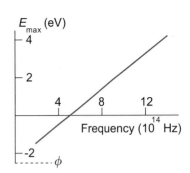

Fig. 9.12 The variation of the maximum photoelectron energy emitted from a sodium surface as a function of wavelength. These results, which were obtained by Millikan in 1914, were a key piece of evidence in favour of the quantum theory.

● *Photoelectrons need enough energy to overcome the work function*

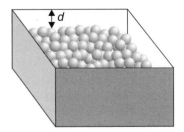

Fig. 9.13 To escape from a metal, electrons must overcome an attractive force. This is like the balls in the box in this diagram, which are held down by the gravitational force: a ball must have at least enough kinetic energy to reach the top of the box before it can escape.

Stating the relation in words:

The energy of a single photon equals the frequency of the light multiplied by Planck's constant h.

The equation that fits the results in Fig 9.12 is

$$E_{max} = h\nu - \phi. \tag{9.9}$$

Expressing this equation in words, we can say that the maximum energy of electrons liberated from a surface as a result of the photoelectric effect is equal to the photon energy minus the work function.

Exercise 9.3 An experiment is performed with clean nickel surfaces using an apparatus like the one in Fig 9.11. The work function of nickel is too high for any photoelectrons to be emitted with visible light, and ultraviolet (UV) wavelengths must be used. For UV of wavelength 150 nm, a retarding potential of 3.3 V is needed to cut off all photoelectrons, while for 75 nm the cut-off voltage is 11.6 V. Use these results to estimate Planck's constant and the work function of nickel.

Answer $h = 6.7 \times 10^{-34}$ J s and the work function of nickel is about 5.0 eV. It is difficult to obtain consistent results for the work function of a metal, because it depends critically on the state of the metallic surface. However, 5 eV is a typical value for nickel, which has one of the highest work functions. Sodium and the other alkali metals have the lowest work functions.

Photoelectric cells and Solar power

Photoelectrons can be emitted from other surfaces besides those of metals. For example, the element silicon has a work function of about $4\frac{1}{2}$ eV, which is found by observing photoelectrons in just the same way as for metals. Silicon is a *semiconductor*, which means that it is a very poor electrical conductor compared to a metal. Electric current is carried by electrons moving about within the silicon, and normally hardly any of the electrons from the silicon atoms are free to move. However, when photons with sufficient energy illuminate a silicon surface, some electrons may leave the atoms to which they were bound, without having enough energy to escape completely from the surface. These electrons can generate electric currents and voltages in a *photoelectric cell* (or *photocell*), which converts light energy into electrical energy.

Photocells are used to measure light intensity, for example, in the light meter of a camera. They can also generate significant amounts of power from sunlight. Huge amounts of power reach the Earth's surface even on cloudy days: however, because of the cost of fabricating silicon photocells, the power they generate is more expensive than electrical power derived from fossil fuels. Solar power is competitive in remote locations where the cost of transporting fuel or transmitting electricity is high. Large panels made up of solar cells are always used to power satellites and space probes.

Solar power uses up no resources as do fossil fuel power stations, and produces no harmful waste such as greenhouse gases. The cost of making solar panels is continually decreasing, and the efficiency of the photocells is increasing; in future solar power generation will surely become much more important than it is at present.

The momentum of a photon and Compton scattering

When the photoelectric effect occurs in metals, energy is transferred from a photon to an electron according to eqn (9.9). This equation is simply a statement of the conservation of energy, since the maximum energy of a photoelectron is the photon energy minus the minimum energy needed to remove the electron from the metal.

Nothing has been mentioned so far about the conservation of momentum. In the apparatus in Fig 9.11, the electrons leave the metal moving outwards: their momentum is pointing outwards. The photons also carry momentum in the direction they are moving, i.e. inwards. Since momentum is a vector, momentum cannot be conserved by the photon and the photoelectron on their own if they are moving in opposite directions. The missing momentum is taken up by the whole piece of metal, which suffers a small recoil after being struck by the photon. Because the metal is so massive, the energy of recoil is very small indeed and is justifiably neglected in eqn (9.9). This is rather like bouncing a ball off a wall—the wall absorbs the ball's momentum but, after it has bounced, the ball retains practically all of its initial kinetic energy.

If a photon interacts with an electron that is free to move on its own, but is unable to transfer momentum to a massive body, we must consider the consequences of energy and momentum conservation more carefully. Energy and momentum cannot both be conserved unless there is a third object as well as the incident photon and electron. In **Compton scattering** the third object is another photon. In this process an incoming photon transfers some of its energy to a free electron, and the remaining energy is carried away by a scattered photon. The net momentum of the electron and scattered photon must be the same as the momentum of the incoming photon, and as a result the relation between the energies of the incident and scattered photons depends on the angle of scatter.

● *Momentum and energy must both be conserved in Compton scattering*

How are the energy and momentum of a photon related? We can get a hint by looking at the energy–momentum relation for the electron. The electron's kinetic energy is $E_K = \frac{1}{2} m_e v^2$ and its momentum is $p_e = m_e v$, where m_e is the electron's mass and v its velocity. The energy–momentum relation can be written as

$$p_e = 2 \frac{E_K}{v}.$$

We might guess that, since the photon moves at a speed c, the equivalent relation for a photon of energy $h\nu$ would be $2h\nu/c$. This is not correct, because the expressions for the electron energy and momentum are only valid provided that the electron is moving at a speed much less than c, a condition that is not satisfied by the photon! The theory of relativity, which is the subject of Chapter 19, is needed for objects moving at speeds

● *The relation between energy and momentum for photons*

comparable to the speed of light, and indeed the expression $E_K = \frac{1}{2}m_e v^2$ that we have used for the electron is only valid if $v \ll c$. But our guess for the energy–momentum relation for the photon has the right dependence on energy and speed, though it is wrong by a factor of two. The correct expression for the momentum p_v of a photon with energy $h\nu$ is

$$p_v = \frac{h\nu}{c} \qquad (9.10a)$$

or, equivalently,

$$p_v = \frac{h}{\lambda}. \qquad (9.10b)$$

The important eqn (9.10a) may be expressed in words by saying that

The momentum of a photon is equal to its energy divided by the speed of light.

Now that we have established the relation between the energy and the momentum of a photon, we can go back to considering Compton scattering. To make the algebra as easy as possible, consider a collision in which the photon is scattered backwards through 180° and the electron moves forwards at a speed v in the same direction as the incident photon as shown in Fig 9.14. The energy of the incident photon is $h\nu$ and that of the scattered one is $h\nu'$. We shall again use the expressions $E_K = \frac{1}{2}m_e v^2$ and $p_e = m_e v$ for the kinetic energy and the momentum, respectively, of the electron, which are only valid if its speed $v \ll c$. The energies and momenta before and after the collision are the same, i.e.

$$h\nu = h\nu' + E_K = h\nu' + \tfrac{1}{2}m_e v^2 \quad \text{or} \quad h\nu - h\nu' = \tfrac{1}{2}m_e v^2 \qquad (9.11)$$

and

$$\frac{h\nu}{c} = p_e - \frac{h\nu'}{c} \quad \text{or} \quad h\nu + h\nu' = p_e c = m_e v c. \qquad (9.12)$$

Now $(h\nu + h\nu')^2 - (h\nu - h\nu')^2 = 4h^2 \nu \nu'$ and, from eqns (9.11) and (9.12),

$$4h^2 \nu \nu' = m_e^2 v^2 c^2 - \tfrac{1}{4}m_e^2 v^4.$$

The second term on the right-hand side of this equation can be neglected since the expressions for the electron energy and momentum are only valid if $v \ll c$ and, using eqn (9.11), we have

$$4h^2 \nu \nu' = m_e^2 v^2 c^2 = 2m_e c^2 \times \tfrac{1}{2}m_e v^2 = 2m_e c^2 h(\nu - \nu'). \qquad (9.13)$$

The wavelength corresponding to the frequency ν is $\lambda = c/\nu$, and eqn (9.13) can be rearranged as

$$\frac{c}{\nu'} - \frac{c}{\nu} = \lambda - \lambda' = \frac{2h}{m_e c}. \qquad (9.14)$$

$h\nu$

Before scattering

$h\nu'$ v

After scattering

Fig. 9.14 Compton scattering of a photon in the backward direction. The electron is knocked forwards, and the energy of the scattered photon is determined by the requirement that both energy and momentum are conserved.

Although eqn (9.14) was derived using approximate expressions for the electron energy and momentum, the equation is exact and applies even when the recoil electron has a speed near to c.

The diagram in Fig 9.15 represents a Compton scattering event in which the photon is scattered at an angle θ to the forward direction. In this case the components of momentum must be conserved both in the direction of the incident photon and perpendicular to it. The result for a photon scattered through an angle θ is that the wavelength change is

$$\lambda' - \lambda = \frac{h}{m_e c}(1 - \cos\theta). \tag{9.15}$$

This equation is again exact, applying even when the speed of the recoil electron is not small compared to c.

The quantity $h/m_e c$ is called the **Compton wavelength** of the electron. It has the value 2.43×10^{-12} m, a very small length compared to the wavelength of visible light. Yellow light has a wavelength of about 500 nm $\equiv 5 \times 10^{-7}$ m. According to eqn (9.15) the maximum wavelength shift is two Compton wavelengths, and for yellow light this is only a change of one part in 100 000. For much shorter wavelengths, i.e. more energetic photons, the relative change between the incident and scattered photons becomes larger. Rewriting eqn (9.15) again in terms of photon energies,

$$\frac{1}{h\nu'} - \frac{1}{h\nu} = \frac{(1 - \cos\theta)}{m_e c^2}. \tag{9.16}$$

Here $m_e c^2$ is a quantity with the dimensions of an energy, and its value is 511 keV.

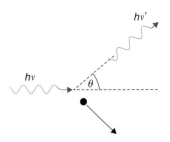

Fig. 9.15 In Compton scattering the electron need not be knocked in the forward direction. The scattered photon then must move as shown in the figure so that the momentum in the direction perpendicular to the incident direction is zero. The angle θ between the incident and scattered photons is the angle occurring in eqn (9.11).

● *Wavelength shifts are very small for Compton scattering of visible light*

Worked Example 9.5 A photon with energy 511 keV is scattered through 180° by a stationary electron. What is the energy of the scattered photon?

Answer Putting $\theta = 180°$ in eqn (9.16),

$$h\nu' = \left[\frac{(1 - \cos\theta)}{m_e c^2} + \frac{1}{m_e c^2}\right]^{-1} = \left[\frac{2}{m_e c^2} + \frac{1}{m_e c^2}\right]^{-1} = \frac{m_e c^2}{3}.$$

Since $m_e c^2$ is 511 keV, the energy of the scattered photon is $511/3 \approx 170$ keV.

Exercise 9.5 X-rays have wavelengths that are comparable to the size of an atom, which is about 0.1 nm. For incident X-rays of wavelength 0.1 nm, calculate the wavelength of photons Compton-scattered through 90°.

Answer 0.1024 nm.

Photons with the energy of 511 keV discussed in Worked example 9.5 are observed when the positively charged counterpart of the electron called the *positron* meets an ordinary negatively charged electron. Both particles disappear, leaving two photons, each of energy 511 keV, travelling in opposite directions. The 511 keV photons do indeed lead to back-scattered photons with an energy 170 keV, showing that eqn (9.10) gives the correct value for the photon momentum.

The energies we have been considering in this example are much larger than the energies needed to remove electrons from atoms. Typically 5–10 electron volts is enough energy to remove an outer electron from an atom—much the same as the work function of a metal. When a photon with an energy of tens or hundreds of keV interacts with an outer atomic electron, the energy transferred to the electron is huge compared to its binding energy, and the electron behaves almost as though it were free, rather than bound in an atom. Photon scattering occurs in all materials, and the energy loss is given to a good accuracy by eqn (9.16).

9.4 Wave–particle duality for photons

The photoelectric effect and Compton scattering show that light can behave as individual particles, each one having a definite energy and momentum as given by eqns (9.8) and (9.10). How is this to be reconciled with the fact that light exhibits interference, a property of waves? Look again at an interference pattern produced by a monochromatic plane wave of visible light passing through two slits. Figure 9.16(a) is a photograph showing how the intensity of light varies across a screen.

The photographic emulsion used in black and white photography consists of tiny particles of silver bromide suspended in gelatine. When the emulsion is developed, silver is deposited around active centres in the silver bromide particles. The developed photograph in Fig 9.16(a) is not smoothly changing from dark to light, as shown in Fig 9.16(b), which is part of the same photograph at a much greater enlargement. The graininess of the photograph is now visible. A spot on the photograph occurs where silver bromide has been activated by the light.

This is a quantum phenomenon and we have to talk about photons again. If the photon energy is too low, no activation occurs in the emulsion: this is why it is safe to look at ordinary film in red light. If the photons are energetic enough, the number of activated sites is proportional to the number of photons that have arrived. More photons will arrive at places where the wave intensity is high than where it is low. The *probability* of arrival of a photon is all that the wave theory can tell us. Where the first photon will activate a silver atom on a photographic emulsion cannot be known, nor if one photon has been detected is it

(a)

(b)

Fig. 9.16 Two photographs of an interference pattern detected in photographic emulsions. The only difference is that the exposure in (b) is much less than in (a), showing that the varying intensity across the pattern is determined solely by the density of developed grains.

● *Waves predict probabilities*

known where the next one will come. But as more and more photons arrive, their probability distribution—the density of spots on the photograph—will approach more and more closely to the wave intensity.

The probability distribution does not arise through photons interacting with each other, ganging up as it were to make up an interference pattern. This can be shown by photographing an interference pattern in such a faint light that there is only a small chance that more than one photon is in the whole apparatus at any time. After a long enough exposure, the same pattern emerges as from a strong light source. The wave pattern is carried by each separate photon, which has its own probability distribution independently of other photons.

The property of the photon of behaving as both a particle and a wave is called **wave–particle duality**. Although at first sight it may seem paradoxical, it is not really very mysterious. After all, once it is recognized that matter is made up of atoms, it is only possible for *individual* atoms to be excited by light. The wave and the particle aspect are describing different parts of a system's response to light. The wave theory gives advance information of the probability that light will reach different places; the particle picture is no good for this because it cannot account for interference. But once an interaction happens, as for example in a Compton scattering, the scattering is determined by the energy and momentum of a particular photon.

● *Photons interact with matter*

Whether the wave-like or the particle-like properties of light are more important in a particular case does depend to some extent on the wavelength, i.e. on the photon energy. For visible light both wave and particle aspects are easily demonstrated. Electromagnetic waves cover a huge range, however, from low-frequency radio waves to energetic γ-rays. All travel at the same speed c *in vacuo*, and all are made up of photons each with energy $h\nu$ and momentum $h\nu/c$. But for low-frequency radio waves these quantities are so minute that the effect of an individual photon is negligible. A wavelength of 1500 m is used for broadcasting in Britain, for example. The energy of one photon at this wavelength is 10^{-9} eV, far too small to cause any effect in a single atom. Huge numbers of photons at this energy can cause electrons to move in a wire and generate a current, but this process is better described by saying that the current is brought about by a changing electric field that moves along with the wave.

At the other extreme, the 511 keV photon we used as an example when discussing the Compton effect has a wavelegth 2.4×10^{-14} m. No objects can ever be positioned with an accuracy approaching such a small distance, and it is not possible to observe any interference effects. Wave theory is not usually needed when discussing the interaction of 511 keV photons with matter. For a beam of photons, wave–particle duality is of practical significance for wavelengths from about 10^{-5} m (infrared) to about 10^{-11} m (X-rays).

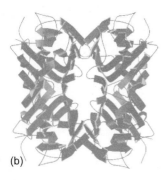

Fig. 9.17 (a) An X-ray diffraction pattern of a complex biological molecule. This pattern was obtained with the very intense X-radiation generated by circulating beams of high-energy electrons in a synchrotron radiation source. (b) The structure of the molecule, which has been derived from the positions and intensities of the spots in (a).

Although X-ray wavelengths are near to the limit where interference is not observed, X-ray diffraction is of the greatest practical importance. The separation of neighbouring atoms in a solid is typically about 0.1 nm, and strong interference occurs when X-rays with wavelength smaller than this spacing pass through solid materials. The X-rays scattered from the regularly spaced atoms in a solid are concentrated into spots in well-defined directions. From the intensities and directions of the spots it is possible to work backwards and deduce the arrangement of atoms in the solid. Figure 9.17 shows the pattern of a complicated biological molecule taken with radiation of wavelength 0.05 nm. This beautiful pattern has been used to work out the structure of the molecule, which is also illustrated.

9.5 Wave–particle duality for all particles

Since photons exhibit wave–particle duality, do ordinary particles like electrons have any wave-like properties? This question was posed in the 1920s by Louis de Broglie, who suggested that wave–particle duality should apply to all particles. He proposed equations for the wave frequency and wavelength in terms of the particle energy and momentum. A few years later de Broglie's equations were shown to be correct for electrons, and it is now known that they apply to *all* particles. Brief accounts of electron and neutron diffraction, i.e. of the wave-like properties of these particles, are given below. Electron and neutron diffraction are chosen not only because they are good examples of wave—particle duality, but also because both are important tools in the study of the structure of condensed matter.

The wavelength of matter waves

Two questions arise when we think about de Broglie's idea that wave—particle duality applies to particles like electrons, which have a mass when they are at rest. We can talk about 'matter waves', but what is it that is undergoing a wave motion? This is an important and difficult question, which is discussed in Chapter 10. The second question is: how can we detect the presence of matter waves? The answer to this question must be to look for interference. If a beam of electrons is wave-like, it will spread out after passing an obstacle, and both constructive and destructive interference effects will be observable. For visible light passing through two slits, the spacing of the interference bands, given by eqn (8.50), is proportional to the ratio λ/a of the wavelength λ to the slit spacing a. Interference is most easily seen if the spacing is not too much larger than the wavelength of the light. Similarly, the wavelength of the X-rays used to study crystal structure is similar to atomic spacings in crystals.

The same applies to matter waves. If we are to observe interference, we must choose structures that have spacings comparable to the wavelength of the matter waves. How can we guess that wavelength? When de Broglie made his proposal, both the photoelectric effect and Compton scattering were understood. The momentum of photons was known in terms of their wavelength according to eqn (9.10b). What de Broglie suggested was the simplest possible assumption, namely, that the wavelengths λ of particles like electrons are given by the *same* equation as for photons. Taking the wavelength from eqn (9.10b) to apply to the electron momentum p_e,

● *The de Broglie relation between wavelength and momentum*

$$\lambda = \frac{h}{p_e}. \tag{9.17}$$

Equation (9.17) is called the **de Broglie relation**. In words the de Broglie relation states that

The wavelength of a particle such as an electron is equal to Planck's constant divided by the particle's momentum.

The de Broglie relation was proposed in 1924: within a few years it was confirmed in independent observations of diffraction of electrons from crystals by Davisson and Germer in the United States and by Thomson in Britain.

Applications of the wave properties of electrons

The wave behaviour of photons is used in many applications. X-ray diffraction patterns such as the one in Fig 9.17 allow the structure of molecules to be worked out. An example with visible light is the formation of images by changing the shape of wavefronts with lenses. Both examples have counterparts using electrons instead of photons.

● *Electron diffraction*

Information about the internal structure of materials is best obtained with X-rays having wavelengths smaller than the separation of individual atoms, namely, about 0.1 nm. The same applies to electrons. It is not very difficult to prepare a beam of electrons all having very nearly the same energy. How much energy is needed for a wavelength around 0.1 nm? This can be calculated using eqn (9.17) by expressing the kinetic energy $E_K = \frac{1}{2} m_e v^2$ of the electron in terms of its momentum p_e:

$$p_e = m_e v = \sqrt{2 m_e E_K} \tag{9.18}$$

where m_e is the mass of the electron. From eqn (9.17)

$$\lambda = \frac{h}{p_e} = \frac{h}{\sqrt{2 m_e E_K}} \tag{9.19}$$

so that, if we want to have an electron wave with wavelength λ, the kinetic energy of the electron must be

$$E_K = \frac{h^2}{2 m_e \lambda^2}. \tag{9.20}$$

Worked Example 9.6 A typical atomic spacing in crystals is 0.1 nm. What is the kinetic energy, in eV, of electrons with wavelength 0.1 nm?

Answer Substituting in eqn (9.20), the energy in SI units is

$$\frac{(6.6 \times 10^{-34})^2}{2 \times 9.1 \times 10^{-31} \times 10^{-20}} = 2.4 \times 10^{-17} \text{ joules.}$$

Since $1 \text{ eV} \equiv 1.6 \times 10^{-19}$ J, this energy is about 150 eV. Shorter wavelengths occur at higher energies, as indicated by eqn (9.19). The kinetic energy of 150 eV is thus an estimate of the minimum energy useful for electron diffraction.

Fig. 9.18 An electron diffraction pattern of a metallurgical sample.

In practice electron diffraction is observed with electrons of energy higher than 150 eV, typically about 100 keV. As well as having shorter wavelengths, the more energetic electrons penetrate matter more easily. A modern electron diffraction pattern is shown in Fig 9.18. It shows light spots where there is constructive interference, just like the X-ray diffraction pattern in Fig 9.17. Given the photographs on their own, there is no way of telling which has been taken with electrons and which with X-rays. Wave–particle duality is interpreted in exactly the same way for electrons as for photons. The light spots on Fig 9.18 are places where there is a high probability that an electron will arrive. To calculate this probability we need to know the wavelength of the electrons (from the de Broglie relation) and the structure of the crystal. This calculation is done in terms of matter waves. But the electrons nevertheless act as particles that can be detected one by one. The arrival of an electron at a particular position has no influence on where the next one will come. Each electron has the same probability distribution.

In a microscope using visible light, a magnified image is formed by passing the light through lenses. Electron microscopes can be constructed on the same principles, although electron lenses are completely different from the glass lenses used to focus light. Electrons are stopped very quickly in solid materials, so they must be in a vacuum when passing through a lens. It is nevertheless possible to make electron lenses using electric and magnetic fields.

Electron microscopes are cumbersome and very expensive by comparison with an optical microscope. They are used in material science because, as we have seen, electrons have a much shorter wavelength than visible light. It may not be obvious why this is advantageous, since we usually think of lenses as bringing a bundle of light rays to a focus, rather than using a wave description. It is fine to assume that light travels in straight lines—i.e. that the ray picture used in Section 8.8 applies—if the sizes of apertures and obstacles and of the light beam are large compared with the wavelength.

● *Electron microscopes*

Lenses for visible light are normally enormous compared with the wavelength, and that is why it makes sense to draw ray diagrams to explain how they work. However, according to a ray diagram, parallel light falling on a perfect lens will be brought to a focus at a single point. This is not strictly correct when the wave properties of light are taken into account. The best possible focus is fuzzy and is spread out over a distance of about a wavelength. Similarly it is not possible to make a clear image of an object that is smaller than one wavelength. Since atomic spacings are typically several thousand times smaller than the wavelength of visible light, it is not possible to use an ordinary microscope to view structures on an atomic scale, no matter how large a magnification is used. X-rays do have wavelengths in the right range, of course, but it is very difficult to make lenses for X-rays. Primitive X-ray microscopes have been made, but for most purposes it is much more satisfactory to use an electron microscope.

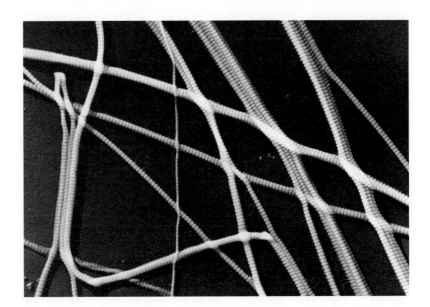

Fig. 9.19 An electron microscope picture of collagen molecules. The distance between neighbouring turns of the helical molecule is 67 nm.

Figure 9.19 is an electron microscope picture of a fibre of collagen, the molecule that occurs in the connective tissue of mammals. The molecule has a helical structure, and the distance between turns of the helix is 67 nm, as shown in the figure. This distance on the page is 1 mm, corresponding to a magnification of about 15 000. The energy of the electrons used in the microscope was 60 keV, corresponding to a wavelength 5×10^{-3} nm. The smallest detail visible in the figure has a size of a few nm, much greater than the wavelength. In practice, instrumental aberrations always limit the resolution attainable in electron microscopes. However, for electron microscopes as well as microscopes using visible light, even a perfect lens system will lead to a fuzzy focus instead of a point, and the focus will be about one wavelength across. We shall return to this unavoidable fuzziness and its relation to wavelength in Section 9.6.

Neutron diffraction

The properties of neutrons have already been briefly mentioned in Section 9.1. Neutrons and protons are constituents of atomic nuclei, and they account for most of the mass of an atom. Neutrons and protons have nearly equal masses, and the main difference between them is that protons carry positive charge while neutrons are uncharged. Because they are uncharged, neutrons do not experience electrostatic forces, and their only interaction with electrons is through a very weak magnetic force. The interaction between neutrons and nuclei is strong, and neutrons are scattered or captured when they collide with a nucleus. However, neutrons travel many times farther than the separation between neighbouring nuclei before making a collision. Energetic neutrons lose energy in scattering collisions and gradually reach thermal energies determined by the temperature T of the material in which they are moving. Thermal energy is discussed in Chapter 12, and a typical value is around $k_B T$ where k_B is a constant called Boltzmann's constant. The whole process of thermalization and eventual capture by a nucleus only takes a few milliseconds, so free neutrons occur only rarely in nature. Intense beams of thermal neutrons are however available in a few specialized laboratories, and these beams may be used to study diffraction of the neutron waves.

The mass m_N of a neutron is 1.675×10^{-27} kg, almost the same as the mass of a hydrogen atom, and nearly 2000 times the mass of an electron. Looking again at eqn (9.20), we see that the energy needed to achieve a particular wavelength gets smaller as the mass increases. The de Broglie relation $\lambda_N = h/p_N$ between neutron wavelength and momentum has been experimentally tested for neutrons with a high precision. At room temperature, the typical energy of a neutron in thermal equilibrium is about 0.025 eV, leading to a wavelength of about 0.18 nm. Thermal neutrons are just what is required to demonstrate diffraction from solid materials. Neutron sources are even more expensive and difficult to operate than electron microscopes. However, the information obtained from neutron diffraction is sufficiently different from what can be learned from electron or X-ray diffraction that it is worthwhile to build a few intense neutron sources mainly to be used for the study of solids and liquids. The diffraction pattern shown in Fig 9.20 was obtained using a neutron beam.

Two features distinguish neutron diffraction from electron and X-ray diffraction.

1. The energy and speed of thermal neutrons are of similar magnitude to those of the atoms in the material being studied. Information about the energy of the atoms and about how they are moving relative to one another can be found by using neutron diffraction.

2. Both electrons and X-rays interact with the electric charge within atoms. This means that a much bigger signal is seen from a heavy element that has a high nuclear charge and many electrons, than from a light one. It is very difficult to observe signals from hydrogen, an important constituent of biological molecules. This does not apply to neutrons, for which signals vary rather randomly from one element to another. This means that data from neutron diffraction are often complementary to data from electron and X-ray diffraction. Hydrogen, for example, gives a strong signal with neutrons.

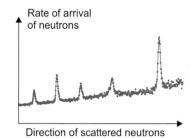

Rate of arrival of neutrons

Direction of scattered neutrons

Fig. 9.20 A neutron diffraction spectrum of a sample of H_2S at a pressure of 60 000 atmospheres. At this huge pressure the substance is a solid, and its structure can be deduced from the spectrum.

● *The special features of neutron diffraction*

As with electrons, it is not only in diffraction patterns that we can see analogies between matter waves and light waves. Neutrons can be prepared in a beam with a very low energy, much less than the typical value $0.025\,\text{eV}$ for neutrons in thermal equilibrium. These neutrons have a wavelength that is long compared to atomic spacings. Since the neutron is fuzzy over a distance of about a wavelength, such a neutron wave interacts with a large group of atoms, as does visible light. The combined effect of many atoms can be described in terms of a refractive index and, as for visible light, refraction and total reflection can be observed with neutron waves.

9.6 The uncertainty principle

In the discussion of electron microscopes in the previous section it was mentioned that no lens can focus light to a point. The diffraction of light waves always limits the smallest size of an image that can be achieved. It is important for optical instruments to be able to distinguish two images that are close together. The measure of this is called the **resolving power** of the instrument. For microscopes the resolving power is defined as the smallest distance between two objects that gives images that can be distinguished as separate. When viewing distant objects, as with an astronomical telescope, it is the angular separation that matters, and the resolving power of a telescope is defined to be the smallest angular separation of two objects that gives distinct images.

Consider a lens used to make an image of a distant object emitting light of wavelength λ. Figure 9.21 is a ray diagram illustrating the lens action for an object on the lens axis. Because the object is very far away, the rays reaching the lens are very nearly parallel. After passing through the lens the rays are bent towards the axis and cross it at the focal point. In the ray picture the focal point can be indefinitely small. But when we think of the light as waves, the aperture of the lens causes the light to spread out by diffraction so that the focus becomes fuzzy.

For a circular lens, the focal point is smeared out into a circular spot surrounded by a dark ring. Outside this dark ring is a series of bright rings, which get fainter and fainter the further they are from the focal point, as is shown in Fig 9.22. The intensity falls to zero at the dark rings, where the incoming waves interfere destructively. The radius of the innermost dark ring turns out to be about $\lambda/\sin\theta$, where the angle θ is the half-angle subtended by the lens at the focus. The largest possible value of $\sin\theta$ is one, so the smallest radius achievable for the image is about λ, the wavelength of the light passing through the lens.

How do we describe the smeared-out focus when talking about photons? Just as for Young's slits or an X-ray diffraction pattern, the

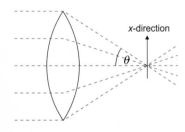

Fig. 9.21 The ray diagram shows light converging towards a focus. Because of the wave nature of light, the focus is not a point, but a blurred spot surrounded by faint rings.

changing intensity across the spot represents the changing probability of arrival. When a photon from the distant object reaches the telescope, it is diffracted by the lens and there is an uncertainty in its position when it arrives at the focal plane. Neglecting the faint outer rings, the photon will pass within the innermost dark ring, and its position in the x-direction (as labelled on Fig 9.21) has an uncertainty of $\delta x \approx 2\lambda/\sin\theta$.

There is also an uncertainty in the direction the photon is travelling— it may not be going along the axis, but upwards or downwards on the diagram in Fig 9.21. The greatest deviation from the axis is θ, for photons passing close to the edge of the lens. The momentum of the photons is $h\nu/c$, and the x-component of photon momentum varies between $\pm h\nu\sin\theta/c$. The x-component of momentum thus has an uncertainty $\delta p_x \approx h\nu\sin\theta$. This can be reduced by using a long focus lens, so reducing $\sin\theta$. But if we do this, the uncertainty δx in the photon's position at the focal point increases, since it varies as $1/\sin\theta$. The *product* of the momentum and position uncertainties is roughly

Fig. 9.22 The intensity of light at the focus of the lens in Fig 9.21. The intensity is greatest where the diagram is white, and the intensity is zero in the series of dark rings where destructive interference occurs.

$$\delta x \times \delta p_x \approx \frac{\lambda}{\sin\theta} \times \frac{h\nu\sin\theta}{c} = \frac{h\nu\lambda}{c} = h.$$

The product of the two uncertainties is determined only by the universal constant h, and is independent of the wavelength of the light and of the lens aperture and focal length. Any attempt to make a more accurate measurement of the two quantities simultaneously is always frustrated, and an uncertainty governed by the magnitude of h always remains.

The value of the uncertainty depends on the properties of the beam of light reaching the lens. However, it can be proved that, if Δx is defined to be the mean square deviation of the position x from its average value, and Δp_x the mean square deviation of the x-component of momentum, then there is a minimum value of $\Delta x\Delta p_x$ equal to $h/4\pi$. We can say that $\Delta x\Delta p_x$ is always greater than or equal to $h/4\pi$. The mathematical symbol for 'greater than or equal to' is $\geq$, and the equation equivalent to this statement is

$$\Delta x\Delta p_x \geq h/4\pi$$

or, defining $\hbar = h/2\pi$ ($\hbar$ is pronounced h-bar),

$$\Delta x\Delta p_x \geq \tfrac{1}{2}\hbar. \qquad (9.21)$$

The argument leading to eqn (9.21) has been derived by considering photons. However, the momentum of any particle is related to its

● *The uncertainty principle*
for momentum and position

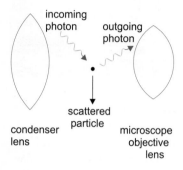

Fig. 9.23 Heisenberg's microscope. The uncertainty in direction of the photon implies a corresponding uncertainty in the momentum of the particle being viewed. Since the position of the particle is also uncertain to within about a wavelength of the photon, the uncertainty principle applies to the particle in the same way as it does to a photon.

wavelength in the same way as for a photon, and eqn (9.21) also applies to any object whatsoever. Equation (9.21) is called the **uncertainty principle**, and it embodies the limitations to the precision of measurement caused by the wave nature of matter. It applies equally to all the coordinate directions and, similarly, $\Delta y \Delta p_y \geq \frac{1}{2}\hbar$; $\Delta z \Delta p_z \geq \frac{1}{2}\hbar$. The uncertainty principle was formulated by Heisenberg in the 1920s and is still often referred to as Heisenberg's uncertainty principle.

Without actually making any measurements, Heisenberg considered a number of 'thought experiments' in the same way as we have considered the fuzzy focus produced by a lens. A famous example is the uncertainty in x and p_x of a particle when viewed in a microscope, as in Fig 9.23. The argument is very similar to the one we have already used for the lens. The object in the microscope is brightly illuminated by light focused by the condenser lens at the left-hand side of the diagram. If the object is to be observed, a photon must be scattered from it and detected after passing through the objective lens. The position and momentum of the scattered photon have an accuracy limited by the uncertainty principle. This means that the same uncertainty applies to the object, since its position is only known as precisely as that of the photon, and conservation of momentum requires that it must have a momentum uncertainty at least as great as that of the photon. The uncertainty principle applies to particles as well as to photons.

Exercise 9.6 Electrons with a kinetic energy of 10 keV pass through a slit with a width 0.1 mm in the x-direction. Estimate the uncertainty in the x-component of momentum, and the ratio of this uncertainty to the momentum in the forward direction.

Answer The momentum uncertainty is about 5×10^{-30} N s and the ratio to the forward momentum is about 10^{-7}. This very small number is also the diffraction angle given in eqn (8.43). The wavelength of the electrons is so small compared to the width of the slit that diffraction is negligible.

● *The uncertainty principle*
for energy and time

The attainable precision of other measurements besides those of position and momentum is also restricted by the uncertainty principle. The units of h are joule seconds; energy and time are linked quantities that must also obey the uncertainty principle in the form

$$\Delta E \Delta t \geq \frac{1}{2}\hbar. \tag{9.22}$$

This has the important consequence that anything that has a limited average lifetime—for example, a radioactive nucleus that undergoes decay—has energies spread out over a range of values: the shorter the lifetime, the more uncertain the energy.

Because Planck's constant is very small, the uncertainty principle is only of significance for measurements on microscopic objects such as electrons or single atoms. In measurements on large-scale objects, other sources of error are always dominant. However, as well as being important in the domain of quantum mechanics, the uncertainty principle has also had a philosophical impact. The mechanistic view, that if only we knew everything about a system we could predict its future behaviour exactly, is no longer tenable. Physicists now have to be more humble about their knowledge than were their predecessors in the nineteenth century.

Problems

Level 1

9.1 The density of ^{27}Al (aluminium of mass number 27) is 2.7×10^3 kg m^{-3} and the density of ^{197}Au (gold) is 17.0×10^3 kg m^{-3}. Calculate the number densities of the atoms of these two metals.

9.2 The caesium isotope ^{137}Cs was widely spread over Europe in the fallout following the Chernobyl accident. The half-life of ^{137}Cs is about 30 years. Estimate the activity in becquerels of 1 μg of ^{137}Cs.

9.3 The decay scheme of ^{232}Th, which has a half-life of 1.4×10^{10} years, is shown in Fig 9.8. All the daughter products have half-lives very much shorter than that of ^{232}Th. How many atoms of ^{224}Ra are present in one gram of ^{232}Th? The half-life of ^{224}Ra is 3.7 days.

9.4 Calculate the rate of energy loss from a 1 mg sample of pure ^{60}Co (cobalt), assuming that all the energy from radioactivity escapes from the sample. Use the following approximate data: the density of cobalt is 9 g cm^{-3}; the half-life of ^{60}Co is 5 years; the energy released by the decay of one ^{60}Co nucleus is 3 MeV.

9.5 An energy of about 1 eV is required to activate a particle of silver bromide in a photographic emulsion so that it will darken on development. Estimate the longest wavelength of light that can be recorded by the emulsion.

9.6 The work function of silver is 4.4 eV. A silver surface is illuminated by UV light with a wavelngth of 200 nm. What is the maximum kinetic energy of the photoelectrons emitted from the silver surface?

9.7 A photon with wavelength 350 nm is Compton-scattered through 90°. Calculate the wavelength difference between the incident and the scattered photon.

9.8 Calculate the number of photons in one wavelength along the beam: (a) for a 1 mW beam of visible light with wavelength 500 nm; (b) for a 1 μW beam of X-rays with wavelength 0.1 nm.

9.9 The ionization potentials (i.e. the energy needed to remove one electron) of neutral atoms of helium (He) and caesium (Cs) are 24.5 and 3.47 eV, respectively. Calculate the maximum wavelength of light that can ionize (a) helium and (b) caesium.

9.10 Calculate the de Broglie wavelength: (a) for a 20 eV electron; (b) for a lead atom with mass number 208 at an energy 0.056 eV (this is a typical energy for a lead atom evaporated at the melting point).

9.11 The diameter of a lead nucleus is about 1.5×10^{-14} m. What is the energy of a neutron that has a wavelength equal to this diameter? The mass of a neutron is almost the same as one atomic mass unit.

9.12 Estimate the speed of electrons and neutrons if each type of radiation has a wavelength of 0.1 nm.

9.13 The Z_0 particle was discovered at CERN in 1983. The root mean square deviation of the total energy of its decay products is about 3 GeV. Estimate the lifetime of the Z_0.

Level 2

9.14 The radioactive radon isotope ^{222}Rn is an inert gas. In the decay chain of the uranium isotope ^{238}U, ^{222}Rn is released and may reach significant concentrations in buildings situated on uranium-bearing rocks such as granite. Indeed, in the UK ^{222}Rn contributes on average about half of the population's total exposure to ionizing radiation. What is the source strength, in Bq, of all the radon released from one tonne (one tonne $\equiv$ 1000 kg) of ^{238}U? The half-life of ^{238}U is 4.5×10^9 years and the half-lives of all the daughter products are much shorter than this.

9.15 When the human eye is fully dark adapted, it can detect the presence of a light source so faint that only about 10 photons reach the retina during the time of about 1/20 s for which an image persists. Estimate the distance at which a 0.1 mW light source can be seen in ideal dark conditions. Assume that 20% of the electrical power is converted into visible light.

9.16 An apparatus like the one in Fig 9.11 is set up to measure the work function of magnesium. Using UV light of wavelength 250 nm, the maximum energy of photoelectrons is measured to be 1.36 eV. What is the work function of magnesium? What is the maximum energy of photoelectrons emitted when the magnesium surface is illuminated with UV at a wavelength of 200 nm?

9.17 An X-ray of energy 30 keV undergoes Compton scattering through an angle of 60°. What is the energy of the scattered X-ray?

9.18 Calculate the angle, with respect to the incident X-ray, of the electron recoiling from the Compton scattering in Problem 9.17.

9.19 A neutron with kinetic energy 0.1 eV is moving horizontally. How far does it move in the horizontal direction before falling through a height equal to one hundred times its de Broglie wavelength?

9.20 Show that the energy E_K of electrons that have the same wavelength as X-rays of energy E_X is $E_X^2/2m_ec^2$, making the assumption that the electrons are non-relativistic, having kinetic energy $\frac{1}{2}mv^2$. X-rays

from tungsten have an energy of 60 keV. What is their wavelength, and what is the energy of electrons with the same wavelength?

9.21 The radioactive isotope ^{99m}Tc is frequently used in medicine. The superscript m indicates that the isotope is not in its lowest energy state. It emits only γ-rays, changing neither the atomic number nor the mass number of the nucleus. The γ-ray has momentum, and to satisfy conservation of momentum the Tc residual nucleus must recoil. When the isotope emits a 140 keV γ-ray, what is the energy of the recoil nucleus?

9.22 An excited atomic state has a lifetime of 10^{-8} seconds, and decays by emitting a photon with wavelength 590 nm. Estimate the energy uncertainty of the photon and the uncertainty of its wavelength.

Level 3

9.23 A source of strength $1 \text{ MBq} \equiv 10^6$ Bq of the oxygen isotope ^{20}O is prepared. The half-life of ^{20}O is 13.5 s. It decays into ^{20}F (fluorine), which in turn decays with a half-life of 11.0 s to the stable neon isotope ^{20}Ne. Calculate the time after the 1 Mbq source is prepared when the daughter ^{20}F activity is strongest, and the fluorine source strength in Bq at that time.

9.24 A radioactive isotope with decay constant λ_1 has a single radioactive daughter, which has a decay constant λ_2. Starting with a pure sample of the parent isotope, obtain an expression for the time taken for the number of parent and daughter atoms to be equal. Hence show that this equality can only be achieved if $2 < \lambda_2/\lambda_1 < 3$.

9.25 A photon of wavelength λ is scattered through 180° by a stationary atom of mass m. Obtain an expression for the recoil velocity v of the atom, neglecting the difference between λ and the wavelength λ' of the scattered photon. Calculate v for a photon of wavelength 589 nm scattered through 180° by a stationary ^{23}Na atom. The scattered photon may be regarded as being emitted from an atom moving at a speed v, and thus has a Doppler-shifted wavelength. Use the wavelength shift from the Compton scattering formula to show that the Doppler shift has the same form as eqn (8.32), which was derived for sound waves.

9.26 A photon of energy 511 keV is scattered through 90°. By considering the conservation of momentum in the forward direction and in the direction of the scattered photon, show that the angle θ between the incident photon and the recoil electron satisfies the equation $\sqrt{5/4}\sin\theta\cos\theta = (\cos\theta - \sin\theta)$. To solve this problem, you must use eqns (19.42), (19.43), and (19.44) from the chapter on relativity. These equations give the relation between kinetic energy and momentum for objects moving at speeds comparable to the speed of light.

9.27 A beam of 10 keV electrons is collimated by passing through a long slit. Make a rough estimate of the width of slit that will give the narrowest beam at a distance of 1 m behind the slit.

Some solutions and answers to Chapter 9 problems

9.1 The number density of atoms in Al is 6.0×10^{28} m^{-3} and in Au is 5.2×10^{28} m^{-3}. The space occupied by a single atom of each of these two elements near opposite ends of the periodic table is nearly the same.

9.3 Since the half-life of ^{232}Th is very much longer than that of any of its radioactive daughter products, the daughters are all in radioactive equilibrium, decaying with the same decay rate as ^{232}Th. Thus $\lambda(\text{Th}) \times N(\text{Th}) = \lambda(\text{Ra}) \times N(\text{Ra})$ and

$$N(\text{Ra}) = \frac{\lambda(\text{Th})}{\lambda(\text{Ra})}N(\text{Th}) = \frac{\tau_{1/2}(\text{Ra})}{\tau_{1/2}(\text{Th})}N(\text{Th})$$

$$= \frac{3.7}{1.4 \times 10^{10} \times 365} \times \frac{6.02 \times 10^{23}}{232}$$

$$= 1.9 \times 10^9 \text{ atoms.}$$

9.7 The change in wavelength is 2.4×10^{12} m, less than 10^{-5} of the incident wavelength.

9.8 Since light travels at a speed c, for a beam power W, the energy in one wavelength is $W\lambda/c$. Each photon has an energy $h\nu = hc/\lambda$, and the number of photons in one wavelength is

$$N = \frac{W\lambda}{c} \times \frac{1}{h\nu} = \frac{W}{h}\left(\frac{\lambda}{c}\right)^2.$$

Thus: (a) for 1 mW of visible light at 500 nm, $N = 4$; (b) for 1 μW of X-rays at 0.1 nm, $N = 1.7 \times 10^{-10}$.

For the visible light there are several photons in one wavelength, and photons are more likely to arrive at a detector at times corresponding to peaks of a classical wave. For the X-rays, however, there are so many wavelengths between each photon that their arrival times are essentially random.

9.9 The maximum wavelength of light that can ionize He is 50 nm. This is ultraviolet light. The maximum

for Cs is 357 nm, which is blue light in the visible spectrum.

9.11 The momentum p of a neutron of mass m_N with kinetic energy E is $\sqrt{2m_N E}$ and, from the de Broglie relation, $\lambda = h/\sqrt{2m_N E}$. Hence

$$E = \frac{h^2}{2m\lambda^2} = \frac{(6.6 \times 10^{-34})^2}{2 \times 1.6 \times 10^{-27} \times (1.5 \times 10^{-14})^2}$$

$$= 5.9 \times 10^{-13} \text{ J}$$

or 3.7 MeV.

9.12 The speed of the electrons is 7.3×10^6 m s^{-1} and of the neutrons 4000 m s^{-1}.

9.13 The energy spread Γ and the lifetime τ satisfy the uncertainty relation

$$\tau \approx \frac{\hbar}{2\Gamma} \approx \frac{10^{-34}}{2 \times 3 \times 10^9 \times 1.6 \times 10^{-19}} \approx 10^{-25} \text{ s.}$$

For almost all purposes this lifetime is so short that the decay of the Z_0 may be regarded as prompt—in 10^{-25} s light can only travel a distance much shorter than the size of a nucleus.

9.14 The activity of the radon is 1.2×10^{10} Bq.

9.15 Since the energy of a single photon is hc/λ, if the efficiency of a lamp is ϵ the number of photons per second at a power W is $\epsilon W\lambda/hc$. At a distance R, the number N passing through a pupil of diameter r in a time τ is

$$N = \frac{\epsilon W\lambda\tau}{hc} \times \frac{\pi r^2}{4\pi R^2}.$$

Hence

$$R^2 = \frac{\epsilon W\lambda\tau r^2}{4Nhc}.$$

For $N = 10$, $\tau = 0.05\,\text{s}$, $\epsilon = 0.2$, $\lambda \approx 500\,\text{nm}$, $W = 0.1\,\text{mW}$, and $r \approx 0.002\,\text{m}$, $R \approx 500\,\text{m}$.

9.18 The electron recoils at at angle 58.5° to the direction of the incident X-ray.

9.19 0.19 m.

9.21 The momentum $h\nu/c$ of the γ-ray equals the recoil momentum p of the Tc nucleus, which thus has energy

$$\frac{p^2}{2m_{Tc}} = \frac{h\nu^2}{2c^2 m_{Tc}} = \frac{(1.4 \times 10^5 \times 1.60 \times 10^{-19})^2}{2 \times 99 \times 0.166 \times 10^{-27}}$$

$$= 1.70 \times 10^{-20}\,\text{J or } 0.106\,\text{eV}.$$

This energy is only a few times the thermal energy $k_B T$, and the recoiling nucleus does no damage to surrounding living tissue.

9.23 The maximum activity of the ^{20}F is 4.1×10^5 Bq, which is reached after 17.5 s.

9.25 The recoil momentum of the atom, of mass m, is $mv = 2h/\lambda$, and $v = 2h/m\lambda = 0.059\,\text{m s}^{-1}$. From the Compton scattering formula (9.17), $\lambda' - \lambda = 2h/mc$. The Doppler shift according to eqn (8.32) gives $\lambda' = \lambda(1 + v/c)$. For $v = 2h/m\lambda$ this is the same as the Compton formula.

9.26 Let ν and ν' be the frequencies of the incident and scattered photons, and p_e the momentum of the electron. The energy $h\nu$ of the incident photon equals $m_e c^2$, where m_e is the mass of the electron. Conservation of momentum gives the equations $h\nu/c = p_e \cos\theta$ in the forward direction and $h\nu'/c = p_e \sin\theta$ in the direction of the scattered photon.

Equation (9.18) for Compton scattering at 90° is

$$\frac{1}{h\nu'} = \frac{1}{h\nu} + \frac{1}{m_e c^2} = \frac{2}{m_e c^2} \quad \text{or} \quad h\nu' = \tfrac{1}{2}m_e^2.$$

Substituting for ν and ν',

$$\frac{1}{p_e \sin\theta} = \frac{1}{p_e \cos\theta} + \frac{1}{m_e c},$$

or

$$p_e = m_e c \left\{ \frac{\cos\theta - \sin\theta}{\cos\theta \sin\theta} \right\}.$$

By conservation of energy, the kinetic energy of the electron after scattering equals the energy difference $h\nu - h\nu'$ of the photons. Using the relativistic eqns (19.42), (19.43), and (19.44) to express the kinetic energy of the electron in terms of its momentum,

$$(h\nu - h\nu' + m_e c^2)^2 = \tfrac{3}{2}m_e c^2 = p_e^2 c^2 + m_e^2 c^4,$$

which gives $p_e^2 c^2 = \tfrac{5}{4}m_e^2 c^4$.
Substituting in the earlier expression for p_e leads to the required equation in θ. The solution, obtained numerically, is $\theta \approx 26.6°$.

9.27 Very close to the slit the width of the electron beam is the same as the slit width d. Further away the beam spreads out because of diffraction: for a wavelength λ the first minimum of the diffraction pattern is at the angle $\theta = \lambda/d$ (eqn (8.43)). At a distance ℓ behind the slit, the width of the beam is at least $\lambda\ell/d$. However, the width cannot be less than d, and is a minimum if $\lambda\ell/d \sim d$, or $d \sim \sqrt{\lambda\ell}$. The wavelength of 10 keV electrons is $\lambda = h/\sqrt{2m_e E} \sim 4 \times 10^{-11}$ m, and the slit width to give a minimum beam size at 10 m is about 2×10^{-5} m.

Chapter 10

The Schrödinger equation

This chapter considers the equation that describes matter waves and how we interpret the wave functions provided by its solutions.

In the last chapter we saw that in the microscopic world particles behave as waves. Particles of mass m moving at speed v, and hence having momentum $p = mv$, are associated with matter waves of wavelength λ given by

$$\lambda = h/p$$

where h is Planck's constant. When considering the motions of electrons in atoms or molecules, or in solid crystals, or when looking at how protons and neutrons move inside atomic nuclei, we have to conduct the discussion in terms of waves in order to reproduce as fully as possible the experimental results on those systems. Chapter 7 was concerned with waves that correspond to various physical processes. For example, the mathematical expression (7.8) gives the transverse displacement of a long string down which energy is being propagated. At a fixed point on the string the displacement at any time can be calculated if the frequency and wavelength of the wave are known. The speed with which the wave travels can also be deduced from the differential equation that the wave motion on the string must obey.

Two questions immediately arise in connection with matter waves.

1. How do we interpret the matter wave if it is known at any point in space and time?

2. What is the differential equation to be solved in any particular situation in order to predict the matter wave?

These questions are considered in this chapter. The precise form of the differential equation varies for each special microscopic situation we wish to describe. However, all equations have a common structure and, as the title of this chapter suggests, the non-relativistic matter wave equation is called the **Schrödinger equation**.

The matter wave describing a given microscopic situation is usually called the **wave function**. Once this has been determined from the appropriate Schrödinger equation, many other properties of the quantum system can also be determined. This involves the interpretation of the wave function, question (1) posed above, together with a set of rules telling us how to predict those properties of the system that can be measured. The rules are often called the **postulates of quantum mechanics** or of **wave mechanics**, where the terms quantum mechanics and wave mechanics are both used for the same thing—the mathematical theory of the quantum, microscopic world introduced in the last chapter.

To obtain the wave function from the Schrödinger equation is usually a tedious mathematical exercise and in almost all realistic situations the solutions cannot be given in closed analytical form, i.e. given in terms of an algebraic formula or sensibly converging series. In real problems computers are used to give approximate wave functions, employing numerical analysis techniques that give numbers on a space–time grid. Here we will be content to limit the discussion of wave mechanics to simple abstract, one-dimensional examples in order to illustrate the principles of the subject. These examples will have analytical (but sometimes complicated) mathematical solutions that can be written out exactly.

10.1 The wave function for a freely moving particle

Before thinking about wave functions for freely moving particles, let us first summarize some important parts of Chapters 7 and 8. A wave pattern generated by an object moving sinusoidally up and down in the middle of a pond spreads out over the surface of the pond in all directions. If the horizontal surface when undisturbed is in the xy-plane at $z = 0$, the vertical displacement of the surface at any point specified by coordinates (x, y) varies sinusoidally with time. The wavefronts are continuous lines along which the displacements are equal. They are concentric circles with the centre at the source of the oscillations.

In a big pond, far away from the source, the wavefronts in any localized region are lines that are nearly straight and perpendicular to the line joining the localized region to the source, i.e. perpendicular to the direction in which the wave is travelling. To a good approximation the wave is then a plane wave, and there is a constant flow of energy in a straight line. If we choose the x-direction to be the direction of propagation[1] and use the symbol ξ for displacement of the surface from its undisturbed position, the mathematical form of the plane wave of a single frequency, analogous to eqn (8.10), is

$$\xi = \xi_0 \cos(\omega t - kx) \qquad (10.1)$$

with ω the angular frequency of the wave, k its wavenumber, and ψ_0 its amplitude.

The above discussion leads to the result that, far away from the source of a three-dimensional wave, the wavefronts are planes perpendicular to the direction of propagation. If energy is moving in the x-direction, the displacement is independent of the coordinates y and z, and the expression for a plane wave continues to be given by eqn (10.1) above in which ξ represents the physical quantity oscillating. As observed in Chapter 7,

---

[1] In the earlier chapters on waves we choose the z-direction to be the direction of propagation. In treatments of wave mechanics in one dimension it is reasonable for us to choose x as the coordinate as is usually done in other texts.

waves of a single frequency do not exist in nature; the purest waves contain a narrow band of frequencies clustered around a central frequency and constitute a narrow wave group. The wave (10.1) must be considered as a first approximation to a wave group consisting of a small range of angular frequencies centred on ω and wavenumbers centred on k.

Now let us move on to consider the form of the wave function ψ that represents a particle of mass m moving freely at speed v, a situation in which energy is also being propagated unchanged in a straight line. We will try an expression somewhat similar to eqn (10.1) but, since a complex number will give more flexibility, we choose the complex expression

$$\psi = \psi_0 e^{j(\omega t - kx)}. \tag{10.2}$$

Here we have used the symbol j for $\sqrt{(-1)}$, the purely imaginary unit number. The real part of the above expression has the same form as expression (10.1), which corresponds to a real displacement. This is the construction presented in Section 8.6, where the real part of a complex expression similar to eqn (10.2) was used to represent an electromagnetic wave.

The dispersion relation

For eqn (10.2) to be the wave function we seek it must be consistent with the physics of photons and particles discussed in Chapter 9. There we saw that a photon of energy E corresponds to an electromagnetic wave of angular frequency ω given by

$$E = h\nu = \hbar\omega \tag{10.3}$$

with $\hbar$ equal to Planck's constant divided by 2π. Photons have zero mass, but we assume that eqn (10.3) is also true for massive particles. The energy E of the moving mass is simply its kinetic energy $mv^2/2$, and application of eqn (10.3) gives the angular frequency of the wave function as

$$\omega = \frac{mv^2}{2\hbar}.$$

The wavelength λ of the moving particle is given by Planck's constant divided by its momentum; this tells us that the wavenumber k of the wave function is

$$k = \frac{mv}{\hbar}.$$

Combining the expressions for ω and k gives the relation between these two quantities

$$\omega = \frac{\hbar k^2}{2m}.$$

This equation gives the dispersion relation (discussed in Section 7.5) for the wave function describing a freely moving particle, and from it we can derive the speed with which the energy in the wave travels. This speed is the group velocity v_g of the waves given by eqn (7.31). If our description is to be valid, v_g must equal the speed v with which the mass is moving. This is indeed so, since

$$v_g = \frac{d\omega}{dk} = \frac{\hbar k}{m} = v.$$

Thus the postulated wave function (10.2) for a particle freely moving at speed v is a wave of frequency ω and wavenumber k, and the wave does indeed have group velocity equal to the particle's speed.

Exercise 10.1 Compare the phase and group velocities of an electron freely moving with a kinetic energy of 20 eV.

Answer The phase velocity, ω/k, is equal to $1.33 \times 10^6\,\mathrm{m\,s^{-1}}$; the group velocity is equal to the speed of the electron, and has the value $2.65 \times 10^6\,\mathrm{m\,s^{-1}}$. ($v_p = v_g/2$.)

10.2 Interpretation of the wave function

At this point question (1) posed in the introduction to this chapter concerning the interpretation of the wave function must be addressed. The interpretation that gives results in agreement with experiments relates the wave function to the probability of finding the particle at particular places.

The probability at a time t of finding the particle in the vanishingly small interval between x and $x + dx$ is proportional to dx and to the square of the modulus of the wave function evaluated at x and t.

● *The probability interpretation*

Let the function $P(x, t)dx$ be the probability of finding the particle between x and $x + dx$. This probability is proportional to the product of ψ and its complex conjugate ψ^*, since this product equals the square of the modulus of the wave function, and

$$P(x, t) \propto \psi^*(x, t) \cdot \psi(x, t),$$

so that

$$P(x, t)\, dx \propto \psi^*(x, t) \cdot \psi(x, t)\, dx. \qquad (10.4)$$

The complex conjugate of a complex number $z = (a + jb)$, where a and b are real numbers, is $z^* = (a - jb)$. Complex numbers may appear in a more complicated form than $(a + jb)$, for example, $(a_1 + jb_1)/(a_2 + jb_2)$ is also a complex number. Although the latter expression can always be reduced to the simple form $(a + jb)$ it is not necessary to do such a reduction in order to determine its complex conjugate. This is obtained by changing the sign of j everywhere it occurs. The product of a complex number with its complex conjugate is a real number, as can readily be verified with any example. The square of the modulus $|z|^2$ of a complex number $(a + jb)$ is the product of z with its complex conjugate z^*. A complex number can be imagined as a point on a two-dimensional diagram. The x-coordinate of the point is given by the value of the real part a of the complex number $(a + jb)$, and the y-coordinate is given by the imaginary part b. The length of the line joining the origin to the point $(a + jb)$ on this diagram is equal to $(a^2 + b^2)^{1/2}$. It can quickly be seen that the square of the magnitude of a complex number is equal to the product of the number with its complex conjugate, $|z|^2 = z \cdot z^*$. Hence eqn (10.4) is often written as $P(x, t)\, dx \propto |\psi(x, t)|^2\, dx$.

So far we have shown that the wave function (10.2) satisfactorily describes a freely moving particle. Does it also satisfy another fundamental principle of quantum mechanics, Heisenberg's uncertainty principle?

Let us remind ourselves of the last section of Chapter 9. The position and momentum of a particle cannot both be specified exactly at the same time. Both can be specified within limits, but the uncertainties Δp in momentum in the x-direction and Δx in position must satisfy the condition given in eqn (9.21) that their product be at least as big as Planck's constant divided by 4π,

$$\Delta x \cdot \Delta p \geq \hbar/2.$$

Can the wave function (10.2) satisfy Heisenberg's uncertainty principle?

The function $P(x, t)$ is constant for the freely moving particle as shown by substitution of the wave function (10.2) into eqn (10.4), which gives

$$P(x, t)\, dx \propto \psi_0^2\, dx. \tag{10.5}$$

There is equal probability that the particle is within a very small interval dx anywhere from $x = -\infty$ to $x = +\infty$. This corresponds to $\Delta x = \infty$, infinite uncertainty in the position of the freely moving particle. This in turn is consistent with the uncertainty principle because expression (10.2) represents a particle with precisely fixed speed and momentum, i.e. $\Delta p = 0$. As the momentum becomes more and more precise and Δp

● *The wave function for a freely moving particle and Heisenberg's uncertainty principle*

tends to zero, Δx tends to infinity in such a way as to make the uncertainty product $\Delta x \times \Delta p$ tend to a finite value greater than $\hbar/2$.

With the probability interpretation of the wave function given above we thus conclude that eqn (10.2) satisfies the uncertainty principle. It also satisfies the considerations of Section 10.1 and so we adopt it as the wave function we are seeking for a freely moving particle.

The full curve shown on the top half of Fig 10.1 is a suitable probability distribution $P(x, t)$ at a time t for particle moving at speed v but now localized within a region of length $\sim \Delta x$. The uncertainty Δx has to be accompanied by an uncertainty $\Delta p \geq \hbar/2\Delta x$. Since $p = h/\lambda = \hbar k$ the spread Δp corresponds to a spread Δk in wavenumber, and the wave function becomes a group of waves with wavenumbers centred on $k = mv/\hbar$ and spread $\Delta k = \Delta p/\hbar$. As the particle's position is less strictly localized, the function $P(x, t)$ becomes wider and the wavenumber and momentum spread become narrower, as shown by the dotted lines on Fig 10.1. With less and less localization the wave group approaches very closely the single-frequency wave given by eqn (10.2).

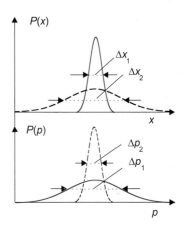

Fig. 10.1 The full line on the top half of the figure is the probability distribution for a particle localized within a region of length $\sim \Delta x_1$, and the full line on the bottom half shows the corresponding spread $\sim \Delta p_1$ on the momentum of the particle. As Δx increases to Δx_2. the spread in momentum decreases to $\sim \Delta p_2$ as indicated by the dotted lines on the drawings. The function $P(x)$ has the dimension of inverse length and units that are the inverse of the units of x. The units of momentum p are $\hbar$ divided by the units of x and the function $P(p)$ is measured in inverse momentum units.

10.3 **Normalization**

If eqn (10.2) is to represent a single particle, the probability of finding the particle somewhere over the complete range $x = \infty$ to $x = -\infty$ must be unity at any time t. This is also true for a wave function that represents a single particle in any state of motion. In mathematical language,

$$\int_{-\infty}^{\infty} P(x, t)\, \mathrm{d}x = \int_{-\infty}^{\infty} \psi^* \psi\, \mathrm{d}x = 1. \tag{10.6}$$

Imposing this condition is called **normalizing** the wave function. It arranges that the wave function is multiplied by the correct constant to ensure that the total probability is unity. We now see that normalizing eqn (10.2) leads to a difficulty; it gives

$$\int_{-\infty}^{\infty} P(x, t)\, \mathrm{d}x = \psi_0^2 \int_{-\infty}^{\infty} \mathrm{d}x = 1.$$

This requires that ψ_0 be zero, since the integral is infinite, and eqn (10.2) becomes meaningless. This problem can be removed by allowing eqn (10.2) to be the approximation to a very narrow wave group that exists not over the complete range of x but only over the range between $-L$ and $+L$ where L is very large. Normalization then requires that ψ_0 be equal to $(2L)^{-1/2}$. The one-dimensional wave function has the dimensions of [length]$^{-1/2}$, and the function $P(x, t)$ has the dimensions of [length]$^{-1}$, as required by the definition of $P(x, t)dx$ as a probability.

Worked Example 10.1 At a certain time the wave function of a particle moving along the x-axis has the triangular form given by

$$\psi = x + \beta, \quad \text{for } -\beta < x < 0,$$

$$= -x + \beta, \quad \text{for } 0 < x < \beta,$$

and zero elsewhere. What is the value of β, and what is the probability that the particle at that time has position between $\beta/2$ and β?

Answer The total probability is unity and normalization requires that

$$\int_{-\infty}^{+\infty} \psi^2 \, dx = 1.$$

Hence,

$$\int_{-\beta}^{0} (x + \beta)^2 \, dx + \int_{0}^{+\beta} (-x + \beta)^2 \, dx = 1.$$

Evaluation of these integrals gives the result $2\beta^3/3$, and hence β is equal to the cube root of 1.5.

Now that the wave function is normalized, the probability that the particle lies between x and $x + dx$ is given by eqn (10.4), with a constant of proportionality equal to unity. The probability P that it lies between $\beta/2$ and β is the sum of the probabilities for all the very small intervals between those limits:

$$P = \int_{\beta/2}^{\beta} (-x + \beta)^2 \, dx.$$

Evaluation of the integral and substitution of the value previously determined for β gives $P = 1/16$.

10.4 The Schrödinger equation

Now that we have agreed on a suitable wave function for a freely moving particle we can proceed to the second of the questions posed in the introduction to this chapter and ask, 'What is the Schrödinger equation?' We will try to find it by first determining the differential equation that the wave function (10.2) satisfies.

The wave function (10.2) for the freely moving particle is a function of the two variables x and t. Differentiation with respect to x at constant t gives the partial differential of ψ with respect to x as

$$\frac{\partial \psi}{\partial x} = -jk\psi_0 e^{j(\omega t - kx)}.$$

Differentiating once more gives the double partial differential

$$\frac{\partial^2 \psi}{\partial x^2} = -k^2 \psi_0 e^{j(\omega t - kx)}.$$

Differentiation with respect to t gives

$$\frac{\partial \psi}{\partial t} = j\omega \psi_0 e^{j(\omega t - kx)}.$$

Using the relationship $\omega = \hbar k^2 / 2m$ between ω and k, it is seen that the wave function satisfies the equation

$$-\frac{\hbar^2}{2m}\frac{\partial^2 \psi}{\partial x^2} = -j\hbar\frac{\partial \psi}{\partial t}. \tag{10.7}$$

This is the Schrödinger equation for a freely moving particle. It has a double partial differential of the function ψ with respect to position x on the left-hand side, and a single partial differential with respect to time on the right-hand side.

Partial differentials were discussed first in Chapter 7. In the above equations the function ψ depends on two independent variables: x and t. The partial differential $\partial \psi / \partial t$ is the function of x and t obtained by assuming that x is constant and differentiating ψ with respect to t. It is more exactly written $(\partial \psi / \partial t)_x$ to indicate that the variable x is to be held constant while the differentiation with respect to t is performed. As an easy example consider a function given by $f = x^2 t^2$. The partial differential $\partial f / \partial t$ of f with respect to time is the function $2x^2 t$. The double partial differential $\partial^2 f / \partial t^2$ of f with respect to time is the function $2x^2$.

Motion in a time-independent potential

Equation (10.7) is correct for a freely moving particle. What happens if the particle is acted upon by a force? What is the Schrödinger equation now, the equivalent of Newton's law of force, eqn (2.5)?

Let us consider a simple force, independent of time and depending only on position x. In this chapter, for simplicity, we are concerned only with one-dimensional problems where the position is specified by the

x-coordinate alone. Let the force be a conservative one, which as explained in Section 3.6 means that the force depends only on the position x and can be derived from a potential $V(x)$ via the relation $F = -\mathrm{d}V/\mathrm{d}x$. (Note that here we use the total differential of V with respect to x because the function V depends only on the single variable x.)

Using eqn (10.2) for a freely moving particle, evaluation of the left-hand side of eqn (10.7) gives the kinetic energy of the particle times the wave function (10.2). Evaluation of the right-hand side gives the total energy of the particle times the wave function, and the equation is consistent in that in this example of a freely moving particle the total energy is simply the kinetic energy. A particle moving in a conservative potential has additional potential energy V, and classically its total energy equals the sum of its kinetic and potential energies. Hence it is possible that the wave equation for a particle in a conservative potential V may be obtained by modifying the left-hand side of eqn (10.7) to include the factor $V(x)\psi$, the potential energy times the wave function.

This is indeed the case, and the result is the Schrödinger equation we are seeking for motion of a mass m in a time-independent potential,

$$-\frac{\hbar^2}{2m}\frac{\partial^2 \psi}{\partial x^2} + V(x)\psi = -\mathrm{j}\hbar\frac{\partial \psi}{\partial t}. \tag{10.8}$$

The Schrödinger equation, when applied in three dimensions to real problems and when the potential V is known with reasonable accuracy, gives an excellent description of observations on microscopic systems in the same way as Newton's laws give an excellent description of macroscopic phenomena. Both theories are approximations and need modifying for particles moving at speeds comparable with the speed of light—$3 \times 10^8\,\mathrm{m\,s^{-1}}$. This speed regime requires theories that take account of relativity, and some of the consequences of relativity for classical mechanics are discussed in Chapter 19. The consequences of relativity for quantum mechanics will not be discussed in this book.

It should be stressed that the above introduction of the Schrödinger equation is not a derivation. The reader has been persuaded to accept it by being eased along a route that has introduced many of the basic ideas of wave mechanics. The equation cannot be derived. It was formulated in the 1920s and was accepted as correct because it gave agreement with all the microscopic experimental results known at the time.

Equation (10.8) may look formidable to those with little mathematical experience. Note that the variables x and t on which the wave function ψ depends have not been displayed explicitly. They are usually omitted, as are the variables on which the potential V depends. The equation is complex; hence solutions will be complex and the value of a wave function at any position x and time t will be a complex number. It is worth repeating that solving eqn (10.8) is in most cases a difficult task, even when working with hypothetical one-dimensional problems. The form of the Schrödinger equation, however, makes it a linear differential equation, and this helps in many applications as well as resulting in far-reaching physical consequences.

Although linear differential equations were discussed in Chapter 6 we repeat some of their important features. They have the property that, if two functions ψ_1 and ψ_2 are separately solutions, the function $\Psi = \psi_1 + \psi_2$ is also a solution. This can readily be verified by substitution in the Schrödinger equation. Also, if ψ is a solution, the function $a\psi$ is a solution, where a is a constant. Hence the function $\Psi = \sum_{n=1}^{n=N} a_n \psi_n$ is a solution to the Schrödinger equation if the functions ψ_n are also solutions.

Stationary states

The solution to the Schrödinger eqn (10.8) for a potential independent of time simplifies greatly. Let us try a solution $\psi(x,t)$ that factorizes into a product of two functions, one dependent on position alone, $\phi(x)$, and the other, $\rho(t)$, on time alone, and put

$$\psi(x,t) = \phi(x)\rho(t). \tag{10.9}$$

Substitution in eqn (10.8) gives

$$-\rho \frac{\hbar^2}{2m} \frac{d^2\phi}{dx^2} + V(x)\rho\phi = -j\phi\hbar \frac{d\rho}{dt}. \tag{10.10}$$

The partial differentials of the Schrödinger equation have now become total differentials because

$$\frac{\partial[\phi(x)\rho(t)]}{\partial t} = \phi(x)\frac{d\rho(t)}{dt},$$

since the operator $\partial/\partial t$ does not affect $\phi(x)$, a function of x alone.

Dividing eqn (10.10) by the product $\rho\phi$ gives the result

$$-\frac{\hbar^2}{2m\phi} \frac{d^2\phi}{dx^2} + V(x) = -j\frac{\hbar}{\rho} \frac{d\rho}{dt}. \tag{10.11}$$

The left-hand side of this equation is a function of the variable x alone; the right-hand side depends only on t. The left-hand side thus does not change with time although the right-hand side does, and the right-hand side does not change as x varies although the left-hand side does. The only way the equation can be true is for both sides always to equal the same constant value irrespective of the values of x or t. If we call this constant E we obtain the following two equations involving ϕ and ρ separately:

$$-\frac{\hbar^2}{2m}\frac{d^2\phi}{dx^2} + V(x)\phi = E\phi \tag{10.12}$$

and

$$-j\hbar\frac{d\rho}{dt} = E\rho. \tag{10.13}$$

Equation (10.13) can immediately be solved. Since

$$\int d\rho/\rho = \ln\rho - \ln(1/A)$$

where $\ln(1/A)$ is a constant,

$$\rho(t) = Ae^{jEt/\hbar}.$$

Choosing the constant A to be unity,

$$\rho(t) = e^{jEt/\hbar}, \tag{10.14}$$

and $\rho\rho^*$ is unity at all times.

The function $\psi(x, t)$ of eqn (10.9) can now be normalized by normalizing $\phi(x)$. It can be seen that the probability distribution function $P(x, t)$ introduced in Section 10.1 is independent of time for the wave function (10.9), as it must be for a particle in a time-independent potential.

We have obtained the result that, if the potential in which a particle moves is independent of time, the state of motion can be described by wave functions that factorize as in eqn (10.9) and for these states of motion the probability distribution does not change with time. Such allowed quantum states are called **stationary states**.

The existence of stationary states is not unexpected. It is similar to those situations in classical physics where only conservative forces act. For example, the Earth would go around the Sun in the same orbit forever if there were absolutely no frictional forces or other effects to remove energy from the Earth–Sun system, and no other processes that changed the potential energy of the Earth in the gravitational field of the Sun.

Equation (10.12) is called the **time-independent Schrödinger equation**, or TISE for short. Its solutions are functions of position alone. For a given potential there are usually several solutions that correspond to different allowed energies E. These solutions are stationary states with different wavefunctions and hence different probability distributions.

The wave function (10.2) for a freely moving particle is a simple example of a particle moving in a time-independent potential. For a freely moving particle, the potential is zero or constant everywhere and the wave function factorizes as in eqn (10.9). It may be written

$$\psi(x, t) = \psi_0 e^{j\omega t}e^{-jkx} \tag{10.15}$$

where ω is equal to $E/\hbar$, with E the total energy of the moving particle. Substituting the solution (10.14) for $\rho(t)$ into eqn (10.9) gives

$$\psi(x, t) = \phi(x)e^{jEt/\hbar}. \tag{10.16}$$

The function ϕ is determined by solving eqn (10.12) and it must satisfy the normalization condition

$$\int_{-L}^{+L} \phi^*(x)\phi(x)\,\mathrm{d}x = 1 \tag{10.17}$$

where L is very large. The problem of normalization of the wave function for a freely moving particle was discussed in Section 10.3.

Comparison of eqns (10.15) and (10.16) indicates that the constant E introduced during the procedure used to obtain eqns (10.12) and (10.13) can also be identified with the total energy (kinetic plus potential) of the particle moving in the potential V, which is, of course, the reason it was given the symbol E in the first place.

Exercise 10.2 A stationary state of a particle of mass m in a one-dimensional system has a wave function $\phi(x) = Ae^{-bx^2}$. The potential V is known to be zero at $x = 0$. What is the energy E of the state?

Answer $b\hbar^2/m$.

The next two sections of this chapter contain examples of wave functions that are solutions to Schrödinger's equation for different time-independent potentials, but first it is necessary to make one more generalization to complete our preliminary discussion of the equation. The TISE has been introduced by consideration of motion in time-independent potentials. For potentials that vary with time, the full Schrödinger eqn (10.8) has to be used. Solutions to this equation cannot be factorized as in eqn (10.9). Equation (10.8) is sometimes called the **time-dependent Schrödinger equation, TDSE**, which we record again, for completeness,

$$-\frac{\hbar^2}{2m}\frac{\partial^2 \psi}{\partial x^2} + V(x, t)\psi = -j\hbar\frac{\partial \psi}{\partial t}. \tag{10.18}$$

10.5 Particles in one-dimensional potentials

In this section we consider in more detail the motion of a particle of mass m under the influence of a conservative force that can be derived from a potential $V(x)$. If the particle is constrained by the force to move within a given region, the simple potential well shown in Fig 10.2 may be taken as an approximation to a more realistic one. The potential has magnitude

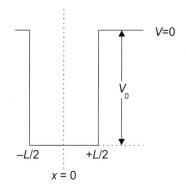

Fig. 10.2 A potential well of depth V_0 with sharp walls at $x = -L/2$ and $x = +L/2$.

$-V_0$ over a region extending from $x = -L/2$ to $x = +L/2$. Outside of this region the potential is zero. If the total energy of the particle E is less than zero, the particle is said to be bound in the potential well. For a negative value of E, an amount of energy $|E|$ would have to be given to the particle in order for it to appear outside the well. Hence a bound state with energy $-E$ corresponds to a state with a positive binding energy $E_B = +E$.

The determination of the complete wave function proceeds by solving the TISE inside and outside the well. This gives functions involving constants of integration, and these are found by using conditions that the complete wave functions must obey. The general conditions that any wave function has to obey are that the wave function itself is continuous and single-valued and that its differential is also continuous and single-valued. The first condition stems from the requirement that there can only be one probability for a particle to be in a particular very small element of space. The second comes from the realization that, if the differential of the wave function is discontinuous, the double differential becomes infinite, thus making the TISE meaningless for all problems involving finite potentials.

When there are sharp boundaries where the potential changes, as in Fig 10.2, the wave function and its derivative must satisfy the above conditions at the boundaries. The algebra involved in determining general solutions and then fitting them to boundary conditions can be tedious, and we will not go through the details of the matching procedures used to determine the wave function for a particle in the well of Fig 10.2. However, some important results can be obtained from the analogy between a particle in the well shown in Fig 10.2 and the situation discussed in Section 8.4 when a wave is reflected at the junction of two different strings. If an incident wave cannot propagate on the second string, an evanescent wave is produced.

From the analogy between an evanescent wave on a string and the bound-state wave function in a non-allowed region, we may anticipate that the wave function decays exponentially to zero outside the well as distance from the potential step increases. An acceptable bound-state wave function for the particle may look rather like that drawn on Fig 10.3, which schematically shows a wave obeying the boundary conditions. Unlike in classical mechanics, the particle now has a nonzero probability of being in the classically forbidden region, although the probability is small and exponentially decreases as the distance from the edge of the well increases.

● Boundary conditions on wave functions

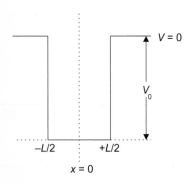

Fig. 10.3 An acceptable wave function for a bound state with negative energy E but $|E| < V_0$ in the potential well of Fig. 10.2. The wave function penetrates into the classically forbidden region where its kinetic energy is negative and thereafter decays exponentially.

Tunnelling

The penetration into classically forbidden regions is an important feature of quantum mechanics. Figure 10.4 shows a potential distribution

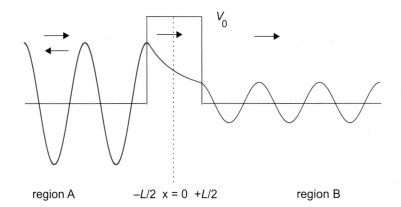

Fig. 10.4 A particle with positive
energy E approaches a potential
barrier of width L between
$x = -L/2$ and $x = +L/2$. The barrier
has height greater than E. The
particle tunnels through the barrier
and its wave function has nonzero
amplitude at $x = L/2$ and in the
region to the right.

involving a barrier between two regions labelled A and B. If a particle has
total energy E, which is positive but less than the barrier height V_0, on
approaching the barrier in region A it is partly reflected and partly
transmitted into the barrier. The solution to the Schrödinger equation in
region A thus corresponds to the sum of wave functions with different
amplitudes for a freely moving forward-going and a backward-going
particle of energy E. Inside the barrier, the wave function decreases
exponentially but at the barrier's end at $x = L/2$ some of the wave
function leaks through into region B. In region B the wave function is
that of a forward-going, freely moving particle of energy E. The result
is that, if a beam of particles is incident on a barrier through which
classically the particles cannot pass, quantum mechanics allows some of
the particles to tunnel through the barrier and emerge on the other side.
The fraction that tunnel through depends on the difference $(V_0 - E)$ and
on the width L of the barrier.

⬤ *Tunnelling through barriers*

In region A on Fig 10.4 the wave function ψ_A is the sum of an incident
wave of amplitude ψ_0 and a reflected wave of amplitude $R\psi_0$, where R is
the amplitude reflection coefficient.

$$\psi_A = \psi_0 e^{j(\omega t - kx)} + R\psi_0 e^{j(\omega t + kx)}$$

with $\omega = E/\hbar$ and wavenumber $k = 2\pi/\lambda = \sqrt{2mE}/\hbar$.

In region B the wave function ψ_B has only a forward-going part with
amplitude $T\psi_0$, T being the transmission amplitude coefficient.

$$\psi_B = T\psi_0 e^{j(\omega t - kx)}.$$

Inside the barrier the TISE is

$$-\frac{\hbar^2}{2m}\frac{\mathrm{d}^2\phi}{\mathrm{d}x^2} + V_0\phi = E\phi.$$

Hence

$$-\frac{\hbar^2}{2m}\frac{d^2\phi}{dx^2} + (V_0 - E)\phi = 0$$

with $(V_0 - E)$ positive, and the general solution is

$$\phi = Ce^{\alpha x} + De^{-\alpha x}$$

with C and D constants to be determined by the boundary conditions, and $\alpha = \sqrt{2m(V_0 - E)}/\hbar$. The total wave function ψ_I inside the barrier when the time-dependent part $\exp(j\omega t)$ is included is thus

$$\psi_I = [Ce^{\alpha x} + De^{-\alpha x}]e^{j\omega t}.$$

There are four unknowns to be determined in this problem: R, T, C, and D. These may be complex numbers to allow for phase changes of the waves at the boundaries. They can be determined using the four equations obtained from the two boundary conditions that are obeyed at $x = -L/2$ and $x = +L/2$, namely, that the wave function and its derivative be continuous. The algebra for determining the unknowns is laborious and will not be done here. If $L \gg 1/\alpha$, the tunnelling probability is small, and it can be shown that

$$T \approx -\frac{4jk\alpha e^{-(\alpha+jk)L}}{(\alpha - jk)^2}.$$

In this approximation, the tunnelling probability, $T^2 = T^*T$, often called the penetrability, is given by the formula

$$T^*T \approx \frac{16\alpha^2 k^2 e^{-2\alpha L}}{(\alpha^2 + k^2)^2}. \tag{10.19}$$

The behaviour of the transmission is dominated by the exponential term $\exp(-2\alpha L)$ and the transmission decreases very rapidly as the width of the barrier increases and also as the factor α, which is proportional to the square root of the difference between the barrier height and the energy, increases. Examples of transmission through barriers are provided by the tunnelling of electrons through thin insulating layers between conductors and, in three dimensions, by the alpha-particle decay of some heavy nuclei when alpha-particles tunnel through the barriers provided by the repulsive Coulomb potentials.

Worked Example 10.2 Estimate the probability for an alpha-particle of mass 6.64×10^{-27} kg and energy 6 MeV to tunnel through a barrier of height 8 MeV and width 4×10^{-14} m? This example is a rough one-dimensional approximation to the tunnelling of an alpha-particle through a barrier in the radioactive alpha-particle decay of certain unstable heavy elements.

Answer The wavenumber in region A is $k = 2\pi/\lambda = \sqrt{2mE}/\hbar$ and substituting the values $m = 6.64 \times 10^{-27}$ kg and energy $E = 6$ MeV gives $k^2 = 1.15 \times 10^{30}$ m^{-2}. The parameter for the wave function inside the barrier is given by $\alpha = \sqrt{2m(V_0 - E)}/\hbar$ and substituting $V_0 = 8$ MeV, $E = 6$ MeV, and $m = 6.64 \times 10^{-27}$ kg gives $\alpha^2 = 0.38 \times 10^{30}$ m^{-2}. The width L of the barrier is 10^{-14} m, and using eqn (10.19) for TT^* gives the penetrability estimate of $3.0 \times \exp(-49.2) = 1.3 \times 10^{-21}$.

Electrons between fixed points

As the simplest example of solving the TISE in a potential well we will treat the hypothetical one-dimensional problem of an electron constrained to move along the x-axis within the limits $x = -L/2$ and $x = +L/2$. The potential well is shown in Fig 10.5. The constraint corresponds to steps in the potential of infinite height at the points $x = \pm L/2$. Within this range the potential is constant and the electron experiences no force. Its wave function has the form of that of a freely moving particle with energy E, but it also has to satisfy certain boundary conditions.

Consider the boundary conditions that acceptable wave functions have to obey. The electron is not allowed outside the range $x = -L/2$ and $x = +L/2$. If the wave function were not zero outside this range, the TISE would involve infinities since the potential is infinite. The square of the modulus of the wave function gives the probability density distribution $P(x)$. A particle cannot have two different probabilities of being in the same region of space and hence the function $P(x)$ must be continuous. Since the probability is zero of the electron being within a very small

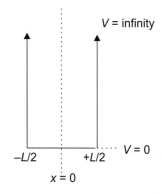

Fig. 10.5 A potential well with walls of infinite height at $x = -L/2$ and $x = +L/2$.

There are no exactly analogous examples in nature of an electron or other particle constrained to move between two points on a line and experiencing a constant potential in that part of the line between the points. Nevertheless, the problem dealt with in this section illustrates many of the basic ideas of quantum mechanics. A semiconductor heterojunction consists of a very thin layer of one material sandwiched between layers of a second. The motions of the electrons in the thin layer are determined predominantly by the constraint imposed by the very short dimension. Their behaviour in this dimension follows closely the description given in this section.

Another example of a real system in which electrons are effectively constrained to one-dimensional motion is provided by certain organic materials that exhibit metal-like conduction at low temperatures. This phenomenon is presently being actively studied. The organic conductors are crystals of an organic salt with layers of flat, elongated molecules oriented along certain directions. The molecules interact weakly and at low temperatures retain their flat shape and aligned distributions. Some of the electrons of the molecules are free to roam. The strong directionality of the intermolecular bonds within the packed layers of molecules then forces these electrons to travel along one-dimensional trajectories.

interval vanishingly close to the points $x = \pm L/2$ outside the allowed range, the wave function must always be zero at these points.

Using only this condition allows us to determine the solutions. However, although we will not prove it, the boundary condition on the differential of the wave function being continuous at $x = -L/2$ and $x = +L/2$ is also satisfied by the solutions to the problem involving a finite well such as that shown in Fig 10.2, in the limit as the potential walls tend to infinite height.

The zero amplitude at the walls is similar to the condition that gave standing waves on a string fixed at both ends, discussed in Section 7.7, and we can guess that the simplest wave is one for which a half wavelength fits exactly into the allowed interval of length L. This would suggest that the wavelength λ for the simplest wave function is equal to $2L$ and hence that for this wave function the electron's energy is $E = p^2/2m = h^2/8mL^2$, since $\lambda = h/p$. However, we will proceed with the formal solution of the TISE and come back to comparisons with standing waves on strings later.

Choosing the constant potential within the allowed region to be zero simplifies the notation. If the potential is put equal to V_0 rather than zero, the effect is simply to increase the energies of the resulting quantum states by the amount V_0 without altering any other aspects of the problem. The TISE now to be solved is eqn (10.12) with the potential $V = 0$ for $-L/2 < x < L/2$, and $V = \infty$ for x outside this range and with m the mass of the electron.

Inside the allowed region the equation reduces to

$$-\frac{\hbar^2}{2m}\frac{d^2\phi}{dx^2} = E\phi. \tag{10.20}$$

This can be written

$$\frac{d^2\phi}{dx^2} + \alpha^2\phi = 0 \tag{10.21}$$

where

$$\alpha^2 = 2mE/\hbar^2. \tag{10.22}$$

Allowing ϕ to be complex the general solution to eqn (10.20) is

$$\phi = Ae^{j\alpha x} + Be^{-j\alpha x} \tag{10.23}$$

where A and B are the two constants in the solution to the second-order differential eqn (10.20) that are determined by the boundary conditions.

In this problem the wave function must be zero at $x = \pm L/2$. Hence $\phi(\pm L/2) = 0$ and substituting in turn in eqn (10.23) gives

$$0 = Ae^{j\alpha L/2} + Be^{-j\alpha L/2}$$

and

$$0 = Ae^{-j\alpha L/2} + Be^{j\alpha L/2}.$$

These two relations can be put in a more useful form by rewriting the exponentials in terms of sines and cosines:

$$(A + B)\cos(\alpha L/2) + j(A - B)\sin(\alpha L/2) = 0$$

and

$$(A + B)\cos(\alpha L/2) - j(A - B)\sin(\alpha L/2) = 0.$$

There are two ways in which these equations can be satisfied. The first is for A to be equal to B and for $\cos(\alpha L/2)$ to be zero; the second is for A to be equal to $-B$ and for $\sin(\alpha L/2)$ to be zero. Let us consider the first alternative. If $\cos(\alpha L/2)$ is zero,

$$\alpha L/2 = n\pi/2 \tag{10.24}$$

with n an odd integer; $n = 1, 3, 5, \ldots$. This gives different values for α for each n and hence different energies E_n depending on n. For a given n the wave function and energy are, from eqns (10.23) and (10.22), respectively,

$$\phi_n = A(e^{jn\pi x/L} + e^{-jn\pi x/L}) = 2A\cos(n\pi x/L) \tag{10.25}$$

and

$$E_n = \frac{\hbar^2}{2m}\alpha^2 = \frac{\hbar^2 n^2 \pi^2}{2mL^2}. \tag{10.26}$$

If $\sin(\alpha L/2)$ is zero.

$$\alpha L/2 = n\pi/2 \tag{10.27}$$

with n an even integer, $n = 2, 4, 6, \ldots$. Now the wave functions are

$$\phi_n = A(e^{jn\pi x/L} - e^{-jn\pi x/L}) = 2A\sin(n\pi x/L) \tag{10.28}$$

with the energies given by the same formula (10.26) but with n even.

Energies and wave functions

We see from eqn (10.26) that each nonzero integer n defines a different wave function corresponding to a quantum state of different energy. The energy increases as the square of n, and depends on the square of the length L and on fundamental constants $\hbar$, π, and m.

● *The lowest energy state*

The lowest energy state is that with $n = 1$, for which $E = \pi^2\hbar^2/2mL^2$. Accordingly, the lowest allowed momentum for the electron is $p = \sqrt{(2mE)} = \pm\pi\hbar/L = \pm h/2L$. The electron has a sharp value for the magnitude of its lowest allowed momentum but it has equal probability of moving in either direction on the x-axis. While it can be seen that the average momentum is zero, there is an uncertainty in momentum that we may roughly equate to the spread between the plus and minus values. The minimum momentum uncertainty the electron can have, Δp, on this basis, is about h/L. Hence, for the electron localized within the region of length L and thus having position uncertainty Δx about equal to L, the minimum uncertainty product $\Delta p\Delta x$ is roughly equal to Planck's constant h and is consistent with the uncertainty principle.

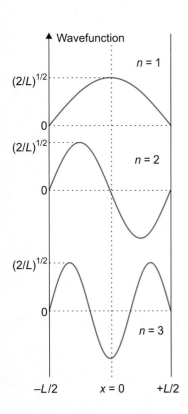

Exercise 10.3 A thin channel, along which electrons are constrained to move back and forth, is fabricated on the surface of a solid material. What is the maximum channel length if an electron in the lowest energy state has to have an energy greater than $1/160$ eV? (1 eV $\equiv 1.6 \times 10^{-19}$ J.)

Answer 7.8 nm.

Quantum numbers

The wave functions for the three lowest values of the integers n are shown on Fig 10.6, and the corresponding probability distributions on Fig 10.7. These figures are indeed similar to the patterns produced by standing waves on strings discussed in Chapter 7. The wave function is simplest for $n = 1$, and increasing n results in wave functions with an increasing number of nodes. The number of nodes is equal to $(n+1)$, including those at the end points. The integers n play a dominant role in the wave-mechanical description of the problem. They define the form of the wave function and give the energy of the corresponding state. They are examples of **quantum numbers**, which occur in the quantum description of all microscopic phenomena. Quantum numbers are associated with definite physical observables that have sharp, constant values. In the present problem the energy E for a particular state with given n is

Fig. 10.6 Normalized wave functions for the $n = 1$, 2, and 3 states for an electron moving between fixed and impermeable walls at $x = \pm L/2$.

constant. The energy is called a constant of the motion and its value depends on the value of the quantum number n.

We end this section with the reminder that the constant A in eqns (10.25) and (10.28) is determined by normalization of the wave functions. Each different wave function may in principle have a normalization constant that depends on n. However, in this problem the constant multiplying all sine and cosine wave functions is the same.

Worked Example 10.3 Show that the normalization constant for any wave function describing the motion of a particle constrained to move between two points separated by a distance L is $\sqrt{2/L}$.

Answer One set of allowed stationary-state wave functions is given by eqn (10.25) with the integer n odd, $n = 1, 3, 5, \ldots$. Normalization requires eqn (10.6) to be satisfied, and for stationary states this reduces to the condition

$$\int_{-\infty}^{+\infty} \phi^* \phi \, dx = 1.$$

Applying this to any of the wave functions with an odd value of n gives

$$4A^2 \int_{-L/2}^{+L/2} \cos^2(n\pi x/L) \, dx = 2A^2 \int_{-L/2}^{+L/2} (\cos(2n\pi x/L) + 1) \, dx = 1.$$

The first term in the last integrand integrates to zero independently of the value of n, leaving

$$2A^2 L = 1$$

and the normalization constant $2A$ is equal to $\sqrt{2/L}$.

The same procedure gives the same normalization constant for all of the allowed wave functions (10.28).

10.6 The quantum harmonic oscillator

We now consider a more realistic problem that has much in common with a naturally occurring phenomenon. Diatomic molecules such as oxygen or hydrogen chloride consist of the nuclei of two atoms held together by the attractive force provided by the surrounding electrons.

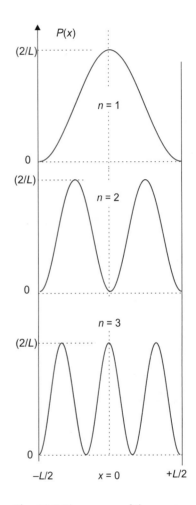

Fig. 10.7 The squares of the wave functions shown in Fig 10.6. The probabilities of the electron being in a narrow position interval of width dx at any point are equal to these squares multiplied by dx.

These are distributed about the nuclei in a way determined by the electron wave functions. The two nuclei may vibrate with respect to each other along the line joining them. If their separation exceeds the equilibrium separation, the electron cloud provides a force that tends to reduce the separation; if their separation becomes less than the equilibrium separation, there is an electrical repulsion between the nearly point-like nuclei that pushes them apart again.

There is no dissipation or exchange of energy to the surroundings for an isolated molecule and the force F between the nuclei is a conservative force. It can thus be derived from a potential V using the relation $F = -dV/dx_s$, where x_s is the separation of the nuclei. The form of this potential is illustrated in Fig 10.8; the shape is very much the same for all diatomic molecules but the depth and equilibrium separation x_0 vary somewhat. For small displacements from equilibrium, the restoring force F is proportional to displacement $(x_s - x_0)$, since the variation of the potential close to its minimum value at x_0 where the force is zero is the same as that of the potential discussed in Section 6.1 for the simple harmonic oscillator. (See Problem 10.9.) The displacement from equilibrium thus obeys the equation for simple harmonic motion and varies with angular frequency $(k/m)^{1/2}$ which depends on the spring constant k of the restoring force and the mass parameter m associated with the vibrations.

The constant k depends on details of the electron distributions in the molecule that are beyond the scope of this chapter. The mass parameter m is the reduced mass of the two oscillating nuclei. If the latter are m_1 and m_2 the reduced mass m is given by

$$m = \frac{m_1 m_2}{(m_1 + m_2)}. \tag{10.29}$$

The description of the vibrational motion in terms of one coordinate and one mass results from disentangling two separate motions. The first is that of the centre of mass of the molecule as a whole. There is no external force on the molecule and so by Newton's second law, eqn (2.5), the centre of mass is at rest or moving with constant velocity, and we have no further interest in it. The second motion is the vibration of the two nuclei in the molecule. Problem 10.22 and the mathematical insert opposite deal with the mathematics of the separation of the two motions and the classical derivation of the equation of relative motion. The procedure follows that given in Section 6.7, which treated the normal modes of vibration of the atoms within the carbon dioxide molecule.

The relative motion of the two nuclei corresponds classically to simple harmonic motion of the variable $(x_s - x_0)$. The angular frequency of the vibration is given by eqn (6.6) involving the parameters k and m, and the

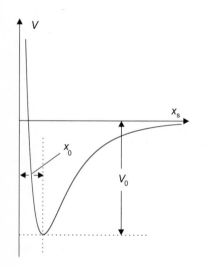

Fig. 10.8 The interatomic potential as a function of the distance x_s between the two atoms of a diatomic molecule. The depth V_0 of the potential and the equilibrium separation x_0 depend on the particular molecule. For oxygen $x_0 \sim 1.21 \times 10^{-10}$ m and $V_0 \sim 5.2$ eV; for HCl $x_0 \sim 1.28 \times 10^{-10}$ m and $V_0 \sim 4.4$ eV.

energy in the vibration depends upon the amplitude, which can vary continuously in classical mechanics. In quantum mechanics, only certain states of relative motion of the nuclei with different discrete energies are allowed. The allowed states are described by wave functions that depend on the separation $(x_s - x_0)$ and time t. In this section we will determine the allowed states and their energies by setting up the appropriate Schrödinger equation and solving it.

In the centre of mass system the vibration of the two nuclei reduces to a problem in which the reduced mass m oscillates on the end of a spring of force constant k and unstretched length x_0. The potential V to which the mass is subject is given by

● *The equation for vibrations in the centre of mass system*

$$V = \tfrac{1}{2} k(x_s - x_0)^2.$$

(10.30)

With $x = (x_s - x_0)$ used for the coordinate giving the displacement of the mass from the equilibrium position, the time-independent potential of eqn (10.30) becomes

$$V = \tfrac{1}{2} k x^2$$

(10.31)

and the TISE to be solved is

$$-\frac{\hbar^2}{2m} \frac{d^2\phi}{dx^2} + \frac{1}{2} k x^2 \phi = E\phi.$$

(10.32)

This equation must be solved to give functions ϕ that are subject to the boundary conditions that they vanish at large separations, that they are continuous (in order that the probability distributions are single-valued), and that their derivatives are continuous (in order that the first term in the TISE remains finite).

The mass m in eqn (10.32) is the reduced mass of the two vibrating nuclei. The rigorous way of deriving eqn (10.32) may be more satisfactory for many students, although it requires more mathematical effort. Two bodies vibrating along the line joining them require two coordinates, x_1 and x_2, to specify their positions with respect to a chosen origin of coordinates. The Schrödinger equation now has two kinetic energy terms, one for each nucleus, and a potential energy term that depends on the separation $(x_1 - x_2)$, for which the symbol x_s was used above. The equation is

$$-\frac{\hbar^2}{2m_1} \frac{d^2\Phi}{dx_1^2} - \frac{\hbar^2}{2m_2} \frac{d^2\Phi}{dx_2^2} + V(x_1, x_2)\Phi = E\Phi.$$

Here the wave function Φ depends on the two variables x_1 and x_2, and the potential V depends on the separation x_s and is as shown in Fig 10.8. The function Φ describes both the motion of the centre of mass of the two-nucleus system and the relative motion of the nuclei. We wish to disentangle these to determine the relative motion alone. This is done by transforming to two new coordinates: one that of the centre of mass, call it X, the other a coordinate x giving the displacement of the separation of the nuclei from the equilibrium separation, i.e. the coordinate $(x_s - x_0)$.

$$X = \frac{(m_1 x_1 + m_2 x_2)}{(m_1 + m_2)}$$

and

$$x = (x_s - x_0) = [(x_1 - x_2) - x_0].$$

When the transformation is made the wave function Φ factorizes into a product of a function ϕ depending on x alone, and a function that depends on X alone describing the centre of mass moving at constant speed. The equation for the function ϕ reduces precisely to eqn (10.32).

Energies and wave functions

It is not necessary at this point to know the details of how the solutions to eqn (10.32) are obtained. The solutions do constitute a discrete set and, as in the example of the last section, are characterized by a quantum number n that can have any integer value from zero to infinity. The energies of the states increase as n increases, and the corresponding stationary state wave functions ϕ_n become more complicated by acquiring more nodes.

● *The lowest energy state*

The lowest energy state has $n = 0$ and the wave function is

$$\phi_0 = A_0 e^{-a^2 x^2 / 2} \tag{10.33}$$

with

$$a^2 = m\omega/\hbar \tag{10.34}$$

and $\omega = (k/m)^{1/2}$. A_0 is the normalization constant, which can be determined using eqn (10.6). If the function (10.33) is substituted into the TISE (10.32) it is found that it is indeed a solution for an energy E given by

$$E_0 = \tfrac{1}{2}\hbar\omega. \tag{10.35}$$

Equation (10.33) is the wave function for the ground state of the oscillator and the ground-state energy, often called the zero-point energy, of the vibrator is $\hbar\omega/2$.

The first excited state has a wave function

$$\phi_1 = A_1 x e^{-a^2 x^2/2} \tag{10.36}$$

where A_1 is the normalization constant for $n = 1$. Substitution of this wave function into the TISE gives the energy of this state as

● *The first excited state*

$$E_1 = \tfrac{3}{2}\hbar\omega. \tag{10.37}$$

If substitution into the Schrödinger equation is used to determine the energies of the more complicated wave functions it is found that

● *Higher states*

$$E_n = \left(n + \tfrac{1}{2}\right)\hbar\omega. \tag{10.38}$$

Figure 10.9 shows normalized wave functions for the lowest two states in an oscillator potential. The wave functions are shown for a value of the parameter a equal to $10^{11}\ \mathrm{m}^{-1}$, a reasonable number for a typical diatomic molecule and close to the value for HCl. The excursion of the separation of the nuclei from its equilibrium value is about $1/a$, which is approximately $10^{-11}\ \mathrm{m}$. This may be compared with the equilibrium separation for HCl of $\sim 1.3 \times 10^{-10}\ \mathrm{m}$. The mean square displacement of the separation of the nuclei from its equilibrium value is a few per cent of the equilibrium separation. Figure 10.10 shows the probability distributions of the displacement in the two lowest states whose wave functions are given in Fig 10.9.

The energies of the allowed stationary states of the one-dimensional harmonic oscillator are equally spaced with separation $\hbar\omega$. We expect to find energy levels in diatomic molecules that closely follow this example and correspond to vibrations of the nuclei along the line joining them. This is indeed the case, and Table 10.1 gives the energies of vibrational levels found experimentally for oxygen and hydrogen chloride. The spacings of the energy levels are not quite equal due to the inadequacy of the assumption (10.31) about the form of the potential V.

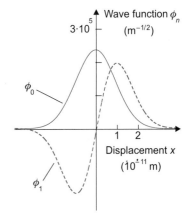

Fig. 10.9 Normalized wave functions for the two lowest vibrational states of a typical diatomic molecule. The first two states of the one-dimensional quantum oscillator are plotted for the value $a = 10^{11}\ \mathrm{m}^{-1}$ for the parameter in eqns (10.33) and (10.36). The position x is the displacement of the separation of the two atoms from their equilibrium separation.

Table 10.1 Energies of vibrational energy levels of the hydrogen chloride and oxygen molecules. The first column is the vibrational quantum number n. Columns 2 and 3 for the vibrational energy and the energy difference between successive levels refer to hydrogen chloride; columns 4 and 5 to oxygen. Energies are given in electron volts (eV) and data are taken from F.J. Lovas and E. Tieman, *Journal of Physical and Chemical Reference Data,* **3** (1974).

n	E_n	$E_n - E_{n-1}$	E_n	$E_n - E_{n-1}$
0	0.184		0.098	
1	0.542	0.358	0.291	0.193
2	0.887	0.345	0.481	0.190
3	1.218	0.331	0.668	0.187
4	1.536	0.318	0.852	0.184
5	1.842	0.306	1.034	0.182
6	2.134	0.292	1.214	0.180

Exercise 10.4 Use the data in Table 10.1 to estimate values of the spring constants k for hydrogen chloride and oxygen molecules.

Answer For hydrogen chloride, $k \sim 490\,\mathrm{N\,m^{-1}}$; for oxygen, $k \sim 1.2 \times 10^3\ N\ m^{-1}$.

Worked Example 10.4 Sodium iodide crystals absorb electromagnetic radiation at a frequency of 4×10^{12} Hz. Assuming that this absorption line is due to transitions between vibrational levels of simple harmonic oscillations of sodium atoms of atomic weight 23, calculate the potential energy of a sodium atom as a function of small displacements x from its equilibrium position.

Calculate $\sqrt{\overline{x^2}}$, the square root of the average of the square of the displacement x from its equilibrium position. This is usually called the root mean square or rms displacement.

Answer The absorption frequency times Planck's constant corresponds to the difference in energy between the vibrational levels of the oscillating sodium atoms. This difference equals $\hbar \omega$ from eqn (10.38), where the angular frequency $\omega = \sqrt{k/m}$ with m the mass of the sodium atom and k the spring constant. The constant k, which gives the potential $kx^2/2$ in which the atoms move, is

$$k = m\omega^2 = 24.1\,\mathrm{kg\,s^{-2}}$$

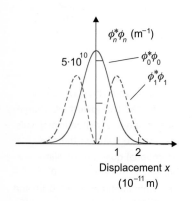

$\phi_n^* \phi_n$ (m^{-1})

$5 \cdot 10^{10}$

$\phi_0^* \phi_0$

$\phi_1^* \phi_1$

1   2
Displacement x
(10^{-11} m)

Fig. 10.10 Squares of the wave functions shown in Fig 10.9 for the two lowest vibrational states of a typical diatomic molecule.

with the mass m equal to $23 \times 1.66 \times 10^{27}$ kg. The potential is

$$V(x) = \tfrac{1}{2}kx^2 = 12.1x^2$$

with $V(x)$ in joules and x in metres.

The probability that x lies between x and $x + dx$ is given by eqn (10.5) and the stationary state wave function (10.33). The factor a^2 in the wave function is given by eqn (10.34) and is $a^2 = m\omega/\hbar = 9.1 \times 10^{21}$ m^{-2}. The probability that the displacement lies between x and $x + dx$ is

$$P(x)\,dx = A_0^2 e^{-a^2 x^2}\,dx$$

where A_0 is the normalization constant. This probability distribution gives the mean of the square of the displacement

$$\overline{x^2} = \frac{\int_{-\infty}^{+\infty} x^2 e^{-a^2 x^2}\,dx}{\int_{-\infty}^{+\infty} e^{-a^2 x^2}\,dx},$$

using the usual formula for the average of x^2. The integrals required to evaluate this expression are

$$\int_{-\infty}^{+\infty} e^{-a^2 x^2}\,dx = \left(\frac{\pi}{a^2}\right)^{1/2}$$

and

$$\int_{-\infty}^{+\infty} x^2 e^{a^2 - x^2}\,dx = \frac{1}{2}\left(\frac{\pi}{a^6}\right)^{1/2}.$$

These give

$$\sqrt{\overline{x^2}} = \frac{1}{\sqrt{2}a}$$

and $\sqrt{\overline{x^2}} = 7.4 \times 10^{-12}$ m.

10.7 The correspondence principle

The probability distributions shown in Fig 10.10 for the lowest two oscillator states differ markedly from that in a classical harmonic oscillator. For the latter the displacement x is given by eqn (6.7). With the phase angle ϕ equal to zero we have

$$x = x_0 \cos(\omega t),$$

and the probability distribution $P(x)\,dx$ when the amplitude is x_0 is worked out by determining what fraction of the period of the motion the particle spends between x and $x + dx$. This gives the curve shown on Fig 10.11; the probability is much larger near the extremes of the motion, $\pm x_0$, because the particle spends more time near these points, slowing

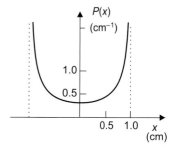

Fig. 10.11 The probability of finding a classical harmonic oscillator between points x and $x + dx$ is given by the function $P(x)$ shown, multiplied by dx. The curve shows the function $P(x)$ for an oscillator with amplitude 1 cm.

down to a stop and reversing. We know classical physics works, and so, in the classical limit, quantum mechanics must in all problems give the same predictions for observable quantities as classical mechanics. This is called the **correspondence principle** and was of great value in establishing the rules or postulates of quantum mechanics.

How do we go to the classical limit when considering the quantum oscillator and show, for example, that we do indeed obtain the same result as classical physics for the probability distribution? The classical oscillator has a continuous distribution of possible energies corresponding to the continuous allowed range of amplitudes. When does a similar situation effectively exist with the quantum oscillator? The answer is at very large values of the quantum number n, when the spacing $\hbar\omega$ between adjacent energy levels becomes very small compared with the oscillator energy $(n + \frac{1}{2})\hbar\omega$.

For large n the wave functions ϕ_n become highly oscillatory and give probability distributions that, averaged over small intervals, approach closer and closer to the classical distribution as n tends to infinity. Figure 10.12 shows the square of the $n = 11$ wave function, together with the classical probability distribution for an oscillator of the same total energy. Although $n = 11$ is not a high value, the figure serves to illustrate the convergence of the quantal and classical distributions. If the equivalent classical oscillator has an amplitude x_0 its total energy is $kx_0^2/2$, which for equivalence must be set equal to the energy $11.5\hbar\omega$ of the quantum oscillator. From this it can be deduced that the amplitude of the classical oscillator is equal to $\sqrt{23}/a \approx 4.80/a$, where a is the parameter defined by eqn (10.34). For this reason the dimensionless parameter (ax) is plotted along the horizontal axis on Fig 10.12 and the classical turning points occur at $ax = \pm 4.80$.

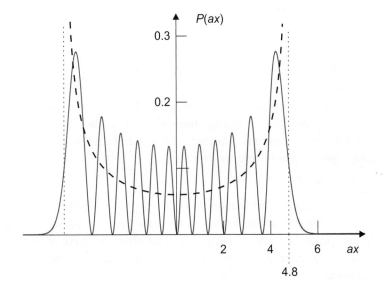

Fig. 10.12 The full curve shows the square of the wave function for the $n = 11$ state of the quantum harmonic oscillator plotted against the dimensionless parameter ax. The dashed curve shows the probability distribution for a classical harmonic oscillator with the same total energy.

Problems

Level 1

10.1 Billiard balls of mass 200 g and speed 100 cm s^{-1} are fired at a slit of width 15 cm. Roughly what value would Planck's constant have to have if the distribution of the balls the other side of the slit were to show diffraction effects?

10.2 Electrons may suffer refraction on entering a region where their wavelength changes. This can happen in some materials in which the effective kinetic energies of moving electrons are increased over their energies before entry of the material. If a beam of electrons of energy 25 eV passes through an interface at an angle of incidence of 65°, what is the angle of refraction if their kinetic energies are increased to 30 eV?

10.3 Determine the normalization constant for the wave function of the ground state of the quantum oscillator given by eqn (10.33). The integral $\int_{-\infty}^{+\infty} e^{-x^2} dx = \pi^{1/2}$.

10.4 The third standing wave on a string of length L fixed at both ends has wavelength $2L/3$. Use this to determine the energy of the $n = 3$ wave function for the electron in the potential discussed in Section 10.3 and show that it is the same value as given by eqn (10.26).

10.5 Instead of moving in zero potential between points $\pm L/2$ in Section 10.3 the electron experiences a constant potential V_0. Show that this simply adds the amount V_0 to the allowed energies (10.26) without changing the wave functions (10.25) and (10.28).

10.6 All atoms have roughly the same radius of about 10^{-10}. Use eqn (10.26) to estimate the lowest kinetic energy of the electrons.

10.7 Show that, if a particle is moving in one dimension under the influence of a time-independent real potential V that has the property that $V(-x) = V(x)$, the solutions of the Schrödinger equation have the property that $\phi(-x) = \pm\phi(x)$.

10.8 An electron is constrained to move on a line between two points separated by 2×10^{-10} m. What is its energy in the lowest state? What wavelength of radiation is emitted in the transition between the first excited state and the ground state?

10.9 Use the Taylor series expansion of a function $V(x)$ about the point x_0 (see Section 20.7) to show that, if $V(x)$ has a minimum at x_0, for points x close to x_0 the potential, to a good approximation, has the form

$$V(x) = V(x_0) - s(x - x_0)^2/2$$

where $s = d^2V/dx^2$ evaluated at the point x_0.

10.10 The wave functions $\psi_1 = \phi_1(x)e^{-jE_1t/\hbar}$ and $\psi_2 = \phi_2(x)e^{-jE_2t/\hbar}$ are two solutions of a particular time-dependent Schrödinger equation (TDSE). Which of the following statements are true?

(a) The wave function $a_1\psi_1 + a_2\psi_2$ satisfies the TDSE.

(b) The modulus of ψ_1 is independent of time.

(c) The modulus of the wave function given in (a) is independent of time.

(d) The wave function $a_1\psi_1$ satisfies the time-independent Schrödinger equation (TISE).

(e) The wave function given in (a) satisfies the time-independent Schrödinger equation.

10.11 Recalling that $|\phi|^2 dx$ is the probability of finding the particle between x and $x + dx$, the mean displacement of the particle is given by $\bar{x} = \int_{-\infty}^{+\infty} x|\phi|^2 dx$. Show that the mean displacements for a particle in the ground state when constrained to move on a line at constant potential between fixed points, is zero if the origin is chosen to be midway between the fixed points.

10.12 A particle is constrained to move on a line between the points $x = -L/2$ and $x = +L/2$. It is in the ground state of the possible states of motion. Determine the probability that the particle lies between $-L/4 < x < +L/4$.

Level 2

10.13 If the stationary state wave function of a particle of mass m moving in one dimension is of the form $A/(b^2 + x^2)$, with A and b^2 constants, and the potential in which the particle moves is known to be zero when $x = 0$, determine the energy of the state.

10.14 Show that, for an electron moving in the uniform potential discussed in Section 10.5 and described by a number of neighbouring wave functions with very large n values, the probability distribution approaches the probability distribution for a classical particle moving in a straight line back and forth between rigid, elastic walls, and hence that the correspondence principle is satisfied.

10.15 Show that the wave function for a particle in the ground state of a simple harmonic oscillator potential gives a description of the particle that satisfies the uncertainty principle.

10.16 Verify that substitution of the wave function (10.33) into the TISE (10.32) gives the energy (10.35). Verify that the technique used in Problem 10.13 gives the same answer. The latter trick will not work for the $n = 1$ state. You will have to verify the statement leading to eqn (10.37) by the slightly less direct method.

10.17 For the simple harmonic oscillator potential $V(x) = m\omega^2 x^2/2$, show that the wave function

$$\phi(x) = (4a^2 x^2 - 2)e^{-a^2 x^2/2},$$

where $a = \sqrt{(m\omega/\hbar)}$, is a solution to the time-independent Schrödinger equation corresponding to energy $5\hbar\omega/2$.

10.18 Calculate the average values of the squares of the positions x for a particle in the ground state when constrained to move on a line between fixed points, and for the ground state of the one-dimensional oscillator. The integral $\int_{-L/2}^{+L/2} x^2 \cos^2(\pi x/L)\,dx = L^3(1/24 - 1/4\pi^2)$. The integral $\int_{-\infty}^{+\infty} x^2 e^{-\alpha x^2}\,dx = \sqrt{\pi/4\alpha^3}$.

10.19 Show that the wave functions of two different stationary states of a particle constrained to move on a line between fixed points are orthogonal, i.e. that the integral of their product over the whole of the range of x from $-\infty$ to $+\infty$ is zero.

Level 3

10.20 The energies of an electron in an infinite one-dimensional well of width L are given by eqn (10.26). An electron moves in a *finite* well of depth $1\,\text{eV}$ and width $3 \times 10^{-9}\,\text{m}$. Explain why the ground-state and first excited state energies approximate to those of an infinite well of the same width.

Electrons in a layered semiconductor structure behave as though they are in a one-dimensional well of depth $1\,\text{eV}$. The electrons strongly absorb radiation of

wavelength $20\,\mu\text{m}$. By assuming the absorption is due to transitions between the ground and first excited states, deduce a value for the width of the well.

10.21 A particle is constrained to move on a line between fixed points. Calculate the average value of the square of the position x when the quantum number n in eqn (10.25) is very large. The integral $\int x^2 \cos^2 x\,dx = x^3/6 + (x^2/4 - 1/8)\sin 2x + x\cos 2x/4$.

10.22 Choose an origin of coordinates and write down the equations of motion for two masses m_1 and m_2, connected by a light spring that has a spring constant k and an unstretched length r_0, in terms of the coordinates x_1 and x_2 of the masses. Rewrite these equations in terms of two new coordinates: one that of the centre of mass of the system; the other the separation r of the two masses. Obtain the equation

$$m\frac{d^2 x}{dt^2} = -kx$$

for the coordinate x, which equals $(r - r_0)$.

10.23 Show that the parameter a, given by eqn (10.34) and used in the wave functions for the quantum oscillator, is equal to the reciprocal of the maximum displacement of a classical harmonic oscillator of energy $\hbar\omega/2$.

10.24 For what fraction of time does a quantum simple harmonic oscillator with the zero-point energy reside in the classically forbidden region? The definite integral $\int_1^\infty e^{-x^2}\,dx = 0.0787\sqrt{\pi}$.

10.25 Calculate the zero-point energy of a mass of $10^{-3}\,\text{kg}$ oscillating on the end of a spring of force constant $10\,\text{N}\,\text{m}^{-1}$. When the energy in the oscillating system is equal to $k_B T$, with k_B Boltzmann's constant and T the temperature equal to $293\,\text{K}$, what is the quantum number giving the vibrational energy?

10.26 A particle of mass m moves in one dimension as in Section 10.5. At time zero, the wave function is

$$\psi(x, 0) = \left(\frac{1}{2L}\right)^{1/2}(\cos(\pi x/L) + \sin(\pi x/L))$$

where L is the length over which the particle is constrained to move. Calculate the wave function at time t.

10.27 Write down the Schrödinger equation for two non-interacting particles of mass m moving in a conservative potential V. Show that the wave function of the two-particle system factorizes into a product of wave functions, each involving the coordinates of one particle only, and that the energy of the whole system is the sum of the energies of the individual particles. (You will find later that there are new rules that identical particles have to obey, so for now we will consider the particles in this example as non-identical. The results derived are, however, still valid when the particles are of the same type.)

10.28 A particle of mass m moves in the potential $V(x) = m\omega^2 x^2/2$ for $x \geq 0$, and $V(x) = \infty$ for $x < 0$. By considering the boundary conditions that must be satisfied by the wave functions and by using standard results from the simple harmonic oscillator potential, show that the possible energy levels are

$$E_n = \hbar\omega(n + \tfrac{1}{2})$$

where n is an odd integer, $n = 1, 3, 5, \ldots$. Obtain the normalized wave function for the lowest energy state.

10.29 The Schrödinger equation applies to non-relativistic particles. For a relativistic particle of mass m, the relation between energy E and momentum p is

$$E^2 = p^2 c^2 + m^2 c^4$$

where c is the speed of light in free space. This is discussed in Chapter 19 on Relativity. Use this equation and the de Broglie relations $E = \hbar\omega$ and $p = \hbar k$ to derive a dispersion relation for the waves of angular frequency ω and wavenumber k associated with the particle. From this show that the group velocity v_g of the waves is equal to the particle velocity and that the phase velocity is given by $v_p = c^2/v_g = E/p$. Make a guess at a possible relativistic wave equation.

Some solutions and answers to Chapter 10 problems

10.2 Outside the medium, the electron's wave function is that of freely moving particles of energy $E = 25$ eV. Outside, their speed is thus $v_1 = \sqrt{2E/m} = 2.96 \times 10^6$ m s^{-1}. Inside the medium, the wave function is that of freely moving particles of energy 30 eV. Inside, their speed is thus $v_2 = 3.25 \times 10^6$ m s^{-1}. The refractive index of the medium for the waves is $n = v_1/v_2 = 0.913$ and the normal to the wavefront is refracted away from the normal to the boundary on entering the medium. If the angle of refraction, for an angle of incidence of $65°$, is ϕ, $\sin\phi/\sin 65° = 1/0.913$ and $\phi = 83.1°$.

10.3 The normalization constant $A_0 = \sqrt{a/\sqrt{\pi}}$.

10.5 If there is a constant potential V_0 between the points $x = \pm L/2$ between which the electron is constrained to move, eqn (10.20) becomes

$$-\frac{\hbar^2}{2m}\frac{d^2\phi}{dx^2} = (E - V_0)\phi.$$

The general solution to this equation is still given by (10.22) but with $\alpha^2 = 2m(E - V_0)/\hbar^2$. The allowed values of α, as before, are given by the conditions that the wave function be zero at $x = \pm L/2$; thus the wave functions remain as given by eqn (10.25) with n an odd

integer, or by eqn (10.28) with n an even integer. Substitution of these into the Schrödinger equation gives

$$(E - V_0) = \frac{\hbar^2 n^2 \pi^2}{2mL^2},$$

and the energy is V_0 plus the allowed energies for $V_0 = 0$.

10.8 The ground-state energy is $E_1 = \pi^2\hbar^2/2mL^2 = 9.4$ eV. The first excited state with $n = 2$ has $E_2 = 37.6$ eV excitation energy. The energy of the photon de-exciting the $n = 2$ level to the ground state is thus 28.2 eV.

10.10 (a) Yes; (b) yes; (c) no; (d) yes; (e) no.

10.12 The probability that the particle lies between $\pm L/4$ is

$$\int_{-L/4}^{L/4} \phi^* \phi \, dx = \frac{2}{L} \int_{-L/4}^{L/4} \cos^2(\pi x/L) \, dx$$

$$= \frac{1}{L} \int_{-L/4}^{L/4} (2\cos(2\pi x/L) - 1) \, dx = \tfrac{1}{2}.$$

10.17 The TISE for the oscillator is

$$-\frac{\hbar^2}{2m}\frac{d^2\phi}{dx^2} + \tfrac{1}{2}m\omega^2 x^2 \phi = E\phi.$$

If the given function ϕ is a solution it must satisfy this equation. The first differential of ϕ is

$$\frac{d}{dx}A(4a^2x^2 - 2)e^{-a^2x^2/2} = A(10a^2x - 4a^4x^3)e^{-a^2x^2/2},$$

and the second differential is

$$\frac{d^2\phi}{dx^2} = A(10a^2 - 22a^4x^2 + 4a^6x^4)e^{-a^2x^2/2}.$$

For the above TISE to be satisfied we thus require

$$\frac{-\hbar^2}{2m}(10a^2 - 22a^4x^2 + 4a^6x^4) + \tfrac{1}{2}m\omega^2x^2(4a^2x^2 - 2)$$

$$= E(4a^2x^2 - 2).$$

For this to be true at all x the constant terms on both sides of the equation must be equal as must the coefficients of the terms in x^2 and of those in x^4.

The coefficient of the term in x^4 on the left-hand side of the equation is

$$-2a^6\hbar^2/m + 2m\omega^2a^2 = -\frac{2m^2\omega^3}{\hbar} + \frac{2m^2\omega^3}{\hbar} = 0,$$

the same as coefficient on the right-hand side. The coefficient of the term in x^2 on the left-hand side is $11\hbar^2a^4/2m - m\omega^2 = 10m\omega^2$. The coefficient of the term in x^2 on the right-hand side is $4m\omega E/\hbar = 10m\omega^2$ also. Finally, taking the constant terms,

$$\frac{-\hbar^2}{2m}10a^2 = -2E,$$

giving the result that the original function ϕ is a solution if $E = \tfrac{5}{2}\hbar\omega$.

10.19 Consider two stationary states that are cosine solutions to the Schrödinger equation corresponding to two odd integers n and m. The integral over all space reduces to an integral from $-L/2$ to $+L/2$, and

$$\int_{-L/2}^{+L/2} \phi^* x^2 \phi dx$$

$$= \frac{2}{L}\int_{-L/2}^{+L/2} \cos(n\pi x/L)\cos(m\pi x/L)dx.$$

The integrand can be written as the sum of two cosine terms by using eqn (20.37), and

$$\frac{2}{L}\int_{-L/2}^{+L/2} \cos(n\pi x/L)\cos(m\pi x/L)dx$$

$$= \frac{2}{L}\int_{-L/2}^{+L/2}\left(\cos\frac{(n+m)\pi x}{L} + \cos\frac{(n-m)\pi x}{L}\right)dx$$

$$= \frac{1}{(n+m)\pi}\left[\sin\frac{(n+m)\pi x}{L}\right]_{-L/2}^{+L/2}$$

$$+ \frac{1}{(n-m)\pi}\left[\sin\frac{(n-m)\pi x}{L}\right]_{-L/2}^{+L/2}.$$

Since both n and m are odd $(n+m)\pi/2$ and $(n-m)\pi/2$ are both integral multiples of π making all terms in the integral zero.

Arguments along the same lines hold if two wave functions are chosen that are both sine functions and thus have even n and m, or if one of the wave functions is a sine function and one a cosine.

The orthogonality has been proved for any two different stationary state solutions of eqn (10.20), but it is generally true for any two different solutions of the same TISE.

10.21 $L^2/12$.

10.24 $0.0787 \times 2 = 0.157$.

10.28 One boundary condition is that, for $x \leq 0$ in the region where the potential is infinite, the wave function is zero. The other is that for $x > 0$ the wave function must tend to zero as x tends to infinity, where again the potential is infinite. The wave functions for $x > 0$ satisfy the same potential and the same boundary condition at large x as those of the simple harmonic oscillator. The allowed wave functions for the problem given are the quantum oscillator wave functions, which satisfy the new condition that they be zero at $x = 0$. This subset consists of the odd n members of the wave functions for the simple oscillator, such as $n = 1$ given in eqn (10.36).

Chapter 11

The microscopic world of quantum mechanics

The principles of quantum mechanics are applied to atoms, molecules, and nuclei, explaining phenomena that cannot be understood in terms of classical physics.

Wave–particle duality is the name given to the behaviour of very small objects such as electrons, which must be regarded both as waves and particles. When they are diffracted by obstacles, beams of electrons are acting like waves. Yet when they are detected, for example, in a photograpic emulsion, they register arrival at a particular position in space. In Chapter 10 it was explained how the wave and particle aspects of matter are reconciled by the Schrödinger equation. The wave function of a particle is a complex function that, as its name implies, has wave-like properties: it has an amplitude and a phase and may show constructive or destructive interference. The square of the modulus of the wave function is everywhere positive, and when normalized is interpreted as being a probability density: the probability of finding the particle in a volume dV is the probability density times dV.

The probability interpretation highlights one of the important features of quantum mechanics—it is not always possible to predict with certainty the result of a measurement. Often a distribution of probabilities is the best that can be done. Another feature emerged when we looked at solutions of the Schrödinger equation. Physical quantities such as energy or momentum may be restricted to certain discrete values instead of having a continuous range. The examples given in Chapter 10 were chosen to be mathematically simple. They illustrated how one sets about solving the Schrödinger equation and the essential features of wave functions, but they gave only a glimpse of how quantum mechanics affects the behaviour of matter in the real world. This chapter endeavours to draw a wider picture, giving quantum mechanical explanations of observations on atoms, molecules, and nuclei.

11.1 **The hydrogen atom**

The simplest atom of all is hydrogen, which consists of a single proton and a single electron. The force that holds the hydrogen atom together is the electrical attraction between the negatively charged electron and the positively charged proton. This force, like the force of gravity, varies as the inverse square of the distance between the two charges. If electrons were to obey the laws of classical mechanics, they would move in orbits around nuclei analogous to the orbits of planets around the Sun, or of artificial satellites around the Earth. Planetary and satellite orbits are discussed in Chapter 5, but we cannot use any of the mathematical detail given there in our present discussion. Information about the orbit of a satellite, for example, may be obtained by direct observation, since its position can be determined by sending radio signals to the satellite and receiving signals back from it. We also know that a continuous range of energies is accessible to the satellite. Any energy within the range may be selected by firing rockets on board the satellite. However, the classical theory of well-defined orbits cannot be applicable to electrons in an atom, since we already know that it is not possible to specify the exact position of an electron on an atomic scale. Nevertheless, we can learn something about the internal structure of an atom by observing how the atom emits or absorbs light.

Atomic spectra

Whenever an atom emits or absorbs a photon, the energy of the atom changes by an amount equal to the energy of the photon. In Section 9.3 it was explained that the energy E of a photon is given by eqn (9.8) as $E = h\nu$, where h is Planck's constant and ν is the frequency. The wavelength λ, which can be measured, is related to ν and to the speed of light c through eqn (7.14): $\nu = c/\lambda$. The wavelength of an absorbed or emitted photon tells us the frequency of the photon and hence the change in energy of the atom. Figure 11.1 shows the spectrum of light emitted by the filament of a light bulb as a function of frequency. Each photon emitted reduces the energy of the filament by an amount equal to the photon energy and, since the spectrum is continuous, we infer that a continuous range of energies occurs within the filament.

What is the spectrum of light emitted by an assembly of separate atoms? This can be investigated by heating atoms in a gas discharge. An electric current is passed between two electrodes in a tube containing the gas as shown in Fig 11.2. The apparatus is just like a fluorescent tube except that it is made of transparent glass: the original source of light in a fluorescent tube is also a gas discharge, but the fluorescent coating on the tube absorbs and re-emits the light to give a distribution of frequencies more acceptable to the human eye.

When a high voltage is applied between the metal electrodes at the ends of the tube, an electric current flows from one end to the other. The current is mostly made up of electrons. Let us consider what happens in the discharge tube if it is filled with hydrogen. Hydrogen is normally in the form of H_2 molecules but, when the electrons collide with hydrogen, the molecules dissociate into separate hydrogen atoms and electrons within the atoms are excited to orbits of higher energy.

The light from a gas discharge tube filled with hydrogen may be viewed in a **spectrometer**, an instrument that separates light of different frequencies. Part of the resulting spectrum, corresponding to visible light, is shown in Fig 11.3. The figure represents images of a slit facing the discharge tube. The position of the image on the horizontal axis depends on the frequency, and the darkness of the image depends on the intensity of the light. There is a series of dark lines that get closer together and merge on the right-hand side of the spectrum. Since photon energies are proportional to frequency, the series of lines shows that photons are only

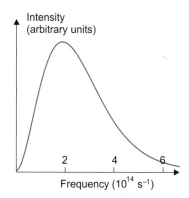

Fig. 11.1 The curve represents the variation with frequency of the intensity of light emitted from a hot filament at 3000°C. The intensity varies smoothly with frequency. All visible frequencies are present: there are no gaps in the spectrum where no light is emitted.

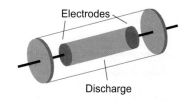

Fig. 11.2 In the gas discharge tube light is emitted by excited atoms in the region between the electrodes.

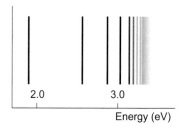

Fig. 11.3 The spectrum of visible light from hydrogen atoms.

emitted at certain definite energies. Each photon comes from a particular hydrogen atom, and conservation of energy requires that the atom changes its energy by the same amount as is carried away by the photon. The line spectrum implies that, when emitting a photon, the hydrogen atom can only change its energy by the discrete amounts corresponding to the photon energies seen in the spectrum.

The Schrödinger equation for the hydrogen atom

To apply quantum mechanics to the hydrogen atom we must calculate a wave function that gives the probability of finding the electron at different distances from the proton. To do this we must solve the Schrödinger equation. In Chapter 10 the Schrödinger equation was solved for a particle moving in one dimension between two fixed points or in a simple harmonic potential. Both examples led to a discrete series of allowed energy levels because the wave function must satisfy certain boundary conditions as well as being a solution of the Schrödinger equation. We may anticipate that electrons in a hydrogen atom are similarly only able to occupy certain discrete energy levels, and that this is why the hydrogen gas discharge produces a line spectrum rather than a continuous spectrum.

● *The time-independent Schrödinger equation may be used for the bound states of the hydrogen atom*

In this section we shall restrict discussion to the time-independent Schrödinger equation. This is not strictly correct, since the excited states of the hydrogen atom are not in a steady state, but after some time they lose energy by emitting a photon. On an atomic scale, however, the lifetime of the excited states is long. According to the uncertainty principle as stated in eqn (9.22), the uncertainty in energy ΔE for a state with lifetime Δt is at least $\hbar/2\Delta t$. Typically, for the lines in the hydrogen spectrum in Fig 11.3, the spread in photon energies ΔE is less than one-millionth of the corresponding photon energy E. For the purposes of learning about the wave function and evaluating energies, it is an extremely good approximation to assume a steady state, which may be described by the time-independent Schrödinger equation.

Worked Example 11.1 Yellow light of wavelength 589 nm is emitted from an excited state of the sodium atom. The lifetime of this state is 1.6×10^{-8} s, and after emission of a photon the atom is in its ground state. What is the spread in energy of the emitted photons?

Answer The spread in energy of the photons is then determined almost entirely by the uncertainty $\Delta E = \hbar/2\Delta t$ in the energy of the excited state. For $\Delta t = 1.6 \times 10^{-8}$ s, $\Delta E = 3.3 \times 10^{-27}$ J, or 2×10^{-8} eV.

Exercise 11.1 Calculate (a) the ratio of the energy to the energy spread of the photons emitted by the sodium atoms and (b) the number of wavelengths emitted in one lifetime Δt.

Answer (a) 10^8; (b) 8×10^6. These two numbers are the same except for a factor 4π, and either represents a rough measure of how good is the approximation of using the steady state Schrödinger equation to work out the wave function.

If the internal energy of an object is not changing, then in quantum mechanics, just as in classical mechanics, the internal motion can be considered in the centre of mass frame of reference, independently of the motion of the object as a whole. This was discussed in the context of the one-dimensional harmonic oscillator in Section 10.6, but it also applies in three dimensions. When we want to find the wave function of an electron in a hydrogen atom, we don't need to know whether or not the atom as a whole is moving. The internal wave function depends only on the relative position of the electron and the proton—the vector **r** in Fig 11.4. The Schrödinger equation is still three-dimensional, since three coordinates are required to specify **r**. The three-dimensional Schrödinger equation can be solved and, as for the one-dimensional harmonic oscillator, solutions are only found for certain discrete values of the atom's energy. It is not the purpose of this book to give the full mathematical detail of how these solutions are obtained. Our aim is to discuss the properties of the discrete states that lead to the line spectrum of emitted light shown in Fig 11.3.

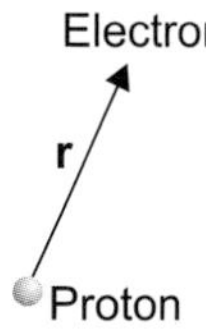

Fig. 11.4 The wave function describing the internal motion of the hydrogen atom depends only on the position vector **r** of the electron with respect to an origin located at the centre of the proton.

Spherically symmetric states of hydrogen

The attractive force between the electron and the proton in hydrogen acts in the direction of the line joining them, and its magnitude depends on the magnitude r of their relative position vector **r** but not on the direction of **r**. Such a force is called a **central** force. For a central force it is not convenient to express the wave function in terms of coordinates (x, y, z), but better to use instead the **spherical polar coordinates** (r, θ, ϕ) which are illustrated in Fig 11.5, which shows their relation to a set of Cartesian axes (x, y, z). In this figure, the coordinate r is the distance OP from the origin at O to a point P and θ is the angle between OP and the z-axis. A line PQ perpendicular to the xy-plane meets this plane at Q, and the third coordinate ϕ is the angle between OQ and the x-axis. Choosing the origin to be at the centre of the proton, the separation r of the proton and electron is the same as the coordinate r.

● *Central forces*

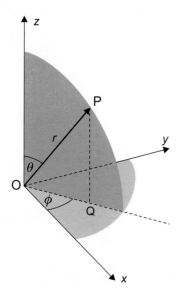

Fig. 11.5 This diagram shows the relation between the Cartesian co-ordinates x, y, z and the spherical polar coordinates r, θ, ϕ. The point P has a position vector $\mathbf{r}$, which has a length r and makes an angle θ with the z-axis. The z-axis and $\mathbf{r}$ lie in the plane that cuts the xy-plane at an angle θ to the x-axis.

⬤ *The Schrödinger equation*
for spherically symmetric states

Cylindrical polar coordinates labelled (r, θ, z) are used in Section 4.2 in a discussion of moments of inertia. You must be careful to distinguish spherical and cylindrical polar coordinates, since the coordinates r and θ have different meanings in the two systems. The coordinate z for a point P is the same in cylindrical polar coordinates and in Cartesian coordinates and r is the perpendicular distance from P to the z-axis. The cylindrical polar coordinate θ is the same as the spherical polar coordinate ϕ: it is the angle between the x-axis and the line OQ between the origin O and the point Q where the perpendicular from P meets the xy-plane, just as in Fig 11.5.

In general, wave functions depend on the angular variables θ and ϕ as well as the radial distance r. However, there are wave functions that are spherically symmetrical, i.e. they are the same in all directions and depend only on r and not on θ or ϕ. We shall write these wave functions[1] as $\phi(r)$. The wave function $\phi(r)$, like the wave functions we have already met, is a probability amplitude, and the probability of finding an electron within an infinitesimal volume dV at r is proportional to $\phi^*(r)\phi(r)dV$.

The volume of a spherical shell lying between r and $r + dr$ is $dV = 4\pi r^2 dr$, and the probability of finding an electron within this shell is therefore proportional to $4\pi r^2 \phi^*(r)\phi(r)dr$. The *total* probability of finding an electron anywhere is one, that is to say, the wave functions must be normalized. The normalization now covers the whole of space, i.e. all values of r between 0 and ∞. This means that $\phi(r)$ must satisfy the equation

$$\int_0^\infty 4\pi r^2 \phi^*(r)\phi(r)\, dr = \int_0^\infty 4\pi [r\phi^*(r)][r\phi(r)]\, dr = 1. \tag{11.1}$$

This equation has the same form as eqn (10.17) for normalizing a one-dimensional wave function, except that there is an extra factor 4π, and the wave function $\phi(r)$ has been replaced by $r\phi(r)$. In terms of the function $u(r) = r\phi(r)$, it can be shown that the three-dimensional Schrödinger equation for spherically symmetric states reduces to an equation with the same form as the one-dimensional eqn (10.12):

$$-\frac{\hbar^2}{2m}\frac{d^2 u}{dr^2} + V(r)u = Eu \tag{11.2}$$

where $V(r)$ is the potential energy of the electron and proton at a separation r, and m is the reduced mass of the electron and the proton. The reduced mass is defined in the same way as in eqn (5.10) and, since

---

[1] It is a common practice to use the symbol ϕ for both the angular variable and the time-independent wave function. This should not cause confusion, since it is usually clear from the context whether ϕ is a function or a variable.

the mass m_p of the proton is much greater than the mass m_e of the electron, $m = m_e m_p / (m_e + m_p)$ is almost the same as m_e.

In Section 9.1 it was explained that the proton and electron carry charges of the same magnitude e coulombs but of opposite sign. Electrical forces between charges obey an inverse square law, like the gravitational force, and they are discussed in detail in Chapter 15. All we need to know now is that the force attracting an electron with charge $-e$ coulombs towards a proton carrying a charge $+e$ coulombs at a distance r metres is

$$F(r) = \frac{1}{4\pi\epsilon_0} \frac{e^2}{r^2}. \tag{11.3}$$

The constant factor $1/4\pi\epsilon_0$ is required to ensure that the force is in newtons: by including 4π in the denominator of eqn (11.3) some other equations in electromagnetism are made simpler. The value of ϵ_0 is, to four significant figures,

$$\epsilon_0 = 8.854 \times 10^{-12} \text{ SI units.} \tag{11.4}$$

Potential energy, not force, occurs in Schrödinger's equation. Potential energy, as explained in Section 3.6, is energy stored by a system. For the electron in the hydrogen atom the potential energy is negative, since work must be done to pull the electron and the proton apart. Work in quantum mechanics is (force × distance) just as in classical physics. Using eqn (11.3), the work done to increase the distance between the proton and electron from r to $r + dr$ is

● *The potential energy in the hydrogen atom*

$$F(r)dr = \frac{e^2}{4\pi\epsilon_0 r^2} dr.$$

The work done to change the electron–proton distance from r to ∞ is

$$W_\infty(r) = \int_r^\infty \frac{e^2}{4\pi\epsilon_0 r^2} dr = \frac{e^2}{4\pi\epsilon_0} \left[\frac{1}{r}\right]_r^\infty = \frac{e^2}{4\pi\epsilon_0 r}.$$

The potential energy $V(r)$ of the electron is $-W_\infty(r)$, and finally

$$V(r) = -\frac{e^2}{4\pi\epsilon_0 r}. \tag{11.5}$$

This equation shows the same dependence on distance as the gravitational potential of a massive object near the Earth, given by eqn (5.5).

Exercise 11.2 The size of an atom is roughly 10^{-10} m $\equiv 0.1$ nm, about 5000 times smaller than the wavelength of yellow light. Calculate the force between an electron and a proton when they are separated by 10^{-10} m, and their potential energy at this distance.

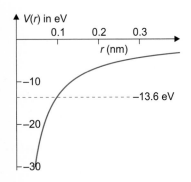

Fig. 11.6 The potential energy $V(r)$ of an electron as a function of its distance r from the hydrogen nucleus. At large distances $V(r)$ is very small compared with the binding energy E_B. The dashed line at -13.6 eV shows the binding energy of the lowest state of hydrogen.

⬤ *The behaviour of the wave function at large* r

Answer The force is 2.3×10^{-8} N and the potential energy is -14.4 eV. As we shall see, 14.4 eV is close to the energy needed to remove an electron from a hydrogen atom.

Strictly speaking, eqn (11.5) does not apply when r is very small, since even the hydrogen nucleus consisting of a single proton has a nonzero radius. However, the size of the nucleus is so small compared with the size of the atom that we can assume that the charge is concentrated in a point. The potential energy is then infinite at $r = 0$ but, fortunately, this does not lead to any mathematical difficulty.

Substituting eqn (11.5) in (11.2), the equation for u becomes

$$-\frac{\hbar^2}{2m}\frac{d^2 u}{dr^2} - \frac{e^2}{4\pi\epsilon_0 r}u = Eu. \tag{11.6}$$

We must emphasize that, for a system like the hydrogen atom, from which the electron does not escape, the total energy E is negative. The electron is **bound** in the atom and work must be done to release it and take it a long way from the proton. This positive amount of work is called the *binding energy*, which we shall write as E_B. This is the same notation as is used in Section 5.1 for planets that are bound within the solar system. When the electron and the proton are far apart, the total energy is zero, and hence the energy E in eqn (11.6) is

$$E = -E_B. \tag{11.7}$$

The variable r is the distance between the electron and the proton, which can only have positive values. The potential energy is sketched in Fig 11.6. It tends to $-\infty$ at $r = 0$ and becomes very small at large r. In order to find solutions of the Schrödinger equation in this potential, we must give boundary conditions, i.e. specify the behaviour of the wave function at $r = 0$ and $r = \infty$. The wave function $\phi(r)$ cannot be infinite at the origin, and $u(r) = r\phi(r)$ must be zero at $r = 0$. At large r the potential becomes very small and does not have much influence on u. Neglecting the potential for the moment, and using eqn (11.7), at large r the equation for u is

$$-\frac{\hbar^2}{2m}\frac{d^2 u}{dr^2} = -E_B u$$

or

$$\frac{d^2 u}{dr^2} - \alpha^2 u = 0, \tag{11.8}$$

where $\alpha^2 = 2mE_B/\hbar^2$. The general solution of eqn (11.8) is

$$u = Ae^{+\alpha r} + Be^{-\alpha r}. \tag{11.9}$$

(Comparing this equation with eqn (10.22), you will see that changing from a positive to a negative total energy gives solutions of the same form except that the exponent changes from imaginary to real.) The coefficient A can immediately be put equal to zero, since the positive exponential gets bigger and bigger as r increases. Only the second term in eqn (11.9) remains, and $u(r)$ tends to zero at large r.

We now have both of our boundary conditions: $u(r)$ must be zero at $r = 0$ and at $r = \infty$. The simplest function that satisfies these requirements, and is normalized according to eqn (11.1), is

$$u_1(r) = \left(\frac{1}{\pi}\right)^{1/2}\left(\frac{1}{a_0}\right)^{3/2} r e^{-r/a_0}. \tag{11.10}$$

● **The boundary conditions**

As may be verified by substitution, this function satisfies eqn (11.6) with

$$a_0 = \frac{4\pi\epsilon_0\hbar^2}{me^2} \tag{11.11}$$

and

$$E = -\frac{e^2}{8\pi\epsilon_0 a_0}. \tag{11.12}$$

The distance a_0 is about 5.3×10^{-11} m. It is called the **Bohr radius**. The normalized wave function corresponding to $u_1(r)$ is

$$\phi_1(r) = \frac{u_1(r)}{r} = \left(\frac{1}{\pi}\right)^{1/2}\left(\frac{1}{a_0}\right)^{3/2} e^{-r/a_0}. \tag{11.13}$$

The functions $u_1(r)$ and $\phi_1(r)$ are plotted in Fig 11.7 (a) and (b), respectively. The maximum value of $u_1(r)$ occurs at the Bohr radius a_0 and, since the probability of finding the electron at a distance between r and $r + dr$ from the proton is $4\pi r^2|\phi_1(r)|^2 dr = 4\pi u_1^2(r)dr$, the probability also has its maximum value at a_0. The wave function $\phi_1(r)$ has its largest value at the origin, which means that the electron cloud is densest close to the proton even though the electron is most likely to be found at distances of about a_0 from the proton.

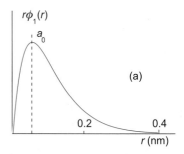

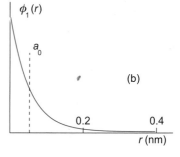

Fig. 11.7 (a) The function $u_1(r) = r\phi_1(r)$ for an electron in the ground state of hydrogen; (b) the wave function $\phi_1(r)$. Both functions are real, and have no imaginary parts.

Exercise 11.3 Show by substitution that the wave function $u_1(r)$ in eqn (11.10) does satisfy eqn (11.6), and verify that the constants a_0 and E have the values given in eqns (11.11) and (11.12).

The function $u_1(r)$ in eqn (11.10) has a single maximum, as shown in Fig 11.7. It is the solution that has the lowest energy, i.e. the most negative value of E, and it represents the *ground state* of the hydrogen atom. The value of E for the ground state is about -2.18×10^{-18} J, or equivalently -13.6 eV (the electron volt (eV)), the unit of energy used in atomic physics, was defined in Section 9.1).

As well as the ground state, there are also excited spherically symmetric states, corresponding to other solutions of eqn (11.6) for which $u(r)$ is

● **Excited states**

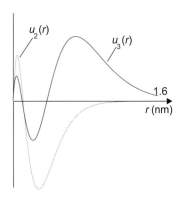

Fig. 11.8 The functions $u_2(r)$ and $u_3(r)$ describing spherically symmetric excited states of the hydrogen atom.

⬤ *All energies are possible for unbound states*

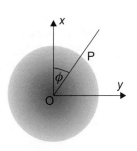

Fig. 11.9 A view of a spherically symmetrical state of the hydrogen atom. The variation of the electron density does not depend on the angle ϕ.

zero at $r = 0$ and $r = \infty$. Like $u_1(r)$, these solutions are real, and have no imaginary part. They have extra maxima and minima, as do the wave functions shown in Figs 10.6 and 10.9 for the excited states in the one-dimensional potential well and the harmonic oscillator. We denote the solution that has n maxima or minima by $u_n(r)$.

The functions $u_2(r)$ and $u_3(r)$ are sketched in Fig 11.8, using a different distance scale from the graph of $u_1(r)$ in Fig 11.7. You will notice that, as n becomes larger, $u_n(r)$ stretches further away from the origin. Remembering that $4\pi[u(r)]^2\mathrm{d}r$ is the probability of finding the electron at a distance between r and $r + \mathrm{d}r$ from the proton, this means that an electron in a state with high n is likely to be far away from the proton where the potential $V(r)$, is small. Since the attraction between the electron and the proton is then weak, the binding energy is small. The energy of the nth state is in fact

$$E_n = -\frac{1}{n^2}\left(\frac{e^2}{8\pi\epsilon_0 a_0}\right) = \frac{1}{n^2}E_1 \tag{11.14}$$

where $E_1 = -13.6\,\mathrm{eV}$ is the energy of the ground state.

All values of n are in principle possible, but eqn (11.14) shows that for high n the binding energy $(-E_n)$ is very small: the electron is very weakly bound to the proton. No atom can be completely isolated, and a weakly bound electron far from its parent nucleus is easily removed, becoming *unbound*. An unbound electron has a positive energy, which is just its kinetic energy when it is far enough from the nucleus for the potential energy to be negligible. The kinetic energy can have any value. As well as the spectrum of discrete bound states, there is a continuum of unbound states representing electrons that are scattered by a hydrogen nucleus (a proton) that has no bound electron.

Exercise 11.4 Calculate the distance at which the potential energy of the electron in hydrogen is equal to the energy of a state with $n = 100$.

Answer $10^{-6}\,\mathrm{m}$, comparable to the wavelength of visible light.

11.2 **Quantum numbers**

Quantum numbers were introduced in Section 10.5 to distinguish between different solutions of the Schrödinger equation. The number n in eqn (11.14) specifies a particular spherically symmetric bound state of the hydrogen atom. For such a state, n tells you the number of maxima and minima between $r = 0$ and $r = \infty$. The force between the electron and the proton is known, and the wave function can be worked out for any n.

It is an important idea that different spherically symmetric states can be *labelled* by a quantum number. The states can still be labelled even if you do not know exactly how the force varies with distance. You can say 'I cannot solve the Schrödinger equation for this force, but I have got to fit n maxima between $r = 0$ and $r = \infty$, so I can label each state by n although I don't know its exact wave function'.

Wave functions that depend on angle

So far we have only discussed hydrogen wave functions that are spherically symmetric. There are others in which the wave function varies with direction as well as with the distance between the electron and the proton. In spherical polar coordinates the direction is given by the two angular coordinates θ and ϕ. Boundary conditions have to be applied to the wave function as θ and ϕ vary, and only special discrete functions will satisfy these conditions. The functions can be labelled by two new quantum numbers, one each for θ and ϕ. Let's look in more detail at how these quantum numbers arise.

Suppose that you have a marvellous instrument that allows you to see inside atoms, and that you are looking at a hydrogen atom in its ground state. The atom is spherically symmetrical, and what you would see is something like Fig 11.9. Choosing the origin of coordinates at the centre of the atom and the z-axis directed into the paper along the line of sight, the angle ϕ in Fig 11.9 is the same as in the coordinate system illustrated in Fig 11.5. Along a line OP the electron density has the same variation with distance from the centre whatever may be the angle ϕ.

If the electron density is not spherically symmetrical, you can imagine that the atom might look like Fig 11.10(a), with a density that changes around a circle at fixed r but varying ϕ. If you go right round the circle, so that ϕ changes by 2π radians, you get back to where you started from. The wave function and the electron density must return to their original values. In other words, the wave function is periodic in ϕ with period 2π, just like the functions $\cos\phi$ and $\sin\phi$.

The solutions of the Schrödinger equation we are looking for do in fact have a sinusoidal variation in ϕ. In Fig 11.10(a) the electron density has a maximum value at $\phi = 0$, corresponding to the wave function varying as $\cos\phi$ and the electron density as $\cos^2\phi$. Figure 11.10(b) is a similar illustration for a wave function varying as $\sin\phi$. Obviously you can get from Fig 11.10(a) to Fig 11.10(b) simply by rotating the whole atom through a right angle about the z-axis. The two states are not independent of one another.

Only two independent wave functions are needed to describe all possible rotations of Fig 11.10(a) about the z-axis. Because the Schrödinger equation is linear, the sum of two solutions is also a solution, and the wave functions obey the *principle of superposition* which is described in more detail in Section 7.4. For example, a wave function varying as $\sin(\phi - \phi_0)$ is zero at the angle $\phi = \phi_0$. Since

● *Boundary conditions for angular variables*

Fig. 11.10(a) A projection of the electron density along the z-axis, and plots of the wave function and the electron probability density as functions of the angle ϕ for a wave function varying as $\cos\phi$.

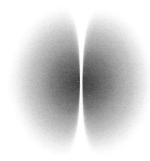

Fig. 11.10(b) (b) A wave function with a $\sin\phi$ dependence is obtained by rotating the wave function in (a) through 90°.

$\sin(\phi - \phi_0) = (\sin\phi\cos\phi_0 - \cos\phi\sin\phi_0)$, the variation with ϕ is the same as for the sum of the original functions $\cos\phi$ and $\sin\phi$ with amplitudes $(-\sin\phi_0)$ and $(\cos\phi_0)$. The coefficients of $\cos\phi$ and $\sin\phi$ may be complex numbers, and the best pair of independent functions to choose is actually $(\cos\phi \pm j\sin\phi)$, These functions can be expressed alternatively in terms of complex exponential functions as $e^{\pm j\phi}$.

The boundary condition that the wave function returns to the same value when ϕ increases by 2π is not only satisfied by $e^{\pm j\phi}$, but also by the functions $e^{jm_\ell\phi}$, provided that m_ℓ is a positive or negative integer or zero. The number m_ℓ is a new quantum number, labelling the members of a set of discrete functions that can describe the ϕ variation of solutions of the Schrödinger equation for the hydrogen atom.

● *The quantum number associated with ϕ variation*

To define the direction of a point P in spherical polar coordinates, the angle θ between the position vector and the z-axis is required as well as the angle ϕ (Fig 11.5). The wave function must obey the same boundary condition for θ as for ϕ, since we also return to the starting point if θ is increased by 2π. The θ variation of the wave function is much more complicated than the ϕ variation, and we shall not give the mathematical form of the θ variation here. All we need to know is that there is a new quantum number ℓ associated with the θ variation of the wave function.

● *The quantum number associated with θ variation*

The quantum numbers n, ℓ, and m_ℓ are not independent of one another. The θ variation of the wavefunction depends on ℓ and m_ℓ, and the radial variation depends on ℓ as well as on n. The relations between the quantum numbers, which follow mathematically from the full solution of the three-dimensional Schrödinger equation, are as follows.

● *Rules for quantum numbers*

1. n **may be any positive integer.**

2. ℓ **may be zero or any positive integer up to and including** $(n - 1)$.

3. m_ℓ **takes integral values from** $-\ell$ **to** ℓ.

The labelling of the solutions to the Schrödinger equation is now complete. The variation of the wave function with position is determined by the three quantum numbers n, ℓ, and m_ℓ, one each for the three-dimensional variables r, θ, and ϕ. The energy of a state of the hydrogen atom is determined by the quantum number n, which is called the **principal quantum number**. In the next section it is explained that the quantum numbers ℓ and m_ℓ are related to angular momentum: ℓ is called the **orbital quantum number** and m_ℓ the **magnetic quantum number**.

11.3 **Angular momentum**

The quantum numbers have been introduced as a mathematical consequence of the requirements that the wave function should obey

For any central force the angular variations of the wave functions are the same

We have talked about the r, θ, and ϕ variation of the wave function separately. This is allowable, because the potential energy $V(r)$ is independent of direction. The angular variation of the wave function therefore does not depend on how the potential varies with the distance between the electron and the proton. The variations with θ and ϕ are also independent of one another, and the three-dimensional wave function $\phi(\mathbf{r})$ can be written as a product of separate functions of r, θ, and ϕ. The ϕ variation is given by $e^{jm_\ell\phi}$, and we shall call the functions describing the r and θ variations $R_{n\ell}(r)$ and $\Theta_{\ell m_\ell}(\theta)$. The full wave function $\phi(\mathbf{r})$ is then

$$\phi(\mathbf{r}) = \phi(r,\theta,\phi) = R_{n\ell}(r) \times \Theta_{\ell m_\ell}(\theta) \times e^{jm_\ell\phi}.$$

$$(11.15)$$

The wave function may be written in the form of eqn (11.15) for any central potential $V(r)$, i.e. one that depends only on r. Since the angular variation of the wave function is independent of $V(r)$, the function $\Theta_{\ell m_\ell}(\theta)$ is the same for all central potentials. The numbers n, ℓ, and m_ℓ are the quantum numbers of the state of the hydrogen atom that has the wave function given by eqn (11.15)—one quantum number for each coordinate of the three-dimensional problem.

The universal functions of direction described here are known as **spherical harmonics**, by analogy with the harmonics in a Fourier series. Spherical harmonics, which are functions of both θ and ϕ, are often written in the notation

$$Y_{\ell m_\ell}(\theta,\phi) = \Theta_{\ell m_\ell}(\theta)e^{jm_\ell\phi}.$$

$$(11.16)$$

The infinitesimal volume enclosed in the space between r and $(r+dr)$, θ and $(\theta+d\theta)$, ϕ and $(\phi+d\phi)$ is $r^2 dr \sin\theta\, d\theta\, d\phi$. The spherical harmonics $Y_{\ell m_\ell}$ are normalized over the angular variables by satisfying the integral

$$\int_{\phi=0}^{2\pi} \int_{\theta=0}^{\pi} Y_{\ell m_\ell}^*(\theta,\phi) Y_{\ell' m_\ell'}(\theta,\phi) \sin\theta\, d\theta\, d\phi = 1$$

if $\ell = \ell'$ and $m_\ell = m_\ell'$. The integral is zero if $\ell \neq \ell'$ or $m_\ell \neq m_\ell'$. This is very similar to the integral relations that were given for sines and cosines in the discussion of Fourier series (Section 6.6). The analogy goes further and, in the same way that any periodic function of time can be expressed as a Fourier series, any function of θ and ϕ that returns to the same value when θ or ϕ advances by 2π can be expressed as a sum of spherical harmonics.

certain boundary conditions. But the quantum numbers ℓ and m_ℓ associated with the angular variation of the wave function have a physical significance too. They are related to angular momentum.

For an isolated hydrogen atom with no external forces or torques acting on it, angular momentum must be conserved, as explained in Section 4.5. This means that the angular momentum, like the energy, does not change with time—it is a constant of the internal motion of the atom. In quantum mechanics the unit of angular momentum is Planck's constant divided by 2π—usually written as $\hbar$.

Worked Example 11.2 Express the units of angular momentum in terms of the basic SI units: metres, kilograms, and seconds.

Answer The dimensions of linear momentum are $[\text{mass}] \times [\text{velocity}]$, i.e. $[M][L][T]^{-1}$. The dimensions of angular momentum are $[\text{length}] \times [\text{linear momentum}]$, i.e. $[M][L]^{2}[T]^{-1}$, and the SI units are $\text{kg m}^2 \text{ s}^{-1}$.

Exercise 11.5 Remembering that the energy of a photon with frequency ν is $h\nu$, check that h, and therefore also $\hbar$, have the same units as angular momentum.

In Section 4.6 it is proved that the internal angular momentum of any system of particles about their centre of mass can be separated from the angular momentum associated with motion of the centre of mass. When thinking about the internal states of a hydrogen atom, we do not have to worry if the atom as a whole is moving, so long as no forces or torques are acting on it. All the information it is possible to know about the internal motion of an atom in a steady state is given by its wave function, which is in turn labelled by the quantum numbers. The magnitude of the internal angular momentum of the hydrogen atom can be calculated from its wave function. The results of this calculation are that:

For a state with quantum number ℓ the magnitude of the angular momentum has the value $\sqrt{\ell(\ell+1)}\hbar$, and the component of angular momentum along the z-axis is $m_\ell \hbar$.

As mentioned before, the quantum number m_ℓ is restricted to integral values between $-\ell$ and ℓ (relation 3 from the previous section). We can see why this is so. The z-component of angular momentum may be positive or negative, but the magnitude of the z-component cannot possibly be greater than the total magnitude. In classical physics, if a vector is pointing along the z-axis, then its z-component is equal to the magnitude of the vector. This leads one to expect the magnitude of the angular momentum of a state with quantum number ℓ to be $\ell\hbar$. The fact that the magnitude of the angular momentum is not $\ell\hbar$ but $\sqrt{\ell(\ell+1)}\hbar$ may be understood as a consequence of the uncertainty principle. In quantum mechanics, if the total angular momentum is not zero, we cannot simultaneously give exact values for its x-, y-, and z-components. If the quantum number m_ℓ has the value ℓ, uncertainties must occur for the x- and y-components. The angular momentum may be pictured as a vector of length $\sqrt{\ell(\ell+1)}\hbar$, with projection $\ell\hbar$ along the z-axis, as in Fig 11.11.

In quantum mechanics it is always necessary to consider what may be the consequences of the uncertainty principle. We have already met the uncertainty principle in the form $\Delta E \Delta t \geq \frac{1}{2}h$ (eqn (9.18), which told us that only steady states, for which Δt is indefinitely large, can have exactly defined energies. The dimensions of $\hbar$ are the same as those of angular momentum, and the z-component of angular momentum $(m\hbar)$ is

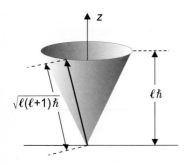

Fig. 11.11 The length of the vector representing angular momentum is $\sqrt{\ell(\ell+1)}\hbar$, but its maximum component along the z-axis is $\ell\hbar$. The vector is equally likely to be at any point on the cone with projection $\ell\hbar$ on to the z-axis.

coupled in the uncertainty principle with the dimensionless angular variable ϕ: $\Delta(m\hbar)\Delta\phi \sim \hbar$. Because m is an integer there is no uncertainty in $m\hbar$, i.e. $\Delta(m\hbar)$ is zero, so that $\Delta\phi$ must be infinitely large—we do not know anything about ϕ. In Fig 11.11, the vector representing the angular momentum may lie anywhere on the coloured cone and all angles ϕ are equally likely.

This schematic picture of the angular momentum in the hydrogen atom is extended in Fig 11.12 to show all the states for $\ell = 3$. There are $(2\ell + 1) = 7$ possible values of $\ell_z = m_\ell \hbar$, with $m_\ell = +3, +2, +1, 0, -1, -2, -3$. For clarity, the vectors representing the angular momentum are all shown at the same angle on the surface of the cone for each value of m_ℓ, but each vector should be allowed to lie at any angle around the cone.

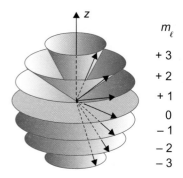

Fig. 11.12 For $\ell = 3$ there are 7 possible values of z-component of angular momentum.

Degeneracy

Figure 11.13 shows the energies of some of the different states of hydrogen. Each state is labelled with the quantum numbers n and ℓ. States with the same n but different ℓ do have slightly different energies, but the differences are too small to show up on the diagram.

The quantum number m_ℓ has been left out of Fig 11.13 because states with the same n and ℓ but different m_ℓ all have *exactly* the same energy. This is because $m_\ell \hbar$ represents the z-component of angular momentum, and we can choose any direction we please for the z-axis. The energy of a state cannot depend on our choice of the mathematical coordinate system: states with the same n and ℓ but different values of m_ℓ must therefore have the same energy. They are said to be **degenerate**. States with the orbital quantum number ℓ may have $(2\ell + 1)$ different values of m_ℓ, and have a **degeneracy** of $(2\ell + 1)$. Thus the degeneracy of the $\ell = 3$ states in Fig 11.12 is seven.

For central forces, the energy depends only on n and ℓ, whatever the form of the potential $V(r)$. Under the approximation that the nucleus is a point charge, and neglecting magnetic and relativistic corrections, the energy of the hydrogen atom is given by eqn (11.14) for all possible values of the quantum number ℓ. This is only true for an exact inverse square force law: it is a fluke, or in more formal language an **accidental degeneracy**. The real hydrogen atom, in which there are small corrections to a pure inverse square law, has tiny differences between the energies of states with the same n but different ℓ.

States with the same m_ℓ are not degenerate when there is an external force acting on the atom. A magnetic field has a definite direction, for example, and, when the atom is placed in a magnetic field, the degenerate states split up. If the z-axis is chosen to lie along the direction of the magnetic field, each m_ℓ-value corresponds to a state with a slightly different energy. This is why m_ℓ is referred to as the *magnetic* quantum number.

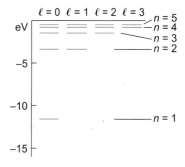

Fig. 11.13 The low-lying energy states of the hydrogen atom, labelled by their quantum numbers.

Transitions between states with different quantum numbers

As mentioned above, the 'accidental degeneracy' of the states of hydrogen with different ℓ only applies for an exact inverse square attraction between the electron and the proton. There are small deviations to this law in a real hydrogen atom, but to a very good approximation the energy of a state with principal quantum number n is given by eqn (11.14) as $E_n = E_1/n^2$, where $E_1 = -13.6$ eV is the energy of the ground state.

When a photon is emitted by a hydrogen atom, the energy of the photon is equal to the difference of two of the discrete states of the atom. The energy $E_{n_1 n_2}$ of a photon emitted in a transition between a state with energy E_{n_2} and one with lower energy E_{n_1} is

$$E_{n_1 n_2} = E_1 \left(\frac{1}{n_1^2} - \frac{1}{n_2^2} \right). \tag{11.17}$$

The energy E_n becomes less negative as n increases, and n_2 must be larger than n_1. For each value of the quantum number n_1 of the lower state, there is a series of transitions with energies becoming larger as n_2 becomes larger, with a limiting value at E_{n_1}. The series with $n_1 = 1$ and $n_2 = 2, 3, 4, \ldots$ is called the Lyman series. The lowest energy of photons in this series is $13.6 \times (1 - \frac{1}{4}) = 10.2$ eV, and the other members of the series are between 10.2 and 13.6 eV, all in the ultraviolet region of the spectrum. The next series, the Balmer series, with $n_1 = 2$ has energies ranging from 1.9 eV to 3.4 eV, in the visible and near-ultraviolet region. This is the series shown in Fig 11.14.

● **The hydrogen spectrum**

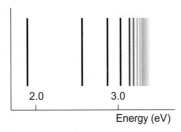

Fig. 11.14 This figure, which is the same as Fig 11.3, shows the spectrum of visible lines emitted by hydrogen atoms making transitions to the $n = 2$ states.

The Rydberg constant

The relation between the wavelength λ of a photon and its energy E is

$$\frac{1}{\lambda} = \frac{\nu}{c} = \frac{E}{hc}.$$

From eqns (11.11) and (11.14) the wavelength $\lambda_{n_1 n_2}$ of a transition between states with principal quantum numbers n_2 and n_1 may be rewritten as

$$\frac{1}{\lambda_{n_1 n_2}} = \frac{me^4}{8\epsilon_0^2 h^3 c} \left(\frac{1}{n_1^2} - \frac{1}{n_2^2} \right) = R_{\mathrm{H}} \left(\frac{1}{n_1^2} - \frac{1}{n_2^2} \right).$$

The constant R_{H} is called the **Rydberg constant for hydrogen**. To four significant figures, R_{H} has the value 1.097×10^7 m^{-1}. The quantities c and ϵ_0 occurring in R_{H} are defined *exactly*. The reduced mass m can be expressed accurately in terms of the actual mass of the electron m_{e}, so that a measurement of the wavelength of hydrogen transitions is connected to the important constants m_{e}, e, and h.

The difference in energy between $n = 1$ and $n = 2$ states can be measured with great precision—about 1 part in 10^{11}. There are other very accurate measurements that are related to different combinations of atomic constants, and the best values of these constants are derived from these very accurate measurements taken together, and not by direct measurement of each constant separately.

Warning. The values of $1/\lambda$ are sometimes quoted in units of m^{-1} or cm^{-1}, and referred to as *wavenumbers*. The usual definition of wavenumber in physics is the one given in eqn (7.1), which differs by a factor 2π.

Intrinsic spin

The main features of the hydrogen spectrum are well explained by the transition energies given in eqn (11.17). However, when the spectrum is viewed with high resolution (i.e. with good discrimination between photons having nearly equal energies), each line in Fig 11.14 is found to consist of a group of narrow lines with very nearly the same energy. A group of lines makes up the **fine structure** of a transition between pairs of states with the same quantum numbers n_1 and n_2. Once the energies of the fine structure lines are known, a new level diagram can be deduced that includes the small energy differences between states with the same principal quantum number n but different values of the orbital quantum number ℓ. The new diagram contains a surprise. For each n, *two* states are found for all $\ell > 0$.

Why are there two states when solving the Schrödinger equation had led us to expect one? The answer is that the electron has **intrinsic spin**. Intrinsic spin is angular momentum possessed by the electron in addition to its orbital angular momentum. We shall describe intrinsic spin by a new quantum number s. The presence of intrinsic spin appears clearly when the atoms are placed in a magnetic field. The states with $\ell = 0$ then split into two components. The magnetic field causes an energy difference to occur between states with different magnetic quantum number m_ℓ: for the $\ell = 0$ states this cannot be associated with the orbital motion of the electron, since m_ℓ is zero for these states.

A state with orbital angular momentum quantum number ℓ has $(2\ell + 1)$ different values of m_ℓ according to the rule derived from the boundary conditions on the wave function. Since the $\ell = 0$ states split into two components with different m in a magnetic field, this implies that the intrinsic spin quantum number s satisfies $(2s + 1) = 2$, or $s = \frac{1}{2}$. This result may seem strange, since the other quantum numbers are all integers. It is nevertheless correct. Intrinsic spin has the same properties as orbital angular momentum—intrinsic spin angular momentum has a magnitude $\sqrt{s(s + 1)}\hbar = \sqrt{\frac{3}{4}}\hbar$ and z-component $\pm\frac{1}{2}\hbar$.

Intrinsic spin is a consequence of the theory of relativity. A fully relativistic quantum mechanical theory of the electron, which was first formulated by Paul Dirac in 1928, *requires* the electron to have an intrinsic spin with the observed properties. When the interactions between electric and magnetic fields and the electron are included, the theory is called quantum electrodynamics. This theory has been tested to a remarkably high precision, and no deviation between theory and experiment has ever been found.

According to the Dirac theory, for states with $\ell > 0$ the intrinsic spin may add to or subtract from the orbital angular momentum. The resultant

● *States with $\ell = 0$ have a degeneracy of two*

● *Intrinsic spin adds to or subtracts from orbital angular momentum*

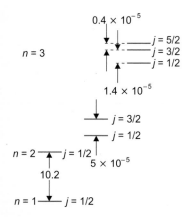

Fig. 11.15 Low-lying energy levels in hydrogen showing the pairs of fine structure states with almost the same energy. Energy differences are marked in eV.

states have total angular momentum labelled by a new quantum number $j = \ell \pm \frac{1}{2}$. Since ℓ is an integer, j is always half-way between two integers: j is described as being **half-integral**. Total angular momentum may be pictured using a diagram like Fig 11.12, which applied to orbital angular momentum on its own. The maximum and minimum values of the z-component of the total angular momentum are $(+j\hbar)$ and $(-j\hbar)$. Because of the uncertainty principle, the magnitude of the total angular momentum is greater than $j\hbar$ and has the value $\sqrt{j(j+1)}\hbar$. Only certain discrete values $m_j\hbar$ of the z-component are allowed, labelled by a magnetic quantum number m_j. Like j, m_j is half-integral, and it has $(2j+1)$ values from $(+j)$ to $(-j)$. For example, a state with $\ell = 1$ has $j = \frac{1}{2}$ or $j = \frac{3}{2}$. For the $j = \frac{1}{2}$ state, m_j may be $\pm\frac{1}{2}$ and for the $j = \frac{3}{2}$ state m_j may be $+\frac{3}{2}, +\frac{1}{2}, -\frac{1}{2}$, or $-\frac{3}{2}$. Since $m_j\hbar$ is the z-component of the total angular momentum, including the contributions of both orbital angular momentum and intrinsic spin, m_j is the only magnetic quantum number that needs to be specified when labelling the state of an electron in the hydrogen atom.

The story is now complete. There are no more hidden states of the hydrogen atom. A new quantum number has been introduced for intrinsic spin, which always has the value $s = \frac{1}{2}$, so there is no need to mention it when labelling a state. The four quantum numbers, n, ℓ, j, and m_j are sufficient to determine unambiguously the state of an electron in hydrogen. A few of the low-lying states in hydrogen are shown again in Fig 11.15 and their energy differences are also shown (not to scale!) to indicate just how small is the fine structure separation between states with the same ℓ but different j.

11.4 Many-electron atoms

How do electrons behave in atoms with atomic number $Z > 1$? The Z electrons in the atom are all bound, and you have to do work to remove any one of them a long way from the atom. As for hydrogen, bound electrons can only exist in discrete states, each with a sharply defined energy. For example, sodium has $Z = 11$. The spectrum of yellow light emitted from a sodium discharge tube—familiar since it is often used for street lighting—is mostly concentrated at two wavelengths, 589.0 and 589.6 nm. You would be right to guess that the separation of about one part in a thousand of these wavelengths is fine structure due to the presence of states with the same ℓ but different j. It still makes sense to assign quantum numbers to each electron wave function.

However, it is not possible to write down an exact solution to the Schrödinger equation for an element with a large value of Z. When there are many electrons, each one is attracted to the nucleus and repelled from the other electrons. The wave function for each electron depends not only on its own coordinates (r, θ, ϕ) but on the coordinates of all the other electrons as well. In classical mechanics many-body problems of this sort are appallingly difficult. It is only possible to calculate accurately the orbits of planets in the solar system, for example, because the Sun is so gigantic—its mass is ten thousand times bigger than the mass of the largest planet, Jupiter. Each planet orbits around the Sun, and its orbit is only slightly affected by the presence of the other planets.

There is no dominant force for the electrons in an atom. The largest force on an individual electron is indeed towards the nucleus with its charge $+Ze$. But there are $(Z - 1)$ other electrons each with a charge $-e$, so their net effect is comparable to that of the nucleus. In classical mechanics this is an impossible problem. Strangely, it is much *easier* in quantum mechanics. This is because the exact position of an electron cannot be specified. It is smeared out like a cloud with a density given by the square of its wave function.

Atoms are more or less spherical, as a microscope picture like Fig 9.1 suggests. For one particular electron, the effect of all the others is like an almost spherical cloud. The potential energy of the electron depends on the distance r from the centre of mass, but scarcely at all on the angular variables θ or ϕ. To a very good first approximation each electron is therefore moving in a central potential $V(r)$ and its wave function can be evaluated independently of all the others. The variation with r of the potential $V(r)$ depends on the wave functions of all the other electrons, so accurate calculations are still very difficult. Nevertheless each electron has to satisfy its own set of boundary conditions, and its wave function can be labelled by the same quantum numbers n, ℓ, j, and m_j used in hydrogen.

⬤ *Exact wave functions are not obtainable for many-electron atoms*

The total number of states is not affected by mixing states with different quantum numbers

So far in this section we have been discussing electrons in a many-electron atom as though we can pick out a particular electron and write down its own wave function. In fact all the electrons are *identical* and there is no way of choosing one electron to be called 1, another 2, and so on. It is not possible to say, for example, that electron 1 has $m_j = \frac{1}{2}$ and electron 2 has $m_j = -\frac{1}{2}$. It is just as likely to be the other way round. The consequence of this is that the wave functions for individual electrons get all mixed up. Exactly how the wave functions are mixed up and how the orbital and spin angular momenta of the electrons are combined to give the angular momentum of the atom as a whole, are subjects outside the scope of this book. However, the *number* of different states that exists is not changed by mixing them. If we are wanting to find out how many different states there are we can make a list of different sets of the quantum numbers n, ℓ, j, and m_j. Furthermore, the energy of a many-electron atom is close to the sum of the energies of separate electrons each moving in a central potential with their own quantum numbers n, ℓ, j, and m_j. In the rest of this section we shall use this approach to describe the states available in atoms of different elements. Individual quantum numbers will be assigned to each electron in turn and qualitative explanations given of the binding energies found in atoms.

The Pauli exclusion principle

In a real many-electron atom with atomic number Z, which states do the Z electrons occupy? If all the electrons were in their individual ground states with $n = 1$ and $\ell = 0$, the atom as a whole would obviously also have the lowest possible energy. But this does not happen. All elements with high Z would have nearly the same properties if all their electrons were in the same state. Chlorine, argon, and calcium would behave in the same way—there would be no chemistry, you would not be here, and the world would be a very uninteresting place. This disastrous scenario is prevented by another feature of quantum mechanics, which we have not yet met. This feature is the **Pauli exclusion principle**, which forbids more than one electron to occupy the same state. The states are labelled by their quantum numbers, and a formal statement of the Pauli principle is as follows.

No two electrons in an atom can have the same set of quantum numbers n, ℓ, j, and m_j.

Now that we know that electrons occupy states with different quantum numbers we may consider the structure of many-electron atoms. Let's start with the next simplest atom after hydrogen, which is helium (chemical symbol He) with $Z = 2$. If we have a helium ion, which has only one electron and a net charge of $+e$, the Schrödinger equation is exactly the same as for hydrogen except that the potential energy becomes $2e^2/(4\pi\epsilon_0 r)$ instead of $(e^2/(4\pi\epsilon_0 r))$, because the attraction of the nucleus with charge $+2e$ is just twice the attraction of the hydrogen nucleus with charge $+e$. The stronger force causes the electron to spend more time close to the nucleus, and the net result is that the binding energy of the electron in its ground state is four times the binding energy for hydrogen, i.e. $4 \times 13.6 \simeq 54\,\text{eV}$. For an ion consisting of a single electron and a nucleus with atomic number Z, the energy of the ground state is Z^2 times the energy given in eqn (11.12) for hydrogen:

● *The helium ion*

$$E_1(Z) = -\frac{Z^2 e^2}{8\pi\epsilon_0 a_0}. \tag{11.18}$$

The binding energy of a single-electron ion of a heavy element like lead, with $Z = 82$, is thus very large: $82^2 \times 13.6 \simeq 91\,\text{keV}$.

What happens when a second electron is added to make a neutral helium atom? The first electron was in a state with quantum numbers $n = 1$, $\ell = 0$, $j = \frac{1}{2}$, and $m_j = \pm\frac{1}{2}$. The second electron can also go into a state that has $n = 1$, $\ell = 0$, and $j = \frac{1}{2}$. But now there is no choice for m_j. One electron must have $m_j = +\frac{1}{2}$ and the other $m_j = -\frac{1}{2}$. The intrinsic spins of the two electrons point in opposite directions and the total angular momentum of the pair is zero. There is only one way of putting

● *The helium atom*

the electrons in these two states, because the two electrons are *identical* and it is not possible to tell which has $m_j = +\frac{1}{2}$ and which $m_j = -\frac{1}{2}$.

The two equivalent electrons have the same energy and, since they repel one another, the binding energy of each is less than the binding energy $E_B \approx 54\,\text{eV}$ of the single electron in the helium ion. The minimum energy required to remove one electron from a helium atom to form an ion is called the **first ionization energy** E_I. The first ionization energy is about 25 eV, less than the binding energy of each electron in the atom, because, when one electron is removed, the remaining one is more tightly bound. The total energy needed to remove *both* electrons is about

$$E_I + E_B \approx (25 + 54) = 79\,\text{eV}$$

and, since this energy is shared equally, the binding energy of each electron in the neutral helium atom is about 78/2 = 39 eV.

After helium comes lithium (Li) with atomic number 3. The third electron in lithium cannot go into a state with principal quantum number $n = 1$, because there are only two of these and both are already occupied. Only $n = 2$ states are available, which are much less tightly bound than the $n = 1$ states. The $n = 2$ wave functions have large amplitudes far away from the nucleus, outside the region where the $n = 1$ amplitudes are concentrated. At large distances two of the three positive charges on the nucleus are almost cancelled out by the negative charge on the two electrons with $n = 1$. The $n = 2$ electron is therefore shielded from the nucleus and the force on it is much less than the force on the $n = 1$ electrons. The $n = 2, \ell = 0$ state does penetrate closer to the nucleus than the $n = 2, \ell = 1$ state and is more tightly bound: it is the $n = 2, \ell = 0$ state that is occupied in the ground state of lithium. The ionization energy of lithium—the energy needed to remove the outermost electron—is only 5.4 eV, very much less than the value of 24.4 eV for helium. The energies of the occupied levels in neutral hydrogen, helium, and lithium atoms are shown in Fig 11.16.

● *Occupied states in the ground state of lithium*

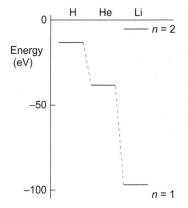

Fig. 11.16 The energies of the states occupied in the ground states of hydrogen, helium, and lithium atoms. All of the states have $\ell = 0$. The $n = 1$ states become more tightly bound as Z increases. After the closed shell at He, the $n = 2$ state in Li has a binding energy of only about 5 eV.

Worked Example 11.3 How many electrons are allowed in states with principal quantum number $n = 2$?

Answer For states with $n = 2$, the orbital quantum number ℓ may be 0 or 1. If $\ell = 0$, the total angular momentum j is the same as the intrinsic spin, i.e. $j = \frac{1}{2}$. There are two degenerate states with $m_j = \pm\frac{1}{2}$. For states with $n = 2$ and $\ell = 1$, j may be either $\ell + \frac{1}{2} = \frac{3}{2}$ or $\ell - \frac{1}{2} = \frac{1}{2}$. These states have degeneracy $2j + 1 = 4$ and 2, respectively, and there are thus 6 states with $\ell = 1$. The total number of electrons that can occupy $n = 2$ states is $2 + 6 = 8$.

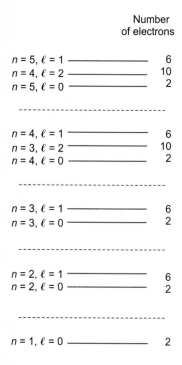

Number of electrons

$n = 5, \ell = 1$ ———————— 6
$n = 4, \ell = 2$ ———————— 10
$n = 5, \ell = 0$ ———————— 2

$n = 4, \ell = 1$ ———————— 6
$n = 3, \ell = 2$ ———————— 10
$n = 4, \ell = 0$ ———————— 2

$n = 3, \ell = 1$ ———————— 6
$n = 3, \ell = 0$ ———————— 2

$n = 2, \ell = 1$ ———————— 6
$n = 2, \ell = 0$ ———————— 2

$n = 1, \ell = 0$ ———————— 2

Fig. 11.17 This figure shows schematically how states of many-electron atoms are grouped together in shells. The states are labelled with the quantum numbers n and ℓ. The number of electrons that can occupy each state is given at the right of the diagram. The energies of the states are not shown, because they change with the atomic number Z.

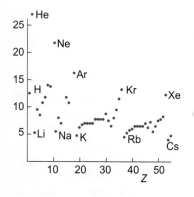

Fig. 11.18 Ionization energies of elements up to $Z = 56$.

As Z increases beyond 3, more electrons are added to $n = 2$ states. In the ground state of neon, with $Z = 10$, all the $n = 2$ states are filled. Each electron is paired off with another that has opposite angular momentum; an electron with $m_j = +\frac{3}{2}$ is matched by one with $m_j = -\frac{3}{2}$, and one with $m_j = +\frac{1}{2}$ by one with $m_j = -\frac{1}{2}$. The total angular momentum is therefore zero. All the $n = 2, \ell = 1$ electrons have wave functions with the same dependence on the radial distance from the nucleus, and none is completely shielded from the positive nuclear charge of $+10e$ by other electrons. The ionization potential of neon is 21.5 eV. Once again we have a tightly bound atom with zero angular momentum, as we did with helium when both $n = 1$ states were occupied.

Groups of electrons with similar binding energies are called **shells**. Electrons in the same shell are located in roughly the same region of space. When all the states in a shell are occupied the atom is said to have a closed shell.

Electrons in the next higher shell are on average farther away from the nucleus, as though they were excluded by a hard inner shell. A closed shell is also *strong* in the sense that the electrons are tightly bound so that the shell can not easily be disturbed by the removal of an electron.

After $n = 2$ the states with the same principal quantum number n are no longer all in the same shell. The electrons in states with $\ell = 0$ are on average closer to the nucleus and are therefore more tightly bound than electrons in states with higher ℓ. The groups of states forming the next shells are shown in Fig 11.17, and the ionization energies of all atoms with atomic numbers up to $Z = 56$ in Fig 11.18. The ionization energy is very large for the closed shell atoms and small for the next one—the extra electron has had to go into a weakly bound state in the next shell.

Exercise 11.6 Show that the Pauli principle allows 18 electrons to occupy the shell above the shell that closes at $Z = 18$.

Chemical bonds

So far in this chapter we have been discussing individual atoms that are not interacting with any other atoms. In nature it is rare to find atoms on their own. Usually they are grouped with other atoms. These may be of the same kind, but more often the atoms of one element are attached to atoms of different elements in chemical compounds. The energy of stable chemical compounds is less than that of the separate energies of the atoms making up the compound: if this were not so, the compound would break up. The following paragraphs give a qualitative description of the two main types of chemical bond. Without attempting the difficult task of solving the Schrödinger equation for interacting atoms, we can gain some insight into the nature of chemical bonds by considering how an electron's density is modified by the presence of a neighbouring atom.

Differences in ionization energy profoundly affect the way in which atoms interact with each other. Let us consider the behaviour of atoms near the closed shell at the element argon (chemical symbol Ar), which has $Z = 18$. Potassium (K, $Z = 19$), has an electron in the next shell, and an energy of only 4.3 eV is needed to remove one electron to form a potassium ion with charge $+e$. This ion is labelled K^+. If the spare electron is now given to a chlorine atom (Cl, $Z = 17$), it can occupy an empty state in the lower shell to form a negatively charged ion Cl^- with the same electronic configuration as Ar. Although the nuclear charge is one less than for Ar, this ion is actually stable: 3.7 eV is required to remove the extra electron and recover the neutral atom.

The net energy needed to form a K^+ and a Cl^- ion from the neutral atoms is only $(4.3 - 3.7) = 0.6$ eV. The two oppositely charged ions attract one another when they are far apart but, when they come close enough for there to be a significant overlap of their electron wave functions, they repel one another strongly. The net result is that, at a distance of 0.2 nm, the potential energy of the two ions has a minimum value of about -7 eV. The energy of the molecule potassium chloride (KCl) is thus lower by $\sim 7 - 0.6 = 6.4$ eV than the energy of the K and Cl atoms separately. The molecule is therefore stable. Because ions have been formed by transferring an electron from one atom to the other, the molecule is described as being held together by an **ionic bond**.

● *Electrons are exchanged to form an ionic bond*

Similarly, other atoms with nearly full shells readily accept electrons, and atoms with only a few electrons in a new shell readily give up electrons. For example, calcium, which is the element after potassium, has two electrons in the shell above argon. Both are easily removed and may be donated to two chlorine atoms to form calcium chloride with the chemical formula $CaCl_2$.

What about the closed-shell atom argon itself? The electrons in argon are already tightly bound, and there is no gain in sharing electrons with atoms of any other element. Argon, like the other closed-shell elements (helium, neon, krypton, xenon, and radon), does not form any chemical compounds except under very unusual conditions. These elements are all gaseous at ordinary temperatures and pressures, and are called inert gases, or noble gases.

Ionic bonds are formed in KCl because so little energy is needed to transfer an electron from one atom to the other. In a molecule containing only one type of atom, the energy required to remove an electron from one atom is always much greater than the energy gained by adding an electron to another atom. Nevertheless many elements, like oxygen and nitrogen, form diatomic molecules in the gaseous state. Neither atom in

● *Electrons are shared in covalent bonds*

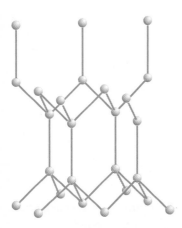

Fig. 11.19 The covalent bonds between the carbon atoms in a diamond crystal act in preferred directions. Each atom has four nearest neighbours placed at the corners of a tetrahedron.

Table 11.1 Electron affinities for a number of elements

Element	Electron affinity (eV)
H	+0.74
C	+1.26
O	+1.46
Cl	+3.62
Ar	−0.36
K	+0.50
Ca	−1.62

the molecule has gained or lost an electron, but the electron density in the space between the atoms is enhanced—each atom contributes to this density and they are in effect *sharing* electrons. As for the ionic bond, the energy has a minimum for a particular separation of the atomic nuclei.

A chemical bond formed by sharing electrons is called a **covalent** bond. A single atom may have several bonds with other atoms. In chemical parlance, the number of bonds is called the *valence* of the atom. The valence of a particular element often depends on the number of electrons outside a closed shell. For example, the carbon atom has four electrons outside a closed helium shell. Covalent bonds are directional: for an atom with more than one covalent bond, the bonds are strongest when they have certain preferred relative orientations. The four covalent bonds in carbon are strongest when the angles between each pair of bonds is the same. Diamond is pure carbon and, in a diamond crystal, all four outer electrons in each atom are shared, and form covalent bonds by occupying the space between neighbouring carbon atoms. In diamond each atom is at the centre of a tetrahedron with another atom at each corner as shown schematically in Fig 11.19. The atoms fill most of the space in the real crystal, but they have been drawn small in order to illustrate the three-dimensional arrangement of the bonds, which are represented by the grey lines.

Worked Example 11.2 The energy required to dissociate 12 g of diamond into separate neutral carbon atoms is 719 kJ. What is the binding energy of the covalent bond between two carbon atoms? (The number of atoms in 12 g of carbon is Avogadro's number $N_A = 6.02 \times 10^{23}$.)

Answer Each carbon atom has four bonds and each bond connects two atoms, so the number of bonds per mole is $2N_A$. The binding energy of the covalent bond is therefore

$$\frac{719 \times 10^3}{2N_A} \text{ J} \quad \text{or} \quad \frac{719 \times 10^3}{2 \times 6.02 \times 10^{23} \times 1.60 \times 10^{-19}} = 3.73 \text{ eV}.$$

Many chemical bonds are neither purely ionic nor purely covalent. One atom contributes more to the shared electron density than the other, but the directional nature of the bond is retained. An example of this mixed type of bond is found in water. The oxygen atom in H_2O shares electrons with the two hydrogen atoms. Both hydrogen and oxygen atoms form stable negative ions with an extra electron: the energy gained in forming the ion is called the electron **affinity**. Table 11.1 lists some electron affinities, and the value for oxygen (1.47 eV) is greater than for hydrogen (0.77 eV). In the competition for the shared electrons in the O−H bond, the oxygen wins, and more of the density of the shared electrons is close to the oxygen atom than to the hydrogen atom. The water molecule therefore has excess negative charge around the oxygen and excess positive charge around the hydrogen. The hydrogen atoms repel each other because of their positive charge, tending to push them to opposite sides of the oxygen atom. The angle between the two bonds is not 180°, however, as would be expected for a purely ionic bond in which the bond energy does not depend on direction. The bond retains some directional, covalent, character, and the angle between the two bonds is 105°, as shown in Fig 11.20.

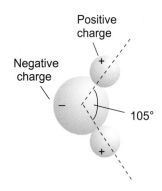

Fig. 11.20 The bonds between hydrogen and oxygen in water are partly ionic and partly covalent: the centres of positive and negative charge do not coincide.

The periodic table

At each closed shell the atoms with one electron in the next shell, which are called alkali metals, have similar properties. In a chemical compound containing potassium, the potassium atoms can usually be replaced by atoms of other alkali metals to form another compound with similar properties. Sodium (Na) is the alkali metal with one more electron than the inert gas neon, and molecules of sodium chloride, which is common salt, form in exactly the same way as molecules of KCl. As the atomic number increases, chemical behaviour repeats itself at each closed shell. The repetitions were well known long before their explanation was understood, and elements were ordered in the **periodic table** on the basis of their chemical affinities. The periodic table is set out in Fig 11.21. The name periodic table is a bit of a misnomer, since the number of elements in each shell is not the same. However, the name has stuck, and it does remind us that there is a regular pattern of chemical behaviour in moving from one shell to another.

It is not the purpose of this book to give any details of the physical basis of chemical interactions. This is a large branch of science in its own right, known as physical chemistry. What this section seeks to emphasize is that the interaction between atoms is governed by the quantum numbers of their outermost electrons. The Pauli exclusion principle ensures that all the electrons have different quantum numbers. Because of the way that wave functions change with the quantum numbers n and ℓ, the inner shells that are full are compact. They are scarcely affected by the presence of other atoms in the neighbourhood. Only the electrons in the outermost shell take part in chemical interactions.

11.5 Molecules

So far our discussion of the consequences of quantum mechanics has concentrated entirely on the behaviour of electrons. However, atoms as a

The Periodic Table

1	2	3	4	5	6	7	8	9	10	11	12	13	14	15	16	17	18
1 H 1.0079																	2 He 4.003
3 Li 6.941	4 Be 9.012											5 B 10.81	6 C 12.01	7 N 14.01	8 O 16.00	9 F 19.00	10 Ne 20.18
11 Na 22.99	12 Mg 24.30											13 Al 26.98	14 Si 28.09	15 P 30.97	16 S 32.07	17 Cl 35.45	18 Ar 39.95
19 K 39.10	20 Ca 40.08	21 Sc 44.96	22 Ti 47.87	23 V 50.94	24 Cr 52.00	25 Mn 54.94	26 Fe 55.85	27 Co 58.93	28 Ni 58.69	29 Cu 63.55	30 Zn 65.39	31 Ga 69.72	32 Ge 72.61	33 As 74.92	34 Se 78.96	35 Br 79.90	36 Kr 83.80
37 Rb 85.47	38 Sr 87.62	39 Y 88.91	40 Zr 91.22	41 Nb 92.91	42 Mo 95.94	43 Tc 98.91	44 Ru 101.1	45 Rh 102.9	46 Pd 106.4	47 Ag 107.9	48 Cd 112.4	49 In 114.8	50 Sn 118.7	51 Sb 121.8	52 Te 127.6	53 I 126.9	54 Xe 131.3
55 Cs 132.9	56 Ba 137.3	57 La- 71 Lu	72 Hf 178.5	73 Ta 180.9	74 W 183.8	75 Re 186.2	76 Os 190.2	77 Ir 192.2	78 Pt 195.1	79 Au 197.0	80 Hg 200.6	81 Tl 204.4	82 Pb 207.2	83 Bi 209.0	84 Po 210.0	85 At 210.0	86 Rn 222.0
87 Fr 223.0	88 Ra 226.0	89 Ac- 103 Lr	104 Rf	105 Db	106 Sg	107 Bh	108 Hs	109 Mt	110 Uun	111 Uuu	112 Uub	113 Uut					

Lanthanides

57 La 138.9	58 Ce 140.1	59 Pr 140.9	60 Nd 144.2	61 Pm 146.9	62 Sm 150.4	63 Eu 152.0	64 Gd 157.2	65 Tb 158.9	66 Dy 162.5	67 Ho 164.9	68 Er 167.3	69 Tm 168.9	70 Yb 173.0	71 Lu 175.0

Actinides

89 Ac 227.0	90 Th 232.0	91 Pa 231.0	92 U 238.0	93 Np 237.0	94 Pu 239.1	95 Am 241.1	96 Cm 244.1	97 Bk 249.1	98 Cf 252.1	99 Es 252.1	100 Fm 257.1	101 Md 256.1	102 No 259.1	103 Lr 260.1

Fig. 11.21 The periodic table. In each box the number above the chemical symbol for the element is its atomic number Z. The number below the chemical symbol is the atomic weight, which applies to the mixture of stable isotopes found in nature. The chemical properties of an element are determined by the structure of the electrons in the outermost shell. For elements with atomic number greater than $Z = 20$, the chemically active electrons are not necessarily changed as Z increases. Chains of elements with different Z may then have very similar chemical properties. This applies particularly for the *lanthanides* from lanthanum (La) with $Z = 57$ to lutecium (Lu) with $Z = 71$ and to the *actinides* from actinium (Ac) with $Z = 89$ to lawrencium (Lr) with $Z = 103$. The naming of elements with Z greater than 100 has caused some controversy. However, in 1997 the names of all the elements up to $Z = 108$ were agreed. Only a few atoms of elements 109–112 have been made to date, and these elements have not yet been named.

whole are also very small objects, and classical laws are not adequate to describe their motion. The simplest possible combination of atoms is a diatomic molecule. Each molecule is a bound system—energy is needed to separate the molecule into its component atoms—and, like the other bound systems we have considered, an isolated molecule has discrete energy states labelled by quantum numbers. If diatomic molecules are in a gaseous form, they are not completely isolated, but from time to time they collide with other molecules. Strictly speaking, their wave functions are not time-independent. However, provided that the time between collisions is long enough, the wave functions are almost the same as the time-independent wave functions that would apply if there were no interactions with the rest of the gas.

This section deals only with the gaseous state of simple diatomic molecules such as the potassium chloride molecule we discussed in Section 11.4. Even for diatomic molecules the exact wave functions cannot be calculated, since the motions of the electrons and the atomic nuclei should be considered together. Larger combinations of atoms are of course very important: biological molecules, for example, may include many thousands of atoms.

The approximation made in discussing diatomic molecules is that the motions of the electrons and the nuclei may be considered separately. Electrons move much faster than nuclei and, as the nuclei slowly adjust their positions, the electrons remain in states with fixed quantum numbers. Restricting attention to the nuclei, there are two types of internal motion relative to their common centre of mass. They may vibrate along the line joining their centres, or thay may rotate about the centre of mass. Both motions occur: we shall consider them in turn and then explain how the spectra of photons emitted by diatomic molecules depend on their vibrational and rotational motion.

Vibrations and rotations

The oscillation of diatomic molecules about their mean separation was discussed in Section 10.6. If the nuclei are displaced by a small amount from their mean separation, the restoring force is proportional to the displacement, just like the restoring force in a classical simple harmonic oscillator. The energies of the states found by solving the time-independent Schrödinger equation are given in eqn (10.38):

$$E_n = (n + \tfrac{1}{2})\hbar\omega.$$

Here n is the quantum number for the oscillator, representing the number of nodes in the wave function. The energy is referred to the potential energy at the mean position of the nuclei. Remember that even in the ground state with $n = 0$ there is a zero-point motion $\tfrac{1}{2}\hbar\omega$.

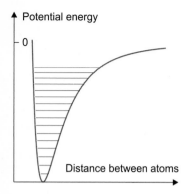

Fig. 11.22 The interatomic potential and the vibrational energy levels for a diatomic molecule.

⬤ *Rotational energy is linked to angular momentum*

The uncertainty principle does not allow the nuclei to be stationary and located at their mean position—if this were to occur, both the positions and the momenta of the nuclei would be exactly defined.

The energy level diagram for the vibrations of the molecule are shown in Fig 11.22. The levels are equally spaced, at least for small n. As the energy increases the approximation that the restoring force is proportional to displacement is no longer valid. The restoring force pulling the two atoms together becomes weak at large distances, and the potential energy $V(r)$ flattens out and approaches zero where the atoms are so far apart that their interaction is negligible. As for the states with high n in the hydrogen atom, the wave functions of the high-n molecular states are very small except at large distances, and these states are weakly bound. The spacing of neighbouring states with high n gets smaller, as indicated in the figure.

As well as oscillating about its mean position, a molecule may also rotate about an axis through the centre of mass. What does quantum mechanics tell us about the rotation of a molecule? For electrons in atoms, the Schrödinger equation requires that orbital angular momentum has values determined by a quantum number ℓ. The magnitude of the angular momentum was given in Section 11.3 as $\sqrt{\ell(\ell+1)}\hbar$, where ℓ could be zero or any positive integer. The same applies to a rotating diatomic molecule. Only those energies occur that lead to allowed values of angular momentum. Rotations in classical physics are discussed in chapter 4, and rotational energy is given by eqn (4.4) as

$$E_K = \tfrac{1}{2} I \omega^2 \tag{11.19}$$

where I is the moment of inertia and ω the angular frequency. Here we need to express the energy not in terms of angular speed but in terms of angular momentum. In classical mechanics the angular momentum has a magnitude $L = I\omega$, from eqn (4.18). Substituting for ω in eqn (11.19) the rotational energy may be written as

$$E_K = \frac{L^2}{2I}.$$

If we assume that the same relation applies on the atomic scale, then, for a state with quantum number ℓ, L^2 has the value $\ell(\ell+1)\hbar^2$ and the rotational energy is

$$E_\ell = \frac{\ell(\ell+1)\hbar^2}{2I}. \tag{11.20}$$

We have derived this result from classical physics, and can only be sure that it is correct for very large values of ℓ (invoking the correspondence principle of Section 10.7 that requires quantum physics to reproduce classical physics in the limit of large quantum numbers). However, the result is actually correct for all ℓ.

The energy level diagram for a rotating diatomic molecule is shown in Fig 11.23. The levels get further apart as ℓ increases. As for electrons, each level is degenerate, and there are $(2\ell + 1)$ states for each ℓ, all having exactly the same rotational energy E_ℓ.

The lowest energy given by eqn (11.20), for $\ell = 0$, is zero. So far we have not specified whether or not the molecule has any other motion besides its rotation. If the molecule is in its lowest vibrational and lowest electronic state, the set of levels in Fig 11.23 is called the **ground state rotational band**. But there is no reason why it should not rotate even if it is in one of the vibrational states with quantum number $n \geq 1$. This can be investigated by studying the spectrum of photons emitted by gaseous KCl. Just as for atomic spectra, each photon corresponds to a transition between two discrete states of the molecule. The energy levels of KCl derived from the spectrum are shown in Fig 11.24. For KCl and for other diatomic molecules, the rotational levels are closer together than the vibrational levels. A rotational band is built on each successive vibrational level.

The energy differences between neighbouring rotational or vibrational states are much smaller than the energy differences of several eV we found for different electronic states of the hydrogen atom. Since the time variation of a steady-state wave function for a state with energy E is given by an angular frequency $E/\hbar$ (eqn (10.14)), the angular frequency of the electrons is much larger than that of the nuclei. This justifies our assumption that the electrons adapt to the instantaneous position of the nuclei, so that we don't need to worry about electrons when thinking about the motion of the nuclei.

Worked Example 11.4 Taking the reduced mass of the KCl molecule to be 18.6 amu, use the rotational energies in Fig 11.24 to estimate the separation of the K and Cl atoms.

Answer The moment of inertia I of the KCl molecule about its centre of mass is mr^2, where m is its reduced mass and r the average separation of the atoms. Substituting for I in eqn (11.20) and rearranging,

$$r^2 = \frac{\ell(\ell+1)\hbar^2}{2mE_\ell}.$$

In Fig 11.24 the level in the ground state band with $\ell = 5$ has an energy close to $5 \times 10^{-4}\,\text{eV} \approx 8 \times 10^{-23}\,\text{J}$. The reduced mass is

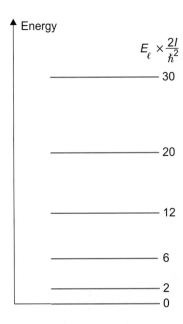

Fig. 11.23 The ground state rotational band for a diatomic molecule.

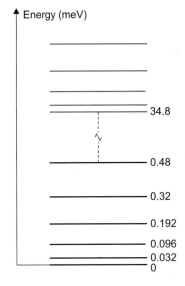

Fig. 11.24 The level scheme for the KCl molecule, with an energy scale in meV (milli-electron volts $\equiv 10^{-3}$ eV). The levels shown in black are part of the ground state rotational band. A similar rotational structure, in blue, is built on the first vibrational state at 34.8 meV.

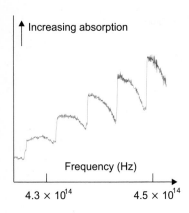

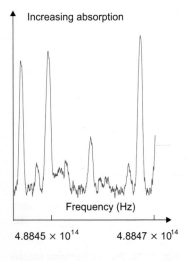

Increasing absorption

Frequency (Hz)

4.3 × 10¹⁴   4.5 × 10¹⁴

Fig. 11.25 Part of the absorption spectrum of iodine vapour, taken with an instrument of low resolution. Individual rotational states cannot be distinguished, but each sharp edge indicates the start of a new rotational band built on a vibrational excitation.

Increasing absorption

Frequency (Hz)

4.8845 × 10¹⁴   4.8847 × 10¹⁴

Fig. 11.26 Part of the absorption spectrum of iodine in the visible region, measured with very high resolution. Note the extremely small frequency interval that covers the whole of this section of the spectrum.

18.6 amu $\approx 3 \times 10^{-26}$ kg; hence

$$r^2 = \frac{5 \times 6 \times (1.06 \times 10^{-34})^2}{2 \times 3 \times 10^{-26} \times 8 \times 10^{-23}} \text{ m}^2,$$

leading to $r \approx 2.6 \times 10^{-10}$ m.

The level structure of molecules may be investigated by studying *absorption spectra*. A molecular absorption spectrum measures the attenuation of a beam of photons as a function of wavelength as it passes through a sample of the molecules under investigation. The absorbing molecules are in their ground state to start with, and only transitions from the ground state to higher states are observed. The spectrum is a map of the higher states as a function of energy. Even for simple diatomic molecules, the spectrum is tremendously complicated.

Figure 11.25 is an absorption spectrum of gaseous iodine (I_2), measured with an instrument that cannot resolve neighbouring rotational states. Successive rotational bands are seen separated by a regular frequency spacing that represents the energy difference $\hbar\omega$ of the vibrational states. This is the information that leads to vibrational level schemes like the one shown in Fig 11.22.

At very high resolution, as in Fig 11.26, the spectrum is found to be a thicket of closely spaced levels. The figure shows part of the absorption spectrum for visible photons, where electronic states as well as rotational and vibrational states may change. Although the iodine absorption spectrum is too complicated to understand in detail, it has been accurately mapped, and can be used to give very accurate calibrations of the frequency of the light delivered by lasers.

11.6 Nuclei

The nucleus of an atom with atomic number Z and mass number A contains Z protons and $(A - Z)$ neutrons, as described at the beginning of this chapter. The masses of protons and neutrons are roughly the same and, because electrons have much smaller mass, the mass of an atom resides almost entirely in the nucleus. Nuclei are much smaller than atoms: for a light element like oxygen, for example, the diameter of the nucleus is about 10^{-14} m, roughly ten thousand times smaller than the diameter of the atom.

If the position of a proton or neutron is to be specified within such a small distance as 10^{-14} m, the uncertainty principle tells us that there must be a large uncertainty in momentum. From eqn (9.21), if the uncertainty in position along the x-axis is Δx, the uncertainty in the

The nuclear force is strong for protons and neutrons that are close together, but it falls off very rapidly with distance. Once outside the nuclear volume, a proton or a neutron is not attracted towards the nucleus. Protons are charged and, because they interact with atomic electrons, they cannot penetrate ordinary matter. Neutrons, however, are uncharged. They are practically unaffected by electrons and pass freely through matter until they

chance to collide with a nucleus. A nuclear reaction may then occur, releasing a large amount of energy. Neutrons are therefore very damaging to living things. They can penetrate deep inside a plant or animal, and then cause severe local damage. When neutrons are present at high intensity, as in a nuclear reactor, great care must be taken to ensure that personnel are adequately shielded from the neutrons.

---

x-component of momentum Δp_x is

$$\Delta p_x \geq \frac{\hbar}{2\Delta x}.$$

To meet this requirement p_x itself must be at least Δp_x. For a proton with mass m_p moving along the x-axis at a speed v, the momentum is $p_x = m_\mathrm{p} v$ and the kinetic energy

$$\frac{1}{2} m_\mathrm{p} v^2 = \frac{p_x^2}{2m_\mathrm{p}} \geq \left(\frac{1}{2m_\mathrm{p}}\right)\left(\frac{\hbar}{2\Delta x}\right)^2.$$

● The uncertainty principle gives a lower limit to the energy of a proton confined within a nucleus

Putting $\Delta x = 10^{-14}$ m, the kinetic energy must be at least about 10^{-13} J, or equivalently about one MeV.

Since the protons and neutrons are bound together inside the nucleus, they must attract one another. To ensure that the protons and neutrons are bound, the magnitude of the attractive potential energy must be greater than the kinetic energy, which the uncertainty principle requires to be at least one MeV. The protons all have positive charge, and the electrical force between them is repulsive. However, there is an additional attractive force, specific to the constituents of nuclei, which has no effect on the electrons moving outside. This nuclear force is called the **strong force**. The strong force is indeed strong compared to the electrical forces acting on protons and electrons. The magnitude of the potential energy binding the nucleus together is typically several tens of MeV. By comparison, the binding energies of electrons in atoms vary from a few eV to about 100 keV.

The nuclear potential

Like electrons, protons and neutrons have wave functions that are solutions of the Schrödinger equation. The wave functions spread throughout the nucleus, and each proton and each neutron is moving in a potential that is made up of the average effect of all the others: as for

Richard Feynman (1918–88)

Most Nobel prizes in physics are awarded to scientists who have started a new branch of the subject or who have developed a new theory. Feynman, together with Schwinger and Tomonaga, unusually received the prize in 1965 for their work *completing* a theory. The theory, which is called *quantum electrodynamics*, describes the interaction between charged particles and electromagnetic fields.

Quantum electrodynamics, or QED for short, was completed in 1948. It is an extraordinarily successful theory: some quantities can be calculated with a precision of better than one part in a billion, yet no discrepancy between theory and experiment has ever been found.

A great contribution made by Feynman to QED was to express the theory in a simple way. He liked to have a visual picture of even the most abstract theories, and he described QED using what are now known as Feynman diagrams. Calculating the results of QED is extremely difficult, but the diagrams indicate clearly which parts of a calculation are important and which may safely be neglected. They also make the basic ideas of the theory understandable to undergraduate physics students.

Richard Feynman was born in New York state in 1918. He was an exceptional student at the Massachussetts Institute of Technology, and by 1942—still only 24 years old—he was already well known to the leading physicists. He was invited to join the Manhattan Project at Los Alamos to develop the atomic bomb. At that time there were no electronic computers and progress on the Project was being held up by the long wait for numerical results obtained from mechanical calculators. Feynman was put in charge of the calculating division. Soon calculations which had previously taken three months were appearing at the rate of three every month. This was partly due to Feynman's great skill at devising the most efficient way to perform a calculation, but also to his ability to motivate people and to make them think that the problems they tackled together were exciting.

After his Nobel prize work on QED, Feynman made important contributions to particle physics and to superconductivity theory. Perhaps because of his experience in the Manhattan Project, he also had an interest in developments in computing. In the 1950s the introduc-

tion of transistors were making computers both more powerful and smaller. In 1959 Feynman gave a lecture entitled 'There's plenty of room at the bottom', speculating on how much miniaturization might be possible. This talk contained an amazingly accurate prediction of the techniques which have actually been used to achieve miniaturization to date, and the limits of this process which are only now being approached.

The talk ended with the offer of two prizes of $1000 each; one 'to the first guy who makes an operating electric motor which is only 1/64 inch cube' and a second to 'the first guy who can take the information on the page of a book and put it on the area 1/25,000 smaller in linear dimensions in such a manner that it can be read by an electron microscope.' The first prize was won within a year, but the second was a tougher proposition and was only won 26 years later. Feynman's talk is generally regarded as the beginning of *nanotechnology*, which is technology using very small objects, down to molecular size (also mentioned in the talk!).

Feynman died of cancer at the beginning of 1988. He was already a sick man in 1987 when he was a member of the Commission investigating what caused the space shuttle *Challenger* to explode shortly after launch. It was soon obvious that because the weather had been cold at the time of the launch, seals on the fuel tanks, called 'O-rings', lost their elasticity and leaked under the strain of takeoff: engineers had warned that the launch should be postponed for this reason, but their warnings had been ignored. How could this be demonstrated so that everyone recognised the fault and necessary modifications to the design would be made as soon as possible? The Commission's hearings were being televised, and when the O-rings were discussed, Feynman produced a beaker of cold water in which he immersed a small O-ring. When the O-ring was taken out, it was stiff and when distorted would not recover its round shape. Pictures of Feynman and his O-ring were seen on TV screens and in newspapers around the world. For the last time Feynman had shown his skill as a communicator by using a simple and dramatic illustration to win his point.

---

electrons in an atom, it is quite a good approximation to regard the protons and neutrons as moving independently of one another.

The nuclear potential is much the same for protons and for neutrons. The word **nucleon** is used for either a proton or a neutron. Within the nucleus the potential experienced by a nucleon varies with distance from the centre of the nucleus in a rather similar way to the potential of a simple harmonic oscillator. There is a force towards the centre of the nucleus that gets stronger as the nucleon moves away from the centre.

Previously we have only met one-dimensional harmonic oscillators. Nuclei are three-dimensional, but many nuclei are spherical, so that the potential is central, depending only on the distance r from the centre and not on the direction. A three-dimensional central potential can be written in Cartesian coordinates as

$$V(r) = \tfrac{1}{2} kr^2 = \tfrac{1}{2} k(x^2 + y^2 + z^2) \tag{11.21}$$

where k is a constant. The mass of a neutron is almost the same as that of the proton, and each nucleon may be taken to have the same mass m. The motion in the x-, y-, and z-directions can be considered separately. In each direction there is a harmonic oscillator with angular frequency $\omega = (k/m)^{1/2}$. From eqn (10.38) the energy of a one-dimensional oscillator is $E_n = (n + \tfrac{1}{2})\hbar\omega$, where the quantum number n is zero or any positive integer. In three dimensions quantum numbers n_x, n_y, and n_z specify the energy associated with motion in each direction. The total energy of a state with quantum numbers n_x, n_y, n_z is simply

$$E(n_x, n_y, n_z) = E_{n_x} + E_{n_y} + E_{n_z} = (n_x + n_y + n_z + \tfrac{3}{2})\hbar\omega. \tag{11.22}$$

● *Three quantum numbers are needed to describe the wave functions of nucleons in three dimensions*

Nuclear shells

Like electrons, nucleons have an intrinsic spin quantum number $\tfrac{1}{2}$ and each has an additional quantum number m_s, which may take the values $\pm\tfrac{1}{2}$ (referred to as 'spin-up' or 'spin-down'). Nucleons also obey the Pauli exclusion principle, and only two protons and two neutrons may occupy the lowest energy state having only the zero-point energy $\tfrac{3}{2}\hbar\omega$. The next nucleon must occupy a state with energy $\tfrac{5}{2}\hbar\omega$ with one of the quantum

● *The Pauli principle applies to nucleons*

numbers n_x, n_y, or n_z equal to one. For each of these three possibilities the intrinsic spin may be up or down, leading to a total of six states for protons and six for neutrons. There is a shell structure for nucleons in a nucleus, just as there is for the electrons in an atom.

Nuclei of the elements with $Z = 2$ or $Z = 6 + 2 = 8$ have closed shells of protons. The closed shell numbers 2 and 8 are called 'magic numbers'. Other closed shells for protons occur at the magic numbers 20, 28, and 50. Elements that have atomic numbers equal to these magic numbers are, like inert gas atoms, particularly stable. For example, they tend to be more abundant in nature than neighbouring elements, and to have many isotopes.

● *Magic numbers*

Worked Example 11.5 The magic number 20 for nuclei corresponds to a closed shell with 12 nucleons in states with energy $\frac{7}{2}\hbar\omega$. Show that there are six combinations of n_x, n_y, and n_z in eqn (11.22) that lead to this energy, and hence explain the existence of the magic number 20.

Answer Since the zero-point energy for the three-dimensional oscillator is $\frac{3}{2}\hbar\omega$, for a total energy $\frac{7}{2}\hbar\omega$ the sum of the quantum numbers $(n_x + n_y + n_z)$ must be two. This requires one of the quantum numbers to be two, or two of them to be one. Writing the sequence of quantum numbers as (n_x, n_y, n_z), the six combinations are (2,0,0), (0,2,0), (0,0,2), (0,1,1), (1,0,1), and (1,1,0). Both protons and neutrons have intrinsic spin quantum number $\frac{1}{2}$: two protons and two neutrons may therefore occupy each oscillator state, making a total of 12 of each. The previous magic number is 8, so that both protons and neutrons have a magic number 20 corresponding to a full shell of nucleons with energy $\frac{7}{2}\hbar\omega$.

The approximation that the nuclear potential is simple harmonic is not a very good one, especially for nuclei with a high mass number. More complicated forces are then required, but the picture of protons and neutrons filling up successive shells is still valid. The forces are very similar for neutrons and for protons, and neutrons have the same set of magic numbers 2, 8, 20, 28, 50, and 82 as protons. Heavy nuclei have more neutrons than protons, and neutrons have an additional magic number at 126.

● *Neutrons and protons have the same magic numbers*

Many other properties of nuclei can be explained using this theory, which is called the **shell model** of nuclei. However, we shall not explain any more of the successes of the shell model here. The point we wish to emphasize is that, like electrons in atoms, the protons and neutrons in atomic nuclei have wave functions that are governed by the rules of quantum mechanics. The important features are these.

● Bound states are only allowed at certain discrete energies.

● Each state is characterized by a set of quantum numbers.

- No two protons or neutrons may occupy states with the same quantum numbers.
- Nuclei have a shell structure exhibited by the existence of magic numbers.

Subnuclear particles

The protons and neutrons making up atomic nuclei are not themselves the ultimate constituents of matter. We saw that the energies involved in the internal structure of many-nucleon nuclei are in the region of tens of MeV. The investigation of nuclear structure is carried out using accelerators that generate beams of particles with energies of tens or hundreds of MeV. When still more energetic particles are used, with energies of many GeV ($1\,\mathrm{GeV} \equiv 10^9\,\mathrm{eV}$) or even TeV ($1\,\mathrm{TeV} \equiv 10^{12}\,\mathrm{eV}$), new phenomena are observed.

Neutrons and protons themselves are made up of still smaller particles called quarks. Part of the reason why such large energies are required to discover this internal structure of nucleons is that the uncertainty principle implies that a small distance scale is associated with large values of momentum. At first sight, then, the progress from atoms to nuclei to quarks is like opening up a set of nested boxes: open one box and a smaller one is discovered inside. But the innermost box does not look like the outer two. Quarks behave quite differently from electrons and nucleons. In the first place, no one has ever seen a free quark—quarks always stick together in twos or threes. There are six different types of quark and, as well as intrinsic spin, quarks have another quantum number that is whimsically called 'colour': quarks are imagined as being red, green, or blue. Quarks are responsible for the strong interaction that holds atomic nuclei together. The theory applied to the strongly interacting particles is called quantum chromodynamics (QCD). There are no straightforward analytical approximations in QCD, and solutions can only be found by undertaking lengthy calculations on ultrafast computers.

Other particles besides quarks appear at very high energies: the study of all their properties is the science called particle physics. As well as the strong interaction and electrical forces, another force called the *weak force* is needed to understand particle physics. The theory known as the 'standard model' has led to a unification of electrical forces with the weak force. The weak force coexists in nuclei with the strong force. Neutrons are not stable particles. After a mean life of about a quarter of an hour a free neutron turns into a proton as a result of the weak force. This time-scale is enormous compared to the times occurring in strong or electrical interactions—hence the name weak.

The study of particle physics was originally driven by the attempt to achieve a fundamental understanding of the behaviour of matter on a microscopic scale. The highly energetic particles that are now produced in huge accelerators were present in the early stages of the universe. The properties of these particles are now linked with cosmology, and particle physics is relevant to man's curiosity about very large as well as very small features of the world about him.

Problems

Level 1

11.1 The gravitational force between masses and the electrical force between stationary charges both vary as the inverse square of their distance apart, according to eqns (5.1) and (11.3). Calculate the relative strength of the gravitational and electrical forces between an electron and a proton.

11.2 A transition occurs in the caesium atom that is accompanied by the emission of radiation of wavelength 3.261 225 57 cm. The uncertainty in the wavelength of this transition is very small, and it is used to define the standards of both time and length. What is the energy of this transition in eV? This is one calculation where it is justifiable to work to nine or ten significant figures, since such precisions are achieved in national standard laboratories.

11.3 Calculate the wavelength of the line in the Balmer series of hydrogen due to the transition from an upper state with $n = 3$.

11.4 The wavelength of the line in the Balmer series of hydrogen due to the transition between the levels with $n = 3$ and $n = 2$ is 656.5 nm. Use this information to obtain an estimate of the Rydberg constant for hydrogen and of the binding energy of the electron in the ground state of hydrogen.

11.5 *Hydrogenic atoms* consist of a single electron bound to a nucleus. Calculate the energy and the wavelength of the transition between the states $n = 2, \ell = 1$ and $n = 1, \ell = 2$ for hydrogenic copper, using eqns (11.17) and (11.18). The atomic number of copper is 29. The wavelength is in the X-ray region, and is comparable to atomic sizes. X-rays are commonly produced by

ejecting an $n = 1$ electron from a neutral atom: the empty state is then occupied by an electron making a transition from a higher state, such as $n = 2$. The energy of this transition is similar to that of the same transition in the hydrogenic atom, because the wave functions of the high-n electrons are small in the region occupied by low-n electrons. In the case of copper the energy of the X-rays from the $n = 1$ to $n = 2$ transition is about 6% lower than the result you calculate for hydrogenic atom.

11.6 It was explained in Section 11.4 that the wave functions for electrons are labelled by the quantum numbers n, ℓ, j, and m_j. The z-component of intrinsic spin, which we call m_s, has only two possible values, $+\frac{1}{2}$ and $-\frac{1}{2}$. An alternative labelling system is to specify n, ℓ, m_ℓ, and m_s. The number of states does not depend on the way they are labelled. Show that the number of states available to an electron with $\ell = 1$ is the same whether it is labelled by m_ℓ and m_s or by j and m_j.

11.7 How many electrons may occupy states with the quantum numbers $n = 3$ and $\ell = 2$?

11.8 The element lithium (Li) has atomic number $Z = 3$. Write down the quantum numbers of the electrons in the ground state of Li and deduce their total angular momentum.

11.9 Explain qualitatively why two atoms of an inert gas with atomic number Z are able to approach one another more closely than two atoms of the alkali metal with atomic number $(Z + 1)$.

11.10 The separation of the two chlorine nuclei in the diatomic molecule Cl_2 is close to 0.2 nm. Estimate the energy in eV of the rotational state of the molecule with orbital quantum number $\ell = 2$. (The most abundant isotope of chlorine has mass number 35.)

11.11 The radii of the ions Cs^+ ($Z = 55$) and I^- ($Z = 53$) are about 0.17 nm and 2.2 nm, respectively. The ionization potential of Cs is 3.9 eV and the electron affinity of I (i.e. the energy needed to remove an electron from the I^- ion) is 3.1 eV. Estimate the binding energy of the CsI molecule.

11.12 When a uranium nucleus undergoes fission in a nuclear reactor it splits into two fragments each with quite large atomic numbers. The kinetic energy of the fragments is small until they have separated. The potential energy caused by their electrical repulsion is then converted into kinetic energy. For a typical split into barium (Ba, $Z = 56$) and krypton (Kr, $Z = 36$), make an estimate of the energy released in fission, assuming that the fragments are point charges at a distance of 1.2×10^{-14} m when they separate. Roughly what mass of uranium is needed to generate 2000 MW of power for a year?

Level 2

11.13 In the spectrum of light from a distant star, the lines due to the Balmer series in atomic hydrogen are identified. The longest wavelength in this series is observed to be Doppler-shifted to a wavelength of 722 nm. Estimate the speed at which the star is moving relative to the Earth. (The Doppler effect is discussed in Sections 8.5 and 19.5: the formulae given in Chapter 8 are good enough approximations for this problem.)

11.14 The negative muon (μ^-) is a particle that carries a charge ($-e$) and is very similar to the electron except that its mass (1.88×10^{-28} kg) is about 1/9 times that of the proton (1.67×10^{-27} kg). A μ^- particle may replace the electron in hydrogen to form 'muonic hydrogen'. What is the most probable distance between the muon and the proton in the ground state of muonic hydrogen?

11.15 The positron is a particle with the same mass as the electron, but it carries a charge $+e$. Under certain circumstances an electron and a positron can combine together to form *positronium*, which has wave functions of the same form as those of the hydrogen atom. Calculate the binding energy of the ground state of positronium.

11.16 Show, by integrating by parts, that the function $u_1(r)$ in eqn (11.10) is correctly normalized.

11.17 Instead of labelling electron wave functions by the quantum numbers (n, ℓ, j, m_j) as in the text, the labels (n, ℓ, m_ℓ, m_s) may be used ($m_s\hbar$ is the z-component of intrinsic spin and m_s has the values $\pm\frac{1}{2}$). The two sets are not the same; each member of one set is a linear combination of one or more members of the other. However, the total number of states must be the same in both labelling systems. Check that this is so for an electron in a state with $n = 2$ and $\ell = 1$ by enumerating the possible values of j, m_j and of m_ℓ, m_s.

11.18 The single valence electron in the alkali metal sodium has the principal and orbital quantum numbers $n = 3$ and $\ell = 0$. The valence electron is removed and placed in a state with $n = 10$. Estimate the most probable distance from the nucleus and the binding energy of the electron in this state.

11.19 When an excited state of an atom emits a photon, only those transitions are allowed in which the quantum number ℓ changes by 1 and j changes by 0 or 1. These rules are called *selection rules*. When hydrogen atoms in the state with $n = 3$, $\ell = 1$, and $j = \frac{1}{2}$ emit light, how many different wavelengths will be observed?

11.20 If one ball is removed from a bag containing six identical balls, there is only on possible outcome—one ball outside and five inside. However, if the balls are labelled 1 to 6, there are six different possibilities, because we know which ball has been removed. The number of possibilities depends on whether the balls are identical or can be distinguished.

The problem becomes more complicated if there is more than one type of label, as for electrons, which are labelled by their quantum numbers, but are otherwise identical. Work out the number of states (i.e. separate pairs of quantum numbers) for two electrons in states with $\ell = 1$ and $j = 3/2$: (a) if one electron has $n = 2$ and the other $n = 3$; (b) if both electrons have $n = 2$. For part (b) you must remember that the Pauli principle does not allow the same set of quantum numbers n, ℓ, j, and m_j to occur twice.

11.21 The energy of the rotational state with $\ell = 1$ for the HCl molecule is 1.25 eV. Estimate the separation of the hydrogen and chlorine nuclei in the molecule.

11.22 The 'spring constant' k for the HCl molecule was estimated in Exercise 10.4 to be $490\,\mathrm{N\,m^{-1}}$. The spring constant is not altered by replacing the hydrogen atom in HCl by deuterium. What is the energy of the first vibrational state in deuterium chloride ($^2\mathrm{H}^{35}\mathrm{Cl}$).

11.23 When energetic α-particles are allowed to strike a $^{197}\mathrm{Au}$ (gold) target, some of the α-particles are scattered through $180°$ in close collisions with gold nuclei. What is the energy of such a back-scattered α-particle if its energy is $5\,\mathrm{Mev}$ as it approaches the target atom?

11.24 As the energy of the α-particles used in the scattering experiment described in Problem 11.23 is increased above $30\,\mathrm{MeV}$, a sudden fall in the probability of back-scattering occurs. This is because the α-particles have sufficient energy to penetrate the gold nucleus. Assuming that the potential energy of the α-particle and the gold nucleus is the same as for two point charges, and that the radius of the α-particle is very small, estimate the radius of the gold nucleus. (The atomic number of gold is 79.)

Level 3

11.25 The function $u_2(r)$ for the spherically symmetric state of the hydrogen atom with $n = 2$ is

$$u_2(r) = \left(\frac{1}{\pi}\right)^{1/2}\left(\frac{1}{2a_0}\right)^{3/2} r\left(1 - \frac{r}{2a_0}\right)e^{-r/2a_0}.$$

Verify that this function is a solution of eqn (11.6).

11.26 Find the most probable distance of the electron from the nucleus in the state with wave function $u_2(r)/r$.

11.27 The carbon atom has six electrons. In the ground state of the atom the two $n = 1, \ell = 0$ and the two $n = 2, \ell = 0$ electron states are filled: the remaining two electrons are in $n = 2, \ell = 1$ states. Calling the other quantum numbers m_{ℓ_1}, m_{s_1} for the first of these electrons and m_{ℓ_2}, m_{s_2} for the second, list all the combinations of quantum numbers for the two electrons that are allowed

by the Pauli principle. What are the possible values of the total angular momentum of this pair of electrons? (NB. Since the electrons are indistinguishable, the quantum numbers cannot be assigned to one electron or the other. The actual wave functions are mixtures of wave functions in which the quantum numbers of the separate wave functions are interchanged. However, the procedure outlined above is valid for counting the number of states.)

11.28 As well as the selection rules for ℓ and j given in Problem 11.19, there is an additional selection rule restricting the changes in m_j that may occur when the state of an electron changes. This rule is that m_j must change by 0 or 1.

The sodium atom ($Z = 11$) has one electron, called the valence electron, outside a closed shell in which all the $n = 1$ and $n = 2$ states are filled. If the valence electron has $n = 4$ and $\ell = 2$, show that it may be in one of 10 different states labelled by j and m_j. These states can only de-excite to the 6 lower states with $n = 3$ and $\ell = 1$. If all possible transitions between the upper ($n = 3$) and lower ($n = 2$) states were allowed, the total number of transitions would therefore be 60. How many transitions between the $\ell = 2$ and $\ell = 1$ states are allowed by the selection rules?

11.29 The separation of the two oxygen atoms in the $^{16}\mathrm{O}_2$ molecule is about $0.12\,\mathrm{nm}$ and the force constant for vibrational motion of the molecule is about $1200\,\mathrm{N\,m^{-1}}$. Estimate the number of rotational levels of $^{16}\mathrm{O}_2$ below the first vibrational level.

11.30 The neutron separation energy for the nucleus $^{17}\mathrm{O}$ is about $5.0\,\mathrm{MeV}$, while for $^{16}\mathrm{O}$ it is about $16.5\,\mathrm{MeV}$. Assume that the nuclear potential is simple harmonic, that it is the same for the two nuclei, and that the reduced mass for a neutron moving in the potential is the same as its actual mass, which is $1.67 \times 10^{-27}\,\mathrm{kg}$. Find the distance from the centre of the nucleus at which the potential equals the energy of the last neutron in $^{17}\mathrm{O}$: this gives an indication of the size of the nucleus.

Some solutions and answers to Chapter 11 problems

11.1 The ratio of the gravitational force to the electrical force between the electron and the proton is 4.4×10^{-40}. When considering the internal motion of atoms, the gravitational force is completely negligible.

11.2 The energy of the transition is $h\nu = hc/\lambda$. In SI units this is measured in joules; to convert to eV the result must be divided by the electronic charge e. Keeping seven significant figures to match the precision of the fundamental constants h and e, the energy of the caesium atom transition is $3.801\,768 \times 10^{-5}$ eV.

11.5 The energy is 8.58 keV and the wavelength 0.145 nm, about a typical atomic spacing.

11.6 For $\ell = 1$, the allowed values of the z-component of orbital angular momentum are $m_\ell = 1, 0, -1$. Each of these three may be associated with $m_s = \pm\frac{1}{2}$ ('spin up' or 'spin down'). The total number of states with different values for (m_ℓ, m_s) is $3 \times 2 = 6$. According to the other labelling system, for $\ell = 1$, j may be $\frac{3}{2}$ or $\frac{1}{2}$. For $j = \frac{3}{2}$, $m_j = \frac{3}{2}, \frac{1}{2}, -\frac{1}{2}$ or $-\frac{3}{2}$, four states in all. For $j = \frac{1}{2}$, $m_j = \frac{1}{2}$ or $-\frac{1}{2}$, two more states, making a total of six, the same as the number obtained in the (m_ℓ, m_s) labelling scheme.

11.8 Two electrons with $n = 1, \ell = 0, m_s = \pm\frac{1}{2}$ fill the $n = 0$ shell. The third electron has $n = 2, \ell = 0$. The total angular momentum J of the atom is the same as the intrinsic spin of the electron in the $n = 2$ shell, i.e. $J = \frac{1}{2}$.

11.10 The rotational energy E_ℓ for a diatomic molecule with orbital angular momentum quantum number ℓ is given by eqn (11.20) as $E_\ell = \ell(\ell + 1)\hbar^2/2I$. The moment of inertia is $I = mr^2$, where m is the reduced mass and r the separation between the atoms. The mass of a ^{35}Cl atom is approximately 35 amu, and the reduced mass of the diatomic molecule is half as much. The energy of the $\ell = 2$ state is therefore

$$E = \frac{6 \times (1.054 \times 10^{-34})^2}{2 \times \frac{1}{2} \times 35 \times 1.6 \times 10^{-27} \times (2 \times 10^{-10})^2}$$

$$\approx 2.87 \times 10^{-23} \text{J, or about } 1.8 \times 10^{-4} \text{ eV.}$$

11.12 The energy of two charges $36e$ and $56e$ separated by a distance 1.2×10^{-14} m is

$$\frac{36 \times 56 \times e^2}{4\pi\epsilon_0 \times 1.2 \times 10^{-14}} \approx 3.9 \times 10^{-11} \text{ J.}$$

The energy required to generate 2000 MW for one year is $2000 \times 10^6 \times 3600 \times 24 \times 365 \approx 6.3 \times 10^{16}$ J. The number of uranium nuclei that must undergo fission is therefore $(6.3/3.9) \times 10^{27} = 1.62 \times 10^{27}$. The mass of each ^{235}U nucleus is 235 amu or $235 \times 1.66 \times 10^{-27} = 3.90 \times 10^{-25}$ kg, and the mass needed to run the reactor for one year is $1.62 \times 3.90 \times 10^2 \approx 600$ kg, a little more than half a tonne.

11.13 $4.0 \times 10^7 \text{ m s}^{-1}$.

11.14 The mass m that occurs in the Schrödinger equation for the hydrogen atom is the reduced mass of the electron and the proton, which is almost the same as the mass m_e of the electron. The equation for muonic hydrogen is the same except that the reduced mass of the muon and the proton is now required. This is

$$\frac{m_\mu m_p}{m_\mu + m_p} \approx \frac{\frac{1}{9}m_p^2}{\frac{10}{9}m_p} = \frac{m_p}{10} \approx 184 m_e.$$

The Bohr radius a_0 is proportional to $(1/m)$, so the most probable distance between the muon and the proton is about $a_0/184 \approx 2.9 \times 10^{-13}$ m.

11.15 -6.8 eV.

11.18 1.06×10^{-8} m. This result assumes that the wave function is almost the same as for the $n = 10, \ell = 0$ state in hydrogen. This is a good approximation since the wave function is very small within the volume containing the closed shell $n = 1$ and $n = 2$ electrons. The $n = 10$ electron is attracted by the net charge $+e$ of the nucleus and the inner electrons.

11.19 Two.

11.20 For a single electron the number of states with $\ell = 1$ and $j = \frac{3}{2}$ is $(2j + 1) = 4$, whatever the value of n.

(a) If two electrons with $\ell = 1$ and $j = \frac{3}{2}$ have *different* values of n, such as $n = 2$ and $n = 3$, the four substates of each electron are independent and the total number of states is $4 \times 4 = 16$.

(b) When both electrons have $n = 2$, the Pauli principle requires that the values of m_j are different. There are four ways to choose one value, but then only three remain for the second value, making 12 pairs $[m_j, m'_j]$ There is an additional restriction because the electrons are identical: the *order* of the values of m_j and m'_j is not significant—m_j values [1,2] describe the same state as [2,1]. There is double counting in the list of 4×3 pairs $[m_j, m'_j]$, and the total number of distinct states is $4 \times 3/2 = 6$.

11.23 If the speed of the α-particle is v_i before the collision, the speed of the centre of mass is $\frac{4}{197+4} v_i = \frac{4}{201} v_i$, and the speed of the α-particle with respect to the centre of mass is $\frac{197}{202} v_i$. In the centre of mass coordinates the α-particle moves at the same speed $\frac{197}{202} v_i$ after the collision, in the opposite direction. The speed in the laboratory is now $\frac{197}{202} v_i - \frac{4}{197+4} v_i = \frac{193}{201} v_i$ and the energy of the α-particle is $5 \times 193^2/201^2 = 4.61$ MeV.

The energy depends on the mass of the target nucleus and the technique known as 'Rutherford back-scattering' is used to determine the composition of surface layers of materials.

11.24 7.7×10^{-15} m.

11.26 $5.2a_0$.

11.27 Both electrons have the same quantum numbers n, ℓ, and j and must have different values of m_j. For the purposes of counting states we may label the quantum numbers m_{j_1} and m_{j_2} though, because the electrons are identical, the *order* in which the quantum numbers are listed is not significant. The table shows the

m_{j_1}	m_{j_2}	$m_{j_1} + m_{j_2}$
$\frac{3}{2}$	$\frac{1}{2}$	2
$\frac{3}{2}$	$-\frac{1}{2}$	1
$\frac{3}{2}$	$-\frac{3}{2}$	0
$\frac{1}{2}$	$-\frac{1}{2}$	0
$\frac{1}{2}$	$-\frac{3}{2}$	-1
$-\frac{1}{2}$	$-\frac{3}{2}$	-2

six possible pairs of different quantum numbers and the z-component of the total angular momentum associated with each pair. There is one state with $J = 2$ and one with $J = 0$.

11.28 28.

11.30 In the simple harmonic oscillator potential, the energy difference between successive states is $E_{osc} = \hbar\omega$, where ω is the oscillator frequency. The nucleus ^{16}O is a closed-shell nucleus with all the $n = 1$ and $n = 2$ states occupied in the ground state. The extra neutron in ^{17}O is in an $n = 3$ state, at an energy higher by $\hbar\omega$ than the last neutron in ^{16}O: from the data and the assumptions of the problem, $E_{osc} = 11.5$ MeV. The last neutron in ^{17}O is at an energy $(n + \frac{1}{2})\hbar\omega = \frac{7}{2}\hbar\omega$ above the zero of the potential. Writing the potential as $U = \frac{1}{2}kx^2$, where x is the distance from the centre of the potential, the force constant $k = m\omega^2$ from eqn (6.6). The potential equals the energy of the last neutron if $\frac{7}{2}\hbar\omega = \frac{1}{2}kx^2$, or after some manipulation, $x = \hbar\sqrt{(7/mE_{osc})}$. This leads to $x \approx 5 \times 10^{-15}$. Nuclei do not have well-defined radii, and this is a very rough calculation, but the effective radius of a light nucleus like oxygen is indeed a few times 10^{-15} m.

Summary of Part 3
Quantum physics

Important concepts and key equations from Part 3 are briefly explained in this summary

Chapter 9 **The foundations of quantum physics**

An atom consists of a number Z of *electrons* each carrying a negative electric charge $-e$ bound to a positively charged *nucleus*. The charge on the nucleus is $+Ze$ and the atom as a whole is electrically neutral. The *atomic number Z* determines the chemical behaviour of the atom—each chemical element has a different value of Z. The nucleus is made up of Z protons, each with charge $+e$, and a number of uncharged *neutrons*. A proton and a neutron have nearly the same mass (about 2000 times the mass of the electron), and the sum of the number of protons and neutrons is the *mass number* of the atom. Atoms of the element with atomic number Z may have many *isotopes* with different mass numbers.

Only a few of the isotopes of a given element are stable: most are *radioactive*, decaying spontaneously to form an atom of a different element. A sample initially containing N_0 radioactive atoms of a single isotope decays exponentially, and after a time t the number in the sample is

$$N = N_0 \exp(-\lambda t) \tag{9.2}$$

where

$$\lambda = (\ln 2)/\tau_{1/2} \tag{9.3}$$

is the time constant of the isotope with $\tau_{1/2}$ (its *half-life*) the time taken for the size of the sample to reduce to half its initial value. The decay of a radioactive isotope is accompanied by the emission of *ionizing radiation* which consists of α-particles (helium ions), β-particles (electrons or *positrons*), or γ-rays (*photons*). The emission of a photon does not change the atomic number or the mass number of the decaying isotope. There are some long-lived γ-emitting nuclei, but γ-radiation usually immediately follows α- or β-decay. The isotope formed in a radioactive decay may itself be radioactive, and chains of radioactivity are formed, which come to an end at a stable isotope.

The energy of light is carried in packets called *photons* and the energy E of a photon of frequency ν is

$$E = h\nu \tag{9.8}$$

where h is Planck's constant.

UV light can cause electrons to be released from the surface of a metal, a phenomenon known as the *photoelectric effect*. The maximum energy E_{max} of the electrons is given by the relation

$$E_{max} = h\nu - \phi \tag{9.9}$$

where ϕ is the *work function* of the metal surface.

The momentum of a photon also depends on its frequency: the momentum of a photon of frequency ν is

$$p_\nu = h\nu/c \tag{9.10a}$$

where c is the speed of light. Since wavelength $\lambda = c/\nu$, the momentum can equivalently be written as

$$p_\nu = h/\lambda. \tag{9.10b}$$

The validity of this expression is checked by observing the *Compton scattering* of photons by free electrons. Conservation of energy and momentum requires that the scattered photon has a longer wavelength λ' than the incident wavelength λ, such that

$$\lambda' - \lambda = \frac{h}{m_e c}(1 - \cos\theta). \tag{9.15}$$

Observations of scattered photon wavelengths agree with this equation.

Although photons act as discrete packets of energy, light nevertheless exhibits interference, and the distribution of light intensity must be described by a wave theory. Light is thus regarded as made up of waves when working out where the light energy will travel, but as a mass of photons when describing how it interacts with objects such as electrons. This phenomenon is known as wave–particle duality.

Wave–particle duality applies to massive particles as well as to photons. The wavelength λ associated with an electron having a momentum p_e is given by the *de Broglie relation*

$$\lambda = \frac{h}{p_e}, \tag{9.17}$$

which has the same form as the corresponding eqn (9.10b) for a photon.

Diffraction causes waves to spread out so that they cannot be focused to a spot with a diameter less than about one wavelength. This applies to matter waves as well as to light or other types of wave. Since the wavelength and the momentum of a particle are linked by eqn (9.19), it is not possible simultaneously to measure precise values of the position coordinate along a particular axis and the component of momentum of the particle along the same axis. If the uncertainties in position and momentum in the x-direction are Δx and Δp_x, respectively, then the *uncertainty principle* requires that the product

$$\Delta x \Delta p_x \geq \tfrac{1}{2}\hbar, \tag{9.21}$$

i.e. that the product has a minimum value $\tfrac{1}{2}\hbar$, where $\hbar = h/2\pi$. A similar uncertainty principle applies to simultaneous measurement of energy and time, for which uncertainties ΔE and Δt must satisfy the inequality $\Delta E \Delta t \geq \tfrac{1}{2}\hbar$ (eqn (9.22)).

Chapter 10 **The Schrödinger equation**

A freely moving particle with momentum p behaves like a wave of wavelength

$$\lambda = h/p \tag{9.17}$$

where h is Planck's constant. This constant has the dimensions of energy times time, and in the MKS system of units has the value 6.67×10^{-34} J s. The size of this fundamental constant determines the nature of the world we live in. It determines the small distances at which wave-like behaviour may be expected.

The angular frequency ω of the particle wave is determined from the total energy E of the particle. For a mass m freely moving at speed v the total energy (non-relativistically) is equal to its kinetic energy $E = mv^2/2$ and

$$\omega = E/\hbar.$$

The wave function of a particle moving freely along the x-axis is

$$\psi = \psi_0 e^{j(\omega t - kx)} \tag{10.2}$$

where k is the wavenumber, equal to $2\pi/\lambda$.

The interpretation of the wave function is that the square of its modulus at any point is proportional to the probability $P(x,t)dx$ of finding the particle in a vanishingly small interval of length dx at the point x and at time t. For one-dimensional wave functions that are normalized,

$$\int_{-\infty}^{\infty} P(x,t)\, dx = \int_{-\infty}^{\infty} \psi^*\psi\, dx = 1. \tag{10.6}$$

This interpretation makes the wave function for the free particle consistent with Heisenberg's uncertainty principle.

For a particle moving under the influence of a force described by the potential $V(x)$ that is independent of time, the wave function is obtained by solving the Schrödinger equation

$$-\frac{\hbar^2}{2m}\frac{\partial^2 \psi}{\partial x^2} + V(x)\psi = -j\hbar\frac{\partial \psi}{\partial t}. \tag{10.8}$$

For a time-independent potential the wave function factorizes into a product of the form $\phi(x)\rho(t)$. The function $\phi(x)$ is determined by solving the *time-independent Schrödinger equation*, the TISE,

$$-\frac{\hbar^2}{2m}\frac{d^2 \phi}{dx^2} + V\phi = E\phi, \tag{10.12}$$

appropriate to the potential V. The function $\rho(t)$ has the simple form

$$\rho(t) = e^{jEt/\hbar} \tag{10.14}$$

where E is the total energy. The TISE gives stationary wave functions that describe different quantum states characterized by different values for a quantum number n. The states have probability distributions that do not change with time.

The simplest application of Schrödinger's equation is to the problem of a particle of mass m constrained to move between fixed points at $x = \pm L/2$. The allowed wave functions $\phi(x)$ are

$$\phi_n = 2A \cos(n\pi x/L) \tag{10.25}$$

where the integer n is odd ($n = 1, 3, 5, \ldots$), and

$$\phi_n = 2A \sin(n\pi x/L) \tag{10.28}$$

where the integer n is even ($n = 2, 4, 6, \ldots$). The constant A is determined by the normalization condition to be $\sqrt{1/2L}$. The energy levels for integer n have excitation energies

$$E_n = \frac{\hbar^2 n^2 \pi^2}{2mL^2}. \tag{10.26}$$

Diatomic molecules can vibrate along the line joining the nuclei. For small amplitude vibrations the potential in which the reduced mass m moves is $V(x) = kx^2/2$ where x is the displacement of the nuclei from their equilibrium separation. The allowed energy levels for this quantum harmonic oscillator are given by

$$E_n = (n + \tfrac{1}{2})\hbar\omega \tag{10.38}$$

where $\omega = (k/m)^{1/2}$ and n is zero or an integer.

The correspondence principle allows a connection to be made between quantum physics and classical physics.

Chapter 11 **The microscopic world of quantum mechanics**

The spectrum of light emitted by atoms of a particular element has lines of definite frequency corresponding to discrete energies of photons. It follows that the atoms themselves have bound states with discrete energies. The energies of these states can be found by solving the Schrödinger equation. For the hydrogen atom, the equation can be solved analytically if it is assumed that the only force binding the electron to the proton is the electric force proportional to the inverse square of the distance r between the electron and the proton. Wave functions that are spherically symmetric are one-dimensional, depending only on r, and have relatively simple functional forms. The ground state (the most tightly bound state) has the wave function

$$\phi_1(r) = \left(\frac{1}{\pi}\right)^{1/2} \left(\frac{1}{a_0}\right)^{3/2} e^{-r/a_0}. \tag{11.13}$$

The size of the hydrogen atom in its ground state is roughly equal to the *Bohr radius*, $a_0 \sim 5 \times 10^{-11}$ m.

Other states are labelled by three quantum numbers n, ℓ, and m_ℓ, which must have integral values in order to satisfy the boundary conditions applying to the spherical polar coordinates r, θ, and ϕ. The quantum numbers ℓ and m_ℓ determine the angular momentum of the electron, which has a magnitude $\sqrt{\ell(\ell+1)}\hbar$ and a projection $m_\ell\hbar$ along the z-axis. The largest permitted value of ℓ is determined by the principal quantum number n; ℓ cannot be greater than $(n-1)$. Within the approximation used, the energy of a state depends only on n and is $E_n = E_1/n^2$, where $E_1 = -13.6$ eV is the energy of the ground state.

As well as the orbital angular momentum related to the orbital quantum number ℓ, the electron possesses *intrinsic spin* with a magnitude of $\frac{1}{2}\hbar$. The orbital angular momentum and the intrinsic spin may couple together to form states of varying angular momentum, and the total number of states is greater than it would be in the absence of intrinsic spin.

Many-electron atoms have much more complicated wave functions than hydrogen, but the wave functions can still be labelled by quantum numbers. The structure of many-electron atoms is simplified by the Pauli principle, which states that no two electrons may occupy states with the same set of quantum numbers. Electrons in an atom occupy the most tightly bound states first and, when these are filled, must be placed in states with higher quantum numbers and smaller binding energies. Since the quantum number n has the greatest effect on the binding energy, there tends to be a particularly large step in energy after all the states with a particular value of n are filled. An atom with this structure is very stable, and is described as having a *closed shell*.

Chemical bonds are formed between atoms as a result of interactions among their least tightly bound outermost electrons. These may be shared (to form covalent bonds) or exchanged (to form ionic bonds) between atoms, and the ease with which this is done depends on the number of electrons outside a closed shell or needed to complete a closed shell. Chemical behaviour thus repeats itself at each closed shell, giving rise to the *periodic table* of elements, and to the similarity of compounds made up of atoms in the same position in the table with respect to closed shells.

Molecules must obey the rules of quantum mechanics as well as atoms, and like atoms have discrete bound states. Molecules may vibrate and rotate, and each of these kinds of motion gives rise to a characteristic set of energy levels. Vibrations of molecules were discussed in Chapter 10: the energy of a vibrational state with quantum number n is $E_n = (n + \frac{1}{2})\hbar\omega$. The rotational states are labelled by an orbital angular momentum quantum number ℓ which is an integer, and the rotational energy E_ℓ of a state with quantum number ℓ is

$$E_\ell = \frac{\ell(\ell + 1)\hbar^2}{2I} \tag{11.20}$$

where I is the moment of inertia of the molecule about the axis of rotation. Since the rotational and vibrational states have discrete energies like the electronic states,

only discrete energies are allowed for photons emitted in transitions between molecular levels. Much can therefore be learned about the structure of molecules by studying their spectra.

The constituents of atomic nuclei are subjected to much stronger forces than are electrons in atoms. Nevertheless they still have wave functions that must satisfy the Schrödinger equation and have bound states with well-defined discrete energies. Nuclear states, like atomic states, are labelled by quantum numbers and the protons and neutrons within the nucleus are subject to the Pauli principle. Shell structure is evident in nuclei, although the 'magic' numbers of particles corresponding to closed shells are different from those in atoms.

Part 4

Properties of matter

Chapter 12

Thermal physics

This chapter introduces heat and temperature and discusses thermal equilibrium and some of its consequences.

This chapter introduces some physical phenomena associated with heat and temperature. The ways in which heat and temperature are measured, how heat is exchanged to melt solids and evaporate liquids, how temperature influences the properties of matter, and several other questions related to heat and temperature are addressed.

Normal experience gives us a feel for what is meant by heat and temperature. We generate heat by friction; mankind probably first made fire by rubbing sticks together. A wood fire corresponds to the chemical combination of carbon and oxygen. This releases energy in the form of heat. If a pot of water is placed over the fire, heat is absorbed by the water and pot and the temperature of the water rises. The temperature increases to the boiling point, where the water rapidly turns to vapour. Heat and temperature have a close relationship but they are not the same thing. A small volume of water will reach the temperature at which the water boils before a large volume will. More heat has gone into the large volume to get it to the same temperature. Temperature is a measure of the degree of hotness of the water but not of the amount of heat put into different masses.

Matter is atomic and the water consists of water molecules held together by attractive intermolecular forces. The molecules are in motion although the forces between them keep them in close contact with each other. As the temperature increases the thermal motions of the molecules increase and, at the boiling point, the thermal motions overcome the molecular attractions and the liquid turns to gas. In the gas, the molecules' motions are chaotic and the gas can expand to fill any volume it occupies. The distances between gas molecules can be very large.

If, instead of heating it up, heat is taken out of water, its temperature falls and at the freezing point the water becomes solid ice. At temperatures below the freezing point the intermolecular forces are strong enough to fix the mean position of the vast majority of the molecules, and the result is solid ice. The densities of ice and water are very similar so that equal numbers of molecules occupy roughly the same amount of space in both liquid and solid.

To understand these commonly observed properties of water, we relate more quantitatively the temperature of a substance to the motion of its atomic constituents. The forces between molecules at a fixed separation and with fixed relative speed are the same in water and in ice. However, as the temperature increases the average molecular speed increases and the influence of the intermolecular forces decreases. We may thus start out with a working definition of the temperature of an object as a property that is proportional to the average kinetic energy of the object's microscopic parts. For liquids and gases in which the atoms or molecules are not in fixed positions, the average kinetic energy can be easily

● *A working definition of temperature*

visualized. In most solids the constituents vibrate about fixed mean positions, and the average kinetic energy is the average in the oscillatory motion.

The interaction between any isolated pair of molecules or atoms may be represented by a curve that shows how the potential energy varies with separation, as in Fig 12.1. The potential well has a finite depth ϵ equal to the energy needed to pull the pair completely apart when their initial separation is r_0. In a solid, the distance r_0 is the mean separation of the molecules and is the separation about which the molecules oscillate.

An atom or molecule in a solid experiences a potential more complicated than that given in Fig 12.1 since in the interior of the solid it is surrounded by other atoms. Nevertheless, we may use the potential of Fig 12.1 in a very simple model to discuss what happens as the temperature is raised. The average kinetic energy of the oscillations, and hence their average amplitude increases. At a sufficiently high temperature, the atom or molecule is no longer bound to a fixed centre but is free to roam and the solid turns to liquid. In the liquid the molecules continue to experience attractive interactions sufficiently strong for them to remain very close to each other but not strong enough for them to oscillate in a rigid structure. As the temperature increases, kinetic energies become high enough for them to overcome this constraint and the liquid turns to gas.

An individual atom or molecule has a fixed mass, which is usually given in atomic mass units (amu). One atomic mass unit is one-twelfth the mass of an atom of the most abundant carbon isotope, and equals 1.661×10^{-27} kg to four significant figures. Atomic mass units and isotopes were introduced in Section 9.1. The most abundant carbon isotope consists of six electrons around a nucleus of six protons and six neutrons and this atom is defined to have a mass of 12 amu. The **atomic weight** of an atom is the mass of the atom in atomic mass units.

The only other stable isotope of carbon has six electrons orbiting a nucleus consisting of six protons and seven neutrons. The atomic weight of this isotope is very close to 13 but is not exactly 13. There are two reasons for this. First, the neutron's mass is different to that of the proton; second, the total binding energy per nucleon—the total energy needed to separate all the microscopic constituents from each other divided by the number of nucleons—of the less abundant isotope is different to that of the most common carbon atom. As we shall see later when discussing relativity, the latter effect also results in a small change of mass per particle.

The **molecular weight** of a molecule is the mass of the molecule in atomic mass units. The molecular weight of the most abundant water molecule is close to 18. It is not exactly equal to 18 even though

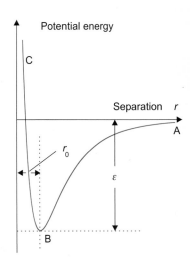

Fig. 12.1 The potential energy between two molecules as a function of the separation r of their centres. An amount of energy ϵ is required to overcome the attractive force and separate the two.

the molecule consists of two hydrogen atoms, each consisting of a proton and an electron, bound to an oxygen atom made of 8 electrons orbiting a nucleus of 8 protons and 8 neutrons. The reasons for this are the same as for the carbon isotopes above: the neutron−proton mass difference and the effect of the binding energy of the microscopic constituents.

The molecular weight of any substance in grams, called a **gram molecule**, contains the same number of molecules. This number is called **Avogadro's number**, symbol N_A, and equals 6.022×10^{23} to four significant figures. For a substance that is monatomic, such as helium gas, a mass equal to the atomic weight in grams contains Avogadro's number of atoms. Avogadro's number is huge since the masses of atoms and molecules are very small. A **mole** of a substance is an amount which contains N_A atoms or molecules.

12.1 **Solids, liquids, and gases**

In this section we make some general remarks on the properties of solids, liquids, and gases. We introduce the definitions of some of the terms used in the discussion of thermal physics and discuss in more detail the changes from the solid to the liquid and gaseous states.

A crystalline solid is made of rather closely packed atoms situated on a *lattice* that has a regular geometric structure. The lattice of a crystal of common salt, sodium chloride, is shown in Fig 12.2. There are solids in which the atoms have no regular structure over distances of several times atomic dimensions. These are called *amorphous* solids and the loose arrangement of their atoms is somewhat similar to that met in liquids. However, many solids are crystalline and the atoms oscillate about fixed positions within lattices. Attractive forces between atoms in solids can arise in various ways, sometimes by sharing of the outer electrons of the atoms between close neighbours, sometimes by donation of electrons to neighbours.

A solid structure with a fixed arrangement of atoms is called a **phase** of the material. If we raise the temperature, the atoms in the crystalline structure vibrate with increased average kinetic energy until a point is reached at which the temperature stays constant and extra input of energy goes into melting the solid until it all has turned to liquid. The liquid state is another phase of the material and the **change of phase** from solid to liquid takes place at a constant temperature. At fixed external conditions—pressure is usually the only one that matters—the constant temperature at which the solid all melts is called the **melting point**. The melting point is the same as the **freezing point**, which is the term used if the process is going the other way

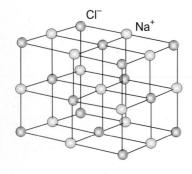

Fig. 12.2 The lattice of a common salt crystal. Each Na⁺ ion has as its nearest neighbour a Cl⁻ ion. To make their positions in the lattice clear, the atoms are not shown closely packed in the figure.

from liquid to solid. Adding more heat energy after the liquefaction of the solid raises the liquid's temperature until we arrive at the **boiling point**, at which the liquid all evaporates as the temperature stays the same. The boiling point also has a well-determined value for a given substance at fixed external pressure. Above the boiling point the phase has changed from liquid to gas and, as the temperature has increased, the substance has gone through the three phases corresponding to solid, liquid, and gas.

Although pressure has not been formally introduced so far, most readers will be familiar with it. Pressure is force per unit area and is usually used in the discussion of gases and liquids. If a liquid is contained in a vessel open to the atmosphere, the pressure in the liquid depends upon distance below the liquid surface. The weight of liquid above increases the pressure below the surface above that at the surface. To a very good approximation, gases exert a constant pressure over the walls of their containing vessels, independent of position, because of their very low density.

● *Pressure*

The SI unit of pressure is called the **pascal** and denoted Pa. One pascal is one newton per square metre. Pressures are often quoted in other units. Low pressures are often given in torr, and high pressures in standard atmospheres. One standard atmosphere pressure is equal to 1.013×10^5 Pa, to four significant figures. The pressure of the atmosphere at sea level and normal temperatures supports a column of mercury about 76 cm high under vacuum. One torr equals 133.3 Pa to four significant figures. A pressure of 1 torr will support a height of about 1 mm of mercury, and low pressures are often simply quoted in mm of mercury. Table 12.1 gives units and conversion factors.

Vapour pressure

At normal temperatures, the liquid and gas phases of many substances coexist. In this situation the gas is called a **vapour**. Consider water alone partially filling a closed container. Vapour at a certain pressure, called the **saturated vapour pressure**, occupies the space in the container above the water. Most of the water exists as liquid and so the temperature is such

Table 12.1 Pressure units and conversion factors

pascal: the SI unit	$1\,\mathrm{Pa} = 1\,\mathrm{N\,m^{-2}}$
bar	$1\,\mathrm{bar} = 10^5\,\mathrm{Pa}$
standard atmosphere	1 standard atmosphere $= 1.01325 \times 10^5\,\mathrm{Pa}$
torr	$1\,\mathrm{torr} = 133.32\,\mathrm{Pa}$
	$760\,\mathrm{torr} = 1$ atmosphere

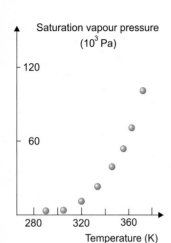

Fig. 12.3 The saturated vapour pressure of water as a function of temperature.

that the average kinetic energy of the molecules is less than the amount needed for escape. However, there is a continuous distribution of kinetic energies and a small number of molecules have kinetic energies sufficiently above the average to allow them to escape the surface. At any temperature the vapour pressure rises to the point where an equilibrium is established between the rate at which molecules leave the surface and the rate at which they return to the liquid via collisions with the surface. This requires further explanation since we invoked the notion that only the faster molecules leave the surface in the first place. However, the interactions of gas molecules with liquid molecules when they collide at the surface cause energy interchanges and all molecules, after an equilibrium has been established, have the same average kinetic energy—the gas is at the same temperature as the liquid. When gas molecules with energies slightly lower than that required for escape strike the liquid surface they re-enter the liquid phase.

In equilibrium, molecules return to the liquid at the same rate as that at which others leave. This dynamic equilibrium for two phases in contact, or for any two bodies in contact after equilibrium has been established, is called **thermal equilibrium**. In equilibrium with its vapour at any given temperature, water has a unique saturated vapour pressure. This increases with temperature, as shown in Fig 12.3, because, as the temperature rises, the fraction of molecules with energies in excess of the escape energy increases.

Evaporation

In contrast to the above section, which considered water and its vapour alone in a closed container, now consider water in a container open to the atmosphere. The water molecules evaporate at a certain rate and escape to the outside world, leading to a slow loss of water from the container. The faster molecules escape from the surface as before but now they escape completely to the large outside environment. This process would cool the water were it not for absorption of heat from the surroundings. The rate at which the water evaporates increases as the temperature is raised because, at the higher temperature, a larger fraction of the molecules have energies in excess of that needed to escape—the same effect as that which causes the rise in saturated vapour pressure shown in Fig 12.3.

As more heat is put in, the water temperature rises until the saturation vapour pressure equals the external atmospheric pressure. At that temperature the vapour pressure stays equal to that of the surroundings and the temperature stays constant. The molecules in the space above the liquid are now almost all molecules of water vapour and, as those on the outside disperse, water turns to vapour more rapidly than at lower temperatures. The constant temperature at which the vapour pressure equals the external pressure is the boiling point. The extra heat put in at the boiling point goes to overcome the molecular attractions in the liquid phase, releasing the molecules to move freely in the atmosphere.

The boiling point changes as the external pressure changes because of the variation of saturated vapour pressure with temperature. For example, on a day when the air pressure is abnormally low during cyclonic weather conditions the boiling point will be slightly lower than during an anti-cyclone. Likewise, water will boil at lower temperatures the higher up one goes from sea level.

The heat required to transform a mole of a liquid into gas is called the **latent heat of evaporation** of the liquid and is usually quoted in joules per mole. It can also be expressed in $J\,g^{-1}$ or $J\,kg^{-1}$. The latent heat is intimately connected with the average binding energy of a molecule in the liquid. This is the energy needed to remove a molecule from the liquid, and can be regarded as made up of contributions from each nearest neighbour.

The attractive interaction between a pair of molecules can be discussed with reference to Fig 12.1. The force as a function of separation r is the negative gradient of the potential. Hence, over the part AB of the curve the force is attractive and arises from distortion of the charge distribution of the molecules as they come close. At very short distances, over the part BC of the curve where the molecules' electron clouds begin to overlap, the force is repulsive corresponding to the difficulty of compressing the electrons' wave functions. On average the separation of molecules in the liquid is close to r_0, and an average energy of ϵ less the average kinetic energy of a molecule is needed to separate one from the other. This is the average binding energy E_B^ℓ of a molecule in the liquid.

If the number of nearest neighbours, called the **coordination number**, in the liquid phase is n_ℓ, the energy needed to release N molecules from the body of the liquid is one-half of the product of NE_B^ℓ and n_ℓ. The factor of one-half arises because the energy E_B^ℓ is shared between two molecules. Hence the latent heat of evaporation per mole is given by

$$L_{\text{evap}} = \tfrac{1}{2} N_A n_\ell E_B^\ell \tag{12.1}$$

where N_A is Avogadro's number.

Melting

Solid and liquid phases do not usually coexist in equilibrium except at the melting point. If we start with ice and heat it, at the melting point more heat turns all the ice into water at constant temperature. A few molecules have too much energy to stay confined in the lattice and transfer to the liquid phase. This would cool the solid but, if heat is supplied to keep the temperature constant, more molecules liquefy, and the heat put in at the melting point goes into turning all the solid to liquid before the average energy can be further increased in the liquid phase. The heat required to transform one mole of solid into liquid is called the **latent heat of fusion** of the solid.

Worked Example 12.1 The density of water at a temperature just above the freezing point is $1\,\mathrm{g\,cm^{-3}}$ and the density of ice just below the freezing point is $0.92\,\mathrm{g\,cm^{-3}}$. Estimate the average separation of the molecules in the two phases.

Answer The molecular weight of water is 18, and 18 g contain 6.02×10^{23} molecules. If we make the gross approximation that each of them occupies a cube of volume $18/(6.02 \times 10^{-23})\,\mathrm{cm^3}$, the side of the cube has length $3.10 \times 10^{-8}\,\mathrm{cm}$. For ice the separation is about $3.19 \times 10^{-8}\,\mathrm{cm}$.

 The binding energy of a molecule in ice or in any solid is reduced when the solid melts. This cannot be explained in terms of changes in the average separation of its constituents. We have to look for a better model than the overly simple one in which atoms occupy individual cubes. The particular arrangement of atoms in a lattice together with the nature of the force between the atoms on the lattice sites provide the crystal binding.

 It is rare that the density of a solid is less than that of the liquid it becomes on melting. Ice forms on the top of ponds because of the unusual properties of water.

Releasing the molecules from their fixed positions in the lattice and allowing them to roam more freely in the liquid requires less energy than releasing them from the liquid and allowing them to roam completely freely in the gas. The average binding energy E_{B}^{ℓ} of a molecule in the liquid arising from the attraction of each nearest neighbour is not much less than its value $E_{\mathrm{B}}^{\mathrm{s}}$ in the solid. If the coordination number in the liquid is n_{ℓ} and in the solid is n_{s}, the latent heat of fusion per mole is given by

$$L_{\mathrm{fus}} = \tfrac{1}{2} N_{\mathrm{A}} \left(n_{\mathrm{s}} E_{\mathrm{B}}^{\mathrm{s}} - n_{\ell} E_{\mathrm{B}}^{\ell} \right). \tag{12.2}$$

Exercise 12.1 The latent heat of fusion of water is about 6.0×10^3 J per mole, and the latent heat of evaporation is about 4.5×10^4 J per mole. If the number of nearest neighbours of an ice molecule is 10 and the number for a water molecule is roughly the same, estimate the average binding energies per pair of ice and water molecules.

Answer Equation (12.1) gives $E_B^\ell = 0.093$ eV, and eqn (12.2) then gives $E_B^s = 0.106$ eV.

In this section we have discussed some particular aspects of gases, liquids, and solids at an introductory level. Real materials show a much wider set of behaviour patterns than those we have described. For example, some materials that have the properties of solids when observed over periods of several years flow like liquids over very long time-scales; glaciers provide a good example. Other materials, such as glasses and glues, exhibit properties of both liquids and solids. Melting and boiling are more complicated when dealt with rigorously, and water, water vapour, and ice, which we have used as examples, have complicated and unusual properties.

12.2 Thermal equilibrium and temperature

In this section we examine more closely the nature of heat and temperature. The hotness of a body is a measure of its temperature and, if one object is placed in contact with a second object that is hotter or colder, the microscopic constituents of each object interact, heat flows between the two, and after a period of time the objects attain thermal equilibrium. The direction of heat flow is determined by the temperatures of the two bodies: heat flows from the hot object to the cold. At thermal equilibrium all microscopic constituents have the same average energy.

In the introduction to this chapter we proposed that the temperature of a body is proportional to the average kinetic energy of the atoms or molecules within the body. This leads to the result that, when two interacting systems are in thermal equilibrium, they are at the same temperature. If they are initially at different temperatures they will always reach an equilibrium in which they are at the same temperature.

At thermal equilibrium the microscopic description of the two systems is that which corresponds to maximum randomness. It is the state for which the probability of the microscopic constituents being distributed amongst the available options has the greatest value. Fluctuations from this state, for macroscopic systems with huge numbers of molecules, are very small and can almost invariably be ignored.

A hot object has a higher temperature than a cold one, but how is temperature measured and what are the units of temperature? We now go on to discuss those matters.

The ideal gas equation and Boltzmann's constant

Consider a gas confined in a large rectangular box. We will take the gas molecules to be featureless spheres of such small size compared to the size of the box that they occupy negligible volume. Furthermore, we will assume that the molecules move independently, making elastic collisions with each other as they move at random in the box. An elastic collision is one in which there is no change in total kinetic energy of the colliding pair. This hypothetical gas is called an **ideal gas**; a monatomic gas such as helium or neon at low pressures when there are relatively few molecules per unit volume behaves approximately like an ideal gas.

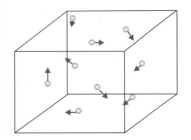

The gas molecules are moving in random directions and with a distribution of speeds, as shown schematically in Fig 12.4. Let us start considering the properties of the gas with a simpler picture, assuming that all the molecules are moving in the positive or negative x-, y-, and z-directions and all at the same speed v, as in Fig 12.5. One-sixth of the molecules are moving in each of the directions shown in the figure.

If the total number of molecules per unit volume is N_0, the number of molecules per unit volume moving in the positive x-direction is $N_0/6$. All the molecules in a column aligned along the x-direction and of length v reach the end of the column in one second. The number that strike a unit area of the wall ABCD shown on Fig 12.5 is therefore $N_0 v/6$ per second. The momentum change of each molecule on reversing direction at the wall is $2mv$ and the momentum change at a unit area of the wall ABCD per second is

$$\Delta p = (2mv) \times \tfrac{1}{6} N_0 v. \tag{12.3}$$

Fig. 12.4 A schematic representation of gas molecules in a container moving independently with different speeds and in random directions.

Force is rate of change of momentum, and the pressure on the wall is the rate of change of momentum on unit area of the wall. Hence, the momentum change Δp given above is the pressure P exerted on the wall by the gas. If the total number of molecules in the box is N, $N_0 = N/V$ where V is the volume occupied by the gas, and the product

$$PV = \tfrac{1}{3} Nmv^2. \tag{12.4}$$

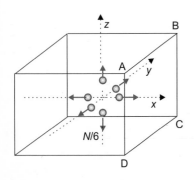

We see that, for a fixed mass of gas that has a fixed number of molecules, the product of pressure and volume is proportional to mv^2, which is twice the kinetic energy of the molecules. But we have defined the temperature of an equilibrated system to be proportional to

Fig. 12.5 The simple model in which all molecules move at the same speed v, and the flux in random directions is replaced by a distribution in which one-sixth of the total molecules move in each of the positive x-, y-, and z-directions, and one-sixth along each of the negative x-, y-, and z-directions.

the average kinetic energy $\overline{E_K}$ of its atoms or molecules. In our simplified example the average is simply $mv^2/2$, and eqn (12.4) shows that

At constant temperature the product of pressure and volume of an ideal gas is constant.

This is **Boyle's law** which is approximately obeyed by real gases at pressures at or below atmospheric pressure.

Since the temperature T is proportional to the average kinetic energy of a molecule, we may write

$$\overline{E_K} = \tfrac{1}{2}m\overline{v^2} = \tfrac{3}{2}k_B T \tag{12.5}$$

where we have chosen the constant of proportionality to be $\tfrac{3}{2}k_B$ for reasons that become apparent later. In the present simplified derivation the molecules all move at the same speed v and $\overline{v^2} = v^2$. Thus $\overline{E_K} = \tfrac{1}{2}mv^2$, and eqn (12.4) becomes

$$PV = Nk_B T. \tag{12.6}$$

If we consider one mole of the gas, N is equal to Avogadro's number N_A and

$$PV = N_A k_B T = RT \tag{12.7}$$

where $R = N_A k_B$ is another constant. This equation is the **ideal gas equation** relating the pressure, volume, and temperature of one mole of an ideal gas. If there are n moles of gas the right-hand side of eqn (12.7) is multiplied by n.

Since Avogadro's number N_A and k_B are constants for any ideal gas, independently of the mass of the ideal gas atom, eqn (12.7) tells us that R is another constant. It is called the **gas constant**. When its value is determined using eqn (12.7) with gases at low pressures and large volumes, when their behaviour approximates closely that of an ideal gas, the values found for different gases are all nearly equal. The constant k_B has a special place in physics, as we shall see. It is called **Boltzmann's constant** after Ludwig Boltzmann, an Austrian physicist who played a major role in the description of the thermal behaviour of matter.

Equation (12.7) can be derived more satisfactorily by taking into account the distribution of speeds and directions of the molecules in a gas. The molecules are moving in random directions with a distribution of speeds.

We take the probability of a molecule having an x-component of velocity between v_x and $v_x + \mathrm{d}v_x$ to be given by $P(v_x)\,\mathrm{d}v_x$ with $P(v_x)$ a fixed function at fixed temperature and normalized to make the total probability that v_x is in the range from $-\infty$ to $+\infty$ equal to unity.

Ludwig Boltzmann

Ludwig Boltzman, pictured here, was a nineteenth century physicist who made pioneering contributions to science at a time when approaches based on the atomistic nature of matter were in their infancy and were the subject of much debate and dispute.

He was born in Vienna on 20 February 1844. His father was a customs official and his grandfather a maker of musical clocks from Berlin. Ludwig's father died in 1859, and his mother used most of the meagre resources left to her to enable Ludwig to continue his education and proceed to study physics at the University of Vienna. There his talents were quickly recognized, and in 1867 the Director of the Physical Institute appointed Ludwig his assistant. Another staff member at the time was Josef Loschmidt, who made one of the first measurements of the number of molecules in a gram molecule. Ludwig rapidly obtained recognition for his early work on the relations between temperature and the kinetic energy of colliding molecules, even though many scientists of the time opposed atomistic notions and considered that the only true physics was that concerned with macroscopic observables. He was appointed Professor of Theoretical Physics at the University of Graz in 1869. From there his career took him to different places over the years: to Munich in 1889, back to Vienna in 1894, on to Leipzig in 1900, and a final return to Vienna in 1902.

Ludwig Boltzmann was extremely near-sighted and suffered head and nerve pains and periods of depression for much of his life. These became worse as he grew older and in 1906 he took his own life at the age of 62. His periods of depression may have been exacerbated by the heavy attacks made on his theories by those who rejected the then hypothetical concept of atoms whose existence had not yet been demonstrated to their satisfaction. It is ironic that Boltzmann's death occurred only a few years before the experiment on Brownian motion by Perrin in 1908 and Millikan's oil-drop experiment in 1909 produced the evidence that placed the atomistic nature of matter beyond doubt.

One of Boltzmann's major achievements was to remove one of the chief impediments to the acceptance of the kinetic theory of heat, which is based on the idea of colliding molecules and the relation between the heat

content of a body and the total kinetic energy of its microscopic constituents. An objection of the time was that mechanical processes were reversible but that processes involving heat were not. If a planet revolved around a star in one direction, another state of motion in which the direction of rotation was reversed was easily imaginable. However, heat flowed only from a hot body to a cold body and never the other way round. The objectors' argument was thus that heat could never be explained mechanistically via moving atoms or molecules. Boltzmann showed that the apparent non-reversibility was a result of the statistical near impossibility of the reverse heat flow. It was not due to any natural law but simply because the probability of it happening for macroscopic systems was negligibly small. An analogy often made uses marbles. If 1000 red marbles and 1000 blue marbles in a box are initially in two separate layers and the box is shaken, the marbles soon become utterly mixed. The probability of more shaking resulting at any future time in two separate layers is minute. If we increase the numbers to 10^{20} the probability becomes vanishingly small.

The distribution functions are the same for all three axes and the averages of the components v_x, v_y, and v_z are zero, since there is no preferred direction. However, we do not need to assume any particular mathematical form for $P(v_x)$ in order to make the derivation.

If the number of molecules per unit volume is N_0, the number with components between v_x and $v_x + dv_x$ striking unit area of the wall per second is $N_0 P(v_x) v_x dv_x$, the number with speed components in the appropriate interval moving along the positive x-axis in a box of length v_x. Each of these suffers a change of momentum $2mv_x$ on reflection at the wall and hence the momentum change at unit area of the wall per second due to molecules with x-components of velocity between v_x and $v_x + dv_x$ is

$$dp = (2mv_x)(N_0 P(v_x) v_x \, dv_x). \tag{12.8}$$

The total momentum change per second is given by summing the contributions from the molecules over the range of values of v_x from zero to $+\infty$, omitting those with negative values of v_x because these do not hit the wall. This is equivalent to integrating over the probability distribution $P(v_x)$,

$$\Delta p = 2mN_0 \int_0^\infty v_x^2 P(v_x) \, dv_x. \tag{12.9}$$

The integral is one-half the average of the square of the x-component of velocity of the molecules, obtained by integrating over the whole positive and negative range. In turn, the average of the square of the speed of the molecules, $\overline{v^2}$, is three times the average of the square of the x-component, since

$$v^2 = v_x^2 + v_y^2 + v_z^2.$$

The pressure P is equal to the total change of momentum per unit area per second, and is finally given by

$$P = \tfrac{1}{3} m N_0 \overline{v^2}. \tag{12.10}$$

If the total number of molecules in the box is N, $N_0 = N/V$. This leads us to eqn (12.7), as before, but in a more rigorous fashion.

The ideal gas scale of temperature

We now have a method of measuring temperature, although its practical application is difficult and requires great care. The product of pressure and volume of an ideal gas is proportional to temperature on the **ideal**

gas scale. Helium gas at low pressure, so that the atoms are well separated, behaves as an ideal gas to a good approximation. The pressure of a fixed volume of helium can thus be used as a measure of the temperature of a system with which the gas is in contact and with which the gas is in thermal equilibrium.

It is necessary to fix the units of temperature, and this is done by defining the numerical temperature value at a fixed point where the temperature is well defined and constant. There is only one temperature and pressure at which the three phases—ice, water, and water vapour—can exist in equilibrium. This state of equilibrium is called the **triple point** and is used to set the temperature scale. The triple point of water is very close to the freezing point at normal pressures. The ideal gas scale is obtained by defining the triple point of water to be $273.16\,\text{K}$.

Temperatures on the ideal gas scale are used in scientific work and are quoted in **kelvins**. One kelvin is given the symbol $1\,\text{K}$ as above. As we shall see, the gas scale of temperature is equivalent to a more fundamental temperature scale, which we will introduce in Section 12.4. On these two scales zero temperature is the minimum attainable; negative temperatures are not possible because it is not possible to have kinetic energy less than zero. At **absolute zero**, the temperature T is zero and matter exists in its lowest possible energy state.

Zero temperature is never precisely reached because the rate at which it is approached becomes slower and slower as T falls. It is now possible to perform experiments at extremely low temperatures, and new phenomena are being revealed that depend on the quantum behaviour of matter close to its coldest possible configurations. A striking example is provided by liquid helium, which becomes superfluid below about $2.2\,\text{K}$, flowing without friction and allowing heat to pass through it with little resistance.

Now that temperature can be measured, the gas constant R can be determined. It has the dimensions of energy divided by temperature and is equal to $8.314\,\text{J}\,\text{K}^{-1}$ to four significant figures.

The constant k_B in eqn (12.6) is determined by N_A and R and is equal to $1.381 \times 10^{-23}\,\text{J}\,\text{K}^{-1}$ to four significant figures.

Both of these numbers are known to better than one part in 10^5. Since R is a constant for an ideal gas of any mass, to the extent that real gases are ideal one mole of all gases occupies the same volume at a given temperature and pressure. Normal atmospheric pressure is close to $10^5\,\text{Pa}$; a standard atmosphere is equal to $101\,325\,\text{Pa}$. At $298.15\,\text{K}$ and a pressure of $10^5\,\text{Pa}$ (called SATP, standard ambient temperature and pressure) the molar volume of an ideal gas equals $24.79 \times 10^3\,\text{cm}^3$. The molar volumes of most common gases are very close to this value at standard pressure and temperature.

Exercise 12.2 What is the value in eV of $k_B T$ at a room temperature of 293 K?

The wavelength of a neutron of momentum p is equal to h/p, where $h = 6.63 \times 10^{-34}$ J s, is Planck's constant. The mass of a neutron is 1.68×10^{-27} kg. What is the wavelength of a neutron when its kinetic energy is $k_B T$ and T is 293 K?

Answer $k_B T = 0.0252$ eV, about 1/40 of an eV; wavelength $= 1.80 \times 10^{-10}$ m. Neutrons with wavelengths in the neighbourhood of this value are produced in nuclear reactors and may be used for diffraction as discussed in Section 9.5.

With the ideal gas as a standard, any physical property that varies with temperature can be calibrated and used as a way of temperature measurement. The most common method is to observe the change in length of a column of liquid, such as mercury, in a tube of small bore. Most materials expand as their temperature increases and for many the expansion varies almost linearly with temperature over limited ranges. Mercury thermometers are convenient and nearly linear over a wide temperature range. Accurate temperature measurements are made with different devices according to the temperature range under examination. For example, the electrical resistance of a platinum wire can be measured accurately and may be used over a broad temperature range from very low temperatures up to over 1400 K.

Temperatures in normal life are usually given in degrees centigrade or degrees Fahrenheit. Assignment of temperatures of zero and 100 degrees to the normal freezing and boiling points of water respectively gives temperature T_C in degrees centigrade. Gabriel Fahrenheit, a German physicist, proposed a different unit by assigning 32 degrees Fahrenheit to the freezing point of water and 212 degrees to the boiling point. Thus, 1.8 degrees Fahrenheit change in temperature is the same as a 1 degree centigrade change. Temperatures are usually given in degrees centigrade in Europe, while degrees Fahrenheit are common in North America. The size of a unit on the gas scale is almost exactly the same as 1 degree centigrade, so that, to sufficient accuracy for most purposes, the ideal gas equation for one mole of gas can be written

$$PV = R(T_C + 273).$$

12.3 Heat, energy, work, and the first law of thermodynamics

Thermodynamics is the subject that deals classically with the thermal behaviour of macroscopic materials.

The first law of thermodynamics deals with the relation between the heat input to a substance and the increase in its internal energy. The total energy of an object is the average energy of one of its microscopic constituents times their number. For now we consider only microscopic constituents or molecules with no internal structure, when the average molecular energy is simply the sum of the average molecular kinetic and potential energies. (The situation in which the constituents may be in excited quantum levels is discussed in Section 12.8.) The potential energies depend upon the forces between the constituents, either as they oscillate about mean positions as in a solid lattice or have more chaotic motions as in liquids or gases. Heat is energy, and as an object is heated its energy increases. Both kinetic and potential energies may change as a result of the increase in energy. The average kinetic energy always increases and the temperature of the object rises.

The energy of a system defined above may include a contribution from the centre of mass of the object. The internal energy of the object, as explained in Section 3.4, is the total energy less the energy associated with the centre of mass. In thermodynamics we are interested only in the internal energy E which is a function of the variables P, T, and V characterizing the system. The internal energy is known precisely if the values of any pair of those variables are specified, since the three are related, and is independent of how those values were obtained.

● *The internal energy of one mole of an ideal gas*

For a dilute gas that approximates an ideal gas, the interactions between the atoms or molecules may be neglected and the potential energies are zero. The internal energy E of one mole of an ideal gas is thus all kinetic energy and from eqn (12.5)

$$E = \tfrac{3}{2} N_A k_B T = \tfrac{3}{2} RT. \tag{12.11}$$

All of the energy given to an object as heat may not go to increasing the internal energy; some may go to do external work. For example, if the object expands, work is done against the external pressure. We will use the symbol Q for heat energy and W for work done. We take work done to be positive if it is done against the external forces on the body. If the internal energy of a system containing a fixed number of particles changes by an infinitesimal amount dE, conservation of energy requires

$$dE = dQ - dW. \tag{12.12}$$

The sign of dQ is positive if heat is put into the system, and negative if it is removed, corresponding to heat being part of the internal energy of the object. Equation (12.12) is the **First law of thermodynamics** which expresses the fact that

Energy is conserved if heat is taken into account.

Specific heat

If a small amount of heat input dQ raises the temperature of an object by an amount dT, the **heat capacity** of the object is defined as dQ/dT and has units of joules per degree. The **specific heat** of a substance is the heat required to raise the temperature of unit mass by 1 degree centigrade, and has units of joules per degree per kg, $JK^{-1}kg^{-1}$. The specific heat is often quoted in JK^{-1} per mole, which is the specific heat per kg multiplied by the mass of a mole of the substance in kg. We will use the symbol C for the specific heat per mole. Like all parameters that describe the macroscopic properties of matter, specific heats can vary with temperature. Figure 12.6 shows how the specific heat of water varies from its freezing point to its boiling point.

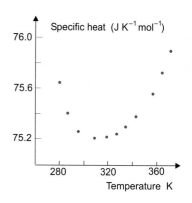

Fig. 12.6 The specific heat per mole of water over the temperature range from freezing to boiling.

Exercise 12.3 An amount of heat equal to 4×10^3 joules is added to 100 g of water, initially at 30 degrees centigrade. Use Fig 12.6 to estimate the rise in temperature of the water.

Answer About 10 K.

The specific heat of a substance depends on the conditions in which the heat input is applied. There may be a magnetic field on the sample; there may be the constraint that the volume of the sample is held constant, or the pressure on the sample is fixed. Thus specific heats are defined under constraints such that certain parameters are held constant.

The simplest example is that of a gas. Physical variables that define the state of a fixed mass of gas are its pressure, volume, and temperature. If we heat one mole of the gas its temperature rises, but it rises by different amounts depending upon whether we keep the volume constant or the pressure constant. If an amount of heat dQ_V input to one mole of gas at constant volume increases its temperature by dT, the specific heat per mole at constant volume is

$$C_V = \left(\frac{dQ_V}{dT}\right). \tag{12.13}$$

If an amount of heat dQ_P input to one mole of gas at constant pressure increases its temperature by dT, the specific heat per mole at constant pressure is

$$C_P = \left(\frac{dQ_P}{dT}\right). \tag{12.14}$$

If the temperature of one mole of gas is changed by an amount dT by adding heat dQ_P at constant pressure P, the gas expands by an amount

dV and does work dW equal to PdV against this pressure. Hence, from eqn (12.12),

$$dQ_P = C_P \, dT = dE + P \, dV.$$

Suppose we change the temperature of the gas by the same amount dT by adding heat dQ_V at constant volume. The internal energy E of the gas changes by the same amount dE as it does when the temperature is raised dT at constant pressure since E depends only on T, but now no work is done against the external pressure. Hence

$$dQ_V = C_V \, dT = dE. \tag{12.15}$$

Combining these two equations we find

$$C_P - C_V = P\left(\frac{\partial V}{\partial T}\right)_P. \tag{12.16}$$

The partial differential $(\partial V/\partial T)_P$ occurs instead of dV/dT because the volume is a function of pressure and temperature and the temperature change was effected at constant pressure.

If we now assume that the gas under consideration behaves approximately as an ideal gas, the relationship between P, V, and T is given by eqn (12.7) which leads to $V = RT/P$. Hence,

$$\left(\frac{\partial V}{\partial T}\right)_P = \frac{R}{P}.$$

Inserting this in eqn (12.16) gives the result that the difference in the specific heats per mole of a gas that behaves like an ideal gas is equal to the gas constant R,

$$C_P - C_V = R. \tag{12.17}$$

The specific heats of different gases vary widely, as shown on Fig 12.7 which shows the specific heats at constant volume of various gases

● *The difference in the specific heats of ideal gases*

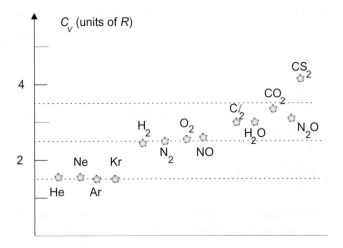

Fig. 12.7 The specific heat per mole at constant volume for various gases at room temperature.

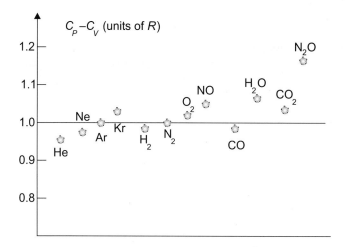

Fig. 12.8 The difference in the specific heat per mole at constant pressure and the specific heat per mole at constant volume for various gases at room temperature.

at a temperature of 273 K. However, the difference in the two specific heats per mole of any gas at normal temperatures is always close to the value R as shown in Fig 12.8. Most solids and liquids expand as their temperature increases, with the result that C_P is larger than C_V, but the difference is usually considerably smaller than the gas constant R.

12.4 The expansion of ideal gases

The ideal gas eqn (12.7) is a relation between the pressure, volume, and temperature of an ideal gas. It is called an **equation of state**. The state of the gas is defined by any two of the three variables P, V, or T, which are sometimes called **functions of state**. Knowing two, the third is fixed, as are other quantities that depend on these variables.

In this section we consider the expansion of ideal gases. If a gas expands, it does work against an external pressure. Let the gas be contained in a cylinder of volume V sealed by a piston as shown in Fig 12.9. At mechanical equilibrium, the force on the piston due to the pressure from the gas inside the cylinder is balanced by an equal external force on the piston. We consider two extreme ways in which the gas expands. The first, called an **isothermal** expansion, occurs when the gas stays at constant temperature. This can be effected by the gas absorbing heat from its surroundings to compensate for the work done by the moving piston against the external force. The second mode of expansion we discuss occurs when no heat at all is exchanged with the surroundings. This is called an **adiabatic** expansion.

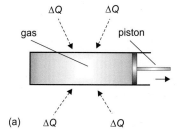

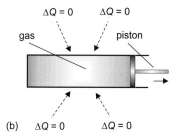

Fig. 12.9 Gas confined to a cylinder by a movable piston exerts a force on the piston. If the force on the piston from the gas inside the cylinder is slightly greater than the force from outside, the gas slowly expands. (a) The situation during an isothermal expansion when the temperature of the gas in the cylinder is maintained constant by heat input ΔQ from an external source or reservoir. (b) An adiabatic expansion when there is no heat interchange.

Isothermal expansion

Consider one mole of gas. Equation (12.12) expresses the conservation of energy when there is an infinitesimal increase in heat input dQ accompanied by an infinitesimal increase in volume dV whilst external work dW is done,

$$dQ = dE + dW.$$

This equation indicates that the heat dQ that must be absorbed from the surroundings during a stage of an isothermal expansion is equal to the external work dW, since there is no change in internal energy E at constant temperature. For an ideal gas, the internal energy is simply the total kinetic energy and, as given in eqn (12.11), this depends on temperature alone.

The concepts of temperature and pressure must remain valid for the whole gas during the expansion in order to discuss the process using those variables, and so the expansion to which our discussion applies must take place very slowly so that at each stage thermal equilibrium is maintained and turbulence does not invalidate the arguments. Such very slow processes are called **quasi-static**. In spite of the academic nature of such exercises they give valuable insight into the properties of gases.

Equation (12.7) relates the initial and final pressures and volumes. Since the temperature T is constant during a quasi-static isothermal expansion, if P_1 and P_2 are the initial and final pressures and V_1 and V_2 the initial and final volumes,

$$P_1 V_1 = P_2 V_2. \tag{12.18}$$

This result was stated earlier in words as Boyle's law.

Adiabatic expansion

The other extreme mode of expansion when no heat at all is exchanged with the surroundings is an adiabatic expansion, for which the heat input dQ is zero during an infinitesimal increase in volume. The external work dW done against the pressure is equal to PdV and conservation of energy ($dQ = dE + dW$) indicates that for each stage of an adiabatic expansion

$$dE = -PdV. \tag{12.19}$$

The work done against the external force has been at the expense of the internal energy of the gas and the temperature falls. We will calculate how much it falls during an adiabatic expansion from volume V_1 to V_2.

Consider one mole of an ideal gas. The change in internal energy dE is equal to $C_V dT$, from eqn (12.15), and substitution in eqn (12.19) gives

$$C_V \, dT = -P \, dV. \tag{12.20}$$

Multiplying both sides of this equation by the gas constant R

$$C_V R \, dT = -RP \, dV,$$

and using the relation $C_P - C_V = R$ given in eqn (12.17)

$$C_V R \, dT = C_V P \, dV - C_P P \, dV. \tag{12.21}$$

At any state of the gas, the ideal gas equation $PV = RT$ holds, and thus for very small changes of P, V, and T

$$P \, dV + V \, dP = R \, dT.$$

Substituting $R \, dT$ from this expression into eqn (12.21) gives

$$C_V V \, dP = -C_P P \, dV.$$

Finally, rearranging and integrating between starting and ending volumes and pressures of V_1, P_1 and V_2, P_2, respectively, we obtain

$$C_V \int_{P_1}^{P_2} \frac{dP}{P} = -C_P \int_{V_1}^{V_2} \frac{dV}{V},$$

leading to the result

$$C_V [\ln P]_{P_1}^{P_2} = -C_P [\ln V]_{V_1}^{V_2}$$

or

$$P_1 V_1^\gamma = P_2 V_2^\gamma \tag{12.22}$$

where γ is the ratio C_P/C_V of the specific heat at constant pressure to the specific heat at constant volume.

Equation (12.17) gives $C_P - C_V = R$ for an ideal gas, and $C_V = (\partial E/\partial T)_V = 3R/2$ from eqn (12.11). Hence, $C_P = 5R/2$ and

● *The ratio of specific heats of an ideal gas*

$$\gamma = \frac{C_P}{C_V} = \frac{5R/2}{3R/2} = \frac{5}{3} = 1.67. \tag{12.23}$$

The ratios of the specific heats of monatomic real gases, which have no rotational or vibrational modes of motion, are very nearly 5/3.

Molecular gases depart considerably from this prediction, especially at high temperatures. This is discussed in Section 12.8.

Worked Example 12.2 Calculate the fractional change in the rms velocity of nitrogen molecules if nitrogen gas is compressed adiabatically from a pressure of one atmosphere to a pressure of 2 atmospheres. Take the ratio of specific heats γ to be constant at 1.4.

Answer During the compression, eqn (12.22) indicates that the product PV^γ is constant. At the start $P_1 V_1 = nRT_1$ where n is the number of moles of gas. Similarly, at the end of the compression $P_2 V_2 = nRT_2$. Putting these relations together, we find

$$\frac{T_2}{T_1} = \left(\frac{P_2}{P_1}\right)^{(\gamma-1)/\gamma}.$$

But

$$\overline{v_2^2}/\overline{v_1^2} = T_2/T_1,$$

and hence

$$\left(\frac{\overline{v_2^2}}{\overline{v_1^2}}\right)^{1/2} = \left(\frac{T_2}{T_1}\right)^{1/2} = \left(\frac{P_2}{P_1}\right)^{1/7}.$$

This ratio is equal to 1.10, and the fractional change in the rms speed of the molecules is 1/10.

Exercise 12.4 An ideal gas expands adiabatically and quasi-statically from pressure P_1 and temperature T_1 to pressure P_2 and temperature T_2. Determine the relationship between initial and final pressures and temperatures.

Helium, which can be taken to behave as an ideal gas, is slowly compressed adiabatically to half its volume. The initial pressure and temperature are 1 bar and 300 K, respectively. Calculate the final temperature and pressure.

Answer Equation (12.22) and the application of the ideal gas equation to initial and final states of the gas give

$$P_1^{(1-\gamma)} T_1^\gamma = P_2^{(1-\gamma)} T_2^\gamma.$$

Final pressure $= 3.18$ bar; final temperature $= 476$ K.

12.5 Heat, energy, work, and the second law of thermodynamics

Heat and work are different forms of energy. Work is converted into heat by frictional or dissipative forces, as when sticks are rubbed together to cause fire. If the sticks do not reach sufficiently high temperature to ignite, they cool down to ambient temperature by giving heat to the surroundings and after cooling are as they were before being rubbed together. This cycle can be repeated continuously and converts work into heat with very high efficiency. How easily can heat be converted into work? In this section we consider this question and briefly discuss the efficiency of engines that transform heat into work, such as steam engines in power stations, which turn heat energy into electrical energy.

The rule that the heat flow between two interacting objects is from the hot to the cold, leading to the concept of thermal equilibrium and equal temperatures, is one of the starting points of thermodynamics. The first law of thermodynamics was met in Section 12.3. A statement of the **second law of thermodynamics** is the following.

If work is to be obtained from a heat engine by removing heat from a hot reservoir, all of the heat cannot be converted into work.

Some of the heat removed from the hot store must be transferred to a cold store and only a fraction is available for work.

Another statement of the second law is that external work must be done to make heat flow from cold to hot, as in a refrigerator.

Steam engines form the basis of electricity generation in most power stations. Sometimes the steam is generated by the heat from chemical reactions as in gas-, oil-, or coal-fired power stations; sometimes the steam is generated by the heat from nuclear reactions as in nuclear reactors. A vital parameter in power generation and all engines is the efficiency with which the heat input can be converted into work output. The efficiency η may be defined as the ratio of those two quantities.

In the same way as a large reservoir of heat (the surroundings) absorbs heat generated by the work done against friction when objects are rubbed together, a large reservoir is needed to absorb heat Q_C not converted into work W when heat Q_H is input to an engine. The efficiency of the engine is then

$$\eta = \frac{W}{Q_H}. \tag{12.24}$$

Conservation of energy requires that $W = Q_H - Q_C$ and hence

$$\eta = 1 - \frac{Q_C}{Q_H}. \tag{12.25}$$

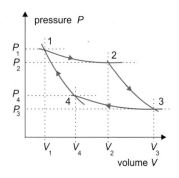

Fig. 12.10 A Carnot cycle.

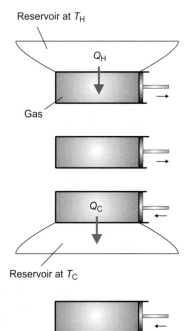

Fig. 12.11 The heat exchanges between gas and hot and cold reservoirs during a Carnot cycle.

The rejected heat Q_C can never be zero and hence the efficiency of a heat engine is always less than unity. We shall see that the efficiency depends upon the temperatures of the reservoirs that supply the input heat and accept the rejected heat, and is usually much less than unity.

The Carnot cycle

A heat engine can be made by taking a gas, which we will assume to be an ideal gas, through a cycle of reversible quasi-static expansions and compressions called a Carnot cycle. Figure 12.10 shows such a cycle. Gas at the point labelled 1 at temperature, pressure, and volume T_H, P_1, and V_1, respectively, is isothermally expanded to point 2 during which time an amount of work is done against the external pressure. The gas is then adiabatically expanded to point 3 during which more work is done at the expense of the internal energy and the temperature falls to T_C. The gas is then isothermally compressed to point 4 and adiabatically compressed back to its starting state at point 1.

The total heat input Q_H to the gas occurs over the path from 1 to 2 and comes from a large reservoir of heat at temperature T_H, and the heat rejected Q_C to the colder reservoir at T_C occurs over the path from 3 to 4. No heat exchanges occur over the adiabatic pathways. The sequence is shown in Fig 12.11.

The cycle of operation of the Carnot engine or of any heat engine is represented schematically in Fig 12.12. An amount of heat Q_H is input to the engine from a reservoir at temperature T_H and an amount Q_C is rejected to a reservoir at T_C while an amount of work W is obtained. The work W is the area of the closed part of the $P-V$ cycle on Fig 12.10 and, since the efficiency $\eta = W/Q_H$ and $W = Q_H - Q_C$, we have merely to determine Q_H and Q_C to determine η.

Over the path from 1 to 2 the heat input to the gas equals the work done against the external pressure.

$$Q_H = \int_{V_1}^{V_2} P \, dV = \int_{V_1}^{V_2} \frac{nRT_H}{V} \, dV = nRT_H \ln\left(\frac{V_2}{V_1}\right) \tag{12.26}$$

where n is the number of moles of gas and R the gas constant. Similarly, over the path from 3 to 4 the heat given out to the cold reservoir at T_C is

$$Q_C = nRT_C \ln\left(\frac{V_3}{V_4}\right). \tag{12.27}$$

The points 2 and 3 are connected by an adiabatic path as are the points 4 and 1. Hence, using eqn (12.22) and the ideal gas equation,

$$T_H V_2^{(\gamma-1)} = T_C V_3^{(\gamma-1)}$$

and

$$T_H V_1^{(\gamma-1)} = T_C V_4^{(\gamma-1)}.$$

Combination of the above equations gives

$$\frac{V_2}{V_1} = \frac{V_3}{V_4},$$

and, substituting into the ratio of eqns (12.26) and (12.27), we find

$$\eta = 1 - \frac{Q_C}{Q_H} = 1 - \frac{T_C}{T_H}. \tag{12.28}$$

It can be shown that the efficiency for the Carnot engine is the best that can be obtained for any heat engine and

Equation (12.28) gives an upper limit to the efficiency of any heat engine operating between temperatures T_H and T_C.

The efficiency depends upon the temperatures T_H and T_C and approaches unity only when the temperature of the cold reservoir approaches absolute zero. A steam engine using steam at 373 K and with the cold reservoir at 273 K has a best possible efficiency of $\eta \sim 1 - (273/373) \sim 27\%$.

Refrigeration

If the Carnot cycle is reversed, work W is done by an external agency to take heat Q_C from the cold reservoir and deposit Q_H in the hot, cooling the cold reservoir and effecting refrigeration. This mode of operation is shown schematically in Fig 12.13. The efficiency is now the heat taken from the cold system divided by the work input to extract it.

$$\eta = \frac{Q_C}{W} = \frac{Q_C}{(Q_H - Q_C)} = \frac{T_C}{(T_H - T_C)}. \tag{12.29}$$

The efficiency now drops as the temperature of the cold system falls, making it more and more difficult to cool down a system as its temperature is lowered.

12.6 The Boltzmann factor

The fact that two systems in contact reach thermal equilibrium has important consequences. The two systems may be separate parts of one large system; they may be two separate volumes of gas within a large

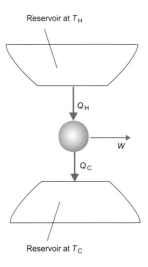

Fig. 12.12 A schematic illustration of the mechanism of the Carnot engine. Some of the heat Q_H taken from a hot reservoir is converted to work W while the rest of the heat ΔQ_C is deposited in a cold reservoir.

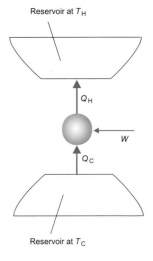

Fig. 12.13 When the Carnot engine is reversed, work W is required in order to transfer heat from the cold reservoir to the hot.

volume, or two liquids mixed together, or two separate solids placed in contact. The thermal equilibrium reached after sufficient time has elapsed means that, if we divide a system up into component parts, any part has the same temperature as the whole.

For a system in thermal equilibrium at a certain temperature, its components are distributed over available energy states to give a total internal energy E. For an isolated system the energy E is constant but it is made up of contributions from any number of subsets into which we care to subdivide the system. For an ideal gas with featureless molecules, the subsystem may be as small as one molecule of the whole. If there is thermal equilibrium, any molecule has the same *average* kinetic energy as any other. At any point in time, different molecules may have quite different speeds and as time changes the speeds of individual molecules change because of their interactions and collisions. However, the probability that a molecule has a speed within given limits is the same for all molecules with the same molecular weight. They all have the same probability that their speed lies in a given interval v to $v + dv$.

Consider a large uniform system, gas, solid, or liquid, in equilibrium at a given temperature and divide the system up into several component parts. Averaged over sufficient time, the energies per unit mass or per unit volume of each of the subdivisions have the same mean value. However, at any instant there is a distribution of the energies of the component parts about that mean. The way the energies are distributed is of fundamental importance to the study of the properties of matter and was clarified by the work of Ludwig Boltzmann in the second half of the nineteenth century.

For any component of a system in thermal equilibrium at a temperature T, the probability that the component has an internal energy in a narrow range at energy ϵ depends upon the number of energy states in that range and the probability of their occupation. The probability decreases as ϵ increases and increases as the temperature T rises. An energy is obtained by multiplying the temperature T with Boltzmann's constant k_B, and Boltzmann showed that

The probability depends exponentially on a combination of ϵ and T and is proportional to the factor

$$e^{-\epsilon/k_B T}. \tag{12.30}$$

This expression is called the **Boltzmann factor**. The temperature T occurring in the Boltzmann factor is a temperature measured on the **absolute scale of temperature**, generated by using the Boltzmann factor to deduce how the properties of physical systems change with the degree of hotness. As seen later this temperature scale is identical to the ideal gas scale.

Worked Example 12.3 A hypothetical system in thermal equilibrium at temperature T is divided into N similar component parts each of which can have one of two values for its internal energy, ϵ or 2ϵ. What is the probability $P(\epsilon)$ that a component has energy ϵ?

Answer The number of components with energy ϵ is proportional to $e^{-\epsilon/k_B T}$, and the number of components with energy 2ϵ is proportional to $e^{-2\epsilon/k_B T}$. If the constant of proportionality is α,

$$N = \alpha e^{-\epsilon/k_B T} + \alpha e^{-2\epsilon/k_B T}.$$

This equation gives the constant α and hence the number $\alpha e^{-\epsilon/k_B T}$ of subdivisions with energy ϵ. The probability that a component has energy ϵ is this number divided by the total number N and is

$$P(\epsilon) = \frac{e^{-\epsilon/k_B T}}{\left(e^{-\epsilon/k_B T} + e^{-2\epsilon/k_B T}\right)} = \frac{1}{1 + e^{-\epsilon/k_B T}}.$$

The distribution of speeds of gas molecules

Let us now be more specific and consider a system consisting of a mass of gas. If we are considering a real gas, the molecules have total energies comprising kinetic energies, potential energies due to the interactions with other molecules, and perhaps internal excitation energies arising from excitations above the normal ground state of the internal structure. However, we consider here the simple case of an ideal gas with constituents of one mass only. These constituents only have kinetic energies.

Subdivide the gas into its smallest components, the individual molecules, and consider the distribution of their kinetic energies, given by their speed distributions, when the gas is in thermal equilibrium at temperature T. The speed v is continuously distributed and is independent of a molecule's position. If the velocity lies in the range with components between v_x and dv_x, v_y and dv_y, and v_z and dv_z, each state of motion of a molecule within these infinitesimal ranges has energy $\epsilon = mv^2/2$. The probability of a molecule occupying a state with this energy is proportional to the Boltzmann factor $\exp(-\epsilon/k_B T)$.

The probability $P(v_x, v_y, v_z)\, dv_x\, dv_y\, dv_z$ that a molecule has velocity components within the ranges v_x to $v_x + dv_x$, etc. obeys the relation

$$P(v_x, v_y, v_z)\, dv_x\, dv_y\, dv_z \propto e^{-\epsilon/k_B T}\, dv_x\, dv_y\, dv_z. \tag{12.31}$$

Since

$$v^2 = v_x^2 + v_y^2 + v_z^2 \tag{12.32}$$

and $\epsilon = mv^2/2$ we find

$$P(v_x, v_y, v_z)\, \mathrm{d}v_x\, \mathrm{d}v_y\, \mathrm{d}v_z \propto e^{-mv_x^2/2k_B T} e^{-mv_y^2/2k_B T} e^{-mv_z^2/2k_B T}\, \mathrm{d}v_x\, \mathrm{d}v_y\, \mathrm{d}v_z.$$
(12.33)

If we are interested in the probability distribution $P(v_x)\mathrm{d}v_x$ of only one component, say that in the x-direction, we integrate over the variables v_y and v_z. Equation (12.33) is equivalent to

$$P(v_x, v_y, v_z)\, \mathrm{d}v_x\, \mathrm{d}v_y\, \mathrm{d}v_z = Ae^{-mv_x^2/2k_B T} e^{-mv_y^2/2k_B T} e^{-mv_z^2/2k_B T}\, \mathrm{d}v_x\, \mathrm{d}v_y\, \mathrm{d}v_z$$
(12.34)

where A is a constant. Since the probability is unity that a molecule has components somewhere in the complete ranges from infinity to minus infinity, the constant A is given by

$$A^{-1} = \int_{-\infty}^{\infty} e^{-mv_x^2/2k_B T}\, \mathrm{d}v_x \int_{-\infty}^{\infty} e^{-mv_y^2/2k_B T}\, \mathrm{d}v_y \int_{-\infty}^{\infty} e^{-mv_z^2/2k_B T}\, \mathrm{d}v_z,$$

and

$$P(v_x)\, \mathrm{d}v_x = \frac{e^{-mv_x^2/2k_B T}\, \mathrm{d}v_x \int_{-\infty}^{\infty} e^{-mv_y^2/2k_B T}\, \mathrm{d}v_y \int_{-\infty}^{\infty} e^{-mv_z^2/2k_B T}\, \mathrm{d}v_z}{\int_{-\infty}^{\infty} e^{-v_x^2/2mk_B T}\, \mathrm{d}v_x \int_{-\infty}^{\infty} e^{-mv_y^2/2k_B T}\, \mathrm{d}v_y \int_{-\infty}^{\infty} e^{-mv_z^2/2k_B T}\, \mathrm{d}v_z}$$

$$= \frac{e^{-mv_x^2/2k_B T}\, \mathrm{d}v_x}{\int_{-\infty}^{\infty} e^{-mv_x^2/2k_B T}\, \mathrm{d}v_x}.$$

The integral in this equation can be looked up to give

$$P(v_x)\, \mathrm{d}v_x = \left(\frac{m}{2k_B T}\right)^{1/2} e^{-mv_x^2/2k_B T}\, \mathrm{d}v_x.$$
(12.35)

A similar function gives the distribution of the y- and z-components of velocity. This function is called a Gaussian function centred on zero and is shown in Fig 12.14.

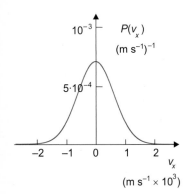

Fig. 12.14 The distribution $P(v_x)$ of the x-component v_x of velocity of sodium atoms in thermal equilibrium at a temperature of 900 K.

Exercise 12.5 Show that the full width at half height of the curve giving the distribution of velocity components equals $2\sqrt{2\ln 2(k_B T/m)}$. What is its value for helium gas at 300 K? Helium is monatomic and the mass of an atom is 4 amu.

Answer Its value for helium at 300 K is $1.86 \times 10^3\ \mathrm{m\,s^{-1}}$.

We now ask for the probability that the molecular speed is in the interval v to $v + \mathrm{d}v$. States of motion in which a molecule has a speed with a given range may be regarded as points on a three-dimensional diagram whose axes correspond to values of the three components of velocity along the x-, y-, and z-directions. The number of states with velocity components between v_x and $v + dv_x$, v_y and $v_y + dv_y$, and v_z and $v_z + dv_z$ is proportional to the volume of the element $dv_x \, dv_y \, dv_z$. The number of states in the speed interval v and $v + dv$ is thus proportional to the volume of the spherical shell between radii v and $v + dv$. This volume equals $4\pi v^2 dv$ and hence

$$P(v) \, dv \propto v^2 \, e^{-mv^2/2k_B T} \, dv. \tag{12.36}$$

The constant of proportionality can be determined using the same ideas as employed in deriving eqn (12.35). Now we have to look up the integral $\int_0^\infty v^2 e^{-v^2/2mk_B T} \, dv$. We find that the probability is given by

$$P(v) \, dv = 4 \left(\frac{m}{2k_B T} \right)^{3/2} v^2 e^{-mv^2/2k_B T} \, dv. \tag{12.37}$$

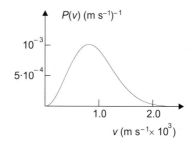

Fig. 12.15 The distribution $P(v)$ of the speed v of sodium atoms in thermal equilibrium at a temperature of 900 K.

Equation (12.37) gives the distribution of speeds of molecules in a gas at thermal equilibrium and is called the **Maxwell–Boltzmann** distribution. It is the expression we have been seeking in this section to show how molecular speeds and hence energies depend upon molecular mass and temperature. Figure 12.15 shows the speed distribution for sodium atoms in thermal equilibrium at 900 K. The most probable speed is a little less than the mean speed (see Problem 12.5) because the distribution is skewed towards high speeds.

The equivalence of the ideal gas and absolute temperature scales

Equation (12.37) can be used to calculate the average of the square of the speeds of gas molecules.

$$\overline{v^2} = \int_0^\infty P(v)v^2 \, dv = 4\pi \left(\frac{m}{2\pi k_B T} \right)^{3/2} \int_0^\infty v^4 e^{-mv^2/2k_B T} \, dv. \tag{12.38}$$

This integral is also found in standard tables and leads to the result

$$\overline{v^2} = 3 \frac{k_B T}{m}$$

or

$$\tfrac{1}{2} m\overline{v^2} = \tfrac{3}{2} k_B T. \tag{12.39}$$

Using the Boltzmann factor (12.30) to define temperature T on the absolute scale, we thus find that the average kinetic energy of a gas molecule is $3k_B T/2$. When the ideal gas scale of temperature was introduced in Section 12.2, we assumed this result in eqn (12.5) in order to proceed to eqn (12.7), which defined the gas scale. The temperature defined by the Boltzmann factor is thus *identical to the ideal gas scale temperature.* However, Boltzmann's constant is a fundamental constant and the definition of absolute temperature via the Boltzmann factor may be regarded as more fundamental than the ideal gas definition.

The absolute temperature determines the distribution of energies of component parts of a system in thermal equilibrium, and is meaningful at all temperatures. A component part may be a featureless molecule as in an ideal (hypothetical) gas; it may be a real molecule or atom with an internal structure. In all cases the Boltzmann factor together with knowledge of the states available to the component determine the average total energy of that component at different temperatures. If the states do not form a continuum but the component also has discrete states available to it, as in the case of a real molecule with discrete quantum states, the Boltzmann factor has to be applied to them individually, rather than as parts of a continuous set. This procedure is part of the subject of statistical mechanics, which we will not discuss. However, Worked example 12.4 is meant to give an indication of how discrete states are dealt with.

Absolute zero corresponds to zero probability of the total energy E of a system being other than zero. This is true in classical physics. In quantum physics the result has to be modified somewhat. We have already seen in Chapter 10 that quantum oscillators have zero-point motions and that the lowest possible states have zero-point energies. Absolute zero in quantum systems corresponds to matter existing in the lowest possible energy state.

Worked Example 12.4 Yellow light of wavelength close to 590 nm is emitted from sodium atoms in a hot sodium vapour in an otherwise evacuated glass bulb. Assume that the only discrete quantum levels of the sodium atoms that have significant occupancy at the temperature of the bulb are the ground state and a single excited level that emits the yellow light. Making the further assumption that the relative populations of these two levels are given by the Boltzmann factor, estimate the number of photons emitted per second from 1 g of sodium at a temperature of 900 K if the decay probability per unit time of the upper state is $6.25 \times 10^7 \, \text{s}^{-1}$.

Answer The molecular weight of sodium is 23 and 1 g of sodium contains $N_A/23 = 2.62 \times 10^{22}$ atoms. We require the number of these that are maintained in the upper state by the thermal equilibrium processes. The number of decays per second, and hence photons per second (assuming all decays proceed by photon emission), is given by the number of atoms in the excited state multiplied by the decay probability per unit time.

If a sodium atom is in the ground state of its electronic structure, its total energy ϵ is its kinetic energy ϵ_K plus the ground-state energy ϵ_0. If an atom is in the excited electronic state, its total energy is its kinetic energy plus ϵ_0 plus the excitation energy ϵ_{exc} of the light-emitting excited state. The excitation energy of the level which emits the yellow line is given in terms of the wavelength λ of the line by the formula $\epsilon_{exc} = hc/\lambda$, where h is Planck's constant and c is the speed of light *in vacuo*. This was discussed in the beginning of Chapter 9; eqn (9.8) gives $\epsilon_{exc} = h\nu$, and the frequency ν and wavelength λ are related via $\lambda\nu = c$. The energy ϵ_{exc} is about 2.10 eV.

Assuming thermal equilibrium, the probability that an atom is in the ground state and thus has total energy $\epsilon_K + \epsilon_0$ is proportional to the Boltzmann factor $e^{-(\epsilon_K+\epsilon_0)/k_B T}$. The probability that an atom is in the excited state and thus has total energy $\epsilon_K + \epsilon_0 + \epsilon_{exc}$ is proportional to the Boltzmann factor $e^{-(\epsilon_K+\epsilon_0+\epsilon_{exc})/k_B T}$. With only two electronic states available, the fraction in the upper state is thus equal to

$$\frac{e^{-(\epsilon_K+\epsilon_0+\epsilon_{exc})/k_B T}}{e^{-(\epsilon_K+\epsilon_0+\epsilon_{exc})/k_B T} + e^{-(\epsilon_K+\epsilon_0)/k_B T}}.$$

This, in turn, equals

$$\frac{e^{-\epsilon_{exc}/k_B T}}{e^{-\epsilon_{exc}/k_B T} + 1}.$$

The factor $e^{-\epsilon_{exc}/k_B T}$ is very small and can be neglected in the denominator of the above expression. Thus the fraction in the excited state is almost exactly equal to $e^{-\epsilon_{exc}/k_B T}$, which has the value 1.18×10^{-12}. Multiplying by the total number of sodium atoms gives 4.50×10^{10} maintained in the excited state by the thermal equilibrium mechanism. The number of photons emitted per second is this number multiplied by the decay probability per second, and is about 2.81×10^{18}.

If all of the photons emerge from the bulb, the radiated power is about $2.81 \times 10^{18} \times 3.4 \times 10^{-19} \sim 1$ watt, since the energy of a single photon is about 3.4×10^{-19} J.

12.7 Equipartition of energy

In this section we examine more closely some implications of the ideas presented in Section 12.6 about energy distributions of component parts of a system in thermal equilibrium, and deduce an important theorem relating to the average energies of the component parts.

We first apply the Boltzmann factor to an atom oscillating in a solid and treat the situation classically. We assume for simplicity that the atom is bound to a centre by a potential that approximates that of the simple harmonic oscillator discussed in Section 6.1. In thermal equilibrium the amplitude of the oscillation and the speed are not constant in time but continually vary as the atom interacts with neighbours. In this dynamic equilibrium the energy ϵ is continuously changing and the probability that the atom has energy ϵ is proportional to the Boltzmann factor $e^{-\epsilon/k_B T}$. We must combine this with the probability that the velocity of the atom has components in the ranges v_x to $v_x + dv_x$, v_y to $v_y + dv_y$, and v_z to $v_z + dv_z$, and with the probability that the position of the atom lies between x and $x + dx$, y and $y + dy$, and z and $z + dz$. This gives

$$P(x, y, z, v_x, v_y, v_z)\, dx\, dy\, dz\, dv_x\, dv_y\, dv_z \propto e^{-\epsilon/k_B T}\, dx\, dy\, dz\, dv_x\, dv_y\, dv_z.$$
(12.40)

The energy ϵ of the oscillator is

$$\epsilon = \tfrac{1}{2} m v^2 + U(r)$$
(12.41)

where m is the associated mass and $U(r)$ is the potential energy due to nearest neighbour interactions when the displacement is r. The simple harmonic oscillation is three-dimensional and, if the force constants have the same value k along x-, y-, and z-directions,

$$U(r) = \tfrac{1}{2} k x^2 + \tfrac{1}{2} k y^2 + \tfrac{1}{2} k z^2,$$
(12.42)

since $r^2 = x^2 + y^2 + z^2$. Substituting eqns (12.41) and (12.42) into (12.40)

$$P(x, y, z, v_x, v_y, v_z)\, dx\, dy\, dz\, dv_x\, dv_y\, dv_z$$
$$\propto e^{-kx^2/2k_B T} e^{-ky^2/2k_B T} e^{-kz^2/2k_B T}\, dx\, dy\, dz$$
$$\times e^{-mv_x^2/2k_B T} e^{-mv_y^2/2k_B T} e^{-mv_z^2/2k_B T}\, dv_x\, dv_y\, dv_z.$$
(12.43)

The constant of proportionality may be evaluated in the same way as in the discussion following eqn (12.34). When this is done the right-hand side of eqn (12.43) is the product of six parts, each of which involves one variable only. For example, that part associated with the variable x is

$$\frac{e^{-k_x x^2/2k_B T}\, dx}{\int_{-\infty}^{\infty} e^{-k_x x^2/2k_B T}\, dx},$$

and the probability that the variable x lies between x and $x + dx$, whatever the values of the other five variables, is given by

$$P(x)\, dx = \frac{e^{-k_x x^2/2k_B T}\, dx}{\int_{-\infty}^{\infty} e^{-k_x x^2/2k_B T}\, dx} = \sqrt{\frac{k_x}{2\pi k_B T}}\, e^{-k_x x^2/2k_B T}\, dx.$$
(12.44)

The average potential energy associated with the variable x is given by

$$\int_{-\infty}^{\infty} \frac{1}{2} k_x x^2 \times P(x)\, dx = \sqrt{\frac{k_x}{2\pi k_B T}} \int_{-\infty}^{\infty} \frac{1}{2} k_x x^2 e^{-k_x x^2/2k_B T}\, dx.$$

When this is worked out by looking up the value of the definite integral, in a way similar to that used for eqn (12.31), it is found that the average energy associated with the variable x is equal to $\frac{1}{2}k_B T$,

$$\frac{1}{2}k_x \overline{x^2} = \frac{1}{2}k_B T. \tag{12.45}$$

Exercise 12.6 Since

$$v^2 = v_x^2 + v_y^2 + v_z^2$$

and the velocity components are independent of each other,

$$\overline{v^2} = \overline{v_x^2} + \overline{v_y^2} + \overline{v_z^2}.$$

Use the relevant distributions to verify this.

Degrees of freedom

The average energies associated with the other five variables, y, z, v_x, v_y, and v_z, of the motion of the oscillating atom discussed above can be determined in a similar fashion to the derivation of eqn (12.45). When this is done, the same result, $\frac{1}{2}k_B T$, is obtained in each case.

Whenever the total energy of a particle in thermal equilibrium with its surroundings can be written as the sum of terms, each of which has the form of a constant multiplying the square of a variable α_i, there is an average contribution of $\frac{1}{2}k_B T$ to the average total energy from each of the separate terms. The variables α_i may be components of a position vector, or of a momentum (or velocity) vector, or any other variables on which the energy of a particle or molecule may depend. This result is known as the theorem of **equipartition of energy** and each variable α_i is called a **degree of freedom**. Associated with each degree of freedom is an average energy $\frac{1}{2}k_B T$.

The derivation of this result depends on the assumption that the allowed states are so close together that they form a continuous set, as well as on the assumption that the total energy of the system can be written as a sum of terms involving the squares of variables. The first condition requires that the phenomena being discussed are adequately described classically.

The equipartition theorem is equally as applicable to the translational degrees of freedom of subsystems of macroscopic size as it is to microscopic subsystems such as molecules. Examples of macroscopic subsystems to which the theorem can usefully be applied are small particles in air or in liquids, or an object suspended in air by a thread. The fluctuating motion of tiny particles suspended in a liquid is called **Brownian motion**, and equipartition indicates that the average kinetic energy of such particles is equal to $\frac{3}{2}k_B T$ corresponding to the three degrees of freedom associated with translational motion in three dimensions.

Exercise 12.7 Particles of mass 3.1×10^{-14} g suspended in a liquid at 300 K have a root mean square speed of $2.0 \, \text{cm s}^{-1}$. Calculate Avogadro's number N_A. Take the gas constant R to be $8.31 \, \text{J mol}^{-1} \, \text{K}^{-1}$.

Answer The data together with the theorem of equipartition give Boltzmann's constant equal to $1.38 \times 10^{-23} \, \text{J K}^{-1}$. Avogadro's number $N_A = R/k_B = 6.02 \times 10^{23}$ molecules per mole.

12.8 Thermal physics and discrete energies

In this section we apply the results of the previous section to examine how the existence of discrete quantum states within the microscopic systems that constitute a macroscopic object modifies the classical predictions. We discuss the energies of real gases over a range of temperature, and thereby highlight the role played by molecular excitations in their specific heats. We also discuss the distribution among frequencies exhibited by the radiation emitted by hot bodies.

Specific heats of real gases

Let us apply equipartition of energy to the molecules of a gas. We assume the pressure is low so that forces between the molecules can be neglected and the gas behaves as an ideal gas. We further assume that the molecules are featureless and have no internal structure. The total energy of a molecule is then the sum of three independent parts corresponding to components of velocity v_x, v_y, and v_z, and each part has the form $mv_x^2/2$. The average total energy of a molecule is thus $3k_B T/2$ and for one mole the total energy depends on temperature alone and is

$$E = \tfrac{3}{2} N_A k_B T$$

as given in eqn (12.11).

The molar specific heat at constant volume is

$$C_V = \frac{\mathrm{d}Q_V}{\mathrm{d}T} = \frac{\mathrm{d}E}{\mathrm{d}T} = \frac{3}{2} k_B T = \frac{3}{2} R. \tag{12.46}$$

Equation (12.16) then gives $C_P = 5R/2$, and the ratio $C_P/C_V = \gamma$ of the specific heats of an ideal gas is equal to 5/3, the same results as obtained earlier in Section 12.4.

We now turn to real gases, whose atoms or molecules are not featureless and which also interact with one another. The atoms and molecules have an internal structure of electrons orbiting nuclei and have discrete excited states of different internal structure lying above the ground states. The probability of a state i at internal energy ϵ_i being occupied is proportional to the Boltzmann factor $e^{-\epsilon_i/k_B T}$. Here, ϵ_i is equal to the ground state energy plus the excitation energy of state i above ground. The internal excitations of a molecule in thermal equilibrium are independent of its translational motion and hence the probability of occupation of state i is given by

$$P_i = \frac{e^{-\epsilon_i/k_B T}}{\sum_i e^{-\epsilon_i/k_B T}} \tag{12.47}$$

where the sum is taken over all the states of internal excitation of the molecule.

If the energy of the lowest excited state above the ground state is large compared with $k_B T$, eqn (12.47) shows that the probability of occupation of any state other than the ground state is very small. When we take account of the total energy E_{int} of the internal structures, the total energy of a mole of gas is given by

$$E = \tfrac{3}{2} N_A k_B T + E_{int}. \tag{12.48}$$

However, if the temperature is such that all molecules are in their ground states, E_{int} does not vary with a small temperature change and the specific heats are unaffected by considerations of internal structure. The ratio γ is equal to 5/3 only if this condition holds.

Now consider a gas of diatomic molecules. Chapter 11 discusses the discrete quantum levels associated with the rotations of these molecules. Rotational levels typically begin at excitation energies above the ground state of a few hundredths of an electron volt and have spacings of about the same amount. If temperatures are such that $k_B T$ is much larger than a few hundredths of an eV (at room temperatures, $k_B T \sim 1/40\,\text{eV}$), these levels have significant occupation probabilities. The occupation probabilities change with temperature, and the average energy of the internal structure changes nonlinearly with temperature. The consequence is that the specific heat is temperature-dependent.

When the temperature is sufficiently high so that $k_B T$ is much greater than a few hundredths of an eV, the rotational levels may be taken to be continuously distributed in energy and the classical approximation made for the translation is now applicable to rotation. The energy in the rotation of a classical dumbell can be expressed as the sum of three terms, with the different terms involving separately the square of an angular speed about three perpendicular axes. However, rotations about the line joining the nuclei of a diatomic molecule are not allowed in quantum mechanics because of the special symmetry of the microscopic dumbell. A diatomic molecule can rotate only about the two axes perpendicular to the line joining the nuclei shown on Fig 12.16. Hence only two degrees of freedom are associated with the rotations of a diatomic molecule.

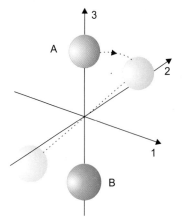

Fig. 12.16 A diatomic molecule can have internal energy in rotations about axes 1 and 2 shown, but symmetry reasons prevent any observable effects due to rotation about axis 3 joining the two nuclei A and B. The figure is schematic. The sizes of the nuclei are very much smaller than the separation drawn.

The moments of inertia for the two rotations perpendicular to the line joining the nuclei have the same value I. The rotational energy can thus be written as a sum of two terms with rotational angular frequencies ω_1 and ω_2, as

$$E_{\text{rot}} = \tfrac{1}{2}I\omega_1^2 + \tfrac{1}{2}I\omega_2^2. \tag{12.49}$$

At temperatures such that $k_B T$ is much greater than the spacings of the rotational levels, the molecules are distributed classically over the rotational motions and each term on the right-hand side of eqn (12.49) has the form required for it to contribute $\tfrac{1}{2}k_B T$ to the average energy. The total energy of a mole of the diatomic gas is then given by

$$E = \tfrac{5}{2}N_A k_B T, \tag{12.50}$$

and the specific heat at constant volume becomes

$$C_V = \tfrac{5}{2}R \tag{12.51}$$

with the ratio of specific heats

$$\gamma = \frac{C_P}{C_V} = \frac{7}{5} = 1.40. \tag{12.52}$$

It is very useful to remember that at room temperatures of $T \sim 300$ K the value of $k_B T \sim 1/40$ eV. At room temperatures we expect some rotational levels to be occupied but perhaps insufficient for the classical assumption to be a good approximation, whereas at $\sim 600{-}700$ K the classical assumption will give good predictions for the specific heats.

Worked Example 12.5 A diatomic nitrogen molecule may be taken to have two nitrogen nuclei each of mass 14 amu separated by 0.3 nm. Such a molecule can rotate about axes that go through the centre of the molecule and are perpendicular to the line joining the nuclei. At 300 K what is the root mean square angular frequency? Assume the spacing of the rotational levels is sufficiently small for the classical approximation to be valid.

Answer The average energy associated with the rotational motion of the molecule about either of the allowed axes on Fig 12.16 is $I\overline{\omega^2}/2$, where I is the moment of inertia and $\overline{\omega^2}$ the average of the square of the rotational angular frequency. The theorem of equipartition tells us that this energy equals $k_B T/2$, with k_B Boltzmann's constant and T the temperature. Hence,

$$I\overline{\omega^2}/2 = k_B T/2.$$

This gives the rms angular frequency

$$\overline{\omega^2}^{1/2} = \sqrt{\frac{k_B T}{I}}.$$

To a good approximation we may take the nuclei to be point masses m and the moment of inertia is

$$I = 2ma^2$$

where $2a$ is the separation of the nuclei. Substitution of the numbers gives $\overline{\omega^2}^{1/2} = 1.99 \times 10^{12}$ radians s^{-1}.

At temperatures higher than a few hundred kelvins a second set of quantum levels available to a diatomic molecule may come into play. The nuclei of a diatomic molecule may vibrate along the line joining their centres. The excitation energies and spacings of these quantum levels are typically a few tenths of an eV. The energy of the vibrations can be written as the sum of two parts, one depending on the square of the displacement, the other on the square of the speed. The vibrations at high enough temperature thus have two degrees of freedom and the energy of a mole of gas becomes

$$E = \tfrac{7}{2}N_A k_B T \tag{12.53}$$

with $C_V = 7R/2$ and

$$\gamma = \tfrac{9}{7} = 1.29. \tag{12.54}$$

The plot of C_V or $C_P = (C_V + R)$ against temperature for a gas that remained gaseous and diatomic over a wide temperature range is thus expected to be somewhat similar to that shown in Fig 12.17. There would be regions over which the specific heat changes little, joined by smooth, more rapidly rising portions where degrees of freedom associated with higher excitations begin to contribute significantly. In reality the various effects overlap to a great extent.

Only hydrogen, of the diatomic molecular gases, is gaseous at the low temperatures where rotations do not contribute. Only for hydrogen and noble gases that are atomic are specific heats observed corresponding to only three degrees of freedom. At the high-temperature end where vibrations would contribute an amount $k_B T$, most diatomic gases have already begun to dissociate into separate atoms. Figure 12.18 shows measured values of the specific heat at constant pressure for hydrogen. The shape of the curve shows the features anticipated and illustrates the validity of the quantal and thermal descriptions of gas behaviour. At room temperatures the contribution to C_P from rotations has already set in to give $C_P \sim (C_V + R) \sim 7R/2$, corresponding to five degrees of freedom. As the temperature rises the curve slopes upwards as vibrations begin to contribute to the internal energy, but near 2500 K the gas has already begun to dissociate.

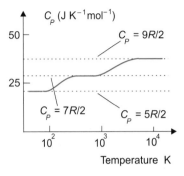

Fig. 12.17 The variation with temperature of the specific heat at constant pressure of a hypothetical gas of diatomic molecules. All real diatomic gases except hydrogen liquefy at temperatures below about 80 K and all dissociate at temperatures well below 10 000 K, so that all aspects of the above behaviour cannot actually be observed. Note the logarithmic temperature scale.

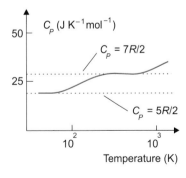

Fig. 12.18 Measured values of the specific heat at constant pressure for hydrogen. Note the logarithmic temperature scale.

Black-body radiation

If a solid surface is maintained at a high temperature it will emit radiation. Different materials have distinct appearances when they are cold—coal, silver, and pottery cannot be confused with one another. However, when they are red-hot, all materials look much the same. The radiation from hot bodies has a wide range of frequencies, and is most intense at a frequency that increases as the temperature increases. For example, a piece of metal at about 1000°C emits mostly infrared and red radiation—it is red-hot. The filament of a light bulb, at nearly 3000°C, emits much more high-frequency blue light, and the broad spread of frequencies makes the light appear white. The variation of intensity with frequency is called the *spectrum* of the radiation: the spectrum of radiation emitted by a suface thus depends on its temperature.

A surface is described as ideally black if it absorbs *all* the radiation that falls on it. When in equilibrium it will emit the same amount of energy as it absorbs within any given fequency interval. If a closed box has ideally black walls, the spectrum of the equilibrium radiation in the box is the same as the spectrum of radiation emitted by an ideal black surface. This equilibrium radiation is therefore called **black-body radiation**.

Although discussed above in terms of ideally black walls, it can be shown that the spectrum of the equilibrium radiation in a box depends *only* on the temperature and not at all on the material of which the box is made. Black-body radiation can therefore be observed by looking through a small hole into a cavity maintained at a constant temperature. The black-body spectrum is shown in Fig 12.19 for a temperature of 3000°C, roughly the temperature of a light-bulb filament.

Classical theory cannot describe black-body radiation; it predicts that the energy emitted should increase steadily with frequency rather than pass through a maximum value as shown in Fig 12.19.

In the introduction to this chapter and in Chapter 14 it is explained that the movement of atoms in a solid is equivalent to a great number of simple harmonic oscillators acting independently. The model used to describe the radiation in a cavity is one in which the cavity supports oscillations rather like a solid box has oscillating atoms. The cavity oscillators acquire energy as a result of thermal motion. Classically, the allowed energies of the oscillators are distributed continuously. The number of oscillators within a given frequency range about a mean frequency can be calculated and according to the equipartition theorem each oscillator has an average energy $k_B T/2$. The number of cavity oscillations per unit frequency interval increases as ν increases, and thus the average energy per unit frequency interval is predicted to increase steadily with frequency and the black-body spectrum is predicted to rise as the frequency increases.

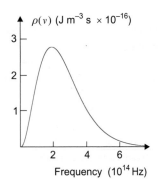

Fig. 12.19 The spectrum of black-body radiation for an object at a temperature of 3000°C. The figure shows the energy density per unit frequency interval $\rho(\nu)$ against frequency. The highest density of radiation is at frequencies near 2×10^{14} Hz.

Quantum mechanics allows an oscillator of frequency ν only to have an energy $nh\nu$ where n is an integer and h is Planck's constant. The probability of an oscillator having an energy $nh\nu$ is proportional to the Boltzmann factor $\exp(-nh\nu/k_B T)$. This enables us to calculate the average energy of an oscillator of frequency ν. The average decreases more rapidly with ν than the number of oscillations per unit frequency interval increases, and the radiated power is now predicted to fall off after reaching a maximum at a frequency dependent on temperature.

The **energy density** of black-body radiation is the amount of radiation energy per unit volume in the box. The energy density, calculated as outlined in the above paragraph, has a definite frequency variation, and we can define an energy density per unit frequency interval $\rho(\nu)$: the energy density at a frequency ν and in the frequency interval $d\nu$ is $\rho(\nu)\,d\nu$. For a black-body at a temperature T,

$$\rho(\nu)\,d\nu = \frac{8\pi\nu^2}{c^3}\,\frac{h\nu}{e^{h\nu/k_B T}-1}\,d\nu. \tag{12.55}$$

This result is called Planck's formula. It accounts perfectly for the observed black-body spectrum illustrated in Fig 12.19 for a temperature of 3000°C. The temperature in Planck's formula must be expressed in *kelvins*, the SI units of temperature.

Worked Example 12.6 Estimate the energy density of radiation at the surface of the Sun, in the frequency range between 4.5 and 5.5×10^{14} Hz. Assume that the temperature of the Sun's surface is about 6000 K.

Answer The average frequency in the range concerned is 5×10^{14} Hz, corresponding to yellow light. At this frequency $h\nu = 3.3 \times 10^{-19}$ J and at $T = 6000$ K, $k_B T = 8.3 \times 10^{-20}$ J, leading to $\exp(h\nu/k_B T) = 54$. Approximate by assuming that $\rho(\nu)$ has a constant value corresponding to this frequency over the whole frequency range $\Delta\nu = 10^{14}$ Hz. In eqn (12.55), the energy density is then

$$\rho(\nu)\Delta\nu = \frac{8\pi \times 25 \times 10^{28}}{27 \times 10^{24}} \times \frac{3.3 \times 10^{-19}}{53} \times 10^{14} \approx 0.15\,\mathrm{J\,m^{-3}}.$$

The total energy density in black-body radiation is obtained by integrating eqn (12.55) over all frequencies from zero to infinity. The result of the integration is

$$\rho_{\text{tot}} = \frac{8\pi^5 k_B^4}{15 h^3 c^3}\,T^4.$$

For $T = 3000°C$, ρ_{tot} is about 0.04 joules per cubic metre. This may seem a very small amount, but of course the radiation is moving at a speed c, so the power emitted by a surface at this temperature is high—only a very small area is needed for the filament in a light bulb. It turns out that the power per unit area J radiated in all directions from a surface at a temperature T is

$$J = \rho_{tot} \times \frac{c}{4} = \frac{2\pi^5 k_B^4}{15 h^3 c^2} T^4 = \sigma T^4. \tag{12.56}$$

Equation (12.56) is **Stefan's law**, and σ is called **Stefan's constant**. The numerical value of Stefan's constant is 5.67×10^{-8} W m^{-2} K^{-4}.

Exercise 12.8 Calculate the power radiated from a 1 mm^2 surface at a temperature of 3000°C.

Answer 6.5 W.

Problems

Level 1

12.1 Helium gas, which is monatomic, is initially at 310 K. It expands slowly and adiabatically to three times its initial volume. Calculate the temperature of the gas after the expansion.

12.2 A Carnot engine is reversed and external work W is done to pump heat from a cold reservoir at temperature T_C to a hot reservoir at T_H. Determine the efficiency of the heat pump.

12.3 Show that the most probable speed of a gas molecule in thermal equilibrium at temperature T is given by $\sqrt{2RT/M}$, where R is the gas constant and M is the mass of one mole. Calculate this speed for oxygen and nitrogen at 300 K.

12.4 At what temperature is the average kinetic energy per atom of neon gas equal to the kinetic energy of a singly charged neon ion that has been accelerated through a potential difference of 5 volts?

12.5 Find the ratio between the most probable speed of a gas molecule and its mean speed, and between the root mean square speed of a gas molecule and its mean speed.

12.6 Show that the mean speed of molecules in a gas at thermal equilibrium at temperature T is $\sqrt{8k_B T/\pi m}$, where m is the mass of a gas molecule. You may use the result that $\int_0^\infty x^3 e^{-\alpha x^2} dx = 1/2\alpha^2$.

12.7 The vibrational levels of carbon monoxide are spaced by 0.27 eV. Would you expect vibrational excitations to make a significant contribution to the specific heat at room temperature?

12.8 The latent heat of vaporization of carbon tetrachloride is 3.2×10^4 J per mole. Estimate the binding energy of a molecule in the liquid state.

12.9 What specific heats are predicted at room temperature by the equipartition theorem for the linear triatomic molecule CO_2 and the nonlinear molecule H_2O?

12.10 A tall building has a height of 300 m. The atmospheric pressure at ground level is 1 bar. If the building is at a uniform temperature of 280 K, estimate the air pressure on the top floor. Take the molecular weight of an air molecule to be 29 and the acceleration due to gravity g to be 9.8 m s^{-2}.

12.11 The specific heat of air at room temperature and pressure is about 20 J K^{-1} mol^{-1}. If a room of floor area

30 m^2 and height 2.5 m is well insulated against heat loss to the outside and to the walls, how long will it take a 2 kW heater to raise the room temperature by 10 K?

12.12 The temperature at the surface of the Sun is about 6000 K. What is the power emitted from 1 m^2 of the Sun's surface?

Level 2

12.13 The coefficient of linear expansion of a metal rod is defined as the change in length per unit length per kelvin temperature change. Most materials expand on heating. The cross-sectional area of a metal rod whose coefficient of linear expansion is $1.2 \times 10^{-6}\,\text{K}^{-1}$ is 500 mm^2. If the temperature changes from 300 K to 350 K what force is required to prevent the rod expanding? Young's modulus for the metal (which was discussed in Section 8.1) is $1.7 \times 10^{11}\,\text{N}\,\text{m}^{-2}$.

12.14 The specific heat per mole of a certain gas at constant pressure varies with temperature according to the equation

$$C = \alpha + \beta T - \frac{\gamma}{T^2}.$$

How much heat is required to change the temperature of n moles of the gas from T_1 to T_2 at constant pressure?

12.15 Sodium is heated to vapour at 900 K in a glass bulb. Yellow light is emitted by atoms that have been thermally excited to a level at an excitation energy of 2.1 eV. If the atoms were at rest the emission line would have a negligible spread in frequency. Since the atoms are moving in random directions with a distribution of speeds the emission line is Doppler broadened. (The Doppler effect was discussed in Section 8.6.) Calculate the full width at half maximum of the line as measured by a stationary observer.

12.16 The electrical resistance R of a platinum wire is used as the basis of a scale of temperature by assuming a linear variation of resistance with temperature over the range from the freezing point of water to the boiling point, these points being assigned the same temperatures of 273 and 373 degrees respectively as on the absolute scale. The actual resistance of the wire varies with absolute temperature T as

$$R(T) = R_0(1 + \alpha T - \beta T^2)$$

where the coefficients α and β are equal to $3.8 \times 10^{-3}\,\text{K}^{-1}$ and $5.6 \times 10^{-7}\,\text{K}^{-2}$, respectively. What is the

absolute temperature in kelvins when the temperature given by the platinum thermometer is 300 degrees?

12.17 A tube of length L and volume V has a uniform cross-section and contains an ideal monatomic gas whose pressure P is constant along the length of the tube. The temperature of the gas varies linearly with position along the length of the tube and T_1 and T_2 are the temperatures at each end. Show that in the steady state the total number of molecules in the tube is given by

$$N = \frac{PV}{k_{\text{B}}(T_2 - T_1)} \log_{\text{e}}\left(\frac{T_2}{T_1}\right)$$

where k_{B} is Boltzmann's constant.

12.18 Mercury of density $13.6 \times 10^3\,\text{kg}\,\text{m}^{-3}$ is poured into a U-shaped tube closed at one end as shown in Fig 12.20. The dimensions of the tube are as shown. Estimate the volume of mercury that can be poured into the tube before it overflows. Assume air behaves as a perfect gas.

12.19 Why is it better to use the expansion of a gas rather than the expansion of a liquid to define a scale of temperature?

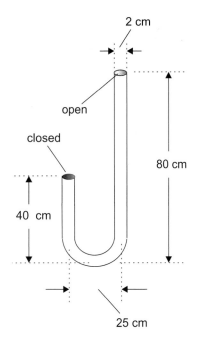

Fig. 12.20 The U-shaped tube in Problem 12.18.

If a gas at the temperature of melting ice is heated at constant pressure to the temperature of boiling water, the volume increases by a certain percentage amount. If the gas is heated over the same temperature range at constant volume the pressure increases by the same percentage amount. Why?

12.20 Particles of mass m are in thermal equilibrium at temperature T in the Earth's gravitational field. Assuming that the potential energy of a particle at height z is mgz, derive expressions for the average height and average energy of a particle.

12.21 A system in thermal equilibrium at temperature T consists of N particles that have two energy states separated by an energy ϵ.

(a) Write down an expression for the number of particles in each state.

(b) Obtain the internal energy of the sytem.

(c) Show that at high temperatures the heat capacity varies as the inverse of the square of the temperature.

12.22 The first excited state of a quantum mechanical system lies at an excitation energy that is much greater than $k_B T$, where T is the temperature and k_B is Boltzmann's constant. Show that the specific heat of the system is much less than k_B.

12.23 By making the approximation that $h\nu/k_B T$ is small in the Planck radiation formula (eqn (9.9)), find the classical expression for the radiation density at frequency ν. By what factor does the classical expression overestimate the radiation density at the frequency for which $h\nu/k_B T = 1$?

Level 3

12.24 The work done on a length of elastic material when it is stretched by an amount dL is $F\,dL$, where F is the tension in the material. A piece of elastic of length L is under a constant tension F. Show that its heat capacity is given by

$$c_F = \left(\frac{\partial E}{\partial T}\right)_F - FL\alpha$$

where

$$\alpha = \frac{1}{L}\left(\frac{\partial L}{\partial T}\right)_F$$

is the coefficient of linear expansion and E is the internal energy.

12.25 The length of a piece of elastic when a tension F is applied is

$$F = KT\left(\frac{L}{L_0} - \frac{L_0^2}{L^2}\right)$$

where K is a constant and L_0 is the length of the elastic at zero tension. The unstretched length is a function of temperature T only. Calculate the work needed to slowly compress the elastic at constant temperature from $L = L_0$ to $L = L_0/2$.

12.26 Show that the isothermal Young's modulus for the piece of elastic of the last problem is given by

$$Y = \frac{F}{S} + \frac{3KTL_0^2}{SL^2}$$

where S is the cross-section of the sample. Show also that the coefficient of linear expansion at tension F is given by

$$\alpha = \alpha_0 - \frac{F}{SYT}$$

where Y is the isothermal Young's modulus at tension F and temperature T, and α_0 is the coefficient of expansion at zero tension.

12.27 The force between the two nuclei in an HCl molecule is given by the potential

$$V(r) = V_0(\exp[-2a(r - r_0)] - 2\exp[-a(r - r_0)])$$

where r is the separation of the H and Cl nuclei, $V_0 = 4.6\,\text{eV}$, $a = 1.9 \times 10^8\,\text{cm}^{-1}$, and $r_0 = 0.1275\,\text{nm}$. Estimate the temperatures at which contributions to the specific heat of gaseous hydrogen chloride become appreciable from the rotational and the vibrational degreees of freedom. The masses of the hydrogen and chlorine nuclei may be taken to be 1 and 37 amu, respectively. (You will need to refer back to Section 10.6 for the vibrational levels of a diatomic molecule and Section 11.5 for the rotational levels to do this problem.)

12.28 Particles of radii $5 \times 10^{-7}\,\text{m}$ and density $1100\,\text{kg}\,\text{m}^{-3}$ are placed in water of density $1000\,\text{kg}\,\text{m}^{-3}$ maintained at a temperature of 300 K and allowed to reach thermal equilibrium. The density of particles in two layers $12 \times 10^{-6}\,\text{m}$ apart is observed to be 220 and 950 per unit volume. Use these numbers to obtain an estimate of Boltzmann's constant.

12.29 A container of height 25 cm holds water at a temperature of 330 K in which are mixed small, spherical, solid particles of density $1.15 \times 10^3 \text{ kg m}^{-3}$. The difference in the volume densities of the particles at the top and bottom of the container is less than 10%. Estimate the maximum diameter of the particles.

12.30 An early method for measuring the ratio of the specific heats of a gas with reasonable accuracy was devised by Rüchardt in 1929. A container is sealed by a steel ball that fits closely in a tube as in Fig 12.21 but can slide up and down with little friction. The ball is given a small displacement z, and executes simple harmonic motion with period τ about its equilibrium position at $z = 0$. The restoring force is provided by the influence of small adiabatic changes in pressure of the gas in the vessel. If V and P are the equilibrium volume of the container and the pressure of the gas, respectively, and S and m are the cross-sectional area of the tube and the mass of the gas molecules, respectively, show that the ratio of specific heats is

$$\gamma = \frac{4\pi^2 mV}{S^2 P\tau^2}.$$

12.31 Sodium atoms at 900 K are contained in an oven at a pressure of 5 Pa. In one wall of the oven there is a

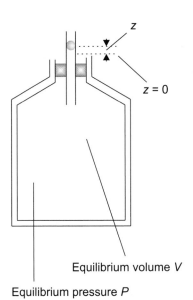

Equilibrium volume V

Equilibrium pressure P

Fig. 12.21 An outline diagram of the apparatus used by Rüchardt to measure the ratio of the specific heat of gases.

narrow slit of size 0.1 cm by 0.1 mm. Outside the oven the pressure is maintained at a very low level by intensive pumping. Because of this we may assume that all atoms hitting the opening pass through. What is the number of atoms emerging per second?

12.32 A thin tube closed at each end is filled with a dilute solution in water of identical molecules with a very large molecular weight. The tube is attached to the flat horizontal surface of a disc with one end of the tube at the disc's centre. The disc rotates at an angular speed ω about a vertical axis passing through its centre. Show that the density of the macromolecules as a function of distance r from the centre of the disc is given by

$$\rho(r) = \rho_0 \exp(r^2/r_0^2)$$

where ρ_0 and r_0 are constants. Estimate r_0 if the mass of each macromolecule is 10^4 times the proton mass, the molecules have a density of $1.2 \times 10^3 \text{ kg m}^{-3}$, the angular speed is 500 rad s^{-1}, and the temperature is 600 K.

12.33 Radio astronomers observe a background of microwave radiation that is visible from the Earth in all directions and that corresponds closely to black-body radiation at a temperature of 3 K. This background radiation is thought to be a remnant of the Big Bang. Calculate the total energy of this radiation in a sphere of radius 100 000 light years, roughly the size of our own galaxy together with its halo. How long would it take the Sun to emit this amount of energy? (The radius of the Sun is 7×10^8 m and its surface temperature is 6000 K.)

12.34 The probability that any one molecule has velocity components between v_x, $v_x + \mathrm{d}v_x$; v_y, $v_y + \mathrm{d}v_y$; and v_z, $v_z + \mathrm{d}v_z$, and a second molecule has velocity components between v_x, $v_x + \mathrm{d}v_x$; v_y, $v_y + \mathrm{d}v_y$; and v_z, $v_z + \mathrm{d}v_z$ is the product of two factors similar to eqn (12.34). By changing variables to those corresponding to relative velocity components and components of the velocity of the centre of mass of the two molecules, show that the mean relative speed of molecules in a gas is equal to $\sqrt{2}$ times the mean speed.

12.35 Estimate the number of molecules in one cubic centimetre of a gas at normal temperature and pressure that have a speed greater than five times the average speed. You will need to estimate the value of the integral involved, either by numerical or graphical means.

Some solutions and answers to Chapter 12 problems

12.4 Neon is a monatomic gas and its behaviour approximates that of an ideal gas. The average kinetic energy of a neon atom at temperature T is thus $\bar{E} \approx \frac{3}{2} k_B T$, as given by eqn (12.11). The energy of a singly charged neon ion after acceleration through 5 volts is 5 electron volts $= 5 \times 1.6 \times 10^{-19}$ J. Hence the required temperature is given by $\frac{3}{2} k_B T = 8 \times 10^{-19}$. The temperature T is 3.86×10^4 K.

12.7 Vibrational excitations make a significant contribution to the specific heats of gases at a given temperature T if vibrational levels are strongly excited at that temperaure. The Boltzmann factor indicates that, if the lowest vibrational energy is much greater than $k_B T$, there is negligible excitation of vibrational levels. At room temperature $k_B T \sim 1/40$ eV. This is much lower than the vibrational spacing given, and hence there is negligible excitation of vibrational levels and negligible contribution from vibrations to the specific heat at room temperature.

12.9 A linear triatomic molecule has three translational degrees of freedom and can rotate independently about only two axes, in a fashion similar to the rotation of a diatomic molecule. It can vibrate but at room temperature the vibrational excitations are usually at an excitation energy too high to be excited and thus do not contribute to the specific heats. The specific heat at constant volume C_V is thus $5R/2$ as in eqn (12.51), and C_P is $7R/2$.

A nonlinear molecule with three atoms, like the water molecule, can rotate about three independent axes and hence has three rotational degrees of freedom. At room temperature, vibrations again do not usually contribute and C_V is equal to $3R$, C_P equal to $4R$.

12.13 5100 N.

12.14
$$C = \alpha + \beta T - \frac{\gamma}{T^2}.$$

The heat capacity of n moles is n times the above. The heat required to raise the temperature from T_1 to T_2 at constant P is

$$\Delta Q = n \int_{T_1}^{T_2} \left(\alpha + \beta T - \frac{\gamma}{T^2} \right) dT = n \left[\alpha T + \frac{1}{2} \beta T^2 + \frac{\gamma}{T} \right]_{T_1}^{T_2}$$

$$= n\alpha \Delta T + \frac{n}{2} \beta \Delta T (T_1 + T_2) - \frac{n\gamma \Delta T}{T_1 T_2}$$

where

$$\Delta T = (T_2 - T_1).$$

12.16 299.7 K.

12.20 $\bar{z} = k_B T/mg$; $\bar{E} = k_B T/2$.

12.21 (a) We may put the energy of the lower state equal to zero since the level populations depend only on the difference in energy. The probability of he level at energy ϵ being occupied is then given by the Boltzmann factor as

$$P(\epsilon) = A e^{-\epsilon/k_B T}$$

where A is a constant. The probability of the ground state being occupied is

$$P(0) = A.$$

The number of particles in the upper level is thus

$$N(\epsilon) = N \frac{e^{-\epsilon/k_B T}}{(1 + e^{-\epsilon/k_B T})}$$

and the number in the ground state is

$$N(0) = N \frac{1}{(1 + e^{-\epsilon/k_B T})}.$$

(b) The total energy E is the number in the upper state times ϵ.

$$E = N\epsilon \frac{e^{-\epsilon/k_B T}}{(1 + e^{-\epsilon/k_B T})}.$$

(c) The heat capacity is the derivative of E with respect to T. We first take the derivative and then in it substitute for the factor $e^{-\epsilon/k_B T}$ the first two terms in the expansion

$$e^{-\epsilon/k_B T} = 1 - \epsilon/k_B T + \cdots.$$

Next, multiply out and retain only the terms that are important at very high T. This gives the heat capacity as $N\epsilon^2/2k_B T^2$.

12.23 The classical expression overestimates the radiation density by a factor 1.72.

12.24 Conservation of energy, expressed in eqn (12.12), tells us that, when a force F stretches a string by an amount dL, doing work $F\,dL$ on the string, the heat input dQ and the change in internal energy dE are related by

$$dQ = dE - F\,dL.$$

Thus

$$c_F = \frac{dQ_F}{dT} = \left(\frac{\partial E}{\partial T}\right)_F - F\left(\frac{\partial L}{\partial T}\right)_F.$$

If

$$\alpha = \frac{1}{L}\left(\frac{\partial L}{\partial T}\right)_F,$$

we have

$$c_F = \left(\frac{\partial E}{\partial T}\right)_F - FL\alpha.$$

12.29 The diameter of the particles is less than 1.4×10^{-8} m.

12.30 In equilibrium the pressure inside and outside the container is the same and the ball is at rest. When the ball has moved up a distance z from its equilibrium position, the gas in the container has expanded adiabatically by an amount $\Delta V = zS$ and the pressure inside is less than that outside by an amount ΔP. There is thus a restoring force on the ball given by $F = \Delta PS$. During the adiabatic expansion PV^γ is constant, and hence

$$V^\gamma \Delta P + \gamma PV^{\gamma-1}\Delta V = 0.$$

This gives

$$\Delta P = -\frac{\gamma P}{V}\Delta V$$

and

$$F = -\frac{\gamma P}{V}S^2\Delta z.$$

The motion of the ball is thus simple harmonic with period

$$\tau = 2\pi\sqrt{\frac{mV}{\gamma PS^2}},$$

and from this the ratio of specific heats is

$$\gamma = \frac{4\pi^2}{\tau^2}\frac{mV}{PS^2}.$$

12.33 The total energy of the microwave background radiation in the galactic halo is about 2×10^{50} J. Although the density of radiation is extremely small, the total is vast. If the Sun were to continue to emit radiation at its present rate, it would take 1.5×10^{15} years to emit this amount of energy.

Chapter 13

Gases and liquids

This chapter discusses some properties of liquids and gases and their description in terms of molecular motions and intermolecular interactions.

This chapter considers some properties of liquids and gases. In both of these fluid phases of matter the microscopic constituents do not oscillate about fixed positions, as in solids, but move around within the vessel which contains them.

In the liquid phase the intermolecular forces are strong enough to keep molecules very close together so that they form a coherent mass. Liquid in a container takes up the shape of the vessel and occupies the lower part because of the gravitational attraction of the Earth on the liquid mass. If a liquid is in a container open to the atmosphere, at the liquid surface the pressure equals atmospheric pressure. However, as the depth below the liquid level increases, the pressure increases above atmospheric pressure by an amount depending on the depth, the density of the liquid, and the acceleration due to gravity g.

Gas alone in a container fills all the vessel and the gas atoms or molecules can be found anywhere therein. The forces between the gas atoms or molecules are on average no longer strong enough to prevent the molecules acting individually. They move at random and, except for very large volumes of gas, such as the atmosphere, can be regarded as unaffected by gravity.

In thermal equilibrium a fixed mass of material can be described in terms of its absolute temperature T, the pressure P it exerts on the walls of a container or on its surroundings, and its volume V. The thermodynamic equilibrium state of the material is defined by any pair of these three variables, since the three are related by an *equation of state*. Different materials and phases each have an equation of state, although it is usually not known. For a gas whose properties approximate those of an ideal gas, however, P, V, and T are related by the ideal gas equation (12.7).

This chapter discusses some properties of bulk matter in the form of liquid or gas. The different behaviours of liquids compared to those of gases are conditioned by the large difference between the average molecular separations. The ease with which they allow the transport through them of objects or of heat, or allow equilibration of density after a density disturbance, are important properties of fluids. Over limited ranges of pressure and temperature the **transport properties** of fluids can usually be characterized by a parameter appropriate to the phenomenon under discussion. For example, the rate at which molecules diffuse throughout the volume occupied by a liquid or gas is given by a diffusivity parameter, and the rate at which heat is conducted is given by a parameter called the thermal conductivity.

In this chapter we determine the transport parameters using appropriate assumptions about the processes being described. One assumption used is that the gas or liquid is in thermal equilibrium. This is not strictly correct. If there is a net transport of molecules or heat from one portion of a gas or liquid to another, the system is clearly not in thermal equilibrium, since transport of material is effected by bulk motion within the fluid and transport of heat requires different parts of an object or fluid to be at different temperatures. Nevertheless, we will assume that the phenomena under discussion are taking place so gradually that results strictly applicable to equilibrium states also apply at all times everywhere in the system.

If the fluid motion is rapid when transport is underway, the assumption of equilibrium is clearly not a good one. The motion may become chaotic, and we have **turbulence** in which the behaviour is unpredictable, as in the patterns made by smoke rising from a fire. Turbulence and its onset are two of the outstanding problems in physics. Although much effort is now directed towards these problems, the nonlinear equations that govern turbulent phenomena are very difficult to interpret and the resulting chaotic phenomena still have no good description.

13.1 **Phase diagrams**

Water may exist in three different phases—solid, liquid, and gas. If water, or any pure substance in thermal equilibrium is contained in a vessel from which all air has been removed, the water may exist in a single phase, in two phases, or under very special circumstances in all three. The relative fractions of the solid, liquid, and gas phases depend upon pressure, volume, and temperature. The manner in which a fixed mass of pure material is distributed between phases as a function of pressure, temperature, and volume may be presented on a **phase diagram** showing corresponding values of temperature and pressure.

Figure 13.1 shows a phase diagram for a typical substance. Consider what happens along the line of constant temperature T_1, shown dotted. At low pressures there is only vapour in the vessel containing the material. The vapour or gas is called an unsaturated vapour. We will use the words gas and vapour interchangeably for the dilute phase of matter in which the molecules are on average far apart and move more or less independently of each other. As the pressure is increased the point A is reached on the line AB. At that pressure solid is in equilibrium with gas, and as the pressure is further increased the substance becomes all solid. The line AB drawn is the line along which gas and solid can exist in equilibrium at different temperatures.

At a particular higher temperature T_{TP}, as the pressure is increased, the point TP is reached, called the **triple point**, first met in Section 12.2, where solid, liquid, and gas coexist together. At a temperature T_2 above the triple point temperature, along the next dotted line shown, at low

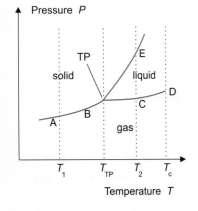

Fig. 13.1 The phase diagram for a typical substance, showing pressure against temperature. The full lines are boundaries between different phases.

pressures there is again unsaturated vapour. As the pressure is increased, the vapour becomes saturated at the point C and gas molecules condense into liquid. The line TP−C−D is the line along which gas and liquid can exist in equilibrium at different temperatures and gives the curve of saturated vapour pressure against temperature. As the pressure is increased further at constant T_2 the point E is reached at which the liquid and solid exist together. More pressure gives all solid.

At a particular temperature T_c called the **critical temperature**, the liquid can no longer exist and the substance is gas or, at sufficiently high pressure, is solid. Above the critical temperature the substance cannot be liquefied by any amount of pressure increase. Common gases such as oxygen or nitrogen have lower values of T_c than substances that are liquid or solid at room temperature. Nitrogen cannot be liquefied at any pressure when its temperature is above $T_c = 126\,\mathrm{K}$, but the critical temperature of sulfur is 1314 K.

Isotherms

Figure 13.2 is another diagram relating to the phase changes of a typical pure substance. The solid green lines on the diagram show the variation of pressure with volume, at different constant temperatures, for a given mass of material. The lines are called **isotherms**, and are lines showing corresponding values of P and V at different constant temperatures such as those drawn dotted on Fig 13.1.

The isotherm corresponding to the temperature T_2 shown on Fig 13.1 is shown on Fig 13.2 as the line ABCD. At the point A the molecules are all in the gas phase in the unsaturated vapour. If the volume is decreased, the vapour pressure increases and the point B is reached at which the vapour becomes saturated and the molecules begin to liquefy. Along the line BC, as the volume continues to be reduced, the pressure stays constant at the saturation vapour pressure appropriate to the temperature T_2. At the point C, the molecules have all turned to liquid. As the volume decreases further, the line CD is followed by a substance that contracts as the pressure increases. This line is very steep since large pressures are required to compress liquids.

If we look at other isotherms on Fig 13.2, as the temperature is increased the region of volume over which the liquid phase exists in equilibrium with the gas phase becomes smaller. The locus of points like B where the vapour starts to coexist with liquid is shown dotted and defines a boundary between all gas and liquid plus gas. This dotted line again gives the saturated vapour pressure as a function of temperature. The locus of points like C, on the dotted line to the left, gives points where the molecules are all in the liquid phase and defines a boundary between all liquid and liquid in equilibrium with gas. The two boundaries

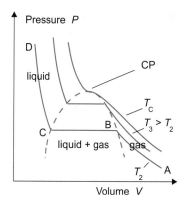

Fig. 13.2 The $P−V$ diagram for a typical pure substance. The full lines are isotherms.

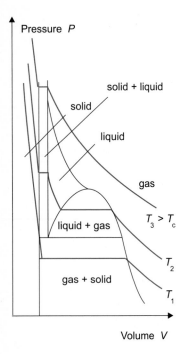

Fig. 13.3 An extended $P-V$ diagram for a typical pure substance. The black lines show phase boundaries. The coloured lines are isotherms.

● *The differences between real gases and an ideal gas*

meet at the **critical point** CP on the isotherm corresponding to a temperature T_c called the critical temperature. Above the critical temperature there is no region where the liquid exists in equilibrium with its vapour; no increase whatever in pressure will cause liquefaction. The isotherm corresponding to the critical temperature is called the **critical isotherm** and at the point CP the pressure is P_c and the volume is V_c.

Figure 13.2 shows isotherms only over pressures and volumes where liquid and gas phases exist or coexist. For normal temperatures and pressures most pure substances exist in the solid phase and substances that are liquids at normal T and P can be solidified by the application of great pressure. If high pressure is applied, there is equilibrium between the solid and liquid along particular isotherms in the $P-V$ diagram. The changes along isotherms in an extended $P-V$ diagram are shown in Fig. 13.3. Where they are the same, the temperature labels on the isotherms correspond to the same temperatures as those on Figs 13.1 and 13.2.

13.2 **Real gases**

The behaviour of real gases only approximates that of an ideal gas at low pressures where the average separation of the atoms or molecules is large. Modifications to the ideal gas eqn (12.7) derived in Section 12.2 must be made in view of two obvious differences between a real gas and an ideal gas.

The first difference is that atoms and molecules have nonzero volumes; the radius of an atom is about 0.1 nm, while gas molecules with more than one atom can have dimensions several times this. The volume V_0 in the gas equation may therefore be replaced by the term $(V_0 - b)$ where b takes account of the fact that part of the molar volume is unavailable because of the nonzero size of the molecules themselves.

The second effect ignored in the ideal gas approximation is the influence of the force between the constituents of the gas. The general features of the force may be obtained from the interatomic potential curve of Fig 13.4. The force is repulsive when atoms or molecules are closer than the separation r_0 since it is very difficult to compress the orbits of the electrons. However, the force is attractive for separations greater than r_0 and the average effect for molecules in a gas is an attraction that reduces the observed gas pressure. The reduction depends upon the average separation of the atoms or molecules and thus increases as the volume decreases; it varies as the inverse of some power of the volume. To obtain a modified equation of state, the observed pressure P should thus have added to it an amount equivalent to the reduction arising from the intermolecular forces.

Exercise 13.1 A gram molecule (the mass of N_A, Avogadro's number, of molecules) of gas at standard ambient temperature and pressure occupies about 25×10^3 cm^3. What fraction of the molar volume is taken up by the molecules themselves if the molecular radius is 3×10^{-8} cm?

Answer About 2.7×10^{-3}.

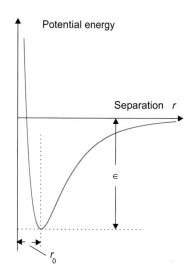

Fig. 13.4 The potential energy of a pair of molecules as a function of their separation r.

The van der Waals equation

A commonly used equation of state for one mole of a real gas and to some extent its liquid is called the **van der Waals equation** and has the form

$$\left(P + \frac{a}{V^2}\right)(V - b) = RT \tag{13.1}$$

with a and b constants depending on the gas. The constant a depends upon the intermolecular force and is thus related to the depth ϵ and the radius r_0 of the intermolecular potential shown in Fig 13.4. The constant b depends upon the volume unavailable because of the nonzero molecular size and the effect of the interaction between molecules at short distances. Molecules cannot approach too closely because their interaction becomes repulsive before touching. The volume unavailable is thus somewhat larger than the total volume of the molecules because the restricted space around any particular molecule is larger than the molecular volume itself.

The van der Waals equation may be rewritten in the form

$$PV^3 - V^2(bP + RT) + aV - ab = 0. \tag{13.2}$$

This is a cubic equation and can be solved for the volume V at given P and T. Over a certain range of temperature and pressure there will be three roots of this equation corresponding to three different volumes. How can this be interpreted in terms of a realistic isotherm such as the isotherm at temperaure T_1 shown in Fig 13.5? This isotherm includes the straight section at constant pressure P_1 which corresponds to the interval between all liquid and all vapour. The dotted curve shown is the isotherm given by eqn (13.2). The points B, C, and D correspond to the three solutions to the van der Waals equation for the volume at the pressure P_1 and temperature T_1. The dotted line and the isotherm may be reconciled to some extent by regarding the area enclosed below the isotherm as corresponding to superheated liquid and the area above as corresponding to supercooled vapour.

As the temperature is raised to the critical point, the three roots converge and become equal at the critical temperature T_c where the distinction between liquid and vapour disappears. There we have $V = V_c$

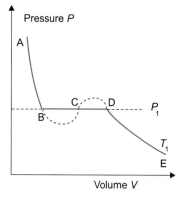

Fig. 13.5 The full line on the pressure–volume diagram is an isotherm for a substance that is liquid over the range AB, liquid and vapour over BD, and is vapour over DE. The dotted line is the isotherm given by the van der Waals equation.

and

$$(V - V_c)^3 = 0 = V^3 - 3V^2 V_c + 3VV_c^2 - V_c^3.$$

Multiplying by P_c gives the new equation

$$P_c V^3 - 3P_c V^2 V_c + 3P_c VV_c^2 - P_c V_c^3 = 0. \qquad (13.3)$$

● The van der Waals equation at the critical point

This equation and eqn (13.2) are equivalent, and comparing coefficients we find

$$V_c = 3b, \qquad (13.4)$$

$$P_c = \frac{a}{27b^2}, \qquad (13.5)$$

and

$$T_c = \frac{8a}{27Rb}. \qquad (13.6)$$

The prediction of the van der Waals equation for the ratio $RT_c/P_c V_c$ is independent of the parameters a and b, which depend upon the gas. Using the above equations, the predicted ratio is equal to 8/3. For many real gases the value of the ratio is indeed not very different from this prediction. However, there are significant variations, since for molecular gases the van der Waals equation is often rather poorly obeyed near the critical temperature. Table 13.1 gives relevant data for some molecular substances.

At temperatures higher than the critical temperature, the cubic eqn (13.2) has only one acceptable solution for the volume at a given temperature and pressure, and gives predictions for isotherms corresponding to states in which the molecules are all in the gas phase.

Table 13.1 Values of the critical constants and the ratio $RT_c/P_c V_c$ for one mole of various gases. In order, the gases for which data are shown are nitrogen, oxygen, hydrogen, chlorine, nitrous oxide, and acetic acid. The notation (Pa×10^5), here and for similar examples elsewhere, means that the number in the table has to be multiplied by 10^5 to give the appropriate value.

Gas	T_c (K)	P_c (Pa×10^5)	V_c (m^3×10^{-6})	$RT_c/P_c V_c$
N_2	126	33.5	90.0	3.40
O_2	155	50.1	78.0	3.29
H_2	33	12.8	64.5	3.32
Cl_2	417	76.1	64.6	7.04
N_2O	310	71.7	95.9	3.74
$C_2H_4O_2$	595	57.1	171.0	5.06

The van der Waals parameters

Estimates of the size of the parameter a in the van der Waals equation can be made in terms of the parameters ϵ and r_0 of the schematic intermolecular potential of Fig 13.4. The reduction in pressure on the walls of a vessel containing gas is due to the attractive intermolecular force. This tends to bunch molecules together and so reduces the frequency with which they hit the walls. The parameter a thus depends upon the depth of the potential and the volume within which one molecule feels the attraction of another.

The dimensions of a are those of [energy] times [volume], since a/V^2 is a pressure. The parameter giving the influence of one molecule on another may be expressed as

$$a \propto \epsilon \frac{4\pi r_0^3}{3},$$

since the volume they occupy in near contact is proportional to the volume $4\pi r_0^3/3$ of a molecule and the energy of interaction is ϵ. For the N_A molecules in a mole there are $N_A(N_A - 1)/2 \sim N_A^2/2$ pairs and we put

$$a \sim \tfrac{1}{2} N_A^2 \epsilon \frac{4\pi r_0^3}{3} \tag{13.7}$$

or, equivalently,

$$a \sim \tfrac{1}{2} N_A \epsilon k b \tag{13.8}$$

where k is a factor of order unity that relates the volume $N_A(4\pi r_0^3/3)$ to the volume b used in the gas equation.

In Section 12.1 we derived the relation (12.1) between the latent heat of evaporation per mole, L_{evap}, the coordination number, n_ℓ, and the binding energy of an atom in the liquid. The latter is approximately equal to the depth of the potential well, ϵ, and we put

$$L_{evap} = \tfrac{1}{2} N_A n_\ell \epsilon.$$

Comparing this with eqn (13.8) we obtain the prediction

$$\frac{a}{b} \sim \left(\frac{k}{n_\ell}\right) L_{evap}. \tag{13.9}$$

This prediction should be roughly in agreement with observation if our proposal (13.7) for the value of a is adequate. Typically, the coordination number n_ℓ is about 8 and the factor k is greater than one. Putting the ratio k/n_ℓ equal to unity gives the prediction

$$\frac{a}{b} \sim L_{evap}. \tag{13.10}$$

The parameters a and b are determined by fitting the variation with pressure and temperature of the volume of the gas to the van der Waals

Table 13.2 Values of the constants *a* and *b* in the van der Waals equation for one mole of various gases together with the latent heat of evaporation

Gas	L_{evap} (J)	a (J m^3)	b (m^{3}10^{-4})	a/b J
N_2	5 600	0.14	0.39	3 550
O_2	10 900	0.14	0.32	4 280
H_2	900	0.024	0.27	920
Cl_2	18 400	0.65	0.56	11 550
N_2O	16 600	0.38	0.44	8 550
$C_2H_2O_2$	23 800	1.76	1.07	16 500

equation. Table 13.2 shows values of these parameters, their ratio, and the latent heat of evaporation of various gases. There is reasonable agreement with the prediction (13.10) considering the approximate nature of our discussion.

Specific heats

In contrast to an ideal gas, the total energy of a real gas contains a term arising from the total potential energy of the molecules in addition to the kinetic energy term. The potential energy depends upon the average separation. At constant volume the average separation is constant and hence the potential energy does not change as a small amount of heat is added while the volume remains the same. The heat input goes solely to increase the total kinetic energy and the specific heat at constant volume per mole of a gas, C_V, is the same as that of an ideal gas.

If heat is added while the pressure remains constant, external work is done against the pressure but in addition the total potential energy changes since the volume of the gas, and hence the average separation of the molecules, changes. The specific heat at constant pressure must include both these energy changes. We will work out the difference $C_P - C_V$ in the specific heats for a gas that obeys the van der Waals equation for comparison with an ideal gas.

Following the argument made above that gave eqns (13.7) and (13.8), we may write the potential energy in terms of the parameter *a* of the equation of state. We make the approximation that the average potential energy between a pair of molecules is roughly ϵ times the ratio of the volume of interaction between them to the total volume. There is an average attractive interaction and the average potential energy is negative. For $N_A^2/2$ pairs, the average potential energy per mole is

● *The difference in the specific heats of a real gas*

$$U = -\frac{1}{2} N_A^2 \epsilon \frac{4\pi r_0^3}{3} \frac{1}{V}. \tag{13.11}$$

Using eqn (13.7) this becomes

$$U = -\frac{a}{V}. \tag{13.12}$$

The total energy E of one mole of a real gas is the sum of the kinetic energy $nRT/2$, where n is the number of degrees of freedom, and the potential energy U. Hence,

$$E = \frac{1}{2}nRT - \frac{a}{V}. \tag{13.13}$$

If an amount of heat dQ is input to one mole of the gas at constant pressure, from eqn (12.12)

$$dQ = dE + PdV,$$

and we have

$$C_P dT = C_V dT + \frac{a}{V^2}dV + PdV. \tag{13.14}$$

For a gas obeying the van der Waals equation, dV and dT are related at constant pressure by

$$\left(P - \frac{a}{V^2} + \frac{2ab}{V^3}\right)dV = RdT.$$

The term $2ab/V^3$ may be neglected compared with $(P - a/V^2)$, and substituting for dV in eqn (13.14) gives

$$C_P - C_V = R\left(\frac{P + a/V^2}{P - a/V^2}\right). \tag{13.15}$$

The term a/V^2 is small compared with the pressure P and $P - a/V^2$ may be expanded using the binomial theorem. Putting $PV = RT$ in the result gives

$$C_P - C_V \approx R\left(1 + \frac{2a}{RVT}\right) \tag{13.16}$$

to sufficient accuracy.

The difference in specific heats is close to that of an ideal gas at normal temperatures and pressures. For example, at a temperature of 300 K and atmospheric pressure, using the value of the parameter a for nitrogen from Table 13.2 gives $C_P - C_V \sim (R + 0.05)\,\mathrm{J\,K^{-1}}$, which is different from R by less than 1 per cent. However, at low temperatures and high pressures, the difference increases.

Sound travelling through air consists of local density variations of very small masses of gas, as described in Section 8.1. The speed of sound, given by eqn (8.9), depends upon the average density ρ and the compressibility χ defined by eqn (8.7). The compressibility of a gas depends upon the equation of state of the gas and on whether the compressions or rarefactions are isothermal or adiabatic. Under most conditions the

● *Compressions in sound waves*

density variations in sound waves in air take place almost adiabatically, since the speed at which heat can pass into and out of the small masses of air undergoing volume changes is less than the speed at which the changes occur.

● *Adiabatic compressibility*

The adiabatic compressibility of a gas can be deduced in terms of the ratio γ of the specific heats C_P and C_V of the gas. For an infinitesimal volume change dV of one mole of gas at pressure P, the heat input dQ is related to the change in internal energy dE and the work done against the pressure, PdV, by eqn (12.12)

$$dQ = dE + PdV.$$

Hence, for adiabatic changes,

$$C_V dT + PdV = 0. \tag{13.17}$$

Let us for the moment consider an ideal gas for which $PV = RT$, and thus

$$PdV + VdP = RdT.$$

Substituting the value of dT from this expression into eqn (13.17) gives

$$C_V(PdV + VdP) + RPdV = 0$$

and thus

$$C_V PdV + C_V VdP + (C_P - C_V)PdV = 0$$

where we have used the relation (12.17) for the difference between the specific heats of an ideal gas to substitute for the gas constant R. This equation can be used directly to give

$$\left(\frac{\partial V}{\partial P}\right)_{\text{adiabatic}} = -\frac{V}{\gamma P} \tag{13.18}$$

and hence the required compressibility

$$\chi = -\frac{1}{V}\left(\frac{\partial V}{\partial P}\right)_{\text{adiabatic}} = \frac{1}{\gamma P}. \tag{13.19}$$

For real gases at normal temperatures and pressures, as we have seen above, the ratio γ differs little from the value predicted using $C_P - C_V = R$ for the difference in specific heats and the value of C_V given by the appropriate number of degrees of freedom. Under normal conditions the ratio for air would thus be close to 7/5, using $C_V = 5R/2$ (corresponding to three translational degrees of freedom for the diatomic molecules nitrogen and oxygen plus two degrees of freedom associated with rotations, as discussed in Section 12.8) and the relation $C_P - C_V = R$. The compressibility predicted at atmospheric pressure is thus $\chi = 5/7P \approx 7 \times 10^{-6}\,\text{m}^2\,\text{N}^{-1}$. This is very close to the value $6.7 \times 10^{-6}\,\text{m}^2\,\text{N}^{-1}$ obtained in Section 8.1 using the observed speed of sound in air at sea level of $340 \, \mathrm{m \, s^{-1}}$ and the density at sea level of $1.29 \, \mathrm{kg \, m^{-3}}$.

Worked Example 13.1 Estimate the factor by which the adiabatic compressibility of chlorine gas at 1 atmosphere pressure and temperature 300 K is different to the adiabatic compressibility at the same temperature but at a pressure of 10 atmospheres.

Answer Equation (13.19) gives the compressibilites after the ratio of specific heats γ has been determined at the two different pressures. The ratios are estimated by using eqn (13.16) to deduce the difference $C_P - C_V$ and the equipartition theorem to deduce C_V.

Equation (13.16) is an approximation and in evaluating the correction term $2a/RVT$ it is good enough to assume that $PV = RT$. Taking the value of the parameter a for chlorine from Table 13.2, eqn (13.16) gives $C_P - C_V \approx 1.21R$ at 10 atmospheres. The molar specific heat at constant volume C_V we take to be $5R/2$, corresponding to the three translational plus two rotational degrees of freedom available to the chlorine molecules. This gives γ at 10 atmospheres as 1.48. The same procedure gives γ at 1 atmosphere as about 1.41. Equation (13.19) then shows that the required ratio of the compressibilities is close to

$$\frac{\chi(1 \text{ atmosphere})}{\chi(10 \text{ atmospheres})} = 10 \frac{\gamma(10)}{\gamma(1)} \approx \frac{14.8}{1.41} = 10.5.$$

13.3 Viscosity

Viscosity is the first example of a transport phenomenon that we discuss. Viscosity is related to the transfer of momentum through a fluid and results in resistance to the passage of objects through the fluid. It requires more force to move an object at a given speed through oil than it does through water. The viscosity of oil at normal temperatures is greater than that of water.

Viscosity is very similar to friction in its origin. Adjacent layers of molecules, which would *on average* be at rest were it not for the object moving through the fluid, resist sliding over each other in an ordered way. If we imagine a flat plate being drawn through a liquid, the molecules next to the plate have an average speed the same as that of the plate. However, before turbulence sets in, in the gentle regime of **streamline flow** in which the velocity gradient is smoothly varying, there is laminar flow in which successive layers more distant from the plate move at slower and slower drift speeds in the direction the plate is moving. The layer in contact with the stationary wall of the vessel containing the fluid is on average at rest, like the wall, and the average molecular drift speed in the direction of movement of the plate varies across the fluid.

Let the average molecular drift speed at perpendicular distance x from the plate be $u(x)$ and the speed gradient be du/dx. For many fluids the force F on a layer of area S at distance x is proportional to the area and the gradient. We may write

$$F = \eta S \frac{du}{dx} \tag{13.20}$$

where the constant of proportionality η is called the **coefficient of viscosity** of the fluid. It has dimensions of [mass] times [length]$^{-1}$ times [time]$^{-1}$, and in the SI system its units are $kg\,m^{-1}\,s^{-1}$. The coefficient of viscosity of water at a temperature of 293 K is about $10^{-3}\,kg\,m^{-1}\,s^{-1}$. Equation (13.20) is obeyed by **Newtonian fluids**, for which the viscosity depends only on temperature. For some fluids the viscosity depends upon the velocity gradient and the shapes and sizes of the restrictions to the fluid flow. For example, blood flows with lower viscosity through narrow veins than measurements of its flow in wider tubes suggest.

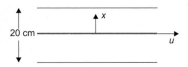

Fig. 13.6 A view looking down on a thin flat plate moving within a fluid contained in a very deep and long box-shaped vessel. The plate is half-way between the walls, parallel to them, and moving with speed u in a direction parallel to the walls.

Exercise 13.2 A long thin flat plate is moving at a steady speed of $1\,m\,s^{-1}$ in the centre of a channel containing water, as shown in Fig 13.6. The depth of the channel is much greater than its width of 20 cm. There is streamline flow of the water along the channel. Determine the viscous force per unit area on the plate, neglecting edge effects. The coefficient of viscosity of the water is $10^{-3}\,kg\,m^{-1}\,s^{-1}$.

Note that the plate has two sides. The water molecules at the surfaces of the plates have an average drift speed u in the direction of motion of the plate. The average drift speed varies with perpendicular distance x from the plate. However, the derivative with respect to x, du/dx, of the average drift speed is constant and equal to $10\,s^{-1}$ on each side of the plate. The derivative is constant because the acceleration of any layer of water is zero in streamline flow. Hence, the force on any layer due to molecules arriving from a layer on one side must be balanced by the force due to those arriving from the other side. Hence the force on one side of a layer is the same for a layer at any distance x. Equation (13.20) then tells us that the derivative with respect to x of the average drift speed is constant.

Answer 0.02 Nm^{-2}.

Consider a cylinder of radius a suspended inside a second coaxial cylinder of larger radius b. Let there be a Newtonian fluid in the space between the cylinders and let the outer cylinder rotate at constant angular speed ω_b, as shown on Fig 13.7. The viscous forces generated by the fluid act on both cylinders, resisting the rotation of the outer cylinder and twisting the suspension of the inner until the restoring torque balances the viscous torque, leaving the inner cylinder at rest.

For low rotational speeds the flow is streamline and molecules in infinitesimally thin cylindrical layers at distance r from the axis of the two cylinders move at steady drift speeds $\omega(r)r$. The layers of fluid molecules rotate in the same sense as the outer cylinder, with the layer in contact with the outer cylinder moving at speed $\omega_b b$ and the layer in contact with the inner cylinder at rest.

To determine the viscous force on a layer we need the derivative with respect to r of the mean drift speed. The argument used in Exercise 13.2 leading to uniform speed gradient is not applicable to the circular motion of the cylindrical layers. The gradient du/dr of the mean drift speed u is

$$\frac{du}{dr} = \frac{d(\omega r)}{dr} = \omega + r\frac{d\omega}{dr}.$$

The first term is the usual term arising from rotation at angular speed ω and is not relevant to the viscous retarding force. The second term arises from layers slipping over one another and is the gradient of the mean drift speed to be used in eqn (13.20) for the viscous force.

The viscous retarding force on a layer at distance r from the axis is everywhere tangential to the surface of the layer. As explained in the discussion of rotation about a fixed axis in Section 4.3, the torque about the axis is the tangential force times the perpendicular distance from the axis. Thus the viscous torque about the axis on the layer over its whole area $2\pi r\ell$ is

$$\tau(r) = r \times (2\pi r\ell) \times \eta \times \left(r\frac{d\omega}{dr}\right)$$

where ℓ is the length of the cylinders. Since the fluid between the cylinders is in a steady state, τ is the same at any distance r and the above relation must hold at all values of r. Integrating between a and b we have

$$\int_a^b \tau\frac{dr}{r^3} = \int_a^b 2\pi\eta\ell d\omega.$$

This leads to

$$\tau = \frac{a^2 b^2}{b^2 - a^2}\, 4\pi\eta\ell\omega_b. \tag{13.21}$$

For constant speed of rotation of the outer cylinder, the suspension of the inner cylinder twists through an angle θ such that the restoring torque it provides is equal and opposite to τ. The angle θ can be measured to give the

● *Drag on rotating cylinder*

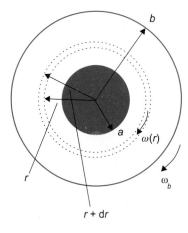

Fig. 13.7 A view looking down on a cylinder of radius b rotating at angular velocity ω outside a coaxial cylinder of radius a. The rotation is opposed by a viscous torque from the fluid between the cylinders.

coefficient of viscosity. Equation (13.21) ignores the torque over the bottom of the outer cylinder. The influence of this and other end effects can be eliminated by repeating the experiment for two different values ℓ_1 and ℓ_2 of the length of the rotating cylinder. The difference in the measured restraining torques can then be attributed to a length $\ell_1 - \ell_2$ of the cylinder.

Exercise 13.3 The inner cylinder of Fig 13.7 has radius 4 cm. The outer cylinder has radius 5 cm and is rotating at angular speed $\omega = 3$ radian s^{-1}. Both cylinders have length 20 cm and the space between them is filled with a liquid that has coefficient of viscosity 3×10^{-3} kg m^{-1} s^{-1}. The inner cylinder is supported by a fibre that has a restoring couple per unit twist of 10^{-3} N m per radian. Through what angle is the inner cylinder twisted from its position when the outer cylinder is at rest? The flow is streamline and you should neglect end effects.

Answer 5.8 degrees.

The viscosity of a liquid usually decreases with temperature: water viscosity decreases by a factor of six as the temperature is raised from that of ice to the boiling point. Sometimes the coefficient of viscosity we have defined above is called the absolute viscosity. The density of a fluid usually changes with temperature, and the coefficient of **kinematic viscosity**, ν, is defined by $\nu = \eta/\rho$, where ρ is the density. The kinematic viscosity is important because it is used in the calculation of the dimensionless **Reynolds number** associated with the flow. This number is given by uL/ν, where u is the speed of flow and L is a length characteristic of the flow situation, such as the diameter of a pipe down which fluid is moving. As the Reynolds number is increased, the flow remains laminar until a certain critical value of the Reynolds number is reached, when the flow becomes turbulent. The critical value varies with the particular example of fluid flow but is typically above several hundred.

The value of the viscosity and its variation with temperature give information on molecular structure. Liquids with molecules having large intermolecular attractive forces have large viscosities. The viscosity at a temperature of 293 K of the organic molecule carbon tetrachloride is lower than that of water. There is a stronger intermolecular force between water molecules than between the organic molecules.

The flow of liquids or gases through pipes or tubes has important practical applications. For non-turbulent and streamline flow, the rate at which fluid emerges from a tube of length ℓ and radius a when a pressure difference ΔP is applied between the tube ends is determined by the fluid viscosity. For most liquids the rate is proportional to $\Delta P a^4/\eta\ell$ if the tube

● *The flow of liquids through pipes*

is very long compared with its diameter. Typically, the ratio of length to diameter must be greater than one or two hundred. For gases, since they are much more compressible than liquids, the rate is given by different formulae depending upon the gas pressure inside the tube as well on the applied pressure difference between the ends.

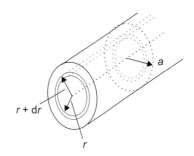

Fig. 13.8 Part of a long tube down which fluid flows uniformly. The length of the tube is very much greater than the radius a. The rate of fluid flow, when this is streamline, is calculated opposite.

Worked Example 13.2 Show that for a Newtonian liquid flowing smoothly and steadily through a tube of length ℓ and radius a, the volume of liquid emerging from the tube per second is equal to $\pi \Delta P a^4 / 8\eta\ell$, where η is the coefficient of viscosity and ΔP the uniform pressure difference between the ends of the tube.

Answer It is instructive to derive the flow by thinking in terms of forces on each side of layers within the liquid. The force per unit area on a layer at distance r from the centre of the tube shown in Fig 13.8 is given by eqn (13.20) as

$$\frac{F}{S} = \eta \left(\frac{du}{dr} \right)$$

with S the area of the layer, equal to $2\pi r\ell$. Hence

$$F(r) = 2\pi r\ell \times \eta \left(\frac{du}{dr} \right).$$

The net viscous retarding force on the annular layer shown in Fig 13.7 between r and $r + dr$ is thus $dF = F(r + dr) - F(r)$.

$$dF = \left\{ 2\pi\ell\eta \left(\frac{du}{dr} \right) + 2\pi r\ell\eta \left(\frac{d^2 u}{dr^2} \right) \right\} dr.$$

For steady flow this force is balanced by the force $\Delta P 2\pi r dr$ due to the pressure difference between the ends of the tube. Hence,

$$2\pi r \Delta P = 2\pi\eta\ell \left(\frac{du}{dr} \right) + 2\pi r\ell\eta \left(\frac{d^2 u}{dr^2} \right).$$

This is a differential equation whose solution gives the speed of layers at distance r. The solution satisfying the boundary conditions of the problem is

$$u = \frac{\Delta P}{4\eta\ell} (a^2 - r^2)$$

as can readily be verified by substitution.

The volume of liquid emerging per second from the fluid between r and $r + dr$ is thus

$$2\pi r dr\, u,$$

and the total volume per second is

$$V = \frac{\pi \, \Delta P}{2\eta \ell} \int_0^a r(a^2 - r^2) \, dr,$$

giving

$$V = \frac{\pi \Delta P a^4}{8\ell\eta}. \tag{13.22}$$

Exercise 13.4 If a pressure of 1/50 of an atmosphere is maintained across the ends of a tube of length 5 m and diameter 4 mm, estimate the volume of water passing through the tube each second assuming streamline flow. The coefficient of viscosity of the water is $10^{-3} \, \mathrm{kg \, m^{-1} \, s^{-1}}$.

Answer About $2.6 \times 10^{-6} \, \mathrm{m^3 \, s^{-1}}$.

Mean free path

The viscosity of gases, in which molecules are moving more or less freely with large separations, is smaller than that of liquids, but not as much smaller as one might guess. For example, at 293 K the viscosity of air is about $1.8 \times 10^{-5} \, \mathrm{kg \, m^{-1} \, s^{-1}}$, about 1/50 that of water.

The coefficient of viscosity for a gas can be estimated using the kinematic theory of colliding molecules introduced in the last chapter to describe gas pressure and velocity distributions. First, however, we must define the **mean free path** of a moving gas molecule. Gas molecules are moving at random and continually make collisions with other gas molecules. The average distance a molecule travels between collisions is the mean free path and is given the symbol λ. The mean free path depends upon the pressure and temperature, which determine the volume of a fixed mass of gas and hence the number of molecules per unit volume, N_0.

Consider molecules of the same type with mass m. We will assume the molecules to be spherical and have diameter d. At thermal equilibrium at temperature T, the molecules have a mean speed $\bar{v}$ equal to $\sqrt{(8k_B T/\pi m)}$, from Problem 12.6. A simple model of the collision processes gives an estimate of the mean free path. Using this picture, in one second a molecule travels on average a distance $\bar{v}$ and collides with all molecules in a cylinder of radius d and length $\bar{v}$. The number of molecules within this cylinder, and hence the number of collisions, is $N_0 \pi d^2 \bar{v}$. The average distance between collisions is thus

$$\lambda = \frac{\bar{v}}{N_0 \pi d^2 \bar{v}},$$

giving

$$\lambda = \frac{1}{N_0 \pi d^2}. \tag{13.23}$$

The above picture is oversimplified. One assumption made is that a molecule continues on a straight line after it collides; a second is that it is a sphere of radius d. All molecules are moving at random and have a mean relative speed $\overline{v_R}$ given in Problem 12.34 as $\sqrt{2}$ times the mean speed $\overline{v}$. We obtain an improved value for the number of collisions made while one molecule travels a distance $\overline{v}$ by using the relative speed to determine the length of the cylinder containing those molecules involved in the collisions. With this change,

$$\lambda = \frac{\overline{v}}{N_0 \pi d^2 \overline{v_R}}$$

and, putting $\overline{v_R}/\overline{v} = \sqrt{2}$, we have

$$\lambda = \frac{1}{\sqrt{2} N_0 \pi d^2}. \tag{13.24}$$

The parameter d in the expression for the mean free path obtained using the simple model is the diameter of a gas molecule. More exactly, the area πd^2 should be replaced by an area that represents the cross-sectional area of the cylinder within which collisions can occur. The collision area varies with the energy of the colliding molecules and usually decreases with temperature. This introduces a small dependence of λ on temperature.

At constant volume the mean free path is independent of temperature if the collision area is constant, since the number of molecules per unit volume does not change. At constant temperature, λ increases as the pressure is reduced because of the consequent reduction in volume and in N_0.

Exercise 13.5 Assuming that nitrogen behaves like an ideal gas, and that the molecules are spherical with diameter 3×10^{-10} m, estimate the mean free path for nitrogen molecules at one atmosphere pressure and a temperature of 293 K. At constant temperature determine how the mean free path increases as the pressure is reduced to: (a) 10^{-3} atmospheres; (b) 10^{-7} atmospheres.

Answer 10^{-7} m. (a) 10^{-4} m; (b) 1.0 m.

Worked Example 13.3 In more sophisticated treatments of transport phenomena in gases it is necessary to know the probability that a molecule will travel a certain distance without collision. Molecules in a gas have a mean free path λ. What is the probability that a molecule travels a distance ℓ without collision?

Answer Let the probability that the molecule travels a distance ℓ without collision be $P(\ell)$. Then the probability that it travels $\ell + \delta\ell$ is $P(\ell + \delta\ell)$. As $\delta\ell$ becomes very small, the probability of a collision in the distance $\delta\ell$

is proportional to $\delta\ell$ whatever the value of ℓ. If the constant of proportionality is β, the probability of no collision in the distance $\delta\ell$ is $(1 - \beta d\ell)$ and the probability of travelling a distance $\ell + \delta\ell$ without collision is given by

$$P(\ell + \delta\ell) = P(\ell)(1 - \beta\delta\ell).$$

As $\delta\ell$ tends to the infinitesimal $d\ell$,

$$P(\ell) + \left(\frac{dP}{d\ell}\right)d\ell = P(\ell) - P(\ell)\beta d\ell$$

and

$$\frac{dP}{d\ell} = -\beta P.$$

Integration gives

$$P(\ell) = Ae^{-\beta\ell}$$

with A a constant. Since $P(\ell)$ is unity if ℓ equals zero, the constant A is equal to one, and

$$P(\ell) = e^{-\beta\ell}.$$

We now have to determine the constant β. The probability that the path length lies between ℓ and $\ell + d\ell$ is equal to the probability that the molecule goes ℓ without collision times the probability of a collision in the distance $d\ell$ and is $P(\ell)\beta d\ell$. Hence the mean value of the path length is given by

$$\lambda = \frac{\int_0^\infty \ell P(\ell)\beta \, d\ell}{\int_0^\infty P(\ell)\beta \, d\ell}.$$

Evaluation of these integrals gives

$$\lambda = \frac{1}{\beta},$$

and hence finally the probability $P(\ell)$ that a molecule travels a distance ℓ without collision is given by

$$P(\ell) = e^{-\ell/\lambda}. \tag{13.25}$$

The distribution of path lengths follows an exponential curve as shown in Fig 13.9.

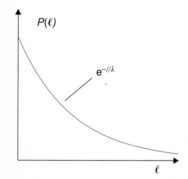

Fig. 13.9 The probability that a molecule travels a distance ℓ without collision decreases exponentially as $e^{-\ell/\lambda}$ with λ the mean free path.

Viscosity of gases

Having discussed mean free path, we may now consider the viscosity of gases. The molecules of a gas at one atmosphere pressure and room temperature are moving randomly with mean speeds of the order of $10^3 \, \mathrm{m\,s^{-1}}$. The average component of velocity in any direction is zero. However, if an object is moving within the gas it drags gas along with it, and at any point the average component of the velocity of gas molecules in the direction of motion of the object is usually not zero.

If there is a slow, steady flow of gas, different layers of the gas move at different speeds and one layer of gas exerts a force per unit area on an adjacent layer given by eqn (13.20). Let us consider the simple example of flow in one direction only, say the z-direction. This simple situation arises to a first approximation when a large flat plate moves in the z-direction with speed u_0 and is separated from a fixed plate by a volume of gas, as shown in Fig 13.10. In that situation the gas molecules in contact with the moving plate have, superimposed on their random velocities, a mean component in the z-direction of average value u_0. Gas molecules in a parallel layer at a perpendicular distance x from the plate still move on average in the z-direction but with a smaller superimposed velocity component $u(x)$. This component of velocity decreases as distance from the plate increases until it becomes zero over the fixed plate. The gradient $\mathrm{d}u/\mathrm{d}x$ is constant, as in the analogous example with liquids.

The drag on a layer of gas parallel to the plate arises because molecules entering the layer from one side have a different average z-component of velocity from molecules entering from the other side. Thus there is a change in momentum over an area of the layer in any time interval and this, by Newton's second law, corresponds to a force on that area. If the layer gains slower molecules and loses faster molecules, the force slows the layer down.

We will calculate the coefficient of viscosity using the same assumptions as in Section 12.2, namely, that at any point in time one-sixth of the molecules are moving with the mean speed $\bar{v}$ in both the positive and negative x-, y-, and z-directions. The mean speed $\bar{v}$ is very much greater than u_0.

Consider any layer parallel to the plates, and let N_0 be the number of molecules per unit volume. With the above assumptions, $N_0 \bar{v} S / 6$ molecules per second arrive at an area S of the layer AB, a distance x from the moving plate, moving in the $(-x)$-direction, and $N_0 \bar{v} S / 6$ molecules per second arrive at S moving in the $(+x)$-direction.

We further assume that molecules moving in the $(-x)$-direction come from the layer at $(x + \lambda)$, as shown on Fig 13.11, and that their average drift speed in the z-direction, $u(x + \lambda)$, is increased to $u(x)$ by collisions with molecules in the layer at x. Similarly, we assume that the $N_0 \bar{v} S / 6$ molecules per second crossing the area S moving in the $(+x)$-direction come from a layer at $(x - \lambda)$ and that their average drift speed in the z-direction is reduced from $u(x - \lambda)$ to $u(x)$. For molecules of mass m, the

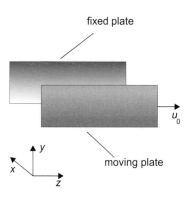

Fig. 13.10 A flat plate parallel to a large fixed plate moves with speed u_0 in the z-direction. Gas between the plates gives a viscous drag on the moving plate. The x-axis is in the direction of the normal to the plates.

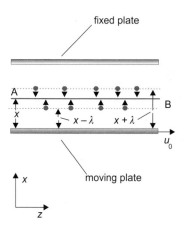

Fig. 13.11 The layer AB is a perpendicular distance x from the moving plate of Fig. 13.10. It is a good approximation at normal pressures to assume that $N_0 \bar{v} / 6$ molecules per second hit unit area of the layer AB and travel perpendicularly from a layer of gas at distance $x + \lambda$. Correspondingly, $N_0 \bar{v} / 6$ molecules per second hit unit area of the layer AB and travel perpendicularly from a layer at distance $x - \lambda$.

net momentum change at the area S per second, dp/dt, is given by

$$\frac{dp}{dt} = \tfrac{1}{6}N_0\bar{v}Sm\{u(x+\lambda) - u(x-\lambda)\}. \tag{13.26}$$

Since λ is very small at normal pressures, we may approximate $u(x \pm \lambda)$ by

$$u(x \pm \lambda) = u(x) \pm \lambda\left(\frac{du}{dx}\right),$$

and substitution in eqn (13.26) gives the force F on the area S,

$$F = \frac{dp}{dt} = \tfrac{1}{3}N_0\bar{v}Sm\lambda\left(\frac{du}{dx}\right). \tag{13.27}$$

Comparing the above equation with eqn (13.20), the coefficient of viscosity is seen to be

$$\eta = \tfrac{1}{3}N_0 m\bar{v}\lambda. \tag{13.28}$$

This equation involves three parameters that depend on the state of the gas, N_0, $\bar{v}$, and λ. Since λ is inversely proportional to N_0, the equation may be rearranged using eqn (13.24) for the mean free path. It becomes

$$\eta = \frac{m\bar{v}}{\sqrt{2} \times 3\pi d^2}.$$

● **Expressions for the coefficient of viscosity**

If we use the result for $\bar{v}$ from Problem 12.6, the above reduces to

$$\eta = \frac{1}{3\pi d^2}\sqrt{\frac{4k_B mT}{\pi}}. \tag{13.29}$$

This shows that the viscosity is independent of pressure and varies as the square root of the temperature. Unlike liquids, the viscosities of gases increase as the temperature is raised.

Worked Example 13.4 A large, thin flat plate is half-way between the walls of a vessel containing nitrogen gas and is moving at a steady speed of $1\,\mathrm{m\,s^{-1}}$ parallel to the walls. The vessel is large and has width 20 cm. The gas is at one standard atmosphere pressure and a temperature of 293 K. Determine the viscous retarding force per unit area of the plate, neglecting edge effects.

Answer We will calculate the viscosity using eqn (13.28), in spite of it being longer to do it that way than to use eqn (13.29), because it will serve to recall several related matters.

1. Exercise 13.5 gives an estimate of the mean free path of the nitrogen molecules as 10^{-7} m.

2. At 293 K, the mean speed $\bar{v}$ in eqn (13.28) is $\sqrt{8k_B T/\pi m}$ where m is the mass of a nitrogen molecule. The molecular weight of N_2 is 28, giving $m = 4.65 \times 10^{-26}$ kg and $\bar{v} = 470$ m s^{-1}.

3. The number of molecules per unit volume can be estimated using the ideal gas equation for one mole, $PV = RT$. At $T = 293$ K and one standard atmosphere pressure, the molar volume is about 0.024 m^3 and $N_0 = 2.5 \times 10^{25}$ m^{-3}.

4. Insertion of these parameter values into eqn (13.28) gives a coefficient of viscosity of $\sim 1.8 \times 10^{-5}$ kg m^{-1} s^{-1}. This is close to the value for air quoted earlier. The force on each side of the plate is given by eqn (13.20), and substituting the value 10 s^{-1} for the gradient du/dx of the average molecular drift speed, the total force per unit area is estimated to be about 3.7×10^{-4} N m^{-2}.

Gases at low pressures

At sufficiently low pressure the mean free path of gas molecules in a vessel becomes larger than the vessel's dimensions. The ideas previously used to calculate the viscosity are then no longer valid since molecules rarely collide with each other, the majority of the collisions being with the walls of the container.

Consider the resistance to the flow of a flat surface moving in the $+z$-direction with speed u_0 parallel to a fixed plane. This is the same situation as shown in Fig 13.10. Now however, the gas between the two planes is at a very low pressure, so that the gas molecules move from surface to surface with very small probability of hitting another molecule, as illustrated in Fig 13.12.

Every time a molecule, of mass m, hits a surface we assume that on release it acquires on average a component of velocity in the $+z$-direction equal to that of the surface which is struck. Since the molecules go from wall to wall, this implies that those hitting the fixed wall on average lose forward momentum mu_0 and those hitting the moving wall on average gain forward momentum mu_0. The rate of loss of momentum from an area S of the moving wall is mu_0 times the number of molecules hitting S in unit time, corresponding to a force

$$F = \tfrac{1}{4}N_0\bar{v}Smu_0. \tag{13.30}$$

In eqn (13.30) we have used $\tfrac{1}{4}N_0\bar{v}S$ for the number of molecules hitting S in unit time. This is the correct expression obtained by averaging over the speeds and directions of the molecules. Previously we simplified the calculations by assuming that molecules are moving in the directions of the axes, and that all collide after travelling a distance equal to the mean free path λ. This procedure leads to the correct functional dependence of

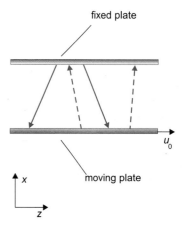

Fig. 13.12 At very low pressures, molecules moving between the plates of Fig 13.10 travel from wall to wall without collisions. When molecules hit the moving wall (shown as solid trajectories) they stick long enough to acquire an average speed in the z-direction equal to the speed of the moving wall u_0, and hence acquire an average momentum $\bar{p}$ in the z-direction. Molecules returning to the fixed wall (shown as dashed trajectories) are brought to rest and give up the average momentum $\bar{p}$.

the transport coefficients, but gives absolute values in error by factors of order unity.

Equation (13.30) for the frictional force may be written in terms of the gas pressure, P, temperature, T, and the mass of a mole of gas, M. The number density of one mole is N_A/V which is equal to PN_A/RT using the ideal gas equation. Also $\bar{v} = \sqrt{8RT/\pi M}$ from Problem 12.6, and $M = N_A m$. Substitution into eqn (13.30) gives

$$F = u_0 PS \sqrt{\left(\frac{M}{2\pi RT}\right)}. \tag{13.31}$$

At constant temperature and very low pressure, the force per unit area becomes dependent on pressure. The force also has a different temperature dependence to that of the viscous force in a gas at higher pressures, decreasing as the temperature is raised instead of increasing. The pressure dependence of the drag on an oscillating object in a gas at very low pressure has been used as the basis of instruments for measuring low pressures.

13.4 Diffusion

The diffusion of gases and liquids is another example of a transport phenomenon. In diffusion, molecules themselves are transported. Imagine that extra molecules are introduced into a small region within a volume of gas, so that within that region the gas density becomes greater than outside. Gas molecules diffuse throughout the whole volume and eventually all regions have on average equal numbers of molecules per unit volume. The rate at which this equilibrium is established depends on the **coefficient of diffusion** of the gas. As with viscosity, the rate can only be simply determined in special circumstances. There must be no bulk motion of the gas; no winds or temperature gradients must cause departures from conditions that approximate thermal equilibrium.

Let the number of molecules per unit volume at time t vary only in the z-direction with gradient $\partial N_0(z, t)/\partial z$, and let the net number of molecules per second crossing a unit area of a plane perpendicular to the z-axis at distance z be $J(z, t)$. The coefficient of diffusion, D, is defined by the equation

$$J(z, t) = -D \frac{\partial N_0(z, t)}{\partial z}. \tag{13.32}$$

This is known as **Fick's law**. The minus sign occurs because the flow J is in the positive z-direction when the number of molecules per unit volume is decreasing in the z-direction, that is, the gradient $\partial N_0/\partial z$ is negative. The coefficient of diffusion has dimensions of [length]^2 times [time]$^{-1}$ and in the SI system is measured in units of m^2 s^{-1}.

For simplicity we will consider only the one-dimensional example of the above introduction. Equilibrium is reached by molecules moving randomly, but with more moving on average in one direction along the z-axis than in the opposite direction. While molecules are diffusing in this way the gas is not in thermal equilibrium, but is very close to it, and as before we assume a Maxwell–Boltzmann distribution of molecular velocities.

Our discussion of diffusion follows closely the discussion of viscosity and also applies when the mean free path is small compared to the dimensions of the vessel containing the fluid. Refer to Fig 13.13, which depicts a similar model to that used with Fig 13.11. We assume as before that at time t, $\frac{1}{6}N_0(z+\lambda, t)\bar{v}$ molecules per second arrive at unit area of the layer AB at distance z having left the layer at $(z+\lambda)$. Similarly, $\frac{1}{6}N_0(z-\lambda, t)\bar{v}$ molecules per second arrive at unit area of the layer AB having left the layer at $(z-\lambda)$. The net rate of increase in number density at the layer at z is thus

$$J(z, t) = \tfrac{1}{6}\bar{v}(N_0(z-\lambda, t) - N_0(z+\lambda, t)).$$

Since λ is very small we may approximate $N_0(z \pm \lambda, t)$ by

$$N_0(z \pm \lambda, t) = N_0(z, t) \pm \lambda\left(\frac{\partial N_0(z, t)}{\partial z}\right)$$

and obtain

$$J(z, t) = -\frac{\bar{v}\lambda}{3}\frac{\partial N_0(z, t)}{\partial z}.$$

Comparison of this equation with eqn (13.32) gives

$$D = \tfrac{1}{3}\bar{v}\lambda. \tag{13.33}$$

The coefficient of diffusion is predicted to be proportional to the mean free path, and thus inversely proportional to the pressure or density, and also predicted to increase as the square root of the temperature.

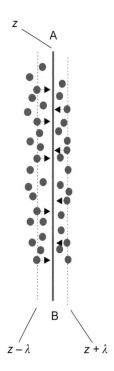

Fig. 13.13 Molecules diffusing in the z-direction when the number per unit volume decreases as z increases, that is, there is a negative concentration gradient $\partial N_0/\partial z$.

● *Expression for the coefficient of diffusion*

Exercise 13.6 Estimate the coefficient of diffusion of nitrogen gas at atmospheric pressure and room temperature.

Answer $1.6 \times 10^{-5}\,\mathrm{m^2\,s^{-1}}$.

The diffusion equation

Equation (13.32) can be used to derive an equation governing the behaviour of the number of molecules per unit volume N_0 as a function of distance and time for the case where the concentration of molecules

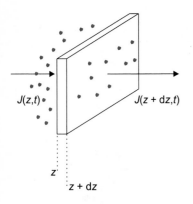

Fig. 13.14 Molecules diffusing in the z-direction through unit area of a large slab of thickness dz.

varies only in one direction. This equation is the one-dimensional **diffusion equation**.

Consider a unit area of a large slab of thickness dz positioned between z and $z + dz$, as shown in Fig 13.14. The number of molecules entering the left-hand side per second at time t is $J(z, t)$, and the number leaving the right-hand side per second is $J(z + dz, t)$. Molecules are not created or destroyed; their number is conserved. Thus, the net gain of molecules per second by unit area of the slab is equal to the rate of increase of molecules within the slab. This rate of increase is given by $(\partial N_0/\partial t)dz$, and thus

$$J(z, t) - J(z + dz, t) = \left(\frac{\partial N_0}{\partial t}\right)dz$$

and hence

$$-\left(\frac{\partial J}{\partial z}\right) = \left(\frac{\partial N_0}{\partial t}\right).$$

The left hand-side of this equation is given by differentiating eqn (13.32), and finally we obtain the diffusion equation

$$D\frac{\partial^2 N_0}{\partial z^2} = \frac{\partial N_0}{\partial t}. \tag{13.34}$$

This is a second-order differential equation and the solution that fits the initial conditions determines how the number density of the molecules changes with time as we go to different distances z. The initial non-uniform density dies away with time and all solutions result in a uniform distribution of the molecules at large times. Problem 13.21 considers the diffusion of gas initially contained within a large, thin slab situated at $z = 0$. The gas spreads out, and at any particular time the distribution has a Gaussian shape centred on $z = 0$, with the width of the Gaussian increasing with time.

The diffusion equation may be compared with the wave equations for unattenuated waves met in Chapter 7. There the equation has double

Diffusion in solids is important in several practical situations. For example, the diffusion into a semiconductor of an evaporated or implanted surface layer is used to manufacture solid-state devices such as transistors and computer chips. Diffusion in solids also obeys the diffusion eqn (13.34), although the mechanism of diffusion in solids differs from that in liquids and gases. In solids, atoms diffuse through the lattice by jumping from one lattice site to a nearby unoccupied site. If the energy required to create a lattice vacancy is ΔE, the probability of a vacancy existing at temperature T at any lattice site is given by the Boltzmann factor $\exp(-\Delta E/k_B T)$. The rate of diffusion is proportional to this probability and thus the diffusion coefficient to be used in the diffusion equation increases exponentially as the temperature increases.

differentials on both sides and their solutions correspond to oscillations of constant amplitude.

13.5 Heat conduction

If the temperature is not uniform throughout a substance, thermal equilibrium will become established by the transport of heat from the hot part to the cold. If the transport is by interactions between molecules without bulk motion of the molecules themselves, the process is called **conduction**.

Consider a large volume of a substance in which there is a temperature gradient $\partial T/\partial z$ in the z-direction. Let $J_Q(z)$ be the amount of heat flowing across unit area of planes perpendicular to this direction each second. For many substances this amount of heat is proportional to the temperature gradient and we may write

$$J_Q(z) = -\kappa \frac{\partial T}{\partial z} \tag{13.35}$$

where the constant of proportionality κ is called the **coefficient of thermal conductivity**. The negative sign arises because the heat flow is in the positive direction of z if the temperature T decreases as z increases, that is, if $\partial T/\partial z$ is negative. The dimensions of the coefficient of thermal conductivity are [mass] times [length] times [time]$^{-3}$ times [temperature]$^{-1}$, which are the dimensions of energy per unit time divided by length and by temperature. In the SI system the coefficient is measured in units of $W\,m^{-1}\,K^{-1}$.

The above description of heat flow applies in gases and liquids only if the flow is smooth and steady and there is no bulk motion of molecules. **Convection** is the name given to the way in which bulk motions within a fluid transfer heat. A hot cup of coffee cools mainly by circulatory motions of masses of liquid, the less dense hot liquid flowing to the top of the cup while the denser cooler liquid drops to the bottom. Experiments on heat flow in gases made to determine the coefficient of conductivity must thus ensure that the mechanism of heat loss is indeed by conduction only. This can be done by observing slow rates of heat loss and requiring consistent results under different conditions. This section is confined to non-convective heat flows in gases, for which eqn (13.35) can be used.

The procedure used to describe conduction in gases follows the same ideas as used in the discussion of viscosity and diffusion. Consider once more a large volume of gas in which there is a positive temperature gradient in the z-direction. Molecules are continually crossing the plane AB at position z shown on Fig 13.15. We make the approximation that $\frac{1}{6}N_0(z+\lambda)\bar{v}(z+\lambda)$ molecules on average cross unit area of AB per

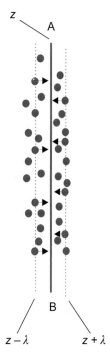

Fig. 13.15 Molecules arriving at the plane AB from the layer at $z + \lambda$ bring in more energy than those from the layer at $z - \lambda$ when there is a positive temperature gradient $\partial T/\partial z$ such that the temperature at $z + \lambda$ is greater than the temperature at $z - \lambda$.

second and come from a plane at position $z + \lambda$, and $\frac{1}{6} N_0(z - \lambda)\bar{v}(z - \lambda)$ cross unit area per second from the plane at $z - \lambda$.

Molecules from different layers carry different mean energies ϵ. This is the most influential aspect of the motion, and we neglect the change in the number of molecules per unit volume N_0 with z in order to obtain an approximate expression for the conductivity and put $N_0(z + \lambda) = N_0(z - \lambda) = N_0$.

The molecules from $(z + \lambda)$ have a higher average energy than those from $(z - \lambda)$ because of the temperature gradient. They carry an amount of energy $\frac{1}{6} N_0 \bar{v}(z + \lambda)\epsilon(z + \lambda)$ per unit area per second. Similarly, molecules from $(z - \lambda)$ carry an amount of energy $\frac{1}{6} N_0 \bar{v}(z - \lambda)\epsilon(z - \lambda)$ per unit area per second. The net transfer of energy per unit area per second in the positive z-direction, J_Q, is then given by

$$J_Q = \tfrac{1}{6} N_0(\bar{v}(z - \lambda)\epsilon(z - \lambda) - \bar{v}(z + \lambda)\epsilon(z + \lambda))$$
$$= (2/m)^{1/2} \times \tfrac{1}{6} N_0(\epsilon(z - \lambda)^{3/2} - \epsilon(z + \lambda)^{3/2}),$$

since $\bar{v} = (2\epsilon/m)^{1/2}$. When the mean free path λ is small

$$\epsilon(z \pm \lambda)^{3/2} \approx \epsilon(z)^{3/2} \pm \lambda \frac{\partial(\epsilon(z)^{3/2})}{\partial z} = \epsilon(z)^{3/2} \pm \tfrac{3}{2}\epsilon(z)^{1/2}\lambda\frac{\partial\epsilon}{\partial z}.$$

Using these approximate values in the equation for J_Q gives

$$J_Q = -\left(\frac{2}{m}\right)^{1/2} \times \tfrac{1}{2} N_0 \epsilon^{1/2}\lambda\left(\frac{\partial\epsilon}{\partial z}\right) = -\tfrac{1}{2} N_0 \bar{v}\lambda\left(\frac{\partial\epsilon}{\partial z}\right).$$

But

$$\left(\frac{\partial\epsilon}{\partial z}\right) = \left(\frac{\partial\epsilon}{\partial T}\right)\left(\frac{\partial T}{\partial z}\right);$$

hence

$$J_Q = -\tfrac{1}{2} N_0 \bar{v}\lambda\left(\frac{\partial\epsilon}{\partial T}\right)\left(\frac{\partial T}{\partial z}\right).$$

Comparing this with eqn (13.35), we see that the coefficient of thermal conductivity

● *Expressions for the coefficient of heat conductivity*

$$\kappa = \tfrac{1}{2} N_0 \bar{v}\lambda\left(\frac{\partial\epsilon}{\partial T}\right). \tag{13.36}$$

The rate of change of mean energy with temperature, $\partial\epsilon/\partial T$, is equal to the specific heat per molecule, c, and depends on the number of degrees of freedom as discussed in Section 12.8. For monatomic gases $c = 3k_B/2$. The specific heat per molecule is reasonably constant over wide temperature ranges, and substituting into the above equation the expression for the mean free path λ given by eqn (13.24)

$$\kappa = \frac{c\bar{v}}{2\pi\sqrt{2}d^2}. \tag{13.37}$$

The parameters c and d are roughly constant, and κ varies only with $\bar{v}$. The coefficient of conductivity is thus independent of pressure and varies with the square root of the temperature T. Putting numbers into eqn (13.37) for a typical gas at normal room conditions gives a value for κ of the order of magnitude of $10^{-2}\,\mathrm{W\,m^{-1}\,K^{-1}}$.

Exercise 13.7 A material has heat conductivity κ, specific heat C per mole, density ρ and molar mass M. Apply eqn (13.35) to a slab of material of thickness dz and unit area. Use the principle of conservation of energy and follow the procedure used to derive the diffusion eqn (13.34) to show that the second-order differential equation giving the variation of temperature with time and distance for heat flow in one direction in a material in which there is a temperature gradient $\partial T/\partial z$ is

$$\kappa \frac{\partial^2 T}{\partial z^2} = \rho \frac{C}{M}\frac{\partial T}{\partial t}. \tag{13.38}$$

As with viscosity, eqn (13.37) for the heat conductivity is valid only when the mean free path of the gas molecules is much less than the dimensions of the containing vessel. The conductivity decreases when this condition is no longer satisfied and at very low pressures gases are good heat insulators. The variation of heat conductivity with pressure at low pressures is the basis of a gauge, called a Pirani gauge, commonly used to measure pressures in the range from about $1\,\mathrm{Pa}$ to $10^{-2}\,\mathrm{Pa}$.

Relations between transport coefficients

The expressions for the coefficients of viscosity, diffusion, and conductivity of gases derived so far are in reasonable agreement with observations for pressures between one-tenth of an atmosphere and a few atmospheres. At high pressures intermolecular forces have to be taken into account and at low pressures, as we have seen, the influence of the long mean free path has to be considered.

It is convenient to repeat here the expressions derived for the three transport coefficients η, D, and κ of gases at normal pressures, eqns (13.28), (13.33), and (13.37).

$$\eta = \tfrac{1}{3}N_0 m\bar{v}\lambda,$$

$$D = \tfrac{1}{3}\bar{v}\lambda,$$

and

$$\kappa = \frac{c\bar{v}}{2\pi\sqrt{2}d^2}.$$

There is a close connection between the coefficients, and the following relation can easily be verified.

$$\frac{D}{\eta} = \frac{1}{N_0 m} = \frac{1}{\rho} \tag{13.39}$$

where ρ is the gas density. Using eqn (13.24), the ratio κ/η becomes

$$\frac{\kappa}{\eta} = \frac{3c}{2m} = \frac{3C_V}{2M} \tag{13.40}$$

where $\partial\epsilon/\partial T = c$, the specific heat per molecule, C_V is the specific heat per mole at constant volume, and M the mass of one mole. These relations predict that $D\rho/\eta$ should be about unity and $\kappa M/\eta C_V$ should be about 1.5. These ratios are reasonably close to the values observed experimentally for many gases.

Exercise 13.8 The coefficient of viscosity of nitrogen gas at a temperature of 300 K and a pressure of one atmosphere is $1.78 \times 10^{-5}\,\mathrm{kg\,m^{-1}\,s^{-1}}$, and the coefficient of thermal conductivity is $0.026\,\mathrm{W\,m^{-1}\,K^{-1}}$. At room temperatures nitrogen molecules have five degrees of freedom. Calculate the ratio $\kappa m/\eta c$, where c is the specific heat per molecule and m the mass of a molecule.

Answer 1.97.

Figure 13.16 shows measured values of the coefficient of viscosity of nitrogen gas, and values of the ratio $\kappa M/\eta C_V$, at various temperatures. The viscosity increases with temperature, roughly as the square root of T, as expected. The ratio $\kappa M/\eta C_V$ is roughly constant at about 2, compared with the predicted value of about 1.5, showing that η and κ vary with temperature in much the same way, again as expected. The difference between the observed ratio and that predicted is not surprising. The numerical factors in the expressions for the transport coefficients are not expected to be accurate to better than a factor of two in view of the approximations made.

13.6 Surface properties

The average force on a liquid molecule at the surface separating a liquid from the gas above it is less than the average force on a molecule within

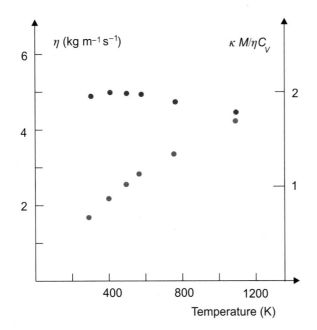

Fig. 13.16 The coloured points show the coefficients of viscosity η of nitrogen gas at various temperatures using the left-hand vertical scale. The black points show the ratios $\kappa M/\eta C_V$ using the right-hand vertical scale.

the liquid. A molecule in the surface has about half the neighbours of one within the liquid since there are very few molecules per unit volume in the space above. Hence, a molecule at the surface experiences only about one-half the attractive intermolecular forces. This effect gives rise to a net force on the surface molecules directed inward and normal to the surface. It is as though the surface were an elastic skin.

A small liquid drop, for which the effects of gravity can be neglected, assumes a spherical shape because of the force normal to the surface. The existence of the inward force gives rise to a **surface energy**. If the surface area of the small sphere is increased, work is done against the inward force and energy is stored in the surface. If the surface energy per unit area is Γ_S, when the area is increased by ΔS the surface energy is increased by $\Gamma_S \Delta S$.

Consider a rectangular area S of liquid surface such as shown on Fig 13.17. Imagine that the liquid is in a flexible container whose area can be increased, when the side AB is pulled, in such a way that the separation $AB = \ell$ stays constant and the surface retains a rectangular shape. The area is increased by a small amount at constant temperature by pulling slowly on the side AB with a force F, moving it a perpendicular distance Δx. The applied force F needed to increase the area by the amount $\Delta x \times \ell$ is opposed by an equal and opposite force that results from the tendency of the molecules not to want to increase the surface energy. This force, per unit length in a surface, is called the **surface tension**, symbol Γ, of the liquid.

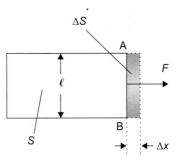

Fig. 13.17 If a force F increases the the surface area of a liquid by an amount ΔS the increase in surface energy is roughly equal to the work done by the force F against the surface tension force.

The force required to extend the surface area is equal and opposite to the force due to surface tension; $F = \Gamma\ell$, and the work done by the force plus any heat exchange ΔQ required to keep the temperature constant equals the total increase in energy of the liquid. Hence,

$$\Gamma_s\Delta S = \Gamma_s\ell\Delta x = \Gamma\ell\Delta x + \Delta Q. \tag{13.41}$$

The second contribution to the energy change can be comparable with the first (see Problem 13.8), although at low temperatures it is usually smaller.

Worked Example 13.5 Show that the excess air pressure inside a spherical soap bubble of radius a is given by $\Delta P = 4\Gamma/a$, where Γ is the surface tension of the liquid.

Answer If the radius of the bubble increases by da, the surface area of each of the bubble's surfaces increases by $8\pi a\,da$. The resultant increase in surface energy of the two surfaces is $16\pi\Gamma_s a\,da$. At constant temperature, some of this increase comes from heat input ΔQ from the surroundings and the remainder from the work done against surface tension by the excess pressure inside the bubble in increasing the volume by $4\pi a^2 da$. Hence

$$16\pi\Gamma_s a\,da = 4\pi a^2 da\Delta P + \Delta Q.$$

But from eqn (13.41) the surface energy per unit area equals the surface tension plus the heat input per unit increase in surface area; hence

$$\Delta P = 4\Gamma/a.$$

The considerations of Worked example 13.5 may be extended to the more general situation of a curved surface separating liquid from gas. At any point on a curved surface the shape is given by two radii of curvature a_1 and a_2 measured in perpendicular directions. Inside a liquid drop where the surface is characterized by radii a_1 and a_2 the excess pressure is given by

$$\Delta P = \Gamma\left(\frac{1}{a_1} + \frac{1}{a_2}\right). \tag{13.42}$$

If the liquid surface is concave towards the body of the liquid, as for a liquid drop between flat plates, both radii are negative and the pressure in the liquid is less than in the gas.

When a liquid is in contact with a solid surface, as at the vertical edge of a container, the surface of the liquid does not usually meet the solid perpendicularly but at an angle less than 90°, either upwards as with most liquids, or downwards as with mercury, for example. The **angle of contact** made by different liquids depends upon the adhesion of vapour molecules to the different surfaces. This angle is defined as the angle between the normal into the liquid surface at the point where the liquid surface touches the solid, and the normal into the solid surface at that point. This is illustrated in Fig 13.18. The angle of contact is less than 90° if the liquid bends upwards at the solid surface, as in Fig 13.18, or greater than 90° if the liquid bends downwards.

A water surface curves upwards as it meets a vertical glass wall and the angle of contact is about 60°. A mercury surface curves downwards at its contact with a vertical glass wall and the angle of contact is about 140° as shown in Fig 13.19.

One commonly observed manifestation of surface tension is the rise of liquids in capillary tubes. A liquid rises or falls in a vertical capillary tube of radius a whose bottom end is submerged. For a liquid with surface shape concave upwards like water, the level rises; for mercury, with shape concave downwards, the level falls.

The height h to which a liquid such as water rises above the level outside the tube can be estimated by ignoring the small amount of liquid absent below the horizontal line at a height h shown on Fig 13.20. With this approximation, the weight of the volume of liquid of height h must be balanced by the upward component of the surface tension force acting around the circle where the liquid contacts the surface of the tube. If the angle of contact is θ, this balance condition gives the result that

$$h = \frac{2\Gamma}{a\rho g}\cos\theta \qquad (13.43)$$

where ρ is the liquid density and a the radius of the capillary tube. If the meniscus is inverted, as for mercury, the surface tension force has a vertical component in the downwards direction pulling the liquid below the level of the surface outside the tube.

It is difficult to make consistent and accurate measurements of surface tension because surface phenomena are very susceptible to impurities in the surfaces. Surface tension decreases as the temperature rises and is rather insensitive to pressure. Techniques for measuring Γ often do not measure the surface tension of the pure liquid. The rise in the height of liquid in a capillary tube, for example, depends upon the interaction of the liquid with the tube surface.

Measurements of the surface tension Γ give information on inter-molecular forces and molecular arrangements on surfaces. Equating the surface tension Γ to the surface energy per unit area for the purpose of making rough estimates, we can relate Γ within the simple model used in Section 12.1 to the depth ϵ of the intermolecular potential shown in Fig 13.4.

The energy needed to remove a number N_A of molecules from the liquid when each molecule has n_ℓ neighbours gives the latent heat of evaporation per mole as $L_{evap} = n_0 N_A \epsilon/2$. If there are N'_0 molecules per

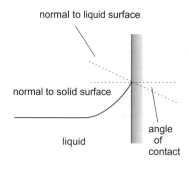

Fig. 13.18 The angle of contact of a liquid surface with a solid surface.

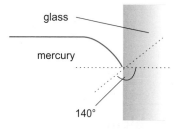

Fig. 13.19 A mercury surface at a vertical plane of glass.

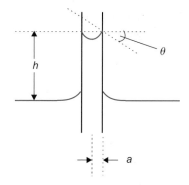

Fig. 13.20 Water in a vertical capillary tube. The water rises to a height h above the surface of the water in the vessel into which the thin-bore tube is immersed.

unit area of the surface these have on average about $n_\ell/2$ nearest neighbours and the surface energy per unit area is $n_\ell N_0' \epsilon/4$. The surface tension thus gives the same sort of information as the latent heat of evaporation and the energy ϵ deduced from the two different sources should be very roughly equal.

Worked Example 13.6 Estimate the depth of the intermolecular potential between water molecules from the surface tension value of $0.074\,\mathrm{N\,m^{-1}}$.

Answer Equating the surface energy per unit area to the surface tension gives $0.074\,\mathrm{J\,m^{-2}}$. Assuming water molecules to be spherical with radius $\sim 3.1 \times 10^{-10}$ m and the surface layer to consist of close-packed spheres, we may estimate the number per $\mathrm{m^2}$ to be about 2.6×10^{18}. Using a value of 10 for the coordination number n_0 gives $\epsilon = 1.14 \times 10^{-20}$ J, or 0.071 electron volts in units more appropriate to atomic phenomena. In view of the crudity of the model and the assumptions made, this value is in reasonable agreement with the value of 0.11 eV obtained from the latent heat of evaporation of water in Exercise 12.1.

Problems

Level 1

13.1 If at a temperature of 293 K and one atmosphere pressure the atoms of helium gas occupy approximately 5×10^{-4} of the actual volume occupied by the gas, give an order of magnitude calculation for the radius of a helium atom.

13.2 The isothermal compressibility of a gas can be defined as

$$\chi = -\frac{1}{V}\left(\frac{\partial V}{\partial P}\right)_T.$$

Derive an expresssion for the isothermal compressibility of an ideal gas.

Discuss the difference in the compressibilities of real gases and ideal gases in terms of the properties of the intermolecular forces.

13.3 If the equation of state of a gas is written in the form

$$\frac{PV}{RT} = 1 + \frac{\alpha}{V} + \frac{\beta}{V^2} + \cdots,$$

the coefficients α, β, etc. are called **virial coefficients**. What is the second virial coefficient of a gas obeying Dieterici's equation

$$P(V - b) = RTe^{-a/RTV}?$$

13.4 The **Boyle temperature**, T_B, is the temperature at which the first virial coefficient α is zero. At that temperature the gas, to a first approximation, obeys Boyle's law and the ideal gas equation. What is the Boyle temperature for a gas that obeys Dieterici's equation?

13.5 Calculate the average time between collisions, sometimes called the mean free time, of nitrogen molecules in nitrogen gas at a temperature of 293 K and a pressure of 10^{-3} atmospheres. Take the mean diameter of a nitrogen molecule to be 3×10^{-10} m.

13.6 Show that the ratio of the pressure to the coefficient of viscosity of a gas is approximately equal to the average number of collisions per unit time experienced by a gas molecule.

13.7 Explain why the heat conductivity of hydrogen gas is greater than that of oxygen gas at the same temperature and pressure.

13.8 If the surface tension is assumed to be a function of temperature T only, it may be shown that the surface energy is equal to $\Gamma - T(d\Gamma/dT)$, where Γ is the surface tension. If the surface tension of water is given approximately by $\Gamma = (650 - T) \times 10^{-3}/5 \, \mathrm{N \, m^{-1}}$, what is the surface energy of water at 290 K?

13.9 A capillary tube of inner radius 0.2 mm is placed vertically in a liquid with surface tension equal to $0.026 \, \mathrm{N \, m^{-1}}$ and density $1.6 \times 10^3 \, \mathrm{kg \, m^{-3}}$. The angle of contact of the liquid with the surface of the tube is 30°. What is the height above the liquid surface to which the liquid rises in the tube? The acceleration due to gravity, g, is equal to $9.8 \, \mathrm{m \, s^{-2}}$.

13.10 Show on a simple kinetic model that the thermal conductivity of a gas should be independent of pressure over a wide range, but that at very low pressures this result no longer applies.

Level 2

13.11 Show that the rate at which gas molecules strike a unit area of the walls of a container is $N_0 \bar{v}/4$, where N_0 is the number of molecules per unit volume and $\bar{v}$ is their average speed.

Gases of two types are contained in a vessel and escape via a small hole. The masses and partial pressures of gases 1 and 2 are m_1, m_2 and P_1, P_2, respectively. Show that the ratio of the masses of the amounts of gases 1 and 2 passing through the hole equals $P_1\sqrt{m_2}/P_2\sqrt{m_1}$.

13.12 The critical point of a gas occurs at a temperature of 310 K and a pressure of 72 atmospheres, the density of the gas being $460 \, \mathrm{kg \, m^{-3}}$. If the van der Waals equation is applicable, what is the molecular weight of the gas? How adequate is the van der Waals equation for this situation?

13.13 Estimate the coefficients a and b in the van der Waals equation for nitrogen using data from Table 13.1. Hence estimate the surface tension of liquid nitrogen far below the critical temperature.

13.14 The coefficient of viscosity of a gas is given by $\eta = \rho\bar{v}\lambda/3$, where ρ is the density, $\bar{v}$ the mean speed of the molecules, and λ the mean free path. Show that η should be independent of density and pressure and

estimate the magnitude of the viscosity of air at a temperature of 293 K.

Explain why the above relationship fails at low pressure and estimate the effective coefficient of viscosity of air in a tube of diameter 5 cm at a temperature of 293 K and a pressure of $10^{-4} \, \mathrm{N \, m^{-2}}$.

13.15 Assuming that the viscosity of a liquid at temperature T is proportional to $k_B T/D$, where k_B is Boltzmann's constant and D is the coefficient of diffusion, show that the viscosity of a liquid should increase as the temperature is decreased.

Why does the viscosity of a gas increase as T is increased whereas the reverse is true of a liquid?

13.16 A straight tube consists of two cylindrical portions joined end to end. Their lengths are ℓ_1 and ℓ_2 and their radii of cross-sections a_1 and a_2, respectively. For both tubes the length is very much greater than the radius. Liquid enters tube 1 at a pressure P_1 and leaves tube 2 at a pressure of P_2. The flow is smooth and steady. Find the pressure at the junction of the two tubes.

13.17 Derive an approximate formula that relates Avogadro's number to the measured values for the surface tension, Γ, the molar latent heat of evaporation, L_{evap}, and the molar volume, V_{mole}. Estimate Avogadro's number from the following data for alcohol at zero degrees centigrade:

$$L_{\mathrm{evap}} = 4.0 \times 10^3 \, \mathrm{J \, mol^{-1}}; \quad \Gamma = 0.024 \, \mathrm{N \, m^{-1}};$$
$$V_{\mathrm{mole}} = 5.7 \times 10^{-5} \, \mathrm{m^3}.$$

13.18 A fish tank is aerated by pumping air through a tube of 1 mm diameter. The end of the tube is 20 cm below the surface of the water in the tank. Estimate the pressure required from the pump to make air bubbles emerge from the submerged end of the tube. Take the surface tension of water to be $0.074 \, \mathrm{N \, m^{-1}}$, and the water density to be $10^3 \, \mathrm{kg \, m^{-3}}$.

13.19 A vessel of volume $10^{-3} \, \mathrm{m^3}$ containing gas at an initial pressure P communicates though an aperture of area $10^{-7} \, \mathrm{m^2}$ with a space in which the pressure is maintained at zero. Find the time taken for the pressure to fall to $P/2$. The density of the gas at unit pressure is $10^{-6} \, \mathrm{kg \, m^{-3}}$.

13.20 The coefficient of thermal conductivity of nitrogen gas is $0.026 \, \mathrm{W \, m^{-1} \, K^{-1}}$ at 300 K. The space between two walls of a cylindrical container is filled with nitrogen.

The inner radius of the cylinder is 12 cm and the gap is 1 cm. The inner wall is at the temperature of melting ice; the outer is at 300 K. Estimate the heat flowing to unit area of the inner wall per second by conduction. Estimate the pressure at which the transfer of heat by conduction begins to decrease.

Level 3

13.21 The solution to the one-dimensional diffusion eqn (13.34) corresponding to the initial condition that $N_0(0, 0)$ gas molecules per unit volume in a thin slab at $z = 0$ are released at $t = 0$ is

$$N_0(z, t) = \frac{N_0(0, 0)}{(4\pi Dt)^{1/2}} e^{-z^2/4Dt}.$$

Discuss the diffusion after release and determine the root mean square displacement of the diffusing particles at time t.

You may use the following integrals,

$$\int_0^\infty e^{-\alpha x^2} dx = \frac{1}{2}\sqrt{\frac{\pi}{\alpha}}$$

and

$$\int_0^\infty x^2 e^{-\alpha x^2} dx = \frac{1}{4\alpha}\sqrt{\frac{\pi}{\alpha}}.$$

13.22 Find the relation that gives the locus of the maxima and minima in the curves of pressure against volume at different temperatures for a gas which obeys the van der Waals equation. The maximum of this curve occurs at the critical point. Use this to determine the critical volume, pressure, and temperature of the gas.

13.23 Use the method of Problem 13.22 to find the critical constants of a gas that obeys Dieterici's equation

$$P(V - b) = RTe^{-a/RTV}.$$

13.24 Calculate the number of oxygen molecules striking 1 cm^2 of the wall of a containing vessel each second if the pressure is 10^{-3} N m^{-2} and the temperature is 293 K.

A clean metal surface is exposed to oxygen at the above pressure and temperature. Assuming that every molecule that strikes the surface sticks there and that a uniform layer is formed, estimate how long it takes for the surface to be covered with a layer of oxygen at least two molecules thick. (Assume oxygen molecules of molecular weight 32 are spheres of diameter 3×10^{-10} m.)

13.25 The viscosity η of a gas depends on the long-range, attractive part of the intermolecular force, which varies with molecular separation r according to

$$F = \mu r^{-n}$$

where n is a number and μ a constant. If η is a function of the mass m of the molecules, their mean speed $\bar{v}$, and the constant μ, use the method of dimensions to show that

$$\eta \propto m^{\frac{n+1}{n-1}} \bar{v}^{\frac{n+3}{n-1}} \mu^{-\frac{2}{n-1}}.$$

If $\eta \propto T^s$, where T is the temperature and s a number, and for helium gas $s = 0.68$, determine n for helium. If $s = 0.98$ for carbon dioxide, determine n for CO_2. Are your answers sensible in terms of the molecular structure of the two gases?

13.26 At position z and time t the fractional concentration $C(z, t)$ and the number of particles crossing unit area per second, the flux, $J(z, t)$ of particles of a solute dissolved in a solvent contained in a long pipe are related by Fick's law

$$J(z, t) = -D\frac{\partial C}{\partial z}$$

where D is the diffusion coefficient. Use the condition that solute particles are conserved to obtain the diffusion equation.

Two large identical vessels at the same temperature are connected by a long narrow tube of length ℓ and cross-sectional area S. Each vessel contains salt solution, and at time zero one has fractional concentration C and the other $C + \Delta C$. ΔC is much smaller than C. Assuming that the concentration gradient varies linearly with distance z down the tube, use Fick's law to show that the approach to equilibrium depends exponentially on time. Calculate the time for the difference in fractional concentrations in the vessels to reduce by a factor of two if $\ell = 1$ m and $D = 10^{-8}$ m^2 s^{-1}.

13.27 A mixture of one mole of helium (ratio of specific heats $\gamma = 5/3$) with 0.2 mole of nitrogen ($\gamma = 7/5$) is initially at 300 K and occupies 4×10^{-3} m^3. It is compressed slowly and adiabatically until its volume is reduced (a) by 1% and (b) by one-quarter. Discuss whether the changes in pressure and temperature of the system can be described adequately in both cases by some average value of γ. Calculate the final pressure and temperature in case (a).

Some solutions and answers to Chapter 13 problems

13.3

$$\left(\frac{a^2}{2R^2T^2} - \frac{ab}{RT} + b^2\right).$$

13.4

$$T_B = \frac{a}{Rb}.$$

13.6

$$\frac{P}{\eta} = \frac{3P}{N_0 m \bar{v} \lambda},$$

from the result (13.28) for the coefficient of viscosity η. The average kinetic energy per molecule is

$$\frac{1}{2} m\overline{v^2} = \frac{3}{2} k_B T$$

from eqn (12.11). If there are n moles of gas, the ideal gas equation is approximately valid and

$$PV = nRT = nN_A k_B T.$$

Hence,

$$\frac{P}{N_0} = \frac{PV}{nN_A} = k_B T$$

and

$$\frac{P}{\eta} = \frac{3k_B T}{m \bar{v} \lambda}.$$

If we now make the further approximation that the mean speed $\bar{v}$ is the same as the rms speed $\sqrt{(\overline{v^2})}$,

$$3k_B T \approx m\bar{v}^2,$$

and

$$\frac{P}{\eta} = \frac{\bar{v}}{\lambda},$$

the average number of collisions per unit time.

13.9 1.4 cm.

13.12 The van der Waals equation for one mole of gas is

$$\left(P + \frac{a}{V^2}\right)(V - b) = RT.$$

Equations (13.4) and (13.5) give the parameters b and a in terms of the critical volume and pressure, respectively. Hence, at the critical point

$$4P_c \times \frac{2V_c}{3} = RT_c.$$

The density at the critical point is

$$\rho = \frac{M}{V_c}$$

where M is the mass of one mole of gas. Combining the last two equations gives

$$\frac{8P_c V_c}{3} = \frac{V_c \rho R T_c}{M}$$

or

$$M = \frac{3\rho R T_c}{8P_c}.$$

Substitution of the numbers into this equation gives M equal to 6.16×10^{-2} kg and the molecular weight equal to 61.6.

13.14 The expression $\eta = \rho \bar{v} \lambda / 3$, derived for gas pressures such that the mean free path is small compared with the size of a containing vessel, reduces to eqn (13.29). The coefficient of viscosity is thus independent of density and pressure and varies as the square root of the temperature.

Using $d = 3 \times 10^{-10}$ m and taking the mass of an air molecule as $30 \times 1.7 \times 10^{-27}$ kg gives an estimate of η of about 3×10^{-5} kg m^{-1} s^{-1}.

At low pressure, when the mean free path becomes comparable with the dimensions of the containing vessel, the ideas used to derive eqn (13.28) no longer apply. To make a rough estimate of the effective coefficient of viscosity in the tube we may start with eqn (13.31) for the force an area S of a moving plate in a gas at low pressure and rewrite it for a tube of radius a moving at speed u_0,

$$F = \frac{u_0}{a} aPS \sqrt{\left(\frac{M}{2\pi RT}\right)}.$$

Comparison of this with eqn (13.20) defining the coefficient of viscosity suggests that at low pressures we may approximate η for the gas in the tube by

$$\eta = aP\sqrt{\left(\frac{M}{2\pi RT}\right)}.$$

Substitution of the appropriate numbers gives $\eta \sim 3.5 \times 10^{-9}\,\text{kg}\,\text{m}^{-1}\,\text{s}^{-1}$.

13.16 The pressure at the junction, P_0, is given by

$$P_0\left(\frac{a_2^4}{\ell_2} + \frac{a_1^4}{\ell_1}\right) = P_2\frac{a_2^4}{\ell_2} + P_1\frac{a_1^4}{\ell_1}.$$

13.18 The pressure inside a bubble of radius a is atmospheric plus the pressure due to 20 cm height of water plus the excess pressure inside the bubble due to surface tension. The latter is $2\Gamma/a$ and, taking atmospheric pressure equal to a standard atmosphere of 1.013×10^5 Pa, the answer is 1.036×10^5 Pa.

13.21 The root mean square displacement is $\sqrt{2Dt}$. The proportionality to the square root of time is characteristic of situations where the motion is random. It occurs, for example, in the random motion of particles suspended in a fluid, called **Brownian motion**. The root mean square displacement of a particle from its position at time zero is proportional to the square root of the elapsed time.

13.23 The locus of the maxima and minima in the curves of pressure against volume at different temperatures for a gas that obeys Dieterici's equation is given by differentiating

$$P = \frac{RT}{(V - b)}e^{-(a/RTV)}$$

with respect to volume and putting the derivative equal to zero.

$$\frac{\partial P}{\partial V} = \frac{RT}{(V - b)}\left(\frac{a}{RTV^2}\right)e^{-(a/RTV)} - \frac{RT}{(V - b)^2}e^{-(a/RTV)},$$

and equating this to zero gives

$$\frac{a}{RTV} = \frac{V}{(V - b)}.$$

The function of volume that gives the locus is then

$$P = \frac{a}{V^2}e^{-V/(V-b)}.$$

The maximum of this curve is the critical point, and the maximum occurs when the gradient is zero, that is, when

$$\frac{\partial P}{\partial V} = -\frac{a}{V^2}e^{-(a/RTV)}$$

$$\times \left(\frac{(V - b) - V}{(V - b)^2}\right) - \frac{2a}{V^3}e^{-(a/RTV)} = 0.$$

This equation reduces to

$$bV = 2(V - b)^2,$$

giving $V_c = 2b$. Substituting this value for the critical volume into the equation for the locus of the maxima and minima gives $P_c = ae^{-2}/4b^2$, and substituting the values of V_c and P_c into Dieterici's equation gives $T_c = a/4Rb$.

13.25 Since $F = \mu r^{-n}$, the dimensions of the constant μ are

$$[\mu] = [\text{mass}] \times [\text{length}]^{1+n} \times [\text{time}]^{-2}.$$

The dimensions of

$$m^{\frac{n+1}{n-1}}v^{-\frac{n+3}{n-1}}\mu^{-\frac{2}{n-1}}$$

are thus

$$[\text{mass}]^{\frac{n+1}{n-1}} \times [\text{length}]^{\frac{n+3}{n-1}} \times [\text{time}]^{-\frac{n+3}{n-1}}$$

$$\times [\text{mass}]^{-\frac{2}{n-1}} \times [\text{length}]^{-\frac{2(n+1)}{n-1}} \times [\text{time}]^{\frac{4}{n-1}}.$$

This expression reduces to

$$[\text{mass}] \times [\text{length}]^{-1} \times [\text{time}]^{-1},$$

the dimensions of the coefficient of viscosity η. Since

$$\eta \propto m^{\frac{n+1}{n-1}}v^{-\frac{n+3}{n-1}}\mu^{-\frac{2}{n-1}}$$

and $\bar{v} \propto T^{1/2}$ with T the temperature,

$$\eta \propto T^{\frac{1}{2}\left(\frac{n+3}{n-1}\right)}.$$

Hence

$$s = \frac{n + 3}{2n - 2}.$$

If $s = 0.68$ for helium, n is thus 12.1, and, if $s = 0.98$ for CO_2, $n = 5.2$. Helium has a closed shell atomic structure and carbon dioxide is a linear molecule. It is plausible that the attractive force between helium atoms will fall off with distance more rapidly than that between carbon dioxide molecules.

13.27 If there are n_1 moles of gas 1 in a vessel of volume V at temperature T, and n_2 moles of gas 2, and if both gases obey the ideal gas equation,

$$P_1 V = n_1 RT$$

and

$$P_2 V = n_2 RT.$$

The total pressure P is the sum of the partial pressures P_1 and P_2, and

$$(P_1 + P_2)V = (n_1 + n_2)RT$$

or

$$PV = (n_1 + n_2)RT = nRT.$$

The compressions apply to each separate gas, and the partial pressure of each component after compression can be calculated separately to give a new pressure P' related to the new volume V' and the new temperature T' by

$$P'V' = nRT'.$$

Chapter 14

Solids

Different kinds of solid material are described in this chapter and their properties explained in terms of their microscopic structure.

We call a material a solid if it more or less retains its shape when it is subjected to a force. If you place one brick on top of another, both bricks look just the same as they did before, though in fact the lower brick has been very slightly squashed by the weight of the upper brick. On the other hand if you bend a plastic ruler, it will bend quite a long way before breaking. Yet it is still a solid. It has 'more or less' the same shape as before it was bent in that it remains a thin sheet of plastic with almost the original length, width, and thickness, even though it is no longer straight along its length. When you let go of the end of the ruler it becomes straight again. However, an aluminium strip with exactly the same shape as the ruler will bend but stay bent when released. These examples show that solids may react in very different ways to the action of a force.

The aim of this chapter is to describe the macroscopic properties of solids, and to explain in an elementary way how these properties are related to the microscopic behaviour of the atoms and molecules making up the solid. An understanding of why solids behave in the way they do is of enormous technological importance. When you understand why a material reacts in the way it does, you may be able to design a new material that will function better. Material science is continuously improving the range of solids. Lighter and stronger materials are developed for construction; ceramics are able to withstand higher and higher temperatures in engines; semiconductors are designed to give better performance in communications and computing.

14.1 The microscopic structure of solids

Solids not only come in all sorts of shapes and sizes, but their appearance varies widely as well. Some solids are shiny, some are dull and some are transparent. If you walk in a rocky countryside, you are very likely to see some crystals of quartz, like the one in Fig 14.1. The crystal has a rather small number of flat faces, and although the arrangement of these faces is not the same in the two parts of the crystal, there is a resemblance between the two parts. The resemblance lies in the orientation of the faces: in each part of the crystal the faces have the same orientation relative to each other. If a crystal is broken, the pieces will also have faces with the same relative orientations. These features of quartz can also be observed in crystals of other crystalline materials, such as common salt and sugar.

Crystal structure

The fact that the external faces of crystals of the same material always have the same relative orientation suggests that they share a common

Fig. 14.1 The angles between the faces of two different quartz crystals are the same.

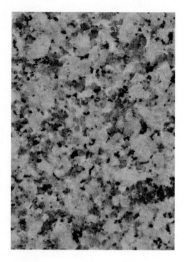

Fig. 14.2 When viewed at high magnification, this piece of rock is seen to consist of a vast number of small crystals.

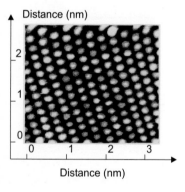

Fig. 14.3 A scanning tunnel electron microsope image of an aluminium surface. Each bright spot represents a single atom. The image is slightly distorted: the atoms lie on a square lattice with a spacing of about 0.2 nm.

internal structure. This is indeed the case, and the modern definition of a crystalline material is one in which the constituent atoms are arranged regularly over distances that are large compared to the interatomic distance. The regularity may not be discernible to the naked eye: many rocks appear to have a rough surface, but when viewed through a microscope are seen to be made up of many tiny crystals, as in the piece of granite shown in Fig 14.2. Such a material is called **polycrystalline**.

Microscopes using visible light cannot distinguish any features smaller than the wavelength of the light, which is much greater than atomic dimensions. However, with scanning tunnel microscopes—these instruments are described in the box opposite—much higher magnification is achievable, and individual atoms can be detected. Figure 14.3 shows the arrangement of atoms on one of the faces of an aluminium crystal. The atoms are spaced regularly; the image is slightly distorted and has different vertical and horizontal appearances, but in fact the distances between each row and each column of atoms are equal. If the whole crystal were displaced so that one of the aluminium atoms moved into a position previously occupied by another one, the figure would look just the same. A pattern like this, which can fill an area of any size by endless repetition, is called a **lattice**. The surface layer of atoms shown in Fig 14.3 makes up a two-dimensional lattice.

Lattices are not always as simple as the one in Fig 14.3, in which all the atoms are in equivalent positions. In the first place, real crystals often contain more than one type of atom. For example, the quartz crystals in Fig 14.1 contain both silicon and oxygen atoms. The chemical formula of quartz is SiO_2, indicating that there are two oxygen atoms for every silicon atom.

Figure 14.4 shows an example of a two-dimensional lattice with two types of atom, represented by the blue and black circles. The atoms are arranged in hexagons, but the positions of the atoms in one hexagon are not all equivalent—the neighbouring atoms around atom A are not in the same direction as those around atom X. However, the atoms A and B are equivalent, as are all the points joined by dotted lines. These points form a two-dimensional lattice, and the pattern of atoms within each rhombus like ABCD is the same.

X-ray diffraction

Scanning tunnel microscopes allow us to see what the surface of a solid is like, but not to investigate the structure of the inside of the material. In a whole crystal, the atoms are regularly arranged in all three dimensions, and they form a three-dimensional lattice. As for a two-dimensional lattice, the arrangement of atoms around each lattice point of a three-dimensional lattice is the same. We need to know the positions of the atoms in three dimensions in order to have a complete picture of the microscopic structure of a crystal.

To find the three-dimensional structure, it is necessary to use a probe that can penetrate the surface without destroying the crystal. X-rays have this property, and most of the information about crystal structure has been obtained by X-ray diffraction. X-rays are energetic photons, with a wavelength of the order of the separation of atoms in a crystal, that is, thousands of times smaller than the wavelength of visible light. The principle of X-ray diffraction is the same as that of the diffraction grating described in Section 8.7. In both cases we have to superimpose the waves from coherent sources. The difference is that, while there are at most a few thousand slits in the grating, each of the vast number of atoms in a crystal acts as a source of scattered X-rays.

To begin with, let us consider what would happen if the X-rays were scattered only from the surface plane of a crystal. Figure 14.6(a) shows an incoming plane wave arriving at this surface layer. The straight blue lines represent wavefronts where the incoming wave amplitude has its maximum value. They are one wavelength apart, and may be pictured as successive crests of the wave. The wave is scattered by atoms in the surface layer, and each atom is the centre for new wavelets setting out in phase with the incoming wave. These wavelets are shown as arcs of circles, which reinforce one another along new wavefronts to form an outgoing plane wave reflected back from the surface.

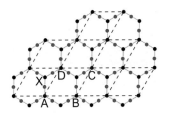

Fig. 14.4 The lattice pattern is repeated when a black atom is moved along a dotted line to the position of the next black atom.

A scanning tunnel microscope is drawn schematically in Fig. 14.5. The instrument is very simple in principle. It relies on the fact that, in the presence of a very strong electric field, electrons may leave the surface of a metal. This process is known as *tunnelling*. Electric fields strong enough to cause tunnelling may be generated close to a sharp point; the tunnelling current varies rapidly with the magnitude of the electric field. In the scanning tunnel microscope a sharp metal point is held about 1 nm from the sample to be scanned, and a voltage is maintained between the point and the sample. The tip of the point has a size comparable to atomic dimensions, and the tunnelling current changes exponentially with the distance between the tip and the sample. The microscope is usually operated in a constant current mode: the tip is moved up and down by a servomechanism to maintain constant current as it scans across the surface of the sample. The vertical and horizontal positioning can be adjusted so delicately that profiles of the surface can be obtained with the extraordinary resolution shown in Fig. 14.3. This requires the distance between the tip and the sample to be controlled to within about 0.01 nm.

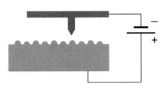

Fig. 14.5 A schematic diagram of a scanning electron microscope

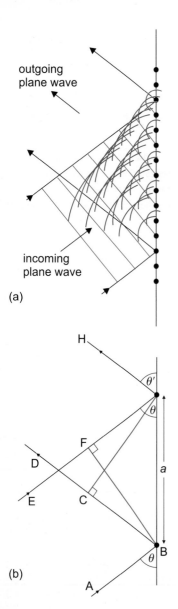

(a)

(b)

Fig. 14.6 (a) An incoming plane wave excites scattered wavelets, which combine to form an outgoing plane wave at large distances. (b) Reflected rays are in phase when the angles of incidence and reflection are the same.

⬤ *X-rays are reflected by planes of atoms*

The same reflection is shown again in Fig 14.6(b). The incoming rays AB and EG travel in the direction of propagation of the incident radiation, perpendicular to the wavefronts, at an angle θ to the surface. The figure is a view along the surface in a direction perpendicular to the rays. Note that θ is defined to be the angle betweeen the ray and the *surface*. This is the usual convention for X-rays, whereas for visible light the convention is that the *angle of incidence* of incoming rays is the angle between the ray and the *normal to the surface*, as in the discussion in Section 8.3.

In Fig 14.6(b) the ray reflected at B moves away from the surface at an angle θ' to the surface. Now look at the ray EG which is scattered by another atom at G, a distance a from the atom at B. The wavefront BF is perpendicular to the rays and the points B and F are at the same phase. The incoming wave arrives at the atoms B and G with a different phase corresponding to the path difference $FG = a\cos\theta$. Similarly, there is a path difference $BC = a\cos\theta'$ for the outgoing rays BD and GH. If $\theta = \theta'$, the path differences BC and FG are the same, and the points C and G are at the same phase. The argument applies to all pairs of atoms in the surface, and the reflected X-ray wavefronts in Fig 14.6(a) have an angle of reflection equal to the angle of incidence, just as for visible light reflected at a smooth surface.

Because the X-rays can penetrate into the crystal, we have to consider what contribution inner layers make to the reflected wave. Two layers separated by a distance d are shown in Fig 14.7. Reflected rays occur for each layer with the angle of reflection and the angle of incidence both equal to θ. There will now be interference between reflections from different layers. The incoming rays AB and FG have the same phase at B and G. If the extra distance $(BC + CD)$ by the ray reflected at C on the second layer is a whole number of wavelengths, the outgoing rays GH and DE will be in phase at G and D. Constructive interference occurs and the amplitudes from the two layers reinforce one another. Both BC and CD equal $d\sin\theta$ and, if the path difference is n wavelengths, then for a wavelength λ

$$n\lambda = 2d\sin\theta. \tag{14.1}$$

This condition, which gives the angles at which the X-ray intensities have maxima, is known as the **Bragg law**, and an angle satisfying eqn (14.1) is called a **Bragg angle**. The integer n is the **order** of the reflection.

When reflections occur from many successive layers, each with the same spacing d, the reflected amplitudes are all in phase with one another at the Bragg angle. The reflections at the Bragg angle are therefore very intense. At a nearby angle the phase difference of the scattered wavelets between the first and third layer is twice the phase difference between the first and second layers, and so on. The deeper layers are progressively further apart in phase and the intensity of the reflected X-rays falls off very rapidly away from the Bragg angle.

If the angle of incidence θ of X-rays on to the surface of a crystal is varied, intense reflections will be observed whenever θ satisfies the Bragg law. The interpretation of X-ray scattering peaks in terms of the Bragg law allows the microscopic structure of crystalline solids to be deduced. The next section outlines how this is done and discusses the microscopic structure of some different types of crystalline solid.

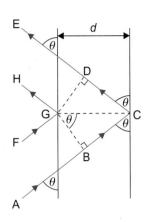

Fig. 14.7 Reflected rays are in phase with one another when θ is the Bragg angle for the two planes.

Exercise 14.1 X-rays of wavelength 0.154 nm illuminate a copper crystal. In this crystal there are planes of atoms separated by 0.18 nm. Calculate the first- and second-order Bragg angles for this set of planes.

Answer 25° and 59°. (When a copper target is bombarded with energetic electrons, the strongest component in the spectrum of X-rays produced has the wavelength used in this example.)

14.2 Crystalline solids

So far we have considered X-ray reflections from layers of atoms that are on the surface, or that are parallel to a surface layer. However, the scattering by an individual atom is small, so that X-rays pass through a small crystal without much attenuation. *All* the atoms in the crystal can take part in reflection, and there may be reflections from any plane that contains a regular array of atoms. Examples of such planes are shown in Fig 14.8, which represents a cubic lattice of common salt (sodium chloride; chemical symbol NaCl), consisting of small cubes of side a stacked together to make a larger cube. An atom is placed at each corner of all the small cubes. Successive horizontal planes of atoms, shaded in the figure, are separated by a distance a. Bragg reflections occur at these planes. Similar planes separated by a distance a are parallel to the vertical faces of the large cube.

Figure 14.9 shows that there are other stacks of successive parallel planes that are not parallel to the sides of the small cubes. In Fig 14.9(a) a shaded plane passes through opposite edges of the large cube. The separation of neighbouring planes parallel to this one, found by applying Pythagoras's theorem to triangles with atoms lying in the planes, is $a/\sqrt{2}$. Figure 14.9(b) is a view looking almost parallel to another set of planes, with separation $a/\sqrt{3}$. Both of these sets of planes can also give rise to

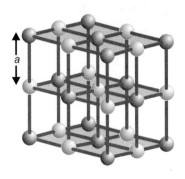

Fig. 14.8 Na^+ ions (blue) and Cl^- ions (grey) alternate in the cubic structure of NaCl. The shaded planes are separated by the nearest neighbour distance a. To illustrate the structure clearly the ions are not to scale—they should be touching.

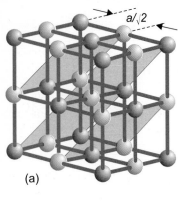

(a)

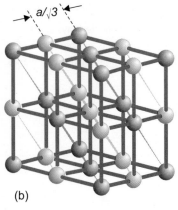

(b)

Fig. 14.9 Two sets of crystal planes that do not lie along the direction of the sides of the cubes. The separation of the planes is $a/\sqrt{2}$ in (a) and $a/\sqrt{3}$ in (b).

Bragg reflections. For different arrangements of atoms in the lattice the ratios of the distances between sets of parallel planes is also different. The Bragg angles at which X-ray maxima occur therefore depend on the three-dimensional structure of the crystal.

The one-dimensional equivalent of the Bragg law is eqn (8.54), which relates the slit separation a in a diffraction grating to the angle θ of a diffraction maximum. For light of wavelength λ incident normally on the grating, the angle of the nth-order maximum is given by $a \sin \theta = n\lambda$. If λ is known, it is a simple matter to deduce a from the observation of a few diffraction maxima.

In spite of the similarity between the Bragg law and eqn (8.54), it is by no means easy to tease out the three-dimensional structure of a crystal from X-ray measurements. The first crystal structures to be determined were those of common salt (NaCl) and the similar mineral potassium chloride (KCl). These were worked out by W.H. Bragg and W.L. Bragg, who were awarded the Nobel prize for their work on crystal structures in 1915: a biographical note on the Braggs is on the facing page.

Ionic and covalent crystals

Figure 14.10 shows a very early X-ray measurement made by the Braggs on KCl. Each spectrum is made by varying the angle of incidence of X-rays on a particular face of a KCl crystal and observing the intensity at the reflection angle. This technique is different from what is usually done nowadays, but these spectra of X-ray intensity versus angle illustrate the similarity between diffraction from a crystal and from a grating. The sharp peaks occur at the Bragg angles for sets of crystal planes parallel to this face. The Braggs did not have a monochromatic source of X-rays, and the peaks picked out in blue and grey are Bragg reflections due to two different wavelengths. The three blue peaks in spectrum 1 are reflections of different orders from the same set of planes. The distance between these planes and the corresponding distances for spectra 2 and 3 are calculated in Worked example 14.1.

Worked Example 14.1 The blue peaks in Fig 14.10 are Bragg reflections due to X-rays of wavelength 0.0587 nm. Estimate the Bragg angles from the diagrams and calculate the spacing between the crystal planes for each spectrum.

Answer In spectrum 1, blue peaks occur close to 11°, 22°, and 33°. These are clearly first-, second-, and third-order reflections from the same set of

William Henry Bragg and William Lawrence Bragg

W.H. Bragg and W.L. Bragg are the only father and son team to have been jointly awarded the Nobel prize. The award, made in 1915, was for research into the structure of crystals using X-rays. The Braggs' work established for the first time the arrangement of atoms within a solid, and opened the way to modern material science.

The father (WH) came from a farming family, but he showed an aptitude for mathematics and won a scholarship to Cambridge where he gained a degree in mathematics. In 1886, at the age of 24, he was appointed professor of mathematics and physics at the University of Adelaide in Australia. In Adelaide there was hardly any equipment, but WH, who had no previous knowledge of physics, designed apparatus for the laboratory and became a skilled experimentalist.

For almost twenty years WH devoted all his time to teaching and to working for improvements in scientific education in Australia. He also gave lectures for the general public on the latest advances in science, which were tremendously popular. In 1904, WH prepared lectures on the recently discovered electron and on radio-activity. This prompted him to obtain some radium and start—at the age of 42—to do his own experimental research. Careful measurements on the distance travelled by the α-particles emitted by the radium proved that the α-particles are not deflected by atoms, as are electrons, but pass straight through them. This was five years before Rutherford (who was awarded the Nobel prize with F. Soddy for their work on radioactivity in 1908) discovered the atomic nucleus. WH was greatly excited by his results, and he wrote to Rutherford about them, starting a lifelong correspondence and friendship.

Letters to Australia went by ship, and WH found the delays in correspondence frustrating. He returned to Britain, to a post at Leeds. Here he worked on the nature of X-rays and their interaction with matter. Again WH was prescient, writing in 1909 to Rutherford (now close at hand in Manchester), 'Supposing the identity of X-rays and light to be established, the supposition is this, I take it: the energy travels from point to point like a corpuscle: the disposition of the lines of travel is governed by wave theory. Seems pretty hard to explain: but that is surely how it stands at present.' The idea of wave–particle duality, discussed in Chapter 9 of this book, was beginning to emerge.

W. H. Bragg

W. L. Bragg

The X-ray tubes designed by WH and made in his laboratory were particularly reliable, and WH's spectrometer was the best instrument available for X-ray studies. The Bragg law was proposed by WH's son W.L. Bragg (by then at Cambridge) and WL first observed Bragg reflections from a single crystal of mica. Working together, WH and WL then obtained the spectra shown in Fig 14.10 and determined the structures of NaCl and KCl, their Nobel prize work.

The fact that both father (WH: 'Willy') and son (WL: 'Billy') were called William caused some confusion. It also led to some chagrin for WL, who felt that credit for his own early work was accorded to his father. But his own career was independently illustrious. During WL's time as head of Cambridge's Cavendish laboratory, Crick and Watson worked out the structure of DNA, Kendrew that of the protein myoglobin, and Perutz that of haemoglobin. Each of these studies earned a Nobel prize, and WL had been closely involved in the analysis, which in each case took many years. WL's laboratory had founded the new science of molecular biology.

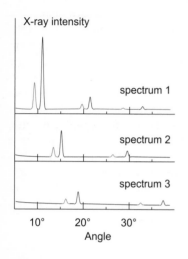

Fig. 14.10 X-ray intensities as a function of scattering angle for three different sets of planes in a KCl crystal.

● *Bragg reflections from different planes are a clue to crystalline structure*

planes. The third-order peak is very nearly half-way between the 30° and 35° markers on the graph. Using a third-order angle of 32.5° in eqn (14.1), the spacing of these planes is

$$d = \frac{3 \times 0.0587}{2 \sin 32.5°} = 0.164 \, \text{nm}.$$

The same calculation for the the second-order peaks at 30° and 37.5° in spectra 2 and 3 give spacings 0.117 and 0.096 nm, respectively.

The ratios of the spacings calculated in Worked example 14.1 are $0.164 : 0.117 : 0.096 \approx 1 : 1/\sqrt{2} : 1/\sqrt{3}$. Figures 14.8, 14.9(a), and 14.9(b) have planes with these relative separations. It follows that KCl has a lattice that is a simple cubic structure and that the side a of one of the small cubes, labelled a in the figures, is 0.164 nm in the KCl crystal.

There is a special simplifying feature in the KCl crystal. In Section 11.4 it was explained that KCl molecules are ionic: an electron is transferred from potassium to chlorine to make K^+ and Cl^- ions. The oppositely charged ions are attracted together strongly to form the ionic bond. The same bonding occurs in a KCl crystal that is composed of K^+ and Cl^- ions each bound to six nearest neighbours of opposite charge. The K^+ and Cl^- ions each have the same number of electrons as the inert gas argon and, since it is the electrons that scatter X-rays, the scattering from both types of ion is almost exactly the same.

The NaCl crystal is also ionic and its ionic separations are similar to those in KCl. But the atomic number Z for sodium is 11, compared to $Z = 17$ for Cl. After transferring an electron, the Na^+ ion has 10 electrons and the Cl^- ion has 18. The X-ray scattering from the two types of ion is quite different. Looking again at Figs 14.8 and 14.9, you can see that the planes lying along the cube faces and the shaded planes in Fig 14.9(a) contain both Na^+ and Cl^- ions. Spectrum 1 and spectrum 2 are nearly the same for NaCl as for KCl. However, the shaded planes in Fig 14.9(b) are alternately Na^+ and Cl^- ions. The separation between two planes made up of the *same* type of ion is $2a/\sqrt{3}$. The larger separation gives a smaller angular difference between orders and spectrum 3 for NaCl has extra Bragg reflections that are not present for KCl: the spectrum thus depends on the type of atoms in the crystal as well as on their positions in the lattice.

The structures of many different crystals have been determined by X-ray diffraction. The way in which atoms are arranged in three dimensions depends on the nature of their chemical bonds. Carbon has four electrons outside a closed shell and, as discussed in Section 11.4, carbon crystallizes as diamond with four strong covalent bonds at equal angles to one another. The shared electrons in the covalent bonds occupy the space between two carbon atoms, and in diamond each atom is at the centre of a tetrahedron with another atom at each corner. This structure is illustrated in Fig 14.11(a). The semiconductors silicon (Si) and germanium (Ge) also have four electrons outside a closed shell and they adopt the same crystal structure as diamond. Silicon crystals in this diamond structure are of great technological importance: large, and nearly perfect, crystals of silicon are used to make the wafers that are the substrate for nearly all the very-large-scale integrated circuits used in computers.

Oxygen atoms share two electrons—in chemical parlance they have a valency of two. They can combine with silicon, with valency four, to form silicon oxide, SiO_2. Quartz is one of the crystalline forms of SiO_2. The structure of the crystal is determined by the number of bonds formed by the atoms of each element, and the angle between the bonds on a particular atom. The Si atoms in quartz are arranged in a tetrahedral array, as in pure silicon. But in quartz an oxygen atom is inserted between each pair of silicon atoms. Each oxygen atom in quartz thus forms a bond with two silicon atoms and each silicon atom is at the centre of a tetrahedron with oxygen atoms at the four corners, as shown in Fig 14.11(b). The geometrical arrangement of the atoms is not easy to visualize; however, the atoms are indeed placed in such a way that atoms do lie in planes that are parallel to the faces of the quartz crystals shown in Fig 14.1.

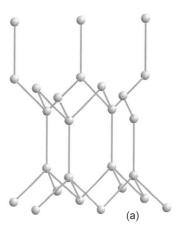
(a)

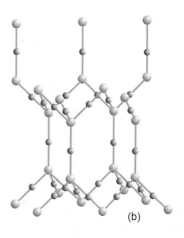
(b)

Fig. 14.11 (a) The tetrahedral structure of diamond. (b) The silicon atoms in quartz also form tetrahedra, but are separated by oxygen atoms.

Metals

Typical metals are aluminium, copper, and silver. All of these metals are good conductors of electrical current. An electric current in a metal consists of moving electrons, and the definition of a metal is that it is a material in which some of the outer electrons are free to move among the atoms of the material. Metals need not necessarily be solid: mercury is liquid at room temperature, and it is a good conductor of electricity. Metals that are solid at room temperature remain metallic when they melt.

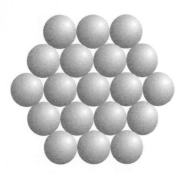

Fig. 14.12 A single layer of atoms can be packed together most closely in a hexagonal array.

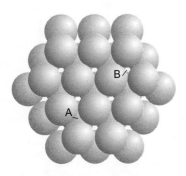

Fig. 14.13 For two close-packed layers of atoms, hollows between the atoms in the upper layer may be above atoms in the bottom layer (as at A) or above hollows in the bottom layer (as at B).

More than 50 of the naturally occurring elements are metals. Alloys formed from mixtures of metallic elements are also usually metallic. As well as being good conductors of electricity, metals are good conductors of heat. Metals also have useful mechanical properties: some are readily machined into almost any shape; some may be drawn into fine wires; others generate strong restoring forces and are used to make springs. Some of the large-scale properties of metals will be discussed later in the chapter, but for the moment we are interested in their microscopic structure.

In the solid phase, metals are normally crystalline. The inner electrons of atoms in a metal crystal are bound to a particular atom, but the outer electrons are free to move. The nucleus and inner electrons are left as ions with a net positive charge. The mobile electrons, called **conduction electrons**, are spread throughout the crystal with a nearly uniform density. The ions are embedded in a sea of negative charge, which has the effect of pulling the ions together until their inner electrons overlap and they repel one another strongly.

Placing the ions in their sea of electrons is rather like packing golf balls in a box: two golf balls also repel one another at a well-defined separation, and the strength of the repulsion is not affected by the whereabouts of other balls. The most efficient way to pack one layer of golf balls is to stagger the rows from a hexagonal array as shown in Fig 14.12. Another layer can be placed on top, with balls from the second layer fitting over the hollows in the first layer, as in Fig 14.13. There are now two possibilities for a third layer. It can be placed on the sites marked A, directly above the atoms in the bottom layer, or on the sites marked B above spaces in the bottom layer. These two structures are not equivalent and have different crystal symmetries. Both structures are commonly found in metal crystals.

Ceramics

Bricks and pottery have been manufactured by mankind for thousands of years. They are examples of **ceramics**, which are materials with properties that have been modified by heat treatment at a high temperature. The raw material for bricks and pottery is simply wet clay of various degrees of fineness, which can easily be moulded to the required shape. On firing in a kiln, water is driven off and any organic material is burnt, leaving a hard and durable polycrystalline inorganic material as the final product.

Many other ceramic materials have been developed in modern times, and all are able to withstand high temperatures since they survived high temperature during production. Ceramics are therefore used as linings for furnaces, or in more extreme conditions such as the re-entry of spacecraft into the Earth's atmosphere, when the spacecraft is protected by a skin of ceramic tiles. Other properties of ceramics may also be valuable. Carbide ceramics such as silicon carbide and tungsten carbide are very hard and therefore suitable for making machine tools and grinding materials. Others are very good electrical insulators, and they are used, for example, to insulate high-voltage cables from their pylons, which are at earth potential.

In contrast, there are some ceramics that are superconducting, which means that they can carry electric current without any dissipation of power. Metallic superconducting materials have been used for many years, but they are only superconducting at very low temperatures and require to be cooled with liquid helium, which boils at 4 K. Ceramic superconductors retain their superconductivity at much higher temperatures. These materials are difficult to fabricate and are at present only used for some specialized purposes. However, future development may well make them economically attractive for more widespread use.

● *Ceramic superconductors*

14.3 **Non-crystalline solids**

Many of the solid materials in everyday use do not have a well-ordered crystal structure. This section decribes the properties of a few important types of non-crystalline solid, with brief discussions of their microscopic structure.

When a crystalline solid melts, its constituent molecules move about and change their relative positions. The bonds between neighbouring molecules may be arranged in much the same way as in the crystal, but the regular order does not persist over a distance of more than a few times the size of the molecules. In other words there is a *short-range* order, but no *long-range* order. If the liquid cools slowly, crystals grow again, and the long-range order is once more established. But if the liquid is cooled rapidly, the molecules do not have time to rearrange into their crystalline order before they have too little energy to escape from neighbours. They become frozen into fixed positions. A solid with no crystal structure is called **amorphous**, which means shapeless: since there is no regular arrangement of molecules within the solid, a large lump of the material can take up any shape, in contrast to the preferred angles between different faces of a crystal.

Glasses

A **glass** is an amorphous solid that is transparent. Common glasses consist mostly of SiO_2, with varying amounts of other materials added to change the properties of the glass. Fused silica is a glass that is useful because it is transparent for ultraviolet light. It consists of pure SiO_2 that

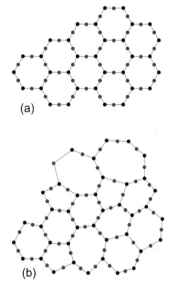

(a)

(b)

Fig. 14.14 This figure illustrates the difference between a crystal and a glass. In (a), Si and O atoms are imagined as forming a crystal in two dimensions. In the glass (b), the long-range order of the crystal is destroyed without greatly distorting the bonds on any one atom.

(a)

$$
\begin{array}{cc}
H & H \\
| & | \\
C & = C \\
| & | \\
H & H
\end{array}
$$

(b)

$$
\begin{array}{ccc}
H & H \\
| & | \\
G-C-C- & \longrightarrow & G-C-C-C-C- \\
| & | \\
H & H
\end{array}
$$

(c)

$$
\begin{array}{cccccccc}
H & H & H & H & H & H & H & H \\
| & | & | & | & | & | & | & | \\
G-C-C-C-C-C-C-C-C-G \\
| & | & | & | & | & | & | & | \\
H & H & H & H & H & H & H & H
\end{array}
$$

Fig. 14.15 Ethylene molecules can combine together to form long-chain polymers.

has been melted and cooled too rapidly to recrystallize. In both the quartz crystal and the glass each Si atom is at the centre of a tetrahedron with an oxygen atom at each corner. The tetrahedra are arranged neatly in the crystal, but in the glass the bonds on the oxygen atoms are bent, so that, although the structure is still made up of tetrahedra touching at their corners, the relative orientation of tetrahedra that are some distance apart is random.

Figures 14.14(a) and (b) illustrate schematically in two dimensions the difference between the crystalline and the glassy forms of SiO_2. In the figures it is assumed that each silicon is connected to three other oxygen atoms lying in a plane. In the crystal the atoms interlock in hexagonal rings that extend in a regular pattern over the whole plane. The ring structure is maintained in the glass, but some of the rings have different numbers of members, and the regular pattern has completely disappeared.

Polymers

So far we have been considering solids containing only one element or made up of very simple molecules such as SiO_2. **Polymers** are individual molecules that have a repeating section and that may have very large molecular weights. A common example is polyethylene. Polyethylene is made by connecting together large numbers of ethylene molecules in a long chain. The ethylene molecule has the chemical formula C_2H_4, and the schematic structure shown in Fig 14.15(a). The valency of carbon is four, and in C_2H_4 each carbon atom forms two bonds with hydrogen. A **double bond** is formed between the two carbon atoms: it is partly because of the ability to form compounds including double bonds that there is such a rich variety of organic compounds based on carbon. The formation of long chains of repeating C_2H_4 units is initiated by the attachment of a group of atoms, labelled G in Fig 14.15(b), to one of the C atoms in C_2H_4 by a single bond. One of the carbon bonds is now unoccupied, resulting in a molecule called a **free radical**. Another C_2H_4 molecule can then be added to form a new free radical as shown in Fig 14.15(b): this is energetically favourable because the only change is that a carbon double bond has been replaced by two single bonds. This process continues until it is terminated when groups are attached to both ends as in Fig 14.15(c). Polymer molecules may also be formed with elements other than carbon in the chain, and with more than one type of repeating unit in the chain.

The length of polymer molecules may be altered by changing the chemical conditions or the duration of the polymerization reaction. The average length of the polymer molecule can be controlled, but there is always a spread in values about the average length. Since different molecules have different lengths, it is not possible to construct a strictly crystalline solid ordered in a three-dimensional lattice. Often there are weak bonds between neighbouring molecules, helping the formation of ordering in two dimensions, like the bundle of wires in a multicore cable. More usually, when a polymer solidifies from the liquid in which the polymerization has been carried out, disorder results. The molecules are not so much like a bundle of wires as a tangled heap of pieces of string of different lengths.

The chemical composition of the repeating units, the length of the molecules, and the conditions of solidification can all be varied in the manufacture of polymers. Polymers can be prepared with a great variety of physical properties, and for this reason they are of commercial importance. Polymers are frequently transparent, and perspex (also called lucite) is often used in place of an inorganic glass. In contrast to metals, polymers do not usually contain any mobile electrons, and they are commonly used as insulators. Their strength and elasticity can be tailored to suit a particular requirement, and polymers are chosen, for example, for ropes, cling-films, and artificial fibres for clothing.

Biological molecules with a very large molecular weight (usually called **biopolymers** or **biomacromolecules**) are similar to non-biological polymers in that they have a backbone consisting of a long repeating structure. However, unlike polyethylene (where the side bonds of the carbon backbone are all attached to hydrogen atoms), the side groups in biopolymers are not all the same. Protein molecules, for example, consist of one or more chains, each with a long central core to which a sequence of side groups called amino acids is attached. Amino acids are organic chemicals essential to life, and each different protein has its own precise sequence of amino acids.

In collagen (the most abundant protein in the body) each molecule is made up of three chains wound together to form a triple helix. The chains, which all have the same length (300 nm), assemble together to form very long fibrils. Assembly into fibrils occurs in a regular way, with all the 300 nm long molecules mutually staggered by a constant distance of 67 nm. This gives rise to a periodic structure in the fibrils, readily seen in electron microscope pictures like the one in Fig 9.18. Like the twisted strands of a rope, collagen fibrils have great strength in tension. Because of this, they are able to provide the body with mechanical support in tissues such as tendon, skin, and cartilage and, after mineralization, in bone.

The study of biological macromolecules is an important and fascinating topic in its own right. The point we wish to emphasize here is that, just as for the inorganic solids we have discussed, the physical properties of the macromolecules are related to the geometrical arrangement of the bonds that hold the material together.

● *The structure of collagen*

Composite materials

All the solids we have considered so far are uniform, having the same structure throughout the material. **Composite solids** are those in which different materials are deliberately mixed, usually in order to achieve better mechanical properties. Concrete, which is a mixture of cement and an aggregate consisting of sand and gravel, is stronger than any of its constituents on its own. Although concrete is immensely strong under compression, it is much weaker in tension. For this reason steel tie rods are added in reinforced concrete to give strength in both compression and tension.

Modern material science has developed many new composite materials, and the properties of these composites can often be designed to be best suited to their particular applications. For example, very hard composites have been developed for use use in grinding tools. Other common examples of composites are glass fibre, in which filaments of glass are embedded in a plastic, and similar but stronger materials using carbon or boron fibres instead of glass.

14.4 Mechanical properties of solids

The discussion in this chapter has so far been about the microscopic structure of solids and how this is related to the interactions between the individual atoms and molecules making up the solid. This information is needed for a proper understanding of the behaviour of particular materials and for suggesting ways in which their properties might be modified. However, when building a bridge or designing a cable to carry large electrical currents, it is not feasible to describe everything at the atomic level. In the rest of this chapter we shall be considering the large-scale behaviour of solids, informed by what we already know about their microscopic structure.

Elasticity

At the beginning of this chapter we pointed out that a characteristic of a solid is that it has a *shape*. A force is required in order to change the shape, and usually the force has to be maintained if the solid is to be kept in its new shape. If you stretch a spring, for example, the force needed to keep the spring stretched does not get less as time goes by. Since there is no motion, the net force is zero, and the spring is exerting an equal and opposite force. This force is called a restoring force, because it is acting in the direction needed to restore the spring to its original length. An **elastic** solid is one that exhibits this behaviour of exerting a restoring force when deformed. If the solid returns to exactly its original shape when the external force is removed, the solid is said to be **perfectly elastic**.

When the shape of a solid is deformed, for example, when a wire is stretched, the magnitude of the restoring force is often proportional to the deformation, at least for small deformations. When a force causes an increase x in the length of a wire, the restoring force F is tending to reduce x. If the force is proportional to x we may write

$$F = -kx \qquad (14.2)$$

where k is a constant. This is Hooke's law, which we met when working out the potential energy of a spring in Section 3.6, and also in the context of simple harmonic motion in Chapter 6. If Hooke's law applies to a material, it is perfectly elastic, since the restoring force only vanishes for $x = 0$.

Stress and strain

When a solid is deformed by an external force, the extent of the deformation depends on the shape of the solid as well as on the strength of the force. For example, if a short fat rod is pulled by equal and opposite forces of magnitude F, as in Fig 14.16, it is in tension but the dimensions of the rod are scarcely changed at all. A long thin rod of the same material, on the other hand, may be considerably elongated by the same forces. In both cases, any elongation occurs because the atoms of the material are pulled further apart by the force. The response must be a property of the material—we should expect the distance between a neighbouring pair of atoms to change by the same amount for the same force between them, whatever the shape of the object being deformed. This is indeed correct, and we can concentrate on the properties of the material by talking not about force and change in length, but about *stress* and *strain*.

For materials under tension or compression, the **stress** is the force per unit area, and the **strain** is the fractional change in length in the direction of the force. In Fig 14.16, the stress on the thin rod is (F/S) and the strain is $\Delta\ell/\ell$.

Remember that the stress here refers to a stationary object. The *net* force acting is zero—equal and opposite forces are always needed to maintain tension or compression in an elastic object.

Young's modulus and the shear modulus

For solids that obey Hooke's law, the ratio of stress to strain is a constant that depends only on the nature of the solid, and not on its shape. If the solid is in compression or in tension, like the rod in Fig 14.16, we may

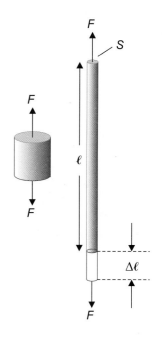

Fig. 14.16 The same force produces a greater stress on the thin rod than on the fat rod, and the thin rod, which is fixed at the top, is more extended.

write

$$\left(\frac{\text{Stress}}{\text{Strain}}\right) = \left(\frac{F}{S} \times \frac{\ell}{\Delta\ell}\right) = Y. \tag{14.3}$$

The constant Y is called **Young's modulus**. Its dimensions are (force per unit area) or equivalently pressure and it is measured in pascals (symbol Pa). We have already met Young's modulus in Chapter 8 when discussing waves travelling down a rod, and eqn (14.3) is a rearrangement of eqn (8.1).

Worked Example 14.2 A transformer that must be operated at high voltage is mounted on a platform supported by four porcelain legs. Each leg is a cylinder that is 50 cm high and has a circular cross-section with radius 10 cm. Young's modulus for the porcelain is 8×10^{10} Pa. The mass of the transformer is 100 kg. By how much is the length of the legs reduced when the transformer is placed on the platform?

Answer The total area of cross-section of the legs is $4 \times \pi(0.1)^2 \, \text{m}^2$, and the force acting on them due to the weight of the transformer is $100g$ newtons. The stress on the legs (i.e. the force per unit area) caused by the transformer is $100 \times 9.8/(4 \times \pi(0.1)^2) \approx 8 \times 10^3$ Pa. The strain is (stress)/(Young's modulus) $= 8 \times 10^3/8 \times 10^{10} = 10^{-7}$. The legs are only compressed by 0.5×10^{-7} m. Porcelain is commonly used for electrical insulators, and in compression it is almost as strong as steel. However, like concrete, porcelain is not very strong when in tension. Even though the compression in the example is so tiny, the dimensions are realistic, and are determined by the stability of the structure against knocks from the side rather than the elastic strength of the porcelain.

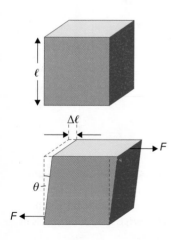

Fig. 14.17 A shear stress twists the cube without altering the area of any of its faces.

Elongation is not the only kind of deformation that a solid can experience. Figure 14.17 shows a solid with faces that are initially rectangular. When equal and opposite horizontal forces are applied to the upper and lower faces, the solid deforms so that the upper face has moved sideways with respect to the lower. The horizontal displacement $\Delta\ell$ of the upper face is perpendicular to the vertical height ℓ. This type of deformation is called **shear**. The strain $\Delta\ell/\ell$ is equal to $\tan\theta$, where θ is the shear angle shown in the figure. The stress is the force F divided by the area S of the horizontal face of the solid. If Hooke's law is obeyed,

$$\left(\frac{\text{Stress}}{\text{Strain}}\right) = \frac{F}{S\tan\theta} = H \tag{14.4a}$$

or

$$H = \frac{F}{S\theta} \qquad (14.4b)$$

if θ is small. The constant H is the **shear modulus** for the solid. Like Young's modulus, the shear modulus is a force per unit area, measured in pascals.

The net force on the cube in Fig 14.17 is zero, but there is a couple acting on it: a couple was defined in Section 4.3 as a system of forces that exert a torque but no net force. The couple has the value $F\ell$, and the stress can also be written as $F\ell/A\ell$, or in words, couple per unit volume.

Worked Example 14.3 A 1.25 cm diameter copper rod is firmly clamped at one end. The length of the rod is 40 cm and a screw thread is cut with a die, starting at the unclamped end. To cut the thread, a force of 10 N is needed at each end of the die at a distance of 15 cm from the axis of the rod, as shown in Fig 14.18. Through what angle is the end of the rod twisted by the die? (The shear modulus H for copper is 4.5×10^{10} Pa.)

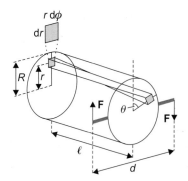

Fig. 14.18 The stress on a small segment of the twisted rod depends on its distance r from the axis.

Answer The rod is twisted by the applied torque, but the shear strain is not uniform throughout a section of the rod at the point where the die is applied. Consider a small segment of the rod at radius r, of thickness dr, and subtending an angle $d\phi$ at the axis of the rod: the area of the segment is $r \, dr \, d\phi$. If the unclamped end of the rod is twisted through an angle θ, the distance moved by a point on the end of the rod at a radius r is $r\theta$. Calling the length of the rod ℓ, the strain on the small segment is $r\theta/\ell$. The stress on the segment is the force per unit area, or, since (stress)/(strain) = H, the force on the segment is

$$H \times \text{strain} \times \text{area of segment} = H \times (r\theta/\ell) \times r \, dr \, d\phi$$

and the torque about the axis of the rod due to this force is

$$\frac{Hr^3\theta \, dr \, d\phi}{\ell} \, .$$

The total torque obtained by integrating over r and θ is equal to the applied torque Fd (see Fig 14.18):

$$\int_{\phi=0}^{2\pi} \int_{r=0}^{R} \frac{Hr^3\theta}{\ell} \, dr \, d\phi = \frac{\pi HR^4\theta}{2\ell} = Fd$$

or

$$\theta = \frac{2Fd\ell}{\pi R^4 H} \, . \qquad (14.5)$$

For the values given in this problem,

$$\theta = \frac{2 \times 10 \times 0.3 \times 0.4}{\pi \times (0.00625)^4 \times 4.5 \times 10^{10}} \approx 0.01 \text{ radians or about } 0.6°.$$

This is quite a small angle and, because of the factor R^4 in eqn (14.5), the angle very rapidly becomes smaller still as the diameter of the rod increases. The same torque on a 2.5 cm (1 inch) diameter rod causes a twist of only about 0.03°.

● *The speed of longitudinal and transverse sound waves*

Both Young's modulus and the shear modulus of a solid can be determined by measuring the speed of waves in a rod. Longitudinal sound waves are excited by exerting a periodic pressure on the end of the rod in the direction of propagation of the wave. The speed v_L of the longitudinal waves is given by eqn (8.6) as

$$v_L = \sqrt{\frac{Y}{\rho}} \tag{14.6}$$

where ρ is the density of the rod. This equation was derived by assuming that stress and strain are proportional, that is, that Hooke's law applies. Hooke's law also applies to transverse waves, excited by moving the end of the rod back and forth in a direction perpendicular to its length. The wave equation for transverse waves is derived in the same way as for longitudinal waves. The restoring force is now given by the shear modulus H instead of Young's modulus, and the speed of the transverse wave v_T is

$$v_T = \sqrt{\frac{H}{\rho}}. \tag{14.7}$$

Poisson's ratio and the bulk modulus

When a length of wire is stretched, its cross-sectional area does not remain constant, but becomes smaller. Associated with the axial strain $\Delta\ell/\ell$ along the length of the wire is a lateral strain in the directions perpendicular to the length. For a wire of circular cross-section, if the radius were r before stretching and $(r - \Delta r)$ after stretching, the strain perpendicular to the length is $\Delta r/r$. The ratio

$$\sigma = \left(\frac{\Delta r}{r}\right) \times \left(\frac{\ell}{\Delta\ell}\right) = \frac{\text{lateral strain}}{\text{axial strain}} \tag{14.8}$$

is called **Poisson's ratio**. Poisson's ratio, like Young's modulus and the shear modulus, is a property of the solid and does not depend on its shape.

Exercise 14.2 A copper wire has a diameter of 1 mm. What is the change in diameter after a mass of 10 kg is suspended by a length of the wire? Young's modulus for copper is 1.1×10^{11} Pa and Poisson's ratio is 0.33. Verify that it is a good approximation to neglect the change in diameter when calculating the stress on the wire.

Answer The diameter is reduced by 0.00038 mm after the weight is attached. The resulting change in area may be neglected, since the elastic constants are only given to two significant figures.

When a solid object is subjected to a uniform pressure P acting in all directions, the object becomes smaller, but its shape is unaltered. For example, if the cube of side a in Fig 14.19 experiences an equal inward force Pa^2 on each of its six faces, there is an equal compression in the x-, y-, and z-directions, resulting in a cube with slightly shorter sides. Suppose the length of the sides is reduced from a to $a - \Delta a$. The volume of the cube reduces from a^3 to $(a - \Delta a)^3 \simeq a^3 - 3a^2 \Delta a$, since Δa is very much smaller than a for practicable pressures, and terms in $(\Delta a)^2$ and $(\Delta a)^3$ may be neglected. The fractional change in volume $\Delta V/V = 3\Delta a/a$ is the strain in response to the stress of the pressure P on all sides of the cube. As usual, (stress)/(strain) is constant for small deformations, and we can define the **bulk modulus** K of the material of the cube as

$$K = \frac{PV}{\Delta V}. \tag{14.9}$$

Since the change in volume has occurred as a result of the application of three separate linear forces in the x-, y-, and z-directions, the bulk modulus is related to Young's modulus. We must remember that, if the forces Pa^2 in the x-, y-, and z-directions are each applied in turn, each causes an extension in the two perpendicular directions by an amount given by Poisson's ratio. When the forces Pa^2 in the $\pm x$-directions are applied to the cube, the length of the sides parallel to the x-axis becomes $a - \Delta a$, and the length of the other sides becomes $a + \sigma \Delta a$. The volume of the cube is now

$$(a - \Delta a) \times (a + \sigma \Delta a)^2 \simeq a^3 - a^2 \Delta a (1 - 2\sigma),$$

and the volume has changed by a fraction

$$(\Delta a(1 - 2\sigma)/a) = P(1 - 2\sigma)/Y.$$

The fractional change in volume when forces Pa^2 are applied to all six faces of the cube is simply three times this amount. The fractional change

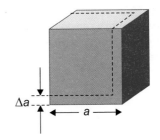

Fig. 14.19 When the cube is subjected to a uniform pressure, all the sides are reduced in length by the same amount.

⬤ *Relations between the elastic moduli*

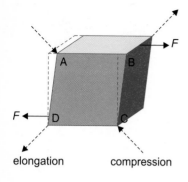

elongation   compression

Fig. 14.20 A shear strain is equivalent to a compression and an elongation in perpendicular directions.

in volume is also given by the bulk modulus, and it follows that

$$K = \frac{PV}{\Delta V} = \frac{Y}{3(1 - 2\sigma)}. \tag{14.10}$$

Figure 14.20 shows a cube that has been subjected to a shear stress. The same deformation can be achieved by forces that squeeze the corners A and C, and pull the corners B and D. An argument similar to the one leading to eqn (14.10) relates the shear modulus to Young's modulus and Poisson's ratio, with the result

$$H = \frac{Y}{2(1 + \sigma)}. \tag{14.11}$$

Exercise 14.3 Young's modulus for aluminium is 6.9×10^{10} Pa and Poisson's ratio is 0.33. Calculate the bulk modulus and the shear modulus of aluminium.

Answer $K = 6.8 \times 10^{10}$ Pa and $H = 2.6 \times 10^{10}$ Pa.

Equation (14.6) for the speed of longitudinal waves is only correct if the measurement is carried out on a rod that is thin compared with the wavelength. There is then no lateral stress on the rod, which contracts by an amount determined by Poisson's ratio, and Young's modulus is the appropriate elastic modulus to use. However, if the lateral dimensions of the solid are not small compared with the wavelength, the contraction cannot occur. The elastic modulus X corresponding to a longitudinal strain with no accompanying lateral strain is given in terms of Young's modulus and Poisson's ratio by the equation

Table 14.1 Young's modulus Y and Poisson's ratio σ for various materials at room temperature. The unit GPa (gigapascal) is 10^9 Pa

$$X = Y \left\{ \frac{(1 - \sigma)}{(1 + \sigma)(1 - 2\sigma)} \right\}. \tag{14.12}$$

This is the modulus that applies to longitudinal waves passing through the Earth. The difference between the arrival time of longitudinal and transverse waves may be used to estimate the position of the epicentre of an earthquake if the elastic moduli X and H are known for the rocks lying between the recording position and the epicentre.

The relations between the elastic moduli given in eqns (14.10), (14.11), and (14.12) mean that only two of the moduli are independent. Measurements of any two enable the other two to be determined by applying these equations. The elastic properties of a number of materials are summarized in Table 14.1, which lists values of Young's modulus and Poisson's ratio.

Material	Y (GPa)	σ
Aluminium	69	0.33
Iron	208	0.29
Copper	110	0.34
Tungsten	407	0.28
Diamond	1040	0.10
Solid argon	7	0.30
Silica glass	64	0.20
Fused silica	73	0.165
Polyethylene	1.1	0.4

As is customary, the elastic moduli have all been defined as positive quantities, without regard to the signs of the stresses and strains. For example, the bulk modulus K is defined in eqn (14.9) as $K = PV/\Delta V$, where the change in volume ΔV is actually a *reduction* in volume for a positive external pressure. The response to changes in pressure of gases was discussed in Chapter 13, and in that context the signs of changes in pressure and volume were taken into account. The compressibility χ of a gas is given in eqn (8.7) as

$$\chi = -\frac{1}{V}\left(\frac{\partial V}{\partial p}\right).$$

This is a positive quantity, as is the compressibility $1/K$ of a solid. Remember also that the elastic moduli for solids have only been defined for the range of stress and strain where Hooke's law applies, that is, where the materials are perfectly elastic. Gases are never elastic, and it is necessary to define the compressibility in terms of a differential coefficient.

How are the elastic moduli related to the bonding between atoms? Consider a polycrystalline solid consisting of atoms all of the same type. If the solid is uniformly expanded or contracted so that the distance r between neighbouring atoms increases or decreases, the potential energy per atom $U(r)$ behaves as shown in Fig 14.21. The energy has a minimum at $r = a$, and rises more steeply for $r < a$ than for $r > a$ because the atoms repel one another very strongly when they are close together. At the minimum dU/dr is zero and, if we make a Taylor expansion of $U(r)$ about $r = a$ (eqn (20.62)), the first terms are

$$U(r) = U(a) + \frac{1}{2}(r-a)^2\frac{d^2U}{dr^2}\bigg|_{r=a} \cdots. \tag{14.13}$$

If $(r - a)$ is small the remaining terms in the expansion may be neglected. In other words, close to the minimum, the energy has the same form as for a simple harmonic oscillator, and the restoring force obeys Hooke's law.

Suppose that a pressure P is applied to a volume V of the solid. According to eqn (14.9) the volume decreases by an amount $\Delta V = PV/K$, where K is the bulk modulus. The work done by the pressure is $P\Delta V = P^2V/K$. This work increases the potential energy of the interatomic bonds. If the sample contains N atoms and each atom has n nearest neighbours, the total number of bonds is $\frac{1}{2}Nn$—the factor $\frac{1}{2}$ arises because each bond is connecting two atoms. The average increase in energy ΔU for each bond is therefore

$$\Delta U = \frac{2P^2V}{KNn}. \tag{14.14}$$

● *Compressibility for elastic solids and for gases*

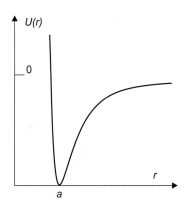

Fig. 14.21 The interaction energy $U(r)$ of an atom as a function of its distance r from a neighbouring atom.

● *The relation between bond energy, bulk modulus, and pressure*

Worked Example 14.4 Aluminium (Al) has a bulk modulus $K = 6.8 \times 10^{10}\,Pa$ and a crystal structure in which each atom has 12 nearest neighbours. The density of aluminium is $2.7\,g\,cm^{-3}$ and its mass number is 27. Calculate the average change in the energy of each interatomic bond when aluminium is subjected to a pressure of $10^8\,N\,m^{-2} \approx 1000$ atmospheres.

Answer Since the density ρ of Al is $2.7\,g\,cm^{-3}$ and its mass number A is 27, one mole occupies a volume of $10\,cm^3$. For a pressure $P = 10^8$ Pa, the increase in bond energy according to eqn (14.14) is

$$\Delta U = \frac{2 \times 10^{16} \times 10^{-5}}{6.8 \times 10^{10} \times 6.0 \times 10^{23} \times 12} = 4.0 \times 10^{-25} \text{ joules,}$$

or 2.5×10^{-6} eV. This is very much less than the energy of the chemical bond between two atoms, which in Section 11.4 was shown to be several eV. Yet the stress of 10^8 Pa on the aluminium is equivalent to the weight of a mass of about one tonne acting on $1\,cm^2$. This illustrates that the macroscopic stresses normally encountered are extremely small when compared to the forces acting between atoms.

In the discussion of elastic properties we have assumed that the relation between stress and strain for the solid is a constant, independent of the direction of the stress. This is justifiable for solids that are amorphous or polycrystalline, as are most structural materials. However, the situation is more complicated for single crystals. Because the interatomic bonds have particular orientations with respect to the faces of the crystal, the magnitude of the strain may depend on the direction of the stress as well as on its magnitude. When this occurs, the solid is said to be **anisotropic**. Different values of the elastic moduli must be specified for different directions of the stress with respect to the crystal axes.

The inelastic behaviour of solids

Hooke's law only applies to solids when they are deformed by a small amount. What happens when the stress becomes very large depends on the nature of the stress. Under compression the restoring force in a solid becomes larger than predicted by Hooke's law. This is because the interaction energy between neighbouring atoms has the form shown in Fig 14.21, rising more steeply than a parabola at short distances. For strains somewhat too large for Hooke's law to apply, a solid may remain elastic under compression, still relaxing back to its undeformed shape when the compressing stress is released. However, eventually all materials become *inelastic*. For example, if a metal rod is clamped too firmly in a vice, when it is removed it will remain flattened where it was gripped by the jaws of the vice. Similar behaviour applies to solids under tension, except that the restoring force becomes weaker for large strains, since the slope of the potential energy curve in Fig 14.21 becomes smaller.

The range of stress and strain over which a solid is elastic is usually fairly constant for different samples of the same material, provided they have been prepared in the same way. Inelastic behaviour sets in at the **elastic limit**. If a material is strained beyond its elastic limit it fails to return to its original shape when the stress is removed and the material is said to be **plastic**.

In the plastic region materials respond in a variety of different ways. Figure 14.22 shows curves of stress versus strain for glass, polyethylene, and aluminium under tension. The magnitudes of stress and strain are realistic for these materials, but the curves are not exact—they are intended to show schematically different kinds of plastic behaviour. In each case Hooke's law only applies for small elongations; this is the linear part of each curve. The elastic limit is close to the end of the linear part of the curve.

For a brittle material like glass, the point of fracture almost coincides with the elastic limit. For polyethylene and aluminium the strain increases much more rapidly with the tension beyond the elastic limit. The maximum possible stress is called the **tensile strength** of the material. After reaching the maximum stress the material may immediately fracture, like polyethylene. In contrast, many metals are ductile, which means that they may elongate to very large strains, though unable to support a tension as large as the tensile strength. Under the right conditions ductile metals can be extended to many times their original length, and drawn out into wires.

All our examples of elastic stresses and strains were in simple geometries, where the stress in the solid sample was uniform or easily calculable. In civil and mechanical engineering the distribution of stress and strain is itself very difficult to calculate. Failure cannot be tolerated in the structural members of a tall building when it is battered by a high wind, or in the ends of a rapidly rotating turbine blade in a jet engine. All the materials must be stressed within their maximum strength, with an adequate safety factor under the most extreme conditions that can be envisaged. On the other hand, cost must be minimized. A compromise between these opposing aims requires reliable information about the materials and stress calculations that, for complicated objects, can only be computed numerically.

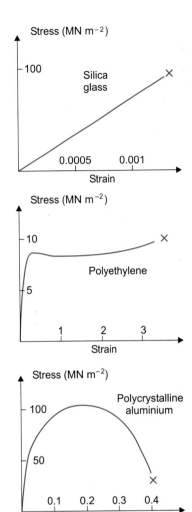

Fig. 14.22 The stress/strain relation for materials under tension. The point of fracture is marked by ×.

14.5 **Thermal and electrical properties of solids**

Specific heat

When the temperature rises in a gas, heat energy is required to increase the translational energy of the molecules and to increase their internal energy by exciting rotational and vibrational degrees of freedom. The molecules in a solid are more restricted. They are not free to rotate or to move from one place to another within the solid. Only vibrational energy remains. According to the principle of equipartition of energy (Section 12.7), in thermal equilibrium there is an energy $\frac{1}{2}k_B T$ for each degree of freedom. Each atom can vibrate in the x-, y-, or z-directions, and each has potential and kinetic energy, giving a total of six degrees of freedom per atom. The total thermal energy of 1 mole is therefore $3N_A k_B T = 3RT$, and the specific heat is $3R$. Notice that this is twice the value $\frac{3}{2}R$ given by eqn (12.46) for the specific heat of an ideal gas at constant volume. This is because the whole of the energy in an ideal gas is kinetic energy. The atoms in an ideal gas exchange kinetic energy with one another, but there is no potential energy, and only three degrees of freedom.

● *Each atom in a solid has six degrees of freedom*

This result, that the specific heat of a solid has the constant value $3R$, is correct provided that $k_B T$ is large compared with the separation of individual vibrational states. The vibration of diatomic gas molecules was considered in Chapter 10, and it was found that vibrational levels at a frequency ν are only found at discrete values separated by an energy $h\nu$, where h is Planck's constant. This energy is much greater than $k_B T$ at room temperature and molecular vibrations are not excited at room temperature in diatomic gases.

In solids an atom is bound to several nearest neighbours, and the restoring force on one individual atom that is displaced is very strong if the neighbours remain fixed. The energy of vibrations of a single atom, restricted by a strong force constant, is large. How, then, do atoms in a solid absorb heat energy? The answer is that the atoms can vibrate collectively in a wave motion. Standing waves can be set up in a solid in which each atom is oscillating at the same frequency, just like the standing waves on a string described in Section 7.7.

● *Modes of oscillation in a solid*

The longest possible wavelength of a standing wave in the string and the corresponding lowest frequency, the fundamental frequency, are determined by the length of the string. The string may also oscillate at higher frequencies that are multiples of the fundamental frequencies. Similarly, standing waves in a solid sample have a longest wavelength determined by the size of the sample. The sample may also oscillate in other modes, but only at particular discrete frequencies. Provided Hooke's law applies, the equations governing all the other modes are linear and each mode is an independent simple harmonic vibration of the sample. Just as for the one-dimensional modes discussed in Section 6.7, any possible vibration is a superposition of the different modes.

Although an enormous number of atoms are moving in each standing wave mode, quantum mechanics requires that each mode can only have certain discrete energy values. The allowed energies are related to frequency in the same way as for the oscillations of the diatomic molecule, and are given by eqn (10.38). For a mode at frequency ν, the allowed energies are $(n + \frac{1}{2})h\nu$, where n is zero or a positive integer.

How do the frequencies of the modes vary with their wavelength? The speeds v_L and v_T of longitudinal and transverse waves in a solid are given in eqns (14.6) and (14.7). In both cases the wave speed v is determined by the density of the solid and an elastic modulus, but is independent of the wavelength λ. The frequency $\nu = v/\lambda$ is therefore small for large λ. The energy $h\nu$ of the low-frequency modes is much less than $k_B T$ at ordinary temperatures, and when solids absorb heat the energy is stored by the vibrating atoms. As the temperature is lowered, the probability of exciting a mode of frquency ν is reduced by the Boltzmann factor $\exp(-h\nu/k_B T)$ and the specific heat becomes smaller.

This section started by pointing out that each atom in a solid can vibrate in the x-, y-, or z-directions, so that the total number of vibrations is simply three times the number of atoms in the sample. This number must also be the total number of modes, since each mode is an independent vibration. There must therefore be an upper limit as well as a lower limit to the frequency of modes of the solid, and a corresponding lower limit to the wavelength.

We can make an estimate of the shortest possible wavelength without enumerating all the different modes. Figure 14.23(a) shows a line of atoms in a crystalline solid with regular spacing a. The blue line represents a standing wave of wavelength λ, and the black circles show the positions of atoms excited by this wave at a particular time. If the wave speed is v, each atom oscillates at a frequency v/λ with an amplitude given by the envelope of the wave.

Now look at the grey line representing the shorter wavelength λ'. This line also passes through the atoms, and it corresponds to a higher frequency v/λ', which is not one of the natural frequencies of the crystal. Of course, the crystal can be forced to oscillate at a higher frequency, or indeed at any frequency whatever, by the imposition of an external force, but forced oscillations do not contribute to the vibrational energy in the crystal due to its thermal motion. When two or more wave patterns are fitted to the line of atoms, the longest wavelength is always more than twice the interatomic spacing a. The highest natural frequency thus corresponds to the wavelength $2a$, which occurs when neighbouring atoms are oscillating in opposite directions, as shown in Fig 14.23(b).

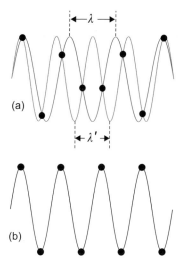

Fig. 14.23 (a) Different wavelengths can be associated with the same atomic displacements. (b) The shortest natural frequency of the crystal occurs when neighbouring atoms oscillate out of phase.

Standing waves in a three-dimensional crystal are more complicated than waves on a line of atoms. However, we can take twice the average separation d of atoms in a solid as a rough measure of the shortest possible wavelength occurring in thermal motion. The speed of longitudinal waves is faster than that of transverse and, using eqn (14.6) for v_L, we estimate the maximum frequency ν_{max} in the solid to be

$$\nu_{max} = \frac{v_L}{2d} = \frac{1}{2d}\sqrt{\frac{Y}{\rho}}. \tag{14.15}$$

Worked Example 14.5 Aluminium has mass number 27, density $2.7\,\text{g cm}^{-3}$ and Young's modulus $6.9 \times 10^{10}\,\text{Pa}$. Estimate the highest frequency ν_{max} of the normal modes of oscillation in aluminium.

Answer One mole of Al occupies $10\,\text{cm}^3 = 10^{-5}\,\text{m}^3$. The average separation of atoms is therefore $\sqrt[3]{10^{-5}/6.0 \times 10^{23}} \approx 2.5 \times 10^{-10}\,\text{m}$. In SI units the density is $2.7 \times 10^3\,\text{kg m}^{-3}$, and in eqn (14.15)

$$\nu_{max} = \frac{1}{5.0 \times 10^{-10}}\sqrt{\frac{6.9 \times 10^{10}}{2.7 \times 10^3}} \approx 1.0 \times 10^{13}\,\text{Hz}.$$

The energy separation of the highest frequency modes is comparable with the energy separation of the vibrations of a diatomic molecule, and at ordinary temperatures these modes do not contribute the full amount $\frac{1}{2}k_B$ per degree of freedom to the specific heat. A more complete, but still approximate, treatment of specific heat uses a simple model to count the permitted modes of oscillation, and fixes the highest frequency by requiring that the total number of modes is $3N_A$ per mole. This model gives a maximum frequency not very different to our rough approximation, and also predicts the variation of specific heat with temperature in terms of a single parameter, which is called the **Debye temperature** Θ_D. The thermal energy $k_B\Theta_D$ at the Debye temperature is equal to the energy $h\nu_{max}$, where ν_{max} is the highest frequency of the normal modes of the solid.

The specific heat is shown in Fig 14.24 as a function of the ratio T/Θ_D of the actual temperature to the Debye temperature. The specific heat is close to $3R$ at high temperatures when $T/\Theta_D > 1$. Figure 14.24 shows that the specific heat becomes smaller and smaller for temperatures well below Θ_D, as more and more modes have vibrational energies above $k_B T$. The high energy modes are then not excited thermally and, like the vibrational levels in diatomic gases, do not contribute much to the specific heat. For Al the Debye temperature, derived from a fit to the low-temperature part of the specific heat versus temperature curve,

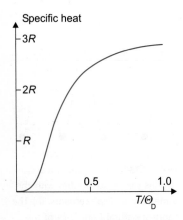

Fig. 14.24 The variation of specific heat with temperature for a solid.

is 428 K: although room temperature is below Θ_D for Al, the specific heat is still more than 90% of the limiting value $3R$.

Exercise 14.4 Use the maximum frequency calculated in Worked example 14.5 to estimate the Debye temperature for aluminium.

Answer 480 K.

The simple model used in eqn (14.15) somewhat overestimates the Debye temperature of aluminium, but leads to the correct qualitative conclusion that the specific heat must fall sharply at temperatures not much less than room temperature. Table 14.2 gives values for the Debye temperature for several solids calculated with eqn (14.15), and compares them with experimental values.

Thermal expansion

So far we have assumed that all the oscillations in a solid are simple harmonic. This is not quite correct, because the potential energy curve between neighbouring atoms, shown again in Fig 14.25, is not simple harmonic. The energy increases faster when the atoms move close together than when they move apart. The horizontal blue lines cover the range of motion of atoms interacting with this potential for a number of different relative energies. As the energy increases, the mean position of the atoms moves a little further apart because of the anharmonic nature of the force. The thermal energy increases with temperature, increasing the average separation of the atoms and thus the size of a solid sample. For a temperature change ΔT that is not too large, the resulting linear strain $\Delta \ell / \ell$ is proportional to ΔT

$$\frac{\Delta \ell}{\ell} = \alpha \Delta T. \tag{14.16}$$

The constant α is called the **linear expansion coefficient** for the material. Since thermal energies are rather small compared with the energies of bonds between atoms the coefficients α are also small. Values are given in Table 14.3 for a number of solids.

Thermal conductivity

When a temperature difference occurs between opposite faces of a slab of solid material, heat is transported from the hotter face to the cooler face.

Table 14.2 The Debye temperature for various solids. The column headed $h\nu_{max}/k_B$ is the estimate of Θ_D based on eqn (14.15)

Material	Θ_D (K)	$h\nu_{max}/k_B$ (K)
Aluminium	430	480
Copper	340	370
Tungsten	400	450
Gold	165	190
Diamond	2230	2400

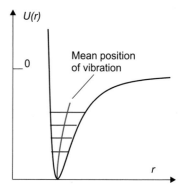

Fig. 14.25 As the thermal energy increases, the average separation of the atoms in a solid increases.

Table 14.3 Linear expansion coefficients α for some metals and for glass. Invar is an iron–nickel alloy with a low value of α

Material	α K^{-1} × 10^{-6}
Aluminium	23
Iron	12
Copper	17
Tungsten	4.5
Invar	1.6
Glass	3.3

The rate of heat transfer per unit area J_Q is proportional to the temperature gradient, and eqn (13.35), which was defined for gases, applies to solids as well. If there is no source of heat in the slab, J_Q is constant in the steady state. Taking the normal to the faces to be the z-direction, we may write

$$J_Q = -\kappa_s \frac{dT}{dz} \tag{14.17}$$

where κ_s is the thermal conductivity of the solid, measured in $W\,m^{-1}\,K^{-1}$.

Exercise 14.5 The area of a window is $2\,m^2$ and its thickness is $5\,mm$. The thermal conductivity of the glass is $1.7\,W\,m^{-1}\,K^{-1}$. When the outside temperature is $0°C$ the inside of the glass is at $10°C$. What is the rate of heat loss through the window?

Answer $6.8\,kW$.

What is the mechanism for transferring heat energy through a solid? In gases the heat is transferred by the gas molecules themselves. On average the molecules travel a distance of one mean free path before colliding: excess thermal energy is thus moved from hot to cool parts of the gas. In solids the thermal energy is not associated with individual atoms or molecules, but is tied up in collective modes of oscillation. In Section 7.6 it was shown that a number of waves in one dimension can be super-imposed to form a wave group that can carry energy through the medium in which the wave is travelling. The same applies in three dimensions, and wave packets may be formed that are localized in space. Because the speed of sound waves in solids is independent of wavelength (i.e. the waves are non-dispersive), such a wave packet travels through the solid at the speed of sound.

In thermal equilibrium the energy density in a solid at a particular moment is not uniform. Fluctuations occur in the vibrational energy within a small volume. Clumps of energy formed by the fluctuations are wave groups called **phonons**, and they move through the solid. In the discussion of specific heat, we assumed that solids have a perfect crystalline structure, allowing waves to travel unimpeded through the entire sample. But crystals are not perfect, and phonons are scattered or change their energy when meeting imperfections in the crystal. The imperfections may, for example, be changes in the crystal orientation in a polycrystalline material or the disturbance caused by other phonons. The result is that the phonons have a mean free path that is much smaller than the size of the solid sample. The movement of phonons

● *Energy is carried from hot to cool parts of a solid by phonons*

from hot to cool regions is responsible for thermal conductivity in a solid in much the same way as the movement of molecules causes heat transfer in a gas.

The thermal conductivity κ_p due to the vibrations may be written in the same form as eqn (13.36) for an ordinary gas. Calling v the speed of the phonons, N their density, and λ_p their mean free path,

$$\kappa_p = \tfrac{1}{3}Nv\lambda_p\left(\frac{\partial\epsilon}{\partial T}\right) = Nk_B v\lambda_p, \tag{14.18}$$

since the thermal energy ϵ associated with each atom is $3k_B T$.

For nonmetallic solids like the glass in Exercise 14.5, the whole of the heat transfer is carried by the phonons. Thermal conductivities are much higher in metals—it is a common observation that, when dipped into a hot liquid, the handle of a metal spoon is much hotter than the handle of a wooden spoon. It was explained in Section 14.1 that in a metal there are some electrons that are not bound to particular atoms, but move through the solid material. The mobile electrons, called conduction electrons, behave rather like an electron gas, travelling considerable distances through the solid. Their contribution κ_e to the thermal conductivity again has the same form as eqn (13.36) for an ordinary gas,

● *Conduction electrons transport heat in metals*

$$\kappa_e = \tfrac{1}{3}N_e\bar{v}_e\lambda_e\left(\frac{\partial\epsilon}{\partial T}\right) \tag{14.19}$$

where N_e is the number density of conduction electrons, $\bar{v}_e$ their mean speed, and λ_e their mean free path.

If we take the rate of change of mean energy with temperature as $\tfrac{3}{2}k_B$ as for the real gas and introduce the mean time between collisions $\tau = \lambda_e/\bar{v}_e$, eqn (14.19) can be rewritten as

$$\kappa_e = \tfrac{1}{2}N_e\bar{v}_e{}^2 k_B\tau = \frac{3N_e k_B^2 T\tau}{2m_e}, \tag{14.20}$$

since the average energy of an electron $\tfrac{1}{2}m_e\bar{v}_e^2 \approx \tfrac{3}{2}k_B T$. For a typical metal each atom contributes about one conduction electron. At ordinary temperatures the thermal conductivity of metals is almost entirely due to the electronic contribution, partly because the speed of electrons is much greater than the speed of sound in the metal.

Worked Example 14.6 The diameter of the base of a copper pan is 20 cm and its thickness is 4 mm. A hotplate delivers 1 kW of heat to the pan, distributed uniformly over its bottom surface. The pan is filled with boiling water. Estimate the temperature of the bottom surface of the pan. The thermal conductivity of copper is $400\,\mathrm{W\,m^{-1}\,K^{-1}}$.

Answer The rate of heat transfer per unit area is $1000/(\pi \times 0.1^2)\,\mathrm{W\,m^{-2}}$. Calling the temperature difference ΔT, eqn (14.17) becomes

$$\frac{1000}{\pi \times 0.1^2} = \frac{400 \times \Delta T}{0.004}, \text{ leading to } \Delta T = \frac{4}{4\pi},$$

about 0.3 K. The copper is an extremely effective heat transfer medium in this situation, and the base of the pan is only just above the water temperature.

Electrical conductivity

Since electrons are charged, a net movement of electrons constitutes an electric current. Conduction electrons can therefore carry electric current in a metal as well as heat.[1] When no electric field is present, the electrons have the same distribution of speeds in all directions and there is no net movement of charge. Suppose that an electric field E pointing in the z-direction is now introduced. The electrons each have a mass m_e and carry a charge $-e$ and they experience a force $(-eE)$ in the z-direction. The mean time between collisions for the electrons is τ and their speed in the z-direction is reduced on average by $eE\tau/m_e$. There is a net movement of electrons in the $(-z)$-direction and a net movement of charge in the z-direction.

For a number density N_e of conduction electrons, a charge $j = e^2 N_e E\tau/m_e$ crosses unit area per second: j is the **electric current density**. For practical electric fields the drift speed $eE\tau/m_e$ is very small compared with the average speed of the electrons, and j is proportional to the electric field E,

$$j = \frac{e^2 N_e E\tau}{m_e} = \sigma E. \tag{14.21}$$

This is Ohm's law, and the constant σ is the **electrical conductivity** of the metal.

---

[1] Electric fields and electric currents are discussed more fully in Part 5. Some definitions and results from Part 5 are anticipated here. Readers who are unfamiliar with this material are referred to Sections 15.2 and 16.2.

Table 14.4 Thermal and electrical conductivities of metals, and insulators at room temperature

Material	κ (W m^{-1} K^{-1})	σ (Ω^{-1} m$^{-1} \times 10^{-7}$)	$\kappa/\sigma T$ (W Ω K$^{-2} \times 10^{-3}$)
Aluminium	240	3.8	2.1
Iron	80	1.0	2.6
Copper	400	5.9	2.3
Lead	35	0.5	2.4
Silica glass	1.7	10^{-20}	
Fused silica	1.4	10^{-25}	

The effect of the Pauli principle on conduction electrons

There is an important difference between an electron gas and an ordinary gas, which should be mentioned. This is that the conduction electrons are restricted by the Pauli exclusion principle, which was discussed in Section 11.5. The exclusion principle requires that no two electrons can have the same wave function. As a result, in a typical metal, even if there is no thermal energy, the conduction electrons occupy states that have a spread of kinetic energies from zero up to a highest value, ϵ_F, that is more than 1 eV. The amount of thermal energy available for exchange between the electrons is about $k_B T = 0.025$ eV at room temperature, much less than ϵ_F. Most of the conduction electrons are unable to change their state, because all the states within an energy range $k_B T$ are already occupied. Only those within the small energy interval $k_B T$ below ϵ_F can be excited into unoccupied states. The average rate of change of energy with temperature $\partial \epsilon / \partial T$ is therefore much less than the value $\frac{3}{2} k_B$ that applies for an ordinary gas. However, this is compensated by the high speed of the electrons engaged in energy exchange, which have energies near ϵ_F. The only change to eqn (14.22) required by a proper treatment of the electron energies is that a factor $\pi^2/3$ appears on the right-hand side.

Because electrons in metals carry both energy and charge, the thermal and electrical conductivities of metals are related. Comparing eqns (14.20) and (14.21), we see that the electron gas model predicts that the ratio κ_e/σ should be proportional to the temperature, and

$$\frac{\kappa_e}{\sigma T} = \left(\frac{k_B}{e}\right)^2. \tag{14.22}$$

Table 14.4 lists $\kappa_e/\sigma T$ for a number of metals at 0°C. The values are remarkably constant, though about three times bigger than predicted by eqn (14.22). This is because our free electron gas model of metals is oversimplified. The discrepancy is explained in the information paragraph below.

Electrical conductivities are given in Table 14.4 for two nonmetallic solids. The range of values is vast: the conductivity of copper is some 10^{20} times greater than that of glass or porcelain. These materials have no

Fig. 14.26 A high-voltage insulator is deeply corrugated along its length to ensure a long path on the surface between the two ends.

conduction electrons and there is effectively no mechanism for transporting electric charge through the bulk of the solid. Porcelain is often used as an electrical insulator in high-voltage equipment. Negligible currents flow through the solid, but it is difficult to avoid 'tracking' of electric current across surfaces: this is why large insulators have the shape shown in Fig 14.26; this ensures that there is no short path on the surface from one end of the insulator to the other.

Silicon and germanium are important examples of materials called semi-conductors. At very low temperatures semiconductors have practically no conduction electrons and are insulators. However, in semiconductors there are states for conduction electrons at energies not much higher than the energy of the highest occupied states. For silicon and germanium the energy gaps E_g between the localized and conduction states are about 1.1 eV and 0.7 eV, respectively. Although these energy gaps are considerably bigger than thermal energies at room temperature, thermal excitation maintains an appreciable population of conduction electrons. The density of conduction electrons, and hence the conductivity, is determined by the Boltzmann factor $\exp(-E_g/k_B T)$, which *increases* as the temperature increases. The conductivity of semiconductors thus increases as T increases, the opposite of the temperature variation observed in metals.

Problems

Level 1

14.1 Each Na^+ ion in an NaCl crystal has six Cl^- ions as its nearest neighbours. How many Na^+ ions are equidistant next nearest neighbours?

14.2 Silicon crystals have a set of planes separated by 0.37 nm. Calculate the Bragg angles for the first-, second-, and third-order reflections from these planes using X-rays of wavelength 0.24 nm.

14.3 If you were to repeat the Braggs' measurement of spectrum 1 in Fig. 14.10 using X-rays of wavelength 0.154 nm (copper X-rays), how many peaks would you see? What would be the angle of the first-order Bragg reflection?

14.4 Polyethylene has a density a little less than that of water—typically 0.95 g cm^{-3}. The length of the C—C bond in the polymer chain is 0.254 nm. Estimate the area

per molecule in the cross-section of a bundle of polymer molecules. What is the average length of a molecule in a sample of polyethylene with average molecular weight 400 000?

14.5 A 10 kg mass is suspended by a nickel wire that has a diameter of 0.5 mm and is 2 m long. The extension of the wire is 5 mm. What is Young's modulus for nickel?

14.6 The edges of an aluminium cube are 10 cm long. One face of the cube is firmly fixed to a vertical wall. A mass of 100 kg is then attached to the opposite face of the cube. The shear modulus for aluminium is 26 GPa. What is the resulting vertical deflection of this face?

14.7 A silica glass rod has a diameter of 1 cm and is 10 cm long. Use the data from Fig 14.22 to estimate the largest mass that could be hung on the rod without breaking it.

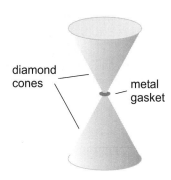

Fig. 14.27 Tapered pieces of diamond are used to subject small samples to extremely high pressure. Metallic gaskets are used to prevent the diamond tips welding together.

14.8 Anvils made of single crystals of diamond, with the shape shown in Fig 14.27, are used to investigate the behaviour of materials under extremely high pressures. The flat faces at the narrow end of the anvil have a diameter of 0.5 mm, and the wide ends have a diameter of 1 cm. The wide ends are subjected to a compressional force of 50 000 N. What is the pressure at the tip of the anvil?

14.9 The width of a rubber band is 5.0 mm when it is not stretched. After stretching to three times its original length, the width of the band has decreased to 3.0 mm. Calculate Poisson's ratio for the rubber, and estimate the volume change accompanying the threefold increase in length.

14.10 The speeds of longitudinal and transverse waves through the Earth's crust are 8 km s^{-1} and 4.4 km s^{-1}, respectively. A seismograph records the P-waves (longitudinal) and S-waves (transverse) from an earthquake. The separation of the P- and S-waves is 20 seconds. How far away is the earthquake?

14.11 Tantalum, which is an easily machinable metal with a melting point of 3000°C, is often used in high-temperature apparatus. The expansion coefficient of tantalum is 6.5×10^{-6} K^{-1}. A tantalum tube is 5 cm long at room temperature. What is its length at 2000°C?

Level 2

14.12 The separation of nearest neighbour Na$^+$ and Cl$^-$ ions in sodium chloride is 0.14 nm. The interaction energy between electric charges is discussed in Chapter 15, and for two charges $\pm e$ at a distance r, is given by eqn (15.24) as $-e^2/(4\pi\epsilon_0 r)$. Estimate the bond energy of the ionic bond between nearest neighbours in sodium chloride. In SI units, $e \sim 1.6 \times 10^{-19}$, $\epsilon_0 \sim 8.8 \times 10^{-12}$.

14.13 Poisson's ratio for nickel is 0.31. Calculate the change in volume of the wire in Problem 14.5 when the 10 kg mass is hung on it.

14.14 A hollow brass cylinder of outer radius 3 cm has walls of thickness 1 mm. When the pressure inside the cylinder is increased by 10^7 Pa, the radius increases by 10^{-6} m. Calculate the bulk modulus of brass.

14.15 An aluminium rod has a diameter of 1 cm and is 10 cm long. Estimate the largest mass that can be hung on it without breaking. Use the data in Fig 14.22, assuming that the cross-section remains constant as the rod elongates and that Young's modulus has a constant value of 0.33.

14.16 The average density of rocks within 20 km of the Earth's surface is 3300 kg m^{-3} and the speeds of longitudinal and transverse waves in these rocks are 8 km s^{-1} and 4.4 km s^{-1}, respectively. Estimate the shear modulus and the bulk modulus of the Earth's crust. (Which is the appropriate modulus to use in calculating the speed of longitudinal waves?)

14.17 A Bragg reflection is observed at 25° in a copper crystal when its temperature is 0°C. The temperature is raised by 110°C, and the reflection angle changes by 0.05°. Is the change an increase or a decrease? What is the linear expansion coefficient of copper?

14.18 A clock with a brass pendulum has been regulated to give the exact time at 18°C. How long will it take to lose one second at 20°C? The thermal expansion coefficient of brass is 2.0×10^{-5} K^{-1}.

14.19 A pond starts freezing on a cold night. After the Sun has gone down the air temperature falls to -10°C. The water temperature is 4°C. The latent heat of melting of ice is 335 J g^{-1}, and its thermal conductivity is 2.3 W m^{-1} K^{-1}. Estimate the thickness of the ice after 12 hours in these conditions.

Level 3

14.20 In the experiment described in Worked example 14.1, spectra for NaCl were measured as well as those for KCl. The spectra look similar, but in spectrum 3 the first two peaks occur at 10.5° and 21°. What is the distance separating neighbouring Na^+ and Cl^- ions in NaCl?

14.21 The binding energy of a single ion in an NaCl crystal is $1.75e^2/4\pi\epsilon_0 a$, where a is the separation between neighbouring Na^+ and Cl^- ions. (The factor 1.75, which arises as a result of summing the positive and negative potential energies due to interaction with all the other ions, is called the Madelung constant for the crystal.) The binding energy per mole for NaCl is 410 kJ. Use this information to estimate Avogadro's number. The electronic charge $e \sim 1.6 \times 10^{-19}$ coulombs and $\epsilon_0 \sim 8.8 \times 10^{-12}$ SI units.

14.22 A torsional pendulum consists of two 1 kg masses at either end of a light horizontal bar. The distance between the centres of the two masses is 50 cm. The bar is suspended at its centre by a fused silica thread which is 1 m long and has a diameter of 0.5 mm. The elastic constants for fused silica are given in Table 14.1. Calculate the period of torsional oscillations of the pendulum.

14.23 The mass in Problem 14.5 is replaced by a circular disc with a mass of 1 kg and a diameter of 5 cm. When the disc is suspended from its centre it undergoes torsional oscillations in a horizontal plane with a period of 13 s. Calculate the shear modulus of nickel.

14.24 The potential energy per atom $U(r)$ in a solid sample varies with the mean separation r of the atoms as shown in Fig 14.21. In terms of the deviation x of the mean separation from the potential minimum at $r = a$, the Taylor series in eqn (14.13) may be written as

$$U(x) = U(0) + \tfrac{1}{2} c_2 x^2 + \tfrac{1}{6} c_3 x^3 + \cdots$$

$$\text{where} \quad c_2 = \left.\frac{d^2 U}{dx^2}\right|_{x=0} \quad \text{and} \quad c_3 = \left.\frac{d^3 U}{dx^3}\right|_{x=0}$$

are the second and third differentials of $U(x)$ evaluated at $x = 0$.

The probability that two nearest-neighbour atoms are displaced by a distance x from the potential minimum is proportional to the Boltzmann factor $\exp(-U(x)/k_B T)$: the Boltzmann factor is discussed in Section 12.6. Assuming that c_3 is small and that higher terms in the Taylor series are negligible, obtain an expression for the mean displacement of any pair of nearest-neighbour atoms at a temperature T. Hence show that the linear expansion coefficient of the solid is $\alpha = -k_B c_3/2ac_2^2$.

14.25 The potential energy of a pair of atoms at a separation r is often approximated by the Lennard-Jones potential

$$U(r) = \frac{a}{r^{12}} - \frac{b}{r^6}.$$

The density of aluminium is 2.7 g cm^{-3}; its binding energy is 326 kJ mol^{-1}. Aluminium has a cubic structure with coordination number six. Use these data to obtain the coefficients a and b for aluminium and then estimate its linear expansion coefficient from the expression given at the end of Problem 14.24.

14.26 A rectangular beam has a length ℓ, a width a, and a depth h as shown in Fig 14.28. The beam is fixed at one end and bent by a mass M at the other end: the mass of the beam is small compared with M. The beam is horizontal at the fixed end, and at a distance x from this end the angle to the horizontal is ϕ. Because of the

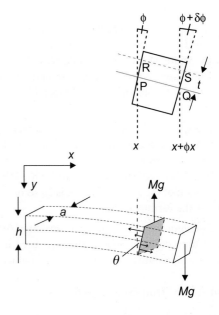

Fig. 14.28 The upper surface of the loaded beam is stretched and the lower surface compressed. The central dashed line has the same length as for the unloaded beam.

bending the top surface of the beam is stretched and the bottom surface compressed. By considering the moments about the point P in the diagram, show that the angle ϕ is given by

$$\frac{d\phi}{dx} = \frac{12Mg(\ell - x)}{Yah^3}$$

where Y is Young's modulus for the beam, and hence that the deflection of the beam at its end is $4Mg\ell^3/Yah^3$. This calculation neglects the shear strain of the beam. Provided that (h/ℓ) is small, this is a good approximation.

14.27 Use the results of the previous problem to estimate the deflection of an aluminium beam of length $\ell = 1$ m, width $a = 10$ cm, and depth $h = 4$ cm when a mass of 100 kg is attached to its free end. Young's modulus for aluminium is $Y = 69$ GPa.

14.28 The load M is removed from the beam in Problem 14.26. Obtain an expression for the deflection at its end due to its own weight. Estimate this deflection for the aluminium beam in Problem 14.27. The density of aluminium is $2.7 \, \mathrm{g\,cm^{-3}}$.

Some solutions and answers to Chapter 14 problems

14.1 12.

14.4 The length taken up by each CH_2 unit in the polyethylene molecule is $\ell = 0.254$ nm. Its volume is ℓS if the cross-sectional area of the molecule is S. Each unit has a mass $m_u = 14$ amu, and the density is $\rho = m_u/\ell S$. Substituting the values given in the problem leads to $S = 1.0 \times 10^{-19} \, \mathrm{m^2}$. The number of units in an average molecule with molecular weight 400 000 is $400\,000/14$ and the average length of a molecule is $400\,000 \times 0.254/14$ nm or 7.3 μm.

14.5 2.0×10^{11} Pa.

14.6 If the cube length is ℓ and a mass m is attached to the free face, the deflection $\Delta\ell$ represents a shear $\Delta\ell/\ell$. The torque is $mg\ell$ and the stress (equals torque per unit volume) is mg/ℓ^2. The shear modulus is

$$H = \frac{\text{Stress}}{\text{Strain}} = \frac{mg}{\ell\Delta\ell};$$

hence

$$\Delta\ell = \frac{mg}{\ell H} = \frac{100 \times 9.8}{0.1 \times 26 \times 10^9} \approx 4 \times 10^{-7} \, \mathrm{m} = 0.4 \, \mu\mathrm{m}.$$

14.8 2.5×10^{11} Pa.

14.9

$$\text{Poisson's ratio } \sigma = \frac{\text{lateral strain}}{\text{axial strain}} = \frac{2/5}{3} = \frac{2}{15}.$$

Assuming that the thickness of the band changes in the same ratio as the width, then

$$\frac{\text{new volume}}{\text{original volume}} = 3 \times \left(\frac{3}{5}\right)^2 = \frac{27}{25},$$

which represents an 8% volume increase.

14.10 The distance to the epicentre of the earthquake is about 200 km.

14.13 The increase in volume is about $1.5 \times 10^{-11} \, \mathrm{m^3}$.

14.14 The bulk modulus is 10^{11} Pa.

14.15 From Fig 14.22 the maximum stress sustainable by the aluminium is 8×10^7 Pa, at a strain of about 0.15. Since Poisson's ratio is 0.33, the lateral strain at this extension is $\Delta r/r = 0.33 \times 0.15$, and the area of the rod is $\pi \times [0.005(1 - 0.33 \times 0.15)^2] = 7.1 \times 10^{-5} \, \mathrm{m^2}$. The stress under these conditions is $mg/(7.1 \times 10^{-5}) \simeq 8 \times 10^7$. Hence

$$m = \frac{8 \times 10^7 \times 7.1 \times 10^{-5}}{9.8} \simeq 580 \, \mathrm{kg}.$$

14.17 Bragg's law is $n\lambda = 2d\sin\theta$. The variables d and θ change when the temperature changes, and by differentiating Bragg's law with respect to these variables we can find the relation between the small changes δd and $\delta\theta$:

$$0 = 2\delta d \sin\theta + 2d\cos\theta \, \delta\theta.$$

Hence θ decreases as d increases, and

$$\frac{\delta d}{d} = -\frac{\delta\theta}{\tan\theta} = \frac{0.05 \times \pi}{180\tan 25°} = 1.87 \times 10^{-3}.$$

Since the temperature change is $110°$, the linear expansion coefficient for copper is $\alpha = 1.87 \times 10^{-3}/110 \approx 1.7 \times 10^{-5}\,\mathrm{K}^{-1}$.

14.18 The clock will lose 1 second in about 14 hours.

14.19 If the thickness of the ice is ℓ and the temperature difference between the water and air is $\Delta\theta$, the rate of energy loss of the water is $\kappa\Delta\theta/\ell\,\mathrm{W\,m^2\,s^{-1}}$. If the latent heat of ice is L and its density ρ, the rate of increase in thickness of the ice is

$$\frac{\mathrm{d}\ell}{\mathrm{d}t} = \frac{\kappa\Delta\theta}{L\rho}\,\mathrm{d}t.$$

Hence

$$\ell\,\mathrm{d}\ell = \frac{\kappa\Delta\theta t}{L\rho} \quad \text{and} \quad \ell^2 = \frac{2\kappa\Delta\theta t}{L\rho}.$$

After 12 hours, the data given in the problem lead to a thickness of ice of about 9 cm.

14.20 The separation of neighbouring Na^+ and Cl^- ions is 0.142 nm.

14.22 From eqn (14.5) the torque exerted by the silica thread when it is twisted through an angle θ is $\pi R^4 H\theta/2\ell$. The moment of inertia of two masses M at a separation d is $2M \times \left(\frac{1}{2}d\right)^2 = \frac{1}{2}Md^2$. The equation of motion for the torsional oscillations is therefore

$$\frac{\pi R^4 H\theta}{2\ell} = \frac{1}{2}Md^2\frac{\mathrm{d}^2\theta}{\mathrm{d}t^2},$$

and the period T is given by

$$T^2 = \frac{4\pi^2}{\omega^2} = \frac{4\pi^2 M\ell d^2}{\pi R^4 H} \quad \text{and}$$

$$T = \sqrt{\frac{4\pi M\ell d^2}{R^4 H}} = \left\{\frac{4\pi \times 0.5^2}{(0.25 \times 10^{-3})^4 \times H}\right\}^{1/2}.$$

For silicon $Y = 7.3 \times 10^{10}\,\mathrm{Pa}$ and $\sigma = 0.165$. From eqn (14.11) the shear modulus is $H = Y/2(1+\sigma) = 3.1 \times 10^{10}\,\mathrm{Pa}$. The oscillation period is about 160 seconds.

14.24 The mean displacement of nearest neighbours in the sample is

$$\bar{x} = \frac{\int_{-\infty}^{\infty} x\exp(-U(x)/k_B T)\,\mathrm{d}x}{\int_{-\infty}^{\infty}\exp(-U(x)/k_B T)\,\mathrm{d}x}$$

$$= \frac{\int_{-\infty}^{\infty} x\exp\{(-U(0) + \frac{1}{2}c_2 x^2 + \frac{1}{6}c_3 x^3)/k_B T\}\,\mathrm{d}x}{\int_{-\infty}^{\infty}\exp\{(-U(0) + \frac{1}{2}c_2 x^2 + \frac{1}{6}c_3 x^3)/k_B T\}\,\mathrm{d}x}.$$

The factor $\exp(-U(0)/k_B T)$ cancels and, since the term in c_3 is small, $\exp(-\frac{1}{6}c_3 x^3/k_B T)$ may be expanded to first order, leading to

$$\frac{\int_{-\infty}^{\infty} x\exp(-\frac{1}{2}c_2 x^2/k_B T)(1 - c_3 x^3/6k_B T)\,\mathrm{d}x}{\int_{-\infty}^{\infty}\exp(-\frac{1}{2}c_2 x^2/k_B T)(1 - c_3 x^3/6k_B T)\,\mathrm{d}x}.$$

The first term in the numerator and the second term in the denominator are odd functions of x and they vanish when integrated from $+\infty$ to $-\infty$. Substituting in the standard integrals gives

$$\bar{x} = -\frac{1}{2}k_B T \times \frac{c_3}{c_2^2} \quad \text{and hence} \quad \alpha = \frac{1}{a}\frac{\mathrm{d}\bar{x}}{\mathrm{d}T} = -\frac{k_B c_3}{2ac_2^2}.$$

14.25 The linear expansion coefficient calculated from the data is 2.04×10^{-5}. This is fortuitously close to the value 2.3×10^{-5} given in Table 14.3, since the potential is approximate. However, the result is not very sensitive to the powers of r used for the attractive and repulsive parts of the potential, and similar calculations will give reasonable estimates for other solids.

14.26 Figure 14.28 shows a small section of the beam between x and $x + \delta x$. The length of the line PQ running through the centre of this section is unaltered by the deflection. This length is $\delta x/\cos\phi = \delta x$ to second order since ϕ is small. The deflections at P and Q are ϕ and $\phi + \delta\phi$. Above PQ the beam is under tension, and the line RS at a distance t from PQ has a length $\delta x + t\delta\phi$, that is, the strain at RS is $t\delta\phi/\delta x$. The force required to cause this strain on a section with area $a\delta t$ is $Yat\delta t(\delta\phi/\delta x)$ and the moment about P of this force is $Yat^2\delta t(\delta\phi/\delta x)$. Taking the limits as δt and δx tend to zero and integrating over the thickness h of the beam, the total moment about P is

$$\int_{-h/2}^{h/2} Yat^2\,\mathrm{d}t\frac{\mathrm{d}\phi}{\mathrm{d}x} = \frac{1}{12}Yah^3\frac{\mathrm{d}\phi}{\mathrm{d}x}.$$

The moment due to the elastic forces balances the moment $Mg(\ell - x)$ of the deflecting mass M; hence

$$\frac{d\phi}{dx} = \frac{12Mg(\ell - x)}{Yah^3} \quad \text{and}$$

$$\phi(x) = \frac{12Mg(\ell x - \frac{1}{2}x^2)}{Yah^3} + \text{constant}.$$

Since $\phi = 0$ at $x = 0$, the constant is zero. Choosing the direction of the y-axis as in Fig 14.28, $\phi = \delta y/\delta x$, and

the deflection at the end of the beam is

$$y(\ell) = \int_0^\ell \frac{12Mg(\ell x - \frac{1}{2}x^2)dx}{Yah^3}$$

$$= \frac{12Mg\left[\frac{1}{2}\ell x^2 - \frac{1}{6}x^3\right]_0^\ell}{Yah^3} = \frac{4Mg\ell^3}{Yah^3}.$$

14.28 The deflection is $3\rho g\ell^4/(2Yh^2)$, which under no load is 0.36 mm for the dimensions given in 14.27.

Summary of Part 4
Properties of matter

Important concepts and key equations from Part 4 are briefly explained in this summary

Chapter 12 **Thermal physics**

One mole of any substance contains Avogadro's number N_A of molecules or atoms if the substance is atomic. The temperature of an object is proportional to the average kinetic energy of its microscopic constituents. At the freezing point a substance changes from liquid to solid at constant temperature. The other way round, the change of phase from solid to liquid occurs at the melting point, which is the same temperature as the freezing point under the same external conditions. Latent heat is given out or taken in when a phase change occurs. The latent heat of evaporation per mole depends upon the average binding energy, E_B^ℓ of a molecule in the liquid, the number of nearest neighbours, n_ℓ, and N_A.

$$L_{evap} = \tfrac{1}{2} N_A n_\ell E_B^\ell. \tag{12.1}$$

Thermal equilibrium is a dynamic equilibrium between different parts of a substance or substances, when the whole system is at the same temperature. For a perfect gas, the pressure P and volume V of N molecules of mass m are related by

$$PV = \tfrac{1}{3} Nmv^2 \tag{12.4}$$

where v is the average molecular speed. For 1 mole of gas, $N = N_A$ and

$$PV = N_A k_B T = RT \tag{12.7}$$

where k_B is Boltzmann's constant and the gas constant R is Boltzmann's constant times N_A. Equation (12.7) is used to define the ideal gas scale of temperature.

The internal energy per mole of an ideal gas is

$$E = \tfrac{3}{2} RT. \tag{12.11}$$

Conservation of energy requires that, when an amount of heat dQ is input to a substance, raising its temperature by dT and doing external work dW,

$$dE = dQ - dW \tag{12.12}$$

where E is the internal energy. This is the first law of thermodynamics.

The specific heat per mole of a substance at constant volume is given by

$$C_V = \left(\frac{dQ_V}{dT} \right). \tag{12.13}$$

An analogous equation, (12.14), defines the specific heat per mole at constant pressure C_P, and for an ideal gas the difference

$$C_P - C_V = R. \tag{12.17}$$

The relation between initial and final pressures and volumes P_1, V_1 and P_2, V_2 for an isothermal expansion of an ideal gas at constant temperature T is

$$P_1 V_1 = P_2 V_2. \tag{12.18}$$

This is Boyle's law.

For an adiabatic expansion,

$$P_1 V_1^\gamma = P_2 V_2^\gamma \tag{12.22}$$

where

$$\gamma = \frac{C_P}{C_V} \tag{12.23}$$

is the ratio of the molar specific heats.

The second law of thermodynamics states that if work is to be obtained from a heat engine by removing heat from a reservoir, all of the heat cannot be converted into work.

The efficiency of a heat engine that takes heat Q_H from a hot reservoir while doing work W and rejecting heat Q_C to a cold reservoir is

$$\eta = \frac{W}{Q_H} = 1 - \frac{Q_C}{Q_H}. \tag{12.25}$$

The efficiency of a heat engine based on the Carnot cycle is

$$\eta = 1 - \frac{T_C}{T_H} \qquad (12.28)$$

where T_H and T_C are the temperatures of the hot and cold reservoirs, respectively. This expression gives the upper limit for the efficiency of any heat engine.

An isolated system in thermal equilibrium has constant total energy. It may be split into any number of equal component parts, to each of which is available a set of continuous or nearly continuous energy states. The different parts may have different energies as long as the total is constant. The probability of a component having energy ϵ is proportional to the Boltzmann factor

$$e^{-\epsilon/k_B T} \qquad (12.30)$$

where T is the temperature on T *the absolute scale.*

This result can be used to determine the speed distribution of gas molecules. The probability $P(v)\,dv$ that a molecule has a speed in the interval v to $v + dv$ is

$$P(v)\,dv = 4\pi \left(\frac{m}{2\pi k_B T}\right)^{3/2} e^{-mv^2/2k_B T} v^2\, dv. \qquad (12.37)$$

This is the Maxwell–Boltzmann speed distribution.

It can be shown that the absolute temperature is the same as the temperature on the ideal gas scale.

Whenever the total energy of a particle in thermal equilibrium with its surroundings can be written as the sum of terms, each of which has the form of a constant multiplying the square of a variable α_i, each term contributes on average $\frac{1}{2} k_B T$ to the average total energy. This result is called the equipartition theorem and each variable α_i is called a degree of freedom. This theorem enables us to deduce the value $C_V = \frac{3}{2} N_A k_B$ for an ideal gas, which together with eqn (12.17) gives the ratio $\gamma = C_P/C_V$ for an ideal gas equal to 1.67.

A real gas has discrete rotational and vibrational energy states as well as translational motion. At sufficiently low temperatures these states are not occupied in molecules and so do not contribute to the specific heats. As the temperature increases rotational and then vibrational degrees of freedom may contribute to the specific heats and change the value of γ. Another area in which the discreteness of microscopic energy levels and the operation of the Boltzmann factor plays a role is in black-body radiation. The energy density between frequencies ν and $\nu + d\nu$ is given by

$$\rho(\nu)\,d\nu = \frac{8\pi \nu^2}{c^3} \frac{h\nu}{e^{h\nu/k_B T} - 1}\,d\nu. \qquad (12.55)$$

Chapter 13 **Gases and liquids**

The values of pressure, temperature, and volume at the boundaries between different phases of a material may be represented on phase diagrams. On a P–V phase diagram, isotherms show corresponding values of P and V at fixed temperatures.

The van der Waals equation for one mole

$$\left(P + \frac{a}{V^2}\right)(V - b) = RT \qquad (13.1)$$

gives a relation between P, V, and T approximately followed by a real gas. The constants a and b are characteristic of the gas and may be related to the microscopic properties of the gas molecules and their interactions, and also to their latent heats of fusion (evaporation). The difference in the specific heats of a real gas is greater than the gas constant R at low temperatures and volumes.

The coefficient of viscosity of a fluid, η, is defined by the equation

$$F = \eta S \frac{du}{dx} \qquad (13.20)$$

where F is the force on a layer of area S at a position x where the fluid speed u has gradient du/dx.

The rate of streamline flow of liquid down a pipe of radius a that is very small compared with its length ℓ is

$$\Delta V = \frac{\pi \Delta P a^4}{8\eta \ell} \qquad (13.22)$$

where ΔP is the pressure difference between the ends of the pipe.

The mean free path of a gas molecule is given by

$$\lambda = \frac{1}{\sqrt{2} N_0 \pi d^2} \qquad (13.24)$$

where d is the molecular diameter and N_0 the number of molecules per unit volume.

The coefficient of viscosity of a gas at pressures such that the mean free path is much smaller than the dimensions of containing vessels is

$$\eta = \tfrac{1}{3} N_0 m \bar{v} \lambda \qquad (13.28)$$

where $\bar{v}$ is the mean speed of the gas molecules of mass m.

The coefficient of diffusion, D, of a gas is defined by the equation

$$J(z, t) = -D \frac{\partial N_0(z, t)}{\partial z} \qquad (13.32)$$

and given by

$$D = \tfrac{1}{3} \bar{v} \lambda. \qquad (13.33)$$

Equation (13.32) is Fick's law.

The coefficient of thermal conductivity, κ, is defined by the equation

$$J_Q(z) = -\kappa \frac{\partial T}{\partial z} \qquad (13.35)$$

and for a gas is given by

$$\kappa = \frac{c \bar{v}}{2 \pi \sqrt{2} d^2} \qquad (13.37)$$

where c is the specific heat per molecule.

Approximate relations between η, D, and κ are

$$\frac{D}{\eta} = \frac{1}{\rho}, \qquad (13.39)$$

and

$$\frac{\kappa}{\eta} = \frac{3 C_V}{2M} \qquad (13.40)$$

where ρ is the gas density, C_V is the specific heat per mole, and M is the mass of one mole.

Molecules at the surface of a liquid experience fewer intramolecular attractions than those in the interior. The liquid thus has a surface energy that is related to its surface tension Γ, the force per unit length on a line drawn in the surface. The direction of the force is perpendicular to the line and in the surface. Surface tension gives rise to spherical bubbles, to the behaviour of liquids in capillary tubes, and to forces due to thin films of liquid between plates. Surface energy and surface tension can be related to the average binding energy per liquid molecule and thence to the latent heat of evaporation.

Chapter 14 **Solids**

Crystalline solids consist of atoms arranged on a regular *lattice* that repeats itself so that if all the atoms are moved by one lattice spacing the lattice is unaltered: each atom has moved to a site previously occupied by an identical atom. The regular structure causes interference effects rather like those produced by a diffraction grating for waves of an appropriate wavelength. X-rays have wavelengths similar to the spacing of atoms in a crystal, and observation of X-ray diffraction by crystals allows the arrangement of atoms to be deduced.

The regular spacings in crystals occur in three dimensions, and X-ray diffraction is much more complicated than the interference due to a one-dimensional grating. However, the interference from atoms lying in a plane leads to reflection, with the angle of incidence equal to the angle of reflection, just as for specular reflection of visible light at a mirror. In a crystal there are many such planes, each separated from a neighbour by a fixed distance. Interference from a stack of planes only gives reflection at certain angles θ determined by the *Bragg law*, which is

$$n\lambda = 2d \sin \theta \qquad (14.1)$$

where λ is the X-ray wavelength, θ the grazing angle of the X-rays to the crystal planes, and n the number of wavelengths in the path difference between reflections from successive planes separated by a distance d.

Many inorganic crystals and metals have simple lattices and it is not too difficult to deduce their structure by working out the expected X-ray pattern from an arrangement of atoms. The structure is then verified when observed and predicted patterns coincide. The structure of large organic molecules such as proteins is extremely complicated, but with advanced techniques and the aid of powerful computers even their structures can be determined.

Not all solids have their atoms arranged in a lattice. Glasses have chemical bonds between neighbouring atoms arranged in such a way that there is some order over a short distance of a few atomic spacings, but distant atoms are randomly placed. Long-chain molecules are ordered in one direction, but may have variable lengths and limited order in directions other than along the length of the chain. Composite materials are deliberately made from more than one type of solid in order to combine the good features of each constituent.

When solids are subjected to external forces their shape is deformed. The external force is countered by an internal restoring force. If the restoring force F obeys Hooke's law

$$F = -kx, \tag{14.2}$$

the solid is *perfectly elastic* and returns to its original shape when the external force is removed. For materials under tension or compression the *stress* is the force per unit area and the *strain* is the fractional change in length. The ratio of stress to strain is a property of the material called *Young's modulus* (Y) and does not depend on the dimensions of the sample.

$$Y = \left(\frac{\text{stress}}{\text{strain}}\right) = \left(\frac{F}{S} \times \frac{\ell}{\Delta\ell}\right). \tag{14.3}$$

Other elastic moduli apply when the stress causes deformations other than tension or compression. *Shear* occurs when the force is perpendicular to the stressed area. The strain is then an angle θ, and the shear modulus H is defined as

$$H = \left(\frac{\text{stress}}{\text{strain}}\right) = \frac{F}{S \tan\theta}. \tag{14.4a}$$

The speed of waves travelling along a solid rod depends on the elastic moduli. For *longitudinal* waves of successive tension and compression the speed is $v_L = \sqrt{Y/\rho}$ (eqn (14.6)). Here ρ is the density of the rod. Transverse waves, which are set up by twisting the rod back and forth, move at a speed $v_T = \sqrt{H/\rho}$ (eqn (14.7)).

When a rod is elongated its cross-sectional area reduces by an amount determined by *Poisson's ratio* σ defined by

$$\sigma = \left(\frac{\Delta r}{r}\right) \times \left(\frac{\ell}{\Delta\ell}\right) = \frac{\text{lateral strain}}{\text{axial strain}}. \tag{14.8}$$

When a solid is compressed by the action of a uniform pressure P the change in volume ΔV is related to the *bulk modulus K*,

$$K = \frac{PV}{\Delta V}. \tag{14.9}$$

The different elastic moduli are not independent of one another, but satisfy the relations

$$K = \frac{PV}{\Delta V} = \frac{Y}{3(1 - 2\sigma)} \tag{14.10}$$

and

$$H = \frac{Y}{2(1 + \sigma)}. \tag{14.11}$$

For large enough stresses all solids cease to be elastic. The maximum stress that they can withstand is called the *tensile strength*, beyond which the material may fracture.

As for gases, the specific heat of solids represents energy used to excite internal degrees of freedom. The atoms in a solid oscillate around fixed positions and in consequence may possess vibrational potential energy as well as kinetic energy. However, since the atoms are in bound states, the vibrations occur as discrete normal modes, and at low temperatures cannot be excited thermally. As the temperature rises, the specific heat increases towards $3R$, twice the value for a perfect gas, since there are twice as many degrees of freedom per molecule in a solid as in the gas, which has no potential energy.

The specific heat of a solid is related to vibrations in the potential well caused by the interaction between neighbouring atoms. The potential is approximately simple harmonic for small displacements but, because the repulsion between atoms when they are pushed close together is stronger than the attraction when they are pulled apart, atoms on average move apart as temperature is raised. The expansion is approximately linear for small temperature changes, and is given by the *linear expansion coefficient* α, which satisfies the equation

$$\frac{\Delta\ell}{\ell} = \alpha\Delta T. \tag{14.16}$$

Energy may be carried through a solid by oscillations of the atoms. Wave packets, called *phonons*, move through the solid with a speed v and a mean free path λ_p, contributing to a thermal conductivity κ_p given by

$$\kappa_p = Nk_B v\lambda_p. \tag{14.18}$$

In a metal there is a much larger contribution κ_e to the specific heat from the *conduction electrons*, which move rather freely through the solid. If the average time between collisions for a conduction electron is τ, the thermal conductivity may be expressed as

$$\kappa_e = \tfrac{3}{2} N_e k_B^2 T \tau / m_e \qquad (14.20)$$

where m_e is the mass of the electron.

As well as carrying energy, the conduction electrons carry charge, and a similar expression to eqn (14.20) applies to electric current, and in an electric field E the current density in the solid is

$$j = e^2 N_e E \tau / m_e = \sigma E \qquad (14.21)$$

where e is the electronic charge and σ the electrical conductivity.

Electricity and magnetism

Part 5

Electricity and
magnetism

Chapter 15

Electrostatics

Electrical forces predominate in the interaction between the atoms and molecules of ordinary matter. This chapter explains the concepts of electric field and electrostatic potential that are needed to understand the behaviour of these forces.

Everyone is familiar with electricity as a source of power. Pressing a switch will turn on a light or heat an oven. Energy is continuously being produced in these processes, energy that is carried by an electric current through the metal wires connected to the electricity supply. The electric current is made up of a flow of moving electrons. We cannot see the movement because the electrons are very small and are able to move through a metal without disturbing the structure of the metal.

Electrons are pushed along a wire by forces that act on them because they carry **electric charge**. These forces are called **electric** forces. The electric force on a charged particle is the same whether the charge is stationary or moving. There are additional forces that act only on moving charges, which are called **magnetic** forces. Moving charges and magnetic forces are discussed in Chapter 16. The subject of this chapter is **electrostatics**, which is the study of the electric forces acting on stationary charges, and of how these forces are modified in the presence of matter.

Like gravitational forces, electrostatic forces act at a distance—there is an electrostatic force between two charged particles even if they are separated by a vacuum. The magnitude of the electrostatic force also has the same inverse square variation with distance as the gravitational force. There are, however, two very important differences between gravitational and electrostatic forces. The first is that the gravitational force between two masses is always attractive, whereas charges may attract or repel one another. The other difference is that, on an atomic scale, electrostatic forces are enormously strong compared with gravitational forces. In the discussion of the internal motion of individual atoms and molecules, which is the subject of Chapter 11, only electrostatic forces were considered and the effects of gravitational forces were completely neglected. This seems paradoxical, because in everyday life we are well aware of gravitational force, but do not often notice electrical forces. The reason is that electrostatic attractions and repulsions tend to cancel out, whereas gravitational forces are always attractive and, in particular, the whole of the Earth attracts everything on its surface.

Although the laws governing the forces between charges are introduced in this chapter in the context of electrostatics, these laws always apply, even when magnetic effects or electromagnetic waves are present. The chapter starts by discussing the forces between very small idealized electric charges in order to explain the concepts of electric field and potential. Later on, in Sections 15.5 and 15.6, we are concerned with objects containing very large numbers of atoms. Only average electrical properties are then of interest: it will be shown how these averages can be obtained without having to consider the electrical forces within each atom in turn.

15.1 Forces between charged particles

Positive and negative charges

It was mentioned above that electrostatic forces are sometimes attractive and sometimes repulsive. This is because there are two different kinds of charge, which are called *positive* and *negative*. Just like positive and negative numbers, positive and negative charges are described as being of opposite sign. Electrons carry a negative charge. Like numbers, positive and negative charges may cancel one another out. For example, as is described more fully in Chapter 9, an atom consists of a number of electrons bound to a positively charged nucleus. The charge on the nucleus has exactly the same magnitude as the charge of all the electrons in the atom. Since the nuclear charge is of opposite sign to the charges on the electrons, the *net* charge carried by the atom, which is the algebraic sum of all its charges, is zero. The atom is said to be **electrically neutral**.

In SI units, charge is measured in **coulombs** (symbol C). The coulomb is defined with reference to the force between wires carrying electric current: this is discussed in Chapter 16. The charge carried by a single electron is written as $-e$, and its magnitude is

$$e = 1.602 \times 10^{-19} \text{ coulombs}$$

to four significant figures. Because the coulomb is a very large unit, charges are often measured in **microcoulombs** (symbol μC: $1\mu\text{C} \equiv 10^{-6}\,\text{C}$).

Ordinary matter, made up of electrons and nuclei, may be electrically neutral or may have a charge that is $\pm e$ times an integer. Other particles besides electrons and atomic nuclei are found in cosmic rays, or may be created in high-energy collisions in accelerators. All these particles also have charges that are zero or $\pm e$ times an integer. Within a nucleus there are thought to be particles called *quarks* that carry an amount of charge that is a fraction of $\pm e$. However, quarks have a property called *confinement*, which means that they are never observed singly but go around in packets that do not have fractional charge. It is thus a universal rule that any object is either electrically neutral or has a positive or negative charge with magnitude that is an integral multiple of e.

● *All observable charges are multiples of the electronic charge*

Coulomb's law

Coulomb's law tells us the strength and direction of the forces acting between two charges. This is the simplest possible case, but on the basis of Coulomb's law it is possible to work out the electric force on any distribution of charges. Consider two charges that we shall label q_1 and q_2: q_1 and q_2 are numbers representing the magnitudes of the charges in coulombs, and these numbers are positive or negative depending on whether the charges are positive or negative. We shall idealize the problem by supposing that q_1 and q_2 occupy such a small volume that they may be treated as *point charges* with no spatial extent at all.

Let us define F_{12} to be the magnitude of the force exerted by a charge q_1 on a charge q_2. According to Coulomb's law, F_{12} depends on the product $q_1 q_2$; doubling the magnitude of either charge doubles the strength of the force. How do the forces between q_1 and q_2 vary with distance? Just like the gravitational force between two masses, the electrostatic force between two charges varies as the inverse square of their distance apart. This inverse square law is known to hold with great precision, not from direct measurement of the force between charges, but from the observation of other phenomena that are deduced from the inverse square law. The magnitude of the force between the charges q_1 and q_2 is expressed mathematically as

$$F_{12} \propto \frac{q_1 q_2}{r_{12}^2}. \tag{15.1}$$

A constant of proportionality is required to give the strength of the force in newtons when q_1 and q_2 are in coulombs and r_{12} in metres. In the SI system the constant is written as $1/4\pi\epsilon_0$—including the factor 4π simplifies other equations in electricity and magnetism. The equation for the magnitude F_{12} of the force becomes

$$F_{12} = \frac{q_1 q_2}{4\pi\epsilon_0 r_{12}^2}. \tag{15.2}$$

The *direction* of the force between the two charges depends on their signs. The force acts on the line joining the charges and, if the charges have opposite sign, like the negatively charged electron and the positively charged nucleus in the hydrogen atom, the force is attractive: it is the electrical attraction that binds an electron to the nucleus and ensures that the atom is stable. On the other hand, if both charges are positive, or both are negative, the force between them is repulsive.

We must be careful to get the direction of the force correct when setting up the final equation for Coulomb's law. To do this we must use the vector notation, and in particular we shall use *unit vectors*, which are vectors pointing in any direction, but which always have unit length. Since we are interested in the direction between the two charges, we introduce the vector $\hat{\mathbf{r}}_{12}$, which is a vector of unit length pointing from an origin at the centre of charge q_1 towards charge q_2.

Unit vectors are used in Section 20.2 (in the mathematical review at the end of the book) to define the directions of the axes of a Cartesian coordinate system. In that context, the unit vectors $\mathbf{i}$, $\mathbf{j}$, and $\mathbf{k}$ (all of unit length) are pointing in the fixed directions chosen for the x-, y-, and z-axes. Any vector **a** that has components a_x, a_y, and a_z along the x-, y-, and z-axes can then be written as $a_x\mathbf{i} + a_y\mathbf{j} + a_z\mathbf{k}$. This expression specifies both the magnitude and direction of the vector **a**. Here it is more convenient to use a different notation, allowing unit vectors to point in any direction. The symbol ˆ is used to indicate that a vector has unit length: thus **â** is a vector of unit length pointing in the same direction as **a**.

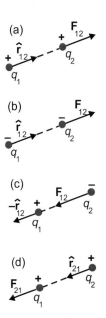

The force exerted by charge q_1 on charge q_2 is denoted by the vector $\mathbf{F}_{12}$. This force points along the line joining the charges, in the direction away from q_1 for a repulsive force (q_1 and q_2 having the same sign) and towards q_1 for an attractive force (q_1 and q_2 having different signs). Different possibilities are illustrated in Fig 15.1, which shows both $\mathbf{F}_{12}$ and $\hat{\mathbf{r}}_{12}$ for different signs of the charges. In Fig 15.1(c), where the signs are different, the force is towards q_1, which is in the opposite direction to $\hat{\mathbf{r}}_{12}$. However, the unit vector $-\hat{\mathbf{r}}_{12}$ is also in the opposite direction to $\hat{\mathbf{r}}_{12}$. The sign required in front of the unit vector $\hat{\mathbf{r}}_{12}$ is thus the same as the sign of the product $q_1 q_2$.

The force between two charges q_1 and q_2 may now be expressed in mathematical terms, using the same notation as in Fig 15.1 for the position vector of q_2 with respect to q_1. The magnitude of the force is given by eqn (15.2) and in SI units Coulomb's law is

$$\mathbf{F}_{12} = \frac{q_1 q_2}{4\pi\epsilon_0 r_{12}^2} \hat{\mathbf{r}}_{12}. \tag{15.3}$$

Similarly, the force $\mathbf{F}_{21}$ exerted by q_2 on q_1 is

$$\mathbf{F}_{21} = \frac{q_1 q_2}{4\pi\epsilon_0 r_{12}^2} \hat{\mathbf{r}}_{21}. \tag{15.4}$$

Since $\hat{\mathbf{r}}_{21}$ is a unit vector pointing from q_2 towards q_1, in the opposite direction to $\hat{\mathbf{r}}_{12}$, the forces exerted on the two charges according to eqns (15.3) and (15.4) are equal and opposite, as they should be (compare Figs 15.1(a) and (d)).

In words, Coulomb's law states that

The force between two charges acts along the line between the charges, and is proportional to the product of the charges and to the inverse square of the distance between them. The force is repulsive for charges of the same sign and attractive for charges of opposite sign.

The dimensions of all the quantities in eqn (15.3) are defined independently of Coulomb's law. The unit of force, the newton, is defined by Newton's second law. The unit of charge, the coulomb, is defined with reference to the magnetic force between wires carrying current. To satisfy eqn (15.3), the units of the constant ϵ_0 are $C^2\,N^{-1}\,m^{-2}$. Its value is related

Fig. 15.1 The force on a charge due to the presence of another charge is in the same direction as the unit vector on the line joining the charges. (a), (b), and (c) show the force on charge q_2 caused by q_1 for different combinations of the sign of the charges. (d) shows the force on q_1 caused by q_2 when both are positive.

to the speed of light, which is the distance light travels in a vacuum in one second. Since the the unit of length is itself *defined* in terms of the time taken for light to travel a distance of one metre, the speed of light is also *defined* to be a particular number of metres per second. Scientists all over the world have agreed that the value of the speed of light is exactly $2.997\,924\,58$ m s^{-1}. Because the constant ϵ_0, which is called the **permittivity of free space**, is derived from the speed of light, it is also in principle known exactly. However, it is not a rational number, but when expressed in decimals it can be calculated to any number of places. To four significant figures, its value is

$$\epsilon_0 = 8.854 \times 10^{-12}\, \text{C}^2\, \text{N}^{-1}\, \text{m}^{-2}. \tag{15.5}$$

Worked Example 15.1 Two small particles of carbon, each weighing 1 mg and each carrying a charge of 10^{-6} C, are one centimetre apart. Calculate the electrostatic force between them.

Answer The force between the particles is found directly by substitution in eqn (15.2). It is

$$\frac{10^{-6} \times 10^{-6}}{4\pi\epsilon_0 \times 10^{-4}} \approx \frac{10^{-8}}{1.1 \times 10^{-10}} \approx 90\, \text{N}.$$

This example illustrates the enormous strength of electrostatic forces. The force of 90 N is nearly equal to the weight of a 10 kg mass, acting between two tiny particles. If the particles were free to move, their initial acceleration would be 90 km s^{-2}. In practice it is not possible to accumulate as much charge as 1 μC on such small pieces of matter, even though only a small fraction of the atoms need to gain or lose an electron to reach this value. The number of atoms in one mole of carbon is Avogadro's number, $N_A \approx 6 \times 10^{23}$, and, since the mass number of carbon is 12, the number of atoms in 1 mg is about 5×10^{19}. The number of electronic charges in 1 μC is $10^{-6}/e \approx 10^{13}/1.6$. The fraction of carbon atoms that must lose one electron to charge the particles with 1 μC is $10^{13}/(5 \times 10^{19} \times 1.6)$, or a little more than one in a million.

15.2 The electric field

Coulomb's law in the form given in eqn (15.3) enables us to work out the forces that two point charges exert on each other. Most practical electrical problems involve not just two charged particles, but vast numbers of them. This section introduces the idea of the *electric field*, which describes the force on a charged particle due to all the other charges in its neighbourhood.

Addition of electric forces

In Fig 15.2 a third positive charge q_3 has been brought close to the two positive charges q_1 and q_2 shown in Fig 15.1(d). There is now an additional force $\mathbf{F}_{31}$ acting on q_1, due to q_3. However, the force $\mathbf{F}_{21}$ exerted on q_1 by q_2 is unchanged, provided that q_1 and q_2 remain at the same positions after q_3 has been introduced. This is described by saying that the electric force between charges is a **two-body force**: the force between two charges is given by Coulomb's law independently of the presence of any other charges. The same applies to q_1 and q_3, and the force $\mathbf{F}_{31}$ is also given by Coulomb's law. The *net* force $\mathbf{F}_1$ on q_1 is simply the vector sum of the forces due to q_3 and q_2 separately,

$$\mathbf{F}_1 = \mathbf{F}_{21} + \mathbf{F}_{31}. \tag{15.6}$$

A very simple application of eqn (15.6) is to two or more charges that are very close together. For example, suppose that in Fig 15.2 q_2 and q_3 both are of magnitude $+e$, and q_3 is moved to the same location as q_2. The force on q_1 then has a magnitude $2eq_1/4\pi\epsilon_0 r_{21}^2$ and is in the direction $\hat{\mathbf{r}}_{21}$. This is, of course, the same as the force due to a single charge $2e$ at the position of q_2. The fact that the force is a two-body force is thus already included in Coulomb's law, which allows q_1 and q_2 to have any values, although we know that in reality the charge in a small volume is always built up of individual charges of magnitude $\pm e$.

As more and more charges are added, each exerts a force on all the others. Labelling the charges one by one as $q_1, q_2, q_3, \ldots q_j, \ldots$, the force $\mathbf{F}_i$ on a particular charge q_i is the vector sum of the forces $\mathbf{F}_{ji}$ due to all the others,

$$\mathbf{F}_i = \sum_{j\neq i} \mathbf{F}_{ji} = \sum_{j\neq i} \frac{q_i q_j}{4\pi\epsilon_0 r_{ji}^2} \hat{\mathbf{r}}_{ji} = \frac{q_i}{4\pi\epsilon_0} \sum_{j\neq i} \frac{q_j}{r_{ji}^2} \hat{\mathbf{r}}_{ji}. \tag{15.7}$$

The caption $j \neq i$ under the summation signs indicates that the sum is taken over all values of j except $j = i$, since the charge q_i is not exerting a force on itself. All the unit vectors $\hat{\mathbf{r}}_{ji}$ in the equation remind us that the force between each pair of charges is pointing along the line joining the charges. However, when doing calculations it is usually convenient to refer all the position vectors to a fixed origin rather than dealing with each pair of charges separately. Figure 15.3 shows two charges q_i and q_j with position vectors $\mathbf{r}_i$ and $\mathbf{r}_j$ referred to an origin at O. The vector from q_j to q_i is $\mathbf{r}_{ji} = \mathbf{r}_i - \mathbf{r}_j$. Writing the length of this vector as $|\mathbf{r}_i - \mathbf{r}_j|$, the unit vector in the direction from q_j to q_i is

$$\hat{\mathbf{r}}_{ji} = \frac{\mathbf{r}_i - \mathbf{r}_j}{|\mathbf{r}_i - \mathbf{r}_j|}. \tag{15.8}$$

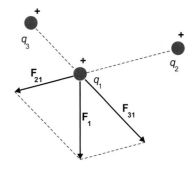

Fig. 15.2 The net force $\mathbf{F}_1$ on the charge q_1 is the vector sum of the forces $\mathbf{F}_{21}$ and $\mathbf{F}_{31}$ caused by q_2 and q_3. In the diagram $\mathbf{F}_1$ is the diagonal in a parallel of forces.

● *The force on a charge is the vector sum of the forces due to all other charges*

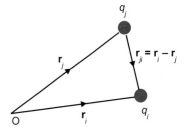

Fig. 15.3 The vector $\mathbf{r}_{ij}$ between the charges q_i and q_j is the difference of the position vectors $\mathbf{r}_i$ and $\mathbf{r}_j$.

Substituting in eqn (15.7), the force on q_i becomes

$$\mathbf{F}_i = \frac{q_i}{4\pi\epsilon_0} \sum_{j\neq i} \frac{q_j(\mathbf{r}_i - \mathbf{r}_j)}{|\mathbf{r}_i - \mathbf{r}_j|^3}. \tag{15.9}$$

Note that, because each term in this expression has a vector of length $|\mathbf{r}_i - \mathbf{r}_j|$ in the numerator, the length appears raised to the power 3 in the denominator, even though the force follows an inverse square law.

A $\quad q$ $\qquad a \qquad$ B $\quad q$

D $\quad q$ $\qquad$ C $\quad q$

Fig. 15.4 The charges at A, B, C, and D lie in the same plane at the corners of a square of side a.

Worked Example 15.2 Four positive charges, each of magnitude q, are situated at the corners of a square of side a, as shown in Fig 15.4. What is the magnitude and direction of the force on the charge at A?

Answer Consider the components of the force in the directions BD and CA. The charges at B and D give rise to equal and opposite components along the direction BD, and each has a component

$$\frac{q^2 \cos(45°)}{4\pi\epsilon_0 a^2} = \frac{q^2}{4\sqrt{2}\pi\epsilon_0 a^2}$$

along the direction CA. The force due to the charge at C is along CA and has a magnitude

$$\frac{q^2}{4\pi\epsilon_0 (\sqrt{2}a)^2} = \frac{q^2}{8\pi\epsilon_0 a^2}.$$

The total force on the charge at A is therefore $q^2(1 + 2\sqrt{2})/(8\pi\epsilon_0 a^2)$ pointing along the direction CA. Each of the other charges has a force of the same magnitude pointing in the direction of the outward diagonal, and the net force on the square is zero.

Exercise 15.1 Calculate the magnitude of the force on the charge at A in Fig 15.4 if the charge at B is replaced by a charge $-q$.

Answer The force due to the charges $+q$ at D and $-q$ at B is now $2\sqrt{2}q^2/(8\pi\epsilon_0 a^2)$ in the direction DB, and the total force has magnitude $3q^2/(8\pi\epsilon_0 a^2)$.

Definition of the electric field

Imagine that you have a test charge q that you can move about in the region near the charges q_i. Suppose also that the positions of the charges q_i are undisturbed by the presence of q. If q is placed at a point with position vector $\mathbf{r}$ with respect to the origin at O, the force on it is, according to eqn (15.9),

$$\mathbf{F} = \frac{q}{4\pi\epsilon_0} \sum_j \frac{q_j(\mathbf{r} - \mathbf{r}_j)}{|\mathbf{r} - \mathbf{r}_j|^3}. \tag{15.10}$$

The sum is over all j now, because the test charge has been excluded from the labelling. Now look at the quantity $\mathbf{F}/q$. It does not depend on the test charge at all. It may be calculated at any location whether or not a test charge is present; the position vector $\mathbf{r}$ is a variable, and $\mathbf{F}/q$ is a **vector function of position**. This function is called the **electric field**, and it is denoted $\mathbf{E}(\mathbf{r})$.

Functions of position are called *fields*. Because $\mathbf{E}(\mathbf{r})$ is itself a vector, it is a *vector field*. In electromagnetism we shall also meet functions of position that are scalar quantities, which have magnitude but not direction. These functions are called *scalar fields*.

● *Vector fields*

By dividing eqn (15.10) by q we find

$$\mathbf{E}(\mathbf{r}) = \frac{1}{4\pi\epsilon_0} \sum_j \frac{q_j(\mathbf{r} - \mathbf{r}_j)}{|\mathbf{r} - \mathbf{r}_j|^3}. \tag{15.11}$$

The dimensions of the electric field are force per unit charge, and it is measured in newtons per coulomb: if a charge of one coulomb were placed in an electric field of strength one newton per coulomb it would experience a force of one newton, and this force would act in the direction of the electric field vector at the position of the charge.

Worked Example 15.3 Calculate the electric field due to a proton at a distance of 0.07 nm (this distance is approximately the separation of the protons in a hydrogen molecule).

Answer There is only a single term in the summation in eqn (15.11) and we can choose the origin to coincide with the proton. At any point a distance $r = 0.07$ nm from the proton, the electric field points in a direction away from the proton, and has a magnitude $e/(4\pi\epsilon_0 r^2)$:

$$\frac{1.602 \times 10^{-19}}{4\pi \times 8.854 \times 10^{-12} \times 0.0049 \times 10^{-18}}$$

$$\approx 2.9 \times 10^{11} \text{ newtons per coulomb.}$$

This is far in excess of any electric field that can be achieved over a distance larger than atomic dimensions, and again illustrates the enormous strength of electric forces within atoms and molecules.

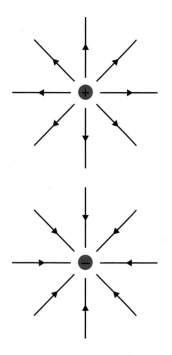

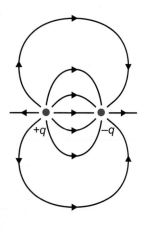

Fig. 15.5 Electric field lines radiate outwards from a positive point charge and inwards towards a negative point charge.

Fig. 15.6 The field lines around an electrostatic dipole start at the positive charge and bend round to end on the negative charge.

● *The electric dipole*

Exercise 15.2 Estimate the repulsive electrostatic force between the two protons in a hydrogen molecule and compare it with their gravitational attraction.

Answer The electrostatic force is about 5×10^{-8} N, and the gravitational force is about 4×10^{-44} N. The electrostatic force is thus more than 10^{36} times larger: this ratio applies at any distance, since both forces obey an inverse square law.

Lines of force

Around an isolated positive point charge q_1 there is only a single term in the summation in eqn (15.11). Choosing the origin to be at the position of the point charge, the electric field is

$$\mathbf{E}(\mathbf{r}) = \frac{q_1}{4\pi\epsilon_0} \frac{\mathbf{r}}{r^3}. \tag{15.12}$$

In this equation $\mathbf{r}$ is the position vector of any point in space with respect to the origin. The electric field $\mathbf{E}(\mathbf{r})$ is everywhere pointing in the same direction as $\mathbf{r}$, directly away from the origin. A positive charge q placed at $\mathbf{r}$ will experience a force $qq_1/(4\pi\epsilon_0 r^2)$ in this direction. This can be visualized by drawing **lines of force** in the direction of the field, as in Fig 15.5. The diagram is only two-dimensional, but it will look the same in any plane passing through q_1. The diagram does not indicate the strength of the force, but notice that close to q_1, where the field is strong, the lines are close together, whereas the lines are far apart at large distances where the field is weak. The figure also shows the field around a negative point charge. The diagram is the same except that the field is in the opposite direction, inwards instead of outwards. Following the arrows, you can see that field lines start from positive charges and end on negative charges.

An instructive diagram of lines of force is shown in Fig 15.6. Here two charges $\pm q$ of the same magnitude but opposite sign are placed not very far apart. Close to each charge, the lines behave in the same way as in Fig 15.5, pointing away from the positive charge and towards the negative charge. But, as the distance from one charge increases, the influence of the other becomes more important. Field lines leaving the positive charge bend round and move towards the negative charge. At larger distances from the charges, the lines are far apart. The electric field has become weak because the contributions from the positive and negative charges almost cancel one another. Once again the diagram is two-dimensional, but it will look the same in any plane passing through both charges.

The pair of equal and opposite charges separated by a small distance is called an **electric dipole**. The field pattern generated by an electric dipole is important in many branches of physics and chemistry; later on we shall evaluate the field due to an electric dipole mathematically. However, it is also very helpful towards understanding the behaviour of electric fields to have a mental picture of the kind given by diagrams of lines of force. As in the example of the electric dipole, these often illustrate important properties of the electric field without the need to do any calculations at all.

Exercise 15.3 Sketch the lines of force in the neighbourhood of two positive charges of equal magnitude.

Superposition of electrostatic fields

The electric field as given by eqn (15.11) is the vector sum of the electric fields generated by each charge q_j separately. If some more charges are added, more terms are added to the summation. However, there is no change to the terms that were already there, provided that the original charges do not move. If we know the electric fields generated by two different sets of charges separately, the electric field generated by both together is simply the vector sum of the two separate fields. The two fields, which each occupy three-dimensional space, are superimposed on one another. Because it has this property, the electric field is said to satisfy the *principle of superposition*.

Superposition is discussed for one-dimensional waves in Section 6.5, where it is shown that different waves may be superposed because the equations governing the wave motion are linear. Similarly here, superposition applies to different electric fields because the field depends linearly on the charges that generate the fields. When superposition is extended to the varying electric and magnetic fields that occur in electromagnetic waves, the phenomena of diffraction and interference described in Sections 8.6 and 8.7 can be explained.

15.3 Gauss's law

The electric field due to any system of charges is found by superposing the fields due to each one separately. This sounds very simple but, since the fields to be summed are vectors, the general expression given by eqn (15.11) may be very difficult to work out.

A completely different way of relating the electric field to the charges is called Gauss's law. It is sometimes much easier to calculate the field from Gauss's law than by summing the fields from all the charges. Gauss's law follows from the inverse square variation of the electric field with distance, and it can be understood by analogy with the spreading out of energy from a light source, which also decreases with the inverse square of distance. If the light source is in an enclosed space such as a room, all the light leaving the source reaches a surface somewhere in the room. Nearby surfaces are brightly illuminated and those that are far away are dimly illuminated. Moving the surfaces will make a difference to their brightness, but not to the total amount of light in the room. We shall prove that a similar result holds for a quantity called the *flux* of the electric field. If a charge is surrounded by a closed surface, the flux over the whole surface has a fixed relation to the amount of charge, independent of the shape or size of the surface.

Fig. 15.7 The vector $\delta\mathbf{S}$ has a magnitude equal to the area δS of the small surface and is perpendicular to it.

Flux

Figure 15.7 shows a small flat surface of area δS placed in an electric field $\mathbf{E}$ so that the normal to the surface is at an angle θ to the direction of $\mathbf{E}$. The projected area of δS viewed along the direction of the field lines of $\mathbf{E}$ is $\delta S \cos\theta$. The **flux** of $\mathbf{E}$ through δS is defined to be $E\delta S \cos\theta$. This may be expressed concisely by associating a vector $\delta\mathbf{S}$ with the area δS, directed along the normal to the surface, as shown in Fig 15.7. The flux through δS can now be written as the scalar product $\mathbf{E} \cdot \delta\mathbf{S}$. Note that $\delta\mathbf{S}$ can be the normal to the surface in either direction from the surface. The sign of the flux depends on whether θ is greater or less than $90°$. If the component of $\delta\mathbf{S}$ in the direction of the field is positive, θ is less than $90°$ and the flux is positive: if this component is opposite to the field, the flux is negative.

If we have a large area S, which is not necessarily plane, we can divide it up as shown in Fig 15.8. If the division is fine enough, the small surfaces like $\delta\mathbf{S}_i$ are practically flat. The flux through $\delta\mathbf{S}_i$ is $\mathbf{E}(\mathbf{r}_i) \cdot \delta\mathbf{S}_i$, where $\mathbf{r}_i$ is the position vector of $\delta\mathbf{S}_i$, the total flux through S is the sum $\sum_i \mathbf{E}(\mathbf{r}_i) \cdot \delta\mathbf{S}_i$ of contributions from all the surfaces $\delta\mathbf{S}_i$. In the limit as the areas of all the surfaces $\delta\mathbf{S}_i$ tend to zero, the summation becomes a two-dimensional *surface integral* over the surface S which is written as

$$\text{Flux through } S = \lim_{\delta S_i \to 0} \sum_i \mathbf{E}(\mathbf{r}_i) \cdot \delta\mathbf{S}_i = \int_S \mathbf{E}(\mathbf{r}) \cdot d\mathbf{S}. \tag{15.13}$$

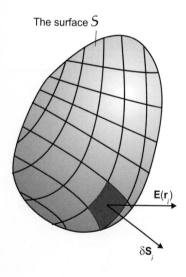

Fig. 15.8 Any surface like the shaded surface S may be divided up into many adjacent surfaces δS_i. In the limit as the δS_i become infinitesimal, each one may be regarded as a plane surface.

How does this equation apply to the electric field around an isolated point charge q_1 located at the origin of coordinates? Choose for the surface S a sphere of radius r centred at the origin. The electric field on the surface of the sphere has a magnitude $q_1/(4\pi\epsilon_0 r^2)$ and is perpendicular to the surface, so that the outward normal to the sphere is everywhere in the same direction as the field. The area of the sphere is

$4\pi r^2$, and the total electric flux out of the surface $\mathcal{S}$ of the sphere is

$$\int_{\mathcal{S}} \mathbf{E}(\mathbf{r}) \cdot d\mathbf{S} = \frac{q_1}{4\pi\epsilon_0 r^2} \times 4\pi r^2 = \frac{q_1}{\epsilon_0}. \tag{15.14}$$

Equation (15.14) relates the flux out of the sphere to the charge inside it. This equation is Gauss's law, though here it has only been derived for the very special case of a point charge at the centre of a sphere.

Surface integrals

In order to generalize Gauss's law to surfaces of any shape, we need to work out the surface integral on the left-hand side of eqn (15.14). Multidimensional integrals are discussed in Section 4.2 in Cartesian and cylindrical polar coordinates. Here it is best to use *spherical polar coordinates*, which are compared with Cartesian coordinates in Fig 15.9. The position vector $\mathbf{r}$ of the point P has Cartesian coordinates (x, y, z). The vector $\mathbf{r}$ can also be specified by the spherical polar coordinates (r, θ, ϕ). The coordinate r is the length OP of $\mathbf{r}$ and θ is the angle between OP and the z-axis. The plane through OP and the z-axis cuts the xy-plane along OQ. The angle between OQ and the x-axis is the coordinate ϕ.

The reason for using spherical polar coordinates here is that the proof of Gauss's law depends on the mathematical concept of *solid angle*, which is best expressed in this coordinate system. For readers unfamiliar with or unsure of the meaning of solid angle, it is described in the box that follows.

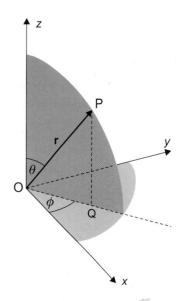

Fig. 15.9 The spherical polar coordinates (r, θ, ϕ) are defined with respect to a set of Cartesian coordinates (x, y, z).

● **Solid angles**

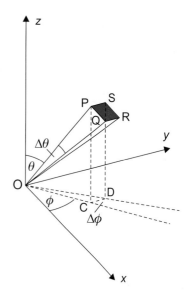

Solid angle is the measure of the angular size of a cone. Figure 15.10 shows part of a sphere with radius r and centre at the origin. The point P with position vector $\mathbf{r}$ has spherical polar coordinates (r, θ, ϕ). Keeping r and ϕ fixed, rotate the position vector $\mathbf{r}$ through an angle $\Delta\theta$. The point P moves to Q along an arc of length $r\Delta\theta$. Next rotate the line OQ through a small angle $\Delta\phi$ while keeping r and θ fixed at the values they have at Q. The point Q moves to R along an arc of length $r\sin\theta\Delta\phi$. When the rotations are performed in the order $\Delta\phi$ followed by $\Delta\theta$, P moves to R via S. Denoting the area of the spherical surface within PQRS by ΔS, the quantity

$$\Delta\Omega = \Delta S/r^2 \tag{15.15}$$

is called the **solid angle** subtended by the area PQRS at the origin O. The area of the whole sphere is $4\pi r^2$, and so $\Delta S/4\pi r^2 = \Delta\Omega/4\pi$ is the fraction of the total area of the sphere covered by the area within PQRS. Equivalently you may think of $\Delta\Omega/4\pi$ as the fraction of the volume of the sphere occupied by the cone that has ΔS as its base. Solid angle is

Fig. 15.10 The area within PQRS is on the surface of a sphere of radius r. The angles at the apex of the cone are $\Delta\theta$ and $\sin\theta\Delta\phi$.

measured in the dimensionless units called **steradians**. The complete sphere subtends a solid angle of 4π steradians at the origin.

Since the area PQRS is not flat, to calculate a solid angle we must perform an integration. If $\Delta\theta$ and $\Delta\phi$ are made smaller and smaller, PQRS gets closer and closer to being a flat rectangle, and in the limit the infinitesimal area $dS = r\,d\theta \times r\sin\theta\,d\phi$ and the infinitesimal solid angle of the cone with base dS is

● *A sphere subtends a solid angle of 4π steradians at its centre*

$$d\Omega = dS/r^2 = \sin\theta\,d\theta\,d\phi. \qquad (15.16)$$

To find the solid angle $\Delta\Omega$ of the cone with angles $\Delta\theta$ and $\Delta\phi$ at the apex, $d\Omega$ must be integrated over both θ and ϕ,

$$\Delta\Omega = \int_{\phi}^{\phi+\Delta\phi} \int_{\theta}^{\theta+\Delta\theta} \sin\theta\,d\theta\,d\phi.$$

To integrate over all directions, the limits are from $\phi = 0$ to 2π, and $\theta = 0$ to π: if θ were allowed to vary from 0 to 2π the whole sphere would be covered twice. The total solid angle subtended by a sphere centred on the origin is thus

$$\int_{\phi=0}^{2\pi} \int_{\theta=0}^{\pi} \sin\theta\,d\theta\,d\phi = 4\pi$$

confirming, by direct integration, the result already derived from the area of the sphere.

In Fig 15.11 a point charge q_1 at the origin is surrounded by a closed surface S. The cone with apex at the origin and solid angle $\delta\Omega = \sin\theta\delta\theta\delta\phi$ cuts through S at the point P with spherical polar coordinates r, θ, and ϕ, and a small area δS of the surface lies within the cone. Because the surface completely encloses the volume within it, a vector normal to δS must point *inwards* or *outwards*. Gauss's law applies to the flux of **E** *out of* a closed surface, and the vector $\delta\mathbf{S}$, of magnitude δS, is chosen to be in the direction of the *outward* normal, as shown in Fig 15.11.

A sphere of radius r centred at the origin also passes through P, and according to eqn (15.16) the area δS_{sphere} of the sphere within the cone is $r^2\delta\Omega = r^2 \sin\theta\delta\theta\delta\phi$. The electric field at P has a magnitude $q_1/(4\pi\epsilon_0 r^2)$ and is perpendicular to the surface of the sphere. The flux through δS is the same as the flux through δS_{sphere} and is

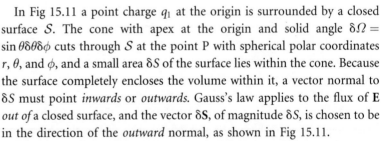

$$\mathbf{E}(\mathbf{r}) \cdot \delta\mathbf{S} = \frac{q_1}{4\pi\epsilon_0 r^2} \times r^2 \sin\theta\delta\theta\delta\phi = \frac{q_1}{4\pi\epsilon_0} \times \sin\theta\delta\theta\delta\phi = \frac{q_1}{4\pi\epsilon_0}\delta\Omega,$$

and in the limit as δS becomes infinitesimal

$$\mathbf{E}(\mathbf{r}) \cdot d\mathbf{S} = \frac{q_1}{4\pi\epsilon_0}\,d\Omega. \qquad (15.17)$$

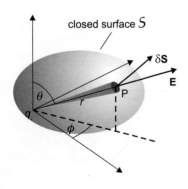

closed surface S

Fig. 15.11 The flux of **E** through the surface δS is determined by the solid angle of the cone and the magnitude of the charge q_1, and does not depend on the orientation of $\delta \mathbf{S}$.

The flux of **E** through δ**S** depends on the solid angle subtended by δ**S** at the charge q_1 but not on the distance of δ**S** or its angle to the position vector **r**. The total flux through $\mathcal{S}$ can now be evaluated using eqn (15.17),

$$\text{Flux through } \mathcal{S} = \int_{\mathcal{S}} \mathbf{E} \cdot d\mathbf{S} = \int_{\phi=0}^{2\pi} \int_{\theta=0}^{\pi} \frac{q_1}{4\pi\epsilon_0} \sin\theta \, d\theta \, d\phi = \frac{q_1}{\epsilon_0}.$$

(15.18)

This result applies for any charge q_1 and any surface $\mathcal{S}$ enclosing it. Any number of charges q_i within $\mathcal{S}$ will each give a contribution q_i/ϵ_0 to the total flux through $\mathcal{S}$.

There may in addition be charges *outside* $\mathcal{S}$. Figure 15.12 shows that such charges make no contribution to the flux over $\mathcal{S}$. The cone from the charge q_1 passes twice through $\mathcal{S}$, once entering and once leaving. The field **E** entering $\mathcal{S}$ makes a negative contribution to the flux out of $\mathcal{S}$, because the *outward* normal makes an angle greater than 90° with **E** at this point. Where the cone leaves $\mathcal{S}$ the contribution is positive and, since the solid angle is the same, the net flux contributed by q_1 is zero. Figure 15.13 shows a more complicated surface $\mathcal{S}$ which is re-entrant. A small cone with apex at a charge q_1 within $\mathcal{S}$ must always pass outwards through the surface one more time than it passes inwards, and the contribution to the flux is just q_1/ϵ_0 as for a sphere. Similarly a cone from a charge outside $\mathcal{S}$ may enter and leave more than once, but the number of entering and leaving fluxes are the same, and the net contribution is zero. The final result, which is **Gauss's law**, is that, for any closed surface $\mathcal{S}$,

$$\int_{\mathcal{S}} \mathbf{E} \cdot d\mathbf{S} = \frac{\sum_i q_i}{\epsilon_0} = \frac{Q}{\epsilon_0}$$

(15.19)

where $Q = \sum_i q_i$ is the sum of all the charges situated within $\mathcal{S}$.

In words, Gauss's law states that

The total flux of the electric field out of any closed surface equals the total charge enclosed within the surface divided by ϵ_0.

Sometimes the electric field possesses a symmetry that may greatly simplify the calculation of the surface integral in Gauss's law. For example, around a point charge q_1 we can say immediately that the electric field must point towards or away from q_1 and that its magnitude must depend only on the distance from q_1. If we place q_1 at the origin of coordinates, the directions of the x-, y-, and z-axes are completely arbitrary. One choice of directions for the axes is as good as any other. If we now use spherical polar coordinates related to the Cartesian coordinates as in Fig 15.10, all values of θ and ϕ, which define a direction, must be equivalent. The field is said to possess *spherical symmetry*, and all points on a sphere centred at the origin are equivalent.

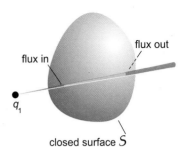

Fig. 15.12 The flux through a closed surface due to a charge outside the surface is zero.

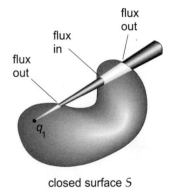

Fig. 15.13 Flux may pass several times in and out of a closed surface, but for a charge located inside the surface the flux always passes outwards one more time than it passes inwards.

● *When a system of charges possesses a simple symmetry, the electric field may be calculated easily using Gauss's law*

There cannot be any component of the field along the surface of the sphere. The magnitude of **E** in the outward direction at a distance r from the origin depends only on r and can be written as $E(r)$. According to Gauss's theorem, the flux through the sphere of radius r is $E(r) \times 4\pi r^2 = q_1/\epsilon_0$, and

$$E(r) = \frac{q_1}{4\pi\epsilon_0 r^2}. \tag{15.20}$$

The argument used to prove Gauss's law from the inverse square law has been turned around by invoking the symmetry of the field. Each law can be derived from the other, and either may be used as the basis of electrostatics.

Worked Example 15.4 A large number of small charges are placed close together along a straight line so that the total charge per unit length is $\eta\,\mathrm{C\,m^{-1}}$ (coulombs per metre). Assuming the line of charges to be infinitely long, calculate the electrostatic field at a perpendicular distance r from the line.

Answer This is an example of *cylindrical symmetry*, because all directions pointing perpendicularly away from the line of charges are equivalent. The electric field must be perpendicular to the line—since there is no way to choose one direction along the line rather than the other, the component of the field along the line must be zero. A cylinder of length ℓ with axis on the line and ends perpendicular to the line, as shown in Fig 15.14, is a suitable surface for the application of Gauss's theorem. The field is everywhere perpendicular to the curved surface of the cylinder and its magnitude $E(r)$ depends only on the distance r. The area of the curved surface of the cylinder is $2\pi r\ell$ and the flux out of this surface is therefore $2\pi r\ell E(r)$. The flux out of the ends of the cylinder is zero, since the field lines do not cross the end surfaces. The total amount of charge within the cylinder is $\eta\ell$ and, applying Gauss's law to the cylinder,

outward flux $= 2\pi r\ell E(r) = $ total charge$/\epsilon_0 = \eta\ell/\epsilon_0$

or

$$E(r) = \frac{\eta}{2\pi\epsilon_0 r}. \tag{15.21}$$

A real line of charges can never be infinitely long. However, eqn (15.21) is a good approximation for the magnitude of the field provided that r is small compared to the distance to the end of the line of charges.

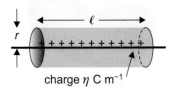

charge $\eta\,\mathrm{C\,m^{-1}}$

Fig. 15.14 The electric field near a line charge may be calculated by applying Gauss's law to an imaginary cylinder of length ℓ and radius r.

15.4 The electrostatic potential

The concepts of work and potential energy are discussed in general terms in Sections 3.3 and 3.6. The work done by a force is defined in eqn (3.12) as (force × the distance moved in the direction of the force). Examples considered in Chapter 3 include the work done against the gravitational force in lifting a mass, and against the restoring force of a spring when it is stretched. In both cases energy must be expended to do the work, but this energy does not disappear. It is stored as potential energy, which may later be released: gravitational potential energy may, for example, be released by allowing an object to fall.

Work done by electric charges

The concepts of work and potential energy also apply when the forces are electrical. Consider the two positive charges q and q_1 shown in Fig 15.15. The charge q_1 is fixed, but q may be moved. There is a repulsive force between the two charges and when they are separated by a distance r the magnitude of the force is given by eqn (15.2) as

$$F = \frac{qq_1}{4\pi\epsilon_0 r^2}.$$

The force on q acts in the direction AB. If q moves a distance dr along the line BC, the work done by the force is $F dr$. The amount of work done when q moves from B to C, changing the separation of the charges from an initial value r_i to a final value r_f is

$$W = \int_{r_i}^{r_f} F dr = \int_{r_i}^{r_f} \frac{qq_1}{4\pi\epsilon_0 r^2} dr = \frac{qq_1}{4\pi\epsilon_0}\left[-\frac{1}{r}\right]_{r_i}^{r_f} = \frac{qq_1}{4\pi\epsilon_0}\left(\frac{1}{r_i} - \frac{1}{r_f}\right).$$

$$(15.22)$$

This work represents the difference in the **electrical potential energy** of the two charges when q moves from B to C. It is natural to choose the potential energy to be zero at $r_f = \infty$, and with this choice the total potential energy U of the two charges when they are at A and B, separated by a distance r_i, is

$$U = W_\infty = \frac{qq_1}{4\pi\epsilon_0 r_i}.$$

$$(15.23)$$

This equation is very similar to eqn (5.5), which gives the gravitational potential energy of two masses, except that the sign is different. The sign change occurs because the gravitational force is attractive, whereas the electrical force between positive charges we have been considering here is repulsive. Work must be done to pull the masses apart, and the gravitational potential energy is therefore negative. The same applies to

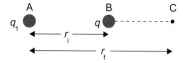

Fig. 15.15 The electric field does work on the charge q when it moves from B to C.

● *Electrical and gravitational potential energies are given by similar expressions*

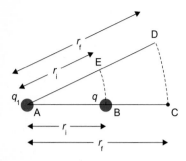

Fig. 15.16 The dashed lines are perpendicular to the electric field due to the charge q_1 and no work is done if q follows the path BCDEB.

⬤ *The electrostatic force is conservative*

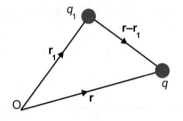

Fig. 15.17 The potential energy of q in the field of q_1 depends on the distance $|\mathbf{r} - \mathbf{r}_1|$ between them.

charges of different sign. Equation (15.23) is still valid, but if q and q_1 have different signs the right-hand side is negative, corresponding, as for the gravitational case, to the fact that work must be done to pull the charges apart.

The potential energy of q depends only on its distance from q_1 and not on the direction. In Fig 15.16 no work is done in moving q from B to E or from C to D, since the electric force is perpendicular to the direction of motion. Furthermore, the loss in potential energy in moving from B to C may be recovered by using an external force to push q back to B. The work done by the external force is also given by eqn (15.22). No work is done if q moves from B to C and back again, nor is there any change in potential energy. The same applies if q is taken round the path BCDEB or any other path starting and finishing at the same point: the potential energy depends only on the position of q and not on the path it took to get there. As explained in Section 3.6, a force that has the property of doing no work around a closed path is called a **conservative** force.

Like the electric field, the potential energy is usually most conveniently expressed in terms of position vectors with respect to a fixed origin. Suppose that q and q_1 are placed at points with position vectors $\mathbf{r}$ and $\mathbf{r}_1$ with respect to an origin at O, as in Fig 15.17,

The potential energy in eqn (15.23) may now be written

$$U = \frac{qq_1}{4\pi\epsilon_0|\mathbf{r} - \mathbf{r}_1|}. \tag{15.24}$$

If more charges $q_2, q_3, \dots$ are now placed at $\mathbf{r}_2, \mathbf{r}_3, \dots$ the potential energy of q with respect to each one is an expression of the form of eqn (15.24). The total potential energy, that is, the energy U_{tot} that is released if q is moved far away while all the other charges remain fixed, is

$$U_{tot} = \sum_j \frac{qq_j}{4\pi\epsilon_0|\mathbf{r} - \mathbf{r}_j|}. \tag{15.25}$$

The charge q has been used as a test charge to sample the potential energy it gains in the neighbourhood of the fixed charges q_j. The potential energy per unit charge U_{tot}/q is determined only by the magnitudes and positions of the fixed charges q_j, just as is the electric field. The quantity U_{tot}/q is called the **electrostatic potential** and it is denoted by $\phi(\mathbf{r})$,

$$\phi(\mathbf{r}) = U_{tot}/q = \sum_j \frac{q_j}{4\pi\epsilon_0|\mathbf{r} - \mathbf{r}_j|}. \tag{15.26}$$

Like the electric field, the electrostatic potential is a function of position. Unlike the electric field, it has a magnitude but no direction: it is a *scalar field*. Since the potential represents energy per unit charge, it may be measured in joules per coulomb. However, the potential is of such great practical importance that it has a special unit called the **volt**, denoted by the symbol V. One volt is the same as one joule per coulomb. One joule of energy is required to move a charge of one coulomb through a potential difference of one volt.

The electrostatic potential depends linearly on the magnitudes of the charges q_j and, like the electric field, the potential obeys the principle of superposition. If the potential is known for two different sets of charges, when both are present the potential is the sum of the potentials for each separately.

Worked Example 15.5 The two electrons in a hydrogen molecule are suddenly removed, leaving two protons separated by about 0.07 nm. The protons then fly apart; calculate their final kinetic energy.

Answer The potential energy of the two protons, given by eqn (15.18), is converted entirely into kinetic energy. They have equal and opposite momenta, and each has kinetic energy E_K equal to half of the initial potential energy,

$$E_K = \frac{e^2}{8\pi\epsilon_0 r} = \frac{(1.602 \times 10^{-19})^2}{8\pi \times 8.854 \times 10^{-12} \times 0.07 \times 10^{-9}} \approx 1.6 \times 10^{-18} \, \text{J}.$$

The unit of energy on the atomic energy scale is the electron volt (eV), which is the work done when a charge of e coulombs is moved through a potential difference of one volt: $1\,\text{eV} \equiv 1.602 \times 10^{-19}\,\text{J}$. E_K is thus about 10 eV.

Equipotential surfaces

For an isolated charge q_1 the electrostatic potential at a distance r from q_1 is $q_1/(4\pi\epsilon_0 r)$. All points on the surface of a sphere of radius r are at the same potential. Spherical surfaces centred on q_1 are **equipotential surfaces**. No work is done in moving a test charge q across the surface from one point to another. The electric field generated by q_1 points radially outwards: electric field lines pointing outwards from an equipotential surface are illustrated in Fig 15.18.

At a point where a field line crosses the equipotential surface the line is perpendicular to the surface. This is obvious for a single charge, for which the field lines are radial and the equipotentials are spherical, but it is in fact a general result. Electric field lines are *always* perpendicular to equipotential surfaces, no matter what the distribution of charge. This is easily proved by considering a small movement of a test charge on an equipotential surface. No work is done, and it follows that the electric field does not have a component lying in the surface, that is, the field is perpendicular to the surface.

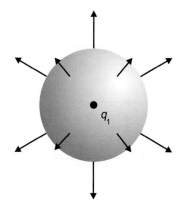

Fig. 15.18 Field lines radiating outwards from the charge q_1 are perpendicular to the spherical equipotential surface.

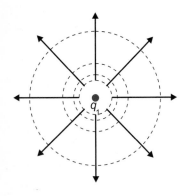

Fig. 15.19 Field lines (solid) and equipotential lines (dashed) in a plane through a charge q_1.

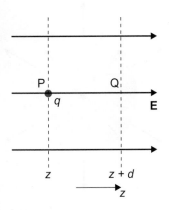

Fig. 15.20 The relation between field E and potential ϕ is found by moving a charge in the field. The equipotentials are the dashed lines.

The electric field and the electrostatic potential are really just different ways of expressing the same information about a system of charges. Figure 15.19 shows the field lines and equipotentials in a plane passing through a point charge q_1. Given a map of the contour lines representing the equipotentials, we could draw field lines cutting them perpendicularly, and vice versa: given the field lines, equipotentials cutting them at right angles would have to be circles.

Up to now we have expressed the electric field in units of newtons per coulomb. Since one volt is the same as one joule per coulomb, newtons per coulomb are equivalent to volts per metre $(V\,m^{-1})$. Note that volts per metre, which is the usual unit for describing electric field, represents the rate of change of potential with distance. In mathematical terms a rate of change is found by differentiation. For a point charge q_1, $\phi(r) = q_1/(4\pi\epsilon_0 r)$. The potential decreases with the distance r from the charge, and to find the rate of change of potential with distance we must differentiate with respect to r. Remembering that for a positive charge the field points outwards, in the direction of decreasing potential, we have

$$-\frac{d\phi}{dr} = \frac{q_1}{4\pi\epsilon_0 r^2},$$

the same as the magnitude of the electric field of a point charge given in eqn (15.20).

Another simple example relating field and potential is illustrated in Fig 15.20, which shows the field lines for a uniform field **E** pointing in the z-direction. Two equipotential surfaces, which are both perpendicular to the z-axis, are a distance d apart, and at potentials $\phi(z)$ and $\phi(z + d)$, respectively. The force on a test charge q is qE and the potential energy at P is $q\phi(z)$. The difference in potential energy of the test charge between the points P and Q equals the work done to move it back from Q to P,

$$q\phi(z) - q\phi(z + d) = qV = qEd. \tag{15.27}$$

The difference in potential V between z and $z + d$ is usually called the **voltage difference** or simply the **voltage** between the two points. Remember that the field points in the direction of *decreasing* potential. In eqn (15.27) V and E are positive if $\phi(z)$ is greater than $\phi(z + d)$.

The uniform electric field may also be represented in differential form by allowing the distance d in eqn (15.27) to become very small. If d is written as δz, then $\phi(z) - \phi(z + \delta z) = E\delta z$ and, in the limit as δz tends to zero,

$$E = -\frac{d\phi}{dz}. \tag{15.28}$$

Notice that there is a minus sign in eqn (15.28), just as in the equation relating field and potential for a point charge.

A differential relation similar to eqn (15.28) applies to any electric field and the argument used here for the uniform field is generalized in the box below.

Figure 15.21 shows two equipotential surfaces very close together, having electrostatic potentials ϕ and $(\phi + \delta\phi)$. The vector $\delta\ell$ is a vector in any direction joining the point P on the surface at potential ϕ to a point Q on the surface at potential $(\phi + \delta\phi)$. The electric field $\mathbf{E}$ at P is perpendicular to the equipotential surfaces. which are a distance PR $= \delta\ell \cos\theta$ apart. The work done on a test charge q when it moves from P to Q is the force qE times the distance PR in the direction of the force, i.e. $qE\delta\ell \cos\theta = q\mathbf{E} \cdot \delta\ell$. This work is equal to the loss in potential energy $(-q\delta\phi)$,

$$q\mathbf{E} \cdot \delta\ell = -q\delta\phi. \tag{15.29}$$

The fact that no work is done on a test charge when it is moved round any closed path that returns to its starting point is expressed mathematically by taking the infinitesimal limit of eqn (15.29) and integrating. The result is that for electrostatic fields

$$\oint \mathbf{E} \cdot d\ell = 0. \tag{15.30}$$

Here the symbol $\oint$ indicates that the line integral is round a closed path made up of infinitesimal segments $d\ell$.

If we choose Cartesian coordinates with unit vectors $\mathbf{i}$, $\mathbf{j}$, and $\mathbf{k}$ in the x-, y-, and z-directions, we may write $\mathbf{E} = E_x\mathbf{i} + E_y\mathbf{j} + E_z\mathbf{k}$ and $\delta\ell = \delta x\mathbf{i} + \delta y\mathbf{j} + \delta z\mathbf{k}$ leading to

$$\mathbf{E} \cdot \delta\ell = E_x\delta x + E_y\delta y + E_z\delta z = -\delta\phi.$$

The partial derivative $\partial\phi/\partial x$ is the rate of change of ϕ with $\mathbf{x}$ when both $\mathbf{y}$ and $\mathbf{z}$ are kept constant. In the limit as δx, δy, and δz tend to zero,

$$E_x = -\frac{\partial\phi}{\partial x}; \quad E_y = -\frac{\partial\phi}{\partial y}; \quad E_z = -\frac{\partial\phi}{\partial z}.$$

The vector with components $\partial\phi/\partial x, \partial\phi/\partial y, \partial\phi/\partial z$ is called *the gradient of ϕ* and is written as grad ϕ. The three eqns above are summarized as

$$\mathbf{E}(\mathbf{r}) = -\text{grad}\phi(\mathbf{r}). \tag{15.31}$$

The function grad$\phi(\mathbf{r})$ is a vector field that has been derived from the scalar field $\phi(\mathbf{r})$. The properties of gradϕ have already been described above in the discussion of the connection between the field and potential: gradϕ is perpendicular to surfaces of constant ϕ and its magnitude is the rate of change of ϕ with distance in that direction. In a two-dimensional contour map the contours are equipotentials of the gravitational potential

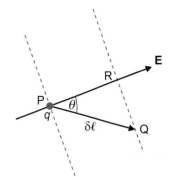

Fig. 15.21 Here the charge is moved in a direction different from the direction of the field.

● *The gradient of a scalar function*

ϕ and grad ϕ at a point on the map is in the direction of the steepest uphill gradient from that point.

Exercise 15.4 An electron is placed in an electric field of magnitude $100 \, \text{V} \, \text{cm}^{-1}$. Calculate the electrostatic force on the electron and compare it with the gravitational force.

Answer $1.6 \times 10^{-15} \, \text{N}$. This is 1.8×10^{14} times greater than the gravitational force on the electron.

The dipole potential and field

Because the electrostatic potential is a scalar field, it is often easier to evaluate the potential than the field for a particular distribution of charges. Once the potential is known, the field can be determined from eqn (15.31). This is the method we shall use to calculate the field in the neighbourhood of an electric dipole consisting of two charges of equal magnitude but opposite sign. The shape of the field lines around a dipole has already been sketched in Fig 15.6, but this figure is only a guess based on the way the field lines must pass from the positive to the negative charge. The figure does not tell us how rapidly the strength of the field falls off with distance from the dipole, or precisely how it varies with direction with respect to the axis of the dipole. It is important to have a mathematical expression for the dipole field, because it is responsible for part of the interaction between molecules and it is also related to electromagnetic radiation.

The dipole in Fig 15.22 has charges $\pm q$ separated by a distance a. Spherical polar coordinates referred to a z-axis along the line joining the charges are the most convenient for discussing the potential of the dipole. The origin is midway between the two charges and the point P has coordinates (r, θ, ψ): the symbol ψ is used for the third coordinate in this section, rather than the usual ϕ to avoid confusion with the potential, which is also usually labelled by ϕ. The dipole has cylindrical symmetry so that all angles ψ are equivalent and the potential depends only on r and θ. The figure represents a plane passing through the z-axis and the point P. The vectors $\mathbf{r}_+$ and $\mathbf{r}_-$ join the charges $+q$ and $-q$ to P. The potential at P is the sum of the contributions from each charge, and is

$$\phi(\mathbf{r}) = \frac{q}{4\pi\epsilon_0} \left(\frac{1}{r_+} - \frac{1}{r_-} \right). \tag{15.32}$$

This expression applies everywhere, but it cannot be expressed simply in terms of the coordinates r and θ. In practice it turns out that what is usually important is the field at distances large compared with a.

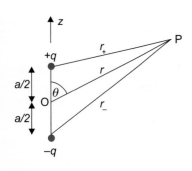

Fig. 15.22 The potential around a dipole is the sum of the potentials of the two charges $\pm q$ separately.

As the distance from the dipole to P increases, the fractional difference between r_+ and r_- becomes smaller. The two terms on the right-hand side of eqn (15.32) therefore cancel each other more and more closely as the distance r from the dipole to P increases, and the potential must diminish faster than $1/r$. When a/r is small, a good approximation to the potential, which is worked out in the following box, is

$$\phi(\mathbf{r}) = \frac{qa \cos \theta}{4\pi\epsilon_0 r^2} = \frac{p \cos \theta}{4\pi\epsilon_0 r^2} \tag{15.33}$$

where $p = qa$.

The magnitudes of the vectors $\mathbf{r}_+$ and $\mathbf{r}_-$ are given in terms of the coordinates r and θ by applying the cosine rule:

$$r_\pm^2 = r^2 + \tfrac{1}{4}a^2 \mp ar \cos\theta = r^2\left(1 + \frac{a^2}{4r^2} \mp \frac{a}{r}\cos\theta\right).$$

To find the potential at distances large compared with a, we must expand it in powers of a/r using the binomial theorem. All terms except the first order in a/r will be neglected, and we may write

$$\frac{1}{r_\pm} = (r_\pm^2)^{-1/2} = \frac{1}{r}\left(1 \mp \frac{a}{r}\cos\theta\right)^{-1/2} = \frac{1}{r}\left(1 \pm \frac{a}{2r}\cos\theta\right).$$

Substituting these expressions for $r_\pm$ in eqn (15.32), the leading terms in $1/r$ cancel and we are left with eqn (15.33).

● *The expansion of $1/r$ in polar coordinates*

The potential given in eqn (15.33) is often called the *dipole potential*, and it represents the potential due to an idealized point dipole in which the distance a has been allowed to tend to zero while $p = qa$ remains nonzero. The dipole potential decreases with distance as $1/r^2$, whatever the angle θ. As we predicted, this decrease is faster than $1/r$ because of the cancellation of the first-order contributions of the positive and negative charges.

We have chosen the z-axis to lie along the line joining the charges of the dipole. The vector $\mathbf{p}$ pointing along the z-axis and with magnitude p is called the **dipole moment**. In terms of this vector, the dipole potential is

● *Expressing the dipole potential in terms of the dipole moment*

$$\phi(\mathbf{r}) = \frac{\mathbf{p} \cdot \mathbf{r}}{4\pi\epsilon_0 r^3}. \tag{15.34}$$

This expression makes no mention of the variable θ and it is in fact correct whatever may be the orientation of the dipole moment with respect to the z-axis.

The general relation between the potential and the electric field is $\mathbf{E}(\mathbf{r}) = -\mathrm{grad}\ \phi(\mathbf{r})$ (eqn (15.31)). In spherical polar coordinates the r-, θ-, and ψ components of $\mathbf{E}(\mathbf{r})$ are in the directions of the unit vectors labelled

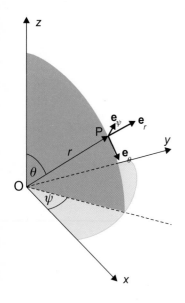

Fig. 15.23 The arrows show the directions of the electric field components E_r, E_θ, and E_ψ at the point P that has spherical polar coordinates (r, θ, ψ).

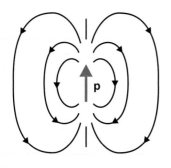

Fig. 15.24 The dipole field due to a very small dipole with dipole moment **p**.

e_r, e_θ, and e_ψ in Fig 15.23. The component E_r points directly away from the origin. The component E_θ is tangential to a circle passing through P having constant r and ψ, in the direction of increasing θ. Similarly, E_ψ is tangential to a circle passing through P having constant r and ψ, in the direction of increasing ψ. In terms of the potential, the components are

$$E_r = -\frac{\partial \phi}{\partial r}; \quad E_\theta = -\frac{1}{r}\frac{\partial \phi}{\partial \theta}; \quad E_\psi = -\frac{1}{r \sin \theta}\frac{\partial \phi}{\partial \psi}. \tag{15.35}$$

Here there is no variation with ψ and the components are

$$E_r = \frac{p \cos \theta}{2\pi\epsilon_0 r^3}; \quad E_\theta = \frac{p \sin \theta}{4\pi\epsilon_0 r^3}; \quad E_\psi = 0. \tag{15.36}$$

The electric field lines given by eqn (15.36) for small a/r are shown in Fig 15.24. This figure applies to any plane that includes the z-axis. Both the outward component E_r of the electric field and the component E_θ following circles of constant r and constant ψ are proportional to $1/r^3$, falling off with distance faster than the field due to a single charge. The terms in higher powers of a/r, which we have neglected, decrease faster still. Electrically neutral molecules may possess a dipole moment and, although their dipole fields may cause important interactions with other molecules, the higher terms are almost always negligible.

15.5 Electric fields in matter

Macroscopic electric fields

Inside a single atom the electric field changes very rapidly with distance. The atomic nucleus is extremely small, even on the atomic scale, and it carries a charge Ze, where Z is the atomic number of the atom (which determines the chemistry of the atom) and e is the electronic charge. Close to the nucleus the positive charge on the nucleus is all that matters and the field is directed away from the nucleus. Further out, the negative charge on the electrons tends to cancel out the effect of the nucleus, and outside the atom the field is very small. On a microscopic scale these changes of field within an atom are extremely important, and indeed in Chapter 11 the attraction given by Coulomb's law between an electron and a proton is used to work out the properties of the hydrogen atom.

In this section we are concerned with the electrical behaviour of pieces of matter made up of an enormous number of atoms. The fields within individual atoms are not of interest: we need to know how the average field varies over volumes large enough to contain very many atoms. Such an average field is called a **macroscopic field**, to distinguish it from the **microscopic field**, which varies rapidly within atoms.

Before discussing the macroscopic field in an assembly of many atoms, consider the average field in a volume containing a single electrically neutral atom such as the inert gas argon. The atom is spherically symmetric, so that the field within it is always pointing away from the centre of the atom. The field has a high value near the centre, like the field around a point charge, but it falls away even faster with distance because of the negative charge on the electrons. To work out the average field, you have to remember that averaging a vector quantity is a bit different from averaging a scalar quantity. Directions as well as magnitudes must be taken into account. For a particular point with position vector $\mathbf{r}$ with respect to an origin at the centre of the argon atom, the field points in the same direction as $\mathbf{r}$. At the point diametrically opposite, which has position vector $-\mathbf{r}$, the field has the same magnitude but is in the opposite direction to the field at $\mathbf{r}$. The sum of the fields at $\mathbf{r}$ and $-\mathbf{r}$ is zero. The same applies to all possible points $\mathbf{r}$, and the average field in a volume including the atom is zero.

The example of an inert gas is a special case because the atoms are spherically symmetric. However, except for some special materials, in the absence of any electric field applied from outside, the macroscopic electric field in electrically neutral matter is zero. When charges are present, it is not necessary to calculate the macroscopic field by adding up the contributions from every single particle carrying a charge $\pm e$ and then find the average—most of the contributions just cancel out. The average charge is determined within a volume small compared with everyday objects, but still large enough to contain many atoms. The electric field caused by this average charge is then calculated.

● *The average field due to electrically neutral atoms is zero*

Suppose that δV_j is a small volume located at a point having a position vector $\mathbf{r}_j$ with respect to the origin. Let the net amount of charge within δV_j be $\rho(\mathbf{r}_j)\delta V_j$: $\rho(\mathbf{r}_j)$ is thus the **charge density**, that is, charge per unit volume, measured in coulombs per cubic metre. Now divide up the whole of the region containing charge into a lot of small volumes δV_j. Each contributes to the macroscopic electric field, which by substitution in eqn (15.11) is

● *Macroscopic fields are calculated from average charge densities*

$$\mathbf{E}(\mathbf{r}) = \frac{1}{4\pi\epsilon_0} \sum_j \frac{\rho(\mathbf{r}_j)(\mathbf{r} - \mathbf{r}_j)\delta V_j}{|\mathbf{r} - \mathbf{r}_j|^3}. \tag{15.37}$$

The average charge density $\rho(\mathbf{r}_j)$ varies smoothly with the position $\mathbf{r}_j$ and it is legitimate to replace the sum in eqn (15.37) with an integral, even though we started with volumes δV_j that are large enough to contain many atoms. The macroscopic field becomes

$$\mathbf{E}(\mathbf{r}) = \frac{1}{4\pi\epsilon_0} \int_{\text{volume}} \frac{\rho(\mathbf{r}')(\mathbf{r} - \mathbf{r}')\mathrm{d}V'}{|\mathbf{r} - \mathbf{r}'|^3} \tag{15.38}$$

where the integral, labelled *volume*, is over all volumes that contain a net charge. From now on, when we refer to an electric field $\mathbf{E}(\mathbf{r})$ without stating whether it is a microscopic or macroscopic field, we shall mean the macroscopic field that has been averaged over many atoms.

As already mentioned, the macroscopic electric field in electrically neutral matter is zero if there is no external electric field. However, if an object is placed in an electric field, this field is modified by the presence of the matter. To investigate how this comes about, we must consider electrical conductors and insulators separately.

Conductors in electric fields

Materials like copper and aluminium that are good electrical conductors are able to carry electric current because some of the electrons in the material are free to move. These electrons, which are called conduction electrons, are not fixed to particular atoms, but are continually moving through the material. In the absence of an electric field there is no net flow of charge, because the electrons are moving at random in all directions. However, if a steady electric field is applied, electrons, each carrying a charge $-e$, experience a force in the opposite direction to the field. There is a net flow in this direction, and the flow may continue for an indefinite time if the conductor is part of a complete electrical circuit. On the other hand, if the conductor is isolated, the electrons cannot continue to move when they reach the boundaries of the conductor. In the slab of conductor shown in Fig 15.25, for example, the electric field pointing to the right causes electrons to migrate from the right-hand side to the left-hand side. Negatively charged electrons accumulate on the left-hand surface, and the deficit on the right-hand side causes a net positive charge to occur there.

The charges appearing on the surface of a conductor are called **induced charges**. The induced charges themselves generate an electric field directed away from the positive charges towards the negative charges, tending to cancel out the external field. Conduction electrons will continue to flow, however small may be the resultant electric field, and they flow until the electric field within the conductor is *zero*. The disposition of surface charges depends on the shape of the conductor and is, in general, very difficult to work out. Whatever the shape, the charges nevertheless arrange themselves so that the field inside the conductor is exactly zero. This applies to any material that contains conduction electrons, and not just to very good conductors like copper. The semiconductors silicon and germanium, for example, have conduction electron densities billions of times smaller than copper at room temperature but, when placed in a steady external field, they also have zero field inside.

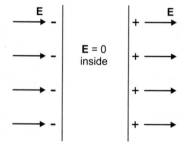

Fig. 15.25 Charges migrate to the surface of a conductor to ensure that the electric field is zero inside the conductor.

● *The electrostatic field inside a conductor is zero*

Because the electric field is zero throughout the conductor, its whole volume is at the same potential. In particular, its surface is an equipotential surface. Since field lines and equipotentials are always perpendicular to one another, the external field is normal to any conducting surface. Using Gauss's law we can relate the magnitude of the electric field to the amount of charge on the conducting surface. In Fig 15.26 the closed surface S is shaped like a pillbox. The curved surface is parallel to the electric field and there is no flux through it. The flat surfaces of the pillbox, each of area δS, are parallel to the conducting surface, one inside and one outside the conductor. The electric field and hence the flux are zero on the inside. The total flux out of S is $E \cdot \delta S$ and, if the charge inside the pillbox is δQ, Gauss's law gives

$$\int_S \mathbf{E} \cdot d\mathbf{S} = E\delta S = \delta Q/\epsilon_0$$

or

$$\epsilon_0 E = \delta Q/\delta S = \sigma \qquad (15.39)$$

where σ is the **surface charge density**, which is measured in coulombs per square metre $(\mathrm{C\,m^{-2}})$. In the simple example of slab geometry illustrated in Fig 15.26 the surface charge density is given directly in terms of the external field by eqn (15.39).

For other shapes of conductor the surface charge density must be distributed in such a way as to ensure that the external field is normal to the conducting surface. This is illustrated schematically in Fig 15.27 which shows a conducting sphere in an electric field that is uniform far from the sphere. Close to the sphere the surface charges modify the field lines so that they curve towards the sphere and meet it normally.

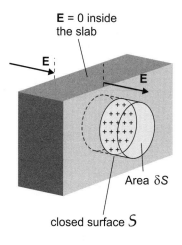

Fig. 15.26 Gauss's theorem relates the induced surface charge to the electric field outside the conductor.

Exercise 15.5 The electric field at the surface of a conductor is $10^4\,\mathrm{V\,cm^{-1}}$. What is the surface charge density on the conductor, and what average area has a charge equal to one electronic charge?

Answer The coulomb is a very large unit, and charge is frequently expressed in microcoulombs $(1\,\mu\mathrm{C} \equiv 10^{-6}\,\mathrm{C})$. The surface charge in this exercise is $8.85\,\mu\mathrm{C\,m^{-2}}$. This is equivalent to one electronic charge on an area $1.8 \times 10^{-14}\,\mathrm{m^2}$ or $1.8 \times 10^4\,\mathrm{nm^2}$, an area large enough to accommodate about one million atoms.

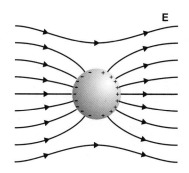

Fig. 15.27 When a conducting sphere is placed in an external electric field $\mathbf{E}$, the field lines bend to meet the conducting surface normally.

The induced charges on the surface of a conductor are located in a very thin layer. The amount of induced charge is given by the surface charge density σ and a surface integral must be added to eqn (15.38) to account

● *Induced charges on the surfaces of conductors contribute to the electric field*

for the contribution of the induced charges to the field. Including the surface charges, the general expression for the macroscopic electric field is

$$\mathbf{E}(\mathbf{r}) = \frac{1}{4\pi\epsilon_0} \int_{\text{volume}} \frac{\rho(\mathbf{r}')(\mathbf{r} - \mathbf{r}')\,\mathrm{d}V'}{|\mathbf{r} - \mathbf{r}'|^3} + \frac{1}{4\pi\epsilon_0} \int_{\text{surface}} \frac{\sigma(\mathbf{r}')(\mathbf{r} - \mathbf{r}')\,\mathrm{d}S'}{|\mathbf{r} - \mathbf{r}'|^3}$$

(15.40)

where the labels *volume* and *surface* indicate that the volume integral is over all volumes containing a volume charge density and the surface integral is over all surfaces on which there is a surface charge density.

Similarly, the potential is

$$\phi(\mathbf{r}) = \frac{1}{4\pi\epsilon_0} \int_{\text{volume}} \frac{\rho(\mathbf{r}')\mathrm{d}V'}{|\mathbf{r} - \mathbf{r}'|} + \frac{1}{4\pi\epsilon_0} \int_{\text{surface}} \frac{\sigma(\mathbf{r}')\mathrm{d}S'}{|\mathbf{r} - \mathbf{r}'|}.$$

(15.41)

Insulators in electric fields

In an insulator, all the electrons are fixed to particular atoms. Over long time periods, practically no migration of charge occurs when an insulating material is placed in an electric field. We can understand how the material responds to the presence of a steady field by considering just one atom.

Imagine that a neutral atom is supported so that it does not fall under gravity, but is free to move horizontally. If a horizontal electric field is switched on, there is no net force on the atom since its charge is zero. However, the nucleus and the electrons experience forces in opposite directions and they tend to move apart, without shifting the centre of mass of the atom. As the centre of the distribution of negatively charged electrons moves away from the positively charged nucleus, the mutual attraction of nucleus and electrons creates a restoring force that balances the force caused by the external field.

Under all conditions that are met in the laboratory, the restoring force is proportional to the distance x between the nucleus and the centre of the electron distribution. Calling the constant of proportionality k, the restoring force is kx. Figure 15.28 shows the forces acting on the nucleus, but greatly exaggerates the relative movement of the electrons and the nucleus: on the scale of the figure, the shift would not be visible for realistic electric fields. For an atom with atomic number Z and nuclear charge Ze, the force due to the external field E is ZeE. This force is balanced by the restoring force when $kx = ZeE$, that is when $x = ZeE/k$.

When the nucleus and the centre of electronic charge do not coincide, the atom is said to be **polarized**. As for the point charges discussed in Section 15.4, the vector in the x-direction and with magnitude equal to

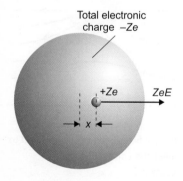

Total electronic charge $-Ze$

$+Ze$ ZeE

x

Fig. 15.28 The force on the nucleus of the atom due to the electric field E is balanced by a restoring force caused by the mutual attraction of the nucleus and the electrons.

the product of the distance x and the charge Ze is called the *dipole moment* of the atom and is measured in coulomb metres (C m). The dipole moment is denoted by the vector **p**: the vectors **x**, **p**, and **E** all point in the same direction and

$$\mathbf{p} = Zex = \frac{(Ze)^2}{k}\mathbf{E} = \alpha\epsilon_0\mathbf{E} \tag{15.42}$$

where the constant α is called the **polarizability** of the atom.

How does polarization affect the macroscopic electric field in an insulator? Let us first consider a slab of uniform insulating material placed in an electric field normal to the faces of the slab. Within the slab the macroscopic electric field must be in the same direction as the field ouside the slab, and we shall assume for the moment that it is constant throughout the slab, having a value $\mathbf{E}_{int}$, say. Each atom of the insulator acquires a dipole momen $\alpha\epsilon_0\mathbf{E}_{int}$ and, according to eqn (15.42), the centre of the electron distribution is displaced a distance $\alpha\epsilon_0 E_{int}/Ze$ from the nucleus.

The nucleus is much more massive than the electrons, and the centre of mass almost coincides with the nucleus. We may picture the polarization as if only the electrons move: this simplifies the argument without altering the results. Figure 15.29 represents a section through a slab of insulator placed in an electric field perpendicular to the sides of the slab. The dashed lines show the boundaries of imaginary closed boxes with faces of area δS perpendicular to the field: we shall apply Gauss's law to these boxes.

When the atoms are polarized, electrons move through both surfaces of the box (b), which is completely inside the insulator. Negative charge has moved out of the left-hand side of the box, but just as much has moved in at the right-hand side. The net charge inside the box is zero, as it was before the atoms were polarized. Gauss's law tells us that the net flux of **E** out of the box is zero. This requires the flux entering the left-hand side to equal the flux leaving the right-hand side, and the assumption that the field is uniform within the slab is justified.

Now look at the box (c), which straddles the right-hand surface of the insulator. Negative charge has moved out of the left-hand side of the box but there are no atoms at the right-hand side and the box has acquired a net positive charge. If the number of atoms per unit volume is N, the charge density of electrons is $-NZe$. All the electrons in the slab have moved the same distance x, and the charge moving out of the area δS on the left of the box is $-NZex\delta S = -Np\delta S$. The box now encloses a net charge $+Np\delta S$, and the surface charge density caused by polarization is $\sigma_p = Np$. The opposite face of the slab acquires a surface charge density $(-Np)$ as electrons move into box (a). The slab as a whole is electrically neutral, as it must be since it is composed entirely of neutral atoms.

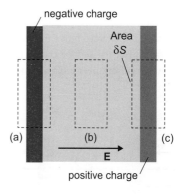

Fig. 15.29 The movement of charge in the slab of insulator builds up charge on the surface but leaves the interior electrically neutral.

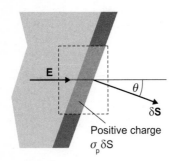

Fig. 15.30 If the field inside the insulator is not perpendicular to its surface, the charges move the same distance, but the surface charge is now spread out over a bigger area.

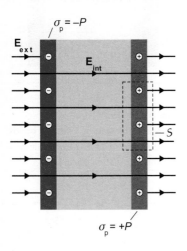

Fig. 15.31 The induced surface charge is related to the electric field inside and outside the insulator.

If the surface of the insulator is at an angle θ to the electric field, as in Fig 15.30, the surface charge density is reduced. If the area of the insulator surface inside the box is δS, the projected area normal to the field is $\delta S \cos \theta$. The net charge inside the box is now $\sigma_p \delta S = Np\delta S \cos \theta$. At a surface where the electric field enters the insulator, σ_p is given by the same expression except for a minus sign. Remembering that the vector $\delta \mathbf{S}$ is the *outward* normal to the surface, we find that both the sign of the surface charge density σ_p and its angular dependence can be expressed concisely by using the vector notation

$$\sigma_p \delta S = N\mathbf{p} \cdot \delta \mathbf{S} = \mathbf{P} \cdot \delta \mathbf{S} \tag{15.43}$$

where the vector $\mathbf{P} = N\mathbf{p}$ is the dipole moment per unit volume of the insulator. The vector $\mathbf{P}$ is called the **polarization density**. The polarization density, like the dipole moment of a single atom, is in the direction from negative to positive polarization surface charge.

The polarization density is useful because it is related to the electric field inside the polarized material. In slab geometry this relation is easily found from Gauss's law. The closed surface $\mathcal{S}$ in Fig 15.31 has surfaces of area δS normal to the field. The polarization charge within $\mathcal{S}$ is $\sigma_p \delta S = P\delta S$. According to Gauss's law the net flux out of $\mathcal{S}$ is therefore $P\delta S/\epsilon_0$. The field entering $\mathcal{S}$ from the left is E_{int} and the field leaving on the right is E_{ext}, and the net flux is $(E_{ext} - E_{int})\delta S = P\delta S/\epsilon_0$. Hence

$$E_{ext} = E_{int} + P/\epsilon_0 \tag{15.44}$$

or, since $P = Np = N\alpha\epsilon_0 E_{int}$,

$$E_{ext} = E_{int}(1 + N\alpha) = E_{int}(1 + \chi_E). \tag{15.45}$$

The dimensionless constant $\chi_E = N\alpha$ is called the **electric susceptibility** of the insulating material. The presence of the polarization charges has *reduced* the field inside the insulator by the factor $(1 + \chi_E)$. This is represented by drawing a reduced density of lines inside: in Fig 15.31 some lines of the external field end on negative polarization charges and start on positive polarization charges.

Worked Example 15.6 At 20°C and one atmosphere pressure helium gas contains 2.7×10^{25} atoms m^{-3}, and the electric susceptibility of the gas is 6.5×10^{-5}. Calculate the separation of the centres of the positive and negative charges in a helium atom when it is placed in an electric field of 10^6 V m^{-1}.

Answer From eqns (15.42) and (15.45) the separation is

$$x = \frac{\alpha\epsilon_0 E}{Ze} = \frac{\chi_E \epsilon_0 E}{ZeN}.$$

Substituting the values given, the separation is

$$\frac{6.5 \times 10^{-5} \times 8.85 \times 10^{-12} \times 10^6}{2 \times 1.6 \times 10^{-19} \times 2.7 \times 10^{25}} = 6.7 \times 10^{-17}\,\text{m} = 6.7 \times 10^{-8}\,\text{nm},$$

a shift of about one-millionth of the radius of the helium atom.

Equation (15.45) has only been proved for slab geometry. The relation between internal and external fields for other shapes of insulator is more complicated, and we shall not discuss it in detail here. The *direction* of the field as well as the magnitude may change at the boundary of an insulator. Figure 15.32 shows the field lines when an insulating sphere is placed in a uniform external field. The external field lines are bent towards the sphere, rather like the field pattern for the conducting sphere. The sphere is a specially simple case that can be solved exactly. The field inside the sphere is uniform: the field inside has a smaller magnitude than the field outside because of polarization charges on the surface of the sphere.

● *Polarization charge density*

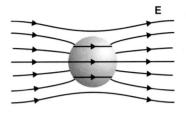

Fig. 15.32 The field lines of an insulating sphere placed in an external electric field E. The field changes direction at the surface of the sphere.

Polarization charge density

Provided that an insulator is uniform, polarization charges appear only on the surface. For a non-uniform insulator there may be polarization charges distributed throughout its volume. For example, if an insulator consists of atoms that all have the same polarizability but a variable density, more charge moves in a more dense region than in a less dense one, and there is a net polarization charge per unit volume, which we shall denote by ρ_p. Once the distribution of polarization charges is known, the electric field and potential are given by eqns (15.40) and (15.41), including polarization charges, induced charges on conductors, and distributions of free charge in the charge densities ρ and σ.

Polar molecules

The atoms in a solid make small vibrations about fixed positions. Each atom is locked in place surrounded by neighbouring atoms, keeping the same set of neighbours over long periods. When the solid is heated, the vibrations become more and more energetic, until at the melting point atoms escape from their fixed positions and the solid turns into a liquid.

In many liquids the atoms do not move independently. They remain as parts of a molecule with a fixed structure. The molecules are the units that change their positions and orientations with respect to their neighbours. The water molecule, for example, consists of one oxygen atom and two hydrogen atoms. As shown in Fig 15.33, the hydrogen and oxygen atoms do not lie on a straight line. (The structure of the water molecule is briefly explained in Section 11.4.) Furthermore, the oxygen atom has more than its share of the electrons in the molecule, so that there is excess negative charge near the oxygen atom and excess positive charge near the hydrogen atoms. The centre of positive charge does not coincide with the centre of negative charge, and the molecule has a dipole moment.

positive charge

negative charge

dipole moment in this direction

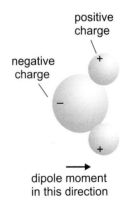

Fig. 15.33 The centres of positive and negative charge do not coincide in the water molecule, and it has a dipole moment.

Molecules like water that possess a dipole moment are called **polar molecules**. Although the water molecule is polar, in the absence of an external electric field the macroscopic electric field inside a volume of water is zero. This is because thermal motion ensures that all directions are equally likely for the dipoles and, on average, their contributions to the electric field cancel out.

When a polar liquid or gas is placed in an electric field, electrons and nuclei are pushed in opposite directions just as in an insulating solid, and as a result the liquid acquires a net dipole moment per unit volume. There is an additional effect for polar molecules that is usually more important. The dipole moments, which were initially randomly oriented, are partially lined up by the field so that they are more likely to be pointing in the direction of the field than opposite to it.

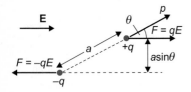

Fig. 15.34 The dipole placed in an electric field experiences a couple.

A polar molecule is represented in Fig 15.34 by positive and negative point charges with the same dipole moment $p = qa$ as the molecule—the dipole moment is all that is needed for working out the effect of a uniform electric field acting on the molecule. The field exerts a couple on the molecule, tending to rotate it so that the dipole moment and the field point in the same direction. The figure shows that, when the dipole is at an angle θ to the field, there is a couple $qaE \sin\theta = pE \sin\theta$ acting on the dipole. This couple is zero for $\theta = 0°$ and for $\theta = 180°$. At $\theta = 0°$ the dipole is in stable equilibrium; when rotated through a small angle the couple will turn the dipole back to $\theta = 0°$. At $\theta = 180°$ the dipole is in a position of unstable equilibrium; after a small deflection it will flip over to $\theta = 0°$.

Work is done by the electric field when it causes the dipole to rotate. The potential energy of the dipole therefore depends on its orientation. In Fig 15.35 equipotential surfaces passing through the charges $+q$ and $-q$ are at potentials ϕ_+ and ϕ_- respectively. From eqn (15.17) the potential energy of the charge $+q$ is $+q\phi_+$ and, similarly, the potential energy of $-q$ is $-q\phi_-$. The potential energy of the dipole in the field is thus $q(\phi_+ - \phi_-)$. The difference between the potential energies is given in terms of the field by eqn (15.29) as $q(\phi_+ - \phi_-) = -\mathbf{E} \cdot \mathbf{a}$, leading to the potential energy of the dipole

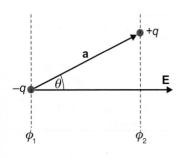

Fig. 15.35 The potential energy of the dipole depends on its orientation in the field.

$$U_{\text{dipole}}(\theta) = q(\phi_+ - \phi_-) = -q\mathbf{E} \cdot \mathbf{a} = -\mathbf{p} \cdot \mathbf{E} = -pE \cos\theta. \quad (15.46)$$

In calculating this potential we have not considered the energy of each charge $\pm q$ due to the presence of the other. For a real molecule this additional energy contributes to the binding energy of the molecule, which does not change as the molecule is rotated in electric fields that can be realized in practice.

If there were no thermal motion, all the dipoles in a polar liquid would line up with the field. But the molecules are continually colliding with their neighbours. There is a conflict between the thermal motion that

● *Thermal motion counteracts the tendency of dipoles to line up along the electric field*

tends to randomize the orientation of the molecules and the couple due to the electric field trying to line them up. The effect of thermal motion is discussed in Section 12.6, where it is explained that the probability of occurrence of states with different energies depends on the comparison of the energy difference with a thermal energy $k_B T$. Here k_B is a universal constant called Boltzmann's constant and T is the absolute temperature measured in kelvins.

In a liquid the electric field acting on each molecule is the internal field $\mathbf{E}_{int}$, and the potential energy is $-\mathbf{p} \cdot \mathbf{E}_{int}$. At ambient temperatures the ratio $pE_{int}/k_B T$ is always small, and the molecular dipoles have only a slight tendency to line up with the field. The molecular dipole moment averaged over many molecules has a magnitude $\frac{1}{3}(pE_{int}/k_B T) \times p$. The average dipole moment is in the direction of the field and, if there are N molecules per unit volume, the polarization density $\mathbf{P}$, which is the dipole moment per unit volume, is

$$\mathbf{P} = \frac{Np^2}{3k_B T}\mathbf{E}_{int}. \tag{15.47}$$

This result is proved in the box that follows.

The polarization density arising from the polarizability of the molecules adds to that arising from the permanent dipole moment. For an isotropic liquid or gas made up of molecules with a dipole moment of magnitude p and polarizability α, combining the results of eqns (15.45) and (15.47) leads to an electric susceptibility

$$\chi_E = N\left(\alpha + \frac{p^2}{3\epsilon_0 k_B T}\right). \tag{15.48}$$

The probability of finding a molecule in a state with energy U_{dipole} is given by the Boltzmann factor $\exp(-U_{dipole}/k_B T)$ (expression (12.30)). Taking the energy U_{dipole} from eqn (15.46), the probability of finding a dipole in a polar liquid at an orientation θ to an external electric field is proportional to

$$\exp\left(\frac{-U_{dipole}(\theta)}{k_B T}\right) = \exp\left(\frac{-pE_{int}\cos\theta}{k_B T}\right).$$

When $(pE_{int}/k_B T)$ is small, the exponential function may be expanded in a Taylor series keeping only the first term,

$$\exp\left(\frac{-pE_{int}\cos\theta}{k_B T}\right) = 1 - \left(\frac{pE_{int}\cos\theta}{k_B T}\right).$$

When the field $\mathbf{E}_{int}$ is zero, all directions are equally probable, and the probability of finding a dipole lying in the range of solid angle $d\Omega$ is $d\Omega/4\pi = \sin\theta d\theta d\phi/4\pi$. Since all values of ϕ are equally probable, only the component $p\cos\theta$ of the dipole moment in the direction of $\mathbf{E}_{int}$ contributes to the average dipole moment, which is

$$\frac{1}{4\pi}\int_{\phi=0}^{2\pi}\int_{\theta=0}^{\pi}p\cos\theta\left[1-\left(\frac{pE_{int}\cos\theta}{k_B T}\right)\right]\sin\theta\,\mathrm{d}\theta\,\mathrm{d}\phi=\frac{p^2 E_{int}}{3k_B T}.$$

If there are N molecules per unit volume, the polarization density $\mathbf{P}$, which is the dipole moment per unit volume, is

$$\mathbf{P}=\frac{Np^2}{3k_B T}\mathbf{E}_{int}.$$

15.6 Capacitors

Capacitors are used to store electric charge. They consist of a pair of conductors with a potential difference maintained between them. Large capacitors installed in oil-filled tanks and operating with very high voltage difference—up to many thousands of volts—may accumulate large amounts of electrical potential energy. At the other end of the scale memory chips incorporate millions of tiny capacitors, each of which represents the number 1 or 0 depending on whether they are charged or uncharged. These memory capacitors operate with a few volts potential difference between conductors separated by silicon oxide insulators. Capacitors also have practical applications in electrical circuits carrying currents that vary with time: this is discussed in Chapter 18.

The conductors in a capacitor are usually close together and separated by a solid insulator. To study the properties of capacitors we shall start by considering a parallel plate capacitor consisting of two parallel conducting plates placed opposite to one another in vacuum as shown in Fig 15.36. Suppose that there is a potential difference V between the plates. In the diagram one plate is at earth potential ($\phi=0$) and the other is at a positive potential V. This choice is only for definiteness and, in fact, the properties of the capacitor depend only on the potential difference and not on the absolute values of the potentials on the plates.

The electric field points in the direction of decreasing potential, and in the centre of the capacitor the field lines are straight from one plate to the other. Near the edges of the plates the field lines are still normal to the conducting surfaces, which are equipotentials, but they bulge out as shown and the electric field does extend a little way outside the region between the plates. However, these edge effects are rather small and, if, as is usually the case, the separation of the plates is very small compared to their length and width, it is a good approximation to assume that the field lines all pass straight across the gap between the plates and that the field outside is zero. The distance between the plates is d and, since their

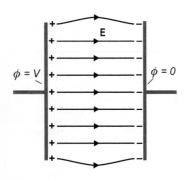

Fig. 15.36 Equal and opposite induced charges occur on the plates of the capacitor when they are held at different potentials.

potential difference is V, the magnitude of the electric field is $E = V/d$, from eqn (15.28). Electric field lines start on positive charges, and positive charges are induced on the plate at potential V. According to eqn (15.39) the surface density of the induced charges is $\sigma = \epsilon_0 E = \epsilon_0 V/d$. Negative charges with the same magnitude of surface charge density are induced on the plate at earth potential where the field lines terminate.

For plates of area S, the total charge on the plate at potential V is $Q = \sigma S = \epsilon_0 VS/d$. Similarly, an induced charge $-Q$ is located on the other plate. The charges $\pm Q$ on the plates are proportional to the potential difference, and the proportionality constant is called the **capacitance** of the capacitor. The capacitance is denoted by the symbol C,

$$Q = CV \tag{15.49}$$

where

$$C = \frac{\epsilon_0 S}{d} \tag{15.50}$$

● *The charge on the plates of a capacitor is proportional to the voltage between them*

for a parallel plate capacitor in vacuum. The unit of capacitance is the **farad** (symbol F). The farad, which has its magnitude fixed by other SI units, is impractically large, and capacitances are usually quoted in **microfarads** ($1\,\mu F \equiv 10^{-6}\,F$), **nanofarads** ($1\,nF \equiv 10^{-9}\,F$), or **picofarads** ($1\,pF \equiv 10^{-12}\,F$).

The capacitance C is determined by the dimensions and geometrical arrangement of the capacitor. Whatever the shape and size of two conductors, it is always true that, when a potential difference is maintained between them, equal and opposite charges are induced on the conducting surface and the magnitude of the charge is proportional to the potential difference. Equation (15.49) applies, with a value of the capacitance determined by the geometry of the two conductors. Since the charges on the conductors are equal and opposite, the total charge on any capacitor is zero. The flux of the electric field through a surface enclosing the capacitor is therefore zero and, apart from the small 'fringing' fields near the edges of the conductors, the field outside the capacitor is everywhere zero.

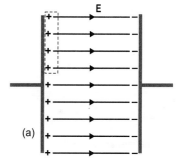

Relative permittivity

The capacitance of a parallel plate capacitor is given by eqn (15.50) only if there is no matter between the plates, that is, the capacitor is in vacuum. Figure 15.37(a) shows such a capacitor with charges $\pm Q$ on its plates. If the capacitor is isolated, so that charge cannot flow on to or away from the plates, the charge remains the same if a slab of insulator is placed between the plates as in Fig 15.37(b). The slab is polarized and the electric field inside the insulator is less than the field outside. Consequently, the

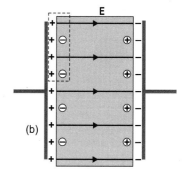

Fig. 15.37 The capacitance is *increased* by inserting dielectric material between the plates of the capacitor.

Table 15.1 Relative permittivities of some materials, measured in steady fields. The value for silicon is included although it is too good a conductor to be used as the dielectric material in a capacitor. However, the relative permittivity of semiconductors in steady fields has an important influence on their behaviour

Material	Relative permittivity ϵ
Mica	7.0
Soda glass	7.5
Polyethylene	2.3
Silicon oxide	3.9
Silicon	11.8
Gas	$10^4(\epsilon - 1)$
Air	5.4
Ne	1.3

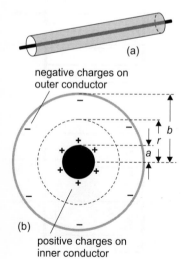

negative charges on outer conductor

positive charges on inner conductor

(b)

Fig. 15.38 (a) A coaxial cable consists of a wire through the centre of a cylinder of insulating material within an outer conductor. A cross-section through the cable is shown in (b).

potential difference between the plates is also reduced, although the charge on them is unaltered. From eqn (15.45), the field inside the insulator is smaller than the field outside by the factor $(1 + \chi_E)$. If the insulator fills the whole of the space between the plates, the field has the reduced value everywhere between the plates, and the potential difference is also reduced by the factor $(1 + \chi_E)$. The charges on the plates have remained the same, and the capacitance $C = Q/V$ (from eqn 15.44)) has increased by this factor,

$$\frac{\text{capacitance with insulator between the plates}}{\text{capacitance in vacuum}} = 1 + \chi_E = \epsilon. \qquad (15.51)$$

The factor ϵ by which the capacitance is increased by the insertion of the insulating material is called the **relative permittivity**. An older name for this factor is *dielectric constant* and, in the context of discussing the behaviour of electric fields, insulating materials are still usually referred to as *dielectric* materials. The relative permittivities of some dielectric materials are given in Table 15.1.

The relative permittivity is simply a number for materials that are isotropic, that is, materials that have no directional properties. In some crystals induced dipole moments are not necessarily in the same direction as the applied field, but depend on the orientation of the field to the crystal axes. The equations we have derived are not then valid. Such crystals have important applications in optics because of their effects on the rapidly varying electric fields in visible light, but only isotropic materials are used in capacitors. Provided that a capacitor is completely filled with a uniform dielectric material, the result that its capacitance is enhanced by the factor ϵ applies to all capacitors and not only to those with slab geometry.

Worked Example 15.7 A coaxial cable consists of a wire with diameter 1 mm, passing through the centre of a polyethylene cylinder of diameter 3.5 mm, which is covered with a conducting coat made by braiding fine wires. The outer conductor is held at earth potential. Calculate the capacitance of a 1 m length of the cable.

Answer The cable has cylindrical symmetry and Gauss's law can be used to determine how the field varies within the cable. First, imagine that there is no dielectric material between the conductors. Suppose that there is a positive charge η per unit length of the inner wire and a charge $-\eta$ per unit length on the outer conductor, as shown in Fig 15.38. If the electric field at a point at a distance r from the axis of the cable has a magnitude $E(r)$, the flux of $\mathbf{E}$ out of a cylinder of radius r and length ℓ centred on the axis is $2\pi r\ell E(r) = \eta\ell/\epsilon_0$. Hence $E(r) = \eta/(2\pi\epsilon_0 r)$. The potential difference between points at r and $r + dr$ is $d\phi = -E(r)dr$ from eqn (15.29). The potential difference between radii a and b is

$$V = \int_a^b d\phi = -\int_a^b \frac{\eta}{2\pi\epsilon_0} \cdot \frac{dr}{r} = \frac{\eta}{2\pi\epsilon_0} \ln\left(\frac{b}{a}\right),$$

and the capacitance of a 1 metre length of cable is

$$C = \frac{\eta}{V} = \frac{2\pi\epsilon_0}{\ln(b/a)}.$$

When the space between the conductors is filled with polyethylene, the capacitance per unit length is increased to $2\pi\epsilon\epsilon_0/\ln(b/a)$. The relative permittivity of polyethylene is listed in Table 15.1 as 2.3. For $a = 0.5\,\text{mm}$ and $b = 1.75\,\text{mm}$, the capacitance of 1 metre is $1.0 \times 10^{-10}\,\text{F}$ or $100\,\text{pF}$. *Coaxial* cables of this kind are frequently used to carry signals from one piece of equipment to another, for example, from a receiving antenna to a television set.

Stored energy

How much potential energy is stored in a capacitor when its plates have a potential difference V? We can work this out by starting with an uncharged capacitor and gradually building up the charge, which is at all times linked to the potential difference by eqn (15.49). Again suppose that one plate is held at earth potential and the other carries a positive charge as in Fig 15.36. When the positive charge has built up to a value $+Q'$ the potential on the left-hand plate is $V' = Q'/C$. If further charge dQ' is moved from a great distance from the capacitor, where the potential is zero, work $V'dQ' = Q'dQ'/C$ must be done. An extra charge $-dQ'$ is induced on the right-hand plate, but no work is required for this since the negative charge does not change its potential. The total work done in building up charges $\pm Q$ on the plates of an initially uncharged capacitor, which is the potential energy stored by the capacitor, is

$$U = \int_0^Q \frac{Q'dQ'}{C} = \frac{1}{2}\left(\frac{Q^2}{C}\right) = \frac{1}{2}CV^2. \tag{15.52}$$

For a 1 μC capacitor with 100 V across the plates, the stored energy is thus $\frac{1}{2} \times 10^{-6} \times 10^4 = 5 \times 10^{-3}\,\text{J}$.

Exercise 15.6 A potential difference of 10 volts is maintained between the two conductors of the coaxial cable in Worked example 15.7. Calculate the energy stored in a 1 metre length.

Answer $5 \times 10^{-9}\,\text{J}$.

Energy density of the electric field

For a parallel plate capacitor with plates of area S separated by a distance d in vacuum, the capacitance is given by eqn (15.50) as $C = \epsilon_0 S/d$. If the potential difference between the plates is V, the energy stored by the capacitor is $\frac{1}{2}CV^2$ (eqn (15.52)). We may equally well express this energy in terms of the field between the plates. Since $V = Ed$, the energy can be written as $\frac{1}{2}C(Ed)^2 = \frac{1}{2}\epsilon_0 E^2 Sd$. Now the volume of the capacitor is Sd and we may think of the amount of stored energy as $\frac{1}{2}\epsilon_0 E^2$ per unit volume. The same expression also applies to capacitors that do not have slab geometry, and we can write the stored energy as

$$U = \tfrac{1}{2}CV^2 = \tfrac{1}{2}\int_V \epsilon_0 E^2 \mathrm{d}V \tag{15.53}$$

where V is the volume where the electric field due to the capacitor is nonzero. For a steady field it is not really possible to locate the energy in a particular region of space and associating the energy with the field does not lead to any advantage in calculations. However, when there are rapidly varying fields, as, for example, in the antenna of a mobile telephone, energy is transmitted from one place to another by radiation. It turns out that the energy density of the electric field that we have calculated for steady fields also applies to radiated energy. Similar expressions, which we shall meet in the next chapter, apply to magnetic fields, and it is energy carried by both electric and magnetic fields that constitutes electromagnetic radiation. The theory of radiation is beyond the scope of this book, but it is important to realize that a thorough grasp of the behaviour of electric and magnetic fields is required before radiation can be understood.

Problems

Level 1

15.1 Calculate the force between two electrons that are 0.1 nm apart.

15.2 Two charges $+q$ and one charge $-q$ are placed at the corners of an equilateral triangle of side a. What is the magnitude and direction of the force on the charge $-q$?

15.3 A conducting sphere of radius R carries a total charge Q. Draw a diagram showing the electric field as a function of distance from the centre of the sphere.

15.4 The sphere in Exercise 15.3 is hollow, having an uncharged conducting concentric sphere of radius R_1 inside the conductor. How much charge is there on the inner surface of the hollow sphere?

15.5 The maximum electric field that can be sustained before breakdown in dry air is about $5 \times 10^6 \,\mathrm{V\,m^{-1}}$. What is the minimum radius of curvature that can be tolerated on the corners of a box that is to be raised to 100 kilovolts?

15.6 A charge q is at a distance d from a thin wire carrying a charge $\eta \, \mathrm{C\,m^{-1}}$. What is the force on the charge q?

15.7 Two parallel thin wires, each carrying a charge per unit length $\eta \, \mathrm{C\,m^{-1}}$, are separated by a distance d. What is the force per unit length between the wires?

15.8 Two parallel wires, each of radius R, are a distance d apart $(d > 2R)$. One of the wires is at earth potential and the other at a potential V. Sketch the lines of the electric field and the equipotentials around the wires. On what part of the wires does the surface charge density have its greatest value?

15.9 Two isolated plates are parallel to each other, One carries a total charge Q and the other a total charge $2Q$. Use Gauss's law to find out how the charges are distributed on the surfaces of the plates, neglecting end effects.

15.10 A slab of material has a uniform charge density ρ throughout its volume. Calculate the electric field as a function of distance from the central plane of the slab.

15.11 The nucleus of a lead atom carries a charge $82e$. It is quite a good approximation to assume that the nucleus is a uniformly charged sphere of radius 7.5×10^{-15} m. Draw a diagram showing the electric field as a function of distance from the centre of the nucleus. What is its greatest value?

Level 2

15.12 Three charges $-q$, $+q$, and $-q$ lie on a line and are separated by a distance a as shown in Fig 15.39. If the negative charges are fixed, calculate the restoring force for small displacements, perpendicular to the line joining the charges, of the positive charge from its equilibrium position half-way between the negative charges.

15.13 Two helium-filled balloons are tied to the same point on the ground by strings 50 cm long. Each balloon

has a lifting force of 1 N, and each has the same charge Q. The angle between the strings is $30°$. What is the magnitude of Q?

15.14 A dipole consists of charges $+q$ and $-q$ separated by a distance a. Derive an expression for the electric field on the axis of the dipole at a distance r from its centre, in powers of (r/a).

15.15 Equal charges $+q$ are situated at the corners of a cube of side a. What force acts on any one of the charges, and what is its direction?

15.16 A charge q is placed at the centre of a cube. What is the flux of the electric field through one of the cube faces? What is the flux through one of the opposite faces if the charge is placed at a corner of the cube?

15.17 A conducting sphere of radius a carries a charge q_a. Outside it are two thin conducting spherical shells, of radii b and c $(a < b < c)$ carrying charges q_b and q_c. The outermost sphere is at earth potential. Obtain expressions for the potentials of the other two spheres.

15.18 Using the same value for the radius of the nucleus of the lead atom as in Problem 15.11, calculate the electrical potential energy of the nucleus. (*Hint.* Calculate the energy needed to build up the nucleus, bringing charge from infinity in infinitesimal steps.)

15.19 Two dipoles, each with dipole moment 6×10^{-30} C m, are placed as shown in Fig 15.40. Their separation a is 0.4 nm. Calculate the potential energy of the dipoles. (These values apply roughly to water molecules.)

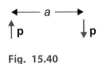

Fig. 15.40

15.20 Calculate the energy of the dipoles in Problem 15.19 at the same separation, but when they are in line as in Fig 15.41.

15.21 A capacitor in a random access memory consists of a layer of SiO_2 0.1 nm thick, sandwiched between

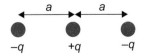

Fig. 15.39

Fig. 15.41

plane conductors each with an area of $0.6 \times 10^{-12}\,\text{m}^2$. The relative permittivity of SiO_2 is 3.9. Estimate the number of electrons on the negatively charged conductor when the voltage across the capacitor plates is 5 V.

15.22 Two spherical conducting surfaces have radii a and b and the space between them is filled with air. Calculate the capacitance of the two conductors.

15.23 A conducting sphere of radius 1 cm is suspended in air. Any other conductors are far away. Estimate the capacitance of the sphere with respect to Earth.

15.24 A parallel plate capacitor has a capacitance of 10 pF when the space between its plates is filled with air. One of the plates is covered with a slab of dielectric material of relative permittivity 7, with a thickness that is half the distance between the plates. What is now the value of the capacitance?

Level 3

15.25 Three charges $-q$, $+2q$, and $-q$ are arranged on a line as shown in Fig 15.42. Calculate the field at a distance $r > a$ on the line, and find the leading term in the expansion in powers of r/a.

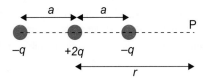

Fig. 15.42

15.26 A slab of dielecric material 2 cm thick with relative permittivity 4.0 has a net charge density $1\,\mu\text{C}\,\text{m}^{-3}$ that is uniformly distributed throughout the material. Calculate the electric field inside and outside the material.

15.27 A line of charges are all separated by a distance 0.1 nm from their nearest neighbours and the magnitude of the charges is alternately $+e$ and $-e$. Calculate the potential energy of one of the charges due to the interaction with all the others, assuming that there is a very large number of charges. (*Hint.* It will help you to find the answer to make a Taylor expansion of the function $\ln(1 + x)$.)

The three-dimensional equivalent of this problem must be solved for different crystal structures to find the electrostatic contribution to the binding energy of ionic crystals.

15.28 Three concentric cylindrical conductors have radii 1 cm, 2 cm, and 3 cm. The space between them is filled with oil with a relative permittivity $\epsilon = 2.2$. If the maximum field that the oil can maintain without breakdown is $5 \times 10^6\,\text{V}\,\text{m}^{-1}$, estimate the highest voltage difference that can be achieved between the inner and outer conductors, and the voltage between the inner and middle conductors under these conditions. (The maximum voltage occurs when both the inner and middle conductors have almost the breakdown field on the outer surfaces.)

15.29 A line charge of strength $\eta\,\text{C}\,\text{m}^{-1}$ is parallel to an earthed conducting plane and at a distance d from it, as shown in Fig 15.43. Calculate the surface density of the induced charge on the conducting plane as a function of y, neglecting end effects. (The solution of this problem requires the use of the uniqueness theorem, which states that if an electric field is known to satisfy the conditions at the boundary of a region of space, it is the only possible field. Here the field must be normal to the conducting surface. Consider the field due to two line charges with strengths $\pm\eta$ and $2d$ apart. This field satisfies the boundary condition and in the region between the line charge and the conductor it is the required solution. The imaginary line charge $-\eta$ is called the image charge of the real line charge $+\eta$.)

15.30 Two parallel plane conductors are a distance d apart. A plane sheet of charge with surface charge density

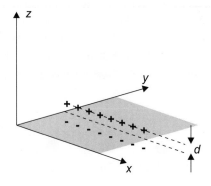

Fig. 15.43

σ lies between the plates at a distance x from one of them. Calculate the induced charge on each plate.

15.31 The relative permittivity of water is 80.36 at 20°C and 60.76 at 80°C when measured in a steady electric field. The density of water is 0.9982 g cm^{-3} at 20°C and 0.9718 g cm^{-3} at 80°C. Use these data to obtain the dipole moment and the polarizability of a water molecule.

Some solutions and answers to Chapter 15 problems

15.1 2.3×10^{-8} N.

15.2 Each of the charges $+q$ exerts a force $q^2/(4\pi\epsilon_0 a^2)$ on the charge $-q$. The horizontal components of these forces are equal and opposite, and the net force on $-q$ is down the page, as shown in Fig 15.44, with a magnitude

$$F = \frac{2q^2}{4\pi\epsilon_0 a^2}\cos 30° = \frac{\sqrt{3}q^2}{4\pi\epsilon_0 a^2}.$$

15.5 Outside a conducting sphere of radius R the potential varies as $1/r$ and may be written $\phi(r) = \phi(R) \times R/r$. The electric field $E(r)$ is

$$E(r) = -\frac{\partial\phi}{\partial r} = \frac{R\phi(R)}{r^2},$$

which has its maximum value $\phi(R)/R$ at $r = R$. If the corners of the box are rounded so that they are portions of spheres of radius R, the field close to the corners is also $\phi(R)/R$. The minimum radius R_{min} that can be tolerated when the box is raised to a potential of 100 kV satisfies

$$\frac{10^5}{R_{min}} = 5 \times 10^6, \quad \text{leading to} \quad R_{min} = 0.02 \text{ m or 2 cm.}$$

15.7 The force is

$$\frac{\eta^2}{2\pi\epsilon_0 r} \text{ N m}^{-1}.$$

15.9 There is a charge $\frac{3}{2}Q$ on the outer surface of each plate. The charges on the inner surfaces are $+\frac{1}{2}Q$ for the plate carrying a total charge $2Q$ and $-\frac{1}{2}Q$ for the plate carrying a total charge $-\frac{1}{2}Q$.

15.10 The electric field points outwards from the centre and at a distance x from the centre, within the slab, its magnitude is $\rho|x|/\epsilon_0$.

15.11 The flux of the electric field at a distance r from the centre of the nucleus is due to the charge within r. If r is less than the radius R of the nucleus, and the charge density is ρ, the total charge within a sphere of radius r is $\frac{4}{3}\pi\rho r^3$. The area of this sphere is $4\pi r^2$ and the flux out of it is

$$4\pi r^2 E(r) = \frac{4\pi r^3 \rho}{3\epsilon_0}, \quad \text{giving} \quad E(r) = \frac{\rho r}{3\epsilon_0}.$$

Outside the nucleus the electric field is the same as for a point charge at the centre, that is, it is proportional to $1/r^2$ and the maximum field is at $r = R$.

For lead the total charge $\frac{4}{3}\pi\rho R^3 = 82e$, and

$$E_{max} = \frac{82e}{4\pi\epsilon_0 R^2} = \frac{(82 \times 1.6 \times 1)^{-19}}{4\pi \times 8.85 \times 10^{-12} \times (7.5 \times 10^{-15})^2}$$
$$\approx 2 \times 10^{21} \text{ V m}^{-1}.$$

This local field that exists close to the nucleus is enormous compared to the largest electric field that can occur even over distances about the size of an atom. As indicated in Problem 15.5, the largest electric field that is sustainable in air is only about 5×10^6 V m^{-1}.

15.13 Assuming that the force between the balloons is the same as if each were a point charge, the charge on each balloon is 1.4 μC.

15.14 The field on the axis is

$$\frac{qa}{4\pi\epsilon_0|r|^3}\left\{1 + \frac{2a}{r^2} + \cdots\right\} = \frac{qa}{4\pi\epsilon_0|r|^3}$$
for small (r/a).

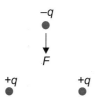

Fig. 15.44

The field acts along the axis in the same direction, from the negative to the positive charge, on both sides of the dipole. For small a/r the magnitude of the field is proportional to $(1/r)^3$: it falls off faster with distance than the field due to a point charge, because to second order in (a/r) the contributions of the positive and negative charges cancel out.

15.15 The three charges at BDE and the three at CFH are symmetrically placed with respect to the diagonal GA in Fig 15.45. By symmetry the force on the charge at A is therefore outwards in the direction GA from the opposite corner. The component of the force due to the charge at B along GA is

$$\frac{q^2}{4\pi\epsilon_0 a^2} \times \frac{1}{\sqrt{3}},$$

the component due to the charge at C is

$$\frac{q^2}{4\pi\epsilon_0 (\sqrt{2}a)^2} \times \frac{\sqrt{2}}{\sqrt{3}},$$

and the force due to the charge at G is

$$\frac{q^2}{4\pi\epsilon_0 (\sqrt{3}a)^2}.$$

The total force is therefore

$$\frac{q^2}{4\pi\epsilon_0 a^2} \left\{ \frac{3}{\sqrt{3}} + \frac{3}{2} \times \frac{\sqrt{2}}{\sqrt{3}} + \frac{1}{3} \right\} = 2.14 \frac{q^2}{4\pi\epsilon_0 a^2}.$$

15.16 For a charge at the centre of the cube the flux of the electric field out of one face is $q/6\epsilon_0$. For a charge at one corner the flux out of an opposite face is $q/24\epsilon_0$.

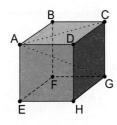

Fig. 15.45

15.18 The electrostatic energy is calculated by building up the charge on the nucleus from the centre, assembling thin spherical shells one by one like the successive layers of an onion. As in Problem 15.11, at a distance r from the centre of the nucleus, less than its radius R, the charge within r is $\frac{4}{3}\pi\rho r^3$. The potential at r is

$$\frac{1}{4\pi\epsilon_0 r} \times \frac{4}{3}\pi\rho r^3.$$

The energy needed to bring from infinity an extra thin shell of radius dr, carrying a charge $4\pi\rho r^2 dr$ is

$$\frac{1}{4\pi\epsilon_0 r} \times \frac{4}{3}\pi\rho r^3 \times 4\pi\rho r^2 dr = \frac{\frac{4}{3}\pi^2\rho^2 r^4 dr}{\epsilon_0},$$

and the electrostatic potential energy of the whole nucleus is

$$\int_0^R \frac{\frac{4}{3}\pi^2\rho^2 r^4 dr}{\epsilon_0} = \frac{4\pi^2\rho^2 R^5}{15\epsilon_0} = \frac{3}{5} \times \frac{Q^2}{4\pi\epsilon_0 R}$$

where $Q = \frac{4}{3}\pi\rho R^3$ is the total charge of the nucleus. For lead, $Q = 82e$, giving a total electrostatic energy of 1.24×10^{-10} J or, equivalently, 775 MeV.

15.19 The two dipoles attract one another, and their potential energy at a distance of 0.4 nm is -0.126 eV.

15.21 The number of electrons is about 1300. Capacitors of about this size are used to store digits in dynamic random access memories (DRAMs).

15.24 When no dielectric is present, the electric field E inside the capacitor is $E = V/d$, where V is the voltage difference between the plates and d their separation. If their area is S, the charge Q on the plates is $\epsilon_0 ES$ and the capacitance $C = Q/V = \epsilon_0 S/d$.

When the space between the plates is half-filled with an insulator, the field inside the insulator is smaller than the field outside by a factor equal to the relative permittivity ϵ. Call the electric field at the surface of the plates E'. The field inside the insulator is E'/ϵ and the voltage between the plates is $V = E'd/2 + E'd/2\epsilon$. The charge on the plates is now $Q = \epsilon_0 E'$ and the capacitance is

$$C' = \frac{2\epsilon_0 S}{d + d/\epsilon} = \frac{2\epsilon\epsilon_0 S}{d(1 + \epsilon)}.$$

The ratio of capacitances with and without insulator is

$$\frac{C'}{C} = \frac{2\epsilon}{1+\epsilon}.$$

For C = 10 pF and $\epsilon = 7$, the capacitance is increased to 17.5 pF.

15.25 The electric field at a point P on the axis is

$$E = \frac{q}{4\pi\epsilon_0}\left\{-\frac{1}{(r-a)^2} + \frac{2}{r^2} - \frac{1}{(r+a)^2}\right\}$$

$$= \frac{q}{4\pi\epsilon_0 r^2}\left\{-\left(1-\left(\frac{a}{r}\right)^2\right)^{-2} + 2 - \left(1-\left(\frac{a}{r}\right)^2\right)^{-2}\right\}.$$

Making a binomial expansion in powers of a/r,

$$E = \frac{q}{4\pi\epsilon_0 r^2}\left\{-1 - \frac{2a}{r} - \frac{(-2)\times(-3)}{1\times 2}\left(\frac{a}{r}\right)^2 - \cdots \right.$$

$$\left. + 2 - 1 + \frac{2a}{r} - 3\left(\frac{a}{r}\right)^2 + \cdots \right\} = \frac{6qa^2}{4\pi\epsilon_0 r^4}$$

to second order in (a/r). For small (a/r) the field from the charges falls off with distance as $1/r^4$, more rapidly than the field of the dipole in Problem 15.14. The charges in this problem may be regarded as two dipoles close together, arranged so that their contributions almost cancel each other at large distances. Such an arrangement of charges is called an **electric quadrupole**.

15.27 The potential energy of one charge due to all the others is $-e^2\ln 2/(2\pi\epsilon_0 d)$. For $d = 0.1$ nm this is -20.0 eV.

15.29 The potential due to the real charges $+\eta$ per unit length and the imagined 'image' charges $-\eta$ per unit length is zero everywhere on the plane midway between them. The potential on the conducting plate is zero, and the induced charge must be distributed on this plate in such a way that the field above the plate is the same as the field due to the charges $\pm\eta$ per unit length. *Inside* the conductor the electric field is actually zero, and the induced charges indeed move to the surface of the plane to ensure that this is so, as illustrated in Fig 15.46.

The electric field at the surface of a conductor is always normal to the surface. To calculate the field due to the line charge and the image charge we only need to consider the component normal to the

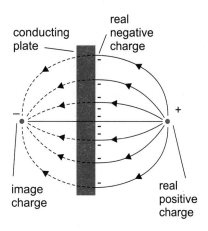

Fig. 15.46

surface. At a distance y from the line joining $\pm\eta$, the normal component is

$$E_\perp = \frac{2\eta}{2\pi\epsilon_0\sqrt{(y^2+d^2)}} \times \frac{d}{\sqrt{y^2+d^2}} = \frac{\eta d}{\pi\epsilon_0(y^2+d^2)}.$$

The surface charge density σ is related to the field by $E_\perp = \sigma/\epsilon_0$ (eqn (15.39)), and the surface charge density on the conductor at a distance y from the line joining the real and image charges is

$$\sigma = \frac{\eta d}{\pi(y^2+d^2)}.$$

15.30 Let the potential at the plane occupied by the positive charge density be ϕ. This plane is at a distance x from one of the conductors; choose the origin of x to be at this plane. The electric field between 0 and x is $-\phi/x$, and the induced charge density on the conductor at $x = 0$ is $-\epsilon_0\phi/x$. Similarly, the induced charge density on the conductor at $x = d$ is $-\epsilon_0\phi/(d-x)$. The total induced charge density must be $-\sigma$ since there is no electric field outside the conductors. Hence

$$\frac{\epsilon_0\phi}{x} + \frac{\epsilon_0\phi}{d-x} = -\sigma \quad \text{and} \quad \epsilon_0\phi = \sigma\frac{x(x-d)}{2x-d}.$$

Hence the charge density on the conductor at $x = 0$ is $-\sigma(x-d)/(2x-d)$ and the charge density on the conductor at $x = d$ is $-\sigma x/(2x-d)$.

For a single charge or any number of charges at a distance x from one of the conductors, the division of the induced charge between the two conductors is the same as it is for a sheet of charge. X-rays and γ-rays cause the formation of electrons and ions when they interact with matter. Many X-ray and γ-ray detectors work by collecting such electrons and ions by placing them in an electric field. The electrons and ions move in opposite directions towards conducting plates. Induced charges appear on these conductors before the ions and electrons arrive, allowing the time of generating the ions and electrons to be determined accurately.

Chapter 16

Magnetism

This chapter is concerned with the magnetic fields produced by steady currents and the forces on charges when moving in the magnetic fields of steady currents.

Static charges are the sources of an electrostatic field that produces a force on a further electric charge. The force on a charge in an electrostatic field is a measure of the field, as discussed in the last chapter.

A stream of charges moving with uniform speed constitutes a steady **electric current**. Currents flow in electrical **conductors**, which are materials through which electric charge can move freely. Energy is normally expended when a current flows and because of this an energy source is required to maintain the flow. A steady current can be maintained in a metal wire if a difference in electrostatic potential is maintained across the ends of the wire. The source maintaining the potential constant must provide the power, which is normally dissipated as the current flows.

If a difference in electrostatic potential is maintained across the ends of a wire, there will usually be an electrostatic field around the wire produced by the source maintaining the current, and this will give an electrostatic force on a stationary test charge close to the wire.

If the test charge outside the wire is not stationary but is moving, a new phenomenon occurs. In addition to any electrostatic force that may be present there is a further force on the moving charge arising from the steady current in the wire. This force is called a *magnetic force* and can be described in terms of a new field generated by the current called the *magnetic field*. This chapter discusses the magnetic fields that arise from simple current distributions and some of the phenomena associated with magnetic fields.

● *The definition of current density*

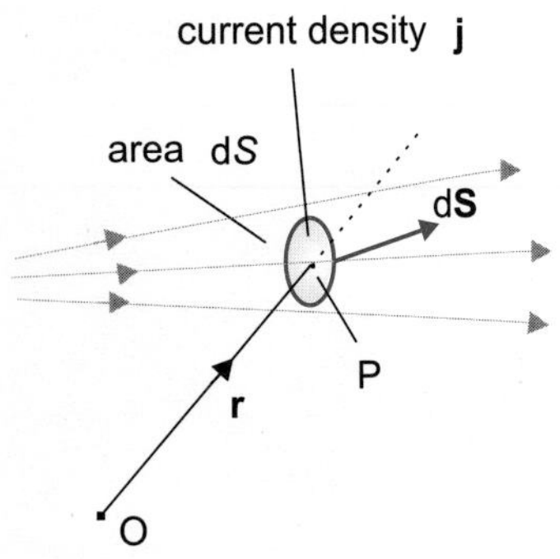

Fig. 16.1 The current through an infinitesimally small element of area dS at the point P is $\mathbf{j} \cdot d\mathbf{S}$, where $\mathbf{j}$ is the current density.

16.1 **Steady currents**

A conductor has free charges that can move under the influence of an electric field. If the field at a point in the conductor is **E**, a charge q at that point experiences a force $q\mathbf{E}$ in the direction of the field. The **current density**, with the symbol $\mathbf{j}$, is a vector in the direction of the electric field. It is a function of position and its magnitude at any point is defined as the positive charge that crosses a unit area per second at that point. Charge in the SI system of units is measured in coulombs, as discussed in Chapter 15, and the current density thus has units of coulombs per square metre per second.

Figure 16.1 shows an infinitesimally small area dS at the point P with position vector **r** with respect to an origin at the point O. The current density at P is $\mathbf{j}(\mathbf{r})$. We will usually omit the position vector on which the current density depends and, as with other vector functions of position, call it simply **j**. We define the vector $d\mathbf{S}$ to have magnitude dS and be in the direction of the normal to the infinitesimal area; the way it points along the normal is associated with a sense of circulation of the perimeter of the area dS given by the right-hand screw rule. This rule relates a direction to a sense of circulation of a loop. If the forefinger of your right hand executes circles in a clockwise sense, the related direction given by the rule is forward. This is also the direction in which a right-handed screw would move if rotated in the same sense as the finger.

● *The right-hand screw rule*

The positive charge passing through the area dS per second in the direction of $d\mathbf{S}$ is equal to $\mathbf{j} \cdot d\mathbf{S}$. The **current**, dI, through the area dS is a scalar equal to the total charge passing through dS per second, and hence

$$dI = \mathbf{j} \cdot d\mathbf{S}.$$

Current has units of coulombs per second and in the SI system this unit is given a special name. It is called an **ampere** and has the symbol A.

The total current I through an area S is the surface integral over the area S of the current density,

● *The definition of current*

$$I = \int_S \mathbf{j} \cdot d\mathbf{S}. \tag{16.1}$$

Electric charge cannot be created or destroyed; it is conserved. All experimental evidence supports the law of conservation of charge. If a steady current is uniformly distributed throughout a straight metal wire of radius a, eqn (16.1) gives $\pi a^2 j$ for the current through a disc perpendicular to the wire. The same answer is obtained for the current through a disc cut at an angle to the wire. The surface integral in eqn (16.1) now has a cosine factor that compensates for the increased area.

● *Conservation of charge*

It is a necessary consequence of charge conservation that the sum of elementary contributions to the surface integral (16.1) is also the same over any surface that the wire passes through. Similarly, if the wire changes its radius from a to a larger value b, the current I is equal on both sides of the join. The magnitude of the current density reduces from j to ja^2/b^2 because otherwise the total charge entering or leaving the join in unit time would not be equal.

The electric field $\mathbf{E}$ is an average, over very small distances on a macroscopic scale, of the atomic electric field. The macroscopic length intervals are large on a microscopic scale and over very small macroscopic distances a charge q experiences resistance to its motion and acquires an average steady drift speed v. Hence, if the macroscopic field is uniform over the length of a conductor such as a straight wire, the moving charges maintain a constant drift velocity $\mathbf{v}$ in the direction of the field.

● *Drift velocity of charges*

For many metals and other materials, and for electric field strengths below the level at which electrical breakdown occurs, the drift velocity is proportional to the field, and the steady motion of charges constitutes a steady electric current. If there are N moving charges per unit volume, each with charge q, the current density is equal to $qN\mathbf{v}$. In metals the charge carriers are electrons, which each have a charge $-e$. The current density is thus

$$\mathbf{j} = -eN\mathbf{v}. \tag{16.2}$$

● *Current density and drift velocity*

Equation (16.2) may be rewritten as

$$\mathbf{j} = eN(-\mathbf{v}),$$

and electrons flowing from right to left in a wire have the same effect as positive charges flowing from left to right. The current is from left to right and, if the cross-sectional area of the wire is S, the current is

$$I = eNvS. \tag{16.3}$$

Since the drift velocity is proportional to the electric field in many materials, the current density in those materials, as given by eqn (16.2), is also proportional to the field and

$$\mathbf{j} = \sigma\mathbf{E}. \tag{16.4}$$

● *Ohmic materials*

The constant of proportionality σ in the above equation is called the **electrical conductivity** of the medium, and media that obey eqn (16.4) over a range of field strength are called **ohmic** over that range. From eqns (16.1) and (16.4) it can be seen that the electrical conductivity is measured in units of amperes per metre per volt.

For a current in a cylindrical wire of cross-sectional area S made from an ohmic metal, the size of the current is

$$I = jS = \sigma ES = eNvS,$$

from eqn (16.3).

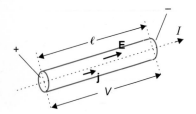

Fig. 16.2 A voltage difference between the ends of a uniform, straight metal wire gives a uniform electric field and a uniform current density in the wire.

Resistance

If a difference in electrostatic potential V is maintained between the ends of a wire of length ℓ, as shown in Fig 16.2, the electric field in the wire is everywhere parallel to the wire and has the constant value V/ℓ.

When currents are being discussed it is usual to refer to such a difference in potential, which is measured in volts, as a voltage difference, and to give the voltage difference the symbol V. The uniform electric field produces a uniform current density and a constant current I in the wire given by

$$I = \int_S \mathbf{j} \cdot d\mathbf{S} = jS. \tag{16.5}$$

Since j is equal to σE and $E = V/\ell$, we have

$$I = \frac{\sigma S}{\ell} V. \tag{16.6}$$

For ohmic materials, the conductivity σ is constant, and we may write

$$I = \frac{V}{R} \tag{16.7}$$

where R, given by

$$R = \frac{\ell}{\sigma S}, \tag{16.8}$$

is called the **resistance** of the wire. Equation (16.7) is **Ohm's law**, which states that

The current I through a resistance R when there is a voltage difference V across it is equal to the voltage divided by the resistance.

The units of resistance are volts per ampere. This unit is also given a special name in the SI system. It is called an **ohm** and has the symbol Ω. The electrical conductivity is usually written in units of $(\Omega\,\mathrm{m})^{-1}$.

Most materials are non-ohmic and no material remains ohmic at all electric field strengths or temperatures. The conductivity often has a marked temperature dependence; conductivities of metals decrease markedly as the temperature increases above room temperature. In addition to being non-ohmic, most materials have very small conductivity indeed and may be classified as **insulators** through which a current will not pass at field strengths normally experienced. At sufficiently high fields insulators suffer electrical breakdown with catastrophic current flow as in a lightning discharge.

Values of electrical conductivities of materials vary over a very wide range. Certain materials have zero resistance at very low temperatures;

these materials are called *superconductors*. Up to now superconductors have only been used in specialized applications, mostly in the research field, because of the need to operate them at low temperatures. Metals are the best conductors under normal conditions, followed by *semiconductors* in which the currents for a given applied voltage, that is, the conductivities, depend strongly on temperature. The proper treatment of conduction in solids is a subject for quantum mechanics and will not be discussed here. A simplified treatment is given in Section 14.5.

The **resistivity** of a material is often quoted instead of the electrical conductivity. The resistivity ρ is the reciprocal of the conductivity and is measured in $\Omega\,\text{m}$.

$$\rho = \frac{1}{\sigma}. \tag{16.9}$$

The conductivities and resistivities of materials vary over a very large range, as Table 14.4 in Section 14.5 shows. Resistivities and conductivities may be quoted for materials that are not usually regarded as ohmic by inserting measurements of the resistance of a regularly shaped body of material, made at a given temperature and electric field strength, into eqn (16.8).

Exercise 16.1 What is the resistance at 293 K of a pure copper wire of diameter 1 mm and length 1 m? The conductivity of copper at 293 K is $5.9 \times 10^7\ (\Omega\,\text{m})^{-1}$.

Answer $2.1 \times 10^{-2}\ \Omega$.

Worked Example 16.1 The resistivity of copper at 293 K is $1.68 \times 10^{-8}\ \Omega\,\text{m}$. The electrical conduction in copper is due to the motion of electrons under the influence of the applied electric field. If there is one electron per atom that can move freely in a copper conductor, what is the mean drift speed of the electrons in the direction opposite to the field when a current of 1 A is flowing in a wire of diameter 1 mm? The density of copper is $8.93 \times 10^3\ \text{kg}\,\text{m}^{-3}$, and the atomic weight is 63.5.

Answer With one free electron per atom, the number of free electrons per unit volume in copper, N, is equal to the number of atoms per unit volume. 63.5 g of copper contains 6.02×10^{23} atoms. Hence 8.93 g has 8.47×10^{22} atoms and $N = 8.47 \times 10^{28}$ free electrons per cubic metre. For a current of 1 A in a wire of cross-sectional area $\pi \times 10^{-6}/4\ \text{m}^2$, the current density j is $1.27 \times 10^6\,\mathrm{A\,m^{-2}}$. The current density is given by eqn (16.2)

$$j = eNv,$$

and substituting into this the values of j, N, and the charge $1.602 \times 10^{-19}\,\mathrm{C}$ of the electron gives $v = 0.094\,\mathrm{mm\,s^{-1}}$.

The drift speed of free electrons in metal wires carrying currrents of the size usually met are of the order of a few millimetres per second. This picture of free electrons moving at constant drift speed under the influence of a field should not be taken too far, although it is a useful model for describing many features of metallic conduction. As already pointed out, the proper description requires quantum mechanics.

Power

If a charge $\mathrm{d}q$ flows down a wire in time $\mathrm{d}t$ the current I is $\mathrm{d}q/\mathrm{d}t$. For a steady current, $\mathrm{d}q/\mathrm{d}t$ is constant and each second a charge I has passed through the wire. If the potential difference across the ends of the wire is V, the charge I has lost potential energy VI in passing from one end to the other. This energy is expended against the forces resisting the motion of the charges and restricting them to a steady average speed v, and is dissipated as heat in the wire. The energy must be provided by the power source that maintains the potential difference V. The source thus does work VI each second. The power P delivered by the source is thus

$$P = VI, \tag{16.10}$$

which, using eqn (16.7), can be written

$$P = I^2 R = \frac{V^2}{R}. \tag{16.11}$$

Exercise 16.2 The resistivity of tungsten at 2500 K is $7.39 \times 10^{-7}\,\Omega\,\mathrm{m}$, about 13 times its value at room temperature. A tungsten wire of diameter 0.1 mm is in an evacuated enclosure and carries a steady current. What must the current be if the wire is to reach a steady temperature of 2500 K, and if the heat loss mechanism at that high temperature is by radiation only?

In Section 12.8 the total energy radiated per second by unit area of a black body at temperature T was discussed. The total power radiated per unit area is equal to Stefan's constant, $\sigma = 5.67 \times 10^{-8}\,\text{W}\,\text{m}^{-2}\,\text{K}^{-4}$, times the fourth power of the temperature. For this question assume tungsten radiates like a black-body.

Answer 2.7 A. The effect highlighted by this exercise is the principle underlying the heated-filament electric light bulb. Electric heating of a resistive wire in a bulb increases the temperature of the wire to the point where it glows and emits significant amounts of radiation in the visible part of the electromagnetic spectrum.

● *Batteries*

Sources or devices that can maintain constant voltage differences across a conductor carrying a current have to deliver power. They are sometimes referred to as sources of **electromotive force**, abbreviated emf. Batteries are a common example of a source of emf. The energy dissipated when a battery drives a current through a resistance comes from chemical changes within the battery.

A car battery consists of separate electrodes of lead and lead oxide in a solution of sulphuric acid in water. The acid dissociates into positive hydrogen, $H^{(+)}$, and negative sulfate $(SO_4)^{(2-)}$ ions. A simplified description of what happens in the cell when the two electrodes drive current through an external circuit is that hydrogen ions move to the lead electrode and combine with electrons from the current in the external circuit. The negative ions in the solution move to the lead oxide electrode where they form lead sulfate and release oxygen, which combines with the hydrogen released at the other end of the cell to form water. This process passes on electrons to the external circuit at the same rate as they are consumed by recombination with positive ions at the lead electrode.

The net result of the above changes is a continuous current, with the direction of positive current from the lead electrode to the lead oxide. The chemical change in the cell is the conversion of sulfuric acid and lead oxide to lead sulfate and water. This reaction gives out energy and the energy released is dissipated as heat in the resistance through which the battery drives current. To a good approximation the battery maintains a constant voltage difference across its terminals, although a practical battery has a small internal resistance across which a voltage drop may occur.

Since the discharge of the battery involves turning acid into water, a common test for when the battery is run down is to measure the acid concentration. When this becomes low, application of a larger voltage across the terminals in opposition to the battery itself reverses the chemical procedure outlined above and restores the dissipated chemical energy.

16.2 Magnetic fields

A steady current gives rise to a force on a nearby moving charge. This force is attributed to a **magnetic field**, symbol **B**. The magnetic force depends on the vector **B**, on the charge q, and on the velocity of the charge **v**. The magnetic force is more complicated than the electric force on a charge q in an electric field **E**. The latter is in the direction of the electric field. The magnetic force is not in the direction of the magnetic field; neither is it in the direction of the velocity of the charge. *It is at right angles to both the field and the velocity.*

First experiences with magnetism often involve permanently magnetized materials, either those found naturally or those man-made. A compass needle points to the Earth's magnetic north pole; the Earth has a magnetic field of its own that produces a torque on the pivoted needle resulting in one end following the direction of the Earth's field. Although not fully understood, it is believed that the Earth's magnetic field arises from large currents circulating in the molten core, and the torque on the compass needle arises from the interaction of the magnetized needle with the field. Only certain materials can be permanently magnetized: there is a continuing research effort in this field because of the immense potential for applications.

● *The Earth's magnetic field*

The force on a moving charge

The magnetic field is defined and measured by the force it produces on a moving charge. Close to a large flat surface of magnetized material there is a uniform magnetic field in a direction normal to the surface, by analogy with the electric field close to an electrified surface. A charge q moving with velocity **v** parallel and close to such a magnetized surface experiences a force. The force on the particle can be measured and experiments performed to determine the law of force.

If the direction of the field and of the moving particle are as shown in Fig 16.3, the force acts perpendicular to both those directions and points as shown for a positive charge q. If the field direction is reversed, the force is reversed. If **B** is the magnetic field, the direction of the force **F** can be determined for arbitrary directions of the velocity with respect to the field. The force is proportional to the vector product of **v** and **B**.

$$\mathbf{F} \propto \mathbf{v} \times \mathbf{B}.$$

Vector products are discussed in Section 20.2.

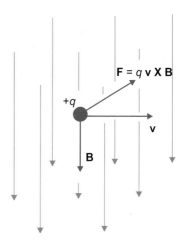

Fig. 16.3 A positive charge q moving with velocity **v** in a magnetic field **B** experiences a force **F** at right angles to both **v** and **B**.

The relative directions of the force, velocity, and magnetic field are given by the **right-hand rule** shown in Fig 16.4. Hold the right hand with the thumb out in the plane of the hand, the first finger pointing forward in the plane of the hand and the middle finger pointing upwards. If the directions of thumb and forefinger are the directions of the velocity and field, respectively, the direction of the force is upwards in the direction of the middle finger.

An alternative way to remember the direction of $\mathbf{v} \times \mathbf{B}$ is the following. Let the vector $\mathbf{v}$ point outwards from your right foot and the vector $\mathbf{B}$ point forwards from your left foot. The vector $\mathbf{v} \times \mathbf{B}$ then points upwards through your head, as illustrated in Fig 16.5.

Experiments with different charges show that the magnitude of the force is also proportional to the charge. Adding this to the observation that the magnitude is proportional to the vector product of the field and the velocity gives

$$\mathbf{F} \propto q\mathbf{v} \times \mathbf{B}.$$

The magnetic field in SI units is now defined by using SI units for q, v, and F in the above equation and putting the constant of proportionality equal to unity.

$$\mathbf{F} = q\mathbf{v} \times \mathbf{B}. \tag{16.12}$$

The dimensions of the magnetic field are thus $[\text{mass}][\text{time}]^{-1}[\text{charge}]^{-1}$, and it is measured in units of $\text{kg s}^{-1}\,\text{C}^{-1}$. The unit of magnetic field is given a special name in the SI system. It is called the **tesla**, symbol T, and a field of one tesla acting perpendicularly to the velocity of a charge of one coulomb moving at a speed of one metre per second produces a force on the charge of one newton.

Equation (16.12) shows that, if a charge is moving parallel to a magnetic field, it experiences no magnetic force. If the angle between the vectors $\mathbf{B}$ and $\mathbf{v}$ is θ, the magnitude of the force is

$$F = qvB\sin\theta. \tag{16.13}$$

● *The right-hand rule*

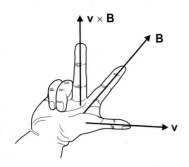

Fig. 16.4 The right-hand rule for giving the relative directions of the velocity $\mathbf{v}$ of a moving positive charge, the magnetic field $\mathbf{B}$ at its position, and the vector product $\mathbf{v} \times \mathbf{B}$.

Fig. 16.5 Another way of remembering the relative directions of velocity, field, and force on a moving positive charge.

Worked Example 16.2 A proton of energy 1 MeV moves in vacuum in a region where there is an electric field of magnitude $10^5\,\text{V m}^{-1}$. Its velocity is perpendicular to the direction of the electric field. Determine the direction and magnitude of an applied magnetic field if the proton is to suffer no deflection.

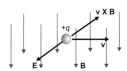

Answer The force from the electric field is in the direction of the electric field and hence at right angles to the proton's velocity. The force from the magnetic field is at right angles to the magnetic field and also at 90° to the proton's velocity. Hence, for the two forces to be equal and opposite the magnetic field must act at right angles to the electric field and to the direction of motion of the proton and be in the direction such that the vector $\mathbf{v} \times \mathbf{B}$ is opposite to the vector $\mathbf{E}$. The relative directions of the three vectors are as shown on Fig 16.6.

For the two forces to be equal

$$qE = qvB,$$

which gives the required magnetic field strength as

$$B = \frac{E}{v},$$

or $B = 7.2 \times 10^{-3}$ T.

Fig. 16.6 The relative directions of the velocity of a moving positive charge and the electric and magnetic fields acting on it, if the forces due to these fields are to be in opposite directions.

Analysis of the orbits that particles follow when traversing known electric and magnetic fields is a technique widely used to determine the mass and charge of atomic and subatomic particles moving at high speed. The worked example above outlines the principle of a filter that selects particles of a given velocity independently of their charge. Different arrangements may be used to provide sensitivity to mass or to charge or to combinations of mass and charge.

The speed of a particle of charge q and mass m moving under the influence of a magnetic field alone remains constant, since the force on the particle from the magnetic field always acts at right angles to its velocity and so cannot increase the particle's kinetic energy. However, since there is a force, the particle suffers an acceleration.

For a field $\mathbf{B}$ we may resolve the instantaneous velocity $\mathbf{v}$ of the particle into two vector components, $\mathbf{v}_\parallel$ parallel to the field, and $\mathbf{v}_\perp$ perpendicular to the field. The former component is unaffected and remains constant. The latter, although constant in magnitude, is continuously changing direction because of the force $q\mathbf{v}_\perp \times \mathbf{B}$ acting in the plane perpendicular to $\mathbf{B}$. In this plane the motion is determined by the force $qv_\perp B$ acting perpendicular to the vector $\mathbf{v}_\perp$. The particle thus has central acceleration $qv_\perp B/m$, and as discussed in Section 4.4 executes circular motion in the plane perpendicular to $\mathbf{B}$ with radius a given by the equation

$$\frac{mv_\perp^2}{a} = qv_\perp B.$$

● The motion of a charged particle in a magnetic field

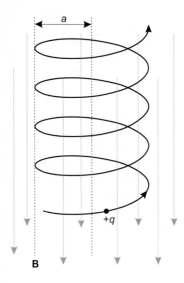

Fig. 16.7 The helical orbit of a moving positive charge in a uniform magnetic field.

⬤ *The difference between magnetic and electric fields*

Hence

$$a = \frac{mv_\perp}{qB}. \qquad (16.14)$$

The resultant path of the particle is a helical orbit, as shown in Fig 16.7. The radius of the helix increases as the angle the initial velocity **v** of the particle makes with the field **B** increases.

Exercise 16.3 An alpha particle, which has charge $2e$ and mass 4 amu, enters a magnetic field of 1 T at an angle of 45°. The alpha particle has energy 2 MeV. (a) How long does it take for it to travel a distance of 10 m in the direction of the field? (b) What is the radius of curvature of the helical motion? (c) What is the total distance travelled by the particle in traversing 10 m in the field direction? (d) How many circles does the particle make in going 10 m in the field direction?

Answer (a) 1.44×10^{-6} s; (b) 0.144 m; (c) 14.1 m; (d) 11.

Magnetic field lines

Stationary charges are the sources of an electrostatic field. In contrast, there are no magnetic charges to act as sources of a magnetic field. No 'magnetic poles' have ever been observed, although some theories of phenomena at very high energies presently unobservable predict their existence. The discussion of magnetism can be pursued without any reference to free magnetic charges. A steady stream of moving charges constituting a current is the source of a magnetic field, and there are important differences between the lines of the electric field and magnetic field lines drawn to represent the magnitude and direction of the magnetic field.

Lines of magnetic field are analogous to the lines of electric field discussed in the last chapter. Their density at any point is proportional to the field strength and their direction, defined by arrows on the lines, gives the field direction. Electric field lines can be mapped by measuring the force on a second small stationary test charge. In the simple example of an isolated electric charge, the electric field lines start on the charge and extend radially outwards. Similarly, magnetic field lines can be mapped by determining the force on a small moving test charge.

When any magnetic field is determined in this way, it is found that the lines of the field have no beginning or end; they always close up in loops. As many lines of the magnetic field leave any volume of space as enter, and there are no sources, no free magnetic charges, or 'poles', within any region. This big difference between the two fields results in a lack of symmetry in their description.

The line integral of the electrostatic field around a closed loop is zero, corresponding to the electrostatic potential at any point having a single value,

$$\oint \mathbf{E} \cdot d\boldsymbol{\ell} = 0.$$

The line integral of the magnetic field around a closed loop need not be zero, and in general we cannot define a magnetostatic potential in magnetism analogous to the electrostatic potential in electrostatics. If a net current passes through the closed loop

$$\oint \mathbf{B} \cdot d\boldsymbol{\ell} \neq 0.$$

The line integral has a nonzero value if a net current passes through the loop.

Ampère's law

Ampère, in the early nineteenth century, performed a series of careful experiments to determine the nature of the magnetic fields and forces when small coils carried currents. His experiments led to the conclusion that, if a closed loop is drawn in any region where there are fields and currents, the line integral of the magnetic field around the loop is proportional to the current passing through the loop,

$$\oint \mathbf{B} \cdot d\boldsymbol{\ell} \propto I.$$

When the magnetic field is measured in teslas and the current in amperes, the constant of proportionality has units of $T\,m\,A^{-1}$ and is defined to be exactly $4\pi/10^7\ T\,m\,A^{-1}$.

With the value of the constant so chosen, the ampere and hence the coulomb can be defined via the force between two current-carrying conductors as discussed in Appendix A. The tesla is then defined by eqn (16.12) and the SI units of B and I have been fixed. We now have

$$\oint \mathbf{B} \cdot d\boldsymbol{\ell} = \mu_0 I. \tag{16.15}$$

This equation is called **Ampère's law** and the constant μ_0 ($= 4\pi/10^7$ $T\,m\,A^{-1}$) is called the **permeability of free space**.

In words, Ampère's law states that

The line integral of the field B around a closed loop is equal to the current through the loop times the permeability of free space.

⬤ *Signs to be used in Ampère's law*

The sign of the current to be used in applying the law has to be inserted carefully. The current to be used is that passing through the area defined by the loop in the direction given by the right-hand screw rule. This is the direction in which a right-handed screw would move if rotated in the same sense as the contour or loop is traversed.

Figure 16.8 shows a loop looked at from the right. The current through the loop is positive from right to left. The right-hand screw rule then indicates that the sense of traversal of the loop when evaluating the line integral of the field has to be clockwise as seen from the right. The elementary vectors dℓ are in the direction shown. If the current density were in the opposite direction to that shown and we insisted on traversing the loop clockwise, the sign of the current to be used in Ampère's law would be negative.

If the total current enclosed by the loop is zero, the line integral of the field is zero. You should note that this does not necessarily mean that the field is zero at all points round the loop.

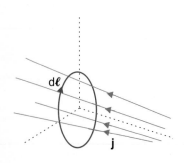

Fig. 16.8 The closed loop is viewed from the right. If the line integral of the magnetic field around the loop is evaluated traversing the loop clockwise, the elementary vectors dℓ are in the direction shown and the current to be used in Ampère' s law is the current flowing in the direction from right to left.

The magnetic dipole

The electric monopole field is the electric field distribution arising from a single isolated charge. The electric dipole field arises from a pair of equal and opposite charges very close together. There is no monopole magnetic field because of the absence of magnetic poles. However, Ampère's experiments showed that a current in a very small loop of wire gives rise to a magnetic field similar to that of an electric dipole, as shown in Fig 16.9. The only difference between the field lines from an electric dipole and the field lines from a magnetic dipole is that the electric field within the electric dipole goes from the real charge $+q$ to the charge $-q$, while there is no reversal of field across the current loop.

A small current loop thus constitutes a magnetic dipole. The dipole moment is in the direction perpendicular to the area of the loop given by the right-hand screw rule and the current. Figure 16.10 shows the relation between current flow and dipole moment.

The strength of the magnetic field from a small loop is found to be proportional to the current I in the loop and its area ΔS. The magnetic dipole moment of the loop is thus proportional to the product of current and area and is defined to be

$$\mathbf{m} = I\Delta\mathbf{S} \tag{16.16}$$

where the vector $\Delta\mathbf{S}$ is in the direction given by the right-hand screw rule. Magnetic dipole moments are measured in units of $A\,m^2$.

As with the electric dipole, provided the distance r from the dipole is large compared with its size, the magnitude of the magnetic field of a magnetic dipole is proportional to the dipole moment and inversely proportional to the cube of the distance.

$$B \propto \frac{I\Delta S}{r^3}.$$

The constant of proportionality has the same dimensions as the permeability of free space μ_0, as can readily be verified. In the SI system of units the constant is equal to $\mu_0/4\pi$.

The magnitudes of the field components at a point P distant r from a magnetic dipole of strength m when r is much greater than the dipole size, are given by formulae analogous to those of eqn (15.36) for an electric dipole. For example, the component B_θ is

$$B_\theta = \frac{\mu_0}{4\pi}\frac{m\sin\theta}{r^3} \tag{16.17}$$

where θ is the angle between the line joining the point P to the dipole and the direction in which the dipole is pointing.

Exercise 16.4 A single electron in the innermost shell of an atom has a spherically symmetric wave function, as discussed in Chapter 11, and so produces no average electric field at the nucleus. However, the electron has an intrinsic magnetic dipole moment of magnitude $9.28 \times 10^{-24}\,A\,m^2$. This moment points perpendicularly to the line joining the electron to the nucleus and gives a nonzero magnetic field at the nuclear site.

A single electron in the innermost shell orbiting a nucleus of atomic number Z is on average at a distance of $5.29 \times 10^{-9}/Z\,cm$. Estimate the field at the nucleus due to a single electron in the innermost shell in (a) a hydrogen atom, and (b) in a highly charged lead ion. The atomic number of lead is 82.

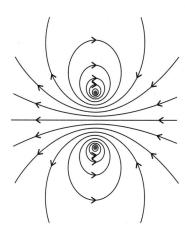

Fig. 16.9 The lines of the magnetic field from a currrent in a small loop of wire. The loop is viewed sideways on. The current flows clockwise in the loop looking at it from the right. The field lines are the same as for a magnetic dipole.

● *Dipole moments*

● *The field of a dipole*

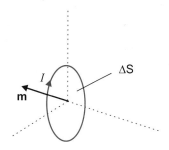

Fig. 16.10 The magnetic dipole moment of a small loop carrying a current. The loop is viewed from the right and the moment is perpendicular to the plane of the small coil in the direction given by the right-hand screw rule and the current.

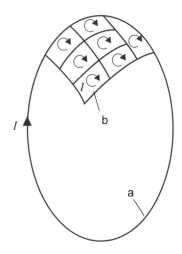

Fig. 16.11 The current in the large loop labelled a is equivalent to a set of currents of the same magnitude in very many small loops like that labelled b. The currents in loops inside the perimeter of the large loop cancel each other, leaving only the current in the perimeter.

● *The magnetic field as a sum of fields from small dipoles*

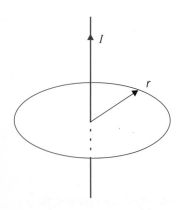

Fig. 16.12 A current in a long straight wire. The field has the same magnitude at all points on the circle drawn, which is centred on the wire, has radius *r*, and lies in a plane perpendicular to the wire. The circle is viewed from above.

Answer (a) 6.3 T; (b) 3.5×10^6 T. The field on the proton in a hydrogen atom in its ground state is a very large field. The field on a lead nucleus when only a single electron is present in the innermost shell is an enormous field, and experiments using such fields provide the most stringent tests of present theories of electricity and magnetism at high magnetic field strengths.

Given the field produced by a very small current loop, we can in principle determine the field due to a current in a loop of macroscopic size. A current I in the large loop shown in Fig 16.11 can be regarded as made up of currents I circulating in very many small loops drawn over the whole area of the large one, as shown. All currents cancel except over the perimeter, and the current distribution of all the small elements reduces to that of a current around the perimeter.

Although we have not yet written it down, the equation satisfied by the magnetic field is a linear equation. Magnetic fields depend linearly on the currents that produce them. Hence, magnetic fields satisfy the principle of superposition, and the total field from separate current distributions is obtained by adding the individual fields. The field from the large loop is accordingly the same as the sum of the fields from the very many small ones and can be evaluated in principle using that equivalence. However, this procedure for calculating magnetic fields is difficult and tedious, and the fields from simple current distributions can be determined more easily using Ampère's law.

16.3 **Fields from simple current distributions**

Currents are often flowing in wires with simple, symmetrical geometrical arrangements. When this is so, Ampère's law is a powerful tool for determining the field strengths and their directions. The two examples below illustrate the way this can be done.

The principle of superposition sometimes enables us to calculate the fields from current distributions that have no simple symmetry. For example, the field that results from currents in two long, parallel wires can be evaluated by superposing the fields from each separate wire.

A straight line current

The first example is a current in an isolated, long, straight wire, as shown in Fig 16.12. Observers at any point on the circumference of a circle lying in a plane perpendicular to the wire and with centre on the wire, such as the circle shown, are indistinguishable. They all see exactly the same physical situation and all points on this circle are equivalent. In mathematical terms, there is cylindrical symmetry about the straight wire and the field is independent of the angle ϕ defined in a cylindrical coordinate system that has its origin on the wire.

● *Cylindrical symmetry of a long, straight wire*

The symmetry requires that the magnitude of the field is the same at all points on the circle of radius r drawn on Fig 16.12. What about the direction? The current I is upwards, and the right-hand screw rule indicates that, looking at the circle from below, a right-handed screw would have to be rotated clockwise to move upwards. Hence the right-hand screw rule tells us that the line integral of the field is positive, going round the circle clockwise looking up. The field lines are continuous, and hence the field direction must everywhere be tangential to the circle in the plane of the circle and in the direction given by the right-hand screw rule and the current.

The field strength can be obtained using Ampère's law. The cylindrical symmetry makes the evaluation of the line integral in eqn (16.15) straightforward. The field at all points on the circumference of the circle shown on Fig 16.12 lying in a plane perpendicular to the wire, has the same magnitude and the field direction at all points on the circle is tangential to the circle, as discussed above. Equation (16.15) thus reduces to

$$\oint \mathbf{B} \cdot \mathrm{d}\boldsymbol{\ell} = 2\pi r B = \mu_0 I, \tag{16.18}$$

and the magnitude of the field is

$$B = \frac{\mu_0 I}{2\pi r}. \tag{16.19}$$

The field at a distance of 10 cm from a current of 1 A in a long wire is 2×10^{-6} T. This is a small field. For orientation, the Earth's field at the surface of the Earth is about 10^{-4} T.

The lines of the field B for a current in a long, straight wire are shown in Fig 16.13.

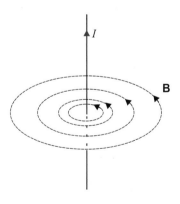

Fig. 16.13 The lines of the magnetic field from a current in a long straight wire, viewed from above.

Worked Example 16.3 A long straight wire of radius 0.5 cm carries a current of 2 A uniformly distributed over the cross-sectional area of the wire. Determine the magnetic field inside the wire at a distance of 1 mm from the axis.

Answer Ampère's law says the line integral of the magnetic field around a closed loop is equal to μ_0 times the current passing through the loop. Any current outside the loop is of no relevance. The current I_r passing through a circle of radius r centred on the wire axis and lying in a plane perpendicular to the wire is the fraction $\pi r^2/\pi a^2$ of the whole current I, where a is the radius of the wire. Hence

$$I_r = I \times \frac{r^2}{a^2}\,.$$

Using the same symmetry argument as above, the field at distance r is of constant magnitude, is tangential to the circle of radius r, and in the same direction as outside. Therefore

$$2\pi rB = \mu_0 I_r = \mu_0 I \frac{r^2}{a^2}$$

or

$$B = r\,\frac{\mu_0 I}{2\pi a^2}\,.$$

This equation shows that inside the wire the field is proportional to distance r from the centre of the wire. At $r = 1\,\text{mm}$, $B = 1.6 \times 10^{-5}\,\text{T}$. Outside the wire the field decreases inversely proportional to distance r according to eqn (16.19).

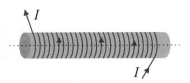

Fig. 16.14 A solenoid consists of wire carrying a current closely wound on a cylindrical former.

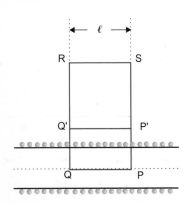

Fig. 16.15 A view through a plane containing the axis of the cylindrical solenoid of Fig 16.14. The field is approximately determined by applying Ampère's law to the rectangular loop PQRS.

The solenoid

Another example of a symmetrical current distribution is provided by the solenoid. This consists of conducting wire closely wound over the outside of a straight, nonmagnetic support as shown in Fig 16.14. If the length over which the wire is wound is much greater than the inner dimension of the support, the current distribution is once more symmetrical about the axis of the support and end effects can be neglected when evaluating the field near the middle of the solenoid.

Consider a cylindrical solenoid that has n_ℓ turns per unit length wound over a tube with the wire carrying a current I. Apply Ampère's law to the loop PQRS drawn on Fig 16.15, which shows a plane containing the axis of the solenoid. Because the solenoid is very long, the field over QR at any particular distance from the axis is the same as the field over PS at that distance. Their contributions to the line integral over the loop PQRS thus cancel each other.

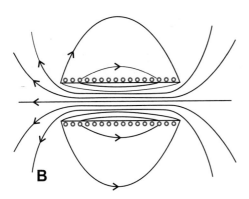

B

Fig. 16.16 The lines of the magnetic field from a solenoid.

The field component in the direction of the line RS outside the solenoid is zero when the lengths QR and PS become very large. By applying Ampère's law to the rectangular loop P'Q'RS which has the line P'Q' parallel to the axis outside the solenoid, we see that the parallel component is zero everywhere outside the solenoid. Thus the contribution to the line integral over the loop PQRS also vanishes.

The only remaining contribution to the line integral is over PQ inside the solenoid. Inside a very long solenoid the symmetry ensures that the field lines are parallel to the axis and that the field has the same magnitude B at all points on the line PQ. The application of Ampère's law to the loop PQRS reduces to

$$B\ell = \mu_0 n_\ell \ell I$$

or

$$B = \mu_0 n_\ell I. \tag{16.20}$$

The direction of the field is given by the right-hand screw rule.

The arguments used above can be applied to any rectangular loop similar to PQRS with PQ parallel to the axis. The current passing through any such loop is the same and the field inside the long solenoid is everywhere parallel to the axis with magnitude given by eqn (16.20). The field strength inside a long solenoid having length 25 cm and a total of 500 turns carrying a current of 2 A is $16\pi \times 10^{-4}$ T.

The field lines for a typical solenoid are shown schematically in Fig 16.16. The end effects cause the field inside to be weaker near the ends than in the idealized case discussed above. However, eqn (16.20) is accurate to a few per cent in the middle of a solenoid if the length exceeds the radius by a factor of about 20.

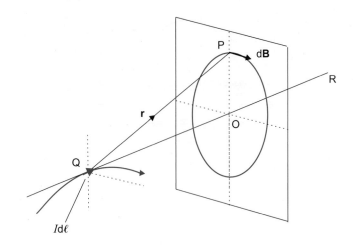

Fig. 16.17 A current element at the point Q is an infinitesimally small part of a complete circuit. The current element at the point Q is the vector $I\mathrm{d}\ell$ and gives rise to a contribution dB to the total field at the point P.

The Biot–Savart law

If we know the contribution to the total field at a point from each infinitesimal element of a current-carrying system of wires, the principle of superposition tells us that the total field can be evaluated as the sum over the whole circuit of the individual contributions. The field produced by a small part of a circuit was deduced by Biot and Savart from the results of experiments they performed early in the nineteenth century.

A wire carrying a current I consists of a very large number of infinitesimally small lengths $\mathrm{d}\ell$, which together make up the complete circuit. We define the **current element** at a point Q in the wire as the vector $I\mathrm{d}\ell$ which points in the direction of the current and the wire at that point.

The total magnetic field at the point P on Fig 16.17 is made up of contributions from all the current elements constituting the current in the wire. Figure 16.17 shows a current element at the point Q in the direction of the line QOR. The contribution dB to the total field at P from the element at Q is given by

$$\mathrm{d}\mathbf{B} = \frac{\mu_0}{4\pi} I \frac{\mathrm{d}\boldsymbol{\ell} \times \mathbf{r}}{r^3} \tag{16.21}$$

where $\mathbf{r}$ is the vector from the current element to the point P.

Since the magnitude of $\mathrm{d}\boldsymbol{\ell} \times \mathbf{r}$ is the same for all points on the circumference of the circle centred on the point O shown on Fig 16.17, the current element at Q gives rise to the same magnitude contribution at all points on the circle. The circle lies in the plane perpendicular to

the line QOR and contains the points P and O. The direction of
the contribution d**B** at any point on the circle is perpendicular to the
vector **r** and perpendicular to the vector representing the current
element. It is thus tangential to the circumference in the direction of
d$\boldsymbol{\ell} \times \mathbf{r}$ as shown.

The total field from the current in the wire is the line integral of the
right-hand side of the above expression around the complete wire.

$$\mathbf{B} = \frac{\mu_0}{4\pi} I \oint \frac{d\boldsymbol{\ell} \times \mathbf{r}}{r^3}. \tag{16.22}$$

Equation (16.22) is the **Biot–Savart law**. In words, the Biot–Savart
law states that

**The field B due to a current I flowing in a closed loop of thin wire is
given by the line integral around the loop of contributions $I d\boldsymbol{\ell} \times \mathbf{r}/r^3$
times the factor $\mu_0/4\pi$.**

Equation (16.21) shows that the contribution to the total field from a
current element falls off as the square of the distance, as does the
contribution to the total electric field of an individual charge. The
Biot–Savart law can be deduced formally from Ampère's law. However,
we will consider that the agreement between expressions for fields
obtained using the Biot–Savart law and using Ampère's law or equivalent
sets of elementary dipoles is satisfactory proof of eqn (16.22).

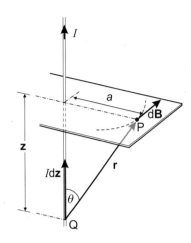

Fig. 16.18 A current element $I d\mathbf{z}$
at the point Q on a long straight
wire gives a contribution d**B** of
magnitude $\mu_0 I dz \sin\theta/4\pi r^2$ to the
magnetic field at the point P.

A straight wire

We will first show how the Biot–Savart law can be used by applying it
to the simple example already treated, that of a current in a long
straight wire. The current distribution is made up of current elements
$I d\mathbf{z}$ like that shown at the point Q in Fig 16.18. The contribution to the
field at the point P from this element lies in the plane perpendicular to
the wire containing the point P and is in the direction shown. Its
magnitude is

$$dB = \frac{\mu_0}{4\pi} I \frac{dz}{r^2} \sin\theta$$

where θ is the angle between the vectors **r** and d**z**.

The contribution at P from every element of the line current is in the
same direction, and so the magnitude of the total field at P is

$$B = \frac{\mu_0}{4\pi} I \int_{-\infty}^{+\infty} \frac{dz}{r^2} \sin\theta = \frac{\mu_0}{4\pi} Ia \int_{-\infty}^{+\infty} \frac{dz}{\left(a^2 + z^2\right)^{3/2}}.$$

The integral of $1/(a^2 + z^2)^{3/2}$ can be evaluated using the substitution $z = a \tan \theta$. The value of the definite integral in the above expression for the field B is $2/a^2$, and the expression reduces to

$$B = \frac{\mu_0 I}{2\pi a},$$

the same relationship as eqn (16.19) which was deduced using Ampère's law.

A current loop

An example that involves a simple current distribution but one whose field is difficult to evaluate without using the Biot−Savart law is a current I in a single circular turn of wire with radius a, as shown in Fig 16.19. We will calculate the field on the line ROS passing through the centre O of the loop and perpendicular to the plane of the loop.

The field at the point P distant x from the centre of the loop is the resultant of contributions from all the current elements comprising the loop. Consider the contribution d**B** from the current element $Id\ell$ shown on the top of the loop at the point Q. This contribution is in the direction shown, at right angles to both the vector **r** joining the current element to the point P and to the vector representing the current element itself. The contribution has magnitude $\mu_0 Idl/4\pi r^2$ and lies in the vertical plane containing the line ROS at an angle θ with that line, the same angle as the vector **r** makes with the normal to the loop.

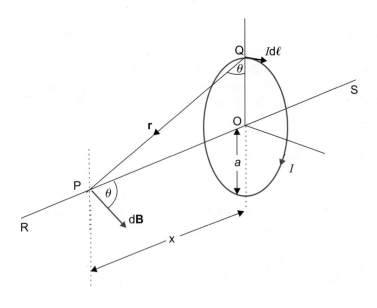

Fig. 16.19 A current element $Id\ell$ at the point Q on a circular loop of wire carrying a current I gives a contribution d**B** of magnitude $\mu_0 Id\ell/4\pi r^2$ to the magnetic field at the point P.

The contribution from any element around the loop has the same magnitude and makes the same angle θ with the line ROS. The contributions from all the elementary current elements around the whole loop thus cover the surface of a cone with half angle θ whose axis lies along the line ROS. Each contribution can be resolved into two components, one along ROS, the other perpendicular to that line. The resultant of the perpendicular components of all the contributions is zero; for every contribution in a certain direction perpendicular to the line ROS, there is another equal and opposite contribution. The resultant field is thus in the direction from R to S and has magnitude B equal to the sum of the components in that direction.

$$B = \frac{\mu_0 I}{4\pi} \oint \frac{\mathrm{d}\ell}{(a^2 + x^2)} \cos\theta = \frac{\mu_0 I}{4\pi} \oint \frac{a\,\mathrm{d}\ell}{(a^2 + x^2)^{3/2}}.$$

The line integral of $\mathrm{d}\ell$ around the loop is simply the circumference of the circle, $2\pi a$, and hence

$$B = \frac{\mu_0 I}{2} \frac{a^2}{(a^2 + x^2)^{3/2}}. \tag{16.23}$$

At the centre of the circle, x is zero and the field is $\mu_0 I / 2a$, a factor π greater than the field at a perpendicular distance a from a straight-line current I. If instead of a single loop there are n turns wound closely together with average radius a, the field on axis, to a good approximation, is n times that given by eqn (16.23).

Exercise 16.5 A circular loop of wire carries a current of 5 A. Its radius is 10 cm. What is the magnetic field on axis a distance of 10 cm from the centre of the loop?

Answer 1.11×10^{-5} T.

Far away from a loop of current, the loop looks like a dipole and the field correspondingly has the same pattern. The field lines near a current loop are shown schematically in Fig 16.20. Near the loop the field has components that are more complicated than those in the dipole field, but these components fall off more rapidly with distance than the dipole component.

Worked Example 16.4 Two coils of the same radius a and number of turns n are placed coaxially. If each coil carries the same current I and if the coils are separated by a distance a, it can be shown that

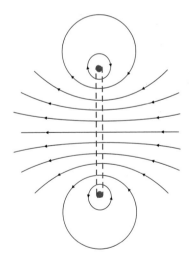

Fig. 16.20 The lines of the magnetic field near the loop shown in Fig 16.19. The field lines are shown in the plane perpendicular to the plane of the loop. The current is out of the page at the bottom of the loop and into the page at the top.

there is a region half-way between the coils on the axis where the field is nearly uniform. This configuration of coils, called Helmholtz coils, is often used to cancel the Earth's field to provide a field-free region.

At a point on the Earth's surface the magnetic field has magnitude 7×10^{-5} T. At this position a current passes in the same direction through two coils each having 50 turns of radius 1 m, their separation being 1 m. How big must the current be to cancel the Earth's field on the axis half-way between the coils?

Answer On axis, half-way between the coils the distance x from each coil is $a/2$, and the addition of two terms given by eqn (16.23), multiplied by the number of turns n, gives the resultant field magnitude as

$$B = \mu_0 n I \ \frac{a^2}{\left(a^2 + a^2/4\right)^{3/2}} = \mu_0 n I \ \frac{a^2}{\left(5a^2/4\right)^{3/2}}.$$

Hence

$$I = \frac{aB}{\mu_0 n (4/5)^{3/2}}.$$

Substitution of the values $a = 1$ m, $B = 7 \times 10^{-5}$ T, $n = 50$, and μ_0 gives $I = 1.56$ A.

16.4 Forces on conductors

A current in a wire gives rise to a magnetic field. This field will exert a force on the moving charges that constitute the current in a nearby current-carrying wire. There are forces between separate current-carrying circuits and there may be forces between different parts of the same circuit.

Consider a long wire in a uniform external field **B**. The wire is thin and straight and carries a current I. The thin wire may be regarded as a linear continuum of elementary current elements $I d\boldsymbol{\ell}$. The field at the position of a current element from any other element is given by eqn (16.21) and is always zero since the vectors $d\boldsymbol{\ell}$ and $\mathbf{r}$ are always in the same direction. Thus no element experiences a force from any of the others. The only force is due to the interaction of the current in the wire with the external field.

A length ℓ of wire of small cross-sectional area S contains $N\ell S$ moving electrons, where N is the number of free electrons per unit volume. The force on these electrons in a constant field of strength B perpendicular to the wire as in Fig 16.21 is in the direction shown and has magnitude

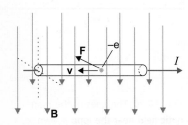

Fig. 16.21 The force on a moving electron in a wire carrying a current and situated in a uniform magnetic field.

$F = N\ell SevB$. But the current in the wire $I = eNvS$ from eqn (16.3), and hence $F = \ell IB$, or

$$\frac{F}{\ell} = IB. \tag{16.24}$$

The force per unit length is equal to the current times the external field, and is in the direction of $\hat{\mathbf{I}} \times \mathbf{B}$ where $\hat{\mathbf{I}}$ is a unit vector in the direction of the current.

There is a constant external field B_2 on a straight wire with a current I_1 if it lies parallel to a second wire that has a current I_2 in it. The field due to I_2 and the force on wire 1 are shown in Fig 16.22 when the currents are in the same direction. In that case the force on wire 1 is towards wire 2 and has magnitude $I_1 B_2$ per unit length; there is an equal and opposite force per unit length $I_2 B_1$ on wire 2. If the wires are 10 cm apart and each carries a current of 10 A the force per unit length is equal to $2 \times 10^{-4} \, \mathrm{N\,m^{-1}}$. This is a small force but readily detectable.

If a plane coil of a single turn or multiple turns carrying a current I is suspended in a uniform magnetic field, there is no net force on the coil but there is a torque. Figure 16.23 shows a rectangular coil PQRS suspended about a vertical axis and with its plane vertical. A uniform magnetic field $\mathbf{B}$ in the horizontal direction covers the coil. The normal NN' to the plane of the coil (pointing in the direction given by the current and the right-hand rule) makes an angle θ with the field direction as shown. The forces $\mathbf{F}_{PQ}$ and $\mathbf{F}_{RS}$ on the top and bottom sides PQ and RS are equal and opposite, and the forces on the vertical sides QR and SP are equal and opposite. Thus, there is no net force on the coil. However, the forces on the vertical sides give rise to a torque about the vertical axis of suspension. These forces have magnitude $IB\ell_2$, where ℓ_2 is the length of the vertical sides, and the torque is

$$\tau = \ell_1 \ell_2 BI \sin \theta = BIS \sin \theta \tag{16.25}$$

where ℓ_1 is the length of the sides PQ and RS, and $S = \ell_1 \ell_2$ is the area of the coil. The torque tends to decrease the angle θ as shown in Fig 16.24.

For a coil with n turns, each of the same area, the right-hand side of eqn (6.25) must be multiplied by n, since the force on each vertical side of the coil is now multiplied by n. The equation holds whatever the shape of the coil, and is the principle behind the moving coil instruments that are sometimes used to measure current. These instruments may have a coil suspended in the field of a permanent magnet by a torsion wire. The

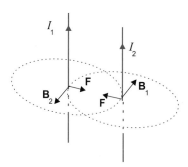

Fig. 16.22 A long, straight wire 1 carries a current I_1 and is parallel to a second long, straight wire 2 carrying a current I_2. The field $\mathbf{B}_2$ at wire 1 from wire 2 has constant strength and is perpendicular to wire 1 at all points on the wire.

⬤ *The torque on a current loop*

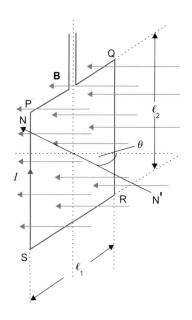

Fig. 16.23 A rectangular coil suspended in a vertical plane in a uniform horizontal magnetic field. The normal NN' to the plane of the coil makes an angle θ with the field direction.

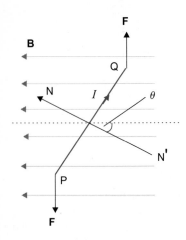

Fig. 16.24 A view looking down on the coil of Fig 16.23 showing the forces on the vertical sides.

deflection of the coil when a current is passed through it is a measure of the current.

Exercise 16.6 A coil consists of 50 turns of wire wound on a circular former of radius 0.5 cm. The coil is suspended in the field of a permanent magnet that gives an average magnetic field of 0.04 T. If the deflection of the coil is 15° degrees when a current of 100 mA passes, what is the magnitude of the restoring torque per radian twist provided by the suspension?

Answer 1.6×10^{-5} N m per radian.

Potential energy

The above derivation of the torque on a coil passing a current in a magnetic field can be used to calculate the energy of a magnetic dipole in an external field and then, in turn, to calculate the forces and torques on coils in general. If the rectangular coil discussed above is very small it constitutes a magnetic dipole of moment equal to the current I times the area ΔS. The torque about the centre of the dipole is given by eqn (16.25). In terms of the dipole moment, which has magnitude $m = I\Delta S$, the torque tending to decrease the angle θ is

$$\tau = mB\sin\theta$$

where the angle θ, as shown in Fig 16.25, is the angle between the direction of the dipole moment and that of the field. In vector notation the equation giving the magnitude and direction of the torque is

$$\boldsymbol{\tau} = \mathbf{m} \times \mathbf{B}. \tag{16.26}$$

If the current in the coil is constant, we may define a potential energy, U_c, of the coil in the external field. The potential energy, when differentiated with respect to position or angular coordinates, gives forces and torques in a manner analogous to the mechanical potential energy. Accordingly, the torque tending to increase θ is

$$\tau = -\frac{\partial U_c}{\partial \theta}$$

and hence

$$U_c = -mB\cos\theta = -\mathbf{m}\cdot\mathbf{B}. \tag{16.27}$$

The potential energy of a large coil carrying a current in a magnetic field can now be determined using the concept of the current sheet introduced at the end of Section 16.1. We fill the area defined by the perimeter of the coil with elementary current loops, each of which is a

● **The torque on a coil**

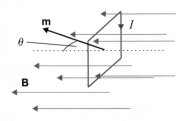

Fig. 16.25 A small rectangular coil in a uniform field **B**. The dipole moment $\mathbf{m} = I d\mathbf{S}$ of the coil makes an angle θ with the field direction.

small dipole with direction given by the right-hand screw rule and the current. The total energy of the large coil in the field is the sum of the energies of the individual small dipoles. If the field at the position of an elementary dipole is **B**, the potential energy of the small dipole is $-I d\mathbf{S} \cdot \mathbf{B}$ and the total potential energy is

● *The force on a dipole*

$$U_{\mathrm{c}} = -I \int_S \mathbf{B} \cdot \mathrm{d}\mathbf{S}. \qquad (16.28)$$

If the field is constant over a coil, and the normal to the plane of the coil in the direction given by the right-hand rule lies at an angle θ to the direction of **B**, eqn (16.28) reduces to

$$U_{\mathrm{c}} = -BIS \cos \theta,$$

and the torque tending to increase θ is

$$\tau = \partial U_{\mathrm{c}}/\partial \theta = BIS \sin \theta,$$

the same result as eqn (16.25). If the field is non-uniform, eqn (16.28) must be used to determine the potential energy and its derivatives taken to determine torques and forces on the coil.

Magnetic flux

Equation (16.28) may be expressed in terms of **magnetic flux** or **flux of the magnetic field** through the coil. This is an important quantity in the physics of electric and magnetic fields. The magnetic flux through an area S is the surface integral over the area of the field **B**. Magnetic flux has the symbol Φ and the flux through an area S is

$$\Phi = \int_S \mathbf{B} \cdot \mathrm{d}\mathbf{S}. \qquad (16.29)$$

This definition is similar to that of the flux of the electric field through a surface, given in eqn (15.13). The area S need not be the area of an actual coil but simply an area enclosed by a continuous line drawn in space, and it need not be a flat surface.

The sign of the magnetic flux through a loop depends on the direction of the field and the way the directions of the vectors d**S** are defined. The vectors d**S** in the flux definition are normal to the small areas $\mathrm{d}S$ in the direction given by the right-hand screw rule and the sense of traversal of the loop. Whatever the shape of the surface, as long as the perimeter stays the same, the flux through any area bounded by that perimeter is the same. This is because of the continuity of magnetic field lines. They have no beginning or end and thus those that pass through the flat area across the loop also pass outwards through any complete area bounded by the perimeter of the loop.

Equation (16.29) applies to a closed loop drawn in space or to a coil of a single turn. The total flux through a coil of n turns, each assumed to have the same area S, is n times the flux through a single turn. Sometimes the total flux through a coil of n turns is called the flux linkage, magnetic flux being reserved for a single loop. We will use the term magnetic flux to cover both cases, since it will be obvious from the context whether the factor n is present or not.

The force on a coil carrying a current in a magnetic field can be expressed in terms of the magnetic flux through the coil. Equation (16.28) gives the potential energy in terms of the flux, and the force in the x-direction, for example, is the negative of the derivative with respect to x of the potential,

$$F_x = -(\partial U_c/\partial x) = I\frac{\partial \Phi}{\partial x}. \tag{16.30}$$

Similar equations give the force components in other directions. The torque on the coil about an axis is given by the negative rate of change of the potential with respect to the angle of twist about the axis. In using eqn (16.30), the current I with the right-hand screw rule defines the directions of the vectors $\mathrm{d}\mathbf{S}$ in the flux Φ, and the resultant sign of F_x determines whether the force component F_x is in the positive or negative x-direction.

Worked Example 16.5 A small circular coil of a single turn carrying a current of 1 A lies with its plane perpendicular to the x-axis. The coil has a radius of 0.5 cm and is in the external magnetic field of a coaxial coil of radius 5 cm supporting a current of 10 A through 100 turns. The currents in the coils are in the same direction. If the separation of the two coils is 10 cm, estimate the force on the small coil.

Answer The small coil can be approximated to a magnetic dipole with a moment of magnitude equal to the current I_1 in the coil times its area S_1, as in eqn (16.16). The direction of the dipole is the same as that of the field due to the large coil. It is along the x-axis as shown in Fig 16.26. If we place the origin at the centre of the large coil, the field B_2 at the small coil is given by eqn (16.23) multiplied by the number of turns of the large coil. The potential energy of the small coil in this field is then given by eqn (16.27) and, since the dipole moment and the field are in the same direction,

$$U_c = -I_1 S_1 B_2 = -I_1 S_1 \frac{\mu_0 n I_2}{2} \frac{a^2}{(a^2 + x^2)^{3/2}},$$

where a is the radius of the large coil, n the number of turns on the large coil, and x the distance between the coils. The force F_x in the x-direction is given by $-(\partial U_c / \partial x)$, and

$$F_x = -\frac{\mu_0 n I_1 I_2 S_1}{2} \frac{3a^2 x}{(a^2 + x^2)^{5/2}}.$$

Substituting the given parameters into this expression gives the estimated force equal to 2.12×10^{-6} N.

We see from Worked example 16.5 and earlier examples that forces between circuits carrying currents of the size normally encountered are small compared with most gravitational forces experienced. However, they are usually large compared with gravitational forces when microscopic phenomena are being considered and the masses involved are very small.

The force on the small coil in the above example is in the negative x-direction. The small coil is being attracted towards the large one. This direction could have been determined by considering the force from the field $\mathbf{B}_2$ on the dipole moment of the small coil. If the dipole is aligned with the field, the force tends to move it to regions of higher field strength; if the direction of the dipole is opposed to that of the field, the force tends to move it to regions of lower field strength.

For exact alignment there is no torque on the dipole and if it is rotated slightly a torque tends to realign it. If the dipole is opposed to the field and rotated slightly, there is a torque that tends to rotate the dipole $180°$ and align it with the field.

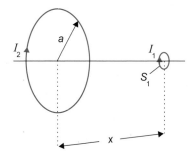

Fig. 16.26 Two coaxial coils, one with radius much larger than the other, separated by a distance x much larger than the radius of the small coil.

16.5 Magnetization

All materials to some extent change a magnetic field into which they are placed, but the influence of most substances is so small that to a good approximation it may be neglected. A rod of copper inside a solenoid that has a steady current will change the field near the end of the solenoid by about one part in 10^4. However, some materials do have a marked influence. A rod of soft iron will increase the field by a factor of a thousand or more.

Before insertion into a solenoid the rods produced no magnetic effects. Since currents are the source of magnetic fields we may thus assume that, in the absence of an external field, the internal microscopic currents due to the motions of the electrons cancel each other. The rods are said to be *unmagnetized*. They become *magnetized* when an external field rearranges the microscopic currents and gives rise to microscopic dipole moments that do produce a net magnetic effect.

The macroscopic magnetic dipole moment per unit volume of a substance arising from the microscopic dipoles in a volume containing very many atoms is called its **magnetization**. The magnetization is a vector function of position and is given the symbol **M**. Its SI units are amperes per metre. The magnetization of a substance gives rise to its own contribution to the total magnetic field, as in the examples above of rods in a solenoid.

Magnetic materials

○ *Ferromagnetism*

Since magnetic fields are caused by currents, we conclude that there are essentially no internal currents induced in a copper rod when placed in an external field but that there are large internal currents induced when an iron rod is inserted. Substances that have large internal currents induced by applied magnetic fields, with the result that the combined field is much larger than the applied field and in the same direction, are called **ferromagnetic**. Examples of common ferromagnetic materials are provided by soft iron, steel (which is predominantly iron but has small quantities of other elements mixed in to give it special properties), nickel, and cobalt. An alloy is a mixture of different metals; brass is an alloy of copper and zinc. Many alloys made of the above ferromagnetic elements are also ferromagnetic.

○ *Diamagnetism and paramagnetism*

Materials that produce very little effect in a magnetic field are non-ferromagnetic, and are divided into two classes depending on whether their very small contributions decrease or increase an external field in which they are placed. The total field near the end of a solenoid into which a copper rod has been introduced is very slightly less than it was before; copper is said to be **diamagnetic**. Insertion of a magnesium or zirconium rod slightly increases the field; zirconium and magnesium are said to be **paramagnetic**. A ferromagnetic material loses its ferromagnetism and becomes paramagnetic above a certain temperature called the **Curie point**. The Curie point for soft iron is about 1040 K.

We will give a brief and simple outline of the microscopic origins of the three different types of behaviour. In ferromagnetic materials, the microscopic magnetic dipoles that arise from individual electron spins or from electron orbits, as discussed in Section 11.3, all line up the same way within very small macroscopic sized regions called domains. Within a domain there are very large numbers of aligned microscopic current loops, equivalent to small dipoles. This gives a large resultant moment of an individual domain capable of giving a measurable macroscopic field by itself. Normally, the very many domains within a piece of soft iron, for example, are oriented at random and their fields average to zero or very small values. In the presence of an applied field, individual domains whose fields are in the same direction as the applied field grow at the expense of others, and the result can be a total field a factor of a thousand or more larger than the original one.

Some ferromagnetic materials, for example, steel, retain the alignment when the applied field is removed. The steel is then said to be *permanently magnetized*, and will remain so unless it suffers special treatment such as heating or application of magnetic fields that alternate in direction while slowly reducing to zero.

The microscopic constituents of diamagnetic materials have no intrinsic magnetic dipole moments. The electrons are all paired in orbits circulating in opposite directions and with opposing intrinsic spins also giving zero net dipole moment to each pair. In the presence of an external field the electron orbits rearrange themselves slightly. This results in a small induced dipole moment that gives a field directed *against* the external field. This contribution is very small for diamagnetic substances, which acquire a small induced magnetization opposite to the external field.

The microscopic constituents of paramagnetic materials do have permanent dipole moments. In the absence of an external field the moments are aligned randomly. When a field is applied, there is a small net alignment in the field direction. This gives a macroscopic dipole moment per unit volume in the *same* direction and an enhancement of the original field, although as in diamagnetism the enhancement is very small. The mechanism opposing alignment of the microscopic dipoles is thermal motion. An equation of the same form as eqn (15.47) for electric dipoles relates the magnetic moments to the applied magnetic field.

The field H

For *isotropic and homogeneous diamagnetic and paramagnetic materials*, the magnetization is proportional to the total field in the material,

$$M \propto B.$$

The constant of proportionality depends upon the material and may depend on other factors such as temperature. The constant is written χ_B/μ_0 where χ_B is called the **magnetic susceptibility**. We then have

● *Magnetic susceptibility*

$$\mathbf{M} = \frac{1}{\mu_0} \chi_B \mathbf{B}. \tag{16.31}$$

The constant μ_0 is introduced in order that the susceptibility be dimensionless. The susceptibility of diamagnetic materials is typically about -1×10^{-4} and that of paramagnetic materials a few times $+1 \times 10^{-4}$.

An existing magnetic field is barely influenced by the presence of non-ferromagnetic materials. However, it is severely affected by iron and other ferromagnetics. The magnetization of the iron usually plays a larger role in determining magnetic effects than the sources of the external field in

Other Magnetic Effects in Materials

The three types of magnetic behaviour outlined in the text do not exhaust the patterns of magnetic effects exhibited by materials. The magnetic fields that arise from magnetizations of materials are all generated by virtue of microscopic currents (which normally average to give zero effect) not cancelling each other when an external field is present. The ordering of very many dipoles gives rise to ferromagnetism. A different type of ordering can occur in certain materials. **Ferrimagnetism** occurs when adjacent microscopic constituents order dipoles in opposite directions. Adjacent dipoles have different strengths and the result is a weaker form of ferromagnetism. **Anti-ferromagnetism** occurs when there is complete ordering within small macroscopic regions as in ferromagnetism. This time, however, the

order is such that adjacent dipoles equal in strength are in opposite directions giving zero magnetization.

In the presence of magnetic fields many materials exhibit unexpected behaviour. Superconductors offer good examples. Superconducting materials have near total loss of resistivity below a critical temperature T_c, which is usually in the range 0.1 to 20 K. The superconductivity is destroyed when the superconductor is placed in a sufficiently large magnetic field.

Superfluids offer another example of dramatic magnetic field effects. Some fluids, such as liquid helium, have near total loss of viscosity below a critical temperature when there is no resistance to flow. Like superconductivity, superfluidity is destroyed by sufficiently large magnetic fields.

which the iron is placed. This makes life difficult because the magnetization vector is discontinuous; it may have a large value within a ferromagnetic material, but immediately outside the material the magnetization is zero. For this reason a third vector field is used in magnetism, and most calculations are performed in terms of this field and the field **B** instead of in terms of **B** and the magnetization.

This new field has the symbol **H** *and we will simply call it the* **field H**. It has the same dimensions as magnetization, and is measured in $A\,m^{-1}$. The fields **B** and **M** have straightforward physical significance. The first gives the force on a moving charge and the second is the dipole moment per unit volume. The field **H** has no simple physical interpretation. It is defined below in terms of both **B** and **M**. Here we regard it as a vector function of position related to currents and introduce it in order to acilitate calculations on magnetism in the presence of materials.

For isotropic and homogeneous paramagnetic and diamagnetic materials, the vectors **B**, **M**, and **H** are simply related. The new field **H** is defined in terms of the total **B** field and the magnetization,

$$\mathbf{B} = \mu_0(\mathbf{H} + \mathbf{M}).$$

(16.32)

The constant μ_0 is introduced because it occurs in the expression for the **B** field from a distribution of dipoles specified by the magnetization.

The magnetization of a substance gives rise to its own contribution to the total magnetic field and corresponds to alignment of microscopic dipoles. These dipoles in turn correspond to circulating microscopic currents and these should be included in Ampère's law, eqn (16.15), which was applied in Section 16.2 only to currents of freely moving charges in wires. (We assumed that the wires had no magnetization. That is true for practical purposes for non-ferromagnetic wires.)

When the circulating microscopic magnetization currents are included in Ampère's law, it is found that the line integral of **H** around a closed loop is related to the current I of *freely moving charges* through the area of the loop and is independent of the magnetization.

$$\oint \mathbf{H} \cdot d\boldsymbol{\ell} = I. \tag{16.33}$$

This is a relation similar to Ampère's law, but it has the advantage that it involves only the current from freely moving charges and not bound currents that arise from magnetization.

Equation (16.32) may be rewritten in terms of a new parameter characteristic of the material called the **relative permeability** and given the symbol μ.

● *Relative permeability*

$$B = \mu \mu_0 H \tag{16.34}$$

where, from eqns (16.32) and (16.34),

$$\mu = 1 + \frac{M}{H}.$$

Using eqns (16.31) and (16.34), this may be written

$$\mu = \frac{1}{1 - \chi_B} \approx 1 + \chi_B, \tag{16.35}$$

since χ_B is small, showing the relation between the relative permeability and the magnetic susceptibility.

For ferromagnetic materials, M is much bigger than H and μ is large. The relative permeability of these materials depends upon the magnetic fields, although over certain ranges of field strengths it may often be assumed to be constant.

Electromagnets

We are now in a position to discuss electromagnets. A simple electromagnet consists of a gap cut in a toroid made of soft iron and overwound with coils that carry a current. Without a gap the closed toroid is similar to a long solenoid overwound on an iron core. The current in the coils generates an external field. This magnetizes the iron and the field generated by the magnetized iron is much larger than the original field.

Figure 16.27 shows a closed toroid. If we ignore any small effects from the leads that bring the current into and take it out of the coil, the toroid has complete symmetry and the lines of the fields **B** and **H** follow circles around the toroid. The magnitudes of the fields are constant around the circles, and are constant across the toroid to a good approximation,

● *The field inside a closed toroid*

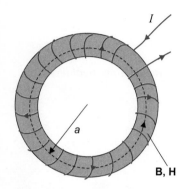

Fig. 16.27 A closed toroid of ferromagnetic material overwound with many turns of a current-carrying coil.

● *The field in a gap cut in a closed toroid*

assuming the width of the toroid to be small compared with its radius. Taking the line integral of the field **H** around the dotted loop shown, eqn (16.33) becomes

$$2\pi aH = nI \tag{16.36}$$

where I is the current and n the total number of turns on the coil. This gives the value of the field H, and in turn the value of the field B is given by eqn (16.34) if the relative permeability at the field strength H is known.

$$B = \frac{\mu\mu_0 nI}{L} \tag{16.37}$$

where $L = 2\pi a$ is the length of the toroid. The field direction is given by the right-hand rule and the current. Typically, the value of μ is around 10^3 and this is the factor by which the field B is amplified by the presence of the ferromagnetic toroidal core.

Exercise 16.7 A toroid of radius 0.3 m is made of material that we assume to have a constant relative permeability of 1000. It is overwound with a coil of 800 turns and a current of 2 A is passed. What are the strengths of the fields B and H in the core?

Answer $B = 1.07\,\text{T}$; $H = 849\,\text{A}\,\text{m}^{-1}$.

Now consider the fields when a gap of width w is cut in the core, as in Fig 16.28. The lines of the fields B and H continue to a good approximation to be circles around the toroid in the same direction as before.

We must now introduce a new property of the field **B**. If the field **B** is perpendicular across a plane boundary it has the same value on either side; *the magnetic field* **B** *is continuous if it is perpendicular to a boundary.*

For the toroid with the gap in it, the field B is normal across the boundaries forming the gap and so has the same value in the gap as in the iron. The field H differs in strength in the two regions. If its value is H_g in the gap and H_i in the iron, eqn (16.33) becomes

$$(2\pi a - w)H_i + wH_g = nI. \tag{16.38}$$

Since B is constant we also have

$$B = \mu_0 H_g = \mu\mu_0 H_i,$$

and combining these relations with eqn (16.38) gives

$$(2\pi a - w)\frac{B}{\mu\mu_0} + w\frac{B}{\mu_0} = nI \qquad (16.39)$$

and

$$B = \frac{\mu\mu_0 nI}{[2\pi a + w(\mu - 1)]} . \qquad (16.40)$$

Exercise 16.8 A toroid of radius 0.3 m is made of material that we assume to have a constant relative permeability of 1000. It is overwound with a coil of 800 turns and a current of 2 A is passed. A small gap of width 2 cm is cut in the toroid. Estimate the strengths of the fields B and H in the gap and in the core.

Answer The magnetic field B has the same strength in the gap and in the core and is equal to 0.092 T. The field H in the gap equals $7.32 \times 10^4 \, \mathrm{A\,m^{-1}}$, and in the core is equal to $73.2 \, \mathrm{A\,m^{-1}}$.

Exercises 16.7 and 16.8 illustrate how cutting a gap in a toroid, for the same current in the coils, reduces the strength of the magnetic field B compared with its strength in the complete toroid. When the second term in the denominator of eqn (16.40) is much greater than the first, the field B in the gap is given to a good approximation by

$$B = \frac{\mu_0 nI}{w} . \qquad (16.41)$$

The ratio of the field B in the gap to the field in the complete toroid is thus $\sim L/\mu w$, from eqns (16.37) and (16.41).

The expressions (16.37) and (16.40) do not give the fields to high accuracy, especially when there is a gap in the toroid. There are fringing fields around the gap that do not follow the circular pattern of field lines used in making the above estimates. The fringing fields reduce the field in the gap compared with the prediction of eqn (16.40).

Hysteresis

The permeability of ferromagnetic materials can only be considered constant over limited ranges of field strengths. The magnetic properties are usually presented in the form of **hysteresis** or **B–H curves**. These show corresponding values of the magnetic field B and the field H in the material as the field H is varied up and down from zero by

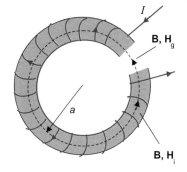

Fig. 16.28 The toroid of Fig 16.27 has a symmetrical gap cut in it in order to produce an accessible region of high magnetic field strength.

● *The hysteresis curve for a*
soft ferromagnetic material

varying the current in coils surrounding a sample such as a continuous toroid.

An example of a hysteresis curve for a typical 'soft' ferromagnetic material such as soft iron is shown in Fig 16.29. Starting at the point O where there is no current in the coils around a core, as H is increased the field B increases along the dashed line, over the part OP, more or less linearly. The value of B increases more slowly at higher fields H over the portion QRS of the curve, eventually saturating at a nearly constant value of the field B characteristic of the material. As the field H is reduced by reducing the current in the coils, the field B decreases along the full line shown. If the current in the coils is reversed, thereby reversing the direction of the field H, only a small field H is needed to reduce the field B to zero at the point C. The $B-H$ curve in the quadrant with negative B and H passes through saturation in the reverse direction. A reduction and second reversal of the field H generates the complete curve shown.

Most ferromagnetic materials have a saturation value of B of about 1 to 1.5 T, and macroscopic fields larger than this are produced in air using superconducting coils. These coils are made of very many super-conducting filaments of small radius and can sustain very large currents without power consumption or heating problems.

Worked Example 16.6 The soft iron used in an electromagnet has the $B-H$ curve shown in Fig 16.29. The magnet consists of a toroid of radius 0.3 m with a gap of width 2 cm cut in it. The toroid is overwound with a coil of 800 turns carrying a current of 2 A. Estimate the field B in the gap.

Answer The field B has the same value both inside the iron and in the gap. The field H has the value H_i inside and H_g in the gap. Using the relation $B = \mu_0 H_g$ and eqn (16.38), the following relation between the two fields in the iron must be obeyed:

$$(2\pi a - w)H_i + \frac{wB}{\mu_0} = nI.$$

With the numbers given in the problem this reduces to

$$1.865 H_i + 1.592 \times 10^4 B = 1600.$$

A second relation between B and H_i that must also be obeyed in the iron is the $B-H$ curve of Fig 16.29. The field strengths present in the iron are thus given by the intersection of the above straight line with the $B-H$ curve. This gives the field B in the iron to be about 0.1 T, and the field B in the gap has the same value.

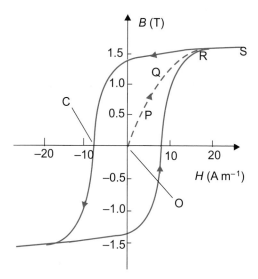

Fig. 16.29 The $B–H$ curve for a typical soft iron.

The bound magnetization currents in the material either side of the gap in an electromagnet are in directions that give a large force tending to close the faces of the gap, called the pole faces of the magnet. The magnet has to be rigid enough to resist these forces. The field produced by a large electromagnet will also induce a large magnetization in a nearby ferromagnetic material that will then suffer a force moving it into regions of large field strength. This force can be very large, and it is wise not to have an iron screwdriver in your pocket when going near a large magnet.

Electromagnets are used for a variety of purposes, from actuating switches to tracking the passage of the products of collisions between very high energy protons and other atomic particles. The radii of curvature of the paths of charged particles of known speed when moving in a known magnetic field determine their mass to charge ratios according to eqn (16.14).

Electromagnets rarely have the shape of the toroids used to simplify our discussions. They vary greatly in size from the very small, as in tape recorders, to the very large. Figure 16.30 is a sketch of a large superconducting magnet used in magnetic resonance imaging (MRI) to give images of parts of the human body. The magnet is in the form of a cylindrical solenoid of large bore. The person being imaged slides into the high field inside the solenoid on the trolley pictured in the foreground. The interaction between the magnetic field and the magnetic moments of the protons in the body's cells aligns the protons along the field direction. The aligned protons can absorb and re-emit energy from radiofrequency electromagnetic waves to give signals that computers are able to analyse to give images.

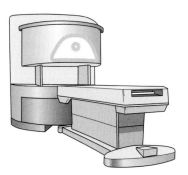

Fig. 16.30 A high-field electromagnet used in medical imaging.

Permanent magnets

Materials that retain magnetization after an external field has been reduced to zero are called *hard ferromagnetic materials*. A typical B–H curve for a hard ferromagnetic material, such as steel, is shown in Fig 16.31. If a toroidal steel specimen is overwound with coils and the current in the coils is increased from zero, the fields B and H in the steel increase over the dashed line shown in much the same way as for a soft ferromagnetic material. At high values of the solenoid current the steel saturates near the point S. However, the behaviour of a hard magnetic material differs from that of a soft material as the current is decreased to zero and reversed. The steel remains magnetized in the original direction for quite large reverse currents.

To reduce the field B to zero requires a relatively large reverse value H_c of the field H. Further increase in H in the reverse direction eventually saturates the field B in the reverse direction. Once more reducing H to zero leaves the steel with a remaining B field in the reverse direction. The full curve is completed as H is again raised to the point where the magnetic field B saturates in the forward direction. The hysteresis curve that takes the steel to saturation is called the *major* hysteresis curve. Minor hysteresis curves correspond to reduced ranges of the back and forward field H.

A permanently magnetized ferromagnetic material is a permanent magnet in which the magnetization is the source of a field B instead of a current in a coil surrounding soft iron, as in an electromagnet. For a given magnetization the fields B and H adjust to fit the geometry and the B–H curve of the magnetized material. Equation (16.33) must be obeyed and, if only permanently magnetized materials are present, the line integral of the field H is zero around any closed loop because there are no free currents anywhere. This does not mean that H is zero everywhere; it means that, if H is nonzero over part of the loop around which the line integral is taken, it must be nonzero in the reverse direction over some other parts of the loop in order for the whole integral to be zero.

As before, we consider a toroid with a gap cut in it, as shown in Fig 16.32, but now the toroid is not overwound with a current-carrying coil. Let the toroid have a permanent magnetization M. As before, we will neglect the small asymmetry caused by the gap and assume the lines of the field B are circles around the toroid. The field B is perpendicular to the pole faces and hence is continuous; it has the same value inside and outside the toroid. In the gap between the pole faces

$$B = \mu_0 H_g, \tag{16.42}$$

● *The hysteresis curve for a hard ferromagnetic material*

● *The field in a gap cut in a permanently magnetized toroid*

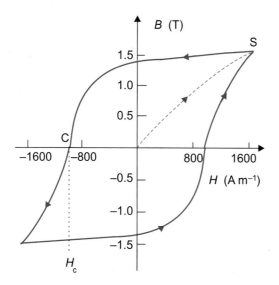

Fig. 16.31 The B–H curve for a typical hard steel.

and B and H_g are in the same direction. Applying eqn (16.35) to the circular dotted loop shown,

$$(2\pi a - w)H_i + wH_g = 0$$

where H_i is the field H in the steel. Substituting from eqn (16.42) gives

$$(2\pi a - w)H_i + \frac{wB}{\mu_0} = 0. \tag{16.43}$$

This relation must be obeyed by the fields B and H within the steel. The fields must also lie on the B–H curve of the material, and the actual fields can be determined by satisfying both relations.

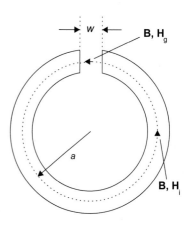

Fig. 16.32 A toroid of permanently magnetized material with a gap cut in it.

Worked Example 16.7 A steel whose hysteresis curve is shown in Fig 16.31 is to be used as a permanent magnet. Estimate the maximum field that can be produced in the gap of a toroidal sample of radius 20 cm and gap width 1 cm.

Answer Taking the steel up to its saturation magnetization and then reducing the current in the coil to zero will leave the steel with the maximum permanent magnetization when used as a permanent magnet. The relation (16.43) between the fields B and H must be obeyed inside

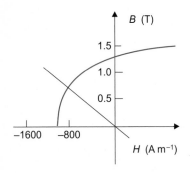

Fig. 16.33 The calculation of the field B in the gap of a permanent magnet made from the steel whose $B-H$ curve is given in Fig 16.31.

the steel toroid in which a gap has been cut. Here we have to take account of the fact that the field H is opposite to the field B inside the material in order to make the line integral of H around the toroid equal to zero. We then have two relations between B and H_i that have to be obeyed inside the steel. One is the eqn

$$(2\pi a - w)H_i + \frac{wB}{\mu_0} = 0.$$

The other is the $B-H$ curve of Fig 16.31. Where these two curves cross gives the fields inside, as illustrated in Fig 16.33. The field B inside is about 0.8 T and the field B outside has the same value.

Permanent magnets have a very wide range of applications. For example, very small magnets are used in tape recorders to write or read information on to magnetic tape, and the tape itself is made from magnetic material. Small magnets are used as door fasteners and somewhat larger magnets are widely used in the power sources in microwave ovens. Permanent magnets are used wherever small size is important or when the power needed to generate magnetic fields from currents is unavailable or difficult to provide.

Permanent magnets are usually made from alloys containing predominantly iron or cobalt. The magnetization of the material of a typical high-quality permanent magnet is close to $10^6 \, A \, m^{-1}$. There is a large ongoing research effort to produce materials that will enable larger fields than presently available to be obtained, and to enable fields of similar strength to be produced using magnets of smaller volume.

Problems

Level 1

16.1 A copper wire of cross-sectional area 1 mm^2 carries a current of 10^{-3} A. How many electrons cross a section of the wire per second? What is the average velocity of the free electrons in the conductor? What is the average time it takes an electron to advance 1 cm along the wire? The atomic weight of copper is 63.5 and its density is $8.9 \times 10^3 \, kg \, m^{-3}$. Assume that there is one free electron per copper atom.

16.2 An electron synchrotron accelerates electrons that travel in an approximately circular path 220 m long at almost the speed of light. If there are 10^{12} electrons on this path during the acceleration cycle, what is the beam current?

16.3 Two square metal plates, each of side 10 cm, are placed parallel and 10 cm apart in air. A potential difference is applied acrosss the plates and as it is increased the current is observed to level off at a value of

10^{-14}A. Assuming that the only ionizing agency is atomic particles from space that penetrate the atmosphere (cosmic rays) and that the mean flux of these particles through any surface is one per cm^2 per minute, integrated over all directions, estimate the mean number of ion pairs formed by a cosmic ray per cm.

16.4 A coaxial cable, similar to a cable that brings a signal from an antenna to a television set, consists of an inner cylindrical wire conductor surrounded by an earthed sheath of larger radius, the two being separated by an insulating material. The inner cylindrical conductor of a very long coaxial cable has a radius a and the earthed sheath is a thin metal outer cylinder of radius b. The two cylinders are separated by non-magnetic spacers. A current I flows in one direction down the inner conductor and an equal and opposite current flows in the opposite direction in the outer conductor. Determine the strength and direction of the magnetic field B between the conductors and outside the cable.

16.5 A horizontally moving, singly charged lead ion of energy 10 keV enters a region in which there is a magnetic field of strength 10^{-3} T in the horizontal direction but at right angles to the ion's motion. Calculate the ratio of the magnetic and gravitational forces on the ion. Take the atomic weight of lead to be 208.

16.6 A toroidal ring of soft iron has a mean diameter of 20 cm and is overwound uniformly with 400 turns of wire. If the relative permeability of iron is assumed to be constant at 500, find the magnitudes of the magnetic fields B and H in the iron when a current of 3 A is passed through the coil. If a gap of 3 mm is cut in the iron, find the directions and magnitudes of the fields B and H in the gap and in the iron.

16.7 An electron moves with 2% of the speed of light at an angle of $60°$ to a magnetic field of 10^{-9} T. How many revolutions does it make while travelling between two stars 6 light years apart along a line of magnetic field?

16.8 Singly charged ions (with net charge equal to the magnitude of the charge on an electron) of the stable sodium atom with mass 23 amu are accelerated to an energy of 18 MeV in a nuclear accelerator. They enter a region in which there is a constant magnetic field B perpendicular to their velocity. What is the magnitude of the field needed to bend the ions in a circle of radius 1 m?

16.9 A very long, hollow, metal tube with inner radius a and outer radius b carries a uniformly distributed current I in the direction along its length. What is the magnetic field outside and inside the tube?

16.10 If the Earth's magnetic field is produced by circulating currents in the core, and the mean radius of such currents is 1000 km, make a rough estimate of the size of the mean current required to produce the field at the North magnetic pole of 6×10^{-5} T. The radius of the Earth is 6.4×10^6 m.

Level 2

16.11 A regular hexagon has sides of length ℓ and carries a current I. Show that the magnitude of the magnetic field B at the centre of the hexagon is $\sqrt{3}\mu_0 I/\pi\ell$.

16.12 The ground state of the stable sodium nucleus has a magnetic dipole moment of 1.1×10^{-26} A m^2. Calculate the difference in energy of the two states that correspond to the dipole moment aligned with, and then against, the magnetic field due to a single electron orbiting the nucleus in the innermost shell. (This field is given in Exercise 16.4.) What frequency of electromagnetic radiation does this energy difference correspond to?

16.13 Two long parallel wires, in a horizontal plane and separated by a distance $2d$, each carry a current I in opposite directions. Derive the magnetic field distribution in a vertical plane midway between the wires. If z is the vertical distance in this plane above the line joining the wires, at what value of z is the field gradient a maximum?

16.14 A thin disc of radius a rotates at angular speed ω. It is made of insulating material and has a charge q per unit area. What are the magnitude and the direction of the magnetic field B produced at the centre of the disc? What are they at a distance x from the disc along the axis of the disc?

16.15 Two thin metal plates, which are very long and wide compared with their separation, carry equal and opposite currents I per unit length. What is the force per unit area on each plate?

16.16 A small coil with magnetic moment 0.1 A m^2 is coaxial with a larger coil of radius 20 cm that carries a current of 10 A. The small coil is 25 cm from the centre of the large coil. What is the force on the small coil?

16.17 A long straight wire carries a current of 1 A. A rectangular coil 2 cm by 4 cm carrying 3 A lies in a plane containing the long wire with the 4 cm sides parallel to the long wire. The nearest of the two 4 cm sides is 2 cm away from the long wire. What is the force on the coil? What is the torque on the coil about an axis through its centre and parallel to the long wire?

16.18 At very large magnetic fields, nickel, which is ferromagnetic at normal temperatures, saturates, that is, as the field H is increased by increasing the current in coils wound round the nickel, there comes a point where the magnetization M no longer increases. At this point for nickel, corresponding values of the fields B and H are 1.3 T and 5×10^5 A m^{-1}, respectively. Estimate the magnetic moment of a nickel atom in Bohr magnetons. The density of nickel is 9×10^3 kg m^{-3}.

16.19 A toroidal ring of soft iron has a mean diameter of 20 cm and is overwound uniformly with 400 turns of wire. A gap of width 3 mm is cut in the iron. If the $B-H$ curve for the iron is given by the table below and a current of 3 A is passed through the coil, what is the magnetic field B in the gap?

H (A m^{-1})	40	80	160	240	320	480	800	1600
B(T)	0.1	0.2	0.6	0.85	1.0	1.2	1.4	1.5

What is the power needed to maintain a field of 1 T in the gap if the resistance of the coil is 0.25 Ω?

16.20 A soft iron electromagnet is in the shape of a circular ring of radius 0.25 m with a gap of width 4×10^{-3} m in it. A current-carrying coil is wound around its length at the rate of 1000 turns per metre. The relation between the fields B and H in the iron is

$$B = 3 \tanh[(H + 180)/400],$$

with B in teslas and H in A m^{-1}. Calculate the magnitude of the field B in the gap when the current has been reduced to 0.8 A.

Level 3

16.21 The inner conductor of one end of a coaxial cable of length ℓ is joined to one terminal of a battery of emf V_0 and the other end is connected to earth by a resistor of value R_1. The battery has internal resistance R_2 and its second terminal is earthed. The cable is faulty and there is a leakage current to the earthed outer conductor at the

rate of $\sigma V(x)$ per unit length at distance x from the battery terminal, with $V(x)$ the voltage at position x. If the resistance per unit length of the inner conductor is ρ, what is the current distribution $I(x)$ along the inner conductor as a function of x?

16.22 A magnet consists of a small cylinder of steel magnetized along its axis. It has magnetic dipole moment **m** and is set spinning about its axis with angular momentum **L**. Show that in a uniform magnetic field **B** it will precess at a rate independent of the angle between **L** and **B**.

16.23 A moving belt that is very long and wide has a thickness d. A charge q per unit length is uniformly distributed throughout its volume. The belt is moving at speed v. Determine the magnetic field inside and outside the belt. How would the field inside and outside change if the charge were carried on the two surfaces?

16.24 A long cylindrical straight wire has radius a_1. A cylindrical cut of radius a_2 is made in the metal wire as shown in Fig 16.34. The distance of the axis of the cut from the axis of the wire is b_1. A current I is passed through the wire and is uniformly distributed throughout the metal. Calculate the magnetic field at the point P, which is a perpendicular distance b_2 from the axis of the wire and on a line passing through the axis of the cut.

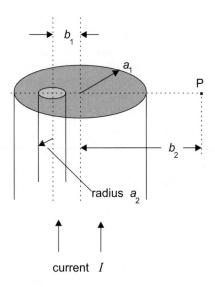

Fig. 16.34 A long, straight wire carrying a uniformly distributed current. The wire has a cylindrical cut in it parallel to the axis.

16.25 A solenoid of length ℓ and radius a has n_ℓ turns per unit length and a current I. Use the Biot–Savart law to determine the field on the axis of the solenoid at a point distant z from an end.

16.26 A circular coil of n turns and radius a carries a current I. A magnetic dipole of strength m lies on the axis of the coil at a distance x. The dipole points towards the coil and the current in the coil circulates clockwise as seen from the dipole. Calculate the force on the coil.

16.27 A spherical nucleus of radius a has a total charge Ze and is rotating at angular speed ω. The charge is uniformly distributed throughout the volume of the nucleus. What is the nuclear magnetic dipole moment?

16.28 A small magnetic dipole of moment m is fixed at the origin of coordinates with its dipole moment pointing in the vertical z-direction. A second small dipole of moment m' and moment of inertia J is freely suspended at a point on the z-axis distant d from the origin. Find the period of small oscillations of the suspended dipole in the vertical plane.

16.29 A square loop of wire of side $2a$ carries a current I. It sits in the xz-plane with its centre at the origin of coordinates. Show that the magnetic field B at the point $(0, b, 0)$ has magnitude

$$\frac{2\mu_0 I a^2}{\pi(a^2 + b^2)\,\sqrt{(2a^2 + b^2)}}$$

and lies in a direction along the y-axis.

16.30 A cylindrical permanent magnet of radius a and length ℓ is uniformly magnetized along its length with a magnetization M. Show that the field B on the axis at a point outside the magnet is

$$B = \tfrac{1}{2}\mu_0 M(\cos\alpha - \cos\beta)$$

where α and β are the semi-angles subtended at P by the circular faces of the magnet.

Some solutions and answers to Chapter 16 problems

16.2 0.22 A.

16.3 When the current has levelled off, the rate at which ion pairs are formed by the cosmic rays is equal to the rate at which they are removed by the current flow. We make the approximation that the only ion pairs collected at the plates arise from interactions of cosmic rays with air molecules inside the square box of which the two plates form two sides. The isotropic distribution of cosmic rays then corresponds to one per minute per square centimetre passing through the surface area of the box. These 10 per second traverse 10 cm of air each on average and, if the mean number of ion pairs formed per cm by a cosmic ray is α, 100α pairs are formed and removed each second. The current of 10^{-14} A corresponds to $10^{-14}/(1.6 \times 10^{-19} \times 2)$ ion pairs, and hence

$$100\alpha = 10^{-14}/(1.6 \times 10^{-19} \times 2),$$

giving $\alpha = 313$.

16.5 4.5×10^6.

16.7 The distance between the stars is 5.67×10^{16} m. The time taken ΔT to go this distance is $5.54 \times 10^{16}/v_{\parallel} = 1.89 \times 10^{10}$ s where $v_{\parallel} = 3 \times 10^6$ m s^{-1}.

The radius a of the orbit in a plane perpendicular to the field is given by eqn (16.14) with $v_{\perp} = 5.20 \times 10^6$ m s^{-1}. The circumference of the orbit is $2\pi a$ and the time $\Delta T'$ for one revolution in a plane perpendicular to the field is $2\pi a/v_{\perp} = 0.036$ s. The number of revolutions is

$$n = \frac{\Delta T}{\Delta T'} = 5.3 \times 10^{11}.$$

16.9 The current in the tube is made up of many currents in thin filaments running along the tube and thus the field at any distance along the cylinder is in the plane perpendicular to the tube at that point. The symmetry of the cylinder tells us that the field is the same at any point on a circle in this plane with centre on the tube axis. Applying Ampére's law to such a circle inside the tube shows that the field inside is everywhere zero since the current through the area of the circle is zero. Outside the tube the same arguments show that the field is the same as the field due to a current I in a thin wire along the axis.

16.13 Figure 16.35 shows a view looking along the wires. We require the field at a point P distant z from the midpoint of the line joining the wires, which are labelled A and B. The magnitudes of the separate fields $\mathbf{B_A}$ and $\mathbf{B_B}$ at P from wires A and B are both equal to $I/2\pi\ell$, where $\ell = AP = BP = \sqrt{d^2 + z^2}$, and their directions are as shown. The resultant field is

$$B = \frac{I}{\pi\ell}\cos(\theta + \phi).$$

The geometry of the figure shows that $\phi = \pi/2 - 2\theta$. Hence $\cos(\theta + \phi) = \sin\theta$, giving

$$B = \frac{\mu_0 I d}{\pi\ell^2} = \frac{\mu_0 I d}{\pi}(d^2 + z^2)^{-1}$$

The maximum field gradient occurs at $z = d$.

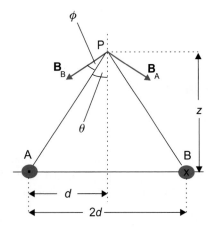

Fig. 16.35 The magnetic field at points equidistant from two long, straight, parallel wires carrying a current I in opposite directions.

16.15 $\mu_0 I^2/2$.

16.17 The force is 6×10^{-7} N perpendicular to the long wire and in the plane containing the long wire and the rectangular loop. The couple on the loop is zero when its plane contains the long wire.

16.19 Equation (16.38) is a relation obeyed by the fields H in the iron and in the gap. In the air gap the field $B_g = \mu_0 H_g$. Also the field B is continuous across the faces of the gap and $B_i = B_g$. Hence one relation obeyed

by the fields B and H inside the iron is

$$(2\pi a - w)H + w\frac{B}{\mu_0} = nI$$

where the symbols a, w, n, and I have the same meaning as in eqn (16.38). The curve of the $B{-}H$ data given is another relation that the fields B and H inside the iron have to obey, and the actual fields are determined by the intersection of the line

$$1.26H + 2390B = 1200$$

with the curve. This is best determined graphically. The result is that the field B in the gap is close to 0.43 T.
    The calculation of the power is left to the student.

16.22 The torque τ on the magnetized cylinder, given by eqn (16.26), is equal to the rate of change of angular momentum $d\mathbf{L}/dt$ of the dipole, from eqn (4.25), and

$$\frac{d\mathbf{L}}{dt} = \mathbf{m} \times \mathbf{B}.$$

The direction of the change in angular momentum $\Delta\mathbf{L}$ is perpendicular to both $\mathbf{m}$ and $\mathbf{B}$, that is, to both $\mathbf{L}$ and $\mathbf{B}$, since $\mathbf{L}$ and $\mathbf{m}$ are parallel. The direction of $\Delta\mathbf{L}$ is thus in the plane of the axes 1 and 2 shown on Fig 16.36, in which the angular momentum $\mathbf{L}$ is drawn in the plane containing the magnetic field $\mathbf{B}$ and axis 1. Resolve the angular momentum into components $L\cos\theta$ along $\mathbf{B}$

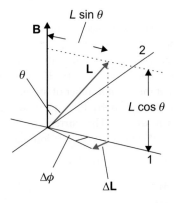

Fig. 16.36 In a time interval Δt, the magnitude of the angular momentum of the magnetized cylinder changes by the amount ΔL.

and $L\sin\theta$ perpendicular to **B** along axis 1. In a time Δt the angular momentum changes by an amount ΔL in the 1−2 plane as shown in Fig 16.36.

The magnitude of the change in angular momentum in time Δt is $\Delta L = L\sin\theta\Delta\phi$, where $\Delta\phi$ is the angle shown on the figure, and is equal to the angle through which the angular momentum has precessed. The above equation for the rate of change of angular momentum now becomes

$$\frac{\Delta L}{\Delta t} = \frac{L\sin\theta\Delta\phi}{\Delta t} = mB\sin\theta.$$

Hence

$$\frac{\Delta\phi}{\Delta t} = \frac{mB}{L},$$

and the rate of precession is independent of the angle between **L** and **B**.

16.25

$$B = \tfrac{1}{2}\mu_0 n_\ell I \left(\frac{z+\ell}{\sqrt{a^2 + (z+\ell)^2}} - \frac{z}{\sqrt{a^2 + z^2}} \right).$$

16.27 Take the z-axis to be the axis of rotation and the origin to be at the centre of the sphere as shown in Fig. 16.37. Consider the disc of thickness $\mathrm{d}z$ cut through the sphere perpendicular to the z-axis at distance z. This disc has radius $\sqrt{a^2 - z^2}$. The dipole moment $\mathrm{d}m$ of the disc is made up from the moments of circulating currents in all the rings of width $\mathrm{d}r$ and radius $2\pi r$ that constitute the disc. The total charge in the ring of radius r is $\rho(2\pi r)\,\mathrm{d}r\,\mathrm{d}z$, where ρ is the volume charge density,

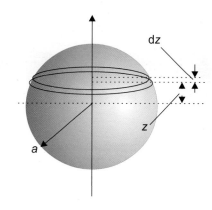

Fig. 16.37 A rotating, uniformly charged sphere.

and the current circulating in the ring of radius r is this charge times $\omega r/2\pi r = \omega r\rho\,\mathrm{d}r\,\mathrm{d}z$. Hence,

$$\mathrm{d}m = \omega\rho\pi \int_0^{\sqrt{a^2-z^2}} r^3\mathrm{d}r\,\mathrm{d}z = \tfrac{1}{4}\omega\rho\pi(a^2 - z^2)^2\mathrm{d}z.$$

The total moment of the sphere m is now obtained by integrating over z.

$$m = \tfrac{1}{4}\omega\rho\pi \int_{-a}^{a} (a^2 - z^2)^2 \,\mathrm{d}z = \frac{\omega a^2}{5}(Ze).$$

16.28

$$T = 2\pi\sqrt{\frac{2\pi d^3 J}{\mu_0 mm'}}.$$

Chapter 17

Induced electric fields

This chapter discusses the electric fields that arise when conductors move in magnetic fields or when magnetic fields change with time, and some effects related to these induced electric fields.

Steady currents produced by constant electric fields give rise to steady magnetic fields, as described in the last chapter. There we were concerned only with situations where both the electric and the magnetic fields do not change with time and the circuits in which the currents flow are stationary. The force on a moving charge in steady fields was described assuming that the time-varying fields produced by the moving charge were negligible compared with the fields in which the charge was moving.

This chapter deals with some phenomena that arise when currents and fields are not constant in time. We consider what happens, first, when conductors move in magnetic fields and, second, when the magnetic field varies with time over stationary conductors. In the former case, when a closed circuit made of conducting wire moves through a magnetic field that is constant in time but changes with position, there are forces on the moving charges due to the magnetic field in which they move and these forces may give rise to currents. In the latter case, the magnetic flux through a closed circuit changes with time and the time-varying magnetic field gives rise to an electric field that exerts forces on charges just like the electrostatic field does.

The electric fields produced when conductors move in magnetic fields or magnetic fields change with time are called **induced electric fields** and the phenomenon is called **electromagnetic induction**.

17.1 **Moving conductors**

If the free electrons in a conductor can be partially separated so that more congregate at one end than at the other, there will be a difference in electric potential, a voltage difference, between the two ends, and an electric field inside the conductor. If a metal rod of length ℓ moves at velocity $\mathbf{v}$ in a steady magnetic field as shown in Fig 17.1, there is a force $\mathbf{F}$ on an electron in the rod given by eqn (16.12) as

$$\mathbf{F} = -e\mathbf{v} \times \mathbf{B}. \qquad (17.1)$$

This acts along the rod in the direction from C to D and pushes negative charge over towards the end D. Negative charge will build up at D, leaving excess positive charge at C, until the electric field $\mathbf{E}$ in the rod produced by the imbalance of charge gives a force on an electron equal and opposite to that given by eqn (17.1). In this steady state,

$$E = vB \qquad (17.2)$$

and the potential difference between the ends of the rod is $vB\ell$.

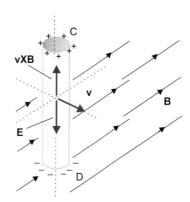

Fig. 17.1 The displacement of charges when a conducting rod moves at right angles to a uniform magnetic field.

Exercise 17.1 An automobile of width 1.5 m is moving in the Northern hemisphere in the direction of magnetic north at a speed of 100 km per hour. The magnitude of the Earth's magnetic field at the position of the car is 6×10^{-5} T. The field lines point into the Earth making an angle of $50°$ to the horizontal. The vehicle's wheels are insulated from the ground. Estimate the voltage difference between the two sides of the metal car? Which side is at the higher voltage? What happens if the car is in the Southern hemisphere?

Answer 1.9×10^{-3} volts. The western side of the car is at a higher potential than the western side. If the car is in the Southern hemisphere, the electric field across it has the same magnitude but is in the opposite direction.

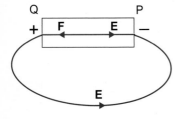

Fig. 17.2 A battery with terminals P and Q. The line integral of the electrostatic field around the closed loop shown is zero.

Electromotive force

With a conductor of resistance R connected to the terminals of a battery of electromotive force V_B, a current $I = V_B/R$ flows. Electromotive force was introduced briefly in Section 16.1. Here we discuss it at more length. Consider an idealized battery that has no internal resistance with terminals at the points P and Q, as in Fig 17.2. The emf of the battery is the integral around a complete circuit, which includes P and Q, of the total force $\mathbf{F}_T$ on a positive charge q, divided by the charge q.

$$\text{emf} = \frac{1}{q} \oint \mathbf{F}_T \cdot d\boldsymbol{\ell}. \tag{17.3}$$

The total force on the charge q is the resultant of the force from the electrostatic field $q\mathbf{E}$ and the force $\mathbf{F}$ due to the mechanism within the battery that converts chemical energy into electrical energy. Hence

$$\text{emf} = \frac{1}{q} \oint (\mathbf{F} + q\mathbf{E}) \cdot d\boldsymbol{\ell}.$$

The force $\mathbf{F}$ exists only within the battery and, since the electric field is electrostatic, its line integral around the closed circuit is zero. The above equation for the emf thus reduces to

$$\text{emf} = \frac{1}{q} \int_P^Q \mathbf{F} \cdot d\boldsymbol{\ell}.$$

This expression is $1/q$ times the work W done to move the charge from P to Q, since the only agency available for work is the battery. However, W is also given by

$$W = \int_Q^P q\mathbf{E} \cdot d\boldsymbol{\ell} = q(V_Q - V_P).$$

The emf of the battery, V_B, is thus equal to the potential difference between the terminals,

$$\text{emf} = V_B = V_Q - V_P. \tag{17.4}$$

Complete circuits

If the moving rod of Fig 17.1 is made part of a complete conducting circuit, the remainder of the circuit being stationary, a flow of charge can be maintained around the circuit under the influence of the magnetic force on the electrons over the side CD. A current flows just as if there were a battery in the loop. This current depletes the charge at the ends of the rod and reduces the electric field in the rod below the value vB given by eqn (17.2).

The current in the complete conducting circuit can be written in terms of an electromotive force. If the instantaneous value of this emf is V and the resistance is R the instantaneous value of the current is

$$I = V/R. \tag{17.5}$$

The emf V in the circuit that contains the moving rod can be obtained from eqn (17.3). The electric field is electrostatic and its line integral around the complete loop is again zero. The only contribution to the line integral around the loop of the total force on a charge q comes from the force due to the magnetic field over the moving rod, and

$$V = \frac{1}{q} \int_C^D \mathbf{F} \cdot d\boldsymbol{\ell}$$

where $\mathbf{F}$ is given by eqn (17.1), after replacing the charge $-e$ by the charge q, and C and D are the ends of the rod. Hence

$$\text{emf} = vB\ell. \tag{17.6}$$

It may be noted that the emf given by this equation is the rate at which magnetic flux is swept out by the rod. It is the area swept out by the rod each second times the magnetic field B.

Homopolar generators

We now discuss a second example of an emf generated by the motion of a conductor in a magnetic field. Consider a metal disc rotating at angular speed ω. Let there be a magnetic field of constant magnitude B over the surface of the disc and in a direction perpendicular to the plane of the disc, as shown in Fig 17.3. If sliding contacts are placed at the centre and edge of the disc and a closed circuit completed between them, a current will flow in the closed circuit. This arrangement is called a homopolar generator.

● *Motional emf*

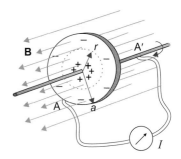

Fig. 17.3 A metal disc of radius a rotates on a thin metal spindle at angular speed ω. A complete electrical circuit through which a current I flows is made by sliding contacts to the disc at A and to the spindle at A'. A uniform magnetic field exists in a direction perpendicular to the plane of the disc.

At a distance r from the centre of the rotating disc the free electrons have an average speed ωr and they experience a force of magnitude $evB = e\omega rB$ from the magnetic field. The direction of the force is from the centre towards the rim. The force varies with distance from the centre and the net effect is for electrons to move to the rim and build up excess negative charge there until the electrostatic repulsion prevents further electron migration. When this happens there is an electric field in the direction from the centre to the rim that at each point balances the force due to the electrons moving in the magnetic field. There is thus a potential difference between the centre and the rim given by

$$V_0 - V_{\text{rim}} = \int_0^a \omega Br \, dr = \tfrac{1}{2}\omega Ba^2. \tag{17.7}$$

● *Drive power*

This potential difference causes a current to flow in an external circuit connected to sliding contacts on the rim and on the spindle on which the disc is rotated. The voltage difference given by eqn (17.7) is the emf that drives the current, and the power dissipated if the complete circuit has resistance R is $(V_0 - V_{\text{rim}})^2/R$. This is the rate at which the agency rotating the disc must do work in order to keep the disc at constant angular speed.

One way in which very large currents can be generated for short times is to build up the angular speed of a disc of large moment of inertia rotating in a magnetic field until the energy stored in the disc is large. If contacts are placed as described above and a complete circuit is formed between them, the energy stored in the disc can rapidly be dissipated as a large current flows.

Worked Example 17.1 A motor drives a large metal flywheel with moment of inertia J to a high angular speed ω_0. A magnetic field B is constant over the area of the flywheel and in a direction perpendicular to the plane of the wheel. At a particular time the motor is disconnected from the wheel and sliding electrical contacts are made to the rim of the wheel and to the metal spindle driving the wheel. The circuit between the contacts is completed with a load of resistance R. What is the initial current through the load and how long does it take for the current to reduce to one-half of the initial value?

Answer The voltage difference between the rim and the spindle is given by eqn (17.7), and at the instant the drive is disconnected and the electrical circuit completed the initial current is

$$I_0 = \frac{1}{2R}\omega_0 Ba^2.$$

A current I through the load R at time t dissipates energy at the rate I^2R. This must be provided at the expense of the rotational energy, E, of the flywheel. Hence

$$\frac{dE}{dt} + I^2R = 0. \tag{17.8}$$

If the rotational speed at time t is ω, the rotational energy is $E = \frac{1}{2}J\omega^2$ and the current is $I = \omega Ba^2/2R$. Thus

$$E = \frac{2JR^2I^2}{B^2a^4},$$

and substitution, after differentiating, in eqn (17.8) gives

$$\left(\frac{4JR}{B^2a^4}\right)\frac{dI}{dt} + I = 0.$$

The solution to this differential equation is

$$I = I_0 e^{-t/\tau}$$

where

$$\tau = \frac{4JR}{B^2a^4}.$$

The current thus falls to one-half its initial value in a time $(4JR/B^2a^4)$ times $\ln 2$.

Large currents can be generated by rapidly spinning a large flywheel. For example, a flywheel with moment of inertia $200\,\text{kg m}^2$ and radius $1.5\,\text{m}$ initially rotating at 50 revolutions per second in a magnetic field of $0.5\,\text{T}$ can give an initial current of nearly $2 \times 10^4\,\text{A}$ through a load of $10^{-2}\,\Omega$. This resistance corresponds to about $50\,\text{m}$ of copper cable of diameter $5\,\text{mm}$, a typical amount of cable used in the windings of a large electromagnet.

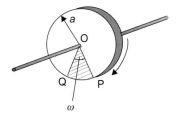

Fig. 17.4 In one second, the point P on the circumference of the disc of Fig 17.3 has rotated from the point P to Q through an angle ω. The radius OP sweeps out the hatched area shown of size $\omega a^2/2$.

⬤ *The motional emf equals the rate of sweeping out flux*

The emfs generated as a rod moves linearly through a magnetic field, or the radius of a wheel rotates in a field, are in both cases equal to the rate at which magnetic flux is swept out by the moving conductor. The emf in a closed loop, given by eqn (17.3), is the line integral around the loop of the total force on a charge q divided by q. For voltage differences produced by motion of conductors in magnetic fields, the line integral reduces to the integral of the force due to the magnetic field over that part of the path occupied by the moving conductor.

In the case of the moving rod of Fig 17.1, the emf is equal to $vB\ell$, as in eqn (17.6). The magnetic flux swept out by the moving rod each second is given by eqn (16.29) and also equals $vB\ell$. Similarly, in one second the rotating radius of the flywheel moves from P to Q on Fig 17.4 through an angle ω, and sweeps out the hatched area shown. This area is equal to $\omega a^2/2$. The rate at which flux is swept out is thus $\omega B a^2/2$, the same as the voltage between the centre and the rim of the wheel.

In the more general situation in which a conductor moves at variable velocity in a non-uniform magnetic field the potential difference between the ends is also equal to the instantaneous rate at which magnetic flux is swept out by the moving conductor.

17.2 Faraday's law

The previous section considered voltage differences produced when conductors move in a magnetic field. If a closed conducting circuit, made of thin uniform metal wire for example, is stationary in a static magnetic field, no current flows; the forces on the randomly moving free electrons in the wire average to zero. If the closed loop moves as a whole, the velocity of the loop is superimposed upon the random motions of the free electrons. There may then be a net effect driving electrons one way or the other round the loop when the magnetic force is integrated around the closed circuit. There will be such an effect giving an induced emf if the spatial variation of the field gives rise to a change of flux through the loop as it moves.

Let us now consider a different situation in which the closed conducting wire is stationary but the magnetic field over it changes with time. *It is an experimental fact that when this happens an emf is induced in the loop and a current flows in the wire.* This can be demonstrated with a large coil of small resistance in series with a device that measures charge flow. If a magnetic field in which the coil is situated is switched off rapidly, charge flows around the circuit. When the plane of the coil is roughly perpendicular to the magnetic field, as in Fig 17.5, the charge flow is larger than when the plane of the coil is inclined to the average field direction, as in Fig 17.6. In the former situation the magnetic flux through the coil is larger than in the latter.

Quantitative experiments on the charge flow under different conditions were first made by Michael Faraday and others in the early 1830s. These experiments resulted in **Faraday's law**, which relates the induced emf in a closed loop to the rate of change with time of the magnetic flux through the loop. The induced emf around a closed loop is defined in analogy with the definition of the emf of a battery given in eqn (17.3). The total force $\mathbf{F}_T$ on a positive charge q now becomes $\mathbf{E}q$ and the line integral in eqn (17.3) becomes the line integral of the electric field around the complete loop.

The induced emf around a closed loop is the line integral of the electric field around the loop.

● *Static field—moving circuit*

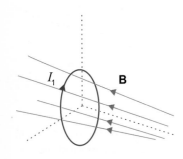

Fig. 17.5 A magnetic field over a fixed coil. When the field is quickly switched off a current I_1 is induced in the coil.

● *Static circuit—changing field*

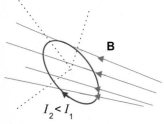

Fig. 17.6 When the coil of Fig 17.5 is inclined to the magnetic field direction, the current I_2 induced when the field is quickly switched off is less than the current I_1 induced with the coil positioned as in Fig 17.5.

This emf produces a current $I = \text{emf}/R$ if the loop is a conductor of resistance R, just as a battery of emf V produces a current V/R in a resistor of resistance R.

The conducting loop is not being moved by an agency that supplies energy, nor is there a mechanism within the circuit for the conversion of chemical or mechanical energy to electrical energy. The force on the moving charges in the loop is due to an electric field induced by the changing magnetic field, and the source of energy is the agency producing the changing magnetic field.

The induced electric field is not an electrostatic field; its line integral around a closed loop is not zero. Faraday's experiments showed that the induced emf in a stationary loop when the magnetic field over the loop changes with time is proportional to the rate of change of the magnetic flux through the loop,

$$\oint \mathbf{E} \cdot d\boldsymbol{\ell} \propto \frac{d\Phi}{dt}. \tag{17.9}$$

In words,

If the magnetic flux through a stationary closed loop changes with time, an emf is induced in the loop that is proportional to the rate of change of the flux through the loop.

The constant of proportionality has unit magnitude in the SI system of units.

Lenz's law

We now consider the direction of the induced emf. The direction predicted by eqn (17.9) depends upon the sign used in that equation. If the loop shown in Fig 17.7 is viewed from the right and is circulated clockwise when evaluating the line integral, the sign of the magnetic flux through the loop is positive if the magnetic field is from right to left and negative if the field is from left to right, as discussed in Section 16.4. Hence, the sign of $d\Phi/dt$ depends on the direction of the field and on whether the field is increasing or decreasing with time. The magnetic flux is positive in the situation shown in Fig 17.7 and, if the field decreases with time, $d\Phi/dt$ is negative.

What is the prediction for the direction of the induced electric field if a negative sign is placed in front of the right-hand side of eqn (17.9)? The clockwise sense of circulation of the loop defines, via the right-hand rule, the direction of the vectors associated with the elementary surface areas comprising the surface area of the loop. This direction and that of the field give the sign of the magnetic flux and, in turn, the time variation of the field gives the sign of $-d\Phi/dt$, which is the sign of the line integral of

● *The sign in eqn (17.9)*

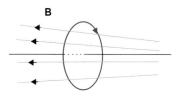

Fig. 17.7 The closed loop shown is viewed from the right. The magnetic flux through the loop due to the field B is positive when the directions of the elementary areas across the surface are defined by circulating the loop clockwise.

the induced field around the loop. If $-\mathrm{d}\Phi/\mathrm{d}t$ is positive, the induced field is clockwise, since we are circulating the loop clockwise. If $-\mathrm{d}\Phi/\mathrm{d}t$ is negative, the induced field is anticlockwise. If a positive sign is placed in front of eqn (17.9), the converse would be true.

Is the inserted sign to be positive or negative? It is an experimental fact that the induced emf in a closed conducting circuit produces a current that generates a magnetic flux opposing the externally imposed change. If the external field is decreasing, thus reducing the flux through the circuit, the induced emf and current are in the direction that gives a magnetic field in the same direction as the original field. If the external field is increasing, the induced emf and current are in the direction that gives a magnetic field in the opposite direction to the original field.

Let us return to Fig 17.7. The sign of Φ is positive and, if the field decreases with time, the rate of change of flux $\mathrm{d}\Phi/\mathrm{d}t$ is negative. The induced field and the current it produces are in a direction that generates a flux opposing the external change. Since the external magnetic field is decreasing, the magnetic field produced by the induced current will be in the same direction as the external field. Thus the induced electric field and current are clockwise and the line integral of the electric field is positive. However, $\mathrm{d}\Phi/\mathrm{d}t$ is negative. We thus require a negative sign on the right-hand side of eqn (17.9) to give the induced field in the observed direction, and Faraday's law can now be written in its complete form as

● **The complete form of Faraday's law**

$$\oint \mathbf{E} \cdot \mathrm{d}\boldsymbol{\ell} = -\frac{\mathrm{d}\Phi}{\mathrm{d}t} \tag{17.10}$$

to give the relation between the induced electric field and the flux change producing it.

The direction of the induced emf is encapsulated in **Lenz's law**, named after Heinrich Lenz who enunciated it in 1834. Lenz's law states in words the result formalized by the negative sign in the above equation.

● **Lenz's law**

An induced emf is in the direction that opposes the magnetic field change producing it.

Lenz's law is consistent with a general principle proposed by Le Chatelier in 1833 that, if a system is in equilibrium and is disturbed, it reacts by trying to restore the original condition. Thus, if the magnetic field over a coil is reduced, the current induced in the coil is in the direction that produces a magnetic field opposing the external reduction.

Equation (17.10) is the mathematical statement of Faraday's law. Faraday's law, together with Gauss's law discussed in Chapter 15, Ampère's law discussed in Chapter 16, and the observation that there are no free magnetic poles, form the basis of the theory of electromagnetism, which is encompassed in a set of equations called **Maxwell's equations**.

Michael Faraday

Michael Faraday was born on 22 September 1791 in a village now subsumed into greater London on the south side of the river Thames. He was the third child of a working blacksmith and a farmer's daughter. He received very little schooling, and in 1804 he went on a 12 month trial as errand-boy to a bookseller and stationer. At the end of the year, having made a very favourable impression with his intelligence and industry, he was taken on as apprentice bookbinder, stationer, and bookseller. He became interested in science and began to teach himself by assiduous reading of books that passed through his place of work. He wrote 'Whilst an apprentice I loved to read scientific books which were under my hands.'

At the end of his seven-year apprenticeship he sought a post in science and was taken on in 1813 as assistant by Humphrey Davy, the distinguished chemist. He accompanied Davy on a journey within Europe and, returning after 18 months, began lecturing at the Royal Institution in London. For over 40 years he lectured there, developing public interest in science, and became the best scientific expositor of his generation in Britain. In 1817 he began research on his own, and until 1830 chemistry was his main area of activity, although during that interval he performed many experiments on the newly discovered magnetic effects of currents. In 1830 he gave up lucrative commissions from the industry of the time, in order to concentrate on his own curiosity-driven research. He began a series of experiments culminating in the demonstration that magnetism could generate electricity, being convinced that Nature had sufficient symmetry to exhibit the inverse effect to electric currents generating magnetism. His discoveries spawned almost all methods used to generate electric power, most of which still depend upon the electric dynamo, and the world was never the same again.

In 1821 Faraday married Sarah Barnard, the sister of a friend, and a member of a small religious body called the Sandemanians that split from the Scottish Presbyterian Church in the middle of the eighteenth century. Faraday's family were members of the same group, and soon after marrying, Faraday himself became a formal member. He retained his religious beliefs throughout his life. The marriage was childless, but the couple were devoted to

each other and almost inseparable. Around 1840 Faraday began to suffer ill-health through overwork. He was forced to stop research for several years, but on returning to his work made more seminal discoveries including in 1845 the effect named after him in which the direction of polarization of light could be affected by the application of a strong magnetic field to the medium in which the light was propagating.

His memory slowly deteriorated from about 1850, and this became a serious handicap to him in the following decade. He gave his last lecture in 1860 and resigned his Professorship in 1861. On 26 August 1867 he died in his chair in his study in the house in Hampton Court that Queen Victoria had given nine years earlier.

Faraday's discoveries had a great influence on the century in which he lived. He was the master experimentalist of his time. He was constantly alert to the practical applications of his work, and his research provides a constant reminder of the benefits that may flow to society through the medium of discoveries brought about by curiosity and a passionate desire to understand natural phenomena.

Exercise 17.2 A circular, flat coil of diameter 4 cm has 50 turns of copper wire of diameter 0.2 mm. A uniform magnetic field of 0.4 T is in a direction perpendicular to the plane of the coil. The field is reduced to zero in 4 seconds at a uniform rate. What is the current in the coil? Neglect the flux through the coil arising from the induced current in the coil itself. The electrical conductivity of copper is $5.9 \times 10^7\,(\Omega\,m)^{-1}$.

Answer 93 mA.

The neglect of the flux through the coil due to the induced current in the coil itself in Exercise 17.2 needs to be justified since the flux Φ to be used in eqn (17.10) is the total flux through the coil from all sources. For an induced current of 93 mA in the coil of Exercise 17.2, the field B through the middle of the coil from this current is about 1.5×10^{-4} T. This is much less than the field from the external source, except when this field has been reduced to very nearly zero, and our neglect is justified for almost all of the winding down process.

Induction in free space

It must be emphasized that induced electric fields are present in any region of space where the field B is changing with time, irrespective of the existence of closed conducting circuits around which induced currents may flow. Induced electric fields exert forces on charged particles whether they are in a solid conductor or in free space. Microscopic charged particles, such as protons or electrons, can be accelerated by the application of an increasing magnetic field.

Consider a proton moving in a vacuum with speed v at a time t in an orbit perpendicular to a magnetic field as shown in Fig 17.8. The protons must be moving in a near-vacuum in order that collisions with gas molecules do not disturb the motion. At the position of the proton the field has magnitude B, but it varies in strength over the plane of the orbit. At time t the proton is moving in an arc of a circle of radius r given by eqn (16.14),

$$r = \frac{mv}{eB} \tag{17.11}$$

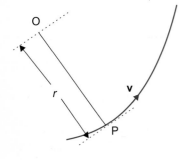

Fig. 17.8 A proton at the point P instantaneously moving in an arc of a circle. The radius of the circle depends upon the speed v of the proton and the magnetic field at the point P.

where m and e are the mass and charge of the proton, respectively.

If the magnetic field is changing with time and it is required that the proton remain in the same orbit of radius r, the above equation must be satisfied at all times. Rearranging the equation and differentiating with respect to time, the condition that the proton remain at radius r may be written

$$\frac{d(mv)}{dt} = er\frac{dB}{dt}. \qquad (17.12)$$

This can be satisfied if the centre of the circular orbit is a centre of symmetry of the magnetic field, as in Fig 17.9, and if the field strength depends only on distance r from the centre.

Now that we have specified that the field strength is a function of r alone, let the field everywhere increase with time while its relative strength from point to point remains the same. The lines of the induced electric field are then circles centred on the centre of the orbit and lying in the plane of motion of the proton, and the induced electric field has constant magnitude E around a circle of radius r.

The induced electric field $\mathbf{E}$ is given by Faraday's law, which says that the line integral of the field around the circle of radius r is equal to the negative time derivative of the magnetic flux through the circle. The magnetic flux through the circle is $\int_S \mathbf{B} \cdot d\mathbf{S}$, which equals $\pi r^2 \times \bar{B}$, where $\bar{B}$ is the average magnetic field over the area enclosed by the orbit. If the magnetic field increases with time, the electric field is in the direction shown and has magnitude E given by

$$2\pi r E = \pi r^2 \frac{d\bar{B}}{dt}.$$

The accelerating force on the proton is thus

$$eE = \frac{er}{2}\frac{d\bar{B}}{dt}.$$

This force is equal to the rate of change of the proton's momentum, which in turn is given by eqn (17.12) for a fixed orbit of radius r. Hence, for a steady orbit,

$$er\frac{dB}{dt} = \frac{er}{2}\frac{d\bar{B}}{dt}$$

or

$$\frac{dB}{dt} = \frac{1}{2}\frac{d\bar{B}}{dt}.$$

Integration of this equation gives the condition that at all times

$$B = \frac{1}{2}\bar{B}$$

where we have put the constant of integration equal to zero, since when $\bar{B}$ is zero there is no field at the orbit. We deduce that for the proton to stay

● *Circular orbit condition*

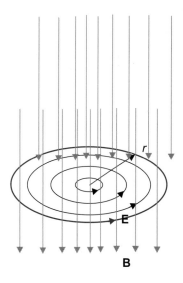

Fig. 17.9 A proton moves in a circular orbit of radius r in a magnetic field perpendicular to the plane of the orbit. The centre of the orbit is a centre of symmetry of the non-uniform field, whose strength depends only on the radial distance from the orbit's centre.

at radius r the average magnetic field over the area enclosed by the orbit must at all times equal twice the field at radius r.

The maximum energy to which a proton can be accelerated in this way is obtained from eqn (17.11) and depends upon the radius r and the maximum field B at distance r.

● *Energy in a circular orbit*

For $r = 0.5$ m and a maximum field of 0.8 T, the maximum energy is close to 7.7 MeV. Protons are not usually accelerated to high speeds in this way. Acceleration by the application of an increasing magnetic field can only be done in bursts and a continuous beam is usually more useful.

Electric motors

Many electric motors produce rotary motion, as in an electric drill. These motors depend upon the torque on a coil in a magnetic field, a topic discussed in Section 16.4. When a flat coil of area S lies in a horizontal magnetic field B with its plane vertical and carries a current I, equation (16.25) gives the torque about a vertical axis through the centre of the coil. If the plane of the coil is perpendicular to the field, the torque is zero and, if the plane is at 90° to the field, the torque has its maximum value.

Consider a plane rectangular coil fixed to a spindle as in Fig 17.10. The spindle is free to rotate and passes through opposite sides of the coil. Let the coil lie in a magnetic field whose direction is perpendicular to the spindle but at an arbitrary angle to the plane of the coil. If a steady current is passed through the coil, there will be a torque on the coil and the coil and spindle will rotate. When the plane of the coil is perpendicular to the field and the direction of the current such that the coil's magnetic moment is in the same direction as the field, the torque is zero. After oscillating, friction will cause the plane of the coil to lie as described above in the position where the torque is zero. The coil will cease to rotate. However, if it is arranged that the field or the current in the coil changes direction each time the plane of the coil is perpendicular to the field, continuous rotation may be possible. By using several coils wound so that their planes intersect at different angles, and with suitable arrangements for passing currents at different times through the coils, a roughly constant torque in the same direction can be applied to the spindle supporting the coils. The spindle rotates at a steady angular speed when a constant retarding torque opposes that produced by the magnetic force. This is the principle of a direct-current electric motor that provides rotary motion, such as used in battery-operated devices.

B

Fig. 17.10 A coil fixed to a spindle that rotates in a magnetic field under the influence of the torque produced when a current is passed through the coil.

In addition to rotary motion, electric motors can also produce linear motion, as in motors designed to drive trains or other track vehicles. These motors depend upon the electric fields generated in a conductor when the magnetic field around the conductor changes. The induced electric fields give rise to currents in the conductor, called **eddy currents**.

These currents can be put to use, as in motors that provide linear motion or in systems that deliberately use eddy currents for heating.

In a rotary motor, the spindle, or rotor, that supports the current-carrying coils rotates in the magnetic field produced by currents in stationary conductors. In one type of linear motor, currents in a stationary conductor produce a magnetic field that induces time-varying electric fields and hence eddy currents in a nearby conductor on the moving object. The linear motion is produced by the forces arising from the interaction of the fields produced by the currents in the coils with the fields produced by the eddy currents.

Another type of linear motor that propels a train has the metal track or guide along which the train runs as the conductor and current-carrying coils on the train. The coils on the train produce a magnetic field that moves along the guide as the train moves and induces large eddy currents in the guide. These currents interact with the field from the coils and the resulting force propels the train. The energy transferred as kinetic energy to the train is provided by the power unit on the train that supplies the current to the coils.

Magnetic levitation

Linear induction motors have been proposed as the power sources for trains and interplanetary vehicles of the future. In order to reduce the frictional resistance to motion and enable faster train speeds with less energy consumption, trains have been built and operated that are raised above the tracks by **magnetic levitation**. An object can be magnetically levitated, that is, suspended in mid-air in a more or less stable position, if the force due to gravity is balanced by an equal and opposite magnetic force.

Nonconducting objects can sometimes be levitated in a non-uniform magnetic field. The magnetic potential energy of a magnetic dipole in a field B is given by eqn (16.27),

$$U_c = -\mathbf{m} \cdot \mathbf{B}.$$

An object with volume V in a magnetic field acquires a magnetization $\mathbf{M}$ given by eqn (16.31) and hence has a magnetic dipole moment

$$\mathbf{m} = \frac{1}{\mu_0} \chi_B V \mathbf{B}$$

where χ_B is the magnetic susceptibility defined by eqn (16.31).

The force on the dipole in a non-uniform field is in the direction in which the derivative of the potential is a maximum, and has magnitude equal to that derivative. When the dipole is placed in a magnetic field that is non-uniform in the z-direction, which we take to be the vertical, it experiences a force of magnitude $-\partial U/\partial z$. If this force cancels out the effect of gravity we require it to be in the upward direction and such that its magnitude

$$-\frac{\partial U}{\partial z} = \frac{1}{\mu_0} \chi_B V \left(\frac{\partial B^2}{\partial z} \right) = \rho V g$$

where ρ is the density of the object, g the acceleration due to gravity, and $(\partial B^2/\partial z)$ the gradient of the square of the field. In the above equation we have used the approximation that the field is in the same direction as its gradient, namely, the z-direction. It can be seen that levitation may result if

$$2B \left(\frac{\partial B}{\partial z} \right) = \frac{\mu_0 \rho g}{\chi_B} \quad \text{or} \quad \frac{\partial B}{\partial z} = \frac{\rho g}{2M}.$$

A diamagnetic object with a negative value of χ_B experiences a force towards regions of low field strength. Thus, if it is arranged that the field increases as height decreases, a diamagnetic object may be in a position of stable equilibrium. Small animals like frogs, which consist chiefly of water which has a negative magnetic susceptibility with magnitude about 9×10^{-6}, can be levitated in a strong field that has a very large gradient.

It is difficult to levitate a paramagnetic or ferromagnetic object in a static magnetic field because of the inherent instability involved in maintaining it stationary in free space. The force on an insulating paramagnetic or ferromagnetic object in a static non-uniform field is in the direction of increasing field, and the object will fall if the force due to gravity exceeds that from the inhomogeneous field, or rise if the reverse is true. A paramagnetic or ferromagnetic object can be picked up by a static field increasing in strength in the upward direction. In that case the ferromagnetic object attaches itself to the pole face of the magnet producing the field. Dead automobiles in scrap yards are sometimes picked up and moved in this way.

If an object is a conductor, levitation can be effected by magnetic fields that change with time, such as those that arise from eddy currents induced when the object is in the presence of external, time-varying magnetic fields. Some designs of levitating trains have coils fixed to the bottom of the train that pass under a raised conducting track along which the train runs. A current through these coils provides a field that moves with the train and induces very large eddy currents in the track. The interaction between the field from the coils and the eddy currents provides the lift to keep the train clear of the guide.

17.3 Inductance

In the last section we saw that, if the magnetic field through a coil changes, there is an induced current in the coil that gives rise to a flux opposing the change. If the flux through the coil is due to a current in the same coil, there is a time variation of flux through the coil when the current through the coil itself changes. This flux variation induces an electric field in the coil in the direction given by Lenz's law. If there are two coils close to each other carrying currents, and the current in one is varied, there is a time variation of flux through the second. An electric field is induced and a current will flow in the second coil in a direction given by Lenz's law, if the second coil is in a complete circuit. These effects are called **inductance**, and have many practical consequences and applications. They are the subject of this section.

Self-inductance

There is a magnetic flux Φ through a coil when a current I is flowing in the coil itself. The magnetic field at all points can in principle be calculated using the Biot–Savart law (eqn (16.22)), which shows that the field everywhere is proportional to the current I. Hence the flux Φ is proportional to I, and we may write

$$\Phi = LI \tag{17.13}$$

where the constant of proportionality L is called the **self-inductance** of the coil. It is the flux through the coil when unit current flows in the coil. The right-hand screw rule, used with the current I to define a sense of circulation for calculating the flux Φ, ensures that the self-inductance L is a positive number.

In the SI system the units of self-inductance are $\mathrm{T\,m^2\,A^{-1}}$ and are given a special name, the **henry** (symbol H). Joseph Henry was a scientist who performed experiments on electromagnetic induction at the same time as Faraday.

The self-inductance of a coil depends upon the size and shape of the coil and can be calculated for simple shapes using the Biot–Savart law. For a circular coil made of thin wire with n turns of the same shape and size, the total flux through the whole coil, sometimes called

● *L depends upon the geometry of the coil*

the flux linkage, is n times that through a single turn. If the current I varies with time, there is an induced emf in the coil equal to the line integral of the induced electric field around the complete loop or set of loops constituting the coil. The emf is given by Faraday's law, eqn (17.10),

$$\oint \mathbf{E} \cdot d\boldsymbol{\ell} = -L\frac{dI}{dt}. \tag{17.14}$$

For an idealized self-inductance with no resistance, which is called an **inductor**, the induced emf is at any instant equal and opposite to the voltage difference V across the inductor from the source driving the current; otherwise the current would be infinite. Hence

● *The voltage across an* *inductor*

$$V = L\frac{dI}{dt}. \tag{17.15}$$

Worked Example 17.2 Estimate the self-inductance of a cylindrical solenoid of radius 1 cm and length 50 cm that has 50 turns per cm? The coils of the solenoid are wound on a nonmagnetic former.

Answer To a good approximation, the magnetic field B inside the solenoid is uniform and parallel to the solenoid axis. When 1 A flows in the coils, the field B, given by eqn (16.20), is $\mu_0 n_\ell$ tesla, where n_ℓ is the number of turns per metre. The self-inductance L of the solenoid is thus the field B times the area of each turn times the total number of turns. Hence

$$L = \mu_0 n_\ell \pi a^2 n_\ell \ell \tag{17.16}$$

where a is the radius of each turn of the solenoid and ℓ its length. Insertion of the numbers gives L just under 5×10^{-3} H.

We see that the henry is a large unit. If the core of the solenoid is ferromagnetic, the flux through it for unit current is greatly augmented by the magnetization of the core and the self-inductance is correspondingly increased.

Mutual inductance

If one coil is situated close to a second that carries a current, there is a flux Φ_1 through the first coil proportional to the current I_2 in the second.

We may write

$$\Phi_1 = M_{1,2} I_2 \qquad (17.17)$$

where the constant of proportionality $M_{1,2}$ is called the **mutual inductance** between coil 1 and coil 2, and is the flux through coil 1 due to unit current in coil 2. If coil 1 has n_1 turns, each of the same size, the total flux Φ_1 is n_1 times the flux through one turn, and the mutual inductance is proportional to the total number of turns on coil 1. The magnetic field from coil 2 for unit current in the coil depends only on its shape and size. Hence the flux through coil 1 in the field of coil 2 depends only on the geometry of coil 2 and the mutual inductance depends only on the shapes, sizes, and positions of the two coils.

⦿ *$M_{1,2}$ depends upon the geometry of the two coils*

The units of mutual inductance are henries, the same as those of self-inductance. The mutual inductance is defined as a positive number. This ensures that the flux Φ_1 is positive when coil 1 is circulated in the sense given by the right-hand rule used with the direction of I_2.

In the steady state, after the current I_2 has been established, there is no current in coil 1. If the current I_2 now changes, the flux Φ_1 changes, and there is an electric field E_1 induced in coil 1 whose value is given by Faraday's law. The line integral of E_1 around the whole coil 1 is

⦿ *Induced emf*

$$\oint \mathbf{E}_1 \cdot d\boldsymbol{\ell} = -M_{1,2}\frac{dI_2}{dt}. \qquad (17.18)$$

The left-hand side of the above equation is the emf induced in the first coil due to current changes in the second. This emf induces a current in the first coil dependent upon the coil's resistance and its self-inductance. The mutual inductance $M_{2,1}$ is the flux through coil 2 due to unit current in coil 1. The mutual inductance between coil 1 and coil 2 is equal to the mutual inductance between coil 2 and coil 1.

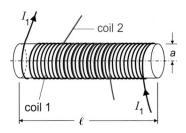

It is easiest to prove this for the simple example of a long solenoid overwound with a second coil 2, as shown in Fig 17.11. When I_1 changes, the flux Φ_2 changes and the emf in coil 2 is

Fig. 17.11 One solenoid wound closely on top of another.

$$\oint_{\text{coil 2}} \mathbf{E} \cdot d\boldsymbol{\ell} = -\frac{d\Phi_2}{dt}$$

where the line integral over coil 2 is taken in the direction that makes the flux Φ_2 positive, as discussed in Section 17.2. The magnetic field due to coil 1 is given by eqn (16.20), and the flux Φ_2 is this field times the area of one turn of coil 2 times its total number of turns n_2. Hence

⦿ *$M_{1,2} = M_{2,1}$*

$$\Phi_2 = \pi a^2 n_2 \mu_0 (n_1/\ell) I_1$$

where a is the radius of coil 2, n_2 is the total number of turns on coil 2, and ℓ is the length of the solenoids. The mutual inductance becomes

$$M_{2,1} = \pi a^2 n_2 \mu_0 (n_1/\ell). \tag{17.19}$$

A similar argument gives the flux through coil 1 due to unit current in coil 2 as

$$M_{1,2} = \pi a^2 n_1 \mu_0 (n_2/\ell).$$

Examination of the above two equations immediately shows that

$$M_{1,2} = M_{2,1} = M. \tag{17.20}$$

This proof is restricted to two coils that constitute two long, overwound solenoids but it can be shown that the result is true quite generally.

The force between the two coils may be expressed in terms of the mutual inductance between them. Equation (16.30) gives the force in the x-direction on coil 1 with current I_1 when there is a flux Φ_1 through it from the field due to coil 2.

$$F_x(1) = I_1 \frac{\partial \Phi_1}{\partial x} = I_1 I_2 \frac{\partial M_{1,2}}{\partial x}. \tag{17.21}$$

● **The force between two coils**

In this expression it must again be remembered that the flux Φ and the mutual inductance are positive, and that this requirement defines the sense of traversal of coil 1. The current I_1 is then positive or negative depending on whether it flows in coil 1 in that same sense or in the opposite sense.

Worked Example 17.3 A solenoid consists of 600 turns of thin wire wound on a nonmagnetic former of radius 1 cm and length 40 cm. A second, closed coil of 1000 turns, made of copper wire of 0.4 mm diameter and insulated from the first coil, is wound over the first. A current in the bottom coil is changing at the uniform rate of $100 \, \text{A s}^{-1}$. Estimate the current in the top coil. The electrical conductivity of copper is $5.9 \times 10^7 \, (\Omega \, \text{m})^{-1}$.

Answer Equation (17.19) gives the mutual inductance of the coils,

$$M = \mu_0 \pi a^2 (n_1 n_2/\ell),$$

where, for estimation, we take the radius of both the top and bottom coils to be equal to the radius of the former. The magnitude of the emf in the top coil 2 is given by eqn (17.18) and, if the resistance of the top coil is R_2, the current through it is

$$I_2 = \frac{1}{R_2} \mu_0 \pi a^2 (n_1 n_2/\ell) \frac{\mathrm{d}I_1}{\mathrm{d}t}.$$

The resistance R_2 is given by

$$R_2 = \frac{n_2 2\pi a}{\pi b^2 \sigma}$$

where b is the radius of the wire of coil 2 and σ is the conductivity of copper. Rearranging gives

$$I_2 = \frac{\pi b^2 \sigma}{2} \mu_0 a (n_1/\ell) \frac{dI_1}{dt},$$

and substitution of $a = 10^{-2}$ m, $b = 2 \times 10^{-4}$ m, $\sigma = 5.9 \times 10^7 \; (\Omega\,\text{m})^{-1}$, $(n_1/\ell) = 3000$ turns per metre, and $dI_1/dt = 100 \; \text{A s}^{-1}$ gives $I_2 = 14$ mA.

Magnetic energy

Energy is required to establish a voltage difference across the plates of a capacitor and, similarly, energy is required to establish a current through an inductor. The energy in a capacitor is stored in the electric field and the energy in an inductor is stored in the magnetic field. The energy in a capacitor may be recovered when the capacitor is discharged. The energy in an inductor may similarly be recovered when the current is reduced to zero.

We will calculate the energy stored in the magnetic field of an inductor that carries a steady current. Consider an external source establishing a final current I through a coil of self-inductance L. We will assume that the resistance of the coil is negligible and ignore the non-recoverable energy loss due to resistive heating of the coil material. At a time t let the current be I'. The voltage across the coil provided by the current source is $V = L\,dI/dt$ from eqn (17.15). In time dt the source passes a charge dq through the coil and does work dW equal to the voltage difference times dq,

● *The energy stored in an inductor*

$$dW = V\,dq = L\frac{dI'}{dt}dq = L\frac{dI'}{dt}\frac{dq}{dt}dt = L\frac{dI'}{dt}I'\,dt.$$

The total work done in the time T taken to establish the final current I is

$$W = \int_0^T LI'\frac{dI'}{dt}\,dt = \int_0^I LI'dI' = \tfrac{1}{2}LI^2. \tag{17.22}$$

Exercise 17.3 Estimate the energy stored in an air-cored superconducting solenoid that has 10^5 turns per metre, length 20 cm, radius 4 cm, and carries a curent of 100 A.

Answer 9.5×10^4 J. This is a very large amount of energy and is enough to turn a significant mass of room-temperature water to steam. If the coils of a superconducting solenoid rapidly become non-superconducting for any reason, the large amount of energy quickly released can cause severe heating problems.

● **Energy in magnetic fields**

The energy stored in an inductor resides in the magnetic field produced by the current. The total energy in the field, given by eqn (17.22), can be written in terms of the magnetic field as an integral over the volume occupied by the field. We may derive this expression using the field due to a current in a long solenoid and the corresponding self-inductance.

The self-inductance of a long solenoid of length ℓ and n_ℓ turns per unit length in which a current I flows is given by eqn (17.16).

$$L = \mu_0 n_\ell^2 \pi a^2 \ell$$

where a is the radius of the solenoid. The energy stored is thus, from eqn(17.22),

$$U = \tfrac{1}{2} L I^2 = \tfrac{1}{2} \mu_0 n_\ell^2 \pi a^2 \ell I^2.$$

The magnetic field B inside the solenoid is given by eqn (16.20),

$$B = \mu_0 n_\ell I,$$

and, writing the energy stored in terms of the field,

$$U = \frac{1}{2\mu_0} B^2 \pi a^2 \ell.$$

Hence the energy per unit volume, the energy density in the magnetic field in the solenoid, U/V, is

$$\frac{U}{V} = \frac{1}{2\mu_0} B^2. \tag{17.23}$$

● **Energy density**

The energy density is often written in terms of the fields B and H, using the relation $B = \mu_0 H$ between B and H in free space given by eqn (16.33) with the relative permeability μ equal to unity.

$$\frac{U}{V} = \tfrac{1}{2} BH. \tag{17.24}$$

It may be shown in a more rigorous treatment that the total energy stored in a magnetic field in the presence of linear, isotropic, and homogeneous materials for which the relative permeability μ is constant

If a magnetic field is produced by a current in coils overwound on a closed, toroidal ferromagnetic core, as in Fig 16.27, the field B is increased by the magnetization of the core. If the core is characterized by a relative permeability μ, the field B is increased by the factor μ over the field present without the core, for the same current through the coils. The total energy in the field is thus a factor μ times greater than it would be without the core. The increase in energy is equal to the gain in potential energy obtained by aligning the dipole moments of individual domains with the field. This potential energy gain is $MB_1/2$ per unit volume, where M is the magnetization, and B_1 the field B with the core absent. Hence

$$\tfrac{1}{2}MB_1 = \tfrac{1}{2}(B_2H - B_1H)$$

where B_2 is the field B with the core and the field H is the same with and without the core. Since $B_2 = \mu B_1$, the

above equation reduces to

$$M = (\mu - 1)H.$$

This result is consistent with eqn (16.32), $B_2 = \mu_0(M + H)$, since substitution for M gives back the identity $B_2 = \mu B_1$ proved in Section 16.5 and used as the starting point for this discussion of the energy in the toroid. The justification for the use of the expression $MB_1/2$ for the work done in producing the magnetization M may not be obvious, since the energy of a small dipole m aligned with a field B is $-mB$. However, the field B_2 in the toroid has a major contribution $(B_2 - B_1)$ from the field due to the magnetization M itself, and there can be no forces on dipoles from the fields to which they themselves give rise.

is given by

$$U = \tfrac{1}{2}\int_V \mathbf{B} \cdot \mathbf{H}\, dV \tag{17.25}$$

where V is the whole volume over which the fields are nonzero.

● *The total field energy*

Exercise 17.4 Compare the energy density in the intense magnetic field in the core of the superconducting solenoid of Exercise 17.3 with the energy density in a very large electrostatic field of $10^7\,\mathrm{V\,m^{-1}}$.

Answer $6.3 \times 10^7\,\mathrm{J\,m^{-3}}$ in the solenoid; $440\,\mathrm{J\,m^{-3}}$ in the electrostatic field.

The field B provided by high-current superconducting solenoids exceeds the saturation field of ferromagnetic materials, and air-cored superconducting magnets provide the largest fields that can presently be generated by macroscopic currents.

17.4 Electric generators

Electric fields are induced in loops through which the magnetic flux is changing or in a circuit part of which is moving in a magnetic field. These fields may induce currents in a conducting circuit and thus generate electric energy. The power available from the electric plug in the wall is

● *A single rotating coil*

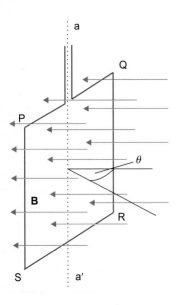

a

Q

P

θ

B

R

S   a'

Fig. 17.12 A plane coil PQRS rotating in a uniform magnetic field. The field is perpendicular to the axis of rotation, aa', of the coil.

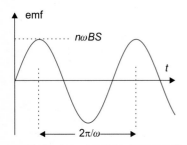

▲ emf

········· $n\omega BS$

t

$2\pi/\omega$

Fig. 17.13 The voltage difference across the ends of the coil as a function of time.

Fig. 17.14 The ends of the plane coil PQRS, which is supported on a spindle running along the line aa', are connected to two rings attached to the spindle. The spindle rotates the coil and rings in a uniform magnetic field that is perpendicular to the axis of rotation aa'.

generated in a power station by moving a coil in a magnetic field. The energy needed to move the coil usually comes from steam turbines with the steam generated either by burning fossil fuel in a boiler or by burning nuclear fuel in a reactor. The primary energy source is natural gas or coal or naturally occurring uranium. The vehicle for converting the energy from such sources into electric energy is electromagnetic induction.

Consider a plane coil PQRS of *n* turns rotating at angular speed ω in a uniform magnetic field as shown in Fig 17.12. The field is perpendicular to the axis aa' about which the coil is rotating. At time *t* let the normal to the plane of the coil make an angle θ with the field. The magnetic flux through the coil at that time is equal to

$$\Phi = nBS \cos\theta$$

where *S* is the area of the coil. If θ is zero at time *t* equal to zero, $\theta = \omega t$, and the emf in the coil, is given by Faraday's law, eqn (17.10):

$$\text{emf} = n\omega BS \sin\omega t. \tag{17.26}$$

This is illustrated in Fig 17.13. If an external load is connected in series with the coil, a current $n\omega BS \sin\omega t/R$ will flow in the load, where *R* is the resistance of load plus coil. The connections of the rotating coil to the load may be made by connecting the load, via carbon brushes over which the rings slip, to two rings that rotate with the coil. The arrangement is shown schematically in Fig 17.14. The above discussion outlines the principles on which the simplest form of alternating current generator, or **alternator**, is based.

Exercise 17.5 A coil of 50 turns, each of area 40 cm², rotates at an angular speed of 500 revolutions per second in a magnetic field of 0.05 T. The magnetic field is perpendicular to the axis of the coil in the plane of

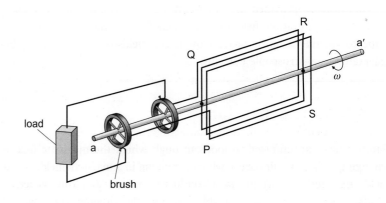

R

a'

Q

ω

load

S

a   P

brush

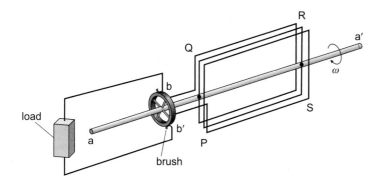

Fig. 17.15 The ends of the rotating coil of Fig 17.14 are now attached to the split halves of a single ring, which is also fixed to the rotating spindle.

the coil. What is the amplitude of the emf generated across the ends of the coil.

Answer 31.4 V.

The average value of the alternating emf of Fig 17.13 is zero, but it becomes an emf with a nonzero average value if the negative-going half-sine waves are converted into positive-going waves. This can be done by connecting the ends of the rotating coil to separate sides of a split ring that rotates with the coil, as shown in Fig 17.15. As the coil rotates, the carbon brush b is always at a higher voltage than brush b' (or instantaneously at an equal voltage to brush b'), and the emf across bb' is as shown in Fig 17.16. The output voltage is similar to that considered in the second part of Problem 6.26, where it was stated that Fourier analysis of the waveform shown in Fig 17.16 gives the average value of the voltage equal to the amplitude $n\omega BS$ times $2/\pi$. The emf delivered depends upon the angular speed at which the armature is rotating.

The split ring on Fig 17.15 is called a **commutator**, and the rotating spindle holding the coil is called an **armature**. In practical alternators, several coils insulated from each other are wound separately on an iron armature and each coil is connected to different segments of a commutator with multiple splits in it. The two brushes make contact with any particular coil when the emf in that coil is near its maximum value. If we consider an armature with four separate coils for simplicity, the output emf is similar to that shown in Fig 17.17. Normally there are more than four coils. As the number of coils increases, the ratios of the average voltage across the load to the amplitudes of the harmonically varying components increases. The ratios can be calculated mathematically by performing a Fourier series analysis on the output.

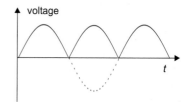

Fig. 17.16 The voltage difference between the two halves of the split ring as a function of time.

● **A multicoiled armature**

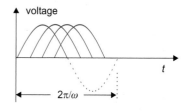

Fig. 17.17 The voltage across the load when there are four coils rotating on the armature with equal angular separation.

Problems

Level 1

17.1 An aeroplane flies horizontally at a speed of 1000 km per hour at a place where the vertical component of the Earth's magnetic field is 4×10^{-5} T. What voltage difference is generated between the wing tips if they are 40 m apart?

17.2 A plane coil of thin wire with 30 turns each of area 20 cm^2 is situated in a uniform magnetic field that varies sinusoidally with time at the frequency of 50 Hz. The amplitude of the field is 0.1 T, and the field direction is perpendicular to the plane of the coil. What is the amplitude of the induced emf in the coil?

17.3 A metal rod of length 0.5 m has its long side perpendicular to a long straight wire that carries a current of 4 A. The end of the rod nearest the wire is 0.1 m from the wire and the other end of the rod is 0.6 m from the wire. If the rod is moving parallel to the wire at a speed of $5 \, \mathrm{m \, s^{-1}}$, what is the voltage difference across its ends?

17.4 A long solenoid of radius small compared with its length and wound with 80 turns per cm is bent to form a toroidal coil enclosing an air volume of 100 cm^3. Estimate its self-inductance.

17.5 An azimuthal current flows around the outside of a thin-walled copper tube as shown in Fig 17.18. Show that the self-inductance of the tube is approximately equal to $\mu_0 \pi a^2 / \ell$.

17.6 Two small circular loops of wire with radii a and b are coaxial and separated by a distance d much greater than both a and b. Estimate their mutual inductance.

17.7 A solenoid 250 cm long and 2 cm diameter is wound with 10 000 turns of wire. A short coil of 100 turns of wire is wound over the centre of the solenoid. Estimate the mutual inductance between the solenoid and the coil.

17.8 A superconducting solenoid wound on a tube 12 cm in diameter and 18 cm long gives a field of 12 T. Estimate the total energy stored in the magnetic field.

17.9 A circular coil is rotated in the Earth's magnetic field at constant angular speed about one of its diameters. An alternating voltage of amplitude 10^{-6} V is observed when the axis of rotation is vertical. When the axis is horizontal and perpendicular to magnetic north, the amplitude is 3×10^{-6} V. What is the angle between the Earth's field and the horizontal in the neighbourhood of the coil?

17.10 What is the maximum emf induced in a circular coil of 1000 turns of average radius 10 cm when it is rotated at 25 revolutions per second about an axis normal to the Earth's magnetic field at a point where the Earth's field has a magnitude of 6×10^{-5} T?

Level 2

17.11 A current I flows in a rectangular bar of semiconductor placed in a uniform magnetic field B as shown in Fig 17.19. The number of free electrons per unit volume in the semiconductor is N. The force on the moving electrons pushes them over to face P of the bar, leaving excess positive charge on face Q, until the electric field set up by the charge distribution gives

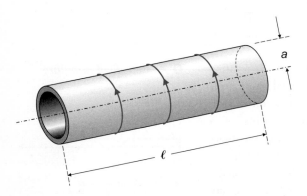

Fig. 17.18 A copper tube has an azimuthal current flowing around its shell.

a force on the electrons cancelling out the magnetic force. What is the potential difference V_{PQ} between faces P and Q?

If $N = 10^{15}\,\text{cm}^{-3}$, $\ell = 4\,\text{mm}$, the current is $10\,\text{mA}$, and the cross-sectional area of the bar is $8 \times 10^{-6}\,\text{m}^3$, what is the voltage difference produced when the bar is in a magnetic field of $0.1\,\text{T}$?

The effect outlined in this problem is known as the **Hall effect**. Magnetic fields are often measured by observing the voltage across two opposite faces of a thin rod of semiconducting material with rectangular cross-section, placed so that the direction of the field is parallel to the faces across which the voltage is measured. Such a device is known as a **Hall probe**.

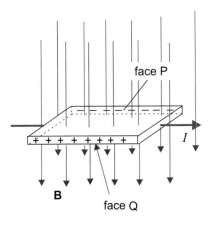

Fig. 17.19 A Hall probe in a magnetic field.

17.12 A small diamagnetic object of density ρ has magnetic susceptibility χ_B. What magnetic field variation is required to levitate the object?

17.13 A coaxial cable consists of an inner cylindrical conductor of radius a and an outer cylindrical conductor of radius b. The two conductors are maintained coaxial by being embedded in a non-conducting material of relative permeability μ. Show that the self-inductance per unit length of the cable is $(\mu\mu_0/2\pi)$ $\ln(b/a)$.

17.14 Determine the mutual inductance of a pair of coaxial, circular loops of thin wire separated by a distance d. The radius of one coil is a and that of the

other b, the former radius being much larger than the latter.

17.15 Determine the mutual inductance between a long straight wire and a rectangular coil with n turns and sides of length a and b. The coil and the wire are in the same plane and the long side of the rectangle, of length a, is parallel to the wire and at a distance d from the wire.

17.16 A small coil has 30 turns of thin wire, each of area $3\,\text{cm}^2$. It is placed in a horizontal magnetic field of strength $1\,\text{T}$ with its plane vertical. The angle between the field direction and a horizontal line drawn in the plane of the coil is $45°$. The coil is connected to an external circuit, the total resistance in the completed circuit being $2\,\Omega$. What is the charge that passes through the coil and external circuit when the coil is removed from the field?

This is a method that can be used to measure magnetic field strengths if a charge-measuring device is available, although it is not capable of high accuracy.

17.17 A steady current I flows in a long, straight wire of radius a. Determine the mean energy stored in the magnetic field inside a unit length of wire.

17.18 Determine the mean magnetic energy stored inside a unit length of a long, straight conducting wire of radius $0.5\,\text{mm}$ carrying a low-frequency sinusoidal current of amplitude $1\,\text{A}$.

At low frequencies the current is uniformly distributed throughout the wire; at sufficiently high frequency the current starts to concentrate at the edge of the wire. This phenomenon is treated in more advanced discussions of electricity and magnetism.

17.19 A loop of wire of area S rotates about a vertical axis with angular frequency ω in the presence of a horizontal magnetic field B. The field B also varies sinusoidally with time at angular frequency ω and has amplitude B_0. Determine the emf in the wire loop.

17.20 A thin conducting disc of thickness d and radius a is stationary in a magnetic field perpendicular to the plane of the disc. The field has amplitude B_0 and varies sinusoidally in time with angular frequency ω. Determine the induced current density as a function of distance from the centre of the disc.

Level 3

17.21 A toroidal tube has a rectangular cross-section with sides Δa and b. The inner radius of the toroid is a and the outer radius $a + \Delta a$. The height of the rectangular cross-section is b. The toroid is overwound with a coil that has a total of n turns. Show that the self-inductance L of the coil is given approximately by

$$L = \frac{\mu_0 b n^2 \Delta a}{2\pi a},$$

if a is much greater than Δa, which in turn is greater than b.

17.22 Two long, straight wires of a hard-wired telephone circuit inside a house are coplanar with and run parallel to a third long, straight conductor that carries an alternating current of amplitude I_0 and angular frequency ω. The telephone wires are distant d_1 and d_2 from the current-carrying wire. What magnetic flux passes through the telephone circuit per unit length? What is the rms voltage noise level introduced into the telephone circuit by the current in the third conductor?

17.23 A circular coil with n_1 turns of radius r_1 is situated with its plane perpendicular to the x-axis and its centre at the origin. It carries a steady current I_1. A second, closed coil of radius r_2 and n_2 turns has its plane perpendicular to the x-axis and its centre on the x-axis at the point $x = d$. The radius of r_2 is much smaller than both r_1 and d. The large coil is now moved sinusoidally along the x-axis such that its position as a function of time is given by $x(t) = x_0 \sin \omega t$, with x_0 much less than d. Show that, to a good approximation, the current I_2 in the small coil is given by

$$I_2 = \frac{3 n_1 n_2 \pi r_2^2 \mu_0 I_1 r_1^2 x_0 \omega}{2 d^4 R} \cos \omega t$$

where R is the resistance of the small coil.

Is the current I_2 as given above in the same sense as the current I_1 at time $t = 0$, or in the opposite sense?

17.24 Find the self-inductance per unit length of two infinite parallel cylinders each of radius r separated by a distance d much larger than r. What is the stored magnetic energy in the field per unit length when a current I is flowing down one cylinder and back along the other?

17.25 Two identical coils are placed close together and the same current passed in the same direction through both. The self-inductance of the combination is found to be $4\,\mu H$. The connections to one of the coils are reversed and the inductance is now found to be $2\,\mu H$. What is the self-inductance of each coil and the mutual inductance between the two?

17.26 A cylindrical metal rod of radius a and length ℓ is placed axially inside a long solenoid of radius b that has n_ℓ turns per unit length. An alternating current $I = I_0 \sin \omega t$ passes through the solenoid. The metal rod has electrical conductivity σ. Show that the average power dissipated by eddy currents in the rod is $(\mu_0 \omega n_\ell I_0)^2 \pi \sigma \ell a^4 / 16$. (At sufficiently high frequency induced currents flow only in a thin layer near the surface of the conducting rod. In this example we assume the frequency to be low enough for this effect to be neglected.)

17.27 A small piece of magnetized material may be regarded as a magnetic dipole of moment **m**. It moves along the x-axis with speed v at a distance d from a circular conducting coil of resistance R. The moment **m** points in the $+x$-direction. The coil consists of a single turn of thin wire and has an area S. The axis of the coil is also along the x-direction and the coil has negligible self-inductance. Determine the force on the small dipole.

17.28 A torsional pendulum consists of two metal spheres of diameter 5 cm at the ends of a thin rod 50 cm long. This is suspended at its midpoint from an earthed support by a conducting fibre and swings in a horizontal plane with an amplitude of 10° and a period of 1 second. A uniform magnetic field of 0.1 T exists in the vertical

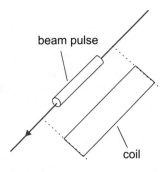

Fig. 17.20 A beam of charged particles in the form of a pulse passes a coil coplanar with the beam.

direction. Find the magnitude of the current in the rod as a function of time if all the components are of negligible resistance. Take the capacitance of each sphere as the value it would have if each sphere were in free space.

17.29 A pulsed beam of sulfur ions of energy 400 MeV, each ion having only two electrons around the sulfur nucleus, is produced by an accelerator. The pulses last for 1 ns, they are separated by 1 μs, and the average beam current is 50 nA. If the beam passes a plane coil of 100 turns each of area 3 cm² situated as shown in Fig 17.20, estimate the current through the coil when a beam pulse passes if the resistance of the coil is 1 Ω?

Some solutions and answers to Chapter 17 problems

17.1 0.42 volts.

17.3 The rod is moving as shown in Fig 17.21 and the magnetic field over the element dx at distance x from the wire is $B = \mu_0 I / 2\pi x$ perpendicular to the rod. There is thus a voltage difference dV between the ends of the element dx of dx, given by eqn (17.6). The voltage difference between the ends of the whole rod is thus

$$V = v\frac{\mu_0 I}{2\pi}\int_a^{a+l}\frac{dx}{x} = v\frac{\mu_0 I}{2\pi}\ln\left(\frac{(a+l)}{a}\right).$$

With the numbers given, the voltage difference is 7.17×10^{-6} V.

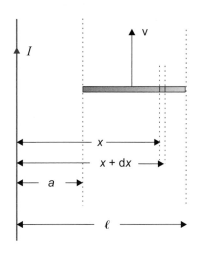

Fig. 17.21 A metal rod moving parallel to a long straight wire that carries a current.

17.6 The magnetic dipole moment of the small loop with radius a when it carries a current I is $m = \pi a^2 I$. The field of this dipole on axis distant d is $B = 2\mu_0 m / 4\pi d^3$, in the direction of the axis. The magnetic flux through the second coil arising from the field of the first is $\Phi = \pi b^2 B = (2\mu_0/4\pi d^3)\pi a^2 I(\pi b^2)$, and the mutual inductance is $M = \mu_0 \pi a^2 b^2 / 2d^3$.

17.9 71.6°.

17.12 The magnetic field must vary in the vertical direction, which we take to be the z-direction, to give a vertical force on the magnetized object. The vertical force is $m\,dB/dz$ where m is the magnetic dipole moment of the object. For this to balance gravity, $m\,dB/dz = V\rho g$, where V is the volume, ρ the density, and g the acceleration due to gravity. But $m = \mu_0\chi_B BV$, from eqn (16.31). Hence

$$B\frac{dB}{dz} = \frac{g\rho}{\mu_0\chi_B},$$

and the required field shape is given by

$$\frac{1}{2}B^2 + A_0 = \frac{g\rho}{\mu_0\chi_B}z$$

with A_0 a constant.

17.14

$$M = \frac{\mu_0 a}{2\pi}\int_d^{d+b}\frac{dx}{x} = \frac{\mu_0 a}{2\pi}\ln\left(\frac{d+b}{d}\right).$$

17.16 The magnetic flux through the coil initially is $\Phi = 30 \times 3 \times 10^{-4} \times \cos 45° = 6.36 \times 10^{-3}$ T m². While the coil is being removed from the field there is an emf in the circuit equal to $d\Phi/dt$ and this causes a current

$$I = \frac{dq}{dt} = \frac{d\Phi}{dt}\frac{1}{R}$$

to flow, where R is the resistance of $2\,\Omega$ in the circuit. Hence

$$q = \int_0^\infty \frac{d\Phi}{R}$$

which equals 3.18×10^{-3} C.

17.19 If the zero of time is chosen so that the coil is parallel with the field B at time $t = 0$, the angle θ that the plane of the coil makes with the field at time t is $\theta = \omega t$. At time t the total magnetic flux through the coil is $\Phi = BS \sin \omega t$ where S is the area of the coil. At that time the field B is given by $B = B_0 \sin(\omega t + \phi)$ where ϕ is a phase angle. Hence

$$\begin{aligned}
\text{emf} &= -\frac{d\Phi}{dt} \\
&= \omega B_0 S(\sin \omega t \cos(\omega t + \phi) \\
&\quad + \cos \omega t \sin(\omega t + \phi)) \\
&= \omega B_0 S \sin(2\omega t + \phi).
\end{aligned}$$

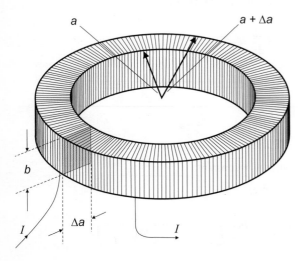

Fig. 17.22 A toroidal tube overwound with a coil that carries a current.

17.21 Figure 17.22 shows the toroid overwound with the coil. The magnetic field B at all points on the perimeter of a circle of radius x is $B(x) = \mu_0 nI/2\pi x$. Since $\Delta a > b$, and $a \gg \Delta a$, we make the approximation that the field is independent of height within the rectangular tube and varies only over the longer side of the tube. The magnetic flux through a small area $b\,dx$ of the n turns is then $d\Phi = b\,dx \mu_0 nI/2\pi xn$. The flux through the whole coil is

$$\Phi = \frac{b\mu_0 n^2 I}{2\pi} \int_a^{a+\Delta a} \frac{dx}{x},$$

and the self-inductance is

$$L = \frac{b\mu_0 n^2}{2\pi} \ln\left(\frac{a + \Delta a}{a}\right).$$

This reduces to

$$L = \frac{b\mu_0 n^2}{2\pi} \frac{\Delta a}{a}$$

when $a \gg \Delta a$.

17.26 Figure 17.23 shows a view looking down the end of the rod inside the solenoid. The magnetic field within the solenoid at time t has the constant magnitude $\mu_0 n_\ell I_0 \sin \omega t$ and is in the axial direction. Around the circular loop with radius r inside the rod the induced emf is $V = \pi r^2 \omega n_\ell I_0 \cos \omega t$. The resistance dR of the thin tube of width dr is $2\pi r/\sigma \ell dr$ and in that tube a

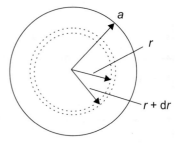

Fig. 17.23 An end-on view of the rod of circular cross-section inside the solenoid.

current I_c flows equal to $V/\mathrm{d}R$. The energy dissipated per second in the tube at time t is

$$I_c^2 \, \mathrm{d}R = (\pi/2)r^3(\omega\mu_0 n_\ell I_0)^2 \sigma\ell \cos^2 \omega t \, \mathrm{d}r.$$

The average total power is given by integrating r from zero to the radius a and putting the average of the square of the cosine term equal to one-half.

17.28 Figure 17.24 shows a view looking down on the suspended rod. The rod executes simple harmonic motion and the angular displacement θ from the equilibrium position at time t is $\theta = \theta_0 \sin \omega t$ where θ_0 is the amplitude of the oscillations and ω their angular frequency. The potential difference V between the ends of the rod at time t is the rate at which the rod sweeps out magnetic flux, and hence

$$V = B\ell^2 \frac{\mathrm{d}\theta}{\mathrm{d}t}$$

where B is the vertical field strength and 2ℓ the length of the rod. If at time t the potential and charge on sphere 1 are V_1 and q_1, respectively, using a similar notation for sphere 2 we have

$$V_1 - V_2 = V = B\ell^2 \frac{\mathrm{d}\theta}{\mathrm{d}t} = \frac{1}{C}(q_1 - q_2),$$

where $C = 4\pi\epsilon_0 a$ is the capacitance of each sphere. The rod is earthed and the total charge is zero; hence $q_1 = -q_2 = q$ and

$$B\ell^2 \frac{\mathrm{d}\theta}{\mathrm{d}t} = \frac{2q}{C}.$$

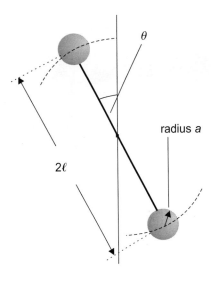

Fig. 17.24 A view in the horizontal plane looking down on the torsional pendulum.

The current $I = \mathrm{d}q/\mathrm{d}t$ and thus

$$I = \frac{1}{2}B\ell^2 C \frac{\mathrm{d}^2\theta}{\mathrm{d}t^2} = \omega^2 \theta_0 B\ell^2 C \sin \omega t.$$

The current varies sinusoidally with period 1 s and amplitude 1.2×10^{-13} A.

Chapter 18

Circuits

The laws of electricity and magnetism are applied to simple direct current and alternating current circuits in this chapter, and a brief introduction is given to some of the concepts needed to understand more complicated electronic devices.

Electrical circuits have the purposes of *transporting power* or of *conveying and processing information.* Circuits have served these purposes for many years. Electric trains and electric light were pioneered near the end of the nineteenth century. The earliest example of the use of electrical circuits for conveying information was the electric telegraph, developed to send messages along the railway lines that were rapidly extending during the 1840s. This was followed by radio broadcasting: the BBC has been broadcasting since 1926. Broadcasting is an example of both power and information delivery, since high-power electromagnetic radiation is beamed from transmitting stations so that anyone may pick up the information it contains. But it is information processing that has more recently had the greatest impact, amounting to a revolution as significant as the Industrial Revolution. With the advent of electronic computers, information may be gathered, stored, and retrieved at previously unattainable speeds. Information may also be processed to perform complex calculations or to control mechanical devices such as the robots in a car factory. Vast numbers of circuits act without human intervention to allocate the routing of calls on world-wide telephone networks and to make conections on the internet.

Even the simplest electrical circuits are too complicated to analyse in complete detail using the concepts introduced in the previous three chapters. It is simply not feasible to calculate the distribution of electric and magnetic fields at all times due to the currents flowing in a circuit. Fortunately, it is not necessary to do this anyway. The performance of circuits can usually be understood more easily and calculated with sufficient precision by dividing them up into **components**, which are connected together by conducting wires or by the electric and magnetic fields between them.

There are two kinds of component, **passive** components and **active** components. Passive components are the resistors, capacitors, and inductors already described in previous chapters. Resistance, capacitance, and inductance are inevitably present in any electrical circuit and it is important to know how electrical signals are affected by their presence. The main aim of this chapter is to explain these effects. Active components have their own source of power and are used to generate or transform signals. Nowadays circuits may contain an enormous number of active components—as many as ten million in a single silicon chip. Only a specialist can know how all the different kinds of component work. What matters for the ordinary physicist is to understand how complicated circuits transform signals, that is, how their output responds to input signals. Towards the end of this chapter we shall give a very brief introduction into the way in which circuits can be analysed in terms of the functions they perform, without knowing anything at all about what goes on inside them.

● *Passive and active circuits*

18.1 **Direct current circuits**

Direct current circuits, usually abbreviated to DC circuits, are those in which steady currents flow. To maintain the steady currents there must be at least one source of emf to provide the power dissipated by resistance. When the currents in a DC circuit are switched on there are transient effects associated with the build-up of energy in inductors and capacitors, but in a very short time a steady state is reached. The analysis of the steady state currents is the subject of this section.

Equivalent circuits

DC currents are only possible in complete circuits in which there is a continuous path through which current may flow. A simple example is given by an electric flashlight (torch), represented diagrammatically by the circuit in Fig 18.1. The components in the circuit are the battery, the bulb, and the switch, which are connected together by wires. The battery and the bulb are idealized. The conventional symbol for the battery indicates a component that has a voltage V across its terminals. The voltage is assumed to have the same value V whether or not the battery is delivering current and the battery is assumed to dissipate no power. Similarly, the bulb is idealized as a pure resistance that obeys Ohm's law.

However, these idealized components do not fully represent the behaviour of the actual battery or the actual bulb. The voltage across the battery terminals, in fact, falls slightly when it is delivering a current, and at the same time some power is dissipated within the battery. Nor is the bulb a simple resistor with a fixed value of resistance. Its resistance actually increases by a considerable factor as the current heats it to the operating temperature. Since it is a finely wound coil, the bulb also has inductance, which does not affect its DC behaviour, but does partly determine how fast the current builds up when the flashlight is switched on. The connecting leads in Fig 18.1 are also idealized and assumed to have zero resistance, although they are really made of copper and have a nonzero resistance.

Although the circuit in Fig 18.1 has so many deficiencies, it does include the most important features of the flashlight. Provided R is close to the value of the resistance of the bulb at its operating temperature, the circuit tells us how the current and power vary with the battery voltage over a small range. The bulb resistance is so much larger than that of the connecting leads that their neglect is unimportant. The great virtue of the circuit in Fig 18.1 is that it can be easily analysed, even though it misses out some of the details occurring in the real world. In this case, the current flowing in the circuit when the switch is closed is $I = V/R$

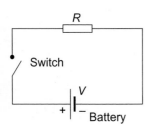

Fig. 18.1 The schematic circuit diagram for a flashlight. A resistor is represented by a small rectangle. In the symbol for a battery the long line is the positive terminal.

(eqn (16.7)) and the power delivered to the bulb is $P = V^2/R$ (eqn (16.11)). This is an approximate result, but it may often be accurate enough. Sometimes more detail is required, and then it is necessary to modify the idealized circuit components. For example, if we are interested in how the voltage across the battery terminals changes with current, Fig 18.1 tells us nothing. In reality some power is dissipated within the battery, and this fact can be included by adding a resistor as shown in Fig 18.2, representing the *internal resistance*, R_i, of the battery. This modification will usually give a good account of both the power loss in the battery and the drop in voltage across its terminals.

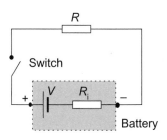

Fig. 18.2 The internal resistance R_i has been added to the circuit to allow for the reduction in battery voltage when the flashlight draws current.

Exercise 18.1 Obtain expressions for the voltage across the battery in Fig 18.2 and the power loss in it when the switch is closed. (From eqn (16.11) the power dissipated in a resistance R is I^2R.)

Answer The voltage is $VR/(R + R_i)$ and the power is $V^2R_i/(R + R_i)^2$.

Both figures, 18.1 and 18.2, are called **equivalent circuits** for the electric flashlight. They represent different degrees of approximation to the real circuit. It must be borne in mind that circuit diagrams are *always* approximate. The skill in the analysis of circuits lies in deciding what is essential and what may be left out in each case. Once an equivalent circuit has been adopted, the currents flowing through any component can be calculated by a systematic procedure, which is described in the rest of this section.

Networks and Kirchhoff's rules

The example of the flashlight is particularly simple because it consists of a single loop of components connected to one source of emf. DC circuits often contain many components and sources of steady emf, joined together by interconnecting wires. Such a circuit is called a **network**. As before, an equivalent circuit may be drawn to represent the network using idealized resistors and idealized sources of emf. Capacitors are not required because steady currents cannot flow through them and, although coils possess both inductance and resistance, only the resistance is needed for a DC equivalent circuit since there is no voltage drop across an ideal inductance with a steady current passing through it.

An arbitrary network of resistors and sources of emf is shown in Fig 18.3. The wires connecting the components are supposed to have zero resistance; if the resistance of a wire is important then it is represented by a resistor with the same value of resistance as the wire. A steady current

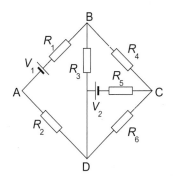

Fig. 18.3 A network of resistors and sources connected by current-carrying wires.

flows through each component, and there is a steady voltage across each resistor as well as each voltage source. Points such as B and C, where three or more wires are joined together, are **junctions** of the network. Paths between junctions, which may contain several components, are called **branches**. For example, the path BAD in Fig 18.3 is a single branch of the network containing three components. A separate branch connects the junctions B and D through the resistor R_3: the total number of branches in the diagram is five.

The values of the currents in any network are determined by the application of **Kirchhoff's rules**, which are:

1. **The algebraic sum of the currents arriving at any junction is zero.**
2. **The sum of the voltages around any loop of the network is zero.**

The first rule follows from the fact that electric charge cannot be created or destroyed or, in other words, that charge is a conserved quantity. In the steady state charge cannot be accumulating or disappearing at a junction, which means that as much current leaves a junction as arrives at it. If current leaving the junction is regarded as a negative current arriving, then the sum of arriving currents is zero as stated in the first rule.

The second rule applies because we are dealing with a steady state with electrostatic fields that satisfy eqn (15.30),

$$\oint \mathbf{E} \cdot d\boldsymbol{\ell} = 0$$

for any closed path such as those following the loops of the network.

Let us illustrate the use of Kirchoff's rules by applying them to the networks in Figs 18.1 and 18.3. In the circuit in Fig 18.1 there is only a single branch and the current I is shown flowing in the conventional direction from the positive terminal to the negative terminal of the battery. If we follow a path around the circuit in the same direction as I, the line element through the resistor R is in the same direction as the electric field $\mathbf{E}$, and the contribution to the line integral $\oint \mathbf{E} \cdot d\boldsymbol{\ell}$ is positive. Inside the battery the field is in the opposite direction to the line element, giving a negative contribution. Kirchhoff's second rule requires that the total line integral is zero, which in this case simply means that the emf of the battery balances the voltage drop across the resistor.

Now consider the more complicated network in Fig 18.3. The voltages V_1 and V_2 of the batteries are given, and the unknown quantities are the currents in each branch of the network and the voltages across each of the resistors. Provided that all the resistors obey Ohm's law there is only one way in which Kirchhoff's rules can be satisfied at all the junctions and round all the loops of the network. The voltage V_i across a resistor R_i carrying a current I_i is $V_i = I_i R_i$, whatever the value of I_i: this is eqn (16.7), the equation which is usually called Ohm's law in the context of

● *Kirchhoff's rules give enough equations to calculate the currents flowing in all branches of a network*

circuit analysis. It is sufficient to regard the currents as the unknowns, since Ohm's law then gives us the voltages across every resistor. Application of Kirchhoff's rules leads to exactly the same number of independent simultaneous equations as there are unknown currents. The equations are linear in the unknown currents and, because the equations fully represent a real physical network, they can always be solved, and there is always only one solution.

Worked Example 18.1 Apply Kirchhoff's rules to the circuit in Fig 18.4 to find the current flowing in each branch.

Answer There are three branches in the network, carrying currents I_1, I_2, and I_3 as shown in the figure. At junction B, Kirchhoff's first rule leads to

$$I_1 = I_2 + I_3. \tag{18.1}$$

Around the loop ABEFA the second rule gives

$$I_1 R_1 + I_2 R_2 = V_1, \tag{18.2}$$

and around the loop BCDEB

$$I_3 R_3 - I_2 R_2 = 0. \tag{18.3}$$

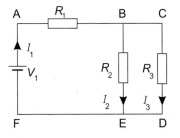

Fig. 18.4 This circuit has three branches and three unknown currents.

Note the minus sign in eqn (18.3). The current I_2 is drawn downwards in the figure, in the opposite direction to the path following the loop BCDEB. For more complicated circuits we may not be able to guess in advance in which direction the current will flow in a particular branch. The voltage contribution in Kirchhoff's second rule must be written in the equation as positive or negative depending on whether the direction chosen for the current is in the same direction or opposite to the path.

There is a third possible loop in the network, namely ACDFA. However, the equation resulting from Kirchhoff's second rule to this loop is not independent of those we have already set up: it is simply the result of adding eqns (18.2) and (18.3).

Equations (18.1)–(18.3) are easily solved. From (18.3) $I_3 = I_2 R_2 / R_3$ and in (18.1), $I_1 = I_2(1 + R_2/R_3)$. Substituting for I_1 in eqn (18.2),

$$I_2 = V_1 \left(\frac{R_3}{R_1 R_2 + R_2 R_3 + R_3 R_1} \right),$$

and hence

$$I_3 = V_1 \left(\frac{R_2}{R_1 R_2 + R_2 R_3 + R_3 R_1} \right),$$

$$I_1 = V_1 \left(\frac{R_2 + R_3}{R_1 R_2 + R_2 R_3 + R_3 R_1} \right). \tag{18.4}$$

Note that the voltage $I_2 R_2$ is the same as $I_3 R_3$ as it should be, and the addition of this voltage to $I_1 R_1$ gives the battery voltage V_1.

Impedance

Equation (18.4), relating the current I_1 delivered by the battery in Fig 18.4 to its voltage V_1, may be written as ($V_1 = I_1 \times$ a constant). This is of the same form as Ohm's law. The constant has the dimensions of resistance and, if, for example, R_1, R_2, and R_3 all have the same value R, then $V_1 = I_1 \times (\frac{3}{2} R)$. If we are primarily interested to know the current delivered by the battery, the network of resistors is equivalent to a single resistor with the resistance $\frac{3}{2} R$ ohms. This equivalent resistance is called the **impedance** of the network of resistors between the points A and F. We shall find that the concept of impedance is particularly useful when extended to AC circuits.

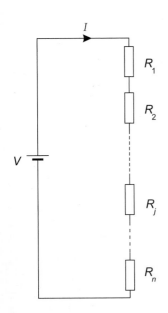

Fig. 18.5 A number of resistors connected in series.

● *Series resistors*

Series and parallel impedances

Figure 18.5 shows n resistors connected together in a string, with a source of emf of voltage V connected to the first and last resistors in the string. The same current I flows through all the resistors. Applying Kirchhoff's second rule to the whole circuit,

$$IR_1 + IR_2 + \cdots + IR_j + \cdots + IR_n = I(R_1 + R_2 + \cdots + R_j + \cdots + R_n)$$

$$= I \sum_{j=1}^{n} R_j = V.$$

Two or more resistors connected together in this way are said to be **in series**. The impedance $R_s = V/I$ of a string of resistors connected in series is simply the sum of the resistances of each resistor separately,

$$R_s = \sum_{j=1}^{n} R_j. \tag{18.5}$$

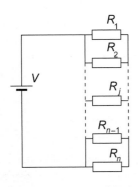

Fig. 18.6 A number of resistors connected in parallel.

● *Parallel resistors*

The resistors in Fig 18.6 are all connected to the same two junctions. They are said to be **in parallel**. The whole of the voltage V of the battery is across each resistor, so that the current through the resistance R_j is $I_j = V/R_j$. The total current I delivered by the battery is found by applying Kirchhoff's first rule to one of the junctions, adding together the currents through all the resistors, with the result

$$I = \frac{V}{R_1} + \frac{V}{R_2} + \cdots + \frac{V}{R_j} + \cdots + \frac{V}{R_n} = V \sum_{j=1}^{n} \frac{1}{R_j}.$$

Calling the impedance of the resistors in parallel R_p, $V = IR_p$, with

$$\frac{1}{R_p} = \sum_{j=1}^{n} \frac{1}{R_j}. \tag{18.6}$$

Frequently, the solution of network problems can be greatly simplified by dividing the network into sections of series or parallel resistors. The impedance of the resistor network in Worked example 18.1 can be found directly in this way, without having to set up equations from Kirchhoff's rules at all. In Fig 18.4, the resistors R_2 and R_3 are in parallel and have an impedance R_p given by $1/R_p = 1/R_2 + 1/R_3$ or $R_p = R_2 R_3/(R_2 + R_3)$. This impedance is in series with R_1, and the impedance of the whole network made up of R_1, R_2, and R_3 is therefore

$$R_1 + \frac{R_2 R_3}{R_2 + R_3} = \frac{R_1 R_2 + R_2 R_3 + R_3 R_1}{R_2 + R_3},$$

the same as the result obtained by solving the simultaneous equations given by Kirchhoff's rules.

Exercise 18.2 What value of R will give an impedance of $50\,\Omega$ for the impedance between the points A and E for the network to the right of the dashed line in Fig 18.7? For this value of R, what are the potentials at points A, B, and C, taking earth as the zero of potential? (The symbol $\doteqdot$ at E indicates that this point is connected to earth.)

Answer The value of R is $50/3\,\Omega$. The potential at A is $V/2$ and at both B and C it is $V/4$.

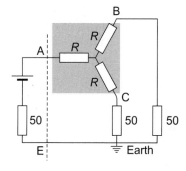

Fig. 18.7 The impedance of the network to the right of A is used to work out the current delivered by the battery. The number placed beside a resistor is its resistance in ohms. The symbol labelled 'Earth' indicates a connection to earth.

The steady signal $V/2$ at A in Fig. 18.7, which is driving an impedance of $50\,\Omega$ is split into two equal signals $V/4$ that also drive $50\,\Omega$ impedances. Non-steady signals are distorted when they are sent along cables unless they drive a particular impedance that depends on the properties of the cable. For scientific equipment this impedance is usually chosen to be $50\,\Omega$. The three equal resistors R act as a simple voltage splitter for AC and pulse signals as well as for DC signals. Notice that signals sent back from B or C towards the three resistors in the shaded area see the same impedance as the signal from A. The network will split signals arriving at any of its terminals A, B, or C.

18.2 Alternating currents

As outlined in Section 17.4, the emf generated at a power station varies sinusoidally with time. The frequency is 50 Hz in Europe and 60 Hz in North America. When a network of passive components are driven by a sinusoidally varying emf, the currents also vary sinusoidally at the same frequency as the emf. Such a network is called an **alternating current** circuit, or AC circuit for short. Alternating current circuits are important, not only for handling mains power at 50 or 60 Hz, but also over a very wide range of other frequencies. Electrical signals are used, for example, to drive loudspeakers at audio frequencies from about 10 Hz to over 10 kHz, and radio waves are broadcast over long distances at frequencies from 10^5 to more than 10^9 Hz. In this section we shall analyse AC circuits in a similar way to DC circuits, but including capacitors and coils with inductance as components in a network.

Sinusoidally varying quantities are described in the context of simple harmonic motion in Chapter 6. A voltage varying sinusoidally with time at an angular frequency ω rad s^{-1} may be written as

$$V(t) = V_0 \cos(\omega t + \phi). \tag{18.7}$$

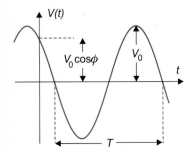

Fig. 18.8 The sinusoidally varying voltage $V(t) = V_0 \cos(\omega t + \phi)$.

The voltage $V(t)$ is sketched in Fig 18.8. The nomenclature for describing AC is the same as is used for simple harmonic motion. The maximum value of the voltage V_0 is called the *amplitude*, and the argument $(\omega t + \phi)$ the *phase*: ϕ is thus the phase at $t = 0$. The repetition frequency of the voltage is $\nu = \omega/2\pi$, and the time $T = 2\pi/\omega$ is called the *period*.

The amplitude V_0 of the mains supply is over 300 volts in Europe and an electric shock from the mains can be dangerous. For safety reasons the metal parts of equipment attached to a mains supply have a connection to earth. In the domestic supply to a house, for example, in Europe an earth connection is provided by the mains supply. One of the two leads carrying the mains current, the *neutral* lead, is attached to a remote earth, and the AC voltage $V_0 \cos \omega t$ is maintained between the *live* and the neutral leads.

In any AC circuit driven by emfs all at the same angular frequency ω, the currents everywhere in the circuit must also vary with the same angular frequency ω. However, voltages and currents in different branches of the circuit are not necessarily at the same phase. The voltage–current phase relationships for resistors, capacitors, and coils are discussed next.

Resistance, capacitance, and inductance in AC circuits

Figure 18.9(a) is the circuit diagram for an alternating emf delivering current to a resistor. If we choose the time zero to be at a moment when

● **Earth connections**

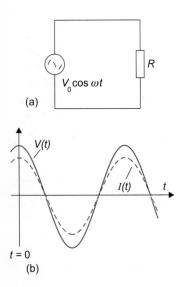

Fig. 18.9 Current and voltage are in phase for a resistor.

the voltage across the resistor has its maximum value V_0, then the voltage as a function of time is

$$V(t) = V_0 \cos \omega t. \tag{18.8}$$

At all times a resistor obeys Ohm's law $V = IR$ where the resistance R is a constant and the current $I(t)$ is

$$I(t) = \frac{V(t)}{R} = \frac{V_0}{R} \cos \omega t = I_0 \cos \omega t \tag{18.9}$$

where $I_0 = V_0/R$ is the amplitude of the current. The current and the voltage both have the same phase at all times—they are said to be *in phase* with one another. The voltage across a resistor and the current through it are given by eqns (18.8) and (18.9) and are drawn in Fig 18.9(b).

In Fig 18.10(a) the resistor has been replaced by a capacitor with capacitance C. If the voltage across the capacitor is V, then there are equal and opposite charges $\pm Q$ on its plates. The magnitude of the charge is given by eqn (15.49) as $Q = CV$. This relation defines capacitance and, although it was introduced in Section 15.6 in the context of steady voltages, it applies for varying voltages as well. If V increases, positive charge must arrive at the positively charged plate and an equal positive charge must leave the negatively charged plate. A current flows through the leads to the capacitor even though no current is flowing in the space between the plates. Current is measured in amperes, which are equivalent to coulombs per second. At any time, the rate of increase of charge Q carried by the capacitor plate is the same as the current $I(t)$ arriving along the lead,

$$I(t) = \frac{dQ}{dt}. \tag{18.10}$$

This equation is correct whatever the rate of change of charge on the capacitor. In particular, if there is an alternating voltage of the same form as in eqn (18.8), $Q = CV_0 \cos \omega t$ and

$$I(t) = \frac{d}{dt}(CV_0 \cos \omega t) = -\omega CV_0 \sin \omega t = \omega CV_0 \cos\left(\omega t + \frac{\pi}{2}\right). \tag{18.11}$$

The current is *in advance* of the voltage by the phase angle $\pi/2$. The voltage and the current for the capacitor are drawn in Fig 18.10(b), which shows the current passing through zero at $t = 0$, before the voltage falls to zero when $\omega t = \pi/2$.

Finally, in Fig 18.11(a) an idealized self-inductance is placed across the source of alternating emf. The self-inductance of a coil, which was

● *Voltage and current are in phase for a resistor*

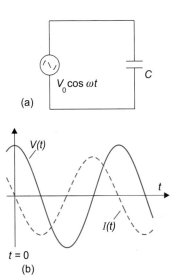

Fig. 18.10 The current leads the voltage for a capacitor.

● *Current leads voltage in a capacitor*

introduced in Section 17.3, gives the strength of the emf induced in the coil when a changing current passes through it.

An idealized coil with no resistance, represented by the coil symbol in Fig 18.11(a), is not realistic, since real coils possess resistance as well as inductance. However, it is very helpful to understand the behaviour of an idealized inductance before incorporating the coil's resistance into the equivalent circuit. Consider the circuit at a moment when the current I is flowing in the direction indicated in Fig 18.11(a) and is increasing. At this moment the upper end of the voltage supply is positive with respect to the lower end. The induced emf in the coil of self-inductance L is given by eqn (17.14) as $-L\mathrm{d}I/\mathrm{d}t$. Since I is increasing, this emf is negative, opposing the change of current as required by Lenz's law, and the upper end of the self-inductance is positive with respect to the lower end. The emfs from the voltage supply and from the self-inductance oppose one another and are equal: as there is no resistance in the circuit the net emf around it must be zero; otherwise an infinite current would flow.

The equation relating voltage and current, which we have already met as eqn (17.15), is

$$V(t) - L\frac{\mathrm{d}I}{\mathrm{d}t} = 0. \tag{18.12}$$

For $V(t) = V_0 \cos \omega t$ eqn (18.12) integrates to give

$$\frac{V_0}{\omega} \sin \omega t = LI + c.$$

The value of the constant of integration c is zero since no steady current is flowing. Rearranging,

$$I(t) = \frac{V_0}{\omega L} \sin \omega t = \frac{V_0}{\omega L} \cos\left(\omega t - \frac{\pi}{2}\right). \tag{18.13}$$

The voltage and current for the idealized self-inductance are drawn in Fig 18.11(b), which shows the current *lagging behind* the voltage by a phase angle $\pi/2$. The current reaches its maximum when $\omega t = +\pi/2$, after the voltage has passed its maximum at $\omega t = 0$.

For AC circuits the ratio of the voltage *amplitude* to the current *amplitude* is called the impedance, by analogy with the impedance of a DC network, which is the ratio of voltage to current for steady currents. From eqns (18.9), (18.11), and (18.13), the impedances of the resistance, capacitance, and self-inductance are R, $1/(\omega C)$, and ωL, respectively. The units of all these quantities are ohms. Capacitors present a high impedance at low frequencies and low impedance at high frequencies, whereas the reverse is true for self-inductance.

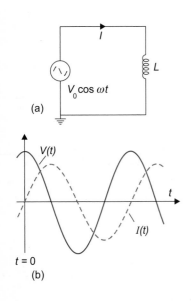

(a)

(b)

$t = 0$

Fig. 18.11 The current lags behind the voltage for self-inductance.

● *Current lags behind voltage in an inductor*

Exercise 18.3 Calculate the impedance, in ohms, of a 1 μF capacitor and a 1 mH inductance at 10 kHz.

Answer 16 Ω and 63 Ω.

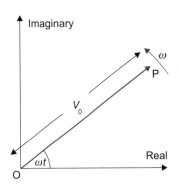

Fig. 18.12 The complex number representing voltage rotates anti-clockwise at a constant angular speed in the Argand diagram.

Complex impedance

The magnitudes of the impedances of the passive components in an AC network are not enough on their own to allow the distribution of currents in the network to be calculated. Phase differences caused by capacitance and inductance must also be taken into account. This can be done in a concise way by including phase information as well as magnitudes in complex numbers. Complex numbers were introduced in Section 6.3 to describe forced mechanical oscillations, and the same technique can be applied to AC circuits. The properties of complex numbers that are required for AC theory are summarized in Section 20.6: you need to be familiar with the contents of this section to be able to work out currents and voltages in AC circuits.

First, we must explain the notation used to represent an alternating voltage by a complex number. According to eqn (20.52) the complex number $e^{j\omega t}$ may be rewritten as

$$e^{j\omega t} = \cos \omega t + j \sin \omega t. \tag{18.14}$$

The magnitude of $e^{j\omega t}$ is unity and its real part is $\cos \omega t$. We can therefore represent an alternating voltage with amplitude V_0 and angular frequency ω by the real part of the complex number $V_0 e^{j\omega t}$.

In Fig 18.12 the complex number $V_0 e^{j\omega t}$ is plotted as the point P on a diagram in which the real part is the horizontal coordinate and the imaginary part is the vertical coordinate. A diagram representing complex numbers in this way is called an *Argand diagram*. The line OP from the origin to P has a length V_0 and the angle between OP and the real axis is ωt, which is the phase of the voltage: OP rotates anticlockwise at a uniform angular speed ω about O. The projection of OP on to the real axis is $V_0 \cos \omega t$, which is the actual alternating voltage. The imaginary part of $V_0 e^{j\omega t}$ has no physical meaning, and it is only introduced as a mathematical device that helps to keep track of phase differences.

We shall denote the sinusoidally varying complex number $V_0 e^{j\omega t}$ by $\widetilde{V}$ to distinguish it from the actual alternating voltage,

$$\widetilde{V} = V_0 e^{j\omega t} = V_0 \cos \omega t + j V_0 \sin \omega t. \tag{18.15}$$

Because the complex number $\widetilde{V}$ contains information about the phase as well as the amplitude of the voltage, it is called a **phasor** and in this context the Argand diagram is referred to as a phasor diagram.

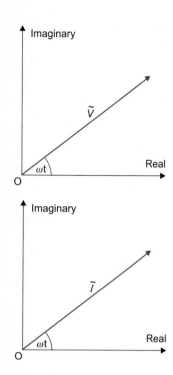

Fig. 18.13 Voltage and current phasors for a resistor.

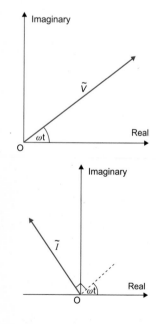

Fig. 18.14 Voltage and current phasors for a capacitor.

Currents may similarly be represented by phasors. The current may not be in phase with the voltage and, if the phase difference is ϕ, we represent the current by

$$\widetilde{I} = I_0 e^{j(\omega t + \phi)} = I_0[\cos(\omega t + \phi) + j\sin(\omega t + \phi)]. \tag{18.16}$$

What is the relation between the phasors $\widetilde{I}$ and $\widetilde{V}$ for resistive, capacitive, and inductive components? First consider resistors. The phase difference ϕ between current and voltage is zero, and from eqn (18.9) the amplitude of the current is $I_0 = V_0/R$. The phasor representing the current is

$$\widetilde{I} = I_0 e^{j\omega t} = \frac{V_0}{R} e^{j\omega t} = \frac{\widetilde{V}}{R}.$$

The phasor equation $\widetilde{V} = \widetilde{I}R$ looks just the same as Ohm's law $V = IR$ for the actual current and voltage. The phasor diagrams for $\widetilde{V}$ and $\widetilde{I}$ are shown in Fig 18.13. Both phasors rotate about the origin at an angular speed ω and both have the same phase at all times.

For a capacitor the actual current is given by eqn (18.11). The current amplitude is $\omega C V_0$ and the current leads the voltage by the phase angle $\pi/2$. The phasor for the current is

$$\widetilde{I} = I_0 e^{j(\omega t + \pi/2)} = \omega C V_0 e^{j(\omega t + \pi/2)}. \tag{18.17}$$

Splitting $e^{j\pi/2}$ into its real and imaginary parts in the same way as for the complex exponential function in eqn (18.14),

$$e^{j\pi/2} = \cos(\pi/2) + j\sin(\pi/2) = j, \tag{18.18}$$

and eqn (18.17) becomes

$$\widetilde{I} = j\omega C V_0 e^{j\omega t} = j\omega C\widetilde{V}$$

or

$$\widetilde{V} = \frac{\widetilde{I}}{j\omega C}. \tag{18.19}$$

This equation again has the form of Ohm's law with the resistance R replaced by the complex number $1/j\omega C$. The phasors $\widetilde{V}$ and $\widetilde{I}$ now point in different directions. At $t = 0$ when $\widetilde{V}$ is real, $\widetilde{I}$ is purely imaginary because of the factor j in eqn (18.19). The two phasors are perpendicular to one another, and remain so at all times, as indicated in Fig 18.14.

The analysis of the current passing through an idealized self-inductance is similar. From eqn (18.13) the phase difference between current and voltage is $(-\pi/2)$ and the amplitude of the current is $V_0/\omega L$. Using

$$e^{-j\pi/2} = \cos(-\pi/2) + j\sin(-\pi/2) = -j, \tag{18.20}$$

the complex current is

$$\tilde{I} = I_0 e^{j(\omega t - \pi/2)} = \frac{-j}{\omega L} V_0 e^{j\omega t}$$

or, since $-j = -j^2/j = 1/j$,

$$\tilde{V} = \tilde{I} \times j\omega L. \tag{18.21}$$

This has the same form as Ohm's law with the resistance R replaced by the complex number $j\omega L$. The phasors $\tilde{V}$ and $\tilde{I}$ are again perpendicular to one another as for the capacitor, but now the current lags behind the voltage (Fig 18.15).

For all three types of passive component the relation between voltage and current phasors is expressed in the same form as Ohm's law for resistors. This is why the use of complex numbers simplifies AC calculations. The ratio $\tilde{V}/\tilde{I}$ is called the **complex impedance** and is denoted by $\tilde{Z}$. The complex impedance satisfies the equation

$$\tilde{V} = \tilde{I}\tilde{Z} \tag{18.22}$$

where

$$\tilde{Z} = R \text{ for resistance } R, \tag{18.23}$$

$$\tilde{Z} = \frac{1}{j\omega C} \text{ for capacitance } C, \tag{18.24}$$

and

$$\tilde{Z} = j\omega L \text{ for inductance } L. \tag{18.25}$$

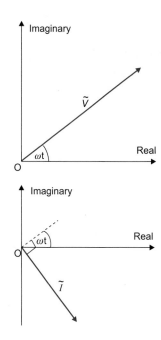

Fig. 18.15 Voltage and current phasors for self-inductance.

Kirchhoff's rules for AC circuits

When an AC network contains capacitance and inductance, the currents in different branches of the network may have different phases. The phase differences have to be taken into account when applying Kirchhoff's first rule at a junction in a circuit. Figure 18.16 shows two alternating currents I_1 and I_2 that have the same amplitude but differ in phase by $\pi/2$. The sum of these currents has an amplitude $\sqrt{2}$ times that of I_1 and I_2, and is not in phase with either. To handle the amplitude and phase changes at a junction using trigonometrical functions leads to some very messy algebra. However, since the real parts of phasors are the same as the actual currents, we can write $\tilde{I}_3 = \tilde{I}_1 + \tilde{I}_2$. The argument may be extended to junctions of any number of connections, and we may say that the algebraic sum of the current phasors arriving at any junction is zero: Kirchhoff's first rule applies to the phasors for AC currents. The current phasors through all the components in any branch of a network must

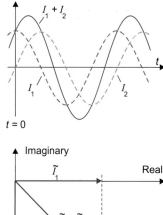

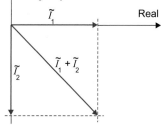

Fig. 18.16 The addition of alternating currents depends on phase as well as amplitude.

therefore be the same, since for any two components connected in series the current leaving one and entering the other must be the same.

Kirchhoff's second rule also applies to AC networks, provided that induced emfs are taken into account. In the presence of changing magnetic fields the electric field is no longer conservative, but Faraday's law holds, and the line integral of the electric field is given by eqn (17.10),

$$\oint \mathbf{E} \cdot d\boldsymbol{\ell} = -\frac{d\Phi}{dt}.$$

The right-hand side of this equation is the rate of change of magnetic flux through the closed circuit around which the line integral is evaluated. For the networks we are representing by equivalent circuits, there are no fields except in the separate components. Induced emfs only occur because of the self-inductance and mutual inductance of coils. The emf due to a self-inductance L, for example, is represented as a complex voltage $j\omega L\tilde{I}$. When induced emfs are included in this way as complex voltages, Kirchhoff's second rule, that the sum of the voltages around any loop of the network is zero, is still true. Worked example 18.2 applies this rule to a real coil that has resistance as well as self-inductance.

● *In AC circuits, Kirchhoff's rules apply to complex voltages and currents*

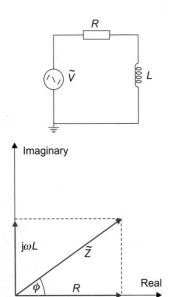

Fig. 18.17 The equivalent circuit for a real coil, including the resistance of the windings. The complex impedance is shown graphically on an Argand diagram.

Worked Example 18.2 A coil with self-inductance L and resistance R is connected to a generator delivering an AC voltage of amplitude V_0 at an angular frequency ω, as represented by the equivalent circuit shown in Fig 18.17. What is the phase difference between the voltage and the current in the coil? In the figure one terminal of the generator is shown connected to earth (the symbol $\doteq$): this does not affect the calculations, but is included to remind you that most real circuits do have earth connections.

Answer The current $\tilde{I}$ through R and L is the same. The complex impedances of R and L are given by eqns (18.23) and (18.24). Equating the voltage of the generator to the voltage drops across R and L,

$$\tilde{V} = \tilde{I}(j\omega L + R) = \tilde{I}\tilde{Z} \tag{18.26}$$

where the complex impedance of the coil, including both its resistance and self-inductance, is

$$\tilde{Z} = j\omega L + R. \tag{18.27}$$

Figure 18.17 includes an Argand diagram showing how the real and imaginary parts of $\tilde{Z}$ are added to give the complex impedance of the whole coil. From this figure we see that the magnitude of the impedance is $Z_0 = \sqrt{(\omega L)^2 + R^2}$ and that it makes an angle ϕ with the real axis given by $\tan\phi = \omega L/R$. The impedance $\tilde{Z}$ may be written alternatively as

$\widetilde{Z} = Z_0 e^{j\phi}$, and eqn (18.26) becomes

$$\widetilde{V} = Z_0 e^{j\phi} \widetilde{I}.$$

The angle $\phi = \tan^{-1}(\omega L/R)$ is thus the phase difference by which the current through the coil lags behind the voltage across it.

This result may be obtained mathematically using equations from the section on complex numbers in the mathematical review Section 20.6. We have started with the complex impedance in the form of eqn (20.46) as $\widetilde{Z} = j\omega L + R$. From eqns (20.52) and (20.47) this may be rewritten as

$$\widetilde{Z} = Z_0 e^{j\phi} = Z_0(\cos\phi + j\sin\phi).$$

The relation between the two forms is given by eqns (20.48) and (20.49) as $Z_0 = \sqrt{(\omega L)^2 + R^2}$ and $\tan\phi = \omega L/R$.

● *The mathematical way of finding impedance and phase difference*

The impedance $\widetilde{Z}$ of the coil given by eqn (18.27) is the sum of the impedances of the components L and R, which are in series in Fig 18.17. Since the voltage and current phasors satisfy Kirchhoff's rules, complex impedances that are in series or in parallel may be combined in the same way as series and parallel resistors. For n impedances $\widetilde{Z}_j$ in series, eqn (18.5) is extended to

$$\widetilde{Z}_s = \sum_{j=1}^{n} \widetilde{Z}_j. \tag{18.28}$$

Similarly for n impedances $\widetilde{Z}_j$ in parallel, eqn (18.6) becomes

$$\frac{1}{\widetilde{Z}_p} = \sum_{j=1}^{n} \frac{1}{\widetilde{Z}_j}. \tag{18.29}$$

The impedance of clusters of parallel and series impedances can be worked out using eqns (18.28) and (18.29) without having to use Kirchhoff's rules to evaluate all the currents in the network.

Exercise 18.4 Find an expression for the complex impedance of the network in Fig 18.18. What are the limiting values of the impedance at high and at low frequencies?

Answer $R + R/(1 + j\omega CR)$; R; $2R$.

Solving network problems for AC circuits is formally the same as for steady currents. If one or more known alternating emfs are delivering current to the network, there is always the same number of linearly

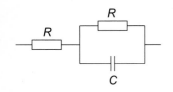

Fig. 18.18 Complex numbers are used to calculate the impedance of a network made up of series and parallel components.

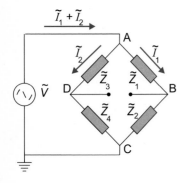

Fig. 18.19 An alternating current bridge.

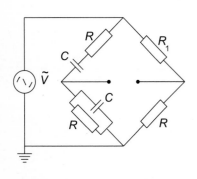

Fig. 18.20 The Wien bridge.

independent equations as the number of unknown currents, and there is only one solution to these equations if all the impedances in the network are fixed. What is different from the steady current networks is that both magnitude and phase must be found for each unknown current. This can be achieved because each complex number equation is really two equations: the real and imaginary parts of the equation must be separately satisfied, leading to the real and imaginary parts of each unknown current.

For example, the circuit shown in Fig 18.19 is called an AC bridge. The bridge is said to be balanced if there is no voltage across the connections to the points B and D. Currents $\tilde{I}_1$ and $\tilde{I}_2$ flow through the parallel arms of the bridge. The voltage across each arm is $\tilde{V}$ and, since each arm is made up of two series impedances,

$$\tilde{V} = \tilde{I}_1(\tilde{Z}_1 + \tilde{Z}_2) = \tilde{I}_2(\tilde{Z}_3 + \tilde{Z}_4).$$

The point C is earthed, and the voltage at B is

$$\tilde{V}_B = \tilde{I}_1\tilde{Z}_2 = \tilde{V}\tilde{Z}_2/(\tilde{Z}_1 + \tilde{Z}_2)$$

and, similarly,

$$\tilde{V}_D = \tilde{I}_2\tilde{Z}_4 = \tilde{V}\tilde{Z}_4/(\tilde{Z}_3 + \tilde{Z}_4).$$

The balance condition is

$$\tilde{V}_B = \tilde{V}_D, \quad \text{i.e.} \quad \tilde{Z}_2/(\tilde{Z}_1 + \tilde{Z}_2) = \tilde{Z}_4/(\tilde{Z}_3 + \tilde{Z}_4).$$

Inverting both sides of this equation,

$$\frac{\tilde{Z}_1 + \tilde{Z}_2}{\tilde{Z}_2} = \frac{\tilde{Z}_3 + \tilde{Z}_4}{\tilde{Z}_4} \quad \text{or} \quad \frac{\tilde{Z}_1}{\tilde{Z}_2} = \frac{\tilde{Z}_3}{\tilde{Z}_4}. \tag{18.30}$$

There is a great variety of AC bridges, which have different combinations of components in the positions of $\tilde{Z}_1$ to $\tilde{Z}_4$. Both the real and imaginary parts of the right and left sides of eqn (18.30) must be equal, giving the two conditions that are required for the magnitude and the phase of the voltages at B and D to be equal. The balance condition for one such bridge is given in Worked example 18.3.

Worked Example 18.3 Find the balance conditions for the AC bridge in Fig 18.20.

Answer Equating the general bridge impedances in Fig 18.19 and the impedances in the Wien bridge in Fig 18.20, $\tilde{Z}_1 = R_1$ and $\tilde{Z}_2 = R$. The impedance of C and R in series, corresponding to $\tilde{Z}_3$, is $(R + 1/j\omega C)$. For the parallel combination of R and C, eqn (18.29) gives $(\tilde{Z}_4)^{-1} = (1/R + j\omega C)$. Substituting these values in eqn (18.30),

$$\frac{R_1}{R + 1/j\omega C} = R(1/R + j\omega C).$$

Multiplying the top and bottom of the left-hand side by $j\omega C$ and then both sides by $(1 + j\omega CR)$ gives

$$j\omega CR_1 = (1 + j\omega CR)^2 = (1 - \omega^2 C^2 R^2) + 2j\omega CR.$$

The real part of this equation is $(1 - \omega^2 C^2 R^2) = 0$, requiring $\omega = 1/CR$. The imaginary part is $j\omega CR_1 = 2j\omega CR$, requiring $R_1 = 2R$. This particular bridge will only balance at one frequency. It is often used as part of circuits that are designed to oscillate at the bridge balance frequency.

Power in AC circuits

The power required to maintain a steady current I in a branch of a network that has a steady voltage V between its ends is given by eqn (16.10) as VI. If the current and voltage are varying with time, having values $I(t)$ and $V(t)$ at time t, the *instantaneous* power required at t, which is the rate of work being done at the time t, is simply $V(t)I(t)$. For an alternating voltage $V_0 \cos \omega t$ applied across a resistor with resistance R, the current is $I(t) = V_0 \cos \omega t/R$ and the instantaneous power is

$$P(t) = \frac{V_0^2 \cos^2 \omega t}{R}. \tag{18.31}$$

The voltage and power are sketched in Fig 18.21: the power reaches its maximum value twice in every cycle of period $T = 2\pi/\omega$, with the maximum power occurring when the voltage is $\pm V_0$. The average power $\overline{P}$ over a complete cycle is found by integrating $P(t)$ from $t = 0$ to $t = T$ and dividing by the period T,

$$\overline{P} = \frac{\omega}{2\pi} \int_0^{2\pi/\omega} \frac{V_0^2 \cos^2 \omega t}{R} dt = \frac{1}{2} \frac{V_0^2}{R}. \tag{18.32}$$

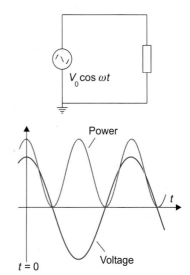

Fig. 18.21 The power dissipated in a resistor by an AC current is always positive.

When discussing power in AC circuits, particularly in the context of mains power, voltage and current are not usually described by their amplitudes. **Root mean square** (rms) values, which are smaller by a factor $1/\sqrt{2}$, are used instead,

$$V_{\text{rms}} = V_0/\sqrt{2}; \quad I_{\text{rms}} = I_0/\sqrt{2}. \tag{18.33}$$

In terms of rms values the power has the same form as for steady currents and the factor $\frac{1}{2}$ in eqn (18.32) is removed,

$$\overline{P} = V_{\text{rms}} I_{\text{rms}} = V_{\text{rms}}^2/R. \tag{18.34}$$

Electricity utilities guarantee to provide their customers with a supply voltage specified in rms volts. Remember that the amplitude of the voltage is larger than this value. For example, if the nominal mains voltage is 240 rms volts, the peak voltage is almost 340 volts.

In an AC circuit power is only dissipated in resistors. Energy is needed to charge a capacitor or to drive current through inductance, and this energy must be provided by a source of emf. However, the energy in a capacitor or an idealized inductance is not lost as heat, but is stored and is released later in the AC cycle. Averaged over a whole cycle, the net energy delivered by the source of emf to a capacitor or an idealized inductance is zero.

For an AC voltage $V_0 \cos \omega t$ across a capacitor with capacitance C, as in Fig 18.22, the current is given by eqn (18.11),

$$I(t) = -\omega C V_0 \sin \omega t.$$

The instantaneous power delivered by the source of emf at a time t is

$$P(t) = V(t)I(t) = -\omega C V_0^2 \cos \omega t \sin \omega t = -\tfrac{1}{2}\omega C V_0^2 \sin 2\omega t. \quad (18.35)$$

The power is sketched in Fig 18.22 below the circuit diagram. Over a whole cycle of length $T = 2\pi/\omega$ the areas above and below the time axis are equal: as much energy is returned by the capacitor as is received. The energy is built up twice in every cycle, once as the capacitor is charged to the positive voltage V_0 and once as it is charged to the negative voltage $(-V_0)$.

A very similar result applies to an idealized inductance L. The current through L is given by eqn (18.13) as

$$I(t) = \frac{V_0}{\omega L} \sin \omega t,$$

and the power delivered is

$$P(t) = \frac{V_0^2}{\omega L} \cos \omega t \sin \omega t = \frac{V_0^2}{2\omega L} \sin 2\omega t. \quad (18.36)$$

Once again the net energy delivered during a complete cycle is zero as the magnetic energy stored by the inductance is built up and dies away.

A word of warning is needed here. In the calculation of power we have not made use of the complex number notation. This is because voltage and current have to be *multiplied* to give the power. Multiplication of the imaginary parts of a complex voltage and a complex current leads to a real contribution that should not be there, since the imaginary parts have no physical meaning. It is a valid procedure to use complex numbers to calculate currents and voltages in a network, but the real parts must be multiplied on their own when calculating the power in any part of the network.

This can be illustrated by considering the power delivered by an alternating emf $V_0 \cos \omega t$ connected to a complex impedance $\widetilde{Z}$, as in Fig 18.23. The current in the circuit is not in phase with the voltage unless $\widetilde{Z}$ is purely resistive. Expressing the complex impedance in the same way as in Worked example 18.2, $\widetilde{Z} = Z_0 e^{j\phi}$. For a voltage $\widetilde{V} = V_0 e^{j\omega t}$, the

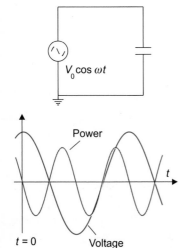

Fig. 18.22 When a capacitor is placed across a source of alternating emf, power is drawn from the source in one half of the cycle and returned in the other half.

● *Complex numbers must not be used for calculating power*

complex current is

$$\tilde{I} = \frac{\tilde{V}}{Z_0 e^{j\phi}} = \frac{V_0}{Z_0} e^{j(\omega t - \phi)}.$$

The real part $(V_0/Z_0)\cos(\omega t - \phi)$ of this complex current corresponds to the actual current flowing in the circuit, and to find the instantaneous power it must be multiplied by the actual voltage $V_0\cos\omega t$ to give

$$P(t) = \frac{V_0^2}{Z_0}\cos\omega t\cos(\omega t - \phi) = \frac{V_0^2}{Z_0}\cos\omega t(\cos\omega t\cos\phi + \sin\omega t\sin\phi)$$

$$= \frac{V_0^2}{Z_0}\left(\cos^2\omega t\cos\phi + \frac{1}{2}\sin 2\omega t\sin\phi\right). \tag{18.37}$$

Averaging over a complete cycle, the term in $\sin 2\omega t$ is zero. As for pure capacitance and pure inductance, this term represents energy stored and returned to the source of emf. Since the average value of $\cos^2\omega t$ is $\frac{1}{2}$, the other term, arising from energy dissipated in the resistance, gives

$$\overline{P} = \frac{V_0^2}{2Z_0}\cos\phi = \tfrac{1}{2}V_0 I_0\cos\phi = V_{\mathrm{rms}}I_{\mathrm{rms}}\cos\phi. \tag{18.38}$$

Because the voltage and current differ by the phase angle ϕ, the average power is reduced by the **power factor** $\cos\phi$.

When large amounts of power are being consumed, it is important that the power factor is not very far from unity. For a given power and a mains supply generating a fixed voltage, if $\cos\phi$ is less than one, I_0 must be increased by $1/\cos\phi$. This causes losses in the transmission cables and it is also inefficient to run the huge alternators at power stations so that the voltage and current are out of phase. For this reason consumers of large amounts of power are charged by a formula that includes a cost for $\cos\phi < 1$, and a heavy penalty if the power factor falls below some value set by the power company.

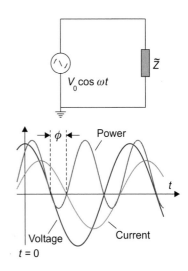

Fig. 18.23 The power dissipated in a resistor is reduced if the total impedance is complex because the voltage and current are not in phase.

Worked Example 18.4 A coil has a self-inductance of 2 mH and a resistance of $10\,\Omega$. Calculate its power factor at 10 kHz.

Answer From Worked example 18.2 the phase angle ϕ between voltage and current in the coil is given by $\tan\phi = \omega L/R$. Here $\omega = 2\pi \times 10^4$ rad s^{-1}, $L = 2 \times 10^{-3}$ mH, and $R = 10\,\Omega$. Hence $\tan\phi = 2\pi \times 10^4 \times 2 \times 10^{-3}/10 = 4\pi$, $\phi = 85.5°$, and $\cos\phi \approx 0.06$. Very little power is dissipated in a small coil at this frequency which is within the audio range.

Resonance

Figure 18.24 shows a circuit consisting of L, R, and C in series across an emf $\widetilde{V}$ and a diagram of the complex impedances, with the real and imaginary axes measured in ohms. The resistance has a magnitude R and points along the real axis. The inductance has impedance $j\omega L$; its magnitude is ωL and it is directed along the positive imaginary axis. The capacitance has impedance $1/j\omega C$ and, since $1/j = -j$, the impedance on the diagram is of magnitude $1/\omega C$ and points downwards, along the negative imaginary axis. The components are in series and their total impedance is the sum of their separate impedances,

$$\widetilde{Z} = R + j\omega L + 1/j\omega C = R + j(\omega L - 1/\omega C). \tag{18.39}$$

The impedance $\widetilde{Z}$ is also shown on Fig 18.24. We can visualize how $\widetilde{Z}$ changes as the angular frequency ω changes. When ω is very small, $1/\omega C$ is much larger than ωL: the imaginary part of the impedance then points downwards and has a much greater magnitude than the real part R. The current leads the voltage by a phase angle which is nearly $\pi/2$, as it is for a capacitor on its own.

As ω increases the phase angle gets smaller and smaller until at $1/\omega C = \omega L$ the impedance is purely real and has a magnitude R. The current and voltage are in phase and the magnitude of the current is the same as if L and C were not present. As ω increases further the phase angle increases, with the current lagging behind the voltage, with a phase angle approaching $\pi/2$ as ω becomes very large. The impedance $j\omega L$ in Fig 18.24 has a bigger magnitude than $1/j\omega C$ and the current lags behind the voltage by the phase angle ϕ. From the geometry of the figure, the amplitude of the current I_0 and the phase ϕ are given by

$$I_0 = \frac{V_0}{|\widetilde{Z}|} = \frac{V_0}{\left\{R^2 + (\omega L - 1/\omega C)^2\right\}^{1/2}}, \tag{18.40}$$

$$\tan \phi = \frac{\omega L - 1/\omega C}{R}. \tag{18.41}$$

The circuit is exhibiting *resonance* behaviour, and for fixed voltage amplitude, the current has its maximum amplitude at the resonance angular frequency ω_0, where $\omega_0 L = 1/\omega_0 C$, that is, $\omega_0 = 1/\sqrt{LC}$. At $\omega = \omega_0$, $I_0 = V_0/R$ and $\phi = 0$: the circuit acts as though only the resistor were present. Electrical resonance is just like the mechanical resonance discussed in Section 6.3. The mathematics of the electrical and mechanical resonances are in fact identical, as we can see by looking at the actual voltages in the circuit in Fig 18.24, omitting the imaginary parts

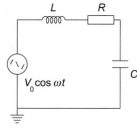

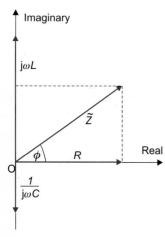

Fig. 18.24 A capacitor and coil connected in series exhibit resonance.

of the complex numbers. The voltages are

$$L\frac{dI}{dt} \quad \text{across} \quad L, \quad IR \quad \text{across} \quad R, \quad \text{and} \quad \frac{Q}{C} \quad \text{across} \quad C.$$

Writing the current I as dQ/dt (eqn 18.10), the voltage balance around the circuit becomes

$$L\frac{d^2Q}{dt^2} + R\frac{dQ}{dt} + \frac{Q}{C} = V_0\cos\omega t$$

or, dividing through by L,

$$\frac{d^2Q}{dt^2} + \frac{R}{L}\frac{dQ}{dt} + \frac{Q}{LC} = \frac{V_0}{L}\cos\omega t. \tag{18.42}$$

This has exactly the same form as eqn (6.22), which is

$$\frac{d^2x}{dt^2} + \gamma\frac{dx}{dt} + \omega_0^2 x = \frac{F_0}{m}\cos\omega t.$$

Comparing electrical and mechanical resonance, we see the following.

- The displacement x is replaced by the charge Q.
- The force amplitude F_0 is replaced by the voltage amplitude V_0.
- The damping constant γ is replaced by R/L.
- The resonant angular frequency ω_0 is replaced by $1/\sqrt{LC}$.
- The mass m is replaced by the inductance L.

With these substitutions almost all the results in Chapter 6 apply to the series LCR circuit. The energy $\frac{1}{2}LI^2$ stored in a coil is analogous to the kinetic energy $\frac{1}{2}mv^2$ (Section 6.1): during the AC cycle this energy is exchanged with the electrostatic energy $\frac{1}{2}Q^2/C$ in the capacitor, which is analogous to potential energy in the mechanical system.

The analogy also applies when there is no alternating voltage in the circuit and the right-hand side of eqn (18.42) is zero. For example, if a capacitor with capacitance C is charged and then allowed to discharge through a coil with self-inductance L and resistance R, the behaviour is similar to that of a spring that has been extended and then released.

If $\gamma/2$ is less than the resonant frequency ω_0, then the displacement of the spring is given as a function of time after release by eqn (6.17),

$$x = x_0\exp(-\gamma t/2)\cos\omega t.$$

By comparison of the electrical and mechanical quantities listed above, we see that, if $R/2L$ is less than the resonant angular frequency $\omega_0 = 1/\sqrt{LC}$, oscillations occur with an amplitude decaying as $e^{-Rt/2L}$. The oscillations disappear for critical damping at $R^2 = 4L/C$ (cf. critical damping in Section 6.2 at $\gamma = 2\omega_0$). For $R^2 > 4L/C$ the currents are the sum of two exponentials with time constants R/L and RC.

If the resistance is small compared with the capacitative and inductive impedances at resonance, the power absorbed by the series *LCR* circuit, when averaged over a whole number of AC cycles, can be shown to be approximately

$$\bar{P} = \frac{V_0^2}{2R} \left\{ \frac{1}{1 + (4L^2/R^2)(\omega - \omega_0)^2} \right\}. \tag{18.43}$$

The sharpness of the resonance is defined by the *Q* factor,

$$Q = \frac{\omega_0 L}{R} = \frac{1}{R} \sqrt{\frac{L}{C}}. \tag{18.44}$$

Much higher *Q* factors can be achieved for electrical than for mechanical systems (for which according to eqn (6.29), $Q = \omega_0/\gamma$) because the resonance frequency may be very much higher. The resonance of an *LCR* circuit is often used for tuning radio and television signals. The larger the value of *Q*, the more completely may signals at close-lying frequencies be separated. Typically, FM radio signals are at frequencies around 100 MHz, and the *Q* value is a few hundred.

The voltages in the circuits we have considered so far depend linearly on the currents, and the principle of superposition therefore applies. The phenomenon of beats occurs (Section 6.5), and signals containing more than one frequency may be subjected to Fourier analysis (Section 6.6). Fourier analysis is particularly important, because impedances, and hence the response of circuits to signals, change with frequency, so that particular frequencies may be enhanced or rejected.

● *Radio and TV signals contain a mixture of different frequencies*

Radio waves broadcast from a transmitter are not continuous waves of fixed amplitude and frequency: the **carrier wave** at the frequency allocated to the particular transmitter is **modulated** by altering its amplitude, frequency, or phase. The information of interest—sound or pictures—is carried by the modulation frequencies. The modulation frequencies are much lower than that of the broadcast wave, and Fourier analysis is essential for analysing the received signals. The receiver must first be tuned to accept only the modulated carrier signals of the transmitter and then sound and pictures are extracted from its modulations.

18.3 Active circuits

So far in this chapter we have considered only circuits that are linear and passive. They are linear because the voltage across each component is proportional to the current through it. They are passive because they cannot independently generate power.

No electrical components are truly linear. All have a voltage range that is limited, by electrical breakdown for capacitors and by overheating for coils and resistors. However, provided that they are operated within their rated values, these components are very nearly linear, and the circuit analysis described in Section 18.2 is a very good approximation.

There are also other components that cannot be represented by fixed impedances because they are deliberately made to be nonlinear. For example, thermistors are devices that have a resistance that changes rapidly with temperature and decreases as the temperature increases. They may be used directly for temperature measurement or for control of current amplitude. Another example is the diode, which passes current easily in one direction but only allows an extremely small current to flow in the opposite direction. This property allows diodes to be used for changing the shape of electrical signals, for example, by rectifying an AC current, only passing current for half the cycle. When nonlinear elements are present there is no systematic procedure for calculating the voltages and currents, and indeed there may be more than one solution to the circuit equations.

● *Nonlinear components*

Active components are those that have the capability of generating an output signal that has more power than the input signal. Conservation of energy requires that power comes from somewhere. Batteries and alternators are power supplies that derive their energy from chemical and mechanical sources, respectively. Power supplies are themselves active components that have no need of input signals to deliver power. All other active components have more than two connections, so it is impossible to represent them by a single impedance.

The most common active element is the transistor, which has three or more connections. Transistors do not themselves generate power, but they can control currents delivered by a power supply and hence amplify the power of an electrical signal. Transistors are the most important components in modern integrated circuits. The number of transistors in a single chip is often very large. It would be almost impossible to make a complete analysis of a circuit containing so many components, especially as the individual transistors are often used in circumstances where their response is nonlinear. Fortunately, this is not necessary. Even a very complicated chip has a limited number of input and output connections. What we need to know about such a circuit is how it responds to signals presented to its inputs, not exactly what is going on inside it.

● *Transistors*

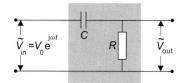

Circuits as input–output devices

Let us first consider the simple linear network in Fig 18.25. We can work out the current and voltages exactly, and here we want to express the result in terms of voltages at input and output connections. The circuit

Fig. 18.25 The input and output terminals are connected inside the shaded area through a network containing a capacitor and a resistor.

consists of a capacitor and a resistor joined in series across an alternating emf $V_0 e^{j\omega t}$. In Fig 18.25 the circuit has been drawn as though it were a 'black box' with a pair of input terminals and a pair of output terminals. The input terminals are connected to the source of emf and the voltage across them is labelled $\widetilde{V}_{in}$.

How does the voltage $\widetilde{V}_{out}$ across the output terminals vary with ω? The impedance of the circuit is $R + 1/j\omega C$, so the current is $V_0 e^{j\omega t}/(R + 1/j\omega C)$ and the voltage across the resistor is $V_0 e^{j\omega t} \times R/(R + 1/j\omega C)$. Hence

$$\widetilde{V}_{out} = \frac{\widetilde{V}_{in}R}{(R + 1/j\omega C)} = \frac{\widetilde{V}_{in}j\omega CR}{(j\omega CR + 1)}.$$

The modulus $|z|$ of a complex number $z = a + jb$ is given by eqn (20.48) as $|z|^2 = a^2 + b^2 = (a + jb)(a - jb)$, and the ratio of the amplitudes of the input and output voltages is therefore

$$\left| \frac{\widetilde{V}_{out}}{\widetilde{V}_{in}} \right| = \left\{ \frac{j\omega CR}{(j\omega CR + 1)} \times \frac{-j\omega CR}{(-j\omega CR + 1)} \right\}^{1/2} = \frac{\omega CR}{\sqrt{1 + \omega^2 C^2 R^2}}. \quad (18.45)$$

The output amplitude is the input amplitude multiplied by the factor $\omega CR/\sqrt{1 + \omega^2 C^2 R^2}$. This factor is small for small ω and tends to unity for large ω.

The network acts as a filter, allowing high-frequency signals to pass, but attenuating low-frequency signals. If we were presented with an actual black box with input and output terminals we could measure the output as a function of input amplitude and frequency. If the result were the same as eqn (18.45) we might guess that what was inside the box was the network in Fig 18.25, but we could not be sure. There might be more than one capacitor in place of C, more than one resistor in place of R, or even some transistors. But we could still use the black box to attenuate low-frequency signals even if we did not know what was inside it. The rest of this chapter introduces some black boxes that contain active circuits. We know how the outputs of the black boxes depend on their inputs, but we do not enquire what is inside them. The circuits we shall consider are basic building blocks for more complicated circuits.

Digital circuits

In digital circuits the signals are short pulses instead of the continuous sinusoidally varying voltages and currents we have been considering so far in this chapter. An example of a digital signal is shown as a function of time in Fig 18.26. The pulses are all the same size, in this case 5 volts, and between the pulses the voltage is at earth potential.

Figure 18.27 shows another digital signal in which each pulse is the same as the corresponding pulse in Fig 18.26, but some of the pulses are

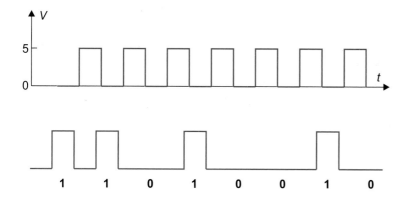

Fig. 18.26 Digital signals consist of a train of identical pulses.

Fig. 18.27 The presence or absence of a pulse corresponds to the binary digits 0 or 1.

missing. The presence or absence of a pulse can be expressed as a digit, either 1 or 0. If the voltage is 'high' (i.e. 5 V) the signal is represented by the digit 1. If the pulse is absent at a time when the voltage is 5 V in the continuous train of pulses (i.e. the voltage in Fig 18.27 is 'low', at 0 V) the signal is represented by the digit 0. These are the numbers below the pulse train in the figure.

Numbers expressed entirely in ones and zeros are called binary numbers. The rules of arithmetic are essentially the same as for decimal arithmetic, but are very simple since long multiplication is done entirely with ones and zeros. As there is no binary digit after 1, the number two is represented by 10. This is just the same as the notation for decimal numbers, where the last digit is 9 and ten is represented as 10. The decimal numbers 0, 1, 2, 3, 4, 5, ... are represented in binary notation as 0, 1, 10, 11, 100, 101, and so on.

In a binary number, each digit is called a **bit**. When writing a binary number on paper it does not matter how many bits are present: 1 is 'one' and 101 is 'five' without any ambiguity. However, when a number is represented by a train of pulses, the length of the pulse train corresponding to a number must be decided in advance and a string of zeros cannot be ignored. For example, if numbers are to be represented by 8 bits, the pulse train corresponding to five is 00000101. Binary numbers are often written in this way, including any leading zeros. Information is usually stored in multiples of 8 bits, and there is a special name for 8 bits: 8 bits $\equiv$ one **byte**.

Exercise 18.5 What decimal number corresponds to the 8-bit binary number 01011001?

Answer 89.

● *Serial and parallel bits*

Binary numbers can be coded as a sequence of pulses as in Fig 18.27 (*serial* bits) or, alternatively, by having separate wires simultaneously carrying a signal for each bit (*parallel* bits). Both methods are of practical importance. Trains of pulses are used to communicate information over long distances, often along optic fibres. Digital signals are increasingly used not only for transmission of numerical data but also other kinds of information such as telephone conversations that are encoded by the sender and decoded by the receiver.

There are two main advantages in using digital technology for these purposes. Firstly, digital signals are very nearly immune to distortion by noise; a pulse only has to be present or absent, and unless the signal is very weak this is a straightforward distinction. Secondly, at the high pulse rates currently used, digital coding is very efficient. Many telephone calls, for example, can be simultaneously transmitted along the same cable. This may sound like a nightmare, since each call has a different origin and a different destination. The world-wide communications network is indeed a miracle of organization as well as of technology since, to make sense of a long series of pulses containing overlapping information, rigorous rules have to be agreed between all users and implemented with standardized circuits and software.

Gates

Within one piece of equipment, such as a personal computer, much information is handled in parallel on several wires. The circuits called **gates** which we shall discuss next have more than one input. Gates perform simple logical operations and, because mathematics is closely related to logic, gates can be combined to do mathematics as well.

The gate in Fig 18.28 is an AND gate. It has two inputs A and B and one output C. The output C at any time is determined only by the inputs A and B at the same time. The symbol in Fig 18.28 is the one always used for AND gates. There are different 'families' of gates that operate with different voltage levels on their inputs and outputs, and Fig 18.27 applies to families for which the output and both inputs nominally have only two values, 0 V or 5 V. But, of course, the voltages do not have to be at exactly these voltages, since all that is required is that the voltage represents 0 or 1. One of the types of gate that operates between 0 and 5 V recognizes a voltage as '1' if it is above 2.0 V and as a zero if it is below 0.8 V.

Since the gate is performing a logical operation it is helpful to use yet another nomenclature: the input or output is called 'true' if it is '1' and 'false' if it is '0'. This nomenclature explains why the gate is called an AND gate. The output C is given for all possible input values in the table in Fig 18.28. The output is 'true' if and only if both A AND B inputs are

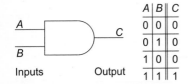

A	B	C
0	0	0
0	1	0
1	0	0
1	1	1

Inputs   Output

Fig. 18.28 A two-input AND gate and its truth table.

'true'. The table that gives the output for each input pair is called the *truth table* for the gate.

A logical operation that applies to a single input signal, is NOT, which is symbolized in circuit diagrams by a small circle. If a signal A is one, the only possible value that is *not A* is zero and vice versa: NOT A is written as $\bar{A}$. Figure 18.29 shows NOT as a triangle followed by a small circle. The triangle stands for the operation of doing nothing at all: this may seem absurd but it is sometimes necessary to show the triangle in circuit diagrams to indicate short delays that are needed to keep signals synchronized.

The other gates in Figs 18.30, 18.31, and 18.32 are NAND, OR, and EOR, respectively, each shown with its circuit symbol and truth table. NAND stands for NOT AND, so its truth table is the opposite of the AND gate truth table, and its symbol is the AND symbol followed by a circle. The output of OR is true if A OR B is true: EOR stands for 'exclusive or' and the output of the EOR gate is true if A OR B is true but not if both are true.

Gates may be connected together in various ways to perform different logical operations. A trivial example is shown in Fig 18.33. The inputs to the NAND gate are either both 0 or both 1. Comparing with the truth table for the NAND gate, we see that the output is the opposite of the input, and the NAND gate has been used to make a NOT gate.

Exercise 18.6 Show that the circuit made up of NAND gates in Fig 18.34 is equivalent to an EOR gate.

One of the important uses of gates is to perform mathematical operations. Addition, for example, is carried out in principle in the same way as the addition of decimal numbers. To add two numbers on paper, you would write them down one below the other and then add each column in turn. When the sum of two numbers exceed ten, the ten is 'carried' to the next column where its value is 1, and 1 is added to the other numbers in this column. The same applies to binary arithmetic, but now the only case when a carry is needed is when both numbers have a 1 in the same column.

Worked Example 18.5 The circuit in Fig 18.35 is called a half-adder. Verify that the circuit gives the correct outputs S and C for binary digits A_1 and A_2 at the inputs, where C is the carry output.

Answer The output S of the EOR ('exclusive or') gate is 1 if either A_1 or A_2 is one, but not if both are. The output C of the AND gate is 1 only if

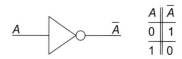

Fig. 18.29 A NOT gate.

A	$\bar{A}$
0	1
1	0

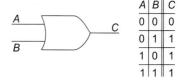

Fig. 18.30 A NAND gate.

A	B	C
0	0	1
0	1	1
1	0	1
1	1	0

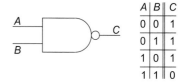

Fig. 18.31 An OR gate.

A	B	C
0	0	0
0	1	1
1	0	1
1	1	1

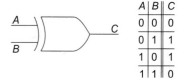

Fig. 18.32 An 'exclusive or' EOR gate.

A	B	C
0	0	0
0	1	1
1	0	1
1	1	0

● *Binary arithmetic*

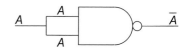

Fig. 18.33 A NAND gate connected as a NOT gate.

Negative numbers in binary notation

Computers have to be able to handle negative numbers as well as positive numbers. The way negative numbers are recorded is to allocate the first digit in a binary number (the *most significant bit*) to the sign. This bit is 0 for a positive number and 1 for a negative number. Note that leading zeros must not be suppressed in this convention—thus, if four bits are being used, plus one is written as 0001. However, negative numbers are not just the same as positive numbers except for the first digit. They are written in a notation called *two's complement*. The two's complement of a positive number has all the digits changed, and then 0001 is added. Thus -1 in two's complement is $(1110 + 0001) = 1111$. Using this notation, if we now add (plus one) and (minus one), we write $0001 + 1111 = 0000$. There is a carry digit from the most significant bit, but this is ignored since there are no spare bits to put it in. The trick of adding two's complements works for all negative numbers. Subtraction has become addition of a negative number, and the computer needs to have an adding circuit but no subtracting circuit!

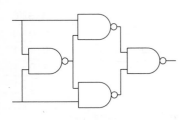

Fig. 18.34 NAND gates connected to make an EOR gate.

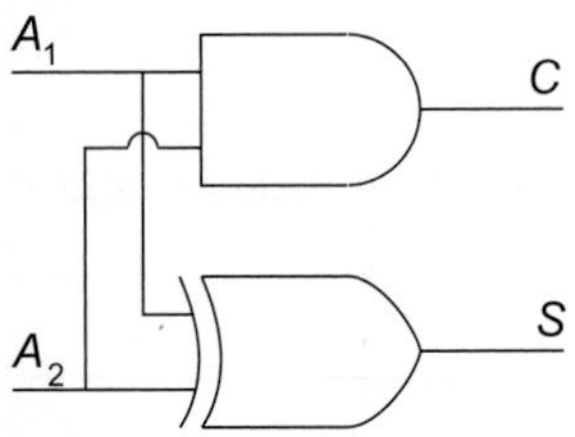

both A_1 and A_2 are 1. The truth table shown for the circuit is thus the same as the binary addition of the two digits A_1 and A_2.

Exercise 18.7 The half-adder is so called because, although it delivers a carry pulse to the next most significant bit, it is not equipped to receive a carry from a less significant bit. Check that the full adder circuit in Fig 18.36 responds correctly to the inputs A_1 and A_2 and the carry pulse C_0.

Gates are not restricted to two input connections, and multiple input gates are part of the internal circuit of chips that perform arithmetic operations on many-bit numbers. All are based on the same principles that we have outlined for simple circuits in this section.

A_1	A_2	C	S
0	0	0	0
0	1	0	1
1	0	0	1
1	1	1	0

Fig. 18.35 A half-adder circuit.

● *Clocks*

Memory

Even if mathematical operations are in parallel, the answer to all except the most trivial of problems does not come out all at once. A calculation has to proceed step by step, and logical circuitry must proceed in a stepwise manner. There must be a signal to determine the moment to move on to the next step. This signal is provided by a *clock*. The clock in a modern personal computer goes extremely fast. The clock signal is a train of pulses like those in Fig 18.26, with a pulse rate of several hundred MHz. This means that the time between successive steps, which is the inverse of the pulse rate, is a few nanoseconds. Not every part of a calculation will be ready to proceed at each step. Some information will have to be stored until it is needed: memory is required.

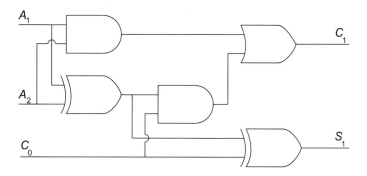

Fig. 18.36 A full adder circuit.

Gate circuits can be configured to store information. The arrangement of gates is quite complicated, and there is no need to go into it here. As usual, what we want to know is how the whole circuit responds to its inputs. A common circuit, made up of gates, is called a *JK flip-flop*, and has the connections shown in Fig 18.37. There are two inputs J and K and also an input for clock pulses. There are actually two outputs, labelled Q and $\bar{Q}$, but these outputs contain only one piece of information, since $\bar{Q}$ is always the opposite of Q: it is provided because it is often convenient to have both signals for later parts of a more complicated circuit. Unlike the gates described above, the output does not change when the inputs J and K change, but only when the clock pulse changes from 5 V to 0 V (gates are also available that operate on the opposite change of the clock pulse). The values of J and K may be set at any time during the previous clock period, ready to alter Q at the moment when the clock changes from 5 V to 0 V. The ouput signals from any number of flip-flops are thus synchronized.

The reason for introducing the *JK* flip-flop here is that its response is different from the simple truth table we have met so far. The output Q following a clock pulse depends not only on the values of J and K, but on what Q was *before* the arrival of the clock pulse. A truth table is not sufficient to describe the operation of the circuit. Clock pulses are arriving regularly, and if the nth pulse has arrived at a time t_n, we want to know the state of the output of the circuit at t_{n+1}. This is given in the *state table* in Fig 18.37, which shows the value of Q at t_{n+1} in terms of J, K, and the value of Q at t_n. There are four possible combinations of J and K, and the four operations performed on Q are the following.

1. Leave it alone.

2. Set it to 0.

3. Set it to 1.

4. Change it to $\bar{Q}$.

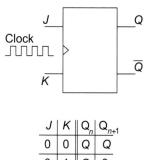

J	K	Q_n	Q_{n+1}
0	0	Q	Q
0	1	Q	0
1	0	Q	1
1	1	Q	$\bar{Q}$

Fig. 18.37 The *JK* flip-flop and its state table.

These are the only possible operations that can be carried out on the single bit Q, and this means that all possible operations on binary numbers can be performed by combinations of JK flip-flops. In particular, Q can be stored without change if $J = 0$ and $K = 0$ until it is required elsewhere—the circuit has memory.

Flip-flops of this kind are important in calculations. An example left for a problem at the end of the chapter is multiplying a binary number by two, which requires shifting all the bits by one column, just like multiplying a decimal number by ten. Flip-flops are also used for organizing a repeating sequence of events, like the cycle that occurs in traffic lights. However, these circuits are complicated and not suitable for storing large quantities of information. It is not necessary to be able to perform all the operations 1 to 4 above on most computer memories; to be able to read each bit and set it to 0 or 1 is sufficient. Very large memories are made with arrays of capacitors, each one with an area less than $(1\,\mu m)^2$. The capacitors do not hold their charge indefinitely, and they are reset at regular intervals. These are the memories known as DRAMs (Dynamic Random Access Memories), random access indicating that any part of the memory can be interrogated when required. Long-term memory usually resides on magnetic discs or tapes, in which opposite directions of magnetization represent zeros and ones.

Operational amplifiers and feedback

The concept of **feedback** is important in all control systems. What it means is best explained with a simple example. Home central heating derives heat from a boiler and, when the boiler is on, the temperature of the house increases. The temperature needs to be controlled so that it remains within a small range that is comfortable for the people in the house. This may be done by having a thermostat that measures the temperature and turns off the boiler when the temperature exceeds a preset value. The output signal—here the temperature—is used to *feed back* information to an earlier stage of the system—here the electrical circuit that turns the boiler on and off. Provided that the boiler is powerful enough, the temperature of the house is determined by the feedback and not by the properties of the boiler.

Feedback is also used within electrical circuits in many applications. A very versatile circuit, which is designed with feedback in mind, is the **operational amplifier**, often shortened to **op-amp**. An op-amp is an active circuit with two inputs and one output: the output amplifies the *difference* of the input values by a very large factor. A schematic diagram of an op-amp is shown in Fig 18.38. The inputs are labelled + and − and the voltages on these inputs are V_+ and V_-, respectively. The output is $G(V_+ - V_-)$, where G is a very large number, typically 10^5 for signals at

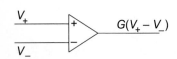

Fig. 18.38 A high-gain operational amplifier.

low frequency. The op-amp is connected to positive and negative power supplies at fairly low voltages such as $\pm15\,\text{V}$. The magnitude of the output voltage cannot exceed that of the power supplies, so the large gain G applies only if $(V_+ - V_-)$ is very small; once the output voltage reaches the supply voltage it is saturated and the gain rapidly falls towards zero. The result is that, if the output voltage is to vary within the limits set by the supply, the difference $(V_+ - V_-)$ must be very small.

We shall start by assuming that the gain is so large that it is a good approximation to assume that the function of the op-amp is to ensure that $(V_+ - V_-)$ is *zero*. We shall make a second assumption, that no current is drawn by the input connections of the amplifier: in practice this is a very good approximation, since the current drawn by the input connections of an op-amp is indeed tiny compared with the currents flowing elsewhere in the circuit.

Let us apply the two approximations to the circuit in Fig 18.39. The feedback network consists of the two resistors R_1 and R_2 joining the output of the op-amp to the $+$ input and to an input voltage V_{in}. Because V_- is at earth potential, so is V_+: the $+$ input is called a 'virtual earth', since it must remain at earth potential even though it is not directly connected to earth. The current through R_1 is therefore V_{in}/R_1. None of this current flows into the op-amp, and the current through R_2 is also V_{in}/R_1 and consequently the voltage drop across R_2 is $V_{in}R_2/R_1$. Since the $+$ input of the op-amp is at earth potential, the output voltage must be $-V_{in}R_2/R_1$ or, in other words, the gain of the whole circuit made up of the op-amp and the feedback network, is $-R_2/R_1$. If, for example, R_1 and R_2 are equal, and provided that the output voltage does not exceed the supply voltage, the op-amp inverts the input signal with a gain of (-1). The behaviour of the circuit is determined by the feedback, and we need to know nothing about the op-amp beyond the fact that it has a very large gain and draws very little current.

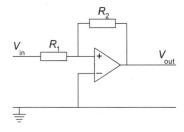

Fig. 18.39 An inverting amplifier with gain determined by feedback.

Exercise 18.8 Calculate the gain of the op-amp circuit in Fig 18.40 using the same approximations as in the above calculation.

Answer $+(R_1 + R_2)/R_1$.

Worked Example 18.6 Calculate the gain of the circuit in Fig 18.40 assuming its gain to be G instead of infinity. Retain the assumption, that no current flows at the $+$ or $-$ inputs.

Answer The current through R_1 is the same as the current through R_2, and hence

$$V_- = \left(\frac{R_1}{R_1 + R_2}\right) V_{out} = \beta V_{out}, \text{ say.}$$

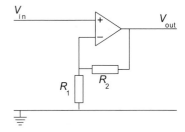

Fig. 18.40 Feeding back part of the output signal to the negative input gives an amplifier with positive overall gain.

The gain of op-amps falls off at high frequency

All op-amps are, in fact, designed so that their gains become smaller at high frequency. This is because there is always some delay in the amplifier, so that high-frequency signals at the output are out of phase with the input. If the gain were still large at frequencies so high that input and output were out of phase by more than π, then negative feedback would become positive feedback. If the gain is greater than one, positive feedback is unstable, since after the short delay in passing through the op-amp the signal would be increased and the output would be driven into saturation. The gain must

therefore fall to values below one at frequencies safely below the frequency at which instability occurs. The response of circuits including op-amps inevitably depends on the frequency of the input signals. Full discussion is beyond the scope of this chapter, which we end by reiterating that the purpose of Section 18.3 has been to demonstrate that circuits such as gates and op-amps can be employed so long as their input–output characteristics are known; a detailed knowledge of their internal mechanisms is not needed.

There is now a difference in the voltages V_{in} and V_{out}, and $V_{out} = G(V_{in} - V_-)$. Substituting for V_- in this equation,

$$V_{out} = G(V_{in} - V_{out}\beta).$$

Rearranging,

$$V_{out} = \frac{GV_{in}}{1 + G\beta}. \tag{18.46}$$

The fraction β is called the **negative feedback fraction**, and it is the fraction of the output signal that is returned to the negative input of the op-amp. If $G\beta$ is large compared to one, then the gain of the circuit is $1/\beta$, the same as the answer to Exercise 18.8. The argument used in this example applies not only to the feedback network in Fig 18.40, but to any network that returns part of the output signal to the negative input, even when this network includes complex impedances. A few examples of complex feedback are included in the problems.

Problems

Level 1

18.1 What is the current delivered by the battery in the circuit in Fig 18.41? In circuit diagrams of this kind, the numbers on resistors represent the resistance in ohms; K stands for one thousand and M for one million. Thus the resistor labelled 2 K has a resistance of 2000 Ω.

18.2 How many independent equations can be set up in Fig 18.3 using Kirchhoff's second rule?

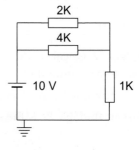

Fig. 18.41

18.3 Calculate the voltage across the shaded resistor in Fig 18.42.

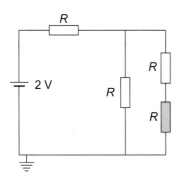

Fig. 18.42

18.4 A 10 V, 1 kHz power supply is connected across a coil with self-inductance 3 mH and resistance 10 Ω. What is the phase difference between the voltage and the current?

18.5 A 1 µF capacitor is connected in parallel with a 1 kΩ resistor across a 240 V (rms), 50 Hz mains supply. Calculate the amplitude of the current drawn from the mains.

18.6 Calculate the amplitude of the output voltage of the network in Fig 18.43 as a function of the angular frequency ω. At what angular frequency is the output amplitude $V_0/\sqrt{2}$?

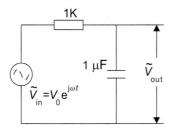

Fig. 18.43

18.7 A coil with self-inductance 1 mH and internal resistance 10 Ω is connected to a capacitor as shown in Fig 18.44. Obtain an expression for the ratio of output to input amplitudes as a function of the angular frequency ω. How does this ratio vary with angular frequency for large ω?

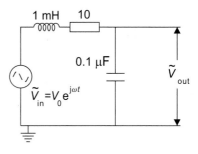

Fig. 18.44

18.8 What values at the inputs A, B, and C will give a 'true' ouput at D in the gate circuit shown in Fig 18.45?

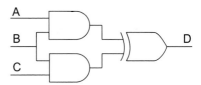

Fig. 18.45

Level 2

18.9 Over the temperature range 0°C to 100°C the resistance of the thermistor in Fig 18.46 may be approximated as $R = 4 \times 10^{(4-T/50)}$ Ω, where T is the temperature in °C. What voltage is measured by the voltmeter V: (a) at 25°C; (b) at 75°C?

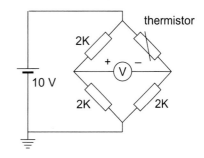

Fig. 18.46

18.10 What is the impedance between A and B of the network shown in Fig 18.47?

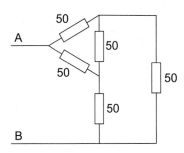

Fig. 18.47

18.11 Calculate the magnitude of the impedance of the network in Fig 18.48 at the angular frequency $\omega_0 = 1/\sqrt{LC}$.

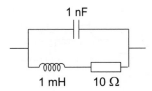

Fig. 18.48

18.12 A coil and a variable resistance are placed in series across an alternating voltage source with a frequency of 10 kHz. The phase difference ϕ between the current and the voltage is measured for a number of different values of the variable resistance. These values are plotted against $\cot \phi$ in Fig 18.49. Estimate the self-inductance and the internal resistance of the coil.

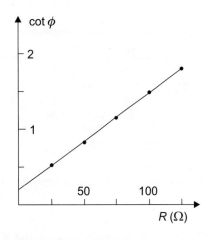

Fig. 18.49

18.13 What is the resonant angular frequency and the Q value for the circuit shown in Fig 18.50? Calculate the angular frequencies at which the current is one-half its value at resonance.

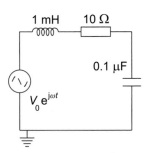

Fig. 18.50

18.14 An AC power supply can be represented by an ideal source of emf $\widetilde{V}_0 e^{j\omega t}$ in series with a resistor R_{out}. The supply is used to drive a load resistance R_1. Calculate the power delivered to the load. What value of R_1 gives the maximum power transfer from the supply to the load?

18.15 Show that the amplitude of the voltage $\widetilde{V}$ between the terminals X and Y in Fig 18.51 is independent of the resistance R. What value of R gives a phase difference of 45° between $\widetilde{V}$ and the 50 Hz power supply?

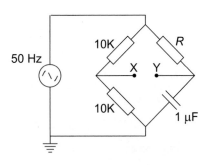

Fig. 18.51

18.16 Set up the differential equation for the circuit in Fig 18.52 and solve it to find the voltage across the capacitor as a function of time after the switch has been closed.

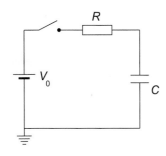

Fig. 18.52

18.17 Calculate the ratio $|\tilde{V}_{out}/\tilde{V}_{in}|$ for the circuit in Fig 18.53. What is the value of this ratio: (a) at low frequency; (b) at high frequency?

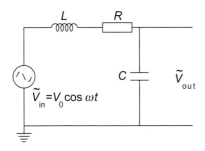

Fig. 18.53

18.18 Show that, when the next clock pulse arrives in the circuit in Fig 18.54, the '1' or '0' at the point A is loaded into Q_1, Q_1 is transferred to Q_2, Q_2 to Q_3, and so on. The circuit is called a shift register. By choosing Q_1 to be the most significant bit in the number stored in $Q_1, Q_2, \ldots$, the number is divided by 2 at the arrival of the clock pulse: conversely, if Q_1 is the least significant bit the number is multiplied by 2.

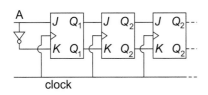

Fig. 18.54

18.19 What is the output of the operational amplifier circuit in Fig 18.55 in the infinite gain, zero current approximation?

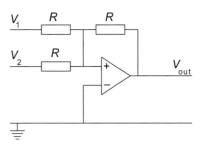

Fig. 18.55

18.20 In the circuit in Fig 18.56 the input signal is $V_0 \cos \omega t$. The resistor R_1 is very large and its function is to ensure that the DC output voltage does not stray too far from earth. Assuming that the feedback is determined only by C and R, calculate the output voltage assuming that $\omega CR = 1$. Use the infinite gain, zero current approximation.

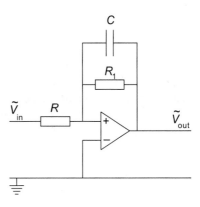

Fig. 18.56

Level 3

18.21 Twelve 1 Ω resistors are arranged to form the edges of a cube, and are connected together at each corner. What is the resistance between opposite corners of the cube?

18.22 Apply Kirchhoff's rules to the circuit in Fig 18.57 and calculate the unknown currents I_1, I_2, and I_3. Show that, if $R_1 = 2R$, then $I_3 = V/3R$.

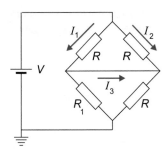

Fig. 18.57

18.23 Find the balance conditions for the bridge in Fig 18.58.

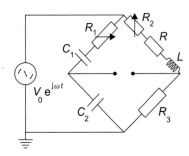

Fig. 18.58

18.24 Obtain an expression for the voltage V_C across the capacitor in Fig 18.59 as a function of time after closing the switch, assuming that the Q factor for the circuit is large. (*Hint.* The differential equation for charge has the same form as eqn (6.14) except that there is an additional constant term. The solution to the

equation with the constant equal to zero is damped simple harmonic. A constant term must be added to this solution to satisfy the initial conditions.)

18.25 Make an expansion of the impedance between the input terminals of the network in Fig 18.60 in powers of the angular frequency ω. For what value of R is the impedance almost purely resistive at low frequencies? Using this value of R, calculate the phase difference between the input and output voltages at low frequencies.

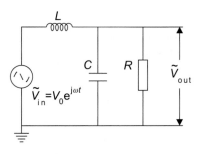

Fig. 18.60

18.26 Figure 18.61 shows a coil and a capacitor connected in parallel to an AC voltage supply. The inverse of the impedance of a circuit is called its *admittance*. Like the series LCR cicuit, the parallel circuit exhibits resonance, nearly at the same angular frequency $\omega_0 = 1/\sqrt{LC}$. Show that, provided $Q = \omega_0 L/R$ is large, the admittance of the parallel circuit at angular frequency ω_0 is almost real, with a value close to $\omega_0 C/Q$. Also show that the magnitude of the admittance increases by a factor $\sqrt{2}$ when the angular frequency differs from ω_0 by $\delta\omega = \pm\frac{1}{2}\omega_0/Q$.

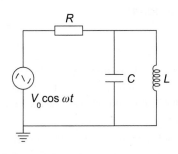

Fig. 18.59

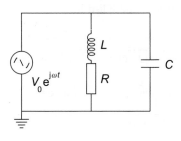

Fig. 18.61

18.27 The frequency of the AC voltage source in Fig 18.62 is 10 kHz. What value of $\tilde{Z}_L$ gives the maximum power transfer to the external circuit?

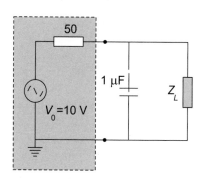

Fig. 18.62

18.28 What are the outputs X and Y for the gate circuit in Fig 18.63 for the inputs $(A, B, C) = (000)$, (001), (011), and (111)?

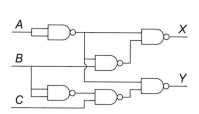

Fig. 18.63

18.29 Derive an expression for the output voltage of the circuit in Fig 18.64 for an AC input.

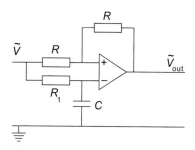

Fig. 18.64

18.30 The op-amp in Fig 18.65 has positive and negative power supplies at ±15 V. When the voltage level at the $(+)$ input is greater than that at the $(-)$ input, the output voltage is $+15$ V and, when it is less, the output voltage is -15 V. The triangular input signal shown in Fig 18.65 varies from $+1$ V to -1 V. At what input voltage does the output flip from $+5$ V to -15 V and vice versa? (This circuit is called a Schmitt trigger.)

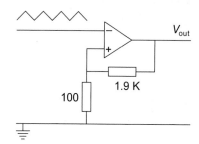

Fig. 18.65

Some solutions and answers to Chapter 18 problems

18.2 There are three independent loops in the circuit.

18.4 The current lags behind the voltage by a phase angle of 62°.

18.5 The impedance $\tilde{Z}$ of the components in parallel is given by

$$\frac{1}{\tilde{Z}} = j\omega C + \frac{1}{R}.$$

Hence

$$\tilde{Z} = \frac{R}{1 + j\omega CR} \quad \text{and} \quad |Z|^2 = \frac{R^2}{1 + \omega^2 C^2 R^2}.$$

The rms current amplitude is $I_{rms} = V_{rms}/|Z|$,

$$\frac{V_{rms}}{|Z|} = \frac{V_{rms}(1 + \omega^2 C^2 R^2)^{1/2}}{R}$$

$$= \frac{240(1 + 4\pi^2 \times 50^2 \times 10^{-12} \times 10^6)^{1/2}}{1000}$$

$$= 0.252 \text{ A}.$$

The amplitude of the current is $I_0 = \sqrt{2}I_{rms} = 356$ mA.

18.7 The same current $\widetilde{I}$ flows through the coil and the resistor. Hence $\widetilde{V}_{out} = \widetilde{I}R_1$ and $\widetilde{V}_{in} = \widetilde{I}(j\omega L + R + R_1)$. Eliminating $\widetilde{I}$,

$$\frac{\widetilde{V}_{out}}{\widetilde{V}_{in}} = \frac{R_1}{j\omega L + R + R_1} \quad \text{and} \quad \left|\frac{\widetilde{V}_{out}}{\widetilde{V}_{in}}\right|$$

$$= \left\{\frac{R_1^2}{\omega^2 L^2 + (R + R_1)^2}\right\}^{1/2}.$$

At low ω, the term $\omega^2 L^2$ may be neglected and the ratio of amplitudes is approximately $(R_1/R + R_1)$, which is nearly one if R is small. At high ω the ratio tends to $R/\omega L$. The circuit thus acts as a low-pass filter, allowing low-frequency signals to pass, but attenuating high-frequency signals.

18.8 $A \cdot B \cdot \overline{C}$ and $\overline{A} \cdot B \cdot C$.

18.9 (a) -3.63 V; (b) 1.13 V.

18.12 1.24 mH and 15.6 Ω.

18.13 At the resonant angular frequency ω_0, the magnitudes of the impedances of the capacitor and the inductance are equal. Thus $1/\omega_0 C = \omega_0 L$ and $\omega_0 = 1/\sqrt{LC}$. For $L = 1$ mH and $C = 0.1$ μF, $\omega_0 = 10^5$ rad s^{-1}. The Q value, given by eqn (18.44), is $Q = \omega_0 L/R = 10$.

18.14 The power delivered to the load is $V_0^2 R/(R + R_{out})^2$. The maximum transfer of power is $\frac{1}{2}V_0^2/R$, when $R = R_{out}$.

18.15 $R = 7688 \ \Omega$.

18.17 The ratio of output to input voltage is

$$\frac{\widetilde{V}_{out}}{\widetilde{V}_{in}} = \frac{(j\omega C)^{-1}}{(j\omega C)^{-1} + j\omega L/R} = \frac{1}{1 - \omega^2 LC + j\omega CR}.$$

$$\left|\frac{\widetilde{V}_{out}}{\widetilde{V}_{in}}\right| = \frac{1}{\{(1 - \omega^2 LC)^2 + \omega^2 C^2 R^2\}^{1/2}}.$$

At small ω this ratio $\rightarrow 1$ and at high ω it $\rightarrow 1/(\omega^2 LC)$. The circuit is resonant at $\omega_0^2 = LC = 10^5$ rad s^{-1}, and, at ω_0, $|\widetilde{V}_{out}/\widetilde{V}| = 1/\omega_0 CR = \sqrt{L/C}/R = 10V_0$ at $\omega = 10^5$ and $0.333V_0$ at $\omega = 2 \times 10^5$. The variation of output voltage with frequency is shown in Fig 18.66. The output voltage at $\omega = 10^5$ is greater than the input voltage. However, all the power is derived from the source of the input voltage: no circuit made up of passive components can produce a gain in power.

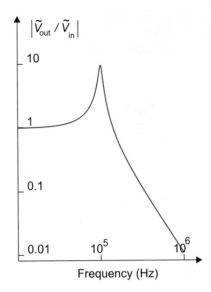

Fig. 18.66

18.19 $-(V_1 + V_2)$.

18.20 The input current through R is $\widetilde{I} = \widetilde{V}_{in}/R$. Neglecting the current through R_1, this current all flows through C. Since the $(-)$ input is at earth potential, the output voltage is $\widetilde{V}_{out} = -1/j\omega C\widetilde{I} = -\widetilde{V}_{in}/j\omega CR = j\widetilde{V}_{in}/\omega CR$. The output is $\pi/2$ ahead of the input—the factor j rotates the phasor through $\pi/2$ in the Argand diagram. The actual input and output voltages are the real parts of $\widetilde{V}_{in}$ and $\widetilde{V}_{out}$. Hence $V_{out}(t) = V_0 \cos(\omega t + \pi/2)/\omega CR = -V_0 \sin \omega t/\omega CR$. Now $\sin \omega t/\omega = \int \cos \omega t$ and we may write

$$V_{out}(t) = -\frac{1}{CR}\int V_{in}(t).$$

This applies for any ω, and the circuit integrates any periodic signal provided that the period is small compared to the time constant CR_1.

18.22 There is no voltage drop across the lead carrying the current I_3, therefore $I_1 = I_2$. Around the loop through R, R_1, and the battery,

$$V = I_1 R + (I_1 - I_3)R. \tag{1}$$

Around the loop through the two right-hand resistors R and the battery,

$$V = I_1 R + (I_1 + I_3)R. \tag{2}$$

If $R_1 = 3R$, eqns (1) and (2) become

$$V = 4I_1R - 3I_3R \quad \text{and} \quad V = 2I_1R + I_3R,$$

leading to $V = 5I_3R$ or $I_3 = V/5R$.

18.23 The balance conditions are $L = C_2R_1R_2$ and $R = R_3C_2/C_1 - R_2$, at any angular frequency ω for the voltage $\tilde{V}$. The self-inductance L of the coil depends on the variable resistance R_1 but not on R_2, and its internal resistance on R_2 but not R_1. These resistors can therefore be calibrated to give direct reading of L in mH and R in ohms. Although the result is independent of ω, the sensitivity of the measurement does depend on ω. At low frequency the voltage drop across L is very small and, conversely, at high frequency almost all the voltage drop across the branch containing the coil occurs across L.

18.24 The equation for the charge Q on the capacitor is the same as eqn (18.42) except that the alternating voltage $V_0 \cos \omega t$ is replaced by the constant battery voltage V_0,

$$\frac{d^2Q}{dt^2} + \frac{R}{L}\frac{dQ}{dt} + \frac{Q}{LC} = \frac{V_0}{L}.$$

By comparison with eqn (6.14), if the right-hand side is zero, the solution is

$$Q = A \exp\left(\frac{-Rt}{2L}\right) \cos(\omega t + \phi),$$

replacing x in eqn (6.14) by Q and with $\omega = (1/LC - R^2/4L^2)^{1/2}$. The initial conditions for the circuit in this problem at the moment when the switch is closed are:

(a) $Q = 0$ at $t = 0$, because the whole voltage is initially across L;

(b) $dQ/dt = 0$, because current cannot suddenly flow through L.

$$\frac{dQ}{dt} = A\left\{-\frac{R}{2L}\exp\left(\frac{-Rt}{2L}\right)\cos(\omega t + \phi)\right.$$
$$\left. - \omega\exp\left(\frac{-Rt}{2L}\right)\sin(\omega t + \phi)\right\}.$$

Initial condition (a) becomes

$$-\left(\frac{R}{2L}\right)\cos\phi - \omega\sin\phi$$

or

$$\tan\phi = -\frac{R}{2\omega L} \approx -\frac{R}{2\omega_0 L},$$

which is very small if the Q factor $= \omega_0 L/R$ is large and $\phi \approx 0$. Hence $Q \approx A \exp(-Rt/2L) \cos\omega t$. After a long time the damped oscillations have died away, and from the differential equation, when $d^2Q/dt^2 = dQ/dt = 0$, $Q = CV_0$. Hence $Q \approx CV_0 + A\exp(-Rt/2L)\cos\omega t$. From initial condition (a) $A = -CV_0$ and, finally, the voltage Q/C on the capacitor is

$$V(t) = V_0\left(1 - \exp\left(\frac{-Rt}{2L}\right)\cos\omega t\right).$$

Figure 18.67 shows how this function fits the damped oscillation to the initial voltage on the capacitor (zero) to the final value (V_0). The functional form of the damped oscillation is always the same, and the response of an LCR circuit to any sudden change can be sketched without doing more mathematics.

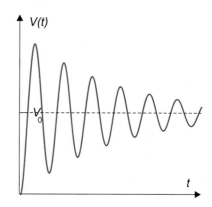

Fig. 18.67

18.25 The impedance of the network is

$$\tilde{Z} = j\omega L + \frac{1}{j\omega C + 1/R} = j\omega L + R(1 + j\omega CR)^{-1}$$
$$= j\omega L + R\{1 - j\omega CR + (j\omega CR)^2 - \cdots\}$$
$$= R + j\omega(L + CR^2) + \text{ terms in higher powers of } \omega.$$

This impedance is purely resistive to first order in ω for $R = \sqrt{L/C}$, and has the value R. The current is then $\tilde{V}_{in}/R = \tilde{V}_{out}(j\omega C + 1/R)$. Hence

$$\tilde{V}_{in}/\tilde{V}_{out} = R((j\omega C + 1/R)) = j\omega\sqrt{LC} + 1,$$

and the phase difference between $\tilde{V}_{in}$ and $\tilde{V}_{out}$ is $\omega\sqrt{LC}$. At an angular frequency ω, the time delay corresponding to a phase difference ϕ is ϕ/ω, and the time delay between $\tilde{V}_{in}$ and $\tilde{V}_{out}$ is therefore $\sqrt{LC}$, independent of ω. Provided that ωCR is small for all the angular frequencies occurring in a signal, the whole signal is delayed by the same time. A number of LC elements may be connected one after the other. Such a network is called a delay line. Coaxial cables possess capacitance and inductance, and they effectively act like delay lines, which can transmit signals without distortion up to very high angular frequency.

18.27 The maximum power transfer occurs if the load is an ideal self-inductance in parallel with a $50\,\Omega$ resistor. The self-inductance must cancel the admittance of the capacitor, which requires $L = 1/\omega^2 C$. For this circuit, $L = 0.25$ mH.

18.28 The outputs for $(A, B, C) = (000)$, (001), (011), and (111) are $(X, Y) = (00)$, (01), (10), and (11), respectively. The (X, Y) values are the binary representations of 0, 1, 2, and 3. C, B, and A can be connected so that they reach the 'true' voltage of the gates when an external voltage reaches 1, 2, or 3 times the 'true' voltage. The circuit thus acts as a converter of analogue signals to digital form.

18.29 At an angular frequency ω,

$$\tilde{V}_{out} = \tilde{V}_{in}\left(\frac{j\omega CR - 1}{j\omega CR + 1}\right).$$

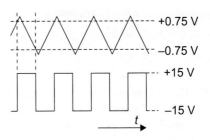

Fig. 18.68

The circuit shifts the phase of the input signal without altering its amplitude. The advantages of the op-amp over the network in Problem 18.15, which also acts as a phase shifter, are that: (a) the input is not attenuated; (b) a load can be placed on the output of this active circuit without significantly changing its response.

18.30 Suppose that the output of the op-amp is at $+15$ V. The $(+)$ input is then at $+15 \times 100/200 = +0.75$ V. The circuit is stable in this state if the signal at the $(-)$ input is less than 0.75 V. If the signal increases to a value just above 0.75 V, the $(+)$ input is less than the $(-)$ input and the circuit flips to a stable state with output at -15 V. Similarly, a decreasing signal passing -0.75 V causes the circuit to flip back again, as shown in Fig 18.68.

Summary of Part 5
Electricity and magnetism

We here summarize important results and equations discussed in Part 5

Chapter 15 **Electrostatics**

The force between two charges is given by Coulomb's law, which states that

The force between two charges acts along the line between the charges, and is proportional to the product of the charges and to the inverse square of the distance between them.

The mathematical form of Coulomb's law for the force between two charges q_1 and q_2 is

$$\mathbf{F}_{12} = \frac{q_1 q_2}{4\pi\epsilon_0 r_{12}^2} \hat{\mathbf{r}}_{12}. \tag{15.3}$$

Here $\mathbf{F}_{12}$ is the force exerted on q_2 by q_1 and $\hat{\mathbf{r}}_{12}$ is the unit vector along the line joining q_1 and q_2 in the direction from q_1 to q_2. The value of the constant ϵ_0 is $8.854 \times 10^{-12}\,\mathrm{C^2\,N^{-1}\,m^{-2}}$. The charges q_1 and q_2 may be positive or negative, and Coulomb's law incorporates the fact that unlike charges attract while like charges repel one another.

When several charges are present the net force $\mathbf{F}_i$ on a charge q_i is the vector sum of the forces due to all the others separately,

$$\mathbf{F}_i = \frac{q_i}{4\pi\epsilon_0} \sum_{j\neq i} \frac{q_j(\mathbf{r}_i - \mathbf{r}_j)}{|\mathbf{r}_i - \mathbf{r}_j|^3}. \tag{15.9}$$

Since the right-hand side of equation (15.9) has the factor q_i outside the summation, the function $\mathbf{F}_i/q_i$ is independent of q_i and defines a *vector function of position* for positions $\mathbf{r}_i$. This function is the *electric field*, and is denoted by $\mathbf{E}(\mathbf{r}_i)$. Dropping the redundant suffix i,

$$\mathbf{E}(\mathbf{r}) = \frac{1}{4\pi\epsilon_0} \sum_j \frac{q_j(\mathbf{r} - \mathbf{r}_j)}{|\mathbf{r} - \mathbf{r}_j|^3}. \tag{15.11}$$

The electric field due to a set of charges is the vector sum of the fields due to each charge separately. Because it has this property, the electric field is said to satisfy the *principle of superposition*.

The *flux* of the electric field through a surface S divided into many small flat surfaces δS_i is defined to be $\sum_i \mathbf{E}(\mathbf{r}_i) \cdot \delta \mathbf{S}_i$, where $\delta \mathbf{S}_i$ is a vector normal to the surface with magnitude equal to the area of the element of surface S_i. In the limit as the areas $\delta \mathbf{S}_i$ all tend to zero

$$\text{Flux of } \mathbf{E} \text{ through } S = \lim_{\delta \mathbf{S}_i \to 0} \sum_i \mathbf{E}(\mathbf{r}_i) \cdot \delta \mathbf{S}_i$$

$$= \int_S \mathbf{E}(\mathbf{r}) \cdot d\mathbf{S}. \tag{15.13}$$

In particular, if S is a closed surface and the infinitesimal vectors $d\mathbf{S}$ are all outward normals, the flux $\int_S \mathbf{E} \cdot d\mathbf{S}$ out of S obeys Gauss's law:

$$\int_S \mathbf{E} \cdot d\mathbf{S} = \frac{\sum_i q_i}{\epsilon_0} = \frac{Q}{\epsilon_0} \tag{15.19}$$

where $Q = \sum_i q_i$ is the sum of all the charges situated within S. In words, Gauss's law states that

The total flux of the electric field out of any closed surface equals the total charge enclosed within the surface divided by ϵ_0.

Work is done in moving charges against electrical forces, and a single charge q brought from infinity into the neighbourhood of a set of fixed charges q_j acquires a potential energy

$$U_{\text{tot}} = \sum_j \frac{qq_j}{4\pi\epsilon_0|\mathbf{r} - \mathbf{r}_j|}. \tag{15.25}$$

The function U_{tot}/q is a *scalar function of position* called the *electrostatic potential* $\phi(\mathbf{r})$,

$$\phi(\mathbf{r}) = U_{\text{tot}}/q = \sum_j \frac{q_j}{4\pi\epsilon_0|\mathbf{r} - \mathbf{r}_j|}. \tag{15.26}$$

The electric field can be deduced from the electrostatic potential and it has components

$$E_x = -\frac{\partial \phi}{\partial x}; \quad E_y = -\frac{\partial \phi}{\partial y}; \quad E_z = -\frac{\partial \phi}{\partial z}. \quad (15.30)$$

Two equal and opposite charges $\pm q$ separated by a distance a are said to have a *dipole moment* $\mathbf{p}$, which is a vector pointing in the direction from the negative to the positive charge and of magnitude qa. The electrostatic potential due to a dipole moment $\mathbf{p}$ situated at the origin is

$$\phi(\mathbf{r}) = \frac{\mathbf{p} \cdot \mathbf{r}}{4\pi\epsilon_0 r^3}. \quad (15.34)$$

Often we are not interested in the electric fields due to individual electrons and atomic nuclei, but need to concentrate on fields averaged over volumes large compared with the size of an atom. Such average fields are called *macroscopic*. In the presence of macroscopic fields charges are moved about and give rise to *induced charges*. Induced charges may appear as a *surface charge density* residing on the surface of a material or as *volume charge density* spread throughout its volume.

In conducting materials, conduction electrons are free to move, and they build up induced surface charge until the electric field within the conductor is zero. If the surface charge density is σ, then there is an electric field pointing in the direction of the outward normal to the conductor with a magnitude $E = \sigma/\epsilon_0$ (eqn (15.39)).

When placed in an electric field, a neutral atom acquires a dipole moment $\mathbf{p}$ with a magnitude proportional to the field,

$$\mathbf{p} = \alpha\epsilon_0 \mathbf{E}. \quad (15.42)$$

where the constant α is called the **polarizability** of the atom. Insulators containing N atoms per unit volume acquire a *polarization density* $\mathbf{P} = N\mathbf{p} = N\alpha\epsilon_0\mathbf{E}$ in an electric field $\mathbf{E}$, leading to induced charges that *reduce* the field within the insulator by a factor $(1 + N\alpha)$ (eqn (15.45)). The product $N\alpha = \chi_E$ is called the *electric susceptibility* of the insulator.

Some materials contain molecules possessing a dipole moment even in the absence of a macroscopic field, though in these circumstances there is no polarization density because the dipoles are randomly oriented. A macroscopic field tends to line up the dipoles, leading to a temperature-dependent electric susceptibility. For a material consisting of a single type of molecule with polarizability α and permanent dipole moment of magnitude p, if the molecular density is N, the electric susceptibility is

$$\chi_E = N\left(\alpha + \frac{p^2}{3\epsilon_0 k_B T}\right). \quad (15.48)$$

Because electric fields induce charge on the surface of conductors, charges are present whenever a votage difference is maintained between two conducting surfaces. Such a pair of conducting surfaces is called a *capacitor*. The charge on the two surfaces has the same magnitude Q, but opposite sign, and is proportional to the voltage between the conductors. The *capacitance* $C = Q/V$ (eqn (15.49)) is a constant depending only on the geometry of the conductors. For two parallel plates of area S separated by a distance d, the capacitance is $C = \epsilon_0 S/d$ (eqn (15.50)). If the capacitor is filled with a material with electric susceptibility χ_E, the capacitance is increased by a factor

$$1 + \chi_E = \epsilon \quad (15.51)$$

where ϵ is called the relative permittivity of the material.

Work must be done to charge a capacitor, and capacitors therefore store energy. When there is a voltage V between the plates of a capacitor, the amount of stored energy is

$$U = \tfrac{1}{2}CV^2. \quad (15.52)$$

This energy can be considered as residing in the electric field, and for a field of magnitude E the energy density is $\tfrac{1}{2}\epsilon_0 E^2$.

Chapter 16 **Magnetism**

A conductor is a material that contains electrons that can move in the presence of an electric field. The movement of electrons gives rise to an electric current density, $\mathbf{j}$, and the current through a surface S is

$$I = \int_S \mathbf{j} \cdot d\mathbf{S}. \quad (16.1)$$

For ohmic conductors, the current density is proportional to the electric field,

$$\mathbf{j} = \sigma\mathbf{E} \quad (16.4)$$

where σ is the eletrical conductivity of the material. If a potential difference V exists between the ends of an ohmic conductor of length ℓ and uniform area of cross-section S, the current through the conductor is

$$I = \frac{V}{R} \qquad (16.7)$$

where R is the resistance of the conductor given by

$$R = \frac{\ell}{\sigma S}. \qquad (16.8)$$

Work must be done to move the electrons from one end of a resistor to the other. This appears as heat in the conductor and the power expended is

$$P = I^2 R = \frac{V^2}{R}. \qquad (16.11)$$

A steady electric current gives rise to a magnetic field that produces a force on a moving charged particle. In a magnetic field $\mathbf{B}$ a particle with charge q moving with velocity $\mathbf{v}$ experiences a force

$$\mathbf{F} = q\mathbf{v} \times \mathbf{B}. \qquad (16.12)$$

This force cannot increase the speed of the particle but can change its velocity. If the mass of the particle is m and the component of the particle's velocity in the plane perpendicular to the field direction is $v_\perp$, the motion of the particle projected on to the plane perpendicular to the field is a circle of radius

$$a = \frac{mv_\perp}{qB}. \qquad (16.14)$$

There are no magnetic charges and the lines of the magnetic field are continuous. The line integral of the field B around a closed loop is related to the current passing through any surface that has the loop as its perimeter. Ampère's law states that

$$\oint \mathbf{B} \cdot d\boldsymbol{\ell} = \mu_0 I \qquad (16.15)$$

where μ_0 is a constant called the permeability of free space and is defined as $4\pi \times 10^{-7} \ \mathrm{T\,m\,A^{-1}}$. The sign of the current is given by the right-hand screw rule with the direction of circulation of the loop. Ampère's law can be used to calculate the magnetic field from symmetric current distributions, for example, a current in a long, straight wire (eqn (16.19)) or a current through a long solenoid (eqn (16.20)).

A very small loop in which flows a current I constitutes a magnetic dipole of moment

$$\mathbf{m} = I\Delta\mathbf{S} \qquad (16.16)$$

where $\Delta\mathbf{S}$ is a vector with size equal to the area of the loop and in the direction given by the right-hand screw rule and the current. The field lines from a magnetic dipole are similar to those from an electric dipole. A large loop of current can be considered to consist of a very large number of small loops and the magnetic field of the large loop is the sum of those due to the small dipoles.

The Biot–Savart law gives the magnetic field from any specified current distribution,

$$\mathbf{B} = \frac{\mu_0}{4\pi} I \oint \frac{d\boldsymbol{\ell} \times \mathbf{r}}{r^3}. \qquad (16.22)$$

An example is the field on the axis of a circular loop of current (eqn (16.23)).

There may be forces or torques on current-carrying wires in a magnetic field arising from another current. The calculation of forces or torques is facilitated by the use of the concept of magnetic flux. The flux through an area S is

$$\Phi = \int_S \mathbf{B} \cdot d\mathbf{S}. \qquad (16.29)$$

Here $d\mathbf{S}$ is a vector normal to the surface and the choice of direction of the normal, together with the direction of the field, define the sign of the flux.

In the presence of a magnetic field, materials become magnetized. They acquire a magnetization $\mathbf{M}$ given by the dipole moment per unit volume. The magnetization contributes to the total field. For ferromagnetic materials, the magnetization is large, and it is usually easier to work with the field $\mathbf{B}$ and a new field, $\mathbf{H}$, defined by

$$\mathbf{B} = \mu_0(\mathbf{H} + \mathbf{M}). \qquad (16.32)$$

In terms of the field H, Ampère's law becomes

$$\oint \mathbf{H} \cdot d\boldsymbol{\ell} = I \qquad (16.33)$$

where I is the current arising from the motion of free charges through the area of the loop and has no contribution from the internal bound currents that correspond to the magnetization. For isotropic, homogeneous materials the fields, if nonzero, are in the same direction at all points and

$$B = \mu\mu_0 H \qquad (16.34)$$

where μ is the relative permeability of the material. For most ferromagnetic materials, μ is not constant and the relation between B and H is given by hysteresis curves.

This expression with the condition on the field B that its component perpendicular to the surface is constant across a boundary enables fields in the gap of an electromagnet to be determined approximately.

Some materials may have a magnetization in the absence of an external magnetizing field and these may be used for permanent magnets.

Chapter 17 **Induced electric fields**

If a conductor moves with velocity **v** in a magnetic field **B** the free electrons move with an average velocity **v** and experience a force

$$\mathbf{F} = -e\mathbf{v} \times \mathbf{B}. \qquad (17.1)$$

This force may produce an induced electric field and an induced electromotive force $\oint \mathbf{E} \cdot d\boldsymbol{\ell}$ in a complete moving circuit such as a closed loop of wire.

The induced electric field is not an electrostatic field and cannot be derived from a scalar potential.

If a circuit is stationary but the magnetic field in which the circuit is situated is changing with time, there is again an induced electric field, and the induced emf around any closed loop is proportional to the rate of change of magnetic flux through the loop. Lenz's law is that the induced emf is in a direction that opposes the magnetic field change producing it. This gives the minus sign in Faraday's law relating the induced emf to the rate of change of flux,

$$\oint \mathbf{E} \cdot d\boldsymbol{\ell} = -\frac{d\Phi}{dt}. \qquad (17.10)$$

In this equation the directions of the vectors used to determine the flux are given by the right-hand screw rule used with the sense of traversal of the loop made when the line integral is evaluated.

If the flux Φ through a closed conducting circuit is due to a current I in the circuit itself,

$$\Phi = LI \qquad (17.13)$$

where L is the self-inductance of the circuit. An emf is induced in the circuit when the current changes. This effect is called self-inductance. The emf is proportional to the rate of change of current, with the constant of proportionality equal to the self-inductance L of the circuit. An idealized circuit with no resistance is called an inductor, and the voltage difference V across the ends of an inductor in which a current is changing is

$$V = L\frac{dI}{dt}. \qquad (17.15)$$

The magnetic flux Φ_1 through a closed circuit 1 due to the field produced by a current I_2 in a second circuit 2 is proportional to I_2. The constant of proportionality, $M_{1,2}$, is called the mutual inductance between the circuits. ($M_{1,2} = M_{2,1}$.) When I_2 changes an emf is induced in circuit 1 given by

$$\oint \mathbf{E}_1 \cdot d\boldsymbol{\ell} = -M_{1,2}\frac{dI_2}{dt}. \qquad (17.18)$$

A capacitor is a store of electrical energy and an inductor is a store of magnetic energy. The energy in a magnetic field that occupies a volume V is

$$U = \tfrac{1}{2}\int_V \mathbf{B} \cdot \mathbf{H}\, dV \qquad (17.25)$$

where **B** and **H** are corresponding field values.

An alternating emf is generated when a coil rotates in a magnetic field. If a coil of n turns, each of area S, rotates at angular frequency ω with the axis of rotation perpendicular to a magnetic field of strength B, the

$$\text{emf} = -n\omega BS \sin \omega t. \qquad (17.26)$$

Rectification, which is the removal of the negative-going parts of this waveform and smoothing the resulting positive-going parts, produces a steady voltage.

Chapter 18 **Circuits**

Circuits are made up of *components* joined together by wires carrying electric current. In circuit diagrams the real components are represented by idealized circuit elements with simplified properties. The resulting *equivalent circuit* can be analysed mathematically to give a good approximation to the behaviour of the real circuit.

In direct current (DC) circuits the elements are sources of emf and resistors, which are assumed to obey

Ohm's law. The equivalent circuit then consists of a network of resistors and emfs with any number of *branches* joined together at *junctions*. The currents in all the branches can be calculated by applying *Kirchhoff's rules*.

1. *The algebraic sum of the currents arriving at any junction is zero.*

2. *The sum of the voltages around any loop of the network is zero.*

The current I between any two points in a network is proportional to the voltage V between the same two points. The ratio V/I is called the *impedance* of the part of the network between the two points. For example, several resistors with resistances R_j connected in a string are said to be in series, and their total impedance R_s is the sum of their separate resistances:

$$R_s = \sum_{j=1}^{n} R_j. \tag{18.5}$$

A number of resistors all connected to the same two junctions are in parallel, and their impedance R_p is given by

$$\frac{1}{R_p} = \sum_{j=1}^{n} \frac{1}{R_j}. \tag{18.6}$$

Alternating current (AC) circuits have voltages and currents that vary sinusoidally but not necessarily in phase with one another. The idealized circuit elements used to build up equivalent circuit diagrams are resistors, capacitors that have no leakage resistance, and inductors that have self-inductance but no resistance. For a voltage $V(t) = V_0 \cos(\omega t)$ with an angular frequency ω and an amplitude V_0, the amplitude and the phase of the current for the three types of element are

Resistor with resistance R:

$$I(t) = \frac{V_0 \cos(\omega t)}{R},$$

Capacitor with capacitance C:

$$I(t) = \omega C V_0 \cos\left(\omega t + \frac{\pi}{2}\right), \tag{18.11}$$

Inductor with self-inductance L:

$$I(t) = \frac{V_0 \cos(\omega t - \pi/2)}{\omega L}. \tag{18.13}$$

The fact that phases are not the same in different parts of an AC circuit can be handled in a convenient way by using complex numbers. An AC voltage $V_0 \cos \omega t$ is represented by a *phasor*

$$\widetilde{V} = V_0 e^{j\omega t} = V_0 \cos \omega t + j V_0 \sin \omega t. \tag{18.15}$$

Only the real part of the phasor has physical meaning, but in the exponential form phase differences are included as part of a complex number that obeys the usual rules of algebra. Since

$$e^{j\pi/2} = \cos(\pi/2) + j \sin(\pi/2) = j, \tag{18.18}$$

a phase change of $\pi/2$ in an AC quantity is equivalent to multiplying the corresponding phasor by j. Equation (18.11) becomes $\widetilde{V} = \widetilde{I}/j\omega C$ (eqn (18.19)). This has the same form as Ohm's law, with the capacitor exhibiting a *complex impedance* $\widetilde{Z} = 1/j\omega C$. The complex forms of the voltage/current relations for resistors, capacitors and inductors are

$$\widetilde{Z} = R \quad \text{for resistance } R, \tag{18.23}$$

$$\widetilde{Z} = \frac{1}{j\omega C} \quad \text{for capacitance } C, \tag{18.24}$$

and

$$\widetilde{Z} = j\omega L \quad \text{for inductance } L. \tag{18.25}$$

Kirchhoff's rules apply to AC circuits in the same way as for DC circuits, and the solutions of the complex network equations give the amplitudes and phases of the currents in all branches of a network.

Complex numbers cannot be used when evaluating the power dissipated in an AC network, because power is (volts) × (amperes), and the products of imaginary components would give false real values. Power is dissipated only in resistors: capacitors and inductors store energy, but they accept and return the same amount of energy twice in every cycle of the alternating emf.

For a purely resistive circuit the peak power is twice the average power, and for that reason it is customary to discuss power in terms of *root mean square* (rms) values. The rms values of voltage and current are related to their amplitudes by the relations

$$V_{\text{rms}} = V_0/\sqrt{2}; \qquad I_{rms} = I_0/\sqrt{2}. \tag{18.33}$$

The average power dissipated in a resistance R is then

$$\bar{P} = V_{\text{rms}} I_{\text{rms}} = V_{\text{rms}}^2 / R. \qquad (18.34)$$

For an impedance $\widetilde{Z}$, which has a magnitude Z_0 and causes a phase difference ϕ between voltage and current, the average power is

$$\bar{P} = \frac{V_0^2}{2Z_0} \cos\phi = \tfrac{1}{2} V_0 I_0 \cos\phi = V_{\text{rms}} I_{\text{rms}} \cos\phi. \quad (18.38)$$

The power is reduced by the *power factor* $\cos\phi$ by comparison with a purely resistive impedance.

Circuits containing both capacitance and inductance may exhibit *resonance* just like the mechanical systems discussed in Chapter 6. For a circuit consisting of L, C, and R in series, the sharpness of the resonance is indicated by the quality factor Q, which is

$$Q = \frac{\omega_0 L}{R} = \frac{1}{R}\sqrt{\frac{L}{C}}. \qquad (18.44)$$

If Q is large the average power absorbed by this circuit, as a function of frequency, is approximately

$$\bar{P} = \frac{V_0^2}{2R} \left\{ \frac{1}{1 + (4L^2/R^2)(\omega - \omega_0)^2} \right\}, \qquad (18.43)$$

which has a narrow maximum peaking when the angular frequency ω equals the resonant angular frequency ω_0.

Unlike circuits containing only resistors, capacitors, and inductors, *active circuits* can generate power and the time variation of a signal may be modified by a single active component. *Gates* are active circuits that perform logical operations on digital signals, Mathematics proceeds in a series of logical steps, and gates are therefore the building blocks of computers. Digital signals are widely used in communication as well as computation, and digital circuits are therefore of great practical importance. Much electronic equipment also generates smoothly varying analogue signals, which are functions of time. The *operational amplifier* is an example of an electronic device that uses *feedback* to obtain different functional relationships between input and output signals. The basic methods of using gates and operational amplifiers are explained in Section 18.3.

Part 6

Relativity

Chapter 19

Relativity

This chapter considers the relationships between measurements of position and time made by observers in uniform relative motion at a speed comparable with the speed of light. The relationships are different to those used in classical physics when the speed is much less than the speed of light. Some of the consequences of this difference are discussed.

Throughout the book we have been stressing that our discussions were restricted to situations where the speeds of moving objects were small compared with the speed of light. When this condition is satisfied, Newtonian mechanics applies and, for example, the mass of an object is a fixed concept and two observers in relative motion measure time in the same way. When the speeds of objects approach the speed of light in vacuum, new ideas, most of them counterintuitive, have to be introduced and classical theories are superseded by a new mechanics.

We know that Newtonian mechanics gives a correct description to an exceedingly good approximation for almost all phenomena normally encountered. This is because moving objects with speeds approaching the speed of light in vacuum, c, are rarely met. When such situations are met, or when observers are moving with uniform velocity with respect to each other and their relative speed is comparable with c, new notions of space and time are used to describe measurements made by the different observers. Newtonian mechanics emerges as a perfectly adequate approximation for speeds small compared with c. The new mechanics is the subject matter of this chapter.

19.1 Introduction

Time and mass at high speeds

If the half-life of an unstable particle is determined when the particle is at rest in the laboratory, different measurements all give the same particular value. A particle called a μ-meson, or muon, has a charge of one electronic unit and a mass when at rest of about 0.11 atomic mass units. It decays at rest with a half-life close to 2.2×10^{-6} s. Muons are formed in the upper atmosphere at mean distances above the surface of the Earth of about 30 miles (about 5×10^{4} m) when cosmic rays hit the residual gas molecules or atoms. The muons travel towards the earth at speeds comparable with the speed of light in vacuum. The time taken for a muon to travel 5×10^{4} m at a speed of 3×10^{8} m s^{-1} is $\sim 10^{-4}$ s, much longer than the muons' lifetime, and it may be anticipated that almost all muons made in the upper atmosphere would have decayed before reaching the Earth's surface. In fact, the majority of muons survive the journey and reach the Earth. The apparent lifetime of a rapidly moving muon is different to that of a stationary muon. We must either postulate that the lifetime depends upon the speed, or agree that time is different for the muon and for the Earth observer. Time has to be reassessed for rapidly moving objects.

If an electron of mass m is accelerated to a speed comparable with the speed of light in vacuum, the speed v can be determined by measuring the

● *Muon lifetime in flight*

time of flight of the electron over a known distance of travel. The momentum mv of the electron can be determined by measuring its radius of curvature in a uniform magnetic field, as discussed in Section 16.2. If the electron's mass is constant, independent of its energy and speed, dividing the momentum so measured by the speed should give the constant value of the mass. When the speed of the electron is comparable with the speed of light in vacuum the mass so measured is not constant but increases as the speed increases. We thus have to reassess the concept of mass as well as of time when objects are moving at high speeds.

The manner in which time, length, mass, and other physical quantities have to be reinterpreted for objects moving at high speeds or for observers moving with high, uniform relative velocity is the subject of this chapter on **special relativity**. Before embarking upon that journey, however, we first recall the details of transformations between space and time coordinates used in classical mechanics.

● *Electron mass in flight*

Galilean transformations

When the speeds of moving objects are small compared with the speed of light in vacuum, Newtonian mechanics is valid and measurements of position and time made in one laboratory can be related in a simple manner to those made in a second laboratory moving uniformly with respect to the first. Measurements of position are made by specifying coordinates in coordinate sytems fixed in the two laboratories and time is the same for observers in either coordinate system. A measuring platform in which position and time are given via coordinates x, y, z, and t is called a **frame of reference** or simply a frame.

There is no absolute frame of reference and no absolute state of rest. Newton's laws are valid in all frames moving with uniform velocity with respect to each other and so all such frames are equivalent. Such frames are called **inertial frames**. For example, the frame in which observations are made by a person stationary on a river bank and the frame used by a person on a boat moving at constant velocity in the river are inertial frames.

Imagine an observer in frame S, on a river bank as in Fig 19.1 watching a boat passing down a river at speed V in the x-direction. An observer on the boat, in frame S′, measures a constant position x' for a particular object at a point on the boat. In the reference frame S, in which the bank is at rest, the position of the object is measured to be $x = x' + Vt$, where t is zero at the instant the observer on the boat is directly opposite the person on the bank. Time is the same for both observers, and also they see no difference in their measurements of height or distance perpendicular to the bank.

The transformations between the x, y, z, and t coordinates used on the bank and x', y', z', and t' used on the boat are those used in classical

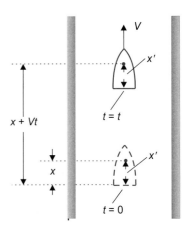

Fig. 19.1 Galilean transformations.

mechanics. They are called Galilean transformations, and are discussed in Section 3.2. They are

$$x' = x - Vt, \tag{19.1}$$

$$y' = y, \tag{19.2}$$

$$z' = z, \tag{19.3}$$

and

$$t' = t. \tag{19.4}$$

If a person on the bank measures a current in the river of speed u_x in the same direction as that in which the boat is moving, a stick in the river is observed by that person to move in the same direction as the boat with speed u_x. If u'_x is the speed of the stick as observed by someone on the boat, $u'_x = (u_x - V)$, the stick is moving backwards with speed u'_x and the bank is moving backwards with speed V. The transformation of speeds, here the transformation between the x-component of velocity measured by one observer and by a second moving at speed V with respect to the first, is thus

$$u'_x = u_x - V. \tag{19.5}$$

Note that this equation gives the x-components with the correct signs. For the boat, V is greater than u_x and the sign of u'_x is negative; the observer on the boat sees the stick move backwards.

Transformations at high relative speeds

Now imagine an observer on Earth and one moving with constant velocity relative to the Earth with the magnitude of the velocity approaching the speed of light. The transformations are no longer simple. Indeed, many of the observations an Earth observer would make about a fast-moving object would be quite different to those that would be made by someone on the object. The way different observers measure position and time when they are moving relative to each other at high uniform speed and the way mass has to be reinterpreted are basic elements of special relativity.

Observations made in frames that are accelerating with respect to each other cannot in general be related using special relativity. Such frames are not inertial frames and the laws of physics stay unaltered only for inertial frames. However, kinematic problems involving only observations of speeds and accelerations can be treated within special relativity, as in Worked example 19.3. Accelerated frames are the province of **general relativity**. Since an acceleration involves a force, an observer in one or both of the non-inertial frames is experiencing a force. Thus general relativity has a particular bearing on gravity. In this chapter we restrict attention to inertial frames.

The axioms of special relativity

Throughout the book so far we have implicitly assumed that the laws of physics are the same for someone performing experiments on the bank of a river in which a boat is moving and someone moving with the boat. This assumption is intuitively acceptable and can be verified by experiment. Does this remain so for observers in uniform high-speed relative motion? The answer is yes. The results of all experiments performed so far indicate that Nature has arranged things so that the laws of physics are indeed the same for all observers in uniform relative motion, however high their relative speed. This is the first axiom, or building block, of special relativity.

We now need a second building block in order to proceed. This is provided by the theory of electromagnetism. We have not developed electricity and magnetism to the point where a complete theory emerges. We have, however, indicated that, when charges in free space are accelerated, an electromagnetic field (first introduced in Chapter 7) is radiated. This field consists of related, time-varying electric and magnetic fields moving outwards from an accelerated charge with a speed c, the speed of light in free space. The theory that describes radiation introduces c in a consistent manner and in free space makes no reference to observer or emitter. The speed of light in free space is the same for all observers irrespective of their state of relative motion. That is the second axiom or building block.

Two fundamental axioms underpinning special relativity are thus:

I   The laws of physics are the same for all observers in uniform relative motion.

II   The speed of light in free space is the same for all observers in uniform relative motion.

Observers in different inertial frames see the same events differently and make different measurements, but the laws they deduce from the measurements are the same. Scientists at the turn of the twentieth century had great difficulty in reconciling axiom II with observations when Galilean transformations were used and the prevalent assumption made that observers in different inertial frames saw time in exactly the same way.

In a paper of 1905, Albert Einstein clarified the situation by taking a profound new look at the meaning of time. This radical step enabled him

Albert Einstein

Albert Einstein, generally accepted as the greatest physicist of his generation and one of the most famous of all time, was born in Ulm, now in Germany, on 31 April 1879. He and sister Maja grew up in Munich, Bavaria, and the family moved to Switzerland in 1894 when Einstein was 15.

At the age of 12 his interest turned to science through reading popular scientific books. Einstein himself says that '...at the age of 12–16 I familiarized myself with the elements of mathematics together with the principles of differential and integral calculus.' This essential preparation for students of physics is not nowadays always completed before commencement of an undergraduate course, and serious students of physics should, like Einstein, ensure for themselves that this groundwork is in place. Einstein was a free-thinker and hated learning by coercion, as he put it, preferring to read deeply and form his own ideas on matters he regarded as essential rather than cramming received material in order to pass examinations. In an autobiographical note made in 1949 he says 'It is...nothing short of a miracle that the modern methods of instruction have not yet entirely strangled the holy curiosity of enquiry....' Many of today's students no doubt have considerable sympathy with that point of view. The technique of almost complete self-instruction works with a genius, in whom it fosters radical departures from conventional wisdom. However, it is probably not such a good method for most of us, although conventional wisdom should always be examined critically.

At the age of 15, his iconoclastic way of educating himself contributed to failure to pass the entrance examination for the Swiss Federal Institute of Technology in Zurich. He returned to school and passed the next year, entering the Institute at age 16 in 1895. On graduation, having alienated many of his teachers by his attitudes, he had difficulty finding a position in which he could continue to pursue his academic interests. He finally joined the Swiss patent office in Berne. There he continued to think deeply about natural phenomena and, while performing his duties as 'technical officer, third class', published papers on heat and thermodynamics, the first in 1901. In 1905 came the astounding trio of publications, each of which had an enormous

impact upon the development of science. The first was on the photoelectric effect and began the acceptance of the corpuscular nature of light. He was awarded the Nobel Prize in 1921 for this work. The second was on Brownian motion and continued the proof of the atomistic nature of matter. The third was 'On the electrodynamics of moving bodies', and placed special relativity on a firm and fruitful basis with the concept of space–time replacing the notions of absolute time and absolute reference frame.

In 1909, at age 30, he obtained a post at the University of Zurich. His first Professorship was at the University of Prague in 1911. He moved back to Zurich in 1912 and went on to Berlin in 1913. In 1915 he published his first paper on general relativity, which gave a geometrical interpretation of gravitation as a curvature in space–time.

Einstein was an avuncular figure, and a kind and humble person. His first wife, Mileva, did not move to Berlin in 1913 and an amicable divorce followed in 1919. He soon married Elsa, a cousin, with whom he was happy, although his life was not without other romantic entanglements. When the Nazis took power in Germany in 1933, and persecution of Jewish people and personal attacks on him as a symbol of Jewish science began, Einstein was in California. He stayed in the USA at Princeton University and worked there for the rest of his life. He died on 18 April 1955.

to produce an elegant and satisfying theory of relativity which describes all observations made up to now. There is a long and interesting history to the evolution of the theory of relativity. Some of it is included in the biographical page on Einstein.

19.2 The Lorentz transformation

How do observers in different inertial frames relate their measurements of position and time? Let us set up two frames, one that we will call S and the other S′ moving at speed V relative to S along the x-axis as shown on Fig 19.2. Let a light signal be emitted in S at time t at the point with coordinates x, y, and z. The four coordinates specify the position of what we will call an *event* in frame S. An observer in S′ may see the signal emitted from the point with different spatial coordinates $x′$, $y′$, and $z′$. An important question is whether the times of events also depend upon which observer is measuring them. The answer is yes. The transformation of time from t to $t′$ depends upon where the event occurred in the frame S, and the observer in S′ measures a time $t′$ for the event different to t.

Our normal perception that time and space differ in a philosophical sense, and should be treated and thought of quite differently, needs revision. There is no absolute time. Observers in different inertial frames measure time differently. We should no longer think in terms of space and time, but in terms of space–time, in which the three space coordinates and one time coordinate are inextricably linked and all four can be transformed from one frame to another. The transforms for time now involve space in the same way as the Galilean transform (19.1) for space involved time.

The two observers measure different time and space coordinates and we require to know how the four coordinates as measured in S transform to the coordinates measured in S′. The transformation between them is known as the **Lorentz transformation**, after Hendrik Lorentz who also played a distinguished part in the evolution of relativity theory.

The Lorentz transformation has the form $x′ = f(x, y, z, t)$, $y′ = g(x, y, z, t)$, $z′ = h(x, y, z, t)$, and $t′ = k(x, y, z, t)$, where f, g, h, and k are functions whose forms are to be determined. The speed of light is the same for observers in either frame and, if a light pulse is emitted at the

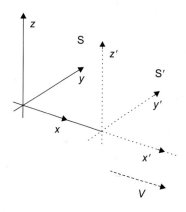

Fig. 19.2 The inertial frame S′ moves at speed V along the x-axis of frame S.

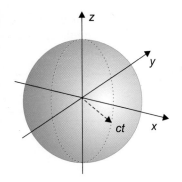

Fig. 19.3 A light signal emitted from the origin of frame S spreads out and at time t reaches points on the surface of a sphere of radius ct.

instant when the origins of the frames coincide, each observer sees the light spreading out in a sphere whose radius increases linearly with time. This requires that the relations between the space and time coordinates in the two frames be linear; no powers of the coordinates higher than the first can be involved in the transforms.

We will present the steps used to derive the Lorentz transforms without going into rigorous mathematical detail. Consider Fig 19.3 in which a light signal is emitted from the origin of system S at time t equal to zero, when the origins of frames S and S′, moving at speed V along the x-axis in S, were coincident. An observer in S sees a spherical wave moving away from the origin at speed c. At time t the wavefront is the surface of a sphere of radius ct whose equation is

$$x^2 + y^2 + z^2 = c^2 t^2.$$

An observer in S′ must also see an outgoing spherical wave moving at speed c, since the same laws of physics hold in both both frames and the speed of light is the same for both. This requires that

$$x^2 + y^2 + z^2 - c^2 t^2 = x'^2 + y'^2 + z'^2 - c^2 t'^2. \tag{19.6}$$

The linear transformations that satisfy this condition are

$$x' = \frac{(x - Vt)}{\sqrt{1 - V^2/c^2}}, \tag{19.7}$$

$$y' = y, \tag{19.8}$$

$$z' = z, \tag{19.9}$$

and

$$t' = \frac{(t - Vx/c^2)}{\sqrt{1 - V^2/c^2}}, \tag{19.10}$$

as can easily be shown by substitution.

Examination of the Lorentz transforms shows that they reduce to the classical forms when the speed V is much less than the speed of light c. Newtonian mechanics is thus the limiting form of the more general mechanics of special relativity. The almost universal validity of Newtonian mechanics, to an excellent approximation, occurs because normal experience does not encompass objects moving at speeds comparable with the speed of light.

The Lorentz transforms are often written in more compact form by introducing the parameter γ given by

$$\gamma = \frac{1}{\sqrt{1 - V^2/c^2}}. \tag{19.11}$$

In terms of γ,

$$x' = \gamma(x - Vt), \tag{19.12}$$

$$y' = y, \tag{19.13}$$

$$z' = z, \tag{19.14}$$

and

$$t' = \gamma(t - Vx/c^2). \tag{19.15}$$

The transformations from coordinates in the S′ frame to those in the S frame are obtained by the substitutions $x \to x'$, $y \to y'$, $z \to z'$, and $t \to t'$ in the above equations while putting $V \to -V$. They are

$$x = \gamma(x' + Vt'), \tag{19.16}$$

$$y = y', \tag{19.17}$$

$$z = z', \tag{19.18}$$

and

$$t = \gamma(t' + Vx'/c^2). \tag{19.19}$$

As the relative speed of the two inertial frames increases, γ increases, and the Lorentz transforms depart more and more from the classical expressions. A second parameter β, equal to the ratio of a speed to the speed of light, is often used in discussions of relativity, and speeds are often quoted in units of c by expressing them in terms of β. The parameter γ also incorporates another important result of relativity. The speed of light in free space is the maximum speed attainable by any object. If V exceeds c, γ becomes imaginary and we are unable to interpret time and space in the moving frame.

The time transform, eqn (19.15), also shows that events at different places seen as simultaneous in one frame are not observed to be coincident by an observer in a second. The times of the two events as seen by an observer in the frame S′ depend upon their positions in frame S.

Worked Example 19.1 An observer on Earth observes two events at x_a and x_b to have occurred simultaneously at time t. What is the time difference between the events measured by someone in a rocket moving at speed V with respect to Earth?

Answer Let t'_a and t'_b be the times of the events as seen by the observer on the rocket. The Lorentz transforms then give

$$t'_a = \gamma(t - Vx_a/c^2)$$

and

$$t_b' = \gamma(t - x_b V/c^2).$$

The time difference is thus

$$t_a' - t_b' = \gamma V(x_b - x_a)/c^2.$$

We may note that, if V is much less than c, $\gamma \sim 1$, V/c^2 tends to zero and the time difference tends to zero, the classical result.

19.3 Kinematics

Kinematics is concerned with motion without reference to the forces that produce motion or to the masses in motion. In this section we pursue the consequences of the Lorentz transforms for length and time intervals as measured by observers in different inertial frames. We also consider how the speed of a third moving body is related in the two frames.

Length contraction

If the length of an object measured in a frame S′ in which the object is at rest is L_0, the length as measured by an observer in a frame S moving with respect to the object is less than L_0. Seen from Earth, an object moving at high speed in space contracts. This effect is known as the Fitzgerald contraction, or the Lorentz–Fitzgerald contraction. George Fitzgerald was another physicist who played an important role in the development of relativity theory.

Consider a rod of length L_0 when measured by an observer in frame S′ in whch the rod is at rest. Let this rod move along the x-axis of frame S with speed V. In frame S, let the positions of the ends a and b of the rod at time t be x_a and x_b. In the frame S′, the positions x_a' and x_b' of the ends of the rod are given by eqn (19.12) as

$$x_a' = \gamma(x_a - Vt)$$

and

$$x_b' = \gamma(x_b - Vt).$$

The length of the rod, L, measured in S is simply $x_b - x_a$, and hence

$$L = (x_b' - x_a')/\gamma = L_0/\gamma = L_0\sqrt{(1 - V^2/c^2)}. \tag{19.20}$$

Once more, it is obvious that the contraction is negligible unless the speed of the moving object is comparable with the speed of light.

Worked Example 19.2 A futuristic train whose length at rest is 200 m passes through a tunnel whose length as measured by an observer at rest is also 200 m. If the driver at the front of the train and a guard at the back of the train think there is a time difference of 10^{-8} s between the front leaving the tunnel and the rear entering it, what is the speed of the train?

Answer Let V be the speed of the train with respect to the Earth. If we look at things in the frame in which the train is at rest, the tunnel is moving past the train and the length of the tunnel is contracted over its rest length by the factor γ. Observers on the train, to whom the train is 200 m long, observe an object of length $200/\gamma$ m passing at speed V. When the back end of the tunnel passes the front of the train, the front end of the tunnel has not yet reached the back of the train. As seen on the train, the time taken for the front of the tunnel to reach the back of the train is $(200 - 200/\gamma)/V$, and this time equals 10^{-8} s. Hence

$$\frac{200 - 200\sqrt{(1 - V^2/c^2)}}{V} = 10^{-8}$$

or

$$1 - \sqrt{(1 - V^2/c^2)} = 5 \times 10^{-11} V.$$

This equation may be expressed conveniently in terms of $\beta = V/c$, when

$$\sqrt{(1 - \beta^2)} = 1 - 0.015\beta.$$

Squaring and solving the resulting equation for β gives $\beta = 0.03$, and $V = 9.0 \times 10^6 \, \mathrm{m\,s^{-1}}$.

Time dilation

If the time interval between two events at the same place in frame S is Δt, the time interval measured by an observer in frame S′ is longer than Δt. This effect is known as time dilation. The time as measured at a fixed point in S is called the *proper* time in S and proper time is ahead of time in a frame moving with respect to S.

An example of time dilation is provided by the radioactive decay of subatomic particles after being created by interactions of cosmic rays from outer space with the Earth's upper atmosphere, as discussed in the introduction to this chapter. Suppose the particles created have a lifetime in their rest frame of τ and that after creation they move towards the surface of the Earth at speed V. The decay curve of the particles is an exponential as discussed in Chapter 9 and the lifetime is a measure of the average time interval over which the particles survive. This interval as observed on Earth is longer than τ because of time dilation and depends on the speed V. Particles with very short lifetimes may still be observed on Earth if the time-dilated lifetimes become long enough for the particles to travel the distance between the point of creation and the Earth's surface.

In the rest frame of a particle, it is created at position x', where it remains, at time t'_a and decays at time t'_b. In the frame in which the Earth is at rest, the particle is moving at speed V along the x-axis and an observer on Earth sees the particle created at x_a at time t_a and decay at x_b at time t_b. The Lorentz transform (19.19) gives

$$t_a = \gamma(t'_a + Vx'/c^2)$$

and

$$t_b = \gamma(t'_b + Vx'/c^2).$$

Hence the time interval between creation and decay, as measured on Earth, is given by

$$t_b - t_a = \Delta t = \gamma(t'_b - t'_a) = \gamma \Delta t'. \tag{19.21}$$

When the decay of many particles with lifetime τ is considered, the average lifetime observed on Earth is increased by the factor γ, and the Earth observer sees particles travelling on average for a time $\gamma \tau$ with a speed V. The distance $(x_b - x_a)$ that particles have moved on average towards the Earth in this time can be determined using the Lorentz transform (19.16), which gives

$$x_a = \gamma(x' + Vt'_a),$$

$$x_b = \gamma(x' + Vt'_b),$$

and

$$x_b - x_a = \gamma V \tau. \tag{19.22}$$

An observer in the rest frame of the particles sees the Earth approaching at speed V, and in the time interval τ the Earth has moved a distance $V\tau$, equal to the length $\gamma V\tau$ shortened by the Lorentz–Fitzgerald contraction.

Worked Example 19.3 An object accelerates from rest in a straight line with uniform acceleration a. What is the time in the object's frame when it is observed to have acquired a speed of $\sqrt{3}c/2$.

Answer Equation (19.21) gives the relation between time $\Delta t'$ and the time Δt measured by the observer in frame S with respect to whom the object is moving. This applies to intervals however short and

$$\frac{dt}{dt'} = \gamma.$$

But $dV/dt = a$, $dt = dV/a$ and hence

$$dt' = \frac{1}{a}\left(1 - \frac{V^2}{c^2}\right)^{1/2} dV.$$

Integrating this equation from $t' = 0, V = 0$ to $t' = T', V = \sqrt{3}c/2$, where T' is the time at which the object is observed in S to have instantaneous speed $\sqrt{3}c/2$,

$$T' = \frac{1}{a}\int_0^{\sqrt{3}c/2}\left(1 - \frac{V^2}{c^2}\right)^{1/2} dV.$$

This integral can be evaluated using the substitution $V = c\sin\theta$, to give $dV = c\cos\theta\,d\theta$ and

$$T' = \frac{c}{2a}\left(\theta + \frac{\sin 2\theta}{2}\right)_0^{\pi/3} = 0.74c/a.$$

Exercise 19.1 Cosmic ray showers produce muons with a speed of $0.988c$. The muons decay in their rest frame with a lifetime of 2.20×10^{-6} s. What lifetime would be measured by an observer on Earth?

Answer 1.42×10^{-5} s.

Transformation of velocity

If an observer on Earth sees an object moving at constant speed u_x in the positive x-direction, what speed u'_x is measured by an observer in a frame S' moving at speed V in the positive x-direction with respect to Earth? In frame S' the motion must also be straight line motion with constant speed, since Newton's laws are obeyed in both frames. In frame S, if the object is at x_a at time t_a and at x_b at time t_b, the speed is given by

$$u_x = \frac{x_b - x_a}{t_b - t_a}. \tag{19.23}$$

The speed u'_x of the object in frame S' is given by

$$u'_x = \frac{x'_b - x'_a}{t'_b - t'_a}. \tag{19.24}$$

Using the Lorentz transforms (19.12) to (19.15) to rewrite this equation in terms of positions and times observed in frame S, we obtain

$$u'_x = \frac{\gamma(x_b - Vt_b) - \gamma(x_a - Vt_a)}{\gamma(t_b - Vx_b/c^2) - \gamma(t_a - Vx_a/c^2)}.$$

Rearranging this expression gives

$$u'_x = \frac{\left(\dfrac{x_b - x_a}{t_b - t_a}\right) - V}{1 - \dfrac{V}{c^2}\left(\dfrac{x_b - x_a}{t_b - t_a}\right)} = \frac{u_x - V}{1 - Vu_x/c^2}, \tag{19.25}$$

from eqn (19.23).

If a particle is moving along the x'-axis with speed u'_x as observed in frame S′, its speed observed in frame S moving at speed V in the negative x'-direction is given by

$$u_x = \frac{u'_x + V}{1 + Vu'_x/c^2}. \tag{19.26}$$

Comparison of eqn (19.25) with expression (19.5) again shows that the relativistic transforms reduce to the classical expression when the speed V is small compared with the speed of light. If the moving object is a pulse of light moving with speed c in frame S, $u_x = c$ and eqn (19.25) shows that the speed of the pulse in frame S′ is also measured to be c. The formula for the transformation of velocities is thus consistent with the second building block of special relativity, which is that the speed of light in free space is the same in whatever inertial frame it is measured.

Worked Example 19.4 A fluid flows with uniform speed V down a hollow tube. A light pulse is passed axially down the tube. The speed of light relative to the tube is measured with the light passing in the same direction as the fluid flow and then in the opposite direction. Determine the ratio of the measured speeds.

Answer The speed of the light pulse is the group velocity v_g of the light in the fluid; v_g is always less than c and is not the same in different inertial frames. The speed v_g is that observed when the medium in which the wave is travelling is at rest relative to the laboratory in which the speed is measured. It is thus the speed of the wave in the frame in which the fluid is at rest.

With frame S the laboratory frame, u'_x in eqn (19.26) is $+v_g$ if the light is travelling in the $+x$-direction, when the speed in the laboratory frame is $u_x(+)$. When the light is travelling in the $-x$-direction, u'_x in eqn (19.26) is $-v_g$ and the speed in the laboratory frame is $u_x(-)$. The frame S′ in which the fluid is at rest is travelling at speed V in the $+x$-direction with respect to the laboratory and, for the light moving in the same direction as the fluid, eqn (19.26) becomes

$$u_x(+) = \frac{v_g + V}{1 + Vv_g/c^2}.$$

For the light moving in the opposite direction,

$$u_x(-) = \frac{-v_g + V}{1 - Vv_g/c^2}.$$

Solving for the ratio $u_x(+)/u_x(-)$ gives

$$\frac{u_x(+)}{u_x(-)} = \frac{(V + v_g)}{(V - v_g)} \frac{(1 - Vv_g/c^2)}{(1 + Vv_g/c^2)}.$$

When the fluid speed is zero, the ratio becomes -1, consistent with measurements in a single frame. When $v_g = c$, as it is for the light pulse travelling in free space, and V is not zero, the ratio is also -1, corresponding to observers in the two frames measuring light going in opposite directions but at the same speed. This is again consistent with the second axiom of special relativity that the speed of light in free space is the same in all inertial frames.

Equation (19.25) gives the transformation of speed when the velocity of an object and that of frame S′ are both along the x-axis of frame S. We will always consider the relative motion of the frames to be along the x- and x'-axes, but we need not restrict the velocity of a particle to lie in any particular direction in frame S for us to construct the velocity components in frame S′. We first note that, although the measurements of y- and z-coordinates in the two frames are the same, the measurements of the time differences $t_b - t_a$ and $t'_b - t'_a$ are different because of the dependence of time in frame S′ on position in frame S. The z-component of the velocity of a particle in frame S is

$$u_z = \frac{z_b - z_a}{t_b - t_a}.$$

The z-component in frame S' is

$$u'_z = \frac{z'_b - z'_a}{t'_b - t'_a}.$$

Using the transforms for the z'- and t'-coordinates given by eqns (19.13) and (19.15), this equation reduces to

$$u'_z = \frac{u_z}{\gamma(1 - Vu_x/c^2)}. \tag{19.27}$$

The same argument leads to a similar formula for the y-component of velocity, u'_y, in S' in terms of the component u_y in S.

19.4 Mechanics

Mechanics considers the motions of massive bodies under the influence of forces. We have seen how the length of an object depends upon the frame in which it is measured, as does the time interval between events. We must ask whether the laws of mechanics and the masses of objects as used classically need to be modified in order to comply with special relativity. We will not discuss the motion of an object under the influence of an arbitrary force, but will consider the application of the principles of conservation of linear momentum and of energy to collisions between two objects.

Collisions between particles moving at speeds comparable with the speed of light, for which relativity must be used, form the essence of high-energy physics research into Nature's fundamental particles. Large accelerators produce electrons, positrons, and protons at speeds little different from c. These particles bombard a fixed target or collide with each other head-on to give new particles whose properties reveal the underlying structure of the matter in the Universe.

Consider two bodies of masses m_A and m_B colliding along a line joining their centres with initial speeds u_A and u_B, as in Fig 19.4. If the internal structures and internal energies of the two objects are the same after the collision as before, the collision is said to be elastic, and only the speeds and directions of the objects change. If A and B move on the line along which they were incident after the collision, the classical expression for conservation of linear momentum is

$$m_A u_A + m_B u_B = m_A v_A + m_B v_B \tag{19.28}$$

where v_A and v_B are the speeds after the collision. The classical expression for the conservation of energy is

$$\tfrac{1}{2}m_A u_A^2 + \tfrac{1}{2}m_B u_B^2 = \tfrac{1}{2}m_A v_A^2 + \tfrac{1}{2}m_B v_B^2. \tag{19.29}$$

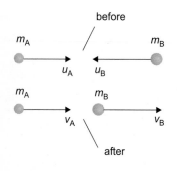

Fig. 19.4 The classical representation of an elastic collision between two masses m_A and m_B that stay on the same line after the collision.

In classical physics these conservation laws must be obeyed by all observers in uniform relative motion. However, when speeds are comparable with c, they are no longer exact. In order to obtain conservation principles that satisfy the principles of relativity, the concept that the mass of a body is fixed and invariant in different inertial frames has to be discarded, just as the concepts of absolute time and fixed length were discarded.

Mass and momentum

In this section we will determine how the conservation laws are reformulated relativistically by determining new expressions for mass and momentum. These expressions will be shown to have transforms similar to the Lorentz transforms.

Since the speed of a moving particle depends on the frame in which it is observed, we begin with the transformation of speeds between inertial frames. If a particle is moving along the x-axis of frame S with speed u_x, its speed observed in frame S′ moving at speed V along the positive x-axis is u'_x, given by eqn (19.26). This equation can be rearranged in the form

$$\frac{u_x}{\sqrt{1 - u_x^2/c^2}} = \frac{(u'_x + V)}{\sqrt{1 - V^2/c^2}\sqrt{1 - u_x'^2/c^2}}. \tag{19.30}$$

It is easiest to show that this expression is true by squaring both sides and showing that the solution for u_x obtained from the resulting equation is the same as given by eqn (19.26).

Multiplication of eqn(19.30) by a quantity m_0 that has the dimensions of mass gives

$$\frac{m_0 u_x}{\sqrt{1 - u_x^2/c^2}} = \frac{m_0(u'_x + V)}{\sqrt{1 - V^2/c^2}\sqrt{1 - u_x'^2/c^2}}. \tag{19.31}$$

In special relativity the masses m and m' to be used in S and S′ are

$$m = \frac{m_0}{\sqrt{1 - u_x^2/c^2}} \tag{19.32}$$

and

$$m' = \frac{m_0}{\sqrt{1 - u_x'^2/c^2}}, \tag{19.33}$$

respectively, and the corresponding momenta p and p' should be

$$p = \frac{m_0 u_x}{\sqrt{1 - u_x^2/c^2}} \tag{19.34}$$

● *Relativistic mass and momentum*

and

$$p' = \frac{m_0 u'_x}{\sqrt{1 - u'^2_x/c^2}}.$$

(19.35)

Equation (19.31) now reads

$$p = \gamma(p' + m'V).$$

(19.36)

Also,

$$\frac{m}{m'} = \frac{\sqrt{1 - u'^2_x/c^2}}{\sqrt{1 - u^2_x/c^2}}$$

and, using eqns (19.25) and (19.30), this reduces to

$$m = \gamma(m' + Vp'/c^2).$$

(19.37)

Comparison of eqns (19.36) and (19.37) with the Lorentz transforms for the coordinates x and t shows that mass and linear momentum transform in exactly the same way as do position x and time t. Substitution of p and p' for x and x', and m and m' for t and t' in the transforms (19.16) and (19.19) gives (19.36) and (19.37).

The mass m_0 is called the **rest mass** of the particle. It is the mass measured in a frame in which the particle is at rest. When the particle is moving, irrespective of the frame in which it is observed, an observer interprets the mass as the rest mass times the factor γ appropriate to the speed of the particle as measured by that observer. The observer interprets the momentum of the particle as the rest mass times the factor γ times the velocity with which the particle is observed to move.

The mass m of an object to be used with the principle of conservation of momentum by an observer with respect to whom the object is moving at speed u is given by

$$m = \gamma m_0$$

(19.38)

where m_0 is the rest mass and $\gamma = (1 - u^2/c^2)^{-1/2}$.

The principle of conservation of linear momentum, instead of eqn (19.28), now becomes

● *Conservation of momentum in special relativity*

$$\frac{m_{0,A}}{\sqrt{1 - u_A/c^2}}u_A + \frac{m_{0,B}}{\sqrt{1 - u_B/c^2}}u_B = \frac{m_{0,A}}{\sqrt{1 - v_A/c^2}}v_A + \frac{m_{0,B}}{\sqrt{1 - v_B/c^2}}v_B$$

(19.39)

for two masses colliding and continuing to move on the same line that before the collision joined their centres, where $m_{0,A}$ and $m_{0,B}$ are the rest masses of particles A and B.

Energy

We have determined how the principle of conservation of momentum is modified to conform with special relativity. How is conservation of energy, as stated by eqn (19.29) for an elastic collision treated classically, to be reconstructed?

A particle of rest mass m_0 moving at speed u has mass given by eqn (19.38). For speeds small compared with the speed of light we may expand this using the binomial theorem to give

$$m = \frac{m_0}{\sqrt{1 - u^2/c^2}} = m_0 + \frac{1}{2} m_0 \frac{u^2}{c^2} + \cdots. \qquad (19.40)$$

If we multiply both sides of this equation by c^2,

$$mc^2 = m_0 c^2 + \tfrac{1}{2} m_0 u^2 + \cdots. \qquad (19.41)$$

The second term on the right-hand side is the classical kinetic energy of the moving particle. We therefore interpret $m_0 c^2$ as the **rest mass energy** of the particle, and mc^2 as the **total energy** E,

$$E = mc^2. \qquad (19.42)$$

$E = mc^2$ is the energy to be used with the principle of conservation of energy for particles at high speeds.

The principle of conservation of energy, instead of eqn (19.29), now becomes

● *Conservation of energy in special relativity*

$$\frac{m_{0,A} c^2}{\sqrt{1 - u_A/c^2}} + \frac{m_{0,B} c^2}{\sqrt{1 - u_B/c^2}} = \frac{m_{0,A} c^2}{\sqrt{1 - v_A/c^2}} + \frac{m_{0,B} c^2}{\sqrt{1 - v_B/c^2}} \qquad (19.43)$$

for the two colliding masses.

We have now reformulated the principles of conservation of momentum and energy in a form that is the same in all inertial frames by changing the definition of mass. That the formulations are indeed correct rests upon a wealth of supporting evidence from data on the collisions of particles at very high energies.

The relation between total energy, mass, and momentum

It should be noted that the kinetic energy of a particle, which is equal to its total energy less its rest mass energy, is given by

● *Kinetic energy in special relativity*

$$\text{Kinetic energy} = mc^2 - m_0 c^2. \qquad (19.44)$$

A useful expression relating the total energy, rest mass, and momentum of a moving particle can be derived using eqn (19.42), which may be rewritten as

$$E^2(1 - u^2/c^2) = m_0^2 c^4.$$

Hence,

$$E^2 = m_0^2 c^4 + E^2 u^2/c^2.$$

Using $E = mc^2$ on the right-hand side gives

$$E^2 = m_0^2 c^4 + c^2 m^2 u^2.$$

The quantity $m^2 u^2$ is the square of the particle's momentum, p, and we obtain

$$E^2 = m_0^2 c^4 + c^2 p^2. \tag{19.45}$$

This equation immediately gives the momentum p of a particle that has energy E but zero rest mass,

$$p = \frac{E}{c}, \tag{19.46}$$

as follows for photons from eqn (9.10a). A photon in the yellow part of the visible spectrum has an energy of about 2 eV and in free space moves at the speed of light. Its momentum is about the same as the momentum of an electron that has a kinetic energy of about 4×10^{-6} eV, an energy at which classical expressions can be used for the electron's momentum and kinetic energy.

A convenient way of determining when the use of classical expressions gives rise to negligible errors is to compare the total energy with the rest mass energy. If the difference is small, mc^2 differs little from $m_0 c^2$ and the factor γ is almost unity. For an electron of kinetic energy 40 eV, the difference between the total energy and the rest mass energy of 511 keV is less than one part in 10^4, and γ differs from unity by the same small factor.

Worked Example 19.5 A fast particle of momentum p_0 and rest mass m_1 hits a stationary particle of rest mass m_2 greater than m_1. The collision is elastic, and the particles retain their initial rest masses. Show that, in the limit of very high energy, the momentum of the mass m_1 when it is scattered through 180° approaches the magnitude $(m_2^2 - m_1^2)c/2m_2$.

Answer Let p_0, E_0 and p_1, E_1 be the initial and final momentum and energy, respectively, of the particle with rest mass m_1, and p_2, E_2 be the momentum and energy after the collision of the particle with rest mass m_2. The problem concerns the situation when the energy E_0 is very high, and the principles of conservation of momentum and energy are best written in terms of energies so that we may take the limit as E_0 tends to infinity. Conservation of momentum, after using eqn (19.45), and energy may be written

$$\sqrt{E_0^2 - m_1^2 c^4} = \sqrt{E_2^2 - m_2^2 c^4} - \sqrt{E_1^2 - m_1^2 c^4} \tag{19.47}$$

and

$$E_0 + m_2 c^2 = E_2 + E_1, \tag{19.48}$$

respectively. The last equation gives

$$E_2^2 = (E_0 + m_2 c^2 - E_1)^2.$$

Using this identity to eliminate E_2 from eqn (19.47), and squaring both sides of the rearranged equation gives

$$E_0^2 + E_1^2 + 2E_0 m_2 c^2 - 2E_1 m_2 c^2 - 2E_0 E_1$$
$$= E_0^2 - m_1^2 c^4 + E_1^2 - m_1^2 c^4 + 2\sqrt{(E_0^2 - m_1^2 c^4)(E_1^2 - m_1^2 c^4)}.$$

We now rearrange this equation and square both sides. After this we need retain only terms in E_0^2 since other terms are negligible when E_0 is very large. This gives

$$E_1 = \frac{c^2(m_1^2 + m_2^2)}{2m_2}.$$

Using eqn (19.45) again,

$$c^2 p_1^2 = E_1^2 - m_1^2 c^4,$$

giving the final result that the magnitude of p_1 is

$$p_1 = \frac{c(m_2^2 - m_1^2)}{2m_2}.$$

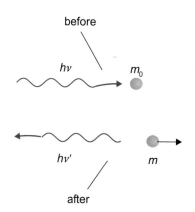

before

$h\nu$

m_0

$h\nu'$

m

after

Fig. 19.5 A high-energy photon back-scattered off an electron.

We note that, if a photon of very high energy collides with a free electron in a Compton scattering process, as in Fig 19.5, and is back-scattered through 180°, m_1 is zero and the momentum $p_1 = h\nu'/c$ where ν' is the frequency and $h\nu'$ the energy of the back-scattered photon. Hence

$$h\nu' = \tfrac{1}{2} m_e c^2 \tag{19.49}$$

where m_e is the rest mass of the electron. The energy of a back-scattered photon approaches half the rest mass energy of an electron, 255 keV, as the incident photon energy becomes very high. This result is consistent with the Compton scattering formula given in eqn (9.16).

The masses of fast-moving particles for which relativistic expressions have to be used are often quoted in units of energy divided by the square of the speed of light. Relativistic mechanics is required in high-energy physics where beams of subatomic particles collide with speeds comparable with the speed of light. It is also necessary to use relativistic mechanics when high precision is required at speeds somewhat less than that of light. In high-energy physics the energy units are MeV, and masses are usually quoted in units of MeV/c^2. The units of momentum are mass times speed, equivalent to energy divided by speed, and momenta are usually quoted in units of MeV/c.

● *Mass in MeV/c^2*

● *Momentum in MeV/c*

Exercise 19.2 A particle has rest mass $300\,MeV/c^2$ and momentum p equal to $400\,MeV/c$. What speed is it moving at?

Answer This exercise is most simply done by using the result that $\gamma = mc^2/m_0 c^2$. The speed is given by $\beta = V/c = 0.8$.

In general, if the conservation of mass is retained, when other forms of energy are also involved we must identify mass and energy as equivalent. If the energy of a system changes by an amount ΔE, its mass changes by an amount

$$\Delta m = \Delta E/c^2. \tag{19.50}$$

Worked Example 19.6 The *solar constant* is the energy per unit time per unit area received from the Sun at a point that is one astronomical unit of distance from the Sun. The astronomical unit of length is the Earth's mean distance from the Sun. The value of the solar constant is 1400 watts per metre squared. This is the energy that would fall on a unit area of the surface of the Earth per second if there were no atmosphere. Estimate the loss of mass per year of the Sun. Take the Sun–Earth distance to be 1.5×10^{11} m.

Answer The solid angle subtended at the Sun by an area of one square metre at a distance of 1.5×10^{11} m is $10^{-22}/1.5^2$ steradians. Solid angles are discussed in Section 15.3. The fraction of a whole sphere subtended is this number divided by 4π. Assuming the Sun radiates energy isotropically, the total energy radiated per second is thus $1400 \times 2.25 \times 10^{22} \times 4\pi \approx 4 \times 10^{26}$ joules. The mass equivalent of this energy is $4 \times 10^{26}/c^2 = 1.1 \times 10^{10}$ kg. In 1 year the Sun thus loses $1.1 \times 10^{10} \times 365 \times 24 \times 3600 = 4.4 \times 10^{16}$ kg.

The mass of the Sun is about 2×10^{30} kg and it can thus continue to radiate at much the same rate for a long while to come as long as the heat-producing mechanisms do not change.

Magnetic deflection

The experimental evidence for the validity of special relativity comes from the fields of particle physics and astronomy, and the most direct evidence from collisions between subatomic particles moving at speeds close to that of light. New subatomic particles are made by converting some of the energy involved in collisions between accelerated electrons and protons into mass. Identification of the masses and charges of the products is a first step in understanding fundamental physical processes that occur at very small distances, and one of the most used methods for determining the mass to charge ratio of a subatomic particle is to observe its deflection in a magnetic field.

In Section 16.2 it was observed that a particle of charge q moving at speed v and momentum p perpendicularly to a magnetic field of magnitude B experienced a force

$$F = qvB$$

in a direction perpendicular to both B and its velocity v. In a very small time interval Δt, the particle thus acquires an increment of momentum in the direction of the force of size

$$\Delta p = F\Delta t = qvB\Delta t,$$

and the direction of its momentum is changed through an angle

$$\Delta\theta = \frac{\Delta p}{p} = \frac{qvB}{p}\Delta t, \tag{19.51}$$

as in Fig 19.6. In the time Δt during which the particle suffers this change of direction it has travelled a distance $\Delta s = v\Delta t$. The angle $\Delta\theta$ and the distance Δs together define a radius of curvature a in which the particle moves,

$$a = \frac{\Delta s}{\Delta\theta} = \frac{v\Delta t}{\Delta\theta}.$$

Substitution of $\Delta\theta$ from eqn (19.51) gives

$$a = \frac{p}{qB}. \tag{19.52}$$

This is the same result as eqn (16.14) but now to be exact the relativistic expression for momentum p must be used. If a particle of rest mass m_0

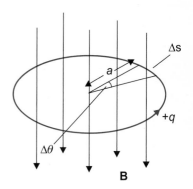

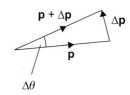

Fig. 19.6 A charged particle moving perpendicularly to a magnetic field experiences a force that gives motion in a circle of radius a, as shown in the top part of the figure. In a small time Δt the momentum vector **p** changes by an amount Δ**p** perpendicular to **p**.

and charge q moves at speed v perpendicularly to a field B its radius of curvature is $\gamma m_0 v / qB$. It may be noted that we have assumed that charge is measured to be the same in all inertial frames. All experimental data indicate that this assumption is correct.

Exercise 19.3 A proton of total energy 2 GeV travels in a plane perpendicular to a magnetic field of strength 1 T. What is its radius of curvature? The rest mass energy of a proton is $938 \, \text{MeV}/c^2$.

Answer 5.9 m.

19.5 **Optics**

Relativity has consequences in all fields of physics when the speeds of objects approach that of light. In the field of optics, the moving objects are light rays themselves. A light ray moving at an angle to the x-axis of frame S appears to an observer in frame S′ to be moving at a different angle to the x'-axis. This phenomenon is called **aberration**.

The frequency of light observed from moving sources depends upon the velocity of the source and the direction of emission of the light with respect to the direction of the source velocity. This phenomenon has already been treated non-relativistically in Chapter 8 and is called the Doppler effect. In this section we discuss aberration and Doppler effects.

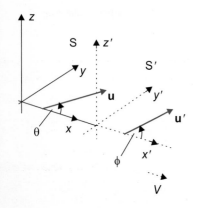

Fig. 19.7 A particle moves with speed u in the zx-plane in frame S at an angle θ to the x-axis. In frame S', the particle moves in the $z'x'$-plane with speed u' at an angle ϕ to the x'-axis.

Aberration

The angle a light ray is observed to make with the x-axis of frame S is different to that observed in a frame S′ moving at speed V along the positive x-axis.

Suppose a particle is moving in frame S in the xz-plane at an angle θ to the x-axis and with speed u. The angle θ is the angle between the positive x-axis and the direction of the velocity vector of the particle, the angle being measured by going anticlockwise from the axis to the velocity vector as shown in Fig 19.7. With this convention for θ the velocity components are

$$u_x = u \cos \theta$$

and

$$u_z = u \sin \theta,$$

and the tangent of the angle θ is equal to u_z/u_x. In frame S' the velocity components are given by eqns (19.25) and (19.27) as

$$u'_x = \frac{u_x - V}{1 - Vu_x/c^2}$$

and

$$u'_z = \frac{u_z}{\gamma(1 - Vu_x/c^2)}.$$

The tangent of the angle ϕ that gives the apparent direction of the particle in frame S' is equal to u'_z/u'_x. The angle ϕ is different to the angle θ and the relation between them can be determined using the above transformations of the velocity components.

We have

$$\begin{aligned}
\tan \phi &= \frac{u'_z}{u'_x} = \frac{u_z}{\gamma(1 - Vu_x/c^2)} \frac{(1 - Vu_x/c^2)}{(u_x - V)} \\
&= \frac{u_z}{\gamma(u_x - V)} \\
&= \frac{u \sin \theta}{\gamma(u \cos \theta - V)}.
\end{aligned} \tag{19.53}$$

If we now make the moving particle a light signal, u becomes equal to the speed of light and

$$\tan \phi = \frac{\sin \theta}{\gamma(\cos \theta - V/c)}. \tag{19.54}$$

A similar argument starting with $\tan \theta = u_z/u_x$ at eqn (19.53) gives, for a light signal,

$$\tan \theta = \frac{\sin \phi}{\gamma(\cos \phi + V/c)}. \tag{19.55}$$

If observers on Earth see light from a star, the light is coming towards them, and the angle θ appearing in the above formulae is greater than 180°. If light from a star appears on Earth to come at an angle of 20° to the horizontal, $\theta = 200°$, as in Fig 19.8. An observer in a frame S' moving at high speed away from the Earth along the positive x-axis sees the star at a smaller angle to the horizontal. If the speed corresponds to $\beta = 0.1$, eqn (19.54) gives $\tan \phi = 0.329$, and the angle ϕ equal to $180° + 18.2°$. The light from the star is seen in S' at 18.2° to the horizontal.

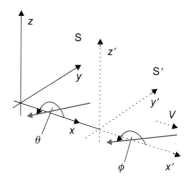

Fig. 19.8 A light ray in the zx-plane from a distant star makes an angle θ with the x-axis of frame S. In the S' frame the ray is seen in the $z'x'$-plane at an angle ϕ to the x'-axis.

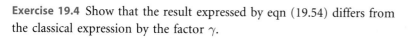

Exercise 19.4 Show that the result expressed by eqn (19.54) differs from the classical expression by the factor γ.

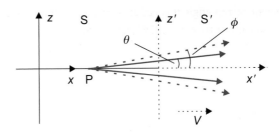

Fig. 19.9 A source of light at the point P on the *x*-axis emits light into a small cone of half-angle θ as measured in frame S. In frame S', the half-angle observed is ϕ.

● *Solid angle ratio*

Solid angle was discussed in Section 15.3. The solid angle subtended at the centre of a sphere by part of the sphere's surface is the area of that part of the surface divided by the square of the radius of the sphere. The solid angle subtended by that part of the surface of a sphere that is the end cap of a cone with half-angle θ is equal to $2\pi(1 - \cos\theta)$.

If light from a source is emitted into a certain solid angle as observed on Earth, what is the solid angle as measured by an observer in a frame S' in uniform relative motion? Let a source of light lie at the point P on the *x*-axis as shown on Fig 19.9 and emit forward along the axis. The light goes into a very small cone of half-angle θ corresponding to solid angle $\Delta\omega$. The observer in S' sees the light emitted into a small cone of half-angle ϕ corresponding to solid angle $\Delta\omega'$. The ratio of the solid angles equals the ratio of the end areas of cones with half-angles θ and ϕ, and

$$\frac{\Delta\omega}{\Delta\omega'} = \frac{1 - \cos\theta}{1 - \cos\phi}.$$

Since the cone angles are small, $\cos\theta \approx (1 - \theta^2/2)$, $\cos\phi \approx (1 - \phi^2/2)$, and

$$\frac{\Delta\omega}{\Delta\omega'} \sim \frac{\theta^2}{\phi^2}.$$

From eqn (19.55), using the approximations $\tan\theta \sim \theta$, $\sin\phi \sim \phi$, and $\cos\phi \sim 1$,

$$\frac{\theta}{\phi} = \frac{1}{\gamma(1 + V/c)}.$$

Squaring gives

$$\frac{\Delta\omega}{\Delta\omega'} = \frac{1 - V/c}{1 + V/c}. \tag{19.56}$$

Equation (19.56) is a relation between elements of solid angle observed by two observers in uniform relative motion. If an observer on Earth observes light travelling along the positive *x*-axis in a small solid angle

$\Delta\omega$, an observer moving at speed V along the positive x-axis observes the light to be emitted into a larger solid angle $\Delta\omega'$.

Electrons can be accelerated to speeds close to the speed of light in accelerators called *synchrotrons*. The paths of the moving electrons are roughly circular and in these paths the electrons suffer central accelerations. Accelerated charged particles emit radiation, and the radiation emitted by the electrons in a synchrotron is called *synchrotron radiation*. The photons typically have energies in the ultraviolet and X-ray parts of the spectrum. This radiation is exceedingly bright and is contained in a very small forward cone whose axis is in the direction of the electron beam at the point of emission of the radiation. The photons in synchrotron radiation are a powerful tool for X-ray diffraction studies of physical, chemical, and biological systems. Equation (19.56) shows why the X-ray beam is contained within a very narrow forward cone, thus making its brightness large and facilitating its use. The ratio of $\Delta\omega$ to $\Delta\omega'$ is large when the electron speed approaches c. Hence, photons emitted forward in a certain solid angle in the system in which the electrons are at rest concentrate in a cone of much smaller solid angle in the laboratory frame.

Doppler effects

Doppler shifts in the frequency and wavelength of waves observed when the source or the observer are moving were discussed in Section 8.5. For a source of waves moving at speed V with respect to a stationary observer, the approach adopted was to determine the time interval between a number, n, of wave crests observed and the time interval, nT_0, between their emission by the source. T_0 is the period of the wave measured when the source is stationary.

The arguments used in Section 8.5 were classical; it was assumed that time was the same for both the observer and the moving source. We now know that is not so. If the period in a frame in which the source is at rest is T_0, the time interval between the emission of wave crests measured in a frame moving at speed V with respect to the source is dilated by the factor γ. For a source moving away from an observer at speed V along the x-axis, eqn (8.33) becomes

$$\nu = \gamma\nu_0\left(1 + \frac{V}{\upsilon}\right)^{-1} \tag{19.57}$$

where υ is the wave velocity of the wave and ν_0 is the frequency observed by the stationary observer. The frequency ν is related to the wavelength λ by the equation

$$\nu = \frac{\upsilon}{\lambda},$$

and we have

$$\lambda = \gamma\lambda_0(1 + V/v) \tag{19.58}$$

where λ_0 is the wavelength observed with the source at rest. The factor γ is the relativistic correction to the classical expression.

If the observer is moving towards the source, which is now stationary in the medium in which the wave is propagating, a similar correction factor has to be applied and the observed wavelength is given by

$$\lambda = \frac{1}{\gamma}\lambda_0 \frac{1}{(1 - V/v)}. \tag{19.59}$$

The corrections made above to the non-relativistic Doppler shift formulae are exceedingly small for acoustic waves. The speeds v of these waves are much less than the speed of light c and the derivation of the formulas in Section 8.5 was valid only for sources or observers moving at speed V less than the wave speed. We introduced the correction in order to discuss Doppler shifts for electromagnetic radiation emitted from moving sources. Consider a light-emitting source moving at speed V in free space along the x-axis in a frame S. The light travels at speed c and eqn (19.58) becomes

$$\lambda = \gamma\lambda_0(1 + V/c).$$

Substitution of the expression for γ gives

● *The Doppler shift for electromagnetic radiation*

$$\lambda = \lambda_0 \left(\frac{1 + V/c}{1 - V/c}\right)^{1/2} \tag{19.60}$$

The wavelength is shifted to a higher value than measured with the source at rest, and the photon energy is decreased. If the source is moving towards the observer, the sign of V changes and the wavelength of the radiation decreases; the photon energy increases.

If the source is moving with velocity $\mathbf{V}$, and the unit vector in the direction of the radiation is $\mathbf{k}_d$, the observed wavelength is given by

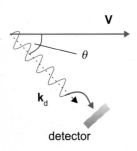

$$\lambda = \gamma\lambda_0 \left(1 - \frac{\mathbf{V} \cdot \mathbf{k}_d}{c}\right). \tag{19.61}$$

In the situation illustrated in Fig 19.10, the wavelength of the radiation observed is

Fig. 19.10 A source moving with velocity $\mathbf{V}$ emits radiation in the direction of the unit vector $\mathbf{k}_d$. The radiation is observed by a detector. The angle between the vectors $\mathbf{V}$ and $\mathbf{k}_d$ is θ.

$$\lambda = \gamma\lambda_0 \left(1 - V\cos\theta/c\right),$$

the same as the non-relativistic expression except for the factor γ. This highlights an important aspect of the relativistic Doppler shift. It is non-vanishing at $\theta = 90°$.

Exercise 19.5 Show that eqns (19.58) and (19.59) both reduce to eqn (19.60) when the wave speed is c. This corresponds to electromagnetic waves in free space when there is no medium and no distinction between moving source and moving observer.

Problems

Level 1

19.1 For an observer moving with the muons of Exercise 19.1, the Earth is approaching at speed $0.988c$. How far would the observer think the Earth had moved during a time equal to the average lifetime of a muon?

19.2 A radioactive source of half-life 2 years emits 10^7 particles per second at a certain point in time. At that time it is sent by rocket at a speed of $0.5c$ on a round trip to a manned space station. It arrives back 2 years later. How many particles per second does it emit at the time of its return?

19.3 Calculate the extra energy in MeV that must be given to a proton to increase its speed from $0.7c$ to $0.8c$. The rest mass of the proton is $938\,\text{MeV}/c^2$.

19.4 Two protons, each with kinetic energy of $1.2\,\text{GeV}$ and rest mass $938\,\text{MeV}/c^2$, collide head-on. What was their relative velocity before the collision?

19.5 A beam of radioactive mesons of rest mass $500\,\text{MeV}/c^2$ and momentum $4\,\text{Gev}/c$ travels a distance of $10\,\text{m}$ before hitting a target. The lifetime of the mesons is $1.2 \times 10^{-8}\,\text{s}$. What fraction of the beam reaches the target? ($1\,\text{GeV} \equiv 10^9\,\text{eV}$.)

19.6 A neutral meson of rest mass equivalent to $135\,\text{MeV}$ is observed to decay in flight into two gamma rays, each with energy of $80\,\text{MeV}$. Calculate the angle between the emitted gamma rays.

19.7 Through what voltages must (a) an electron and (b) a proton be accelerated in order to produce speeds of $0.01c$, $0.1c$, and $0.999c$?

19.8 A radar signal of frequency $30\,\text{GHz}$ is sent to a space ship approaching at speed $0.9c$. What is the frequency of the reflected signal received back on Earth? ($1\,\text{GHz} \equiv 10^9\,\text{Hz}$.)

19.9 Can two unconnected events in a frame S appear to be time-inverted in a frame S′ moving at speed V with respect to frame S?

Level 2

19.10 A positron–electron pair is formed from the annihilation of a γ-ray in the presence of a heavy nucleus. Each particle of the pair has a speed $0.82c$. Estimate the energy of the γ-ray. Neglect the recoil energy given to the nucleus.

19.11 A hypothetical atomic particle of rest mass m_0, moving at speed $0.85c$, decays into two particles, each of rest mass $m_0/4$. The two particles move at the same angle to the direction in which the original particle was moving. What is this angle?

19.12 Find the maximum energy transferred to a stationary electron by a photon of energy $1\,\text{MeV}$ in a Compton scattering process.

19.13 A K-meson of rest mass $494\,\text{MeV}/c^2$ at rest in the laboratory decays into three π-mesons of rest mass $139\,\text{Mev}/c^2$. Why must the directions of emission of the π-mesons lie in a plane? If the π-mesons travel in directions that make angles of $120°$ to each other, find their energy and speed. If the π-meson lifetime is $2.6 \times 10^{-8}\,\text{s}$, how far will they travel on average before decaying?

19.14 A cosmic dust particle moving at one-hundredth the speed of light strikes and sticks to an identical object at rest. If radiation losses are negligible, estimate the temperature of the composite object immediately after the collision. Take the average specific heat of the composite object to be $400\,\text{J}\,\text{kg}^{-1}\,\text{K}^{-1}$.

19.15 A high-energy electron moves in a circle of radius 0.2 m perpendicularly to a uniform magnetic field of 10 T. What is its energy?

19.16 A photon is emitted at an angle θ to the x-axis of a frame S. What angle does it make with the x'-axis of a frame S' moving at speed V along the x-axis? Determine the solid angle into which all the radiation emitted into a hemisphere in frame S is concentrated in frame S' if the frames are approaching each other at speed $\beta = 0.97$.

19.17 A space probe travelling directly away from the Earth contains a transmitter radiating at a constant frequency of 10^9 Hz. The frequency of the signals reaching the Earth can be measured with an accuracy of ± 1 Hz. At what speed relative to the Earth will the relativistic Doppler shift differ measurably from that expected classically?

19.18 A rocket is moving away from the Earth with a speed $0.5c$. Its power source emits a beam of hydrogen atoms backwards with a speed $0.3c$ with respect to the Earth. Equipment on the rocket excites hydrogen atoms, which emit light of wavelength 486 nm in the rest frame of a hydrogen atom. What is the difference in wavelength of the light emitted by this equipment and that emitted by the beam ejected by the rocket, as measured by an observer on Earth?

19.19 An accelerator produces a beam of excited ^{16}O atoms of kinetic energy 160 MeV. Light emitted when the atoms de-excite is observed at $90°$ to the beam direction. Determine the fractional change between the wavelength observed and the wavelength that would be observed if the excited oxygen atoms were at rest. (Take the rest masses of neutron and proton to be equal at $938\,\text{MeV}/c^2$.)

Level 3

19.20 A rigid rod of length ℓ moves with velocity v in the y-direction in frame S with its axis parallel to the x-axis. At time $t = 0$, the end A of the rod passes through the origin of the frame S. In frame S, the motion of the end B of the rod is therefore described by the relations $x = \ell, y = vt$. Determine the equations that describe the motion of the end B in the frame S'. Sketch the situation as it appears to an observer in S' at time $t' = 0$.

This example illustrates the difficulty in passing from the description of the motion of a rigid body in one

frame of reference to its motion in a different frame. The notion of a rigid body cannot be used in relativity theory.

19.21 Two particles with momenta $\mathbf{p}_1$ and $\mathbf{p}_2$ and total energies E_1 and E_2 interact and annihilate into photons. Show that at least two photons must be produced. When two are produced, show that the energies and momenta of the individual photons are not uniquely determined. Also show that, if the two photons are emitted parallel to the vector sum of $\mathbf{p}_1$ and $\mathbf{p}_2$, their energies are $(E_1 + E_2 + c|\mathbf{p}_1 + \mathbf{p}_2|)/2$ and $(E_1 + E_2 - c|\mathbf{p}_1 + \mathbf{p}_2|)/2$.

19.22 A neutral meson of energy E moving at speed V decays into two gamma rays. The maximum and minimum energies of the gamma rays are measured to be E_1 and E_2, respectively. Show that

$$E_1 = \frac{E}{2}\left(1 + \frac{V}{c}\right),$$

and E_2 is given by replacing the positive sign in the above equation with a negative sign.

If $E_1 = 70\,\text{MeV}$ and $E_2 = 60\,\text{MeV}$, deduce the rest mass of the meson.

19.23 A moving proton strikes a second proton at rest and produces two π-mesons of rest mass $139\,\text{MeV}/c^2$ according to the reaction

$$\text{p}^+ + \text{p}^+ \rightarrow \text{p}^+ + \text{p}^+ + \pi^+ + \pi^-.$$

The plus and minus signs indicate the electric charges on the particles in units of the electronic charge. What is the minimum kinetic energy of the moving proton for the reaction to occur? The rest mass of the proton is $938\,\text{MeV}/c^2$.

19.24 A proton collides with a second proton at rest. Show that the angle between the directions of the two protons after an elastic collision is always less than $90°$.

19.25 Determine the minimum energy required of a π-meson incident on a hydrogen target to produce two π-mesons in the reaction

$$\pi + \text{p} \rightarrow \pi + \text{p} + \pi.$$

Take the rest masses of the π-meson and proton to be $139\,\text{MeV}/c^2$ and $938\,\text{MeV}/c^2$ respectively.

19.26 Two light pulses are moving in the positive direction along the x-axis of a frame S, the distance between them being d. The frame S' is moving in the same direction with speed V. Show that the separation of the pulses as measured in S' is

$$d\sqrt{\frac{c+V}{c-V}}$$

where c is the speed of light. Use this result to derive the formula for the longitudinal Doppler effect.

(Remember that the distance between the pulses in frame S is the difference in their x-coordinates observed at the same time t.)

19.27 A plane mirror moves with speed V along the positive x-axis. Its plane is parallel to the yz-plane, as shown in Fig 19.11. A ray of light in the xz-plane from a source at the origin makes an angle θ with the x-axis and is incident on the mirror. Determine a relation between the angle θ and the angle ϕ the reflected ray makes with the x-axis.

19.28 An object outside of our galaxy is receding from Earth at a speed V. The object emits a spectral line that has a wavelength λ_0 in the object's rest frame. As measured on Earth the line is observed at a wavelength λ. The factor $z = (\lambda - \lambda_0)/\lambda_0$ is called the red shift and is used to determine the speeds of receding objects. Show that

$$\beta = \frac{V}{c} = \frac{(1+z)^2 - 1}{(1+z)^2 + 1}.$$

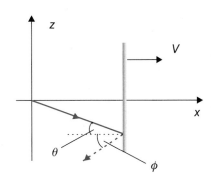

Fig. 19.11 The mirror M moves at speed V along the x-axis and reflects a ray of light incident at angle θ.

An extragalactic object 1 has $z = 3.53$ and another object 2 in the opposite direction from Earth has the same z-value. The rest-spectra of the two objects and that of our galaxy contain a line of 121.6 nm. What is the wavelength of this line emitted by object 1 as seen in our galaxy? What is the wavelength as seen at object 2 of the line emitted by object 1?

Some solutions and answers to Chapter 19 problems

19.2 5.49×10^6.

19.3 When $V = 0.7c$, the factor $\gamma = (1 - V^2/c^2)^{-1/2}$ equals 1.40 and the proton's momentum p_1 is $m_0 \gamma V$ where m_0 is the proton rest mass. Its total energy E_1 is given by

$$E_1^2 = c^2 p_1^2 + m_0^2 c^4 = \gamma^2 m_0^2 c^4$$

and evaluation gives $E_1 = 1.400 m_0 c^2$.

A similar calculation for $V = 0.8c$ gives $E_2 = 1.667 m_0 c^2$.

The extra energy $E_2 - E_1$ is thus $0.267 m_0 c^2 = 250$ MeV.

19.6 The bisector of the angle between the two gamma rays is the direction of the original meson. If the angle between the two is 2θ the sum of the components of the photon momenta along the bisector equals the momentum p of the original meson,

$$2\frac{h\nu}{c}\cos\theta = p$$

or

$$2h\nu\cos\theta = cp.$$

The total energy E of the meson is the sum of the photon energies and is given by

$$E^2 = c^2 p^2 + m_0^2 c^4.$$

Substituting in this expression the meson rest mass m_0 of 135 MeV/c^2 and $E = 160$ MeV gives $cp = 85.9$ MeV. Hence, $160 \cos\theta = 85.9$ and $2\theta = 115.1°$.

19.8 570 GHz.

19.12 796.5 keV.

19.13 For the total linear momentum to be zero after the decay, as it is before, the individual momenta of the three π-mesons must be coplanar. The mesons are identical and if they travel at 120° to each other they must each have the same momentum p and speed V. Thus they each have the same total energy E equal to one-third of the initial rest mass energy of the K-meson. This gives each a total energy of 164.7 MeV.

The speed can be obtained using the expression

$$E = m_\pi \gamma c^2 = \frac{m_\pi c^2}{\sqrt{1 - V^2/c^2}}.$$

This gives $\beta = V/c = 0.536$ corresponding to a speed of $1.61 \times 10^8 \text{ m s}^{-1}$.

The lifetime of the moving meson as seen by a stationary observer on Earth is longer than that observed in the meson's rest frame by the factor γ. Thus the Earth observer sees the mesons decay on average after $2.6 \times 10^{-8}\gamma$ s. The value of γ corresponding to $\beta = 0.536$ is 1.185, giving a 'lifetime' as seen on Earth of 3.08×10^{-8} s. In this time the Earth observer sees the mesons move a distance of $1.61 \times 10^8 \times 3.08 \times 10^{-8} = 4.95$ m.

19.14 2.8×10^5 K.

19.17 About $2 \times 10^5 \text{ m s}^{-1}$.

19.22 The maximum and minimum energy gamma rays occur when one photon is emitted in the same direction as that of the initial meson and one photon in the opposite direction. If V is the speed of the meson and m_0 its rest mass, conservation of linear momentum then indicates that the momentum p of the meson, equal to $m_0\gamma V$, is

$$m_0\gamma V = \frac{(E_1 - E_2)}{c}$$

where $\gamma = (1 - V^2/c^2)^{-1/2}$. Hence

$$cp = (E_1 - E_2),$$

and

$$m_0 = \frac{(E_1 - E_2)}{\gamma c V}.$$

We also have

$$E^2 = c^2 p^2 + m_0^2 c^4$$
$$= (E_1 - E_2)^2 + \frac{(E_1 - E_2)^2 c^4}{\gamma^2 c^2 V^2}$$
$$= (E_1 - E_2)^2 \frac{c^2}{V^2}.$$

Hence

$$E = (E_1 - E_2)\frac{c}{V}.$$

But

$$E = (E_1 + E_2),$$

from conservation of energy, and the last two relations together give

$$E_1 = \frac{E}{2}\left(1 + \frac{V}{c}\right)$$

and

$$E_2 = \frac{E}{2}\left(1 - \frac{V}{c}\right).$$

If $E_1 = 70$ MeV and $E_2 = 60$ MeV, $E = 130$ MeV and $cp = 10$ MeV. The equation $E^2 = c^2 p^2 + m_0^2 c^4$ then gives the value of the rest mass to be 129.6 MeV/c^2.

19.27 $\dfrac{\sin\theta}{(\cos\theta + V/c)} = \dfrac{\sin\phi}{(\cos\phi - V/c)}.$

Summary of Part 6
Relativity

We here summarize important results and equations discussed in Part 6

Chapter 19 Relativity

The Galilean transformations between positions and time used in one frame and those used in a frame moving with constant velocity with respect to the first are no longer valid for relative speeds comparable with the speed of light c.

The theory of special relativity is based on the following two postulates which are verified by experiment:

  I *The laws of physics are the same for any two observers in uniform relative motion.*

  II *The speed of light in free space is the same for all such observers.*

These postulates lead to the Lorentz transforms between position and time measured in two frames with uniform relative velocity. For motion of frame S$'$ at speed $+V$ along the x-axis of frame S,

$$x' = \frac{(x - Vt)}{\sqrt{1 - V^2/c^2}}, \tag{19.7}$$

$$y' = y, \tag{19.8}$$

$$z' = z, \tag{19.9}$$

and

$$t' = \frac{(t - Vx/c^2)}{\sqrt{1 - V^2/c^2}}. \tag{19.10}$$

Two consequences of this transformation are length contraction and time dilation.

1. The length L of an object moving along the x-axis with speed V is measured to be shorter than the length L_0 measured in the frame in which the object

is at rest by the factor γ given by

$$\gamma = \frac{1}{\sqrt{1 - V^2/c^2}}. \tag{19.11}$$

$$L = L_0/\gamma. \tag{19.20}$$

2. Consider two events occurring at the same place but separated by a time $\Delta t'$ in frame S$'$, which is moving along the x-axis at speed V with respect to a frame S. The time difference Δt between the events measured in frame S is increased by the factor γ, and

$$\Delta t = \gamma \Delta t'. \tag{19.21}$$

If a particle is moving along the x'-axis with speed u'_x as observed in frame S, its speed observed in frame S$'$ moving at speed V in the negative x'-direction is given by

$$u'_x = \frac{u_x + V}{1 + Vu_x/c^2}. \tag{19.25}$$

The mass m of an object to be used with the principle of conservation of momentum by an observer with respect to whom the object is moving at speed u is given by

$$m = \gamma m_0 \tag{19.38}$$

where

$$\gamma = (1 - u^2/c^2)^{-1/2}.$$

The mass m_0 is called the rest mass of the object and is the mass as used by a observer with respect to whom the object is at rest.

The total energy of a particle moving at high speed is

$$E = mc^2, \tag{19.42}$$

and its

$$\text{Kinetic energy} = mc^2 - m_0 c^2. \qquad (19.44)$$

The momentum p of the particle is given by

$$E^2 = m_0^2 c^4 + c^2 p^2. \qquad (19.45)$$

In terms of this momentum the radius of curvature of the circular path of a particle of charge q moving perpendicular to a magnetic field B is

$$a = \frac{p}{qB}. \qquad (19.52)$$

The angle θ a light ray makes with the x-axis of frame S is different to the angle ϕ the ray makes with the x'-axis of frame S'. They are related by

$$\tan \phi = \frac{\sin \theta}{\gamma(\cos \theta - V/c)}. \qquad (19.54)$$

If a photon-emitting source is moving with speed V directly away from an observer, the wavelength of the radiation observed is

$$\lambda = \lambda_0 \left(\frac{1 + V/c}{1 - V/c} \right)^{1/2} \qquad (19.60)$$

where λ_0 is the wavelength observed if the source is at rest.

If a photon-emitting source is moving with velocity $\mathbf{V}$ and the unit vector in the direction of the radiation is $\mathbf{k}_d$, the wavelength of the radiation observed is

$$\lambda = \gamma \lambda_0 \left(1 - \frac{\mathbf{V} \cdot \mathbf{k}_d}{c} \right). \qquad (19.61)$$

There is thus a Doppler shift in the frequency observed at right angles to the motion of a fast-moving source.

Chapter 20

Mathematical review

This chapter reviews the mathematics assumed to be known at the start of the use of this book.

In this chapter we give a very brief summary of the basic mathematics used in this book. The treatment presupposes knowledge of algebra at the non-calculus level, and some familiarity with the elements of geometry. This introduction is the start and not the end, by any means. It deals with mathematics that many students will already know, in which case it may only be necessary to make a quick survey to familiarize yourself with the terminology used throughout the text, or to use it for revision or for access to formulae. More advanced mathematical material than appears in this chapter is introduced in the text as it is needed. Definitions of mathematical terms are printed in bold type in this chapter as they have been throughout the book, even though many terms defined here are met in previous chapters.

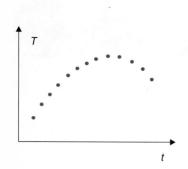

Fig. 20.1 Values of air temperature T measured at different times t.

20.1 **Functions**

The concept of a function is vital for the development of the mathematical language used in physics. A quantity that changes value when the value of a second quantity, called a **variable**, changes is called a **function** of that variable. For example, if the air temperature T is measured at regular intervals during the day it varies with times t at which the measurements are made. Plotting the measurements on a graph, with T on the vertical axis and t on the horizontal axis as in Fig 20.1, gives a series of points equally spaced along the t-axis. If we imagine the measurements to be made at smaller and smaller intervals of time the points become successively closer, and in the limit of a vanishingly small interval the plotted points constitute the smooth curve shown on Fig 20.2. This curve represents the function that gives the dependence of temperature T on time t. The functional dependence of T on t may be written as

$$T = f(t), \tag{20.1}$$

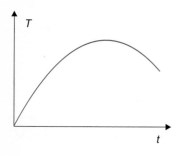

Fig. 20.2 If corresponding values of temperature and time are taken at very small time intervals and plotted, they constitute a smooth curve.

with $f(t)$ shorthand for the function that allows us to tell what the air temperature is at any time t.

In general, functions $f(x)$ of any variable x met in physics have a smooth dependence on x, and often in simple situations $f(x)$ can be written down algebraically in relatively simple terms. The simplest functions can be expressed in powers of x, each power multiplied by a **coefficient**. For example, a **linear** function, such as that shown in Fig 20.3, can be written

$$f(x) = a + bx \tag{20.2}$$

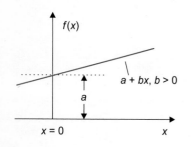

Fig. 20.3 A linear function $f(x) = a + bx$.

where the coefficient a gives the value of f at $x = 0$, and the coefficient b gives the increase in f for unit increase in x.

A **quadratic** function is

$$f(x) = a + bx + cx^2 \tag{20.3}$$

where the size and sign of the coefficient c indicate whether the curve of $f(x)$ versus x bends upwards or downwards, as shown in Fig 20.4, and how rapidly it curves.

More complicated functions have to be represented by adding terms involving higher powers of x, and the expression

$$f(x) = \sum_{n=0}^{n=N} a_n x^n \tag{20.4}$$

is shorthand for

$$f(x) = a_0 + a_1 x + a_2 x^2 + a_3 x^3 + \cdots + a_N x^N. \tag{20.5}$$

The summation symbol $\sum_{n=0}^{n=N}$ means add all terms like $a_n x^n$ for values of n from $n = 0$ to $n = N$.

There are all sorts of functions to be met other than those that can be represented by limited polynomial expansions as in eqn (20.5). Some new functions have very useful properties and it is often helpful to express a complicated function as a sum, not of terms each of which is a power of x multiplied by a coefficient, but as a sum of terms each of which involves one of the helpful functions multiplied by a suitable coefficient.

We have so far restricted attention to functions of a single variable only. However, a function can in principle depend on any number of variables and physical quantities usually depend on more than one. For example, the air temperature depends on time and on the position where it is measured. If we imagine the position to be restricted to points on a line and give position in terms of distance x from some other fixed point on the line, then

$$T = F(t, x). \tag{20.6}$$

Now, if we wish to display graphically the variation of T with both t and x, it is necessary to have two horizontal axes along which t and x are displayed and the measured values of T fall on a surface as shown in Fig 20.5.

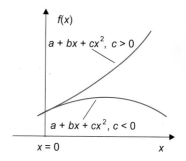

Fig. 20.4 Quadratic functions $f(x) = a + bx + cx^2$ shown for the two signs of the coefficient c.

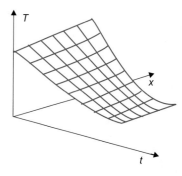

Fig. 20.5 Values of temperature T at times t and positions x constitute a surface.

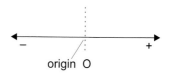

Fig. 20.6 A scalar can have have a positive or negative value, and can be represented by the distance from an origin O of a point on a straight line through the origin. The point corresponding to a negative scalar by convention lies to the left of the origin.

20.2 **Scalars and vectors**

A physical quantity that can be specified by a single number alone is called a **scalar**. The air temperature mentioned above is a scalar. Several other examples immediately come to mind: the mass of an object; the electrical charge on an object; the separation between two points; the speed of an object; the rate of flow of water out of a hole in a bucket; an interval of time; and so on.

A scalar is a one-dimensional quantity. It has a size, or magnitude, and the number giving the size has a sign; scalars can be positive or negative. The electric charge on an object is an example. The size of a scalar can be represented on a drawing by a length interval on a line. If we choose a point on the line to be the point from which the length intervals are displayed, this point may be called the **origin**, and the sign of the scalar can be taken into account by drawing positive to the right and negative to the left, as in Fig 20.6.

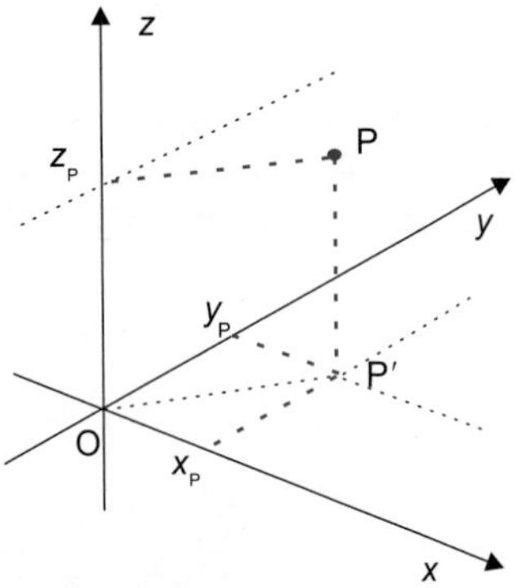

The real world has a three-dimensional space, and to define the position of a point requires a system of coordinates, or **coordinate system**, and an origin of coordinates. The simplest system of coordinates is the **Cartesian** system. In this, three mutually perpendicular axes are drawn through the origin, the point labelled O on Fig 20.7. The axes are labelled x, y, and z, and positive values of these variables are measured along the axes from the origin in the directions shown by the arrows. With this convention the coordinate system shown in Fig 20.7 is called a right-handed coordinate system. The position of a point P is now exactly and unambiguously specified by the values of the three Cartesian coordinates x_P, y_P, and z_P. Note that each of these coordinates is a scalar. The coordinates x_P and y_P are obtained by dropping the perpendicular from the point P to meet the xy-plane at P$'$, and then the perpendiculars from P$'$ to the x- and y-axes. The coordinate z_P is obtained by drawing the perpendicular from P to the z-axis. Other systems of coordinates are more suitable for certain problems and these will be discussed when they arise.

An alternative way to specify the position of the point P with respect to the origin is to use the position vector $\mathbf{r}_P$ shown on Fig 20.8. A **vector** is a physical quantity that requires both a magnitude and a direction for it to be defined completely. Examples of vectors are the velocity of an object, which needs speed and direction to tell which way it is going; force, which needs a number to tell how strong it is and a direction to tell the direction in which it is being applied; and so on. Similarly, the position vector of the point P has size and direction. The position of P is exactly specified by giving the length r_P of the line joining the origin

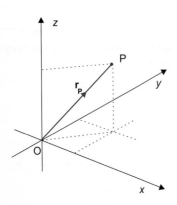

Fig. 20.7 The x-, y-, and z-axes of a Cartesian coordinate system centred on the origin O.

Fig. 20.8 The position of a point P can be represented on a three-dimensional diagram by a position vector $\mathbf{r}_P$, which is the line drawn from the origin to the point P and in the direction from O to P.

to P and the direction of the line joining the origin to P. The position vector may be represented graphically by the line with an arrow on it as shown in Fig 20.8.

Components of vectors

A vector can be resolved into two components along perpendicular directions when the directions and the vector lie in the same plane. Resolving a vector into two components is often useful in the treatment of physics problems, and such a decomposition is shown in Fig 20.9. The components of the vector **a** along the directions of axes 1 and 2 shown are a_1 and a_2, respectively. The components of the position vector **r** of Fig 20.8 along the x-, y-, and z-axes are the scalars x, y, and z, respectively. We can define unit vectors **i**, **j**, and **k** as vectors of unit length along the positive directions of the x-, y-, and z-axes. The position vector **r** is then given in terms of its three component scalars by

$$\mathbf{r} = x\mathbf{i} + y\mathbf{j} + z\mathbf{k}. \tag{20.7}$$

Addition of vectors

The arithmetical manipulation (addition, subtraction, multiplication, and division) of numbers corresponding to scalars works by simple rules that are accepted at an early stage. The manipulation of vectors is more complicated. Addition can be discussed with reference to Fig 20.10. The vector $(\mathbf{a} + \mathbf{b})$ is obtained as the diagonal of the parallelogram that has sides of lengths proportional to the sizes a of vector **a** and b of vector **b**, and directions corresponding to those of **a** and **b**. The size of the diagonal of the parallelogram, denoted by $|(\mathbf{a} + \mathbf{b})|$, can be worked out using the scalar product of two vectors which is defined below.

In terms of the Cartesian components a_x, a_y, a_z and b_x, b_y, b_z of the two vectors, the vector $(\mathbf{a} + \mathbf{b})$ is

$$(\mathbf{a} + \mathbf{b}) = (a_x + b_x)\mathbf{i} + (a_y + b_y)\mathbf{j} + (a_z + b_z)\mathbf{k}. \tag{20.8}$$

The subtraction of a vector **b** from a vector **a**, denoted $(\mathbf{a} - \mathbf{b})$, is obtained by adding to **a** the vector $-\mathbf{b}$, which is simply **b** reversed. Figure 20.11 illustrates this.

In terms of its Cartesian components, the vector $(\mathbf{a} - \mathbf{b})$ is

$$(\mathbf{a} - \mathbf{b}) = (a_x - b_x)\mathbf{i} + (a_y - b_y)\mathbf{j} + (a_z - b_z)\mathbf{k}. \tag{20.9}$$

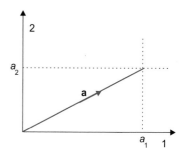

Fig. 20.9 A vector a resolved into two components along axes 1 and 2. If axis 1 is in the direction of a unit vector $\hat{\mathbf{r}}_1$ and axis 2 along the direction of a unit vector $\hat{\mathbf{r}}_2$, we have $\mathbf{a} = a_1\hat{\mathbf{r}}_1 + a_2\hat{\mathbf{r}}_2$.

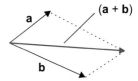

Fig. 20.10 The addition of vectors a and b is the vector $\mathbf{a} + \mathbf{b}$ and is obtained by joining them head to tail, retaining their correct orientations, then completing the triangle.

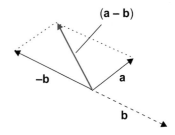

Fig. 20.11 The subtraction of a vector b from a vector a is the addition to a of the vector $(-\mathbf{b})$.

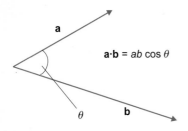

Fig. 20.12 The scalar product of two vectors **a** and **b** is the scalar $ab \cos \theta$.

Multiplication of vectors

There are two ways of multiplying vectors: one gives what is called the **scalar product**, the other the **vector product**. The scalar product of two vectors **a** and **b** is written $\mathbf{a} \cdot \mathbf{b}$. As expected from the name, the scalar product is a scalar, and has magnitude equal to the product of the size of each vector and the cosine of the angle θ between them. (Trigonometric functions such as sine and cosine are considered in Section 20.5. They are often abbreviated as sin and cos, and always abbreviated in that way when they are included in formulae.)

$$\mathbf{a} \cdot \mathbf{b} = ab \cos \theta. \tag{20.10}$$

This is illustrated in Fig 20.12. In terms of the Cartesian components a_x, a_y, a_z and b_x, b_y, b_z of the two vectors

$$\mathbf{a} \cdot \mathbf{b} = a_x b_x + a_y b_y + a_z b_z. \tag{20.11}$$

We may now easily work out the magnitude of the vector $(\mathbf{a} + \mathbf{b})$.

$$(\mathbf{a} + \mathbf{b}) \cdot (\mathbf{a} + \mathbf{b}) = a^2 + b^2 + 2\mathbf{a} \cdot \mathbf{b} = a^2 + b^2 + 2ab \cos \theta$$

where θ is the angle between the two vectors. Hence

$$|(\mathbf{a} + \mathbf{b})| = (a^2 + b^2 + 2ab \cos \theta)^{1/2}.$$

The vector product of two vectors **a** and **b** is in turn a vector. Its direction is perpendicular to the directions of both **a** and **b** and points the way given by the right-hand rule. This rule is demonstrated in Fig 20.13. Another method of remembering which way the vector product points is to stand up and let your right foot point in the direction of **a** and your left foot point in the direction of **b**. The vector product then points upwards through your head. The vector product is written $\mathbf{a} \times \mathbf{b}$, and has magnitude equal to the product of the size of each vector and the sine of the angle between them. This is shown in Fig 20.14. In terms of the Cartesian components of the two vectors

$$\mathbf{a} \times \mathbf{b} = (a_y b_z - a_z b_y)\mathbf{i} + (a_z b_x - a_x b_z)\mathbf{j} + (a_x b_y - a_y b_x)\mathbf{k}. \tag{20.12}$$

There is no such thing as division of one vector by another. The expression $\mathbf{a}/\mathbf{b}$ has no meaning.

Students unfamiliar with the algebra of vectors should now work through the problems at the end of this chapter.

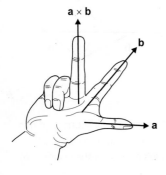

Fig. 20.13 The direction of the vector product of two vectors **a** and **b** is in the direction given by the right-hand rule.

20.3 Differential calculus

The differential calculus tells us what happens to a function as a variable on which the function depends is changed by smaller and smaller amounts, and finally by an infintesimally small amount such as introduced in Section 20.1 when functions were discussed. The notion of the variable x changing from x_1 to x_2 by a nonzero amount $(x_1 - x_2)$ with consequent change in $f(x)$ is easy, as is appreciation of the consequences of an increase in x by a very small amount δx. Suppose the interval δx becomes smaller and smaller. As it approaches zero the change in $f(x)$, which we will write $\delta f(x)$, also approaches zero. However, the ratio $\delta f(x)/\delta x$ can be nonzero. The **differential** of the function $f(x)$ at any point x is the limit of the above ratio as the interval δx becomes vanishingly small, when it is given the symbol dx, and the (vanishingly small) increase in $f(x)$ is given the symbol $df(x)$.

$$\frac{df(x)}{dx} = \text{limit as } \delta x \text{ tends to zero of } \frac{\delta f(x)}{\delta x}. \tag{20.13}$$

Suppose we calculate the differential of a function, and we will see how that may be done in a moment, what does it tell us about the function, and how useful is it for physics? The differential tells us, at any value of the variable on which the function depends, how the function changes when there is a minute change in that variable. For example, Fig 20.15 shows a plot of the distance x travelled by a car as a function of time when the car is going at a constant speed v; the plot is a straight line, and the average speed $\bar{v}$ is, of course, equal to the constant speed. For a small change in time δt at a time t the change δx in the distance travelled is always the same—the change divided by the time interval always equalling the constant speed, and equal to the slope of the straight line.

Now let the car speed up and slow down. The plot of distance against time becomes uneven, as in Fig 20.16, although its trend is always upwards unless the driver stops and puts the car into reverse. The average speed over the time t_1 to t_2 is still the total distance divided by the total time, $(x_2 - x_1)/(t_2 - t_1)$, but the speed at any instant of time varies. The instantaneous speed at time t_1 is obtained by taking the limit as δt becomes infinitesimally small of the ratio $\delta x/\delta t$, that is, the differential of the function giving the dependence of x on time. The limit of this ratio is the slope or gradient of the curve at the time t_1. The differential calculus thus enables a more detailed description to be given of the car's motion than would be obtained by simply discussing average speeds over measurable time intervals.

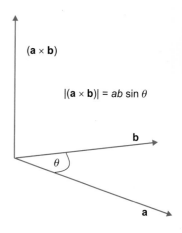

Fig. 20.14 The vector product of two vectors **a** and **b** is the vector of magnitude $ab \sin \theta$ in the direction given by the right-hand rule.

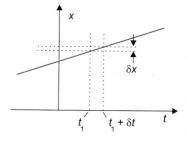

Fig. 20.15 A graph of distance versus time for a car travelling at constant speed.

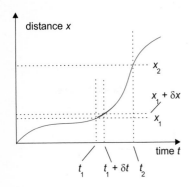

Fig. 20.16 A graph of distance versus time for a car that travels in one direction with variable speeds.

If the car starts from the origin as in Fig 20.17, speeds up, but then slows down to a stop at the point A, where it reverses direction and goes back towards the origin, the curve of distance against time shows a maximum at a time t_1. The rate of change of distance with time at A is the gradient of the curve at A, and there the gradient is zero. Setting the differential of a function equal to zero thus gives a method of finding the maxima of the function (if it has any). However, the gradient is also zero at minima of the function, at times such as t_2 where the curve bends upwards after the car has stopped at point B on the way back to the origin and reversed direction once more. Hence, setting the differential to zero determines maxima and minima at the same time. We leave it as an exercise to show that when the first differential is zero the sign of the **second differential** of a function $f(x)$,

$$\frac{d^2 f}{dx^2} = \frac{d}{dx}\left(\frac{df}{dx}\right), \tag{20.14}$$

determines whether the point is a maximum or a minimum.

Simple differentials

We now proceed to show how to calculate the differentials of simple functions and give prescriptions how to do some complicated cases. The change $df(x)$ in $f(x)$ as x goes from x to $x + dx$ is

$$df = f(x + dx) - f(x)$$

and

$$\frac{df}{dx} = \frac{f(x + dx) - f(x)}{dx}. \tag{20.15}$$

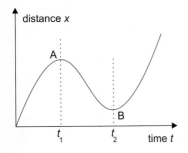

Fig. 20.17 A graph of distance versus time for a car that speeds up, slows down, and then comes to a momentary halt at time t_1. It then moves in the reverse direction before coming to a halt again at time t_2, when it reverses once more and proceeds in the original direction.

The right-hand side of this equality can be evaluated and written down discarding terms involving powers of dx higher than the first. These can be neglected because, if we kept them, division by dx to give $df(x)/dx$ would leave terms involving positive powers of dx. Since dx is vanishingly small these terms tend to zero. The simplest example of differentiation is when the function is linear. (The differential of a constant function is, of course, zero.) Consider the function given by eqn (20.2),

$$f(x + dx) = a + b(x + dx). \tag{20.16}$$

Hence

$$f(x + dx) - f(x) = b\, dx$$

and

$$\frac{df}{dx} = b. \tag{20.17}$$

Again consider the quadratic function (20.3),

$$f(x + dx) = a + b(x + dx) + c(x + dx)^2$$
$$= a + bx + cx^2 + b(dx) + 2xc\,dx + c(dx)^2.$$

Hence

$$\frac{df}{dx} = b + 2cx. \tag{20.18}$$

Similarly, it can be shown that

$$\frac{d(ax^n)}{dx} = anx^{(n-1)} \tag{20.19}$$

where n is any rational number. Also it can be shown that there is a number, e, that when raised to the power x produces a function, called the **exponential function**, e^x, whose differential is the same exponential function. Hence

$$\frac{d}{dx}(e^x) = e^x. \tag{20.20}$$

The numerical value of e is 2.71828 to the nearest 5 decimal places. The logarithm of x to the base e is called the **natural logarithm** of x and is denoted by ln x. It can be shown that

$$\frac{d}{dx}(\ln x) = \frac{1}{x}. \tag{20.21}$$

The technique illustrated above of using eqn (20.15) can in principle be used to determine the differential of any function. Here we simply quote rules about finding the differentials of: (1) products of functions; (2) of functions raised to some power, such as $(2x^3 - 3x)^3$; and (3) of expressions that can be written as one function divided by another.

1. $$\frac{d(f_1 \cdot f_2)}{dx} = f_1 \frac{df_2}{dx} + f_2 \frac{df_1}{dx} \tag{20.22}$$

    where f_1 and f_2 are both functions of x.

2. $$\frac{d(f(x))^n}{dx} = n(f(x))^{n-1} \times \frac{df(x)}{dx}. \tag{20.23}$$

3. $$\frac{d(f_1/f_2)}{dx} = \frac{f_2(df_1/dx) - f_1(df_2/dx)}{f_2^2}. \tag{20.24}$$

The student should now work through the examples on differentiation at the end of the chapter to become familiar with the operation of these rules.

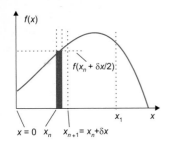

Fig. 20.18 The curve shows a function $f(x)$. The area under the curve from the point x_n to the point $x_n + \delta x/2$ is the strip shown shaded.

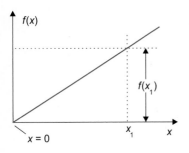

Fig. 20.19 The function $f(x) = x$. The integral of the function from $x = 0$ to x_1 is the area of the triangle with base of length x_1.

20.4 Integral calculus

The **integral** $I(x = 0, x = x_1)$ of a function $f(x)$ over the range $x = 0$ to $x = x_1$ is defined as the limit, as the very small interval δx becomes vanishingly small, that is, becomes dx, of the sum of the areas of strips of width δx such as those shown on Fig 20.18. The thin strips are all adjacent to each other and cover the whole curve from the origin to the point x_1. Thus, the x-coordinate x_n of the beginning of the nth strip is $(n - 1)\delta x$, and the value of the function at the midpoint $x = (n - 1/2)\delta x$ of the nth strip is $f(n\delta x - \delta x/2)$. The integral is then

$$I(x = 0, x = x_1) = \text{Limit as } \delta x \to dx \sum_{n=1}^{n=x_1/x} f(n\delta x - \delta x/2)\delta x.$$

(20.25)

This integral is called a **definite integral** because the beginning and end points, $x = 0$ and $x = x_1$ are specified. As the strip width δx becomes infinitesimally small, the sum of the areas of the strips becomes equal to the area under the curve and the integral is equal to the area under the curve from $x = 0$ to $x = x_1$. It is a number that depends on the function f and the limits, zero and x_1, over which the sum is taken. The function being integrated is called the **integrand**.

The simplest function whose integral can be worked out is $f(x) = x$. This is shown in Fig 20.19, and, from the formula giving the area of a triangle as half the base times the height, the integral of this function over the range zero to x_1 is simply

$$I(x = 0, x = x_1) = x_1^2/2.$$

(20.26)

The integral of a function is given a specific symbol and the definite integral from $x = 0$ to $x = x_1$ is written

$$I(x = 0, x = x_1) = \int_0^{x_1} f(x)dx.$$

(20.27)

If we let the upper limit be regarded as a variable then the integral becomes a function of x and is now called an **indefinite integral**, $\int_0^x f(x)\,dx$. When indefinite integrals are written down the limits are usually omitted. From eqn (20.26)

$$\int x\,dx = \frac{1}{2}x^2,$$

(20.28)

and we see that when this is differentiated we get back x, the function we integrated. This suggests that integration is the reverse of differentiation, and that one way to find the indefinite integral of a function f is to find a function that when differentiated gives back f. This is indeed so and we prove it formally below, although anyone happy just to accept the statement can omit the formal proof.

$$I(x) = \int^x f(x)\,dx = [\text{limit as } \delta x \to dx] \sum^x f(x)\delta x.$$

$$\frac{dI}{dx} = [\text{limit as } \delta x \to dx] \frac{I(x + \delta x) - I(x)}{\delta x}$$

$$= [\text{limit as } \delta x \to dx] \frac{\sum^{x+\delta x} f(x + \delta x)\delta x - \sum^x f(x)\delta x}{\delta x}$$

$$= [\text{limit as } \delta x \to dx] \frac{f(x + \delta x)\delta x}{\delta x}$$

$$= f(x).$$

Simple integrals

The integrals of functions can now be determined, knowing that integration is the reverse of differentiation. It can easily be verified, for example, that

$$\int x^n\,dx = \frac{1}{(n+1)} x^{n+1}. \tag{20.29}$$

The integrals of more complicated functions have been dealt with as they were needed. Here, we write down that

$$\int \frac{1}{x}\,dx = \ln(x) \tag{20.30}$$

where $\ln(x)$ is the natural logarithm of x. We also note that a constant, any constant, may be added to an indefinite integral because its differential is zero. Definite integrals, of course, do not involve such arbitrary constants.

Before we end this section on integrals it is also useful to point out that integrals occur throughout physics, and we have to be familiar with them. It often happens that a physical effect is worked out as the sum of an infinite number of vanishingly small contributions, that is, an integral. Simple examples, such as working out the volume of an object, readily spring to mind. The volume of the square-cut cylindrical rod shown in Fig 20.20 equals the height of the rod multiplied by the area of the base. Although we know the area is πa^2 with a the radius, this area is determined by adding an infinite number of infinitesimally small annular contributions, each of area $2\pi r\,dr$ with r going from zero out to the value a.

$$\text{Area} = \int_0^a 2\pi r\,dr = [\pi r^2]_0^a = \pi a^2$$

where we have used, as is customary in this context, the expression $[\pi r^2]_0^a$ to indicate the quantity obtained by putting $r = a$ inside the square brackets, evaluating, and subtracting from the result the quantity obtained by substituting $r = 0$ inside the square brackets.

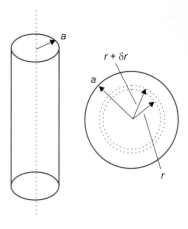

Fig. 20.20 The volume of the cylindrical rod shown is its height times its cross-sectional area.

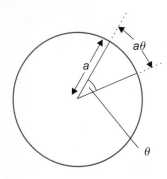

Fig. 20.21 The movement by an angle θ of a line drawn from the centre of a circle to two different points on the circumference defines an arc of length $a\theta$, where a is the radius.

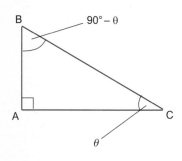

Fig. 20.22 A right-angled triangle used to define the trigonometric functions.

20.5 Trigonometric functions

Consider the line drawn from the centre of the circle in Fig 20.21 to a point on the circumference. If the end point of this line is moved around the circumference to return to its original starting point, the line has swept out an **angle** of 2π radians or, in the units more commonly used, $360°$. If the angle between two such lines is θ radians, the length of the **arc** of the circle defined by the touching points is $a\theta$.

Trigonometry deals with the properties of triangles. Here we are concerned in particular with the properties of right-angled triangles, that is, triangles in which one angle is $90°$, such as that shown in Fig 20.22. The angle between sides AC and BC of the triangle is θ and hence the third angle of the triangle is $(90° - \theta)$. If the angles are kept fixed but the lengths of the sides of the triangle are changed, the six ratios that can be constructed involving the three sides depend only on the angle θ; the ratios are functions of the variable θ and are called **trigonometric functions**. The functions have names and are defined in terms of the following ratios of the lengths of sides.

$$\sin\theta = \frac{AB}{BC}, \tag{20.31}$$

$$\cos\theta = \frac{AC}{BC}, \tag{20.32}$$

$$\tan\theta = \frac{AB}{AC}, \tag{20.33}$$

$$\operatorname{cosec}\theta = \frac{BC}{AB} = \frac{1}{\sin\theta}, \tag{20.34}$$

$$\sec\theta = \frac{BC}{AC} = \frac{1}{\cos\theta}, \tag{20.35}$$

$$\cot\theta = \frac{AC}{AB} = \frac{1}{\tan\theta}. \tag{20.36}$$

Trigonometric relations

Many relationships can be derived between these trigonometric functions. For example, Pythagoras's theorem tells us that the square of the hypotenuse in a right-angled triangle equals the sum of the squares of the other two sides, and this immediately leads to the relation

$$\sin^2\theta + \cos^2\theta = 1 \tag{20.37}$$

where θ is any angle. It can be proved using geometric properties of triangles, although we will not do it here, that

$$\sin(A \pm B) = \sin A \cos B \pm \cos A \sin B, \tag{20.38}$$

$$\cos(A \pm B) = \cos A \cos B \mp \sin A \sin B \tag{20.39}$$

where A and B are any two angles, and $\mp$ means for $\cos(A+B)$ use the minus sign on the right-hand side, and vice versa.

Differentials and integrals

Use can be made of eqns (20.38) and (20.39) to determine the differentials and hence the integrals of the trigonometric functions.

$$\frac{\mathrm{d}}{\mathrm{d}\theta}(\sin\theta) = \frac{1}{\delta\theta}(\sin(\theta + \delta\theta) - \sin\theta),$$

in the limit as $\delta\theta$ becomes vanishingly small. Hence, from eqn (20.38),

$$\frac{\mathrm{d}}{\mathrm{d}\theta}(\sin\theta)\frac{1}{\delta\theta}(\sin\theta\cos\delta\theta + \cos\theta\sin\delta\theta - \sin\theta) = \cos\theta. \qquad (20.40)$$

Here we have used the fact that, as $\delta\theta$ becomes infinitesimally small, $\sin(\delta\theta)$ becomes $\delta\theta$, and $\cos(\delta\theta)$ becomes unity.

In a similar way, using eqn (20.39), it can be shown that

$$\frac{\mathrm{d}}{\mathrm{d}\theta}(\cos\theta) = -\sin\theta \qquad (20.41)$$

and

$$\frac{\mathrm{d}}{\mathrm{d}\theta}(\tan\theta) = \sec^2\theta \qquad (20.42)$$

where the last relation can quickly be derived using eqn (20.24).

The integrals of the trigonometric functions can be obtained as the reverse of the differentials, giving

$$\int \sin\theta\,\mathrm{d}\theta = \cos\theta, \qquad (20.43)$$

$$\int \cos\theta\,\mathrm{d}\theta = -\sin\theta, \qquad (20.44)$$

and

$$\int \tan\theta\,\mathrm{d}\theta = -\ln(|\cos\theta|) = \ln(|\sec\theta|). \qquad (20.45)$$

The last relation is not so obvious as the first two but can readily be verified by differentiation.

Other relationships between trigonometric functions, and differentials and integrals of more complicated expressions involving these functions, will be introduced as required. For now, as before, the appropriate problems at the end of this chapter should be tackled.

20.6 Complex numbers

We are familiar with real numbers. It is easy to think about 7 apples, or even 7.213 apples, where we understand 0.213 of an apple to be a fraction 213/1000 of a whole one. The square root of a positive number is a real number: $\sqrt{4.84} = 2.2$. Minus 4.84 is a real number; what is the square root of -4.84? The concept of complex numbers enables us to deal with this question, and thereafter provides the mathematical tools used to treat many physical problems in a convenient fashion. If we define the square root of -1 to be the purely **imaginary** quantity j then the square root of -4.84 equals 2.2j, or j2.2 as it is usually written. A **complex number** z normally has both a real part a and an imaginary part jb, and is simply an ordered pair of two real numbers a and b. It is written

$$z = (a + jb). \tag{20.46}$$

A complex number can be displayed graphically as a point P on an **Argand diagram**. Consider Fig 20.23 in which the real part of the complex number $(a + jb)$ is plotted along the x-axis and the imaginary part along the y-axis. In terms of the angle θ and the distance r from the origin to the point P,

$$z = r(\cos\theta + j\sin\theta). \tag{20.47}$$

Here r is called the **modulus** or **magnitude** of z, and θ is called its **argument**. From eqns (20.46) and (20.47) we see that

$$r = \sqrt{a^2 + b^2} \tag{20.48}$$

and

$$\tan\theta = b/a. \tag{20.49}$$

Exponential forms

The complex number

$$z = \cos\theta + j\sin\theta \tag{20.50}$$

has unit modulus whatever the value of θ. Other complex numbers are numbers like

$$Ae^{j\theta} \tag{20.51}$$

where A and θ are any real numbers, and it is not surprising, given the special nature of the exponential function, that

$$e^{j\theta} = \cos\theta + j\sin\theta. \tag{20.52}$$

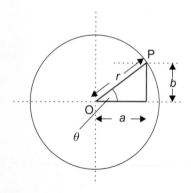

Fig. 20.23 The Argand diagram. The point P represents the complex number $(a + jb)$.

This connection between the exponential function with imaginary argument and sines and cosines is extremely useful for the mathematical description of many phenomena throughout physics and mathematics. Since $\cos(-\theta) = \cos(0 - \theta) = \cos\theta$, from eqn (20.39), and $\sin(-\theta) = -\sin\theta$, from eqn (20.38),

$$e^{-j\theta} = \cos\theta - j\sin\theta. \tag{20.53}$$

These expressions for $e^{j\theta}$ and $e^{-j\theta}$ can be used to verify eqns (20.38) and (20.39), which can be proved geometrically. Verification of the first relation is given below. These procedures may be considered as validation of eqns (20.52) and (20.53).

$$\sin(A + B) = \frac{1}{2j}\left(e^{j(A+B)} - e^{-j(A+B)}\right).$$

$$\sin A \cos B = \frac{1}{4j}\left(e^{jA} - e^{-jA}\right)\left(e^{jB} + e^{-jB}\right)$$

$$= \frac{1}{4j}\left(e^{j(A+B)} + e^{j(A-B)} - e^{-j(A-B)} - e^{-j(A+B)}\right).$$

$$\cos A \sin B = \frac{1}{4j}\left(e^{jA} + e^{-jA}\right)\left(e^{jB} - e^{-jB}\right)$$

$$= \frac{1}{4j}\left(e^{j(A+B)} - e^{j(A-B)} + e^{-j(A-B)} - e^{-j(A+B)}\right).$$

Hence

$$\sin A \cos B + \cos A \sin B = \frac{1}{2j}\left(e^{j(A+B)} - e^{-j(A+B)}\right) = \sin(A + B).$$

Using the equality (20.52) and eqn (20.47) we may now write the complex number (20.46) in its **polar** form

$$z = (a + jb) = re^{j\theta} \tag{20.54}$$

where the modulus r is given by eqn (20.48) and the argument θ is given by eqn (20.49). It is often useful to express complex numbers in polar form when products or expansions have to be made.

20.7 Series

A series is a summation of terms, usually of increasing complexity. The series can have an infinite or a finite number of terms. An infinite series, to be useful in physics, must converge to a finite sum; usually this means that successive terms become smaller and smaller. A simple example of a finite series is the **arithmetic series** consisting of a sum of terms like $A \times n$, where A is a constant and n ranges from n_1 to n_2,

$$\sum_{n=n_1}^{n=n_2} An = A \sum_{n=n_1}^{n=n_2} n = \frac{A}{2}\left(n_2(n_2 + 1) - n_1(n_1 + 1)\right). \tag{20.55}$$

Another example is the **geometric series**, in which successive terms equal the previous term multiplied by a constant,

$$\sum_{n=0}^{n=N-1} ar^n = \frac{a(r^N - 1)}{(r - 1)} \tag{20.56}$$

where a and r are constants.

Exponential series

Some often used functions of a variable x can be expressed as an infinite series. Using the property of the exponential function given by eqn (20.20), it is straightforward to see that e^x can be written

$$e^x = 1 + x + \frac{x^2}{2!} + \frac{x^3}{3!} + \frac{x^4}{4!} + \cdots \qquad (20.57)$$

$$= \sum_{n=0}^{n=\infty} \frac{x^n}{n!} \qquad (20.58)$$

where n is an integer, and $n!$, called **factorial** n, is given by

$$n! = n(n-1)(n-2)(n-3)\cdots 1. \qquad (20.59)$$

Note that $0! = 1$.

Other useful series can be obtained using the same technique of proving that the differential of a function is equal to the differential of a proposed series expansion of that function. For example, using eqn (20.21) and the results of the subsection immediately below for the expansion of $1/(1+x)$, it is straightforward to show that

$$\ln(1+x) = x - \frac{x^2}{2} + \frac{x^3}{3} - \frac{x^4}{4} + \cdots. \qquad (20.60)$$

Binomial expansion

The **binomial expansion** is a valuable series representation of $(1+x)$ raised to a power n, where n is any real number.

$$(1+x)^n = 1 + nx + \frac{n(n-1)}{2!}x^2 + \frac{n(n-1)(n-2)}{3!}x^3 + \cdots. \qquad (20.61)$$

We may satisfy ourselves that this expansion is valid by differentiating both sides and showing the equality of the differentials. This is done below.

$$\frac{d}{dx}((1+x)^n) = n(1+x)^{n-1} = n\left(1 + (n-1)x\right.$$

$$+ \frac{(n-1)(n-2)}{2!}x^2 + \cdots \left.\right) = n + n(n-1)x$$

$$+ \frac{n(n-1)(n-2)}{2!}x^2 + \cdots = \frac{d}{dx}(\text{rhs})$$

where rhs is the right-hand side of eqn (20.61).

The expansion stops at the nth term after the first 'one' if n is a positive integer. The binomial expansion is very useful for values of x small compared with unity. Since the ratio of successive terms decreases roughly as x, an answer to any required accuracy can be obtained by stopping the series at an appropriate point.

Taylor expansion

We end this section by giving a valid series expansion of any well-behaved function such as those usually encountered in physics formulae. This is the **Taylor expansion**. Once more we will not prove it. The proof can be found in mathematics texts for science undergraduates, but it is not essential here—even though it is essential to grasp how the Taylor series is used in physics. Taylor's series gives the value of a function $f(x)$ in terms of the value of the function at $x = a$ and the values of the first and higher differentials of the function evaluated at $x = a$,

$$f(x) = f(a) + (x - a)\frac{df}{dx}\bigg|_a + \frac{(x - a)^2}{2!}\frac{d^2f}{dx^2}\bigg|_a + \cdots. \tag{20.62}$$

In this formula the notation $|_a$ following each differential indicates that, after their determination as functions of x, we evaluate them at $x = a$. The closer a is to x the more rapidly, in general, does the series converge, and as $(x - a)$ tends to the infinitesimally small interval dx we recover the definition of the differential of $f(x)$, eqn (20.15). At the end of the chapter there are examples of the use of Taylor expansions for some functions met in physical situations.

20.8 Differential equations

Most of the physical laws extracted from experimental observations are expressed mathematically in terms of **differential equations**. These equations relate the differentials—first- and/or higher-order differentials—of functions $y = f(x)$ to other functions. Solutions of the differential equations determine the behaviour of the functions y. A very simple example will serve to introduce the way in which they are used.

Suppose we have a regularly shaped container such as that shown in Fig 20.24. It contains water but has a leak and water is lost. We wish to deduce a function that describes how the height of water in the cylinder, h, varies with time after time zero at which the water height is h_0. We make measurements, as time progresses and the height h goes down, of the volumes, δV, of water leaking from the cylinder during equal short time intervals δt when the height is h. We plot those data as shown on Fig 20.25 and deduce, following the argument of Section 20.1, that the rate at which water flows out of the container is proportional to the height of water remaining, h. Expressed mathematically this statement becomes

$$-\frac{dV}{dt} = \alpha h \tag{20.63}$$

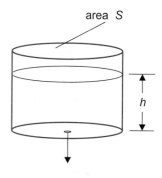

Fig. 20.24 A cylindrical container with cross-section area S from which water leaks.

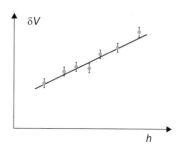

Fig. 20.25 A plot of the volumes of water, δV, flowing out of the cylinder during equal short time intervals δt as a function of the height h at which the measurements were made.

where α is the constant of proportionality and the minus sign arises because dV is negative when V is the volume of fluid in the cylinder. Since $V = Sh$, where S is the constant cross-sectional area of the container, eqn (20.63) may be rewritten as

$$S \frac{dh}{dt} = -\alpha h. \tag{20.64}$$

This is a differential equation relating the differential of the function h and the function itself. The equation can be rewritten

$$\frac{dh}{h} = -\frac{\alpha}{S} dt, \tag{20.65}$$

and this can be solved by integrating both sides between appropriate limits. These limits are that, at $t = 0$, $h = h_0$ and that, at an arbitrary time t, the height is the function $h(t)$ that we are trying to determine,

$$\int_{h_0}^{h} \frac{dh}{h} = -\int_{0}^{t} \frac{\alpha}{S} dt.$$

Performing these integrations we have

$$\ln\left(\frac{h}{h_0}\right) = -\frac{\alpha}{S} t,$$

which finally allows us to write the function h as

$$h = h_0 \exp\left[-\frac{\alpha}{S} t\right]. \tag{20.66}$$

The height of water decays exponentially with time in this hypothetical example, as shown in Fig 20.26.

Equation (20.63) is called a *first-order* differential equation because it involves only the first derivative of the function V whose behaviour the equation describes. First-order equations involve one constant of integration, which is determined by specifying corresponding values of the two variables. In the above discussion of the leaking can, the height h was equal to h_0 at time $t = 0$.

First-order equations may also involve the function itself. An important first-order equation that describes the time variation of the electric current $I(t)$ though a particular circuit has the form

$$L \frac{dI(t)}{dt} + RI(t) = V(t) \tag{20.67}$$

where L, R are parameters of the circuit and $V(t)$ is a time-varying voltage. There is a technique for solving this equation for the current $I(t)$ for any form of the function $V(t)$. However, for the special cases usually considered in elementary physics the solutions can often be obtained by

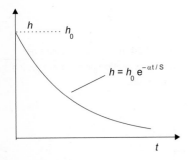

Fig. 20.26 The curve showing how the height of water in the cylinder decays with time.

intelligent guesswork and verification by substitution. For example, if $V(t)$ is a step function of height V_0 applied at time $t = 0$, the current as a function of time approaches the limiting value V_0/R slowly. We may guess that the approach is exponential and try as the solution

$$I(t) = \frac{V_0}{R}(1 - e^{-Rt/L}).$$

This is indeed the solution as can readily be verified by substitution of I and its derivative into eqn (20.67).

More complicated differential equations for a function $y(x)$ may involve double differentials, d^2y/dx^2, as well as the first differential dy/dx. The equation is then a *second-order* differential equation. The form of a commonly met second-order differential equation is

$$a\frac{d^2y}{dx^2} + b\frac{dy}{dx} + cy = f(x) \tag{20.68}$$

where a, b, and c are numbers and $f(x)$ is a function of the variable x. Second-order equations involve two constants of integration, which are determined by specifying two pairs of corresponding values of the two variables.

Much of elementary physics is expressed in terms of second-order differential equations. There are techniques for obtaining solutions but again we need not consider them here. In the text we will often suggest solutions that can be verified by substitution; where particularly useful, more details will be given in mathematical inserts at appropriate places. We note here that any function $g(x)$ that satisfies

$$a\frac{d^2y}{dx^2} + b\frac{dy}{dx} + cy = 0 \tag{20.69}$$

may be added to the solution $y(x)$ of eqn (20.68) to give a function $g(x) + y(x)$ that is also a solution of eqn (20.68). The function $g(x)$ is called the *complementary function* and $y(x)$ is called the *particular integral*. The complementary function contains the two constants of integration determined by the specified values of the pairs of variables.

Problems

20.1 An algebraic expression like $(2x^2 + 7x - 9)/(x^2 - x - 6)$ can be rewritten as $3/(x + 2) + 2x/(x + 3)$. This is called writing $(2x^2 + 7x - 9)/ (x^2 - x - 6)$ as the sum of two partial fractions, $3/(x + 2)$ and $2x/(x + 3)$, and is a useful way of rewriting an expression so that it can more easily be integrated. Express $(3x^2 - 8)/(x^4 - 16)$ as the sum of three partial fractions.

20.2 If $\alpha x^2 + \beta x + \gamma = 0$ is a quadratic equation that has solutions, or roots, a and b, then

$$(x - a)(x - b) = 0,$$

and the solutions are given by

$$2ax = -\beta \pm \sqrt{\beta^2 - 4\alpha\gamma}.$$

If a and b are the roots of the equation $x^2 + px + q = 0$, find the equation whose roots are $a + 1/b$ and $b + 1/a$.

20.3 If a and b are the roots of the equation $x^2 - 5x + 2 = 0$, find the value of $a^3 + b^3$.

20.4 Solve the simultaneous equations

$$y^2 - x^2 - 4 = 0,$$

$$3x + 3y + 1 = 0.$$

20.5 Evaluate the scalar and vector products of the vectors $\mathbf{A} = 3\mathbf{i} + 2\mathbf{j} - 7\mathbf{k}$ and $\mathbf{B} = 3\mathbf{i} - 4\mathbf{j} + 2\mathbf{k}$.

20.6 Two joggers are running in straight lines with positions at time t given by $0.3\mathbf{i} + 250t\mathbf{j} + \mathbf{k}$ and $100t\mathbf{i} + 250t\mathbf{j} + 250t\mathbf{k}$, respectively. Distances are in metres and time in minutes. Calculate the distance between them as a function of time. How far apart are they after 10 minutes? Do they collide?

20.7 Manipulation of geometrical relationships is regularly necessary for the solution of problems in physics. This example considers the relationships between the angles and sides of a triangle.

A triangle has sides a, b, and c. The angle opposite to the side a is A, etc. Show that

$$\frac{a}{\sin A} = \frac{b}{\sin B} = \frac{c}{\sin C}.$$

In a triangle ABC, $a = 8$, $b = 7$, and $A = 43°$. Calculate (a) the side c, (b) the area of the triangle.

20.8 Given $\mathbf{a} = (t^2\mathbf{i}, t\mathbf{j}, 1\mathbf{k})$, $\mathbf{b} = (t\mathbf{i}, t\mathbf{j}, 1\mathbf{k})$, and $\mathbf{c} = (4t\mathbf{i}, 3t^2\mathbf{j}, 1\mathbf{k})$, verify the following theorems involving vector products:

(a) $\dfrac{d}{dt}(\mathbf{a} \times \mathbf{b}) = \dfrac{d\mathbf{a}}{dt} \times \mathbf{b} + \mathbf{a} \times \dfrac{d\mathbf{b}}{dt};$

(b) $\dfrac{d}{dt}(\mathbf{a} \cdot \mathbf{c}) = \dfrac{d\mathbf{a}}{dt} \cdot \mathbf{c} + \mathbf{a} \cdot \dfrac{d\mathbf{c}}{dt}.$

20.9 Differentiate the expressions below with respect to x. It is often useful in differentiation (and integration) to make the substitution $y = f(x)$ and to put

$$\frac{d}{dx} = \frac{d}{dy}\frac{dy}{dx}.$$

(a) $\sin\sqrt{x}$;

(b) $\sqrt{(\sin x)}$;

(c) 10^{2x};

(d) $\sqrt{(ax + b)/(cx + d)}$;

(e) $\sqrt{x/(1 - x)}$.

20.10 Find the maximum and minimum values of $x/(1 + x^2)$, and the points of inflexion of the curve $y(1 + x^2) = x$. A point of inflexion of a curve occurs when the curve's gradient has a maximum or a minimum and thus when the second differential of the curve is zero. The curve continues after a point of inflexion to slope upwards or downwards as it did before.

20.11 Obtain the equation of the tangent at the point $(3t^2, 2t^3)$ to the curve $27y^2 = 4x^3$.

20.12 Evaluate

$$\int_0^4 \frac{x\,dx}{\sqrt{2x + 1}}.$$

By means of the substitution $t = \tan x$ or otherwise find

$$\int \frac{1}{(\cos^2 x + 4\sin^2 x)}\,dx.$$

20.13 Express $2(x + 1)/[(x - 1)(2x - 1)]$ in partial fractions and prove that

$$\int_2^5 \frac{2(x + 1)}{(x - 1)(2x - 1)}\,dx = \ln\frac{256}{27}.$$

20.14 Find the values to three significant figures of

$$\int_0^1 \frac{x}{1 + x^2}\,dx, \quad \int_0^1 \frac{x}{\sqrt{(1 + x^2)}}\,dx, \quad \int_0^1 \frac{\tan^{-1}x}{(1 + x^2)}\,dx.$$

20.15 If $x = a(\theta - \sin\theta)$ and $y = a(1 - \cos\theta)$, where θ is a parameter, prove that

$$\frac{dy}{dx} = \cot\left(\frac{1}{2}\theta\right)$$

and

$$\frac{d^2y}{dx^2} = -\frac{1}{4a}\mathrm{cosec}^4\left(\frac{1}{2}\theta\right).$$

20.16 Prove that

$$\tan^2(45° + \theta) = \frac{1 + \sin 2\theta}{1 - \sin 2\theta}.$$

Prove also that $\tan 22.5° = \sqrt{2} - 1$ and $\tan 67.5° = \sqrt{2} + 1$.

20.17 Draw a rough graph of the curve $y = \sec^2 x$ for values of x between $-\pi/2$ and $\pi/2$.

The area enclosed by the curve, the line $y = 0$, and the lines $x = 0$ and $x = \pi/3$ is rotated about the line $y = 0$. Find the volume of the solid of revolution so obtained.

20.18 Find angles between 0 and 2π such that $\tan x = \tan 4x$.

20.19 The locus of a point whose position is determined by the requirement that it satisfies certain given conditions is the line in space followed by the point. Sketch the locus of the point $(\cos^3 t, \sin^3 t)$ where t is a parameter that can take on any value. Find the area enclosed by the locus.

20.20 In any triangle ABC, prove that

$$\tan\frac{B - C}{2} = \frac{b - c}{b + c}\cot\frac{A}{2}.$$

In one such triangle, $a = 10.78$, $c = 8.42$, and $B = 30° 10'$. Calculate the remaining angles and side of the triangle.

20.21 Evaluate

$$\int_0^{\pi/2} \sin 2x \cos 3x \, dx.$$

Using the substitution $y = e^{-x}$, find

$$\int \frac{dx}{(e^x + 1)^2}.$$

20.22 Prove that

$$\sum_{n=1}^{n=2N-1} n^3 = N^2(2N^2 - 1)$$

where the sum runs only over odd integers n.

20.23 Expand $(2x + 3)^5$ and $(2x - 3)^5$ in descending powers of x and determine the coefficient of x^7 in the product of the two expansions.

20.24 Sketch the regions in the complex plane (the Argand diagram) where (a) the modulus of z is less than 3, and (b) the argument of z lies between 0 and $\pi/4$.

20.25 Simplify the following complex numbers by writing them in the form $a + jb$.

(a) $(1 + 3j)(2 - j)$;

(b) $(2 + 3j)/(2 + 5j)$.

20.26 Express in polar form:

(a) $\sqrt{3} + 2j$;

(b) $(1 + j)^3$;

(c) $\dfrac{(1 + j)^8}{(1 + j\sqrt{3})^4}$.

20.27 If $y = ae^{-2x} \sin 3x$, prove that

$$\frac{d^2y}{dx^2} + 4\frac{dy}{dx} + 13y = 0.$$

This is a differential equation of the sort often met in physics. For example, it is similar to that describing the oscillations of a mass on the end of a spring when the mass suffers a frictional force. This is discussed in Chapter 6.

Some solutions and answers to Chapter 20 problems

20.2 If a and b are the roots of the equation $x^2 + px + q = 0$,

$$(x - a)(x - b) = x^2 - (a + b)x + ab = 0,$$

and hence $ab = q$, $(a + b) = -p$.

The equation that has roots $(a + 1/b)$ and $(b + 1/a)$ is

$$\left(x - \frac{ab + 1}{b}\right)\left(x - \frac{ab + 1}{a}\right)$$

$$= x^2 - \frac{ab + 1}{b}x - \frac{ab + 1}{a}x + \frac{(ab + 1)^2}{ab} = 0.$$

Writing the equation as

$$x^2 + Ax + B = 0,$$

we find

$$A = -\frac{ab^2 + b + a^2b + a}{ab} = \frac{(a+b)(1+ab)}{ab}$$

and

$$B = \frac{(ab+1)^2}{ab}.$$

Hence $A = p(1+q)/q$ and $B = (q^2 + 2q + 1)/q$.

20.3 $a^3 + b^3 = (a+b)^3 - 3ab(a+b) = 95$.

20.5 The scalar product is -13. The vector product is $-24\mathbf{i} - 27\mathbf{j} - 18\mathbf{k}$.

20.7 If the perpendicular from A meets the line BC at D, then $\sin C = AD/b$ and $\sin B = AD/c$. Hence

$$\frac{b}{\sin B} = \frac{c}{\sin C}.$$

Similar constructions show that

$$\frac{a}{\sin A} = \frac{b}{\sin B} = \frac{c}{\sin C}.$$

20.9 (c) $10^{2x} = e^{2ax}$ where $a = \ln(10)$. Hence

$$\frac{d}{dx}(10^{2x}) = \frac{d}{dx}(e^{2ax})$$

$$= 2ae^{2ax} = 2a\,10^{2x} = 2(\ln 10)10^{2x}.$$

20.10 The derivative of the function $y = x/(1+x^2)$, is

$$\frac{dy}{dx} = \frac{1 - x^2}{(1+x^2)^2}.$$

The derivative is zero at $x = \pm 1$, and, at $x = 1$, $y = 1/2$ and, at $x = -1$, $y = -1/2$. The second derivative is

$$\frac{d^2y}{dx^2} = \frac{2x^3 - 6x}{(1+x^2)^3}.$$

This is equal to zero when $x = \pm\sqrt{3}$. The curve thus has points of inflexion when $x = \sqrt{3}$ and $x = -\sqrt{3}$.

20.14 $\int_0^1 \frac{x}{1+x^2}\,dx = 0.347, \quad \int_0^1 \frac{\tan^{-1}x}{(1+x^2)}\,dx = 0.785.$

20.16 $\tan^2(45° + \theta) = \dfrac{\sin^2(45° + \theta)}{\cos^2(45° + \theta)}$

$$= \frac{(\sin 45° \cos\theta + \cos 45° \sin\theta)^2}{(\cos 45° \cos\theta - \sin 45° \sin\theta)^2}$$

$$= \frac{(\cos\theta + \sin\theta)^2}{(\cos\theta - \sin\theta)^2}$$

$$= \frac{1 + 2\sin\theta\cos\theta}{1 - 2\sin\theta\cos\theta}$$

$$= \frac{1 + \sin 2\theta}{1 - \sin 2\theta}.$$

$$\tan^2(22.5°) = \tan^2(45° - 22.5°)$$

$$= \frac{1 - \sin(45°)}{1 + \sin(45°)},$$

from above. But $\sin 45° = 1/\sqrt{2}$; hence

$$\tan^2(22.5°) = \frac{1 - 1/\sqrt{2}}{1 + 1/\sqrt{2}} = \frac{\sqrt{2} - 1}{\sqrt{2} + 1}$$

$$= \frac{(\sqrt{2} - 1)^2}{(\sqrt{2} + 1)(\sqrt{2} - 1)} = (\sqrt{2} - 1)^2.$$

20.18 We require to find angles for which

$$\sin x \cos 4x = \sin 4x \cos x.$$

The left-hand side of this equation reduces to $2\sin x((2\cos^2 x - 1)^2 - 1)$ and the right-hand side to $4\sin x(\cos^4 x - 4\cos^2 x \sin^2 x)$. Reducing these expressions further and equating them gives

$$4\cos^2 x = 1$$

or $\cos x = \sqrt{1/4}$. The angle x thus has $\cos x = \pm 0.5$ giving $x = 60°$, $120°$, $240°$, and $300°$. Angles of zero and 2π also satisfy the requirement and so the complete set are the above four plus zero and $360°$.

20.20 The sum of the angles of a triangle adds up to $180°$; hence

$$\cot\frac{A}{2} = \frac{\cos(\pi/2 - B/2 - C/2)}{\sin(\pi/2 - B/2 - C/2)} = \frac{\sin(B/2 + C/2)}{\cos(B/2 + C/2)}.$$

Since

$$\frac{b}{\sin B} = \frac{c}{\sin C},$$

$$\frac{b-c}{b+c} = \frac{\sin B - \sin C}{\sin B + \sin C}.$$

The identities (20.38) and (20.39) can be manipulated by putting $A = X + Y$ and $B = X - Y$ to show that

$$\sin X + \sin Y = 2\sin(X/2 + Y/2)\cos(X/2 - Y/2)$$

and

$$\sin X - \sin Y = 2\cos(X/2 + Y/2)\sin(X/2 - Y/2).$$

Hence

$$\frac{b-c}{b+c} = \frac{\cos(B/2 + C/2)\sin(B/2 - C/2)}{\sin(B/2 + C/2)\cos(B/2 - C/2)}$$

and thus

$$\frac{b-c}{b+c}\cot\frac{A}{2}$$

$$= \frac{\cos(B/2 + C/2)\sin(B/2 - C/2)\sin(B/2 + C/2)}{\sin(B/2 + C/2)\cos(B/2 - C/2)\cos(B/2 + C/2)}.$$

The right-hand side of this equation reduces to the required $\tan(B/2 - C/2)$.

20.22 If

$$\sum_{n=1}^{n=2N-1} n^3 = N^2(2N^2 - 1),$$

then

$$\sum_{n=1}^{n=2(N+1)-1} n^3 = (N+1)^2(2(N+1)^2 - 1),$$

and the difference between the sum for $n = N$ and $n = N + 1$ should equal the value of the single additional term that has $n = 2N + 1$, namely, $(2N+1)^3$. But

$$(2N+1)^3 = 8N^3 + 12N^2 + 6N + 1$$

and

$$\sum_{n=1}^{n=2(N+1)-1} n^3 - \sum_{n=1}^{n=2N-1} n^3$$

$$= (N+1)^2(2(N+1)^2 - 1) - N^2(2N^2 - 1).$$

This also reduces to $8N^3 + 12N^2 + 6N + 1$. The equality holds for any value of N and thus the proposed identity is true.

20.25
(a) $(1 + 3j)(2 - j) = 2 - j + 6j + 3 = 5 + 5j$;

(b) $\dfrac{(2 + 3j)}{(2 + 5j)} = \dfrac{(2 + 3j)}{(2 + 5j)}\dfrac{(2 - 5j)}{(2 - 5j)} = \dfrac{1}{29}(19 - 4j)$.

20.26 (a) $\sqrt{3} + 2j \equiv Ae^{j\theta} = A\cos\theta + jA\sin\theta$.
Hence $A\cos\theta = \sqrt{3}$ and $A\sin\theta = 2$, and the tangent of the angle θ is equal to $2/\sqrt{3}$. Also

$$A^2\cos^2\theta + A^2\sin^2\theta = 7$$

and, since $\cos^2\theta + \sin^2\theta = 1$, $A = \sqrt{7}$.

Appendix A

Units

This appendix explains the system of units used in physics

In order to be able to convey quantitative information in science, a set of **standards** of measurement must be agreed. Most books and publications in the physical sciences use units of measurement defined by the Système International, which are referred to as **SI units**.

It is not necessary to keep standards for all the SI units. Speed, for example, is measured in metres per second and, provided that the SI standards of length (the metre) and time (the second) are fixed, there is no need for a separate standard of speed. The SI **base units** are defined in Table 1. All the other SI units are then defined in terms of the base units.

There is another base unit, the **candela**, which is a measure of **luminous intensity**, but it is not used in this book and has been omitted from the table.

Note that the metre, which is a standard of length, depends on the agreed standard of time. The metre is defined in such a way that the speed of light is *exactly* 299 792 458 metres per second. The reason for this apparently strange way of defining length is that times can be measured more accurately than any other quantities. The choice of the precise number of metres light must travel per second has been made so that one metre is as nearly as possible unchanged from its value according to an earlier definition.

The definition of length in terms of time does not invalidate the use of length and time as separate dimensions in the dimensional analysis described in Section 1.3. Temperature and electric current are also used in dimensional analysis.

The **mole** is a dimensionless unit referring to substances consisting of atoms of a single element or molecules of a single type. One mole of a substance is thus the amount of the substance having a mass equal to its atomic or molecular weight expressed in grams. Because the ^{12}C atom is

Table 1 The SI base units of measurement

Quantity (symbol)	Definition
Mass (kg)	One **kilogram** is the mass of the standard kilogram at the International Bureau of Weights and Measures at Sèvres, France
Time (s)	One **second** is the time taken for 9 192 631 770 periods of the oscillations of a caesium clock
Length(m)	One **metre** is the distance travelled by light in vacuum in a time 1/299 792 458 seconds
Electric current (A)	One **ampere** is the steady current which, when passed through two infinitely long, thin parallel wires one metre apart, generates a magnetic force of $2\pi \times 10^{-7}$ newtons between them
Temperature (K)	One **kelvin** is 1/273.16 times the absolute temperature difference between absolute zero and the triple point of water
Amount of substance (mol)	One **mole** is the amount of a substance containing the same number of atoms or molecules as 12 grams of ^{12}C

the standard for atomic and molecular weights, the formal definition of the mole is also referred to ^{12}C.

To give the number of moles of a substance in a sample is equivalent to saying how many atoms or molecules are present: one mole contains Avogadro's number N_A of atoms or molecules. Avogadro's number is not known exactly: it is a measurable quantity that has an error. The best value is currently

$$N_A = (6.022\,136\,7 \pm 0.000\,003\,6) \times 10^{23}.$$

The **radian** is another dimensionless unit. It is defined to be the angle between two radii of a circle that cut off an arc on the circumference equal in length to the radius. Angles are mathematical constructions, and the number of radians in a complete circle is known to be 2π independently of any measurements. The unit of solid angle, the **steradian**, is similarly a mathematical unit (steradians are discussed in Section 15.3).

There are other units that are used commonly and have special names of their own, even thought they are defined in terms of the SI base units. They are called **derived units**. For example, force is a derived unit that is defined to be rate of change of momentum according to Newton's second law,

$$\mathbf{F} = \frac{d\mathbf{p}}{dt}. \tag{2.5}$$

Momentum is (mass × velocity), which is measured in $m\,kg\,s^{-1}$, and the units of force are $m\,kg\,s^{-2}$. For such an important quantity as force, 'one metre kilogram per second squared' is a very cumbersome unit, and so the unit of force has been given the name 'newton' and the symbol N. But the newton is defined in terms of the SI base units, and one newton is exactly equivalent to 'one metre kilogram per second squared' or, in symbols,

$$1\,N \equiv 1\,m\,kg\,s^{-2}.$$

Frequently, other derived units are expressed in terms of units that have special names. For example, a torque, which is the moment of a force about a particular point, is usually given the units N m (newton metres) rather than the equivalent expression in base units, which is $m^2\,kg\,s^{-2}$.

The base units in the last column of Table 2 indicate the dimensions to be used in a dimensional analysis of the kind described in Section 1.3. For example, the units of power are $m^2\,kg\,s^{-3}$, and the dimensions associated with power are $[L]^2[M][T]^{-3}$.

Units of *frequency* and *angular frequency* are hertz (cycles/second) and radians/second respectively, which are both acceptable SI units. They differ by a factor 2π, but both have the same dimensions $[T]^{-1}$. This is

Table 2 A list of the SI units with special names that occur in this book

Quantity	Name	Symbol	Base units
Force	newton	N	$m\,kg\,s^{-2}$
Energy	joule	J	$m^2\,kg\,s^{-2}$
Power	watt	W	$m^2\,kg\,s^{-3}$
Frequency	hertz	Hz	s^{-1}
Pressure	pascal	Pa	$m^{-1}\,kg\,s^{-2}$
Electric charge	coulomb	C	$s\,A$
Electric potential	volt	V	$m^2\,kg\,s^{-3}\,A^{-1}$
Capacitance	farad	F	$m^{-2}\,kg^{-1}\,s^4\,A^2$
Resistance	ohm	Ω	$m^2\,kg\,s^{-3}\,A^{-2}$
Magnetic field **B**	tesla	T	$kg\,s^{-2}\,A^{-1}$
Inductance	henry	H	$m^2\,kg\,s^{-2}\,A^{-2}$
Activity of radionuclide	becquerel	Bq	s^{-1}

because the radian, which is the SI unit of angle, is dimensionless. As explained in Section 1.3, dimensional analysis can check the dimensional consistency of equations, but gives no information about constants of proportionality.

Appendix B

Errors

This appendix outlines some simple formulae and procedures adopted when considering errors in experimental measurements

Distribution functions

The Gaussian distribution

Least-squares fitting

Propagation of errors

Systematic errors and quadrature

The Poisson distribution

In the introductory chapter of this book it was emphasized that all experimental measurements have errors. There are random errors that arise from the random inexactitude of the measuring devices. One measurement of a quantity that has a constant value, such as the period of a pendulum, is different from the previous measurement and the deviation of any measurement from the unknown 'true' value may be on either side of that value independently of the sign of the previous deviation. The measured values of the period are distributed about a most probable value and the manner in which they are distributed is described by a **distribution function**. If the measuring inexactitudes, or errors, are small, and many measurements of the period of the pendulum are made, the measurements will be distributed about the most probable value in a narrower curve than if the errors are large.

Distribution functions

We here discuss distribution functions in terms of the distribution of the measurements of a quantity that have measurement errors. Let us retain the example of a pendulum or of a clock that ticks regularly. If the period of the pendulum or the interval between ticks is measured several times, and the number of times an interval t is recorded is plotted against t, the result is similar to Fig B.1. There, we have put the measurements into bins of a given time width Δt to produce a **histogram** of the data. The number of times a period is measured with a value between $(t + \Delta t/2)$ and $(t - \Delta t/2)$ is plotted in a bin centred on the time t.

For a reasonably large number of measurements, the histogram is roughly symmetrical about the bin containing the most probable interval measured. If very many measurements of the period are made, the increased number of data points will allow narrower bins to be used in the histogram. Eventually, we can plot a continuous curve such as that drawn on Fig B.2. The function $f(t)$ describing the continuous curve is the distribution function of the measured periods.

The **moments** of the distribution are properties of the distribution that have important uses. The first moment is the **mean** or the **average**. The average number $\bar{n}$ of apples per box when there are N boxes containing apples is obtained by counting the number in each box and determining the number of times N_i when a box contains n_i apples. The average is defined as

$$\bar{n} = \frac{\sum_i N_i n_i}{\sum_i N_i} = \frac{\sum_i N_i n_i}{N}. \tag{B.1}$$

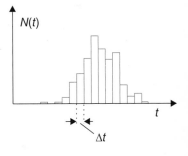

Fig. B.1 A histogram of the number of times, $N(t)$, a period is measured with a value between $(t + \Delta t/2)$ and $(t - \Delta t/2)$.

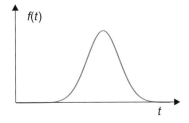

Fig. B.2 The distribution curve of the measured periods.

Similarly, for the continuous distribution $f(t)$ the mean or average is

$$\bar{t} = \frac{\int_{-\infty}^{\infty} tf(t)\,dt}{\int_{-\infty}^{\infty} f(t)\,dt}. \tag{B.2}$$

In eqn (B.2) we have written the lower limit of integration to be $-\infty$ to cover the more general situation in which negative values are possible.

For distributions that are symmetrical about zero, the mean is zero. The first non-vanishing moment of the distribution is then the **mean square** value. If a quantity x has a distribution function $g(x)$, the mean square of x is

$$\bar{x^2} = \frac{\int_{-\infty}^{\infty} x^2 g(x)\,dx}{\int_{-\infty}^{\infty} g(x)\,dx}. \tag{B.3}$$

Higher moments of the distribution may be defined in an analogous manner to the definitions of eqns (B.2) and (B.3).

The **variance** of the distribution of the apples in the boxes is a measure of the spread of the numbers in boxes about the mean, and is defined as

$$V = \frac{\sum_i N_i(n_i - \bar{n})^2}{\sum_i N_i} = \frac{\sum_i N_i(n_i - \bar{n})^2}{N}. \tag{B.4}$$

For a continuous distribution function $g(x)$ the variance is

$$V = \frac{\int_{-\infty}^{\infty} (x - \bar{x})^2 g(x)\,dx}{\int_{-\infty}^{\infty} g(x)\,dx}. \tag{B.5}$$

The Gaussian distribution

Now consider the distribution functions that random errors of measurements actually obey. The deviations of measurements of a quantity from its *true* value are as likely to be positive as negative and, if many measurements are taken, the distribution is symmetrical about the most probable measured value. Their distribution function is a **Gaussian** curve.

A Gaussian curve is shown in Fig B.3 and is symmetrical about its maximum. The mathematical expression for such a curve when it has a maximum at x_0 is

$$y(x) = y_0 \exp\left(-\frac{(x - x_0)^2}{2\sigma^2}\right) \tag{B.6}$$

where y_0 is the maximum value of y and the parameter σ is related to the full width Δx of the curve at one-half its maximum height y_0. It can easily

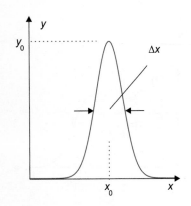

Fig. B.3 A Gaussian curve.

be shown that

$$\Delta x = 2\sigma\sqrt{2\ln 2} = 2.355\sigma \tag{B.7}$$

to four significant figures. The parameter σ is called the **standard deviation** of the distribution; when σ is large the curve is wide and vice versa.

The integral from minus infinity to plus infinity of a Gaussian function with peak height unity is

$$\int_{-\infty}^{\infty} \exp\left(-\frac{(x-x_0)^2}{2\sigma^2}\right) dx = \sqrt{2\pi}\sigma. \tag{B.8}$$

The variance of a Gaussian distribution function can be evaluated using the definition (B.5) and can be shown to equal the square of the standard deviation,

$$V = \sigma^2.$$

The measurements x of a quantity are distributed about the most probable value x_0 (which is also the mean value for symmetrical distributions) according to eqn (B.6). The probability of the occurrence of a measured value between x and $x + dx$ is proportional to $y(x)$. However, the deviations from the 'true' value occur randomly and, for a finite number of measurements in a set or sample, the mean differs from the mean of a similar set taken before or after.

If a sample contains an infinite number of measurements, their distribution is a continuous curve and the mean is known accurately. However, an infinite number of measurements cannot be made and the determination of the 'true' value from the mean of a limited set has an uncertainty that depends on the number of data points taken. Let a sample of N measurements be made and let N_i be the number of times the value n_i is obtained. If N_i is plotted against n_i, the points are distributed around a Gaussian curve that has standard deviation σ. The best estimate of σ from the numbers is equal to the square root of the variance, and this is obtained from

$$\sigma^2 = \frac{\sum_i N_i (n_i - \bar{n})^2}{\sum_i N_i} = \frac{\sum_i N_i (n_i - \bar{n})^2}{N}. \tag{B.9}$$

It can be shown that this expression is equivalent to

$$\sigma = \sqrt{\bar{x^2} - (\bar{x})^2}. \tag{B.10}$$

The ratio

$$\frac{\int_{-2\sigma}^{2\sigma} \exp(-x^2/2\sigma^2)\, dx}{\int_{-\infty}^{\infty} \exp(-x^2/2\sigma^2)\, dx} = 0.955.$$

Hence the probability of a measurement lying two standard deviations or more from the mean of a set of measurements of the same quantity is about 4.5%.

The mean of the measurements is usually quoted as the best value determined, and the error is given as the **standard error on the mean** which is

$$\text{Standard error on the mean} = \sigma/\sqrt{N}. \qquad (\text{B.11})$$

It is important to recognize the distinction between the standard deviation of the distribution function, which represents the spread of measured values, and the accuracy with which the mean is known. If a very large number of measurements is taken, the mean is known with little uncertainty.

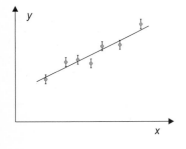

Fig. B.4 Measured values of a quantity y as a function of a variable x when y is expected to vary linearly with x. Errors on x are assumed to be negligible.

Least-squares fitting

Another simple experiment often performed is to measure the value of one quantity y as a second x is varied. For example, y may be the pressure of a fixed volume of gas and x may be the temperature at which the pressure is measured. If we assume that there are only small errors on the measurements of temperature and that these errors can be neglected compared with the errors on the measured pressures, the data may be plotted as shown in Fig B.4.

We may have reason to believe that y varies linearly with x over the temperature range considered and thus that over that range y is given by the formula

$$y = \alpha + \beta x. \qquad (\text{B.12})$$

Because of the errors on the points, a line given by eqn (B.12) will not pass through all the points shown on Fig B.4 whatever line is chosen. The problem is to determine the best line from the data set. The best values of the parameters α and β are given by **fitting** the points to a straight line of the form given by eqn (B.12).

Let us assume, as in plotting Fig B.4, that x is known much more accurately than y and thus that the errors on x may be neglected. Let a measurement y_i of y have an error Δy_i. We define a quantity called the

chi-squared, symbol χ^2, given by

$$\chi^2(\alpha, \beta) = \sum_{i=1,N} \frac{(y_i - y_{\text{th},i}(\alpha, \beta))^2}{\Delta y_i^2} \tag{B.13}$$

where the sum is over the N pairs of (x, y), and $y_{\text{th},i}(\alpha, \beta)$ is the value predicted by eqn (B.12) using parameter values α and β. The best values of α and β are then those that correspond to the minimum value of χ^2.

This technique of estimating the best parameter values is called **least-squares fitting**. The best values may be shown to be given by

$$\alpha = \frac{\overline{x^2}\,\bar{y} - \bar{x}\,\overline{xy}}{\overline{x^2} - (\bar{x})^2} \tag{B.14}$$

and

$$\beta = \frac{\overline{xy} - \bar{x}\,\bar{y}}{\overline{x^2} - (\bar{x})^2}. \tag{B.15}$$

Using the above values for α and β it can be shown that the best fit straight line through a set of points (x, y) passes through the point $(\bar{x}, \bar{y})$.

Propagation of errors

There are errors on the fitted parameters given by eqns (B.14) and (B.15). The procedure for determining them may be illustrated with an example of a more general nature in which a measured quantity z depends upon two other measured quantities x and y through a functional relationship

$$z = f(x, y).$$

A small change dz in z is produced by small changes dx and dy in x and y, with

$$dz = \left(\frac{\partial f}{\partial x}\right) dx + \left(\frac{\partial f}{\partial y}\right) dy.$$

Squaring both sides gives

$$(dz)^2 = \left(\frac{\partial f}{\partial x}\right)^2 (dx)^2 + \left(\frac{\partial f}{\partial y}\right)^2 (dy)^2 + 2\left(\frac{\partial f}{\partial x}\right)\left(\frac{\partial f}{\partial y}\right) dx\,dy.$$

If the small increment dz is the small error introduced by errors dx and dy, and if the latter errors are uncorrelated and equally as likely to be negative as positive, over many measurements the cross-term involving $dxdy$ in the above equation averages to zero. Regarding dx and dy as standard deviations on the measurements of x and y, and dz as the consequent standard deviation in the measured value of z, we then have

$$\sigma_z^2 = \left(\frac{\partial f}{\partial x}\right)^2 \sigma_x^2 + \left(\frac{\partial f}{\partial y}\right)^2 \sigma_y^2. \tag{B.16}$$

Systematic errors and quadrature

There are often systematic errors present in a measurement. For example, a clock used to measure the period of a pendulum may have a calibration error. It may consistently read too high or too low by a given amount but we don't know by how much and in what direction. In that case, the systematic error has to be combined with the random errror on the mean, obtained from the spread in the clock readings using eqns (B.9) and (B.11), to give a total error.

Let the clock's calibration be uncertain to $\pm\sigma_S$. If the mean period measured from the clock readings is $\bar{T}$ and the standard error on the mean is σ_M, the mean value of the period $\bar{T}$ has the additional systematic error σ_S. The answer for the period is usually quoted as

$$T = \bar{T} \pm \sigma_M \pm \sigma_S. \tag{B.17}$$

The Poisson distribution

The Poisson distribution describes the frequency with which rare events occur. For example, if many people pay the same money into a scheme that hands out a limited number of prizes each month and the winners are drawn at random, any particular participant may expect to receive a prize only rarely. If the mean number of prizes received by a participant per month is $\bar{n}$, what is the probability of that participant receiving n prizes in any month?

This probability $P(n)$ is given by the Poisson distribution

$$P(n, \bar{n}) = \frac{e^{-\bar{n}}(\bar{n})^n}{n!}. \tag{B.18}$$

In this expression, $n!$ equals $n(n-1)(n-2)\cdots 1$ and is called **factorial** n. Note that n is an integer but $\bar{n}$ will normally be non-integral. The variance on the distribution is $\bar{n}$ and the standard deviation of the mean is $\sqrt{\bar{n}}$. If the average number of prizes received by a participant per month is 0.5, the probability of receiving one prize in any month is about 0.30 and the probability of receiving two is about 0.08. The maximum probability occurs at n equal to zero, and the probability of receiving no prize at all is about 0.61.

If the mean $\bar{n}$ is large, such as it may be if the events recorded are counts in a detector observing a long-lived radioactive decay over several fixed time intervals, the Poisson distribution becomes closely equal to the Gaussian distribution. If the number of counts observed in any particular one of the time intervals is N, the error on that number is $\sqrt{N}$ and the fractional error on the count is $1/\sqrt{N}$.

Appendix C

Physical constants and other physical quantities

This appendix gives values of physical constants and some useful numbers

Physical constants

Useful physical quantities

Useful conversions and equivalences

Physical constants

Speed of light *in vacuo*, c	$2.997\,924\,58 \times 10^8\,\mathrm{m\,s^{-1}}$
Permeability of vacuum, μ_0	$4\pi \times 10^{-7}\,\mathrm{N\,A^{-2}}$
Permittivity of vacuum, ϵ_0	$8.854\,187 \cdots \times 10^{-12}\,\mathrm{F\,m^{-1}}$

The speed of light *in vacuo* and the permeability of the vacuum are defined to be exactly equal to the above values. The permittivity of the vacuum is related to these two by $\epsilon_0 = 1/\mu_0 c^2$, and hence the permittivity of the vacuum can be calculated to any required degree of accuracy. We give its value above to seven significant figures.

All other constants are measured experimentally and thus have associated errors. We adopt here a common convention for quoting standard deviations on the accepted values of the constants. Numbers in brackets are the uncertainties in the last digits of the accepted values given. For example, 4.321(12) is shorthand for 4.321 ± 0.012.

Gravitational constant, G	$6.672\,59(85) \times 10^{-11}\,\mathrm{m^3\,kg^{-1}\,s^{-2}}$
Planck's constant, h	$6.626\,075\,5(40) \times 10^{-34}\,\mathrm{J\,s}$
	$4.135\,669\,2(28) \times 10^{-15}\,\mathrm{eV\,s}$
Magnitude of charge on electron, e	$1.602\,177\,33(49) \times 10^{-19}\,\mathrm{C}$
Mass of electron, m_e	$9.109\,389\,7(54) \times 10^{-31}\,\mathrm{kg}$
Mass of proton, m_p	$1.672\,623\,1(10) \times 10^{-27}\,\mathrm{kg}$
Mass of neutron, m_n	$1.674\,928\,6(10) \times 10^{-27}\,\mathrm{kg}$
Atomic mass unit	$1.660\,540\,2(10) \times 10^{-27}\,\mathrm{kg}$
Avogadro's constant, N_A	$6.022\,136\,7(36) \times 10^{23}\,\mathrm{mol^{-1}}$
Boltzmann's constant, k_B	$1.380\,658(12) \times 10^{-23}\,\mathrm{J\,K^{-1}}$
	$8.617\,385(73) \times 10^{-5}\,\mathrm{eV\,K^{-1}}$
Gas constant, R	$8.314\,510(70)\,\mathrm{J\,mol^{-1}\,K^{-1}}$

Useful physical quantities

Bohr magneton, $(e\hbar/2m_\mathrm{e})$	$9.274\,015\,4(31) \times 10^{-24}\,\mathrm{J\,T^{-1}}$
	$5.788\,382\,63(52) \times 10^{-5}\,\mathrm{eV\,T^{-1}}$
Nuclear magneton, $(e\hbar/2m_\mathrm{p})$	$5.050\,786\,6(17) \times 10^{-27}\,\mathrm{J\,T^{-1}}$
	$3.152\,451\,66(28) \times 10^{-8}\,\mathrm{eV\,T^{-1}}$
Rydberg constant, $(m_\mathrm{e}e^4/8h^3\epsilon_0^2 c)$	$1.097\,373\,153\,4(13) \times 10^7\,\mathrm{m^{-1}}$
Stefan–Boltzmann constant, $(\pi^2 k_B^4/60\hbar^3 c^2)$	$5.670\,51(19) \times 10^{-8}\,\mathrm{W\,m^{-2}\,K^{-4}}$

Mass of Earth	$5.974 \times 10^{24}\,\text{kg}$
Earth equatorial radius	$6.378 \times 10^{6}\,\text{m}$
Mass of Sun	$1.989 \times 10^{30}\,\text{kg}$
Solar equatorial radius	$6.96 \times 10^{8}\,\text{m}$
Earth–Sun distance	$1.50 \times 10^{11}\,\text{m}$

Useful conversions and equivalences

$1\,\text{eV} = 1.602\,177\,33(49) \times 10^{-19}\,\text{J}$

$1\,\text{eV}/c^2 = 1.782\,662\,70(54) \times 10^{-36}\,\text{kg}$

Mass of electron, $m_e = 0.510\,999\,06(15)\,\text{MeV}/c^2$

Mass of proton, $m_p = 1.007\,276\,470(12)$ atomic mass units

$\qquad\qquad = 938.272\,31(28)\,\text{MeV}/c^2$

1 atomic mass unit $= 931.494\,32(28)\,\text{MeV}/c^2$

$k_B T$ at $300\,\text{K} \approx 0.0258\,\text{eV}$

1 atmosphere $\equiv 760\,\text{torr} \equiv 101\,325\,\text{Pa}$

1 inch $\equiv 0.0254\,\text{m}$

Acknowledgements

We wish to thank the colleagues who have kindly allowed us to use the figures listed below.

Figures 2.20, 2.21, 9.1 and 14.3: Prof F. Besenbacher and Dr K. Hansen, Institute of Physics and Astronomy, University of Aarhus.
Figure 3.27: Dr R. J. Davis, Jodrell Bank, University of Manchester.
Figure 5.17: Photograph © 1997 by H. Mikuz & B. Kambic, Crni Vrh Observatory, Slovenia.
Figure 7.8: Dr Bill Jack Rodgers, Los Alamos National Laboratory.
Figure 9.17: Prof J. R. Helliwell, Department of Chemistry, University of Manchester.
Figure 9.18: Dr J. A. Chapman, Biological Sciences, University of Manchester.
Figure 9.20: Prof R. Nelmes, Department of Physics, University of Edinburgh.
Figures 11.25 and 11.26: Dr P. Campbell, Department of Physics and Astronomy, University of Manchester.
Figure 14.1: Photograph by Mike and Darcy Howard, RockhoundingAR.com.
Figure 14.2: Earth Sciences Library, University of Manchester.

For permission to reproduce pictures of scientists, we thank the University of Vienna (Boltzmann) and the Caltech Archives (Einstein, Feynman).

Index

A page number in bold type indicates the most important reference for the entry in the index.